Handbook of Natural Fibres

The Textile Institute Book Series

Incorporated by Royal Charter in 1925, The Textile Institute was established as the professional body for the textile industry to provide support to businesses, practitioners and academics involved with textiles and to provide routes to professional qualifications through which Institute Members can demonstrate their professional competence. The Institute's aim is to encourage learning, recognise achievement, reward excellence and disseminate information about the textiles, clothing and footwear industries and the associated science, design and technology; it has a global reach with individual and corporate members in over 80 countries.

The Textile Institute Book Series supersedes the former 'Woodhead Publishing Series in Textiles' and represents a collaboration between The Textile Institute and Elsevier aimed at ensuring that Institute Members and the textile industry continue to have access to high calibre titles on textile science and technology.

Books published in The Textile Institute Book Series are offered on the Elsevier web site at: store.elsevier.com and are available to Textile Institute Members at a substantial discount. Textile Institute books still in print are also available directly from the Institute's web site at: www.textileinstitute.org

To place an order, or if you are interested in writing a book for this series, please contact Matthew Deans, Senior Publisher: m.deans@elsevier.com

Recently Published and Upcoming Titles in The Textile Institute Book Series

Advances in Textile Biotechnology, Artur Cavaco-Paulo, 978-0-08-102632-8
Woven Textiles: Principles, Technologies and Applications, 2nd Edition, Kim Gandhi, 978-0-08-102497-3
Auxetic Textiles, Hong Hu, 978-0-08-102211-5
Carbon Nanotube Fibres and Yarns: Production, Properties and Applications in Smart Textiles, Menghe Miao, 978-0-08-102722-6
Sustainable Technologies for Fashion and Textiles, Rajkishore Nayak, 978-0-08-102867-4
Structure and Mechanics of Textile Fibre Assemblies, Peter Schwartz, 978-0-08-102619-9
Silk: Materials, Processes, and Applications, Narendra Reddy, 978-0-12-818495-0
Anthropometry, Apparel Sizing and Design, 2nd Edition, Norsaadah Zakaria, 978-0-08-102604-5
Engineering Textiles: Integrating the Design and Manufacture of Textile Products, 2nd Edition, Yehia Elmogahzy, 978-0-08-102488-1
New Trends in Natural Dyes for Textiles, Padma Vankar Dhara Shukla, 978-0-08-102686-1
Smart Textile Coatings and Laminates, 2nd Edition, William C. Smith, 978-0-08-102428-7
Advanced Textiles for Wound Care, 2nd Edition, S. Rajendran, 978-0-08-102192-7
Manikins for Textile Evaluation, Rajkishore Nayak Rajiv Padhye, 978-0-08-100909-3
Automation in Garment Manufacturing, Rajkishore Nayak and Rajiv Padhye, 978-0-08-101211-6
Sustainable Fibres and Textiles, Subramanian Senthilkannan Muthu, 978-0-08-102041-8
Sustainability in Denim, Subramanian Senthilkannan Muthu, 978-0-08-102043-2
Circular Economy in Textiles and Apparel, Subramanian Senthilkannan Muthu, 978-0-08-102630-4
Nanofinishing of Textile Materials, Majid Montazer Tina Harifi, 978-0-08-101214-7
Nanotechnology in Textiles, Rajesh Mishra Jiri Militky, 978-0-08-102609-0
Inorganic and Composite Fibers, Boris Mahltig Yordan Kyosev, 978-0-08-102228-3
Smart Textiles for In Situ Monitoring of Composites, Vladan Koncar, 978-0-08-102308-2
Handbook of Properties of Textile and Technical Fibres, 2nd Edition, A. R. Bunsell, 978-0-08-101272-7
Silk, 2nd Edition, K. Murugesh Babu, 978-0-08-102540-6

The Textile Institute Book Series

Handbook of Natural Fibres

Volume 1: Types, Properties and Factors Affecting Breeding and Cultivation

Second Edition

Edited by

Ryszard M. Kozłowski

Maria Mackiewicz-Talarczyk

The Textile Institute

ELSEVIER

WP WOODHEAD PUBLISHING
An imprint of Elsevier

Woodhead Publishing is an imprint of Elsevier
The Officers' Mess Business Centre, Royston Road, Duxford, CB22 4QH, United Kingdom
50 Hampshire Street, 5th Floor, Cambridge, MA 02139, United States
The Boulevard, Langford Lane, Kidlington, OX5 1GB, United Kingdom

Notices

Knowledge and best practice in this field are constantly changing. As new research and experience broaden our understanding, changes in research methods, professional practices, or medical treatment may become necessary.

Practitioners and researchers must always rely on their own experience and knowledge in evaluating and using any information, methods, compounds, or experiments described herein. In using such information or methods they should be mindful of their own safety and the safety of others, including parties for whom they have a professional responsibility.

To the fullest extent of the law, neither the Publisher nor the authors, contributors, or editors, assume any liability for any injury and/or damage to persons or property as a matter of products liability, negligence or otherwise, or from any use or operation of any methods, products, instructions, or ideas contained in the material herein.

Library of Congress Cataloging-in-Publication Data
A catalog record for this book is available from the Library of Congress

British Library Cataloguing-in-Publication Data
A catalogue record for this book is available from the British Library

ISBN: 978-0-12-818398-4

For information on all Woodhead Publishing publications visit our website at https://www.elsevier.com/books-and-journals

Publisher: Matthew Deans
Acquisitions Editor: Brian Guerin
Editorial Project Manager: Ana Claudia A. Garcia
Production Project Manager: Sreejith Viswanathan
Cover Designer: Mark Rogers
Partial cover designers: Katarzyna Schmidt-Przewoźna and Jacek Kołodziej, Poland

Typeset by TNQ Technologies

Contents

Contributors

Aifen Tao Fujian Agriculture and Forestry University, Fuzhou, Fujian Provence, China

Anwar M. Allam†

K. Murugesh Babu Bapuji Institute of Engineering and Technology, Davangere, Karnataka, India

Jorge Barriga-Bedoya The ŁUKASIEWICZ Research Network - Industrial Institute of Agricultural Engineering, Poznań, Poland

Gautam Basu ICAR-National Institute of Natural Fibre Engineering and Technology, Kolkata, West Bengal, India

Elmer A. Carvajal Pontificia Universidad Católica del Ecuador Sede Ibarra, Escuela de Diseño, Ibarra, Imbabura, Ecuador

Xingfen Chen National and Local Unified Engineering Research Center for Basalt Fiber Production and Application Technology, Southeast University, Nanjing, Jiangsu Province, China

Abderrezak Djabeur Laboratoire des Productions Valorisations Végétales et Microbiennes, Faculté des Sciences de la Nature et de la Vie, Université des Sciences et de la Technologie d'Oran Mohamed Boudiaf B.P., Oran, Algeria

Umit Halis Erdogan Dokuz Eylul University, Faculty of Engineering, Department of Textile Engineering, Tınaztepe Campus, Buca, Izmir, Turkey

S. Fergusson School of Fashion and Textiles, RMIT University, Brunswick, Australia

M. Griga AGRITEC Plant Research Ltd., Šumperk, Czech Republic

Víctor H. Guerrero Escuela Politécnica Nacional, Departamento de Ingeniería y Ciencias de los Materiales, Quito, Pichincha, Ecuador

R. Harwood Copernicus Textile Solutions Ltd., Desford, Leicestershire, United Kingdom

† Deceased.

Matthew R.L. Horne Shuttleworth College (Bedford College Group) Old Warden, Biggleswade, Bedfordshire, United Kingdom

S. Houshyar School of Fashion and Textiles, RMIT University, Brunswick, Australia

L. Hunter CSIR and Nelson Mandela University (NMU), Port Elizabeth, Eastern Cape, South Africa

Meriem Kaid-Harche Laboratoire des Productions Valorisations Végétales et Microbiennes, Faculté des Sciences de la Nature et de la Vie, Université des Sciences et de la Technologie d'Oran Mohamed Boudiaf B.P., Oran, Algeria

A. Khandual Department of Textile Engineering, CET Bhubaneswar, Bhubaneswar, Orissa, India

Ryszard M. Kozłowski Journal of Natural Fibers Editor-in-chief, coordinator of FAO/ESCORENA European Cooperative Research Network on Flax and other Bast Plants, former director of the Institute of Natural Fibres, Poznań, Poland

Jianxun Liu National and Local Unified Engineering Research Center for Basalt Fiber Production and Application Technology, Southeast University, Nanjing, Jiangsu Province, China

M. Ludvíková AGRITEC Plant Research Ltd., Šumperk, Czech Republic

Latifa Binte Lutfar Former Operations Officer, Dhaka, Bangladesh

Maria Mackiewicz-Talarczyk Institute of Natural Fibres and Medicinal Plants (IWNiRZ), Associate Editor Journal of Natural Fibers, Secretary of FAO/ESCORENA European Cooperative Research Network on Flax and other Bast Plants, Poznań, Poland

Moraima C. Mera Pontificia Universidad Católica del Ecuador Sede Ibarra, Escuela de Ciencias Agrícolas y Ambientales, Ibarra, Imbabura, Ecuador

Leena Mishra ICAR-National Institute of Natural Fibre Engineering and Technology, Kolkata, West Bengal, India

R. Nayak School of Communication and Design, RMIT University Vietnam, Saigon, Vietnam

Mohamed Negm Cotton Research Institute, Giza, Egypt

E. Ondráčková AGRITEC Plant Research Ltd., Šumperk, Czech Republic

Paola E. Ordóñez Pontificia Universidad Católica del Ecuador Sede Ibarra, Escuela de Ciencias Agrícolas y Ambientales, Productos Naturales Bioactivos, Ibarra, Imbabura, Ecuador

Yadira F. Ordóñez Pontificia Universidad Católica del Ecuador Sede Ibarra, Escuela de Ciencias Agrícolas y Ambientales, Productos Naturales Bioactivos, Ibarra, Imbabura, Ecuador

R. Padhye School of Fashion and Textiles, RMIT University, Brunswick, Australia

M. Pavelek AGRITEC Plant Research Ltd., Šumperk, Czech Republic

Richard S. Peigler Department of Biology, University of the Incarnate Word, San Antonio, TX, United States

Marcin Praczyk Institute of Natural Fibres and Medicinal Plants (IWNiRZ), Poznań, Poland

C. Prakash Department of Fashion Technology, Sona College of Technology, Salem, Tamil Nadu, India

Jianmin Qi Fujian Agriculture and Forestry University, Fuzhou city, Fujian Provence, China

S. Roy†

Yasemin Seki Dokuz Eylul University, Faculty of Engineering, Department of Textile Engineering, Tınaztepe Campus, Buca, Izmir, Turkey

Figen Selli Dokuz Eylul University, Faculty of Engineering, Department of Textile Engineering, Tınaztepe Campus, Buca, Izmir, Turkey; Advanced Fibers, Empa Swiss Federal Laboratories for Materials Science and Technology, St. Gallen, Switzerland

Edwin A. Simbaña Pontificia Universidad Católica del Ecuador Sede Ibarra, Escuela de Ciencias Agrícolas y Ambientales, Productos Naturales Bioactivos, Ibarra, Imbabura, Ecuador

E. Smith De Montfort University, Leicester, United Kingdom

Terry Townsend Chair, Discover Natural Fibres Initiative, Consultant Administration, Cotton Analytics, Houston, TX, United States

M. Vrbová - Prokopová AGRITEC Plant Research Ltd., Šumperk, Czech Republic

Judyta Walentowska Institute of Natural Fibres and Medicinal Plants, Poznań, Poland

Xin Wang School of Civil engineering/National and Local Unified Engineering Research Center for Basalt Fiber Production and Application Technology, Southeast University, Nanjing, Jiangsu Province, China

Yufu Wang Institute of Bast Fibres (IBFC), Changsha city, Hunan Province, China

Katarzyna Wielgusz Institute of Natural Fibres and Medicinal Plants (IWNiRZ), Poznań, Poland

Zhishen Wu School of Civil engineering/National and Local Unified Engineering Research Center for Basalt Fiber Production and Application Technology, Southeast University, Nanjing, Jiangsu Province, China

Jiantang Xu Fujian Agriculture and Forestry University, Fuzhou city, Fujian Provence, China

† Deceased.

Introduction to natural textile fibres

1A

Ryszard M. Kozłowski[1], *Maria Mackiewicz-Talarczyk*[2]
[1]Journal of Natural Fibers Editor-in-chief, coordinator of FAO/ESCORENA European Cooperative Research Network on Flax and other Bast Plants, former director of the Institute of Natural Fibres, Poznań, Poland; [2]Institute of Natural Fibres and Medicinal Plants (IWNiRZ), Associate Editor Journal of Natural Fibers, Secretary of FAO/ESCORENA European Cooperative Research Network on Flax and other Bast Plants, Poznań, Poland

1A.1 Historical background of natural fibres

Natural fibres have played an important role in human society since approximately over 7000 BCE (Ryder, 1965; Hillers, 1974). Textile fibres have been used to make a cloth for the last 4000 or 5000 years. Until 1885, when the first man-made (artificial) fibre was produced commercially, fibres were obtained from plants and animals only. The fibres most commonly used were: flax, hemp, silk, wool and cotton.

Rayon, also known as artificial silk, was the first man-made fibre and was produced in filament length until the early 1930s, and an enterprising textile worker discovered that the broken and wasted rayon filaments could be used as a staple fibre.

Many man-made fibres were developed in the first half of the twentieth century and since that time onward tremendous advances have been made in the man-made fibre industry.

Cotton cloth was used in ancient China, Egypt and Peru. Cloth fabrics found in Egypt indicate that cotton may have been used there in 12,000 BC, before flax was discovered (Gordon, 2006).

Cotton spinning and weaving as an industry began in India and fabrics of good quality cotton cloth were being produced as early as in 1500 BCE. The Pima Indians were growing cotton when the Spaniards came to the New World.

One of the items that Columbus took back to Queen Isabella was a hank of cotton yarn, and until eighteenth century Europe was dominated by flax.

It has been argued that flax is the oldest fibre that was used by mankind.

Flax samples, delivered by the British National Museum of Antiques for testing at the Institute of Natural Fibres in Poznań (Poland), were taken from Çatalhöyük (Turkey) and Kerma (Nubia, Sudan) and proved to be flax from 6500 years BCE and 2000 years BCE, respectively (Ryder, 1965).

It is widely known that more than 5000 years old linen mummy cloth has been found in Egyptian tombs. It is fascinating that this fibre is so fine that it has lasted until the present day (Taylor, 1995).

Handbook of Natural Fibres. https://doi.org/10.1016/B978-0-12-818398-4.00001-3

Wool sheep were probably the first animals domesticated by people. Crossbreeding sheep to increase the amount of hairy undercoat began in approximately CE 100. By CE 1400 the Spanish had developed the merino sheep, whose fleece is the precursor of contemporary wool raw materials (Johnson et al., 2008; Simpson et al., 2002).

Silk is the natural protein fibre as well. Silk culture, according to oldest Chinese literature, began when a Chinese Empress observed the life of silkworms and studied their features. Based on this observation, Chinese experts began to breed them and from the filaments the empress gained and produced expensive silk cloth. The Chinese silk industry spread over the years and brought China an extensive income – to the extent that revealing the secret of silk production became punishable by death. For 300 years, therefore, the secret was known only to Chinese experts. During that time China was the only producer of silk in the world and sold it through Persian and Egyptian agents at the price of gold (Franck, 2001).

The elements of the history of natural fibres, their properties and classification have been incorporated also into the *Handbook of Natural Fibres* chapters.

1A.2 Introduction

The United Nations and the *Food and Agriculture Organization of the United Nations (FAO)* declared 2009 as the *International Year of Natural Fibres*. Their objectives were to raise the profile of natural fibres, to stimulate demand for them by promoting the efficiency and sustainability of the natural fibre industries and to encourage appropriate policy responses from governments to the problems faced by these industries. In so doing they hoped to establish an effective and enduring international partnership among the various natural fibre industries (IYNF, 2009).

Some fibrous plants can grow in different climates, from Northern to Southern Arctic Circle and recycle carbon dioxide (CO_2). Now this is especially very important fact, that natural fibrous resources, lignocellulosic and protein ones, are completely sustainable, renewable, completely biodegradable and they and their production does not generally damage the ecosystem, in contrary to man-made fibres, which as the microfibres are the biggest plastic pollution issue you have not heard of yet (Bomgardner, 2017; Nussbaumer, 2009). Unfortunately, these small pieces of plastic and man-made fibres made from polymers (including small particles of polyester fibres) pose a significant risk to aquatic life because they are often mistaken as food and are readily consumed. Moreover, microplastics have a tendency to pick up toxic chemical pollutants from the water. By various processes, these harmful chemicals have the potential to be transferred from ingested microplastics to the tissues of the animals which consume them (Thai Luong, 2017).

The natural fibres are classified as follows:

Bast (flax, hemp, jute, kenaf, ramie, nettle, etc.), **Leaf** (sisal, abaca, etc.), **Seeds** (cotton, kapok, etc.), **Fruit** (coir, African palm, luffa, etc.), **Grass** (bamboo and totora), **Wood** (hardwood and softwood), **Animal Fibres** (wool, silk, hair, etc.), **Mineral Fibres** (Asbestos, basalt, wollastonite, etc.) (Figs. 1A.1–1A.3).

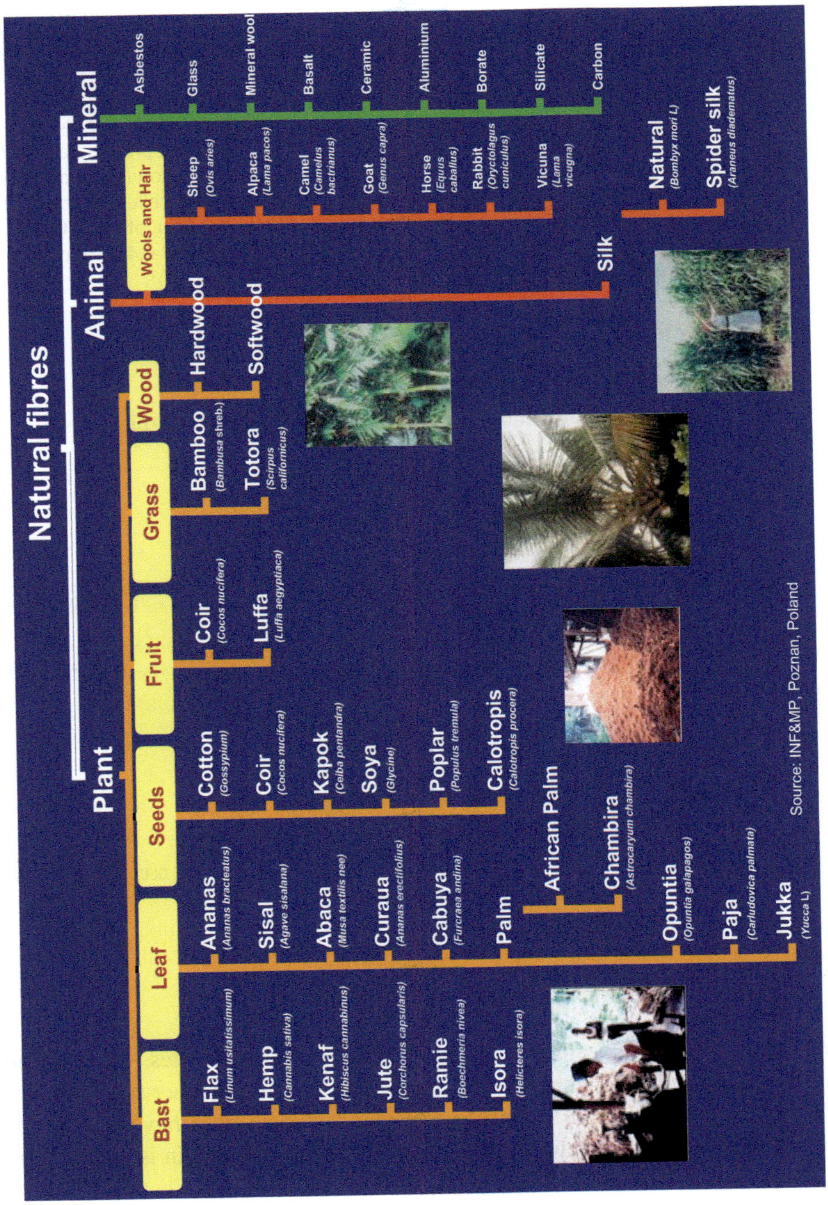

Figure 1A.1 The classification of natural fibres.

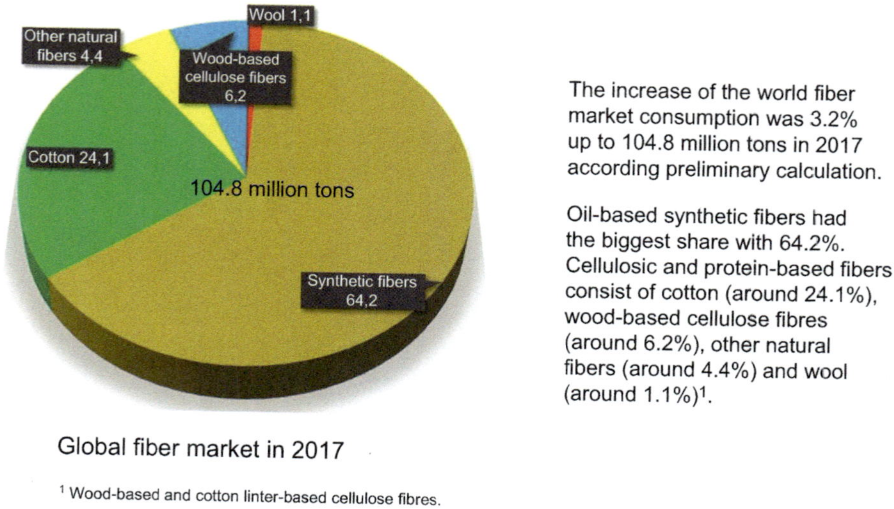

Figure 1A.2 Production of natural fibres in comparison to other fibres, 2017.
Source: ICAC, CIRFS, TFY, FEB, Lenzing Estimates.

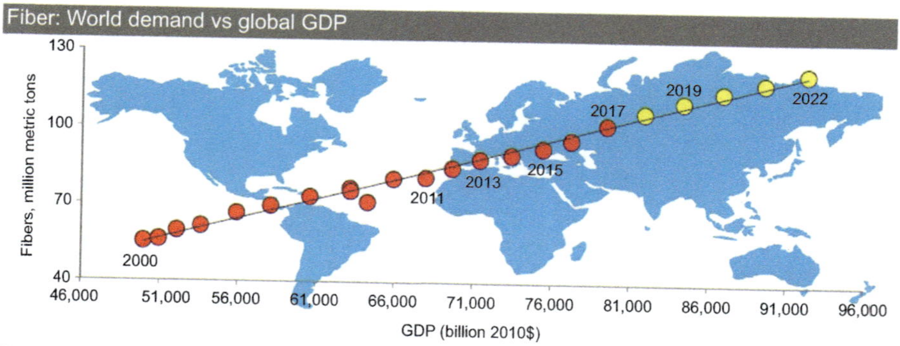

Figure 1A.3 Demand for fibres in the world.
Source: HIS Markit.

Generally natural lignocellulosic fibres are miniature composites, formed from a reinforcement of cellulose embedded in a matrix of lignin and other polysaccharides e.g., hemicellulose. Cellulose molecules of lignocellulosic fibre have a hydrophilic nature due to its hydroxyl group. It is worth to underline that pectin is a natural polymer present in plants. Pectin is a polysaccharide composed of a linear chain of 1–4 linked galacturonic acids which are etherified with methanol at 80% and has ability to biodegradation process (Franck, 2005).

Wool, biosilk and fibres based on polyamino acids and very promising fibres based on fibroin and chitin are also arousing increasing interest.

The most common natural alfa amino acids found in wool are as follows: aspartic acid, glutamic acid, asparagine, glutamine, arginine, lysine, histidine, tryptopfan, serine, threonine, tyrosine, thiocysteine, cysteine, methionine, glycine, alanine, valine, proline, leucine, isoleucine and phenylalanine. More information about wool is in vol. 1, Chapter 9, *Wool fibres* by Umit Halis Erdogan et al.

Silk fibres consist of fibroin (75%–83%) in the inner layer and sericin (17%–25%) in the outer layer, waxes about 1%–5% and other about 1% by weight. More data about silk are provided in vol. 1, Chapter 11 Silk fibres and Vol. 2, chapter *Silk production and the future of natural silk manufacture* (by Murugesh Babu).

The new natural fibres, explored by researchers and industrial people, are described in Chapter 20, vol. 1. *New emerging natural fibres and relevant sources of information.* The growing world population will mean increased demand for and consumption of fibres, including natural. The current population explosion is alarming enough to have necessitated a search for alternative textile resources, mainly for healthy, sustainable, and comfort-providing natural fibres.

Currently there are four main drivers in world wide fibre market: population increase, replacement, wealth, and new applications. The production of these fibres is expected at the level of 35–40 million tons/year in the middle of this century. The demand should be responded by the increase of cotton production up to 26–30 million tons/year, and growth of production of other natural, including modified bast fibres in harmonious coexistence with the man-made fibres (Oerlikon).

Total world production of natural and man-made fibres: polyester 52.7%, cotton 23.8%, polyamide 5.3%, viscose 5.2%, polypropylene 3.2%, jute 3.1%, acrylic 1.8%, wool 1.1%, acetate 0.8%, flax 0.4% and other 2.7% (Gherzi, 2017).

Current data show that fibre demand for natural and man-made fibres increases constantly in line with growing GDP (*Gross Domestic Product).*

More statistical data and survey of the natural fibres market in Vol. 1, Chapter 1B *World natural fibres production and employment* by Terry Townsend.

In 21st century, mankind pays more attention to green, renewable resources, not only for chemical, mainly polymer industry, but also in technology of producing and processing different fibrous resources. Nowadays we have the green light for expanded use of other green fibres in textiles for healthy and comfortable clothing which after lifetime will be fully recyclable and biodegradable.

Most natural fibres can be obtained from lignocellulosic fibrous plants such as flax, jute, hemp, kenaf, sisal, ramie, nettle, abaca, curaua, coir, cabuya, pineapple, bamboo etc. Many others can be extracted, processed, modified and used both in textiles (woven, knitting, nonwoven, technical, and 3D textiles), and for technical purposes, for example, as reinforcements to create more environmentally friendly composites and also in areas such as pulp and paper (from whole plants, shives and fibres) and agro-fine chemicals, including biocosmetics, as well as sources of biopolymers and energy (by combustion or by transforming to bioethanol or biobutanol). For more information see Vol. 2, Chapter 18. *The application of flax and hemp seed in food, nutraceutical and personal care products,* by Anna Bakowska-Barczak, Marie-Anne de Larminat, Paul P. Kolodziejczyk.

Jute, flax, hemp, kenaf and abaca are being used for production of ropes, canvas and other products are also used to reinforce moulded thermoplastic composites in automobile industry and other areas (Kozlowski and Wladyka-Przybylak, 2004).

More information in chapters of Volume 2, namely Chapter 2. *Processing of jute fibre and its applications,* Chapter 4 *Bast fibres: The use of flax and hemp for textile end applications*, Chapter 5. *Bast fibres: Use of flax fibre for high challenge technical textile applications.* Extraction, preparation and requirements for the manufacturing of composite reinforcement fabrics and for geotextiles, Chapter 6. *Bast fibres: Multi-scale mechanical characterization of flax fibres for the reinforcement of composite materials*, Chapter 14. *Natural fibres composites (NFCs) for automotive applications,* Chapter 15. *Natural fibre composites (NFCs) for construction and automotive industries*, Chapter 16. *Natural fibres for geotextiles*, Chapter 19. *Natural fibres for paper and packaging.*

The short core fibres go to insulation products, fibre boards and erosion control mats (geotextiles), while the fibrous core can be blended with lime to make strong and lightweight concrete.

Flax, hemp, kenaf, jute shives are appropriate raw materials for the production of particle boards and excellent bedding materials for animals e.g., for horse stables and chicken farms and pets as well. Also shives can be used for the production of the edible fungi Pleurotus ostreatus.

By-products from the bast fibrous plants, especially seeds and chaffs, are also a source of valuable fatty acids, amino acids, phytoestrogens, cyclolinopeptides, waxes, lignin, pectin, mucilage and other agrochemicals. They are one of the richest plant sources of polyunsaturated fatty acids (omega 3 and omega 6 family), which are effective for prevention of CHD (cardiovascular and heart diseases). More pieces of knowledge in the above mentioned area are provided in the chapter in Vol. 2, Chapter 18. *The application of flax and hemp seed in food, nutraceutical and personal care products.*

The above mentioned positive facts influence now the position of natural fibres and the stable level of their production thanks to the growing area of their application and economic value.

Protein fibres such as wool of sheep, alpaca, camel and yak and silk are used in the production of luxury textiles as well as for the production of sophisticated composites. More information is in Vol. 1, Chapter 10. Hunter L. *Mohair, cashmere and other animal hair fibres.*

In agrotechnology, several species, for example bast fibrous plants, can be used to clean soil polluted by heavy metals, due to their ability to extract cadmium (Cd), lead (Pb), copper (Cu) and zinc (Zn) (Kozłowski et al., 2004; Vandenhove, Van Hees, 2005; Mańkowski et al., 2018; Baraniecki et al., 1995; Kozłowski et al., 1993; Kozłowski and Manys, 1997). More piece of information in terms of remediation of polluted lands is in Vol. 1, Chapter 11. J. Mańkowski, J. Kołodziej, R. Kozłowski R. *The role of hemp (Cannabis sativa* L*) in remediation of degraded lands.*

Natural fibres are characterized by their most important properties such as air permeability, hygroscopicity, their ability to release moisture, the fact that they do not release any substances harmful to health or do not cause allergic reactions (resulting from a higher level of histamine in human blood derived from influence of natural fibres like linen and hemp and by their lower flammability in comparison to man-made fibres (Kozłowski et al., 2010).

Special treatment and functionalization methods such as degumming, enzymatic, ultrasound, plasma, corona, liquid ammonia, flame retardant treatments, and protection against biodeterioration provide new promising features and properties of all natural fibres and derived products (Kozlowski et al., 2017).

More information about flame retardancy in the Chapter 10, Vol. 2, *Improving the flame retardancy of natural fibres*, and about biodeterioration in Vol. 1, Chapter 18, *Prevention of fungi and bacteria growth in natural fibres* and Antimicrobial natural fibres, Chapter 21 in Vol. 2. The enzymatic treatment is described in details in chapter …, vol. … Enzymatic treatment of natural fibres.

Some of natural fibres are susceptible to dyeing by natural dyestuffs, are UV resistant and could be also easily protected against flammability (Schmidt-Przewozna, 2001, 2016).

Natural fibres and fabrics after special functionalization for example by MOF (metal–organic frameworks), POM (Polyoxymetalates) (Kozlowski et al., 2017) and by dendritic polymers will play an important role in the near future not only in defence and military apparels, but in very effective filter materials as well (Akbari, Kozłowski, 2018; Akbari, 2013).

More information is in Vol. 2, Chapter 13. Shirin Khaliliazar, *Chemical and physical characteristic of modified cotton and linen fabrics with amine-terminated dendritic polymer.*

The potential of genetically modified (GM) fibrous plants, called bioengineered in USA, has increased in recent years, due to obtained better yield efficiency of major natural products such as cellulosic fibres and carbohydrates. The possibility now also exists for obtaining polyhydroxyalkanoate (PHA), for example polyhydroxybutyrate – natural polyester in statu nascendi, which is not typically synthesized in a plant (Kozłowski et al., 2012; Wielgus et al., 2012a,b).

GM fibrous plants can be resistant to herbicides and environmental stress including salinity and drought (Jacoby, 2009). In case of bast fibrous plants production, the most important future direction is towards the genetic modification of content of lignin and pectin. GM plants could be used for improving the biomass of natural fibrous resources as well as for obtaining a higher oil content in plants e.g., linseed and hemp.

The potential for novel fibrous plants containing the nanofibres and cellulose modified with phosphate groups to achieve a higher thermal resistance has been the focus of many researchers' and scientists' attention.

Summing up, natural fibres with their long history in the service of mankind have finally been acknowledged by the modern world of 21st century, as a most welcome and user friendly material both in clothing and for other end uses.

Following a period of dynamic development of man-made fibres and a drastically lowered production trend for natural fibres in the last decades, a more stable coexistence between them may now be observed.

The ecological alarm over atmospheric and water pollution dictates the conditions of future coexistence. In the opinion of textile authorities, the present ecological legislation is a pale shadow of what will be needed in the future to come to grips with the ecological problems which the world is facing. Treatments of effluents, disposal of waste and packing material will require enormous capital outlays. The competitive edge in the future will be determined to a great extent by the ability to solve the ecological problems. This is what may be termed as the greatest challenge for textiles industries in this new Millennium (Bomgardner, 2017).

Currently, in 21st century the coexistence and competition between man-made and natural fibres is stabilized-especially in the area of quality, sustainability and economy of their production. Natural and renewable fibres conduct heat, dye well, resist mildew and have natural and antibacterial properties, blocking UV and are easily treated for flame retardancy. This makes them ideal for the production of comfortable healthy clothing that provide UV protection for the body, decrease oxidative stress and muscle tension, increase the level of alpha globulin, thus improving the wellbeing of users (Kozłowski et al., 2004; Kozłowski, Manyś, 1997).

1A.3 Handbook of natural fibres and its content

The *Handbook of Natural Fibres* describes the present state of art in the area of natural fibres. The authors have been chosen from a group of high-ranking, renowned specialists in the area of natural fibres. The Handbook consists of two volumes: Volume 1 is entitled: *Types, properties and factors affecting breeding and cultivation* with 20 chapters, written by 49 authors from 13 countries. Volume 2 entitled *Processing and applications* includes 21 chapters, prepared by 56 authors from 10 countries. Totally 105 authors contributed to the creation of this Handbook. The second edition of the *Handbook of Natural Fibres* (published by Elsevier) is the enlarged, corrected and improved version of the *Handbook of Natural Fibres*, first edition published in 2012 by Woodhead Publishing Inc.

Volume 1 starts with the introduction to natural textile fibres, prepared by the Editors and the survey of the world natural fibre production and employment. Volume 1 consists of two main parts: *Part I: Fundamentals: types of fibres, properties, identification and testing* and *Part II: improving natural fibre production through breeding and cultivation.* Part I provides fundamental knowledge about major plant and animal fibres, their properties, identification and testing. The fibres described are selected bast fibres (jute, ramie, kenaf, flax and fibrous hemp, abaca, bamboo, coir or coconut fibre, wool fibres, mohair, cashmere and other animal hair fibres, as well as silk fibres. Moreover, one chapter describes fibre plants of arid regions of Africa. Editors included also a chapter about such natural mineral fibre as basalt fibre, which are very important for production of composites to perform in extreme conditions. Chapter on identification of natural fibres provides knowledge about contemporary methods in that regard, allowing for unambiguous identification of diversified natural fibres. Chapter on testing of natural textile fibres describes traditional as well as contemporary testing methods.

Part II of Volume 1 is entitled as follows: *Improving natural fibre production through breeding and cultivation.* This part deals with the following subjects: evolution of cotton breeding, developments in fibrous flax breeding and cultivation; prevention of fungi growth in natural fibres, especially mildew; wild silks: entomological aspects and textile applications; new emerging natural fibres and relevant sources of information.

Volume 2: *Processing and applications* consists of two parts: *Part I: Processing techniques for natural fibres* and *Part II: Applications of natural fibres and case studies.* Part I describes current knowledge regarding methods of improving specific properties of natural fibres as well as selected treatments and processing. The following topics and areas are covered: cotton fibre, spinning and weaving; processing of jute fibre and its applications; silk production and the future of natural silk manufacture; the use of flax and hemp for textile end applications; bast fibres: use of flax fibre for high challenge technical textile applications; extraction, preparation and requirements for the manufacturing of composite reinforcement fabrics and for geotextiles; bast fibres: multi-scale mechanical characterization of flax fibres for the reinforcement of composite materials, enzymatic treatment of natural fibres; improving the properties of natural fibres by chemical treatments; electrokinetic properties of natural fibres; improving the flame retardancy of natural fibres. Environmental aspect of bast fibres cultivation is described in the chapter entitled: the role of hemp (*Cannabis Sativa* L.) in remediation of degraded lands. Also the chapter about the new hot topic is included: medical potential of cannabis: an overview. Dendritic polymers and their possibilities of application are gaining recently more attention, the handbook presents specific topic: chemical and physical characteristic of modified cotton and linen fabrics with amine-terminated dendritic polymer. The natural fibres (NFCs) for automotive applications, for construction, for geotextiles, as well as for paper and packaging are described in three separate chapters. In this book, for the first time, are discussed properties and applications of ultrafine powders produced from natural fibres. The application of flax and hemp seed in food, nutraceutical and cosmetic products – is described in the chapter, which has been changed, enlarged and amended. The chapter dealing with the environmental textiles from jute and coir has been updated. Two last chapters are a reprint of the chapters printed in the first edition, namely antimicrobial natural fibres and biomimetics and textile materials.

1A.4 Sources of further information and advice

Blackburn, R.S., ed. 2005. Biodegradable and Sustainable Fibres, Woodhead Publishing, Cambridge.

Buitrago, B., Jaramillo, F., Gómez, M 2015. Some properties of natural fibres. (Sisal, Pineapple, and Banana) in comparison to man-made technical fibers (Aramide, Glass, Carbon), Journal of Natural Fibers, 12 (4), 357–367.

Cook, J.G., ed. 1984. Handbook of Textile Fibres, Vol. 1 - Natural Fibres. Woodhead Publishing, Cambridge.

Denton, M.J., Daniels, P.N., eds. 2002. Textile Terms and Definitions. The Textile Institute, Manchester.

Diversity in Harmony: Asian Textile in the 21st Century. 1999., *Proceedings of the fifth Asian Textile Conference,* Kyoto, Japan, 30 Sept. to 2 Oct., organized by the Society of Fibre Science and Technology.

Debnath Singhal. 2016. Critical parameters affecting mechanical behaviour of natural fibre reinforced plastics. Journal of Natural Fibers, 13 (6), 640–650

Dowgielewicz, S. 1954. Roslinne Surowce Wlokiennicze. Panstwowe Wydawnictwo Naukowe, Warsaw.

Guebitz, G.M., Cavaco-Paulo, A. and Kozłowski, R., eds. 2006., Biotechnology in Textile Processing. Haworth Press, New York.

Gupta S.K., Goswam K.K., Majumdar A. 2015. Durability of handmade wool carpets: A review, Journal of Natural Fibers, 12(5), 399–418

Herzog, R.O. 1930. *Der Flachs, Abtl.1: Botanik, Kultur, Aufbereitung Bleicherei und Wirtschaft des Flachses, Technologie der Textilfasern,* Herausgegeben von Herzog R O, Band 5, Teil 1. Berlin: Ernst Schilling, Botanik und Kultur des Flachs.

Kind, W., Koenig, P., Müller, W., Schilling E., Steinbrinck C. 1930.'Der Flachs. Abteilung 1: Botanik, Kultur, Aufbereitung, Bleicherei und Wirtschaft des Flachses: mit einer Einführung in den Feinbau der Zellulosefaserns', Springer, Berlin, Germany, 427.

Kozłowski, R., Mackiewicz-Talarczyk, M., Demeš, M. 2009. 'Report: The international year of natural fibers and ESCORENA involvement', Journal of Natural Fibers, 6, 347–349.

Kozlowski R., Baraniecki P., Barriga-Bedoya J. 2005. Bast fibres (flax, hemp, jute, ramie, kenaf, abaca). Chapter 2, In Blackburn, R. S. (ed.) *Biodegradable and Sustainable Fibres.* Woodhead Publishing, Cambridge, 36–110.

Kozłowski, R.M., Muzyczek, M., Walentowska, J. 2014. Polymer Green Flame Retardants. Flame Retardancy and Protection against Biodeterioration of Natural Fibers: State-of-Art and Future Prospects, Chapter 23, In *Polymer Green Flame Retardants,* Elsevier B.V., ISBN 978-0-44,453-808-6, Amsterdam, 801–838

Handbook of Natural Fibres. (2012). Ryszard Kozłowski (Ed.), Woodhead Publishing. Series in Textiles, Cambridge, Vol. 1, 1–620:

- Chapter 1. Kozłowski R., Mackiewicz-Talarczyk M. Introduction to natural fibers, one to eight
- Chapter 2. Dochia M., Sirghie C., Kozłowski R., Rozkwitalski Z. Cotton fibres, 1–23
- Chapter 5. Kozłowski R., Mackiewicz-Talarczyk M., Allam A.M. Bast fibers - Flax., 56–113
- Chapter 16. Walentowska J., Kozłowski R. Prevention of fungal growth in natural fibres, 532–549
- Chapter 2. Kozłowski R., Muzyczek M. Improving the flame retardancy of natural fibres, 30–59
- Chapter 5. Konczewicz W., Kozłowski R. Enzymatic treatment, 168–180
- Chapter 9. Leao A., Cherian B.M., Souza S.F., Kozłowski R., Thomas S., Natural fibres for geotextiles, 280–310

Kozłowski R.M., Muzyczek M., Kręgielczak A. Natural Fibers. 2012. Encyclopaedia of Composites. John Willey & Sons, New York, 1–22

Kozlowski R., Muzyczek M. 2016. Chapter 11, Smart environmentally friendly composite coatings for wood protection, in smart composite coatings and membranes, transport, structural, environmental and energy applications. (Ed.) Montemor,M.F., A Volume in Woodhead Publishing Series in Composites Science and Engineering, ISBN: 978-1-78,242-283-9, 293–325

Natural Fibers: Properties, Mechanical Behaviour, Functionalization and Applications, eds. Ryszard M. Kozlowski, Muzyczek M. 2017. Nova Publishers, USA, ISBN: 978-1-53,612-071-4 (available online), Pub. Date: 2017 -third Quarter.

Chapters:

1. Kozlowski R.M., Muzyczek M.: Introduction.
2. Kozlowski R.M., Muzyczek M., Chapter 2. A New Challenge For Flax and Hemp Bast Fibers, 19–44.
3. Kozlowski R.M., Cesarino, I., Leao A.L., Machado I.S. Chapter 4, Curaua Fibers: the Queen of the Fibers, 83–106.
4. Kozlowski R.M., Mackiewicz-Talarczyk M., Barriga Bedoya J., Chapter 6. New Emerging Sources of other Lignocellulosic Fibers, 124–134.
5. Kozlowski R.M., Gavrilas S. Chapter 7: Natural Lignocellulosic Raw Materials as a Source of Nanocellulosic Fibres, 135–148.
6. Kozlowski R.M., Muzyczek M., Chapter 11. Trends in Natural Lignocellulosic Fibre Functionalization and Finishing Processes, 215–236.
7. Jorge Barriga Bedoya, Maria Mackiewicz-Talarczyk, Source of further information about natural fibers, 349–354.

Kozlowski R., Muzyczek M., Mackiewicz-Talarczyk M., Wu, Wei. 2017. Bast Plants – Flax and Hemp Yesterday, Today and Tomorrow, third International Conference on Natural Fibres ICNF 2017, 21–23.06.2017, Braga, Portugal, 9–10, ISBN: 978-972-8600-29-7, Book of Abstracts Ed. R. Fangueiro Title: *Proceedings of the third International Conference on Natural Fibers -Advanced Materials for a Greener World.*

Marshall, G. 1988. *Flax: Breeding and Utilisation (Advances in Agricultural Biotechnology).* Kluwer Academic Publisher, Dordrecht, Boston, London.

Mauersberger, H.R., ed. 1954. *Matthews' Textile Fibers: Their Physical, Microscopic, and Chemical Properties.* New York, John Wiley; London, Chapman & Hall.

Pfäfficon, A.G., ed. 2010. *The Fibre Year 2009/10 (2010), A World Survey on Textile and Nonwovens Industry.* Switzerland: Oerlikon.

Proceedings of the International Conference on Advanced Fibre Materials. 1999. Organized by the Society of Fibre Science and Technology, Ueda, Japan, October 3–5.

Sapuan, S & Jawaid, Mohammad. 2016. Integrated mechanical-economic–environmental quality of performance for natural fibers for polymeric-based composite materials, Journal of Natural Fibers, 13 (6), 651–659

Sharma, H.S.S. and Van Sumere, C.F., eds. 1992. The Biology and Processing of Flax. M Publications, Belfast, Northern Ireland.

Sinha F. 1989. Economic Dev. Associates, New Delhi, for Intermediate Technology Development Group, Rugby, England, India, Allo Nettle processing in Nepal, A preliminary study of its potential for development.

Wallenberger, F.T. and Weston, N., eds. 2004. Natural Fibers, Plastics and Composites. Kluwer Academic Publishers, Dordrecht.

Zaikov G., Rakhimkulov A., Lomakin S., Dubnikova I., Shchegolikhin A., Davidoff E., Kozlowski R. 2014. Chapter 2: Thermal properties of the polypropylene/MWCN composites. Apple Academic Press, Taylor & Francis Group.

References

Akbari, S., 2013. The application of dendritic material in textile engineering. Sc. Bull. Escorena 7 (11–26), 2069-8070.

Akbari, S., Kozłowski, R.M., 2018. A review of application of amine-terminated dendritic materials in textile engineering. J. Text. Inst. 1–8. https://doi.org/10.1080/00405000.2018.1512361.

Baraniecki, P., Grabowska, L., Mańkowski, J., 1995. Len i konopie na terenach zdegradowanych przez przemysł miedziowy. Flax and hemp in areas made derelict by the copper industry. Natural Fibres. XXXIX, 79–86, ISSN 1230-4476.

Bomgardner, M.M., 2017. The great lint migration. C&EN 95 (2), 16–17. January 9.

Franck, R.R. (Ed.), 2001. Silk, Mohair and Other Luxury Fibres. Woodhead Publishing, Cambridge.

Franck, R.R. (Ed.), 2005. Bast and Other Plant Fibres. Woodhead Publishing, Cambridge.

Gherzi, L., 2017. Based on the Fiber Year 2017 (Figures 2016) and Various Sources.

Gordon, S. (Ed.), 2006. Cotton Science and Technology. Woodhead Textiles Series 59. Woodhead Publishing, Cambridge.

Hillers, D.R., 1974. Discovered in the WADI ED DALIVEH. In: Hillers, D.R. (Ed.), The Annual of the American Schools of Oriental Research, vol. 41. American Schools of Oriental Research, Cambridge, Massachusetts.

IYNF, 2009. UN International Year of Natural Fibres. http://www.naturalfibres2009.org/en/.

Jacoby, M., 2009. Getting by on little water. Advances in breeding and basic science confer drought tolerance to crops. C&EN Chicago 87 (39), 60–62. September 28, 2009.

Advances in Wool Technology. In: Johnson, N.A.G., Russell, I. (Eds.), 2008. Woodhead Textiles Series, 72. Woodhead Publishing, Cambridge.

Kozlowski, R., Grabowska, L., Baraniecki, P., Mscisz, J., 1993. Recultivation by flax and hemp culture of soil polluted by heavy metals. In: Proceedings of the Third European Workshop of Flax, 15–17 June, 1993, Bonn, pp. 159–164.

Kozlowski, R.M., Muzyczek, M., 2017. Chapter 11. Trends in natural lignocellulosic fiber functionalization and finishing processes. In: Ryszard, M., Kozlowski, M.M. (Eds.), Natural Fibers: Properties, Mechanical Behavior, Functionalization and Applications. Nova Publishers, USA, pp. 215–236, 978-1-53612-071-4 (available online), Pub. Date: 2017 -3rd Quarter.

Kozlowski, R., Wladyka-Przybylak, M., 2004. Uses of natural fiber reinforced plastics. In: Wallenberger, F.T., Weston, N. (Eds.), Natural Fibers, Plastics and Composites. Kluwer Academic Publishers, Dordrecht, pp. 249–274.

Kozlowski, Kajzar, F., Nyszko, G., Wiertejuk, K., Majewski, K., Mackiewicz-Talarczyk, M., Małkowski, Z., 2017. Special textiles with multifunctional systems of human protection against contamination, fire, cold & UV radiation. Mol. Cryst. Liq. Cryst. 655 (1), 195–203. https://doi.org/10.1080/15421406.2017.1361297. Published online: 07 Dec 2017.

Kozłowski, R., Manys, S., 1997. The coexistence and the competition of natural and man-made fibres. In: Proceedings of the 78th World Conference of the Textile Institute in Association with the 5th Textile Symposium of SEVE and SEPVE, Thessalonica, Greece, vol. 3, pp. 3–52.

Kozłowski, R., Kozłowska, J., Rawluk, M., Barriga, J., 2004. Potential of lignocellulosic fibrous raw materials, their properties and diversified applications. Nonlinear Optics, Quantum Optics 31 (1–4), 61–89.

Kozłowski, R.M., Mackiewicz-Talarczyk, M., Barriga-Bedoya, J., 2010. Natural fibers production, processing, and application: inventory and future prospects. Contem. Sci. Poly. Mat. Am. Chem. Soc. Symp. Ser. 1061 (3), 41–51.

Kozłowski, R.M., Mackiewicz-Talarczyk, M., Muzyczek, M., Barriga-Bedoya, J., 2012. Future of natural fibers, their coexistence and competition with man-made fibers in 21st century. Mol. Cryst. Liq. Cryst. 556 (1), 200–222. Taylor & Francis, MS ID: 635962.

Luong, T., 2017. Microplastic pollutants highlights the dangers posed to the environment and the food chain. In: Chemical&Engineering News, Posted on: February 21, Based on Christopher Blair Crawford, Brian Quinn. 2016. Microplastic Pollutants, first ed. Elsevier Science.

Mańkowski, J., Kołodziej, J., Kubacki, A., Baraniecki, P., Pniewska, I., Pudełko, K., 2018. Wykorzystanie konopi włóknistych do rekultywacji i zwiększenia żyzności gleb zdegradowanych – projekt LIFE11 ENV/PL/445; Remediacja, rekultywacja i rewitalizacja. red. prof. G. Malina; Poznań, pp. 67–84. ISBN: 978-83-64959-05-9.

Nussbaumer, L., 2009. The role of textiles in indoor environmental pollution: problems and solutions. In: Rowe, T. (Ed.), Interior Textiles Design and Development. Woodhead Publishing Ltd, UK, pp. 180–210.

Oerlikon, Analyst & Media Briefing OMF.

Ryder, M.L., 1965. 'Report of textiles from catal Hüyük' *anatolian studies*. J. Br. Inst. Archaeol. 15, 175–176. Ankara.

Schmidt- Przewoźna, K., 2016. YOCOCU 2014. Professional's Experiences in Conservation of Cultural Heritage in America, Europe, Asia, A Short History of Natural Dye Materials Used in Dyeing and Printing of Polish Folk Fabrics. Scholars Publishing, Cambridge, pp. 106–120. ISBN (13): 978-1-4438-8519-5.

Schmidt-Przewozna, K., 2001. Application classes of natural dyes and their usage on animal and vegetable fibres. In: Proceedings of the 2nd Global Workshop of the FAO European Cooperative Research Network on Flax and Other Bast Plants, Bast Plants in the New Millennium, Borovets, Bulgaria.

Wool Science and Technology. In: Simpson, W.S., Crawshaw, G. (Eds.), 2002. Woodhead Textiles Series, 25. Woodhead Publishing, Cambridge.

Taylor, J.H., 1995. Unwrapping a Mummy: The Life, Death and Embalming of Horemkenesi. British Museum Press, London.

Vandenhove, H., Van Hees, M., 2005. Fibre crops as alternative land use for radioactively contaminated arable land. J. Environ. Radioact. 81 (2–3), 131–141.

Wielgus, K., Szalata, M., Słomski, M., 2012a. Genetic engineering and biotechnology of natural textile fibre plants, Chapter 17. In: Kozłowski, R. (Ed.), Handbook of Natural Fibres, first ed., vol. 1. Woodhead Publishing Ltd, Cambridge, ISBN 978-1-84569-697-9, pp. 550–575.

Wielgus, K., Grajek, K., Szalata, M., 2012b. Bioengineered natural textile fibres. Chapter 10. In: Kozłowski, R. (Ed.), Handbook of Natural Fibres, first ed., vol. 1. Woodhead Publishing Ltd, Cambridge, pp. 291–313. 978-1-84569-697-9.

Further reading

Kozlowski, R., Kicińska-Jakubowska, A., Muzyczek, M., 2009. 'Natural fibres for interior textiles'. In: Rowe, T. (Ed.), Interior Textiles Design and Development. Woodhead Publishing Ltd, UK, pp. 1–35.

World natural fibre production and employment

1B

Terry Townsend
Chair, Discover Natural Fibres Initiative, Consultant Administration, Cotton Analytics, Houston, TX, United States

1B.1 Overview

World fibre production equalled approximately 110 million metric tons in 2018 (Table 1B.1), including 32 million tons of natural fibres (Chart 1B.1) and 79 million tons of chemical fibres, both cellulosic (fibres made from the cellulose contained in the pulp of wood, bamboo and other sources) and non-cellulosic (fibres made from petroleum molecules) (Chart 1B.2). The man-made fibre complex accounted for 70% of total fibre production in 2018 (Townsend, Terry, Chair of DNFI, January 2019, Report to the Annual Meeting of the Discover Natural Fibre Initiative, Frankfurt, Germany).

Cotton accounted for 80% of natural fibre production by weight, while polyester, particularly polyester filament, dominated the chemical fibre universe. Synthetic filament alone was nearly half of all fibre production in 2018.

Jute production totalled about three million tons in 2018, while wool and coir each accounted for about one million tons. Specialized natural fibres (abaca, agave, flax, hemp, kapok, ramie, silk, and sisal and the animal fibres other than wool) added another 1.5 million tons.

Since the 1960s when world fibre use was approximately one-eighth of the current level, almost all the growth in consumption has occurred with polyester and cotton, especially polyester (Chart 1B.3). While there have been year-to-year fluctuations in demand for all fibres, consumption of wool, other natural fibres such as silk and flax, and non-cellulosic fibres such as nylon have either been relatively flat or declining.

Prior to the 1900s, all fibres were natural, and in the 1960s, natural fibres still represented 80% of world fibre use. By the 1980s, the natural fibre share had fallen to about half, and while the downward trend abated during the 1980s because of below-average prices for cotton and other natural fibres, the downward slide in market share continued in the 1990 and 2000s (Chart 1B.4).

There has been rapid growth in cellulosic fibres (called rayon or viscose) over the last 2 decades, from two million tons in the early 2000s to seven million tons in 2018. Some of the growth in cellulosic staple fibre production has been offset by a decline in cellulosic filament production.

Handbook of Natural Fibres. https://doi.org/10.1016/B978-0-12-818398-4.00002-5

Table 1B.1 World production, value and employment of natural fibres (2018).

	Production Estimated	Households Millions	Employment Millions	Value of production $Billion
Natural fibres	Metric tons			
Vegetable origin				
Abaca	83,000	0.1	0.5	$0.1
Bast fibres, other	1,90,000	0.25	1	$0.3
Coir	9,70,000	1.28	6	$1.3
Cotton lint	2,61,20,000	45	170	$45.0
Fibre crops not specified elsewhere	2,80,000	0.37	2	$0.4
Flax fibre and tow, ex scutching mill	3,10,000	0.01	0.01	$0.0
Hemp fibre and tow	70,000	0.00	0.002	$0.0
Jute, kenaf and allied fibres	25,00,000	6	33	$6.0
Kapok fibre	96,000	0.13	0.6	$0.1
Ramie	1,00,000	0.13	1	$0.1
Sisal, henequen and similar hard fibres	2,10,000	0.04	0.2	$0.0
Animal origin				
Silk, raw	1,64,000	0.3	0.8	$0.3
Wool, clean	10,80,000	5	10	$5.0
Other, greasy weight[a]	31,000	0.1	0.3	$0.1
Total natural fibres	3,22,00,000	59	226	$59

[a](Includes: Alpaca, Angora Rabbit, Camel hair, Cashmere, Guanaco, Llama, Mohair, Vicuna, Yak hair).
Sources: Estimates for 2016 were developed from data provided by http://faostat3.fao.org/, FAO statistics on hard fibres 2016, ICAC, and IWTO; 2008–14: Bast fibres, other, fiber crops not specified elsewhere, hemp tow waste, kapok and ramie: http://faostat3.fao.org/; 2008–15: Abaca, coir, jute, kenaf and allied fibres, and sisal, henequen and similar fibres: FAO Statistics December 2014, 2015, and 2016; Cotton: International Cotton Advisory Committee; Flax: IWTO Market Information 2015; Silk: http://faostat3.fao.org/ and IWTO Market Information 2018; Wool: International Wool Textile Organization; All other Natural Fibres, Animal origin: Various sources as reported by the International Wool Textile Organization; Man-made Fibres: CIRFS, published in IWTO Market Report 2018.

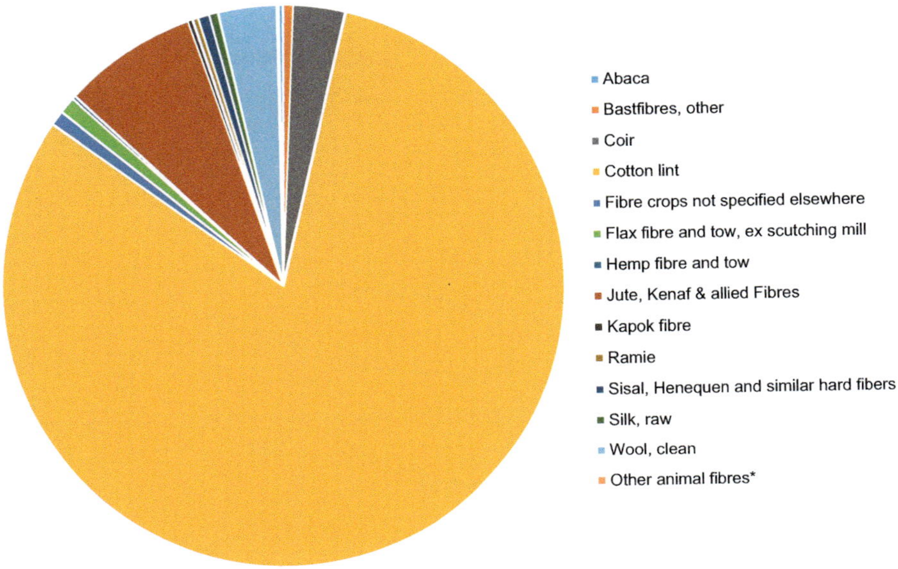

Chart 1B.1 World natural fibre production, 32 million tons in 2018.

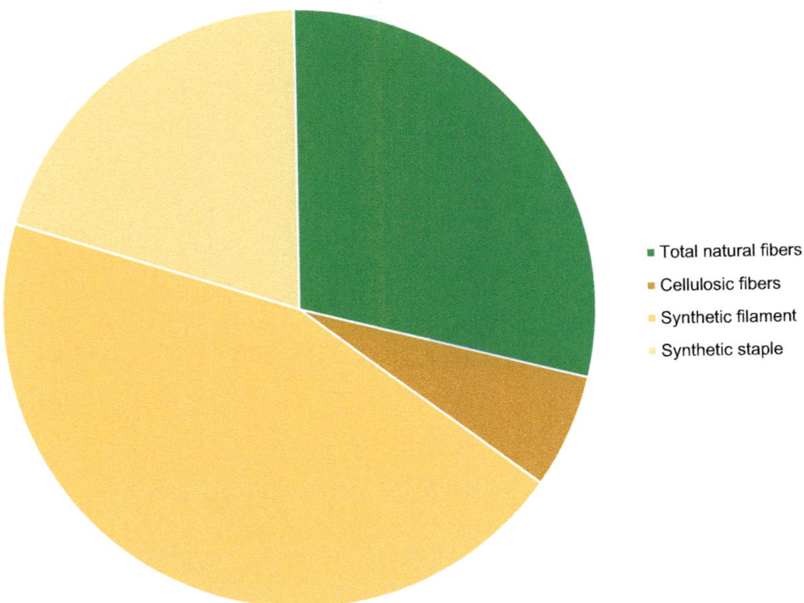

Chart 1B.2 World fibre production, 110 million tons in 2018.

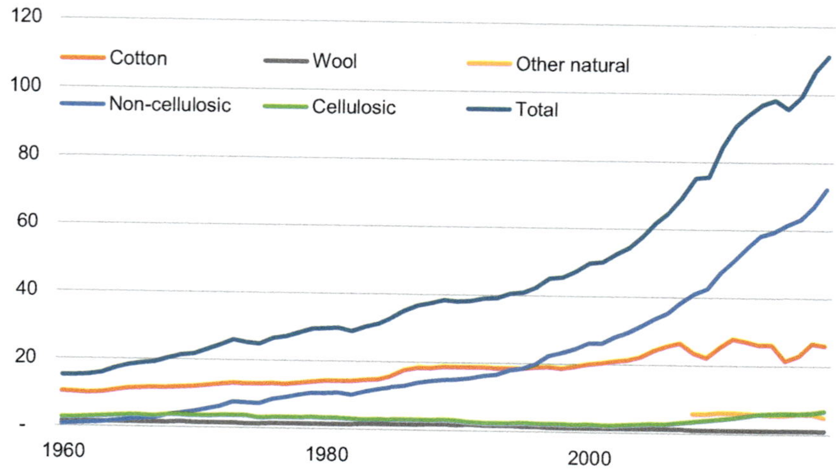

Chart 1B.3 World fibre production, million tons.

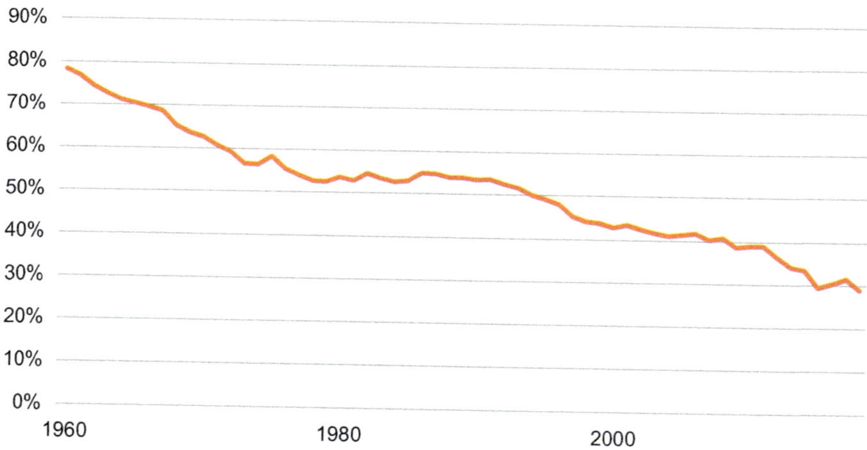

Chart 1B.4 Natural fibres share of world fibre production.

1B.2 Natural fibres connect people to markets

The United Nations General Assembly declared 2009 to be the International Year of Natural Fibres because of the benefits of natural fibre industries to the world economy, environment and consumers.

Even though the share of natural fibres in world fibre production declined during the 2000s, production of natural fibres as a whole has stayed near 30 million tons, and their contributions to the world economy remained very significant.

Natural fibres are grown commercially in around 100 countries, and the area used to grow or raise natural fibres accounts for about 3% of the world's arable land. The value of world natural fibre production is about US$60 billion (Chart 1B.5). Natural fibres serve a unique role in the world economy because they connect people to markets.

Natural fibres can be stored for years without loss of value. They do not require refrigeration, and when baled and stored properly, they do not degrade in quality and are not vulnerable to mould, bacteria, insects or other vermin. A bale of natural fibres packed to international standards is sufficiently dense that oxygen is squeezed from the interior, preventing the establishment of fire or the survival of diseases or insects. A bale will not absorb water, and if the bale ties are not broken will float indefinitely.

Natural fibres can be transported over rough roads for thousands of kilometres without damage. If a bale falls out of the back of a truck or off the side of a rail car, you just pick it up, brush off the dust, put it back and continue on.

Natural fibres have high ratios of value to weight and density, so they can be economically shipped from interior locations. As an example, for a bale of cotton shipped from average location in the United States to an average destination in East Asia, freight represents one-sixth of the landed value. For wheat, corn and soybeans, the analogous calculation is half.

Further, natural fibres contribute to food security. Grain crops are in the grass family and have horizontal roots, good for stabilizing soil but leaving crops vulnerable to water and heat stress. Cotton is a woody perennial with vertical roots that typically descend 1.5 m, allowing it to provide an economic yield in semi-arid and arid regions where food crops would fail. Likewise, other natural fibres such as jute, coir, sisal, and

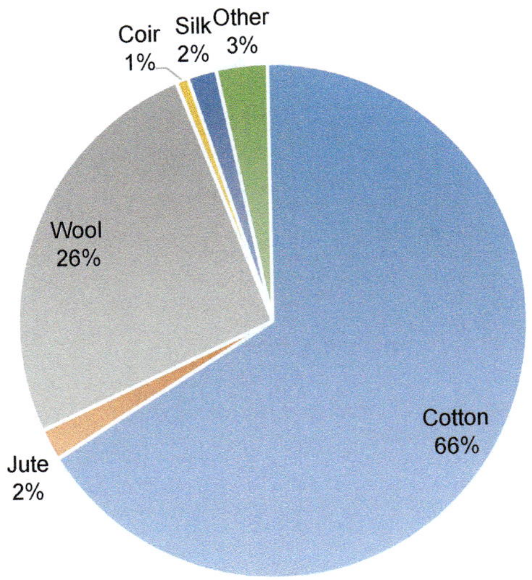

Chart 1B.5 Value of natural fibre production, $60 billion.

hemp, and the animal fibres such as wool and silk are adapted to the regions in which they are grown and provide economic activities that would otherwise be unavailable.

Vegetable fibres are grown in rotation with food crops, and the animal fibres are raised in regions that are not suitable for intensive agricultural activities. As cash crops, natural fibres often serve as collateral against input loans to farmers in developing countries, allowing farmers' access to fertilizer, insecticides and seeds for food crops that would otherwise be unavailable. Consequently, production of food rises in areas where natural fibres are grown or raised.

Natural fibres also provide cash incomes that enable farmers to purchase food, as well as paying for school fees, clothes and health care. The governments of some developing countries, particularly those in West Africa, provide inputs to grow natural fibres to low-income families as a form of social support as an alternative to direct cash assistance.

These are the reasons why natural fibres are grown in landlocked and/or water-scarce regions like Mali and Burkina Faso, Zambia and Zimbabwe, Uzbekistan and Xinjiang, Northern Texas and Central Maharashtra, the interior of Australia or arid regions of South Africa and Russia, thousands of kilometres from ports and weeks from shipping destinations.

Natural fibres are grown on the frontiers of global trade, and in many regions, natural fibres are the only viable economic activity available, providing incomes to millions. Around 60 million family units are engaged directly in natural fibre production (Chart 1B.6). When family labour, hired on-farm labour and workers in ancillary services such as transportation, ginning, baling and storage are considered, total involvement in the natural fibre production sectors is estimated at around 200 million people, or 4% of the world's population.

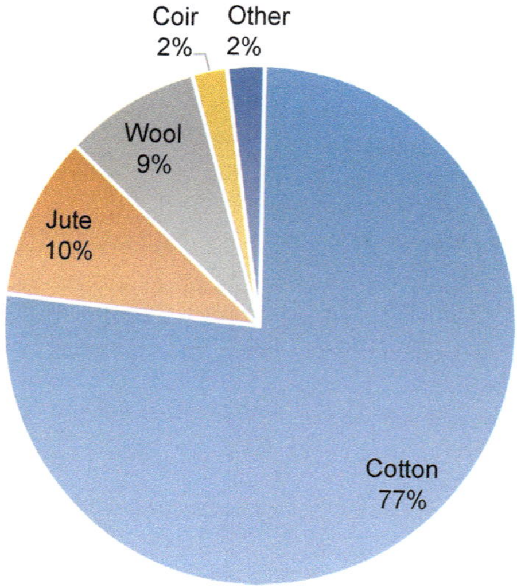

Chart 1B.6 Households involved in natural fibre production, 60 million.

1B.2.1 Cotton: the wearable fibre

The world cotton industry experienced dramatic changes over the last 7 decades. Production nearly quadrupled, rising from seven million tons in 1950–1951 (August to July) to a record of 27 million tons in 2004–2005, and production was still near that level in 2018–2019 (Chart 1B.7). (Ruiz, Lorena, Economist, December 2018, World Textile Demand, Washington, DC, USA, International Cotton Advisory Committee (ICAC)).

The average annual rate of growth in world production over the last 7 decades was 2.5%, or about 290,000 tons. Cotton is the largest of the natural fibres because of its relative ease of growth and wide range of applications in apparel and home furnishings. About 80% of all cotton is used in apparel, with denim being the single biggest end-use category accounting for nearly one-fifth of all cotton use. About 15% of world cotton use is in home furnishings, especially sheets and towels, and the remaining 5% goes into a variety of mostly non-woven applications such as filters and padding.

Growth in cotton production was steady during the 1950 and 1960s but slowed during the 1970s because of slower world economic growth and limited gains in cotton yields. World cotton production exploded from 14 million tons in the early 1980s to 19 million tons in 1984–1985, as political change in China, higher prices, and the widespread use of better seed varieties and better methods of plant protection against pests led to increased yields. World production climbed to a record of nearly 21 million tons in 1991–1992 but levelled off during the 1990s. With the commercial use of biotech cotton varieties beginning in 1996 and the expansion of cotton areas in Francophone Africa, Australia, central Brazil, northwestern China, and Turkey, world production reached 27 million tons in 2004–2005. Production declined during the 2008–2009 world recession to 22 million tons in 2009–2010, still well above the pre-2004–2005 level.

World area dedicated to cotton has fluctuated since the 1950s between 28 million hectares and 36 million hectares, with an average of 33 million hectares (Chart 1B.8).

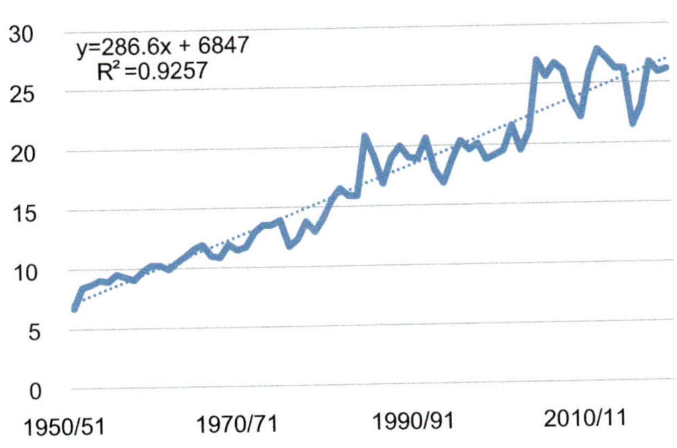

Chart 1B.7 World cotton production, mil tons.

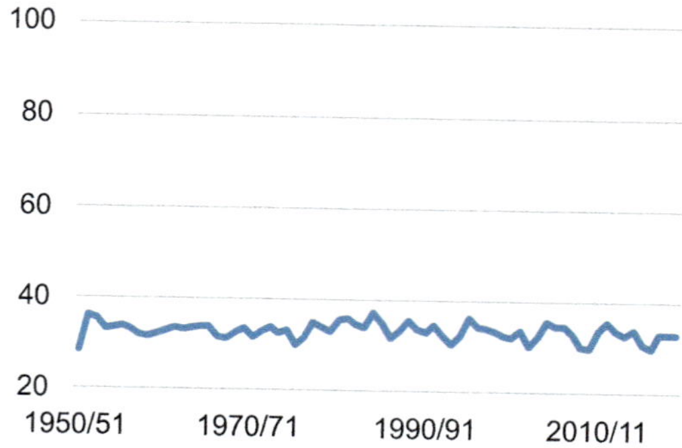

Chart 1B.8 World cotton area, million ha.

While there have been dramatic reductions in cotton area in some regions since the 1950s, particularly in the United States, Central Asia, northern Brazil and North Africa, there have been offsetting increases in Francophone Africa, China, India and Pakistan. With total area showing no tendency to rise, all growth in world cotton production since the 1950s came from improved yields linked to improved technologies.

The world yield has trended higher since the 1950s, with periods of slow growth alternating with periods of rapid growth (Chart 1B.9). The world cotton yield in the early 1950s was 230 kg of lint per hectare. Yields rose steadily at an average rate of more than 2% per year during the 1950 and 1960s, and then more slowly from the mid-1970s until the mid-1980s. During the 1980s the world cotton yield rose dramatically and reached a record of nearly 600 kg per hectare in 1991–1992. However, yields stagnated during the 1990s due to problems associated with diseases, resistance to insecticides, and the disruption of production in the former Soviet Union. Yields began rising again in the late 1990s with improvements in seed varieties and the use of biotech traits, and the world yield in 2007–2008 reached a record 793 kg per hectare. However, the pace of technology development and adoption slowed after 2007–2008, and the world yield in 2018–2019 was still near 800 kg per hectare.

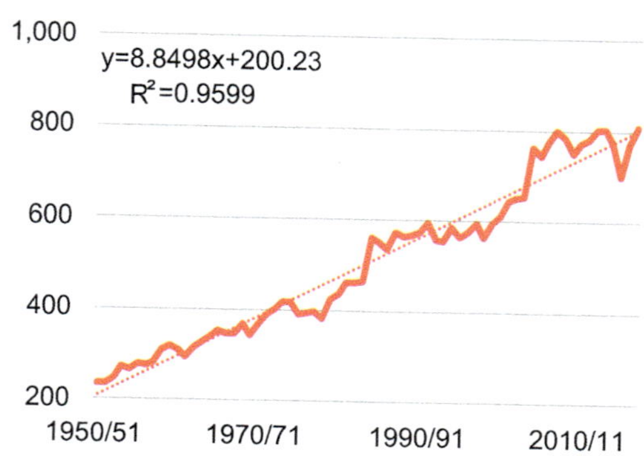

Chart 1B.9 World cotton yield, Kgs/Ha.

The average rate of increase between 1950–1951 and 2018–2019 was 9 kg per hectare per year.

Cotton is produced commercially in about 80 countries. The largest cotton producing countries are India and China, each producing about one-fourth of the world total. The United States and Brazil together account for about one-third of production, and all other countries together produce about one-fifth of the world total (Chart 1B.10).

Between 35 and 60 million family units are engaged directly in cotton production during some portions of each season. When family labour, hired on-farm labour and workers in ancillary services such as transportation, ginning, baling and storage are considered, total involvement in the cotton production sector is estimated at between 150 million and 190 million people, or 2%–3% of the world's population.

World average gross revenue per household from cotton production is about $1000 a year. However, the range around that average is enormous, with a small holder in a developing country growing cotton on one-half a hectare receiving about $400 in gross revenue. In contrast, a large mechanized operation in Australia, Brazil or the United States might gross more than $2 million, and really large operations might gross more than $50 million a year. The mode, or most common experience, would be a small holder receiving between $500 and $1000 a year in gross revenue from cotton production, and net revenue would be between $150 and $300 per year. Such a farmer would also earn additional non-cash income from food production.

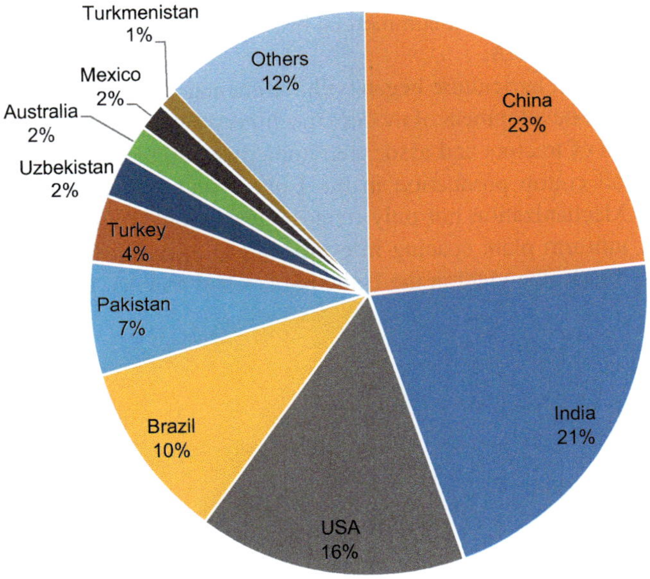

Chart 1B.10 World cotton production, 2018–2019.

1B.2.2 Technology drives yield growth

The role of agricultural science in improvements in natural fibre productivity is apparent in data on cotton yields from the United States. Cotton has been grown for thousands of years, but commercial production began in the 1790s with the invention of the cotton gin to separate lint from seeds, and the United States has been involved in commercial cotton production longer than any other country. Yield data in the United States are available from the 1860s, when all of agriculture was entirely organic (Chart 1B.11) (United States Department of Agriculture, Economic Research Service, various publications. Contact: Mr Leslie Meyer, Economist).

From the 1860s to the 1920s, there were no gains in U.S. cotton yields, which averaged 200 kg per hectare. Small year-to-year variances occurred with weather and pest pressure, but there were no significant structural changes in productivity.

However, since the 1920s, the yield in the United States has risen five-folds to one ton per hectare, mirroring changes in agricultural productivity in all crops. The major technologies that have driven these achievements include directed breeding, mechanization, synthetic fertilizers, plant protection chemicals, and most recently the tools of genetic engineering.

Gregor Mendel was a scientist and an Augustinian Friar who served as the abbot of St. Thomas' Abbey in the city of Brno in present day Czech Republic. He experimented with pea plants in a garden between 1856 and 1863 and established many of the rules of heredity that underlie the modern science of genetics. His work was not understood for many decades, but after World War I, scientists began applying the principles he developed in commercial breeding programs, and agricultural yields began to rise in the 1930s.

Mechanization of agriculture began with rudimentary animal-powered machines in the 1800s, but the development of modern machinery during World War II led to widespread adoption of tractors and associated implements in the 1950 and 1960s. By the early 1970s, all cotton production in the United States, including harvesting, was mechanized. Mechanization not only reduces labour, but it also enhances yields by allowing for uniform plant spacing in even rows to ensure proper nutrient management. By the 1950s, the yield in the United States had doubled to 400 kg per hectare.

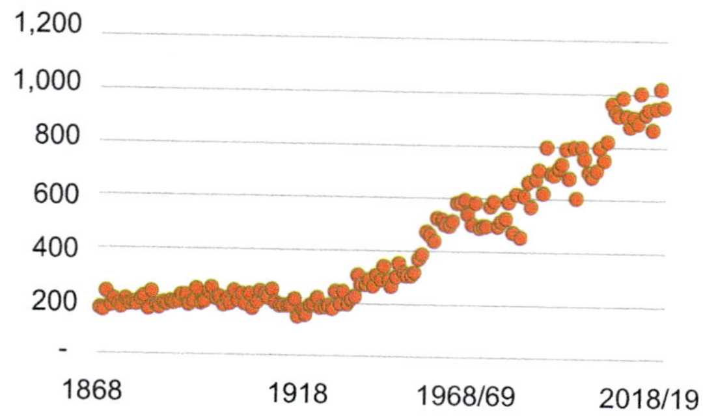

Chart 1B.11 US cotton yields, Kgs/Ha.

In the 1960 and 1970s, scientists and engineers developed synthetic nitrogen for fertilizer and plant protection chemicals to control weeds, disease and insects, and by the 1980s, the United States yield rose to between 600 and 800 kg per hectare.

1B.2.3 Biotechnology: the newest driver of change

And in the 1990s, world agriculture was transformed with the commercial release of varieties containing herbicide tolerant (HT) and insect-resistant (IR) traits conferred by the tools of biotechnology, popularly called GM or GMOs, for genetically modified, and also referred to as biotech varieties. As of 2018, GM cotton was approved for commercial cultivation in 18 countries: Argentina, Australia, Brazil, China, Colombia, Costa Rica, Ethiopia, India, Japan, Mexico, Myanmar, Nigeria, Pakistan, Paraguay, South Africa, Sudan, Swaziland, and USA. Burkina Faso, which approved the use of GM cotton in 2009, discontinued the use of such varieties in 2015. In 2018, GM cotton accounted for about 80% of world cotton area.

Biotech cotton was introduced in 1996 and as of 2019 accounts for about 13% of the world's overall GM crop production (ISAAA, 2018). Biotech cotton is grown from seeds expressing traits conferred by genes that were derived from other organisms. These extraneous genes, known as transgenes, confer economically important traits, primarily insect resistance and herbicide tolerance.

Insect-resistant GM cotton is commonly referred to as *Bt*, short for *Bacillus thuringiensis*, a common soil bacterium. One or more genes from *Bt* are inserted into the cotton genome, producing increasing levels of one or more proteins that cannot by digested by a category of insects called bollworms, what most people would call caterpillars. The expression of the protein increases as the plant matures and becomes increasingly toxic to the bollworm. *Bt* cotton serves as a substitute for insecticides in controlling chewing pests.

Weeding is an extremely labour-intensive process, requiring 30 to 40 man-days per hectare per season if done manually. HT cotton is seen as an essential component of mechanized cotton farming systems in countries where labour is expensive. HT cotton contains genes that protect the crop from the toxic effects of specific broad-spectrum herbicides. When HT cotton is planted, herbicides can be sprayed in the field, resulting in weed control without harming the crop. In addition to weed control, the HT technology is also used to kill cover crops in conservation tillage systems.

1B.2.4 Identity cottons

World cotton production is increasingly disaggregated by the marketing program or initiative under which it is produced. There are many programs to collect data, encourage improvements or assure consumers of responsible production practices; some of the programs are organized by producers in a particular country, some are sponsored by input suppliers, and some are multinational initiatives facilitated by the private sector and governments. Because cotton is increasingly identified in marketing channels by the program under which it was produced, these are called 'identity cottons'.

Production under the four major identity cottons [Certified Organic, Fairtrade, Cotton made in Africa (CmiA) and Better Cotton Initiative (BCI)] is estimated at 5.2 million tons in 2017–2018, up from 3.4 million tons in 2016–2017 and 2.6 million in 2015–2016. The identity cottons as a group accounted for 19% of world cotton production in 2017–2018, with BCI and CmiA together accounting for almost all (Chart 1B.12).

A problem common to all the identity cottons is that less than half is actually sold as an identity product with consumers, retailers and brands paying various fees or price premiums. The majority of cotton produced under the four major identity programs is being sold as regular cotton under commercial terms.

1B.2.5 Organic

Production of certified organic cotton was 117,525 tons on 473,000 ha in 2016–2017, indicating that the average organic yield was 248 kg of lint per hectare. (Textile Exchange, Organic Cotton Market Report, 2018) There were 220,000 farmers involved in organic cotton production in 2016–2017, indicating that production per farmer was about 500 kg of lint on a little more than 2 ha (Chart 1B.13).

India accounted for 50% of organic cotton production in 2016–2017, China 22% and Kyrgyzstan, Tajikistan and Turkey each accounted for about 7% of world production. Another 20 countries accounted for the remaining 11% of world organic cotton production.

1B.2.6 Fairtrade

'Fairtrade cotton' was launched in 2005. Producer organizations are audited against Fairtrade Standards, which provide a framework for a sustainable approach to

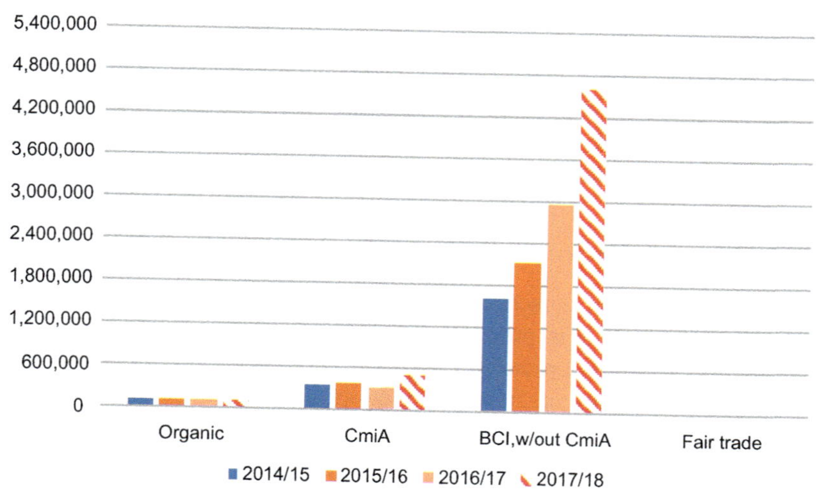

Chart 1B.12 Identity cotton production, tons.

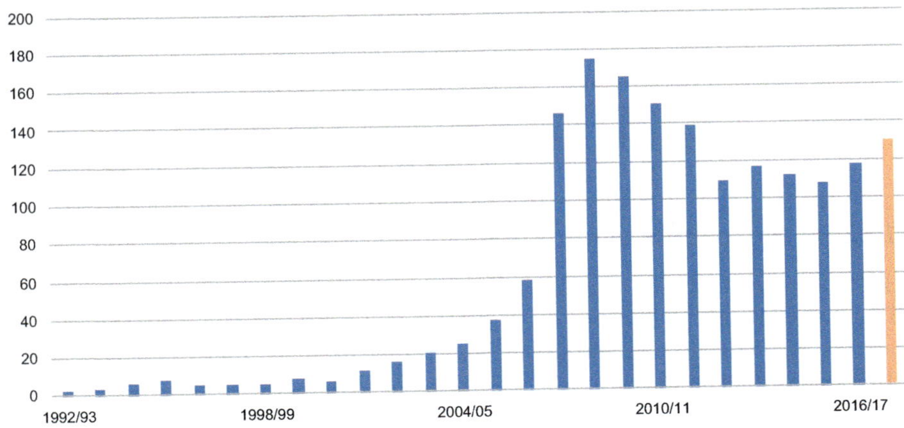

Chart 1B.13 World certified organic cotton production, 000 tons.

production that can have long-term economic, social and environmental benefits for farmers and their communities. The Standards include a guaranteed Fairtrade Minimum Price for seed cotton and the additional Fairtrade Premium for farmers to invest in strengthening their organizations, developing their businesses and improving the infrastructure of their communities (Fairtrade U.K., 2017).

Key Fairtrade products are tea, coffee, cocoa, flowers and bananas. Cotton is classified as a 'smaller, newer category'. There are about 55,000 Fairtrade-certified cotton farmers worldwide, mainly in India and West Africa. 65% of them grow organic cotton. However, producers sold an average of slightly more than 40% of their production volume on Fairtrade terms – meaning they received the Fairtrade Minimum Price and Premium. The lack of market uptake continues to challenge Fairtrade cotton farmers.

About 7000 tons of lint are produced on 61,000 ha in seven countries under Fairtrade (http://www.fairtrade.org.uk/). The average yield among Fairtrade cotton farmers was less than 200 kg of lint per hectare.

1B.2.7 Cotton made in Africa

Cotton made in Africa is an initiative of the Aid by Trade Foundation (AbTF) that helps African smallholder cotton farmers to improve their living condition. (Aid by Trade Foundation, Cotton made in Africa (CmiA), various press releases.)

AbTF works to improve living conditions and promote environmental protection. Farmers are expected to benefit from participation in CmiA through training in improved agricultural practices, not from price premiums.

Production under CmiA reached nearly 500,000 tons in 2017–2018 on 1.6 million hectares. CmiA reports that one million farmers participated in 2017–2018, meaning that production per farmer was about 500 kg of lint, with each farm household managing about 1.5 ha of cotton.

The average yield among farmers participating in CmiA was a little over 300 kg of lint per hectare, compared with almost 400 kg per hectare for other farmers in Sub-Saharan Africa. CmiA accounted for 30% of all cotton grown in Sub-Saharan Africa in 2017–2018 (http://www.cottonmadeinafrica.org/en/english-docs/annual-reports/160-cmia-annual-report/file).

1B.2.8 Better cotton initiative

BCI was created in 2005 from discussions led by the World Wildlife Fund (WWF). The goal of BCI is to encourage sustainable cotton production that benefits farmers and the environment and ensures the future of the sector (Better Cotton Initiative (BCI), 2017 Annual Report.).

The modus operandi of BCI is that farmers maintain careful records of input use and farming practices in order to benchmark against best practices, leading to improvement over time.

BCI was established as an independent entity in 2009, and the first BCI harvests from Brazil, India, Mali and Pakistan were in 2010–2011. As of 2017–2018, producers participating in BCI accounted for 4.6 million tons of cotton from 3.3 million hectares (exempting cotton counted in CmiA). The average yield among producers participating in BCI in 2017–2018 was 1380 kg of lint per hectare. Two-thirds of BCI production occurred in two countries, Brazil and China, where yields are above the world average, explaining the high average yield for BCI producers overall. Across all countries, about 500,000 farmers were participating in BCI in 2016–2017 (the most recent year for such data).

Producers who participated in BCI/CmiA accounted for two-thirds of total production in Brazil in 2016–2017 and one-fifth of production in China and Sub-Saharan Africa. BCI participants also accounted for one-fifth of total production in Pakistan during 2016–2017 (Chart 1B.14).

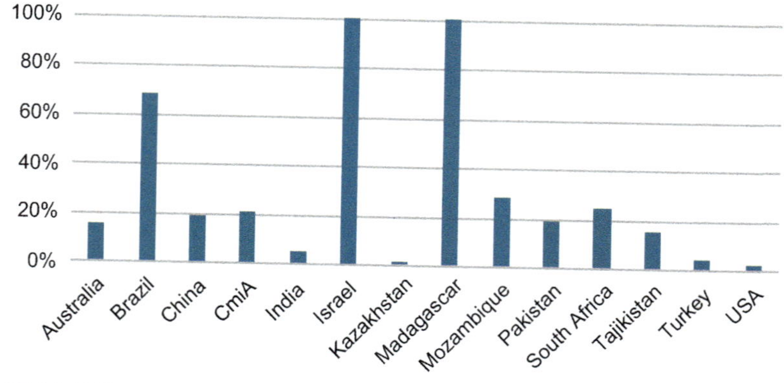

Chart 1B.14 BCI as a share of national production in 2016/17.

All cotton production in Israel and Madagascar is within the BCI system, and around one-third of production in Mozambique and South Africa was BCI. Other countries with producers participating in BCI included Australia, India, Kazakhstan, Tajikistan, Turkey and the United States.

1B.2.9 Wool: the elegant fibre

As of 2018, wool, another major natural apparel fibre along with cotton, accounted for about 1% of world fibre use, or 1.1 million tons, down from 10% or 1.5 million tons in the 1960s (Chart 1B.15). (International Wool Textile Organization (IWTO), Market Information Edition 13, 2018).

There have been some gains in wool production per hectare since the 1960s, but animal husbandry is inherently complicated, resulting in high production costs relative to polyester and cotton. Production has been further limited by drought in animal-producing regions of the Southern Hemisphere since 2017.

Wool production is more evenly dispersed among countries than is cotton production, with the five largest producing countries accounting for half of world wool production, compared with nearly 90% of cotton production (Chart 1B.16). Australia is

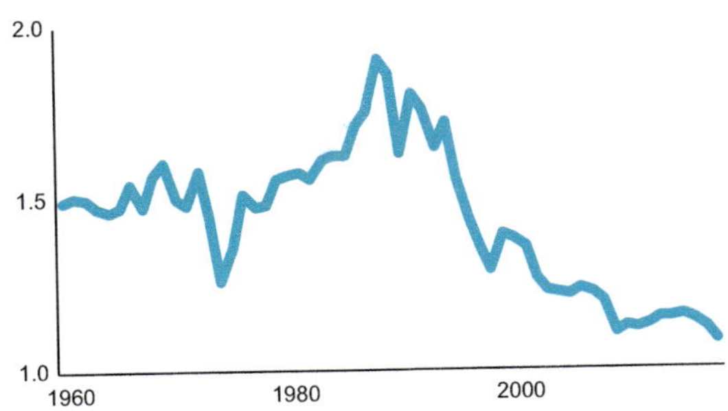

Chart 1B.15 World wool production, million tons.

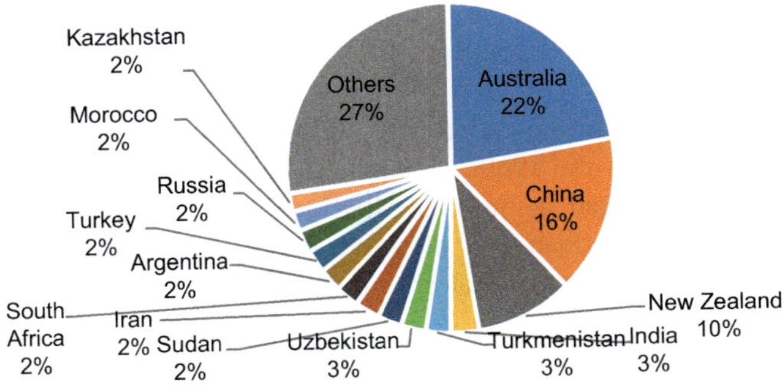

Chart 1B.16 World wool production, 2016, 1.1 million tons.

the largest wool producer, accounting for one-fifth of the world total. China accounts for one-sixth of the total, and New Zealand accounts for one-tenth of world production. More than 40 countries account for the remainder of world wool production.

The world sheep population is about 1.15 billion head, meaning that fleece weights, clean basis, per head average just 1 kilogram. However, there is enormous variation in wool production per head from country to country. Production per sheep averaged 4.1 kg clean basis per year in New Zealand, 3.7 kg in Australia, 2.4 kg in Argentina, 1.1 kg per year per sheep in China, Russia and South Africa, and just 500 g per sheep in India and Sudan. Production per sheep varies with breed and nutrition, with some breeds intended for meat production, while sheep and goats raised in arid conditions naturally produce less wool.

An estimated five million households are engaged in sheep husbandry and wool production, and when total employment, including seasonal labour, is considered, 10 million people are involved in wool production during parts of each season. World wool production is valued at the farm gate at $16 billion in 2018, or about $3 thousand per producing household.

Wool has suffered since the 1960s from shifts in fashion toward more casual appearances and preferences among consumers for easy-care products. The industry also struggles with competition from the oil-based fibres in markets for floor coverings and blankets.

On the other hand, the wool industry is working to develop machine-washable wool products so as to expand the range of product choices available. As of 2019, the wool industry is also making inroads in outdoor apparel markets as consumers recognize the unique properties of natural fibres. Fibre blends involving wool are possible, but most wool products, including suits and carpets, are 100% wool reflecting the unique, elegant, properties of wool.

1B.2.10 Jute, trending upward

Jute is the least expensive of the bast fibres (fibres recovered from the bark or skin of the plant's stem). Jute fibres are much stronger than cotton and have little elasticity, so they are particularly well suited for roping, packaging material, fabrics used in construction and landscaping applications. (FAO, International Year of Natural Fibres, 2009; fibre updates).

World production of jute, known as the golden fibre for its yellowish colour, rose from 2.5 million tons in the 1960s to about 3.5 million as of 2017 (Chart 1B.17). Production is concentrated in the Ganges River Delta of India and Bangladesh, each accounting for more than 40% of world production. Smaller quantities of jute amounting to 5% of the world total are grown in other countries, such as China, Thailand, Myanmar, and Pakistan.

Jute is a traditional crop in India and Bangladesh, and both governments support the industry by mandating the use of jute, or hessian, woven cloth in packaging material. Sacking is still the largest end-use category for jute fibres. India is the largest single consumer of jute, and Bangladesh the largest exporter.

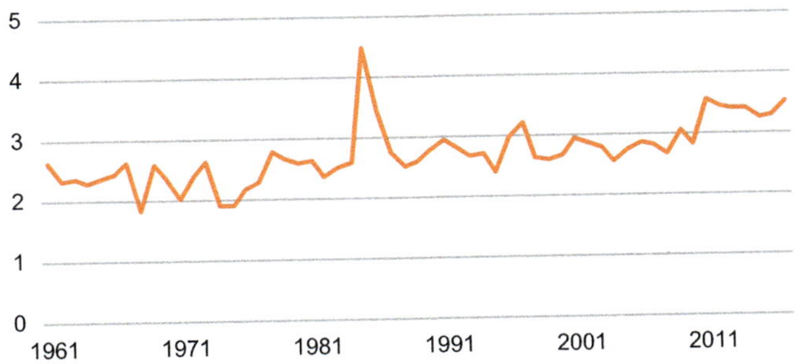

Chart 1B.17 World jute production, million tons.

A major advantage of jute, and other natural fibres, is its biodegradability, and the use of natural fibres in environmental and industrial applications in which biodegradability is an asset is expanding.

Jute requires between 150 and 200 mm of rainfall weekly, humidity above 70% and temperatures between 25° and 30° centigrade. Stalks are harvested by cutting them close to the ground. The stalks are then retted (the process of separating fibres from the stalks) by soaking in water for about 3 weeks to soften the hard tissues. The fibres are then stripped from the stalks, washed, and dried by hanging or laying out on roofs. Retting can also be accomplished mechanically by hammering the stalks, chemically by boiling them and applying chemicals, or by steaming, but these methods tend to be much more expensive. Water retting is by far the most popular method because of the abundance of water in the Ganges Delta.

Jute yields average about two tons of dried fibre per hectare, about four times higher than for cotton in India and Bangladesh. Jute fibres are 1–4 m in length, compared with 3 cm for cotton.

An estimated six million households in India and Bangladesh are involved in jute production, making the industry of national importance, especially in Bangladesh.

Jute has traditionally been used in applications such as cargo sacks, netting and industrial covers. Jute has been hard hit by the growth of alternative roping and wrapping material, such as polypropylene. However, a major advantage of jute over the chemical fibre alternatives is its biodegradability, and future consumption of jute may increase as concerns about polymer pollution mount.

1B.2.11 Coir: production expanding

World production of coir, (fibres extracted from the husks of coconuts) rose from 400,000 tons in 1960 to 1.2 million tons in 2017, and coir production continues to trend upward (Chart 1B.18). Like jute, coir fibres are obtained through a retting process involving either fresh or salt water. However, because of the inherent durability of coir, the retting process takes between 6 and 10 months, compared to about 3 weeks for jute (Food and Agriculture Organization of the United Nations (FAO), FAOStat3).

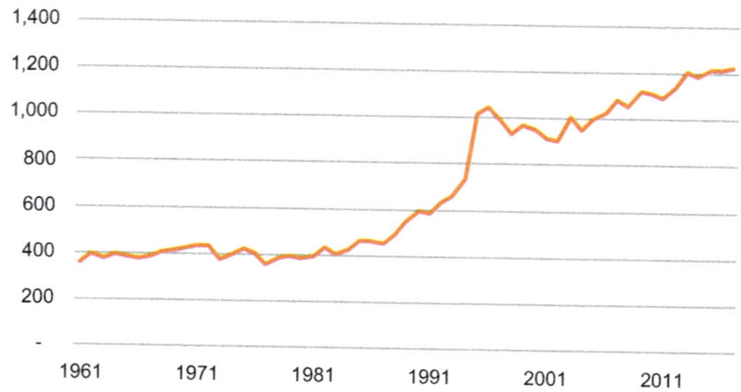

Chart 1B.18 Coir production, 000 Tons.

Coir is used in floor mats, brushes and mattresses. Brown coir, made from ripe coconut, is used in upholstery padding, sacking and horticulture. White coir, harvested from unripe coconuts, is used for making brushes, string, rope and fishing nets. White coir does not absorb water and floats, so it can be used to make long lengths of rope for use on water without dragging down boats and buoys.

Coir is the only natural fibre resistant to degradation in salt water. Therefore, coir is used in fishing nets. Coir is also resistant to abrasion, thus it is used in floor mats and brushes.

India and Sri Lanka account for about 90% of world coir output. India produces about 60% of total white coir, mainly in the Southern regions of Pollachi and Kerala, and Sri Lanka is the largest producer of brown coir.

Approximately one million households are involved in coir production each season, and the annual farm value of production is $500 million.

1B.2.12 Specialized natural fibres

After cotton, wool, jute and coir, world production of specialized fibres totals 1.5 million tons each year. Fibres including abaca, flax, industrial hemp, kapok, ramie, sisal, silk and a range of animal fibres such as Angora rabbit, llama and mohair, have a combined value of $3 billion and provide incomes for more than one million households. While relatively small in world economic terms, each fibre has important impacts in the regions of production and use.

Ramie, silk and flax, which is used to produce linen fabric, are used primarily in apparel and household furnishings. Abaca, hemp and sisal are often used in roping, paper products and as components in automotive composites.

Ramie is a bast fibre that is harvested two to three times a year, and under good growing conditions can be harvested up to six times per year. Unlike other bast crops, ramie requires chemical processing to de-gum the fibre. World production is estimated at 100,000 tons per year. China is the major producer, followed by Japan, Taiwan, the Philippines and Brazil. Most ramie production is used domestically in each producing country, and only a small amount is exported. Japan, Germany, France and the UK are the major importers. Most ramie is used in apparel and is often blended with cotton.

Silk, a filament made by silk worms to form their cocoons, has been used for thousands of years. Production in 2018 is estimated at 160,000 tons. China is the largest producer accounting for 70% of the total, India accounts for about 20%, and Uzbekistan, Thailand, Brazil, Vietnam and numerous additional countries account for the remainder.

Flax fibre production is estimated at 300,000 tons per year. Flax, also known as linseed, is another bast fibre, and it is both a food and fibre crop. The fibre, which is double the strength of cotton, is used to make a fabric known as linen that is used primarily in home furnishings. Flax is grown in cooler climates, ranging from Northern Europe to Canada.

World production of abaca is about 80,000 tons per year, of which about 65,000 comes from the Philippines alone. Eight times stronger than cotton, but very coarse, abaca is used primarily in paper production.

Hemp fibres are derived from varieties of cannabis that are low in THC and have no recreational value. World hemp fibre production is estimated at about 70,000 tons per year. About half is produced in northern Europe and about half in China. The largest single use of hemp is for paper and composite products, but the use of hemp in apparel is expanding.

Sisal comes from the leaves of the *agave* cactus plant common to the semi-desert regions of the Western Hemisphere. Sisal originated in Mexico and was introduced into Tanzania in 1893 and Brazil around 1900. The first commercial production in Brazil started around 1930. Today, Brazil and Tanzania account for the vast majority of the world's sisal production. Sisal is the coarsest of the natural fibres and is used predominantly in floor coverings, ropes and twines, and as a feedstock for paper production.

The animal fibres other than wool and silk are used primarily in apparel, particularly outwear. Annual production of all the exotic animal fibres is estimated at about 25,000 tons per year on a clean basis.

1B.3 The future

Natural fibres have long and romantic histories in the world economy, and despite the growth in synthetic alternatives, natural fibres remain vibrant industries that employ hundreds of millions. Almost every person on earth encounters natural fibres every day of their lives.

However, natural fibres are threatened by the growth of man-made fibre alternatives, and a fundamental question for all those associated with natural fibre industries is how they can survive in competition with polyester, nylon, acrylic, polypropylene and other man-made fibre products.

World production of natural fibres rose from 15 million tons in the early 1960s to more than 30 million by 2004, and production was still at that level in 2018. In contrast,

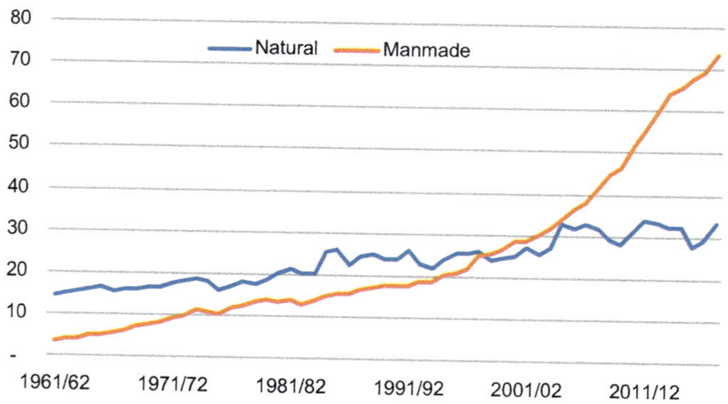

Chart 1B.19 World fibre production, million tons.

world production of man-made fibres grew from a few million tons in the 1960s to more than 70 million by 2018 and was rising sharply (Chart 1B.19).

In the age of sail, sisal, hemp and linen were major industries, employing hundreds of thousands of people in fibre production and processing and many thousands more in the production of ropes and ship's lines, cargo nets, sails and other products with industrial applications. Today, these fibres are relatively minor industries in the world economy, largely supplanted by man-made fibre products.

In contrast to sisal, hemp and linen, cotton and wool remain significant apparel fibres, jute production is still measured in the millions of tons, and production of coir is rising. Cotton, wool, jute and coir show that natural fibre industries can hold their own in competition with man-made fibres. But, to remain competitive, natural fibres must change, and research is the key.

Just as sailing ships have given way to modern cargo carriers and passenger cruise ships to remain competitive with automobiles, trains and aeroplanes, so must natural fibres transform themselves to remain relevant in competition with man-made fibres. Natural fibres must meet consumer demands for fibre performance characteristics at prices that consumers are willing to pay by raising yields, reducing resource use and developing new uses.

For natural fibre industries to remain competitive, the technological progression that started with directed breeding, and continued with mechanization, the use of synthetic fertilizers and plant protection chemicals, and the use of the tools of genetic engineering to improve plant characteristics must continue.

The world cotton yield rose from 200 kg of lint per hectare before World War I to 800 kg today, and cotton fibres today are about 50% longer and stronger than they were 100 years ago. Likewise, yields and qualities of jute are rising with better agronomic practices, wool qualities are improving with better sheep breeding, and palm trees are being bred to yield better coir fibres. Such improvements in yield and quality must continue with all the natural fibres.

In addition to agronomic improvements, natural fibre industries must transform themselves by expanding the range of uses for which they are suited. A major advantage held by the natural fibres over their man-made fibre competitors is biodegradability, rendering natural fibres superior materials for agricultural and landscaping applications and as binders in composite products. Apparel and home furnishings made from natural fibres do not contribute to microfibre pollution in waterways. As consumers and regulators become more aware of the environmental load imposed by non-biodegradable polymers, pressure will build to switch to biodegradable products, and natural fibres will be well-positioned to provide the materials for apparel, home furnishings, construction materials and other products.

It is obvious that natural fibres will never regain their dominant position as the only fibres used by mankind. However, even with rapid growth in man-made fibre production, natural fibre production remained around 30 million tons per year between 2004 and 2018. Even if the share of natural fibres in total fibre production continues to decline in the decades ahead, natural fibres can still remain economically important in absolute terms. And, if environmental concerns result in reduced consumption of man-made fibres, while improvements in technology result in higher yields and better fibre characteristics, natural fibre production could even grow over the next several decades.

Natural fibres have long and romantic histories, and they have optimistic futures as the ingredients for products responsibly produced that consumers want and as the basis for industries that provide employment around the world.

Abbreviations

AbTF	Aid by Trade Foundation
BCI	Better Cotton Initiative
Bt	Bacillus thuringiensis
CmiA	Cotton made in Africa
DNFI	Discover Natural Fibre Initiative
FAO	Food and Agriculture Organization of the United Nations
GM	Genetically Modified
GMO	Genetically Modified Organism
HT	Herbicide Tolerant
ICAC	International Cotton Advisory Committee
IR	Insect Resistant
ISAAA	International Service for the Acquisition of Agri-biotech Applications
IWTO	International Wool Textile Organization
TE	Textile Exchange
WWF	World Wildlife Fund

References

Aid by Trade Foundation, Cotton Made in Africa (CmiA), Various Press Releases.

Better Cotton Initiative (BCI), 2017. Annual Report.

Discover Natural Fibre Initiative, January 2019 (Frankfurt, Germany).
Fairtrade U.K., Cotton Commodity Briefing, 2017. http://www.fairtrade.org.uk.
FAO, International Year of Natural Fibres, Fibre Updates.
Food and Agriculture Organization of the United Nations (FAO), FAOStat3.
International Wool Textile Organization (IWTO), 2018. Market Information Edition 13.
Textile Exchange, 2018. Organic Cotton Market Report.

Part One

Fundamentals: Types of fibre, properties, identification and testing

Bast fibres: jute

2

S. Roy[†,a], *Latifa Binte Lutfar*[1]
[1]Former Operations Officer, Dhaka, Bangladesh

2.1 Introduction to jute

Among the natural fibres, jute ranks next to cotton in terms of production. Jute is a cellulosic fibre under the category of bast fibres and its cultivation is almost as old as human civilization. Jute, an annual herbaceous plant, is mainly cultivated in South and Southeast Asia. Jute was first used as an industrial raw material for making packaging materials, replacing flax and hemp grown in Europe.

2.1.1 *Origin*

Corchorus species are found in warm regions throughout the world, in all continents and in numerous tropical and subtropical regions. *Corchorus* is a pan-tropical genus comprising an uncertain number of species, with estimates ranging from 40 to 100. However, the centre of diversity and origin of the genus appears to be Africa, where the largest number of *Corchorus* species (around 30) has been found with the highest concentration reported from East and South Africa. Of the cultivated species *Corchorus capsularis* is omnipresent in Indo-Myanmar (Burma) and South China, from where it migrated to India and Bangladesh. The primary centre of origin of *Corchorus olitorius* is probably Africa, while India or the Indo-Myanmar region is a secondary centre. Both species are cultivated and naturalized in many parts of the tropics, including Southeast Asia.

2.1.2 *History*

When the jute plants were recognized as a source of fibre and utilized for making ropes and sacking, mainly in the Indian subcontinent, certainly, they are unknown. References to sacking bags made of jute have been traced to the literary works of the region as far back as 1575. Sackcloth made of jute has been referred to as an article of trade in several Bengali poetical works of the 16th and 17th centuries. Rumphius in 1743, one of the earliest workers on Bengal plants, gave an illustrated account of jute plants along with a figure of *C. capsularis* mentioning therein that it was under cultivation in Bengal (India), the Arakans and South China. He even mentioned that the fine white thread made out of this fibre was stronger than that from cotton. Jute is

† Deceased
[a] Dr. Sudripta Roy of the International Jute Study Group (IJSG), Dhaka, Bangladesh

Handbook of Natural Fibres. https://doi.org/10.1016/B978-0-12-818398-4.00003-7

believed to have been traditionally in use in many other parts of Asia and Africa since ancient times to provide cordage and weaving fibres from the stem and vegetables of the leaves.

2.1.3 Adaptation/agroclimatic conditions

Favourable conditions for jute cultivation are found in the floodplains of the great rivers of the tropics and subtropics – the Ganges, the Irrawaddy, the Amazon and the Yangtze – where irrigation, often characterized by extensive flooding, and alluvial soils combine with long day lengths. The crop traditionally thrives very well under rain and hot humid and subtropical conditions in the Bengal Basin in India and Bangladesh, where more than 80% of the world crop is grown.

Jute is mainly grown between 16°N and 27°N, during the hot wet summer season in a hot and humid climate with temperature in the range of 24°C–37°C. Growth is retarded at temperatures below 17°C and above 42°C. The annual rainfall should be 1000–2000 mm of premonsoon showers at sowing time.

2.1.4 Areas of production

Jute is mainly cultivated in the equatorial, the tropical and the subtropical zones. It is extensively cultivated in India and Bangladesh. Other major jute-growing countries are Myanmar, Nepal, China, Vietnam, Thailand and Brazil.

2.1.5 Economic importance

Jute plays a vital socioeconomic role in producing countries. It is produced by hundreds of thousands of small-farm families in Southeast Asia. About 12 million farm families are dependent on this crop. It provides a vital cash flow to the people who live at subsistence level. Besides the fibre, farmers get a substantial quantity of jute sticks, which also have a variety of applications such as fuel, fencing and particle boards.

The importance of jute in the national economy of the producing countries, especially Bangladesh, can hardly be overemphasized. The jute sector in Bangladesh provides about 4% of gross domestic product and has a 4.07% share of the country's national export earnings.

Moreover, jute occupies the land for fibre production for only 4–5 months and is commonly rotated with food crops including rice and potatoes. Multiple cropping gives increased agricultural production, tends to maintain and improve soil fertility and reduces the incidence of weeds and plant diseases. The inclusion of jute in the cropping system enhances soil organic matter through leaf shedding during the growing season and improves soil nutrient availability by breaking the subsoil hard pan caused by continuous cropping of rice.

2.2 Types of jute

Although there are several types of jute plant, which yield fibre, only two species are cultivable types and the others are mostly wild types, bushy and dwarf.

2.2.1 Taxonomy

Jute is an annual plant belonging to the genus *Corchorus* of the family Tiliaceae and also belongs to the suborder Malvineae of the order Malvales. The number of *Corchorus* species is probably around 50–60; however, over 170 *Corchorus* names are given in the *Index Kewensis*. The genus is extremely variable, but all species are apparently highly fibrous.

2.2.2 Main species

Out of over 30 important species belonging to the genus *Corchorus,* only two – *C. capsularis* and *C. olitorius* – are utilized for fibre production on a commercial scale. *C. capsularis* is commonly known as 'white jute', and *C. olitorius* is known as 'tossa jute'.

2.2.3 Botanical description

Jute is an annual herbaceous dicotyledonous plant that grows to a height of 1.5–4.5 m. The stems are about 1–2 cm in diameter with few branches. The colour of the stem, petiole, leaf and pod varies in different forms. Jute fibre is obtained from the bast or phloem layer of the stem. The two species differ in the quality of fibre they yield. Fibres of *C. olitorius* are frequently softer, stronger and more lustrous than those of *C. capsularis.*

2.2.4 Cultivation

Jute is grown in a wide range of soil types, mainly alluviums, laterite and calcareous with soil texture varying from sandy loam to clay loam. Basically, the soil should be well-drained, and its pH should preferably be in the range of 5.5–6.5. White jute is relatively more tolerant to waterlogging especially at later stages of crop growth. Conversely, tossa jute does not tolerate waterlogging and is usually grown on higher lands. In general, both species are more sensitive to waterlogging during the early stage of crop growth.

Jute is propagated by seed. In both India and Bangladesh, sowing is done mostly by broadcasting at seed rates of 7–13 kg/ha for *C. capsularis* and 5–9 kg/ha for *C. olitorius,* though line sowing or planting in lines has many advantages such as 50% lower seed rates, fewer rounds of thinning and weeding, opportunities to reduce costs by utilizing mechanical implements for sowing, thinning and weeding, more convenient harvesting and generally higher fibre yields. Despite these advantages,

Figure 2.1 A view of a jute field.

row cropping is not widely popular and not widely practised by South Asian jute growers (Fig. 2.1).

Organic manure is mainly used to provide nutrients to the jute crop. Usually 6–8 t/ha of farmyard manure are applied to the field during land preparation. Reports indicate that inorganic fertilizers are used by a very small percentage of farmers for jute cultivation, the main reason being that most jute farmers are poor and occupy smallholdings and thus cannot afford to incur the cost of fertilizers. Even when fertilizers are used they are applied in small doses.

Because of the thermosensitive nature of the jute plant, it requires an appropriate time for sowing, matching with the temperature and day length required for optimum growth and development. Soil temperature needs to be 15°C or above for favourable seedling growth.

Raking, thinning and weeding are generally adopted in all jute fields, which are best practised under optimum soil moisture conditions.

Jute is mostly grown in rotation with other crops. It should be emphasized that jute leaves left on the field improve the soil fertility and increase yields of the follow-on crop. Many farmers adopt crop rotation practices with a view to minimizing crop losses due to diseases, insects and pests.

2.2.5 Harvesting

The harvesting time of jute is calculated by taking into account the crop age, height and flowering stage, which varies according to the varieties grown and sowing time, but usually 110–120 days are required for jute to mature for harvesting. Jute is generally harvested when the plants are at early pod stage. The fibre remains weak if it is harvested before flowering. On the other hand, the fibre becomes coarse and lacks lustre if harvesting is delayed.

The plants are usually cut close to the ground by hand. A curved, sickle-type knife is used to cut the plants. The plants are then sorted out according to height and diameter. The assorted plants are tied into bundles. Each bundle weighs 8–10 kg. The bundles are kept standing in the fields for 3–4 days for defoliation. The bundles of jute plants are carried to ponds, canals, ditches or any other water bodies for retting. Harvesting the plants at the correct time is most important and requires long experience.

2.2.6 Retting

After harvesting and defoliation of plants in the fields for 3–4 days, the jute stems are retted in water, and the fibre is extracted. The traditional method is to ret the jute stems for about 15–18 days and extract the fibre manually after retting.

During retting, bacteria break down the soft tissues around the fibre bundles and the fibres. Retting is complete when the bark separates out easily from the core. The end point of retting is a critical stage that largely determines the quality of the fibre. It is difficult to extract the fibre if the plants are taken from the water too early. On the other hand, the fibre is weakened if retting is continued for too long. Before extraction, the farmers make frequent checks to determine the end point of retting. Correct retting is essentially the first step in the production of good quality fibre.

Conventionally, all the operations of retting followed by extraction of the fibres are done manually to date.

An alternative to this traditional retting method is to ret the outer ribbons without the woody stick. This is called ribbon retting and is widely practised in China. Different retting techniques, ribboners/fibre decorticating machines/decorticators have been developed by different R&D organizations in Bangladesh and India. The ribbons, the outer skins of the stem, are extracted mechanically or manually by these machines.

There are also other methods of retting, chemical, enzymatic/microbiological, etc., advocated by different scientists/researchers.

2.2.7 Extraction

After retting, plants are taken out of water, and the fibre is traditionally extracted by hand.

2.2.8 Washing

The washing process consists of holding the extracted fibre bundle by the butt end and jerking it through the water. Clean water is used for washing. The entire dirt, gum, extraneous plant materials and retting residues are removed thoroughly.

2.2.9 Drying

The washed fibre is spread over a bamboo perch or bar for thorough sun drying for 4–7 days before storage. Drying on bare ground is discouraged because it affects quality of fibre by contamination with dirt, sand, dust particles, etc.

2.2.10 Fibre quality and grading

The main parameters of fibre quality include colour, lustre, strength, texture, length and presence or absence of root cuttings, the hardy reeds basal portion of fibre, etc.

The outcome of desired attributes or undesired defects is dependent on various factors such as cultivation practices, diseases and pests and retting. Agronomic manipulation in terms of balanced fertilizer, maintenance of desired plant population and yield, adoption of a new cropping system approach, integrated pest management and improvement of retting facility are suggested for quality improvement of jute.

2.2.11 Grading systems in different countries

In Bangladesh, based on the major quality attributes, fibres are graded by two methods for commercial purposes. One is called the *Kutcha* grading method, where fibre is categorized into five grades: Top, Middle, B-Bottom, C-Bottom and X-Bottom in descending order of quality mainly for the local markets. The other method is called *Pucca* grading, where the basal/root parts of the fibres are cut away as cuttings and are categorized into six grades, namely BW-special, BW-A, BW-B, ..., BW-E for white jute and BT-special, BT-A, BT-B, ..., BT-E for tossa jute in descending order of quality, mainly for the international market.

In India, depending on the six fibre quality attributes – namely strength, fineness, defects, root contents, colour and density – jute is graded into eight white (*C. capsularis*) grades, that is, W_1, W_2, ..., W_8 and eight tossa (*C. olitorius*) grades, that is, TD_1, TD_2, ..., TD_8 in descending order of quality.

Similarly, different grading systems are used in different countries, for example, a grading system with four grades in China, with three grades in Indonesia, with two grades in Nepal and with three grades in Thailand.

2.3 Fibre morphology

Jute fibre, unlike cotton, is a multicellular fibre. In the jute plant, the fibre is formed as a cylindrical sheath made up of single fibres (ultimate cell) joined together in such a way as to form a three-dimensional network from top to bottom of the stem.

The commercial fibres, in the form of fibre bundles of 1.5–3 m long, called *reed*, are held together as a unit by the meshy or network structure of the fibre elements and represents only a very small proportion (4%–6%) of the whole plant.

2.3.1 Macrostructure

Each fibre element of these meshes of a raw jute reed is basically a group of *ultimate cells*, cemented together laterally and longitudinally by means of intercellular materials being chiefly noncellulosic in composition. A single fibre of jute thus comprises a bundle of ultimates. Thus, jute fibre is multicellular.

The ultimate cells are spindle-shaped and of variable size in length and width, being on average 2.5 mm long and 0.02 mm width at the middle. The cells are some 200 times longer than their breadth.

The cross sections of the ultimate cells are found to be polygonal with rounded corners. The layer of natural gum present between the ultimate cells is known as the *middle lamella.* Each ultimate cell has thick cell wall and *lumen,* the central canal, with a more or less oval cross section (Fig. 2.2).

2.3.2 Microstructure and appearance

The cell wall of each ultimate cell is composed of an outer thin *primary wall* and an inner thick *secondary wall,* differing from each other in the molecular architecture. Both these walls of a jute ultimate cell are composed of ultrafine *microfibrils* (Fig. 2.3).

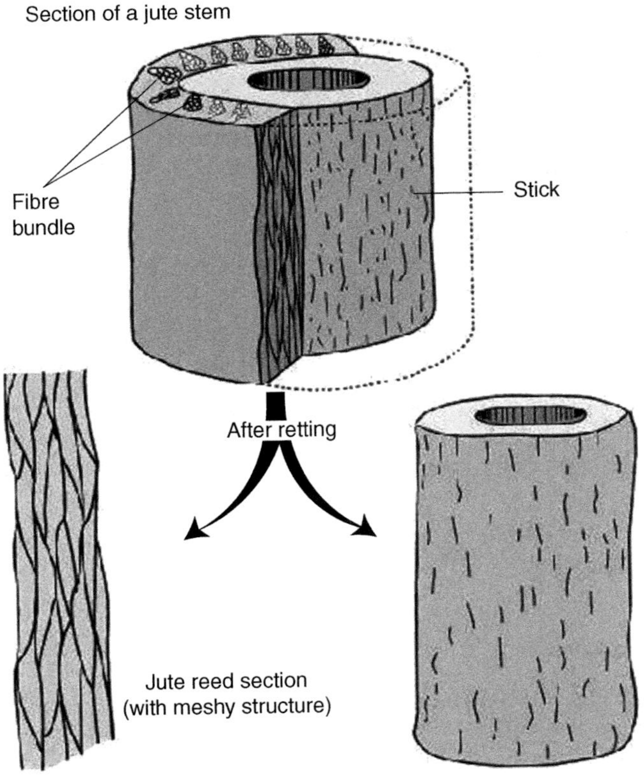

Figure 2.2 Disintegration of a jute stem into a jute reed and a stick after retting.

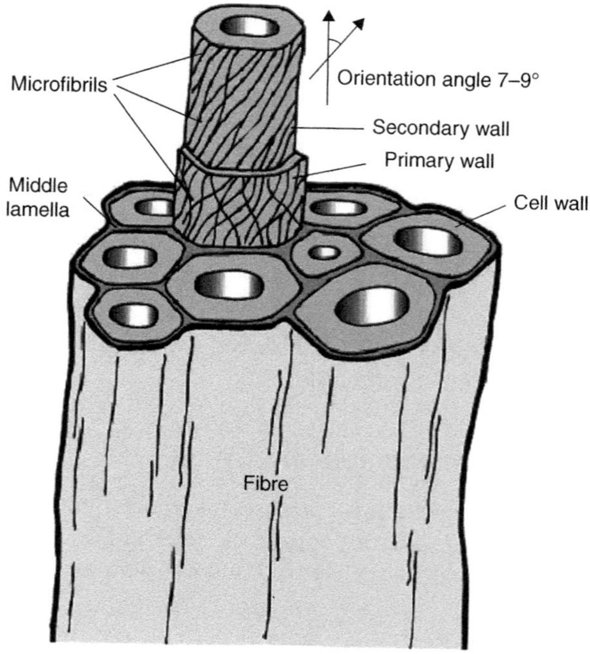

Figure 2.3 Jute fibre morphology: microstructure.

While in the primary wall the fibrils lie in a crisscross manner, the fibrils are arranged almost parallel as right-hand spirals in the secondary wall. The fibrils in the jute cell wall are arranged in a right-handed spiral with angle of orientation of 7–9° in reference to the cell axis.

Within the ultimate cells of a jute fibre, the ultrafine fibrils, being purely cellulosic, are the highly ordered regions, while the interfibrillar regions are less ordered regions, which can make room for the presence of short chain hemicellulose molecules to a larger extent and the bulky lignin molecules to a smaller extent as the bonding material of the middle lamella, providing strong lateral adhesion between the ultimates.

2.4 Chemical composition

The major aspects related to the chemistry of jute and its composition have been described below.

2.4.1 Description

Jute fibre is basically a compound of lignocellulose. It is a complex of organic molecules, which on combustion leaves a little ash consisting of calcium, magnesium, aluminium, iron, etc., that are present either in the free state or bonded with functional groups of cellulosic chain.

The number of ultimate cells in one such bundle constituting a single fibre ranges from a minimum of 8–9 to a maximum of 20–25. This wide variation in the number of cells in a bundle is believed to be a major cause of variation in the physical and mechanical properties of the fibre and its quality.

The spinnable units in jute fibre strands are, like those in most other bast fibre crops, filaments composed of a string of cells bonded together by pectin and hemicelluloses.

2.4.2 Chemical structure

Chemically, jute fibre is mainly composed of polysaccharides and lignin. The fibre also contains smaller amounts of chemical compounds such as fats and waxes, pectin, nitrogenous, colouring and inorganic matters. The polysaccharides are also called carbohydrates (or holocellulose) and are divided into two groups: alphacellulose and hemicellulose.

2.4.2.1 Structure of alphacellulose

Alphacellulose is the major constituent of jute. It forms the skeletal structure of jute fibre and belongs to the family of the compounds of carbohydrates. It contains 44.4% carbon, 6.2% hydrogen and 49.4% oxygen. Its molecular formula is expressed as $(C_6H_{10}O_6)_n$. The chemical structure of alphacellulose is shown in Fig. 2.4.

Alphacellulose is a natural polymer of the polycondensation kind consisting of D-anhydroglucopyranose units linked together in the chain molecule by 1, 4 β-glycosidic bonds. The alphacellulose is a long-chain polymer. The degree of polymerization (DP) of an alphacellulose chain is about 10,000 if we consider the 'glucose unit' as the monomer or about 5000 if the 'cellobiose unit' is considered to be the monomer.

2.4.2.2 Structure of hemicelluose

The hemicellulose in jute is a complex mixture of polysaccharides and polyuronides. It is formed by a number of comparatively low molecular weight polysaccharides of various sugar units, namely xylan in the pentosan, galactan in the hexosan, araban, rhamnosan, mannose, etc. Polysaccharide hemicellulose contains a large number of hydroxyl groups and also carboxyl groups. The hemicelluloses are short chain linear polymers. The molecular weight of hemicellulose is 26,000 (determined osmometrically). The chemical structure of hemicellulose is shown in Fig. 2.5.

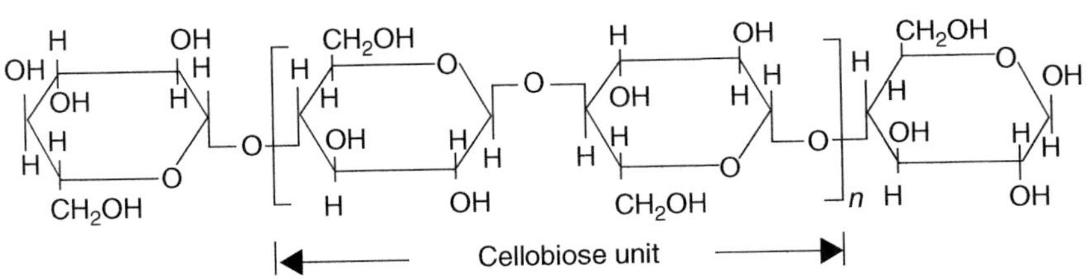

Figure 2.4 Chemical structure of alphacellulose.

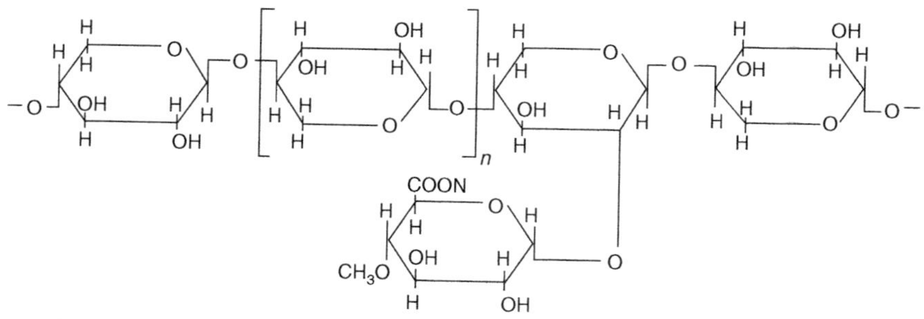

Figure 2.5 Chemical structure of hemicellulose.

2.4.2.3 *Structure of lignin*

In jute, lignin with a DP of 60, the third major constituent, is a long-chain substance of high molecular weight which, like the hemicelluloses, varies in composition from one type of vegetable material to another. Structurally, the molecule of lignin is a polymer of phenylpropane units consisting of phenolic hydroxyl group, methoxyl group, ether linkages and conjugated double bond in the alpha position to the benzene ring. Lignin is a complex noncrystalline polymeric constituent that functions as the structural support material in the plant body.

2.4.2.4 *Structure of pectin*

A small quantity of pectin is present in the jute fibre. The pectin holds the fibre bundles together. Pectin is a large molecule built up of repeating units of hexuronic acid. It undergoes decomposition during retting to water-soluble products by bacteria.

2.4.2.5 *Mineral and colouring matters*

The jute fibres are found to be associated with small amounts of different inorganic metals such as calcium, potassium, magnesium aluminium, iron and silica in the form of their salts. The fibre also contains some pigments and colouring matter. The colouring matter consists of xanthophyll, carotene and tannin. The tannin combines with the iron of the retting water to form iron tannate that imparts a dark grey colour on the fibre surface after retting. They are also present within the cortical cells and in the fibre bundle. The minerals and the colouring matters are removable from the fibre by chemical means such as treatments with dilute mineral acids, bleaching agents, etc.

2.4.3 **Chemical composition of jute**

The retted jute fibres consist of alphacellulose, hemicellulose and lignin along with some minor constituents. The chemical composition of jute fibres is given in Table 2.1.

Table 2.1 Chemical composition of jute (in % of bone dry weight of the fibre).

	Jute	
Constituents	***Corchorus capsularis***	***Corchorus olitorius***
Alphacellulose	60.0–63.0	58.0–59.0
Hemicellulose	21.0–24.0	22.0–25.0
Lignin	12.0–13.0	13.0–14.0
Fats and waxes	0.4–1.0	0.4–0.9
Pectin	0.2–1.5	0.2–0.5
Proteins/nitrogenous matter, etc.	0.80–1.9	0.8–1.6
Ash	0.7–1.2	0.5–1.2

2.5 Properties of jute

Commercial jute ranges from pale cream to golden yellow and from light brown to dirty grey in colour. It possesses a natural silky shine. Jute is a relatively coarse, stiff, inelastic and somewhat rigid fibre that has slightly higher moisture regains (12%–13%) than cotton (7%–8%). Good frictional property, tenacity, very high modulus and low breaking elongation make jute an ideal packaging material.

The use of jute is limited to coarse fabrics because the length/diameter ratio of jute filaments is only 100–120, which is much below the minimum of 1000 required for fine spinning quality. Jute fibre is hygroscopic and wetted filaments may swell up to 23% in diameter.

Other than being of agroorigin (and hence renewable) and biodegradable (and hence environmentally friendly), the major advantageous features of jute are its high strength and initial modulus, moderate moisture regain, good dyeability using different dyes, good heat and sound insulation properties and low cost. However, the major disadvantages of jute are its coarseness, stiffness, low wet strength, moderate wash shrinkage, harsh feel, hairiness and high fibre shedding, photoyellowing and poor crease recovery.

2.5.1 Physical properties

Like other bast fibres, jute is a strong fibre with very low extensibility due to its *'composite-like'* structure with highly oriented long-chain molecules. The jute fibre shows very poor extensibility, the breaking elongation ranging between 1.0% and 1.8%. Tossa jute is found to be stronger than white jute. Tenacity-wise, the jute filament is comparable with steel.

The flexural and torsional rigidities of jute fibre are quite high as compared with cotton or wool due to its coarseness and inelastic structure. For this, jute textile material develops wrinkles and creases easily (Table 2.2).

Table 2.2 Physical properties of jute.

Physical properties (macro- and microstructure)	
Ultimate cell length (L)	
Average	2.50 mm
Range	0.8–6.0 mm
Ultimate cell breadth (B)	
Average	18 μm
Range	10–25 μm
L/B ratio (average)	110
Fibre fineness	1.3–4.0 tex
Fibre length (after carding)	2–50 cm
Density	
True	1.46 g/cc
Apparent	1.10–1.34 g/cc
Bulk density	0.4–0.5 g/cc
Degree of crystallinity (X-ray)	55%–60%
Angle of orientation (X-ray)	7–9°
Moisture absorption	
Moisture regain	
At 65% RH	13.8%
At 100% RH	36.0%
Transverse swelling in water	
Diameter wise	20%
Cross-sectional area wise	45%
Water holding capacity	500%
Mechanical properties	
Tenacity	
Single (gauge length – 1 cm)	30–50 gf/tex; 0.29–0.48 N/tex
Bundle (gauge length – 5 cm)	12–35 gf/tex; 0.18–0.34 N/tex
Elongation at break	1.0%–1.8%
Initial modulus	1700–3000 gf/tex; 17–30 N/tex; $0.9–1.7 \times 10^{11}$ dyn/cm^2
Specific torsional rigidity	0.4–0.5 mN·mm^2/tex^2

Table 2.2 Continued

Physical properties (macro- and microstructure)	
Specific flexural rigidity	0.7–0.8 mN·mm^2/tex^2
Elastic recovery:	
From 3 g/den stress	75%
From 1.5% strain	75%
Specific work of rupture	2.7 mN/tex
Work factor	0.5
Thermal properties	
Specific heat	1.36 × 103 J/kg/K
Thermal conductivity	427.3 mW/m/K
Heat of combustion	17.5 J/g
Ignition temperature	193°C
Heat of wetting	18.2 cal
Electrical properties	
Dielectric constant (at 2 kHz)	
Dry	1.8
At 65% RH	2.4
At 100% RH	3.6
Insulation resistance	10^{14}–10^{17} ohm; depending strongly upon RH
Electric strength	500 kV/cm
Optical properties	
Refractive index	
Parallel to fibre axis	1.577
Perpendicular to fibre axis	1.536
Birefringence	0.041
Frictional properties	
Coefficient of interfibre friction	
Tossa jute	0.45
White jute	0.54

RH, relative humidity.
Source: Sur, D., 2005. Understanding Jute Yarn. Anindita Sur, Kolkata, India.

2.5.2 Thermal properties

Heating at very high temperature, jute fibre chars and burns without melting. With increasing heat, the chain molecules vibrate more increasingly and ultimately disintegrate leading to violent chemical reactions observed as fibre combustion. With a high specific heat value as of 1.36×10^3 J/kg/K, the jute fibre shows good thermal insulation. The ignition temperature of jute, about 193°C, is not high and may be one of the reasons making jute susceptible to catching fire.

Like all cellulosic fibres, jute also loses strength on prolonged exposure to sunlight. Jute loses strength at more than twice the rate for cotton.

2.5.3 Electrical properties

Dry jute exhibits high electrical resistance, but in moist jute, the electrical resistance drops by about 10,000 times. Such variation of electrical resistance with moisture led to the development of an electronic moisture metre. The dielectric constant of jute at a frequency of 2 kHz is 1.8 in the dry state, 2.4 at 65% relative humidity (RH) and 3.6 at 100% RH.

2.5.4 Optical properties

The colour of jute fibres varies widely from pale creams/yellow to dark brown. In general, the tossa variety tends to have a reddish tinge while the colour of white jute varies from pale yellow to yellow. Good quality jute shows excellent lustre. Jute fibres with good molecular orientation show good amount of birefringence, +0.041, the difference between the refractive index along the fibre axis and that perpendicular to it, due to high orientation of chain molecules.

2.5.5 Frictional properties

The higher coefficient of friction in the case of white jute may be attributed to its better fineness and higher meshiness as compared with tossa jute.

Jute fibre shows increasing friction with increase in moisture regain. The removal of fat and wax from the fibre surface also increases the friction. However, the batching oil helps to reduce the friction between jute fibres and metallic pins during carding and drawing stages undertaken for jute yarn spinning. Lower interfibre friction is desirable for reduced drafting-related irregularity, but higher interfibre friction helps to develop good yarn strength.

2.5.6 Chemical properties

Jute fibre is an acid fibre and contains a fairly large amount of hemicellulose (21%–25%) having acidic group and lignin (12%–14%). The fibre is susceptible to the action of alkali. Even a mild treatment with alkali reduces the wet strength of the fibre considerably. The treatment weakens the intercellular cementing materials to hold together the short ultimate cells of the fibre. The acidity of the fibre is due to carboxylic group and phenolic hydroxyl group. By virtue of carboxylic acid, the fibre binds cationic ash minerals and possesses strong affinity for basic dyes such as methylene blue. The chemical properties of jute fibre chiefly arise due to the major three integral constituents alphacellulose, hemicellulose and lignin, which comprise more than 97% of the total jute constituents.

2.5.7 Effect of chemicals on jute

2.5.7.1 Effect of alkali

Jute fibres have poor resistance to alkali due to presence of hemicelluloses. Alkali treatment extracts out hemicellulose from the jute fibre structure making it weak. With treatment of strong (18%) alkali, jute fibres become much weaker but develop crimp due to irregular swelling, a process popularly known as the woollenization of jute.

2.5.7.2 Effect of acid

Jute fibres are weakened and destroyed by acids. The cellulose chains disintegrate due to hydrolysis of the glucoside oxygen atom (which joins glucose units to form a cellulose chain). Inorganic acids have a more adverse effect on jute than organic acids as they hydrolyze cellulose chains more rapidly.

2.5.7.3 Effect of bleaching agent

The most common bleaches used for jute are sodium chlorite and hypochlorite. Sodium chlorite/hypochlorite dissolves out lignin from jute making it whiter as lignin is responsible for the natural yellow to reddish colour of jute. Because of partial removal of lignin, bleaching treatment makes jute weaker as the middle lamella (major location of lignin) between the ultimate cells gets weakened. Because of this, jute fibres also become finer on partial disintegration of the multicellular structure.

2.6 Typical applications

Jute and its products are being used historically in a number of ways. The conventional or traditional uses of jute along with some unconventional and newly developed innovative uses of jute have been described below under the section.

2.6.1 Conventional/traditional applications

The jute industry is about 150 years old, and its products met the global share of packaging successfully though it had ups and downs. Traditionally, jute has been used to manufacture packaging materials such as hessian, sacking, ropes, twines, carpet backing cloth, etc.

The large historical markets for these traditional products of jute have decreased over the years as it has been replaced by synthetics.

Jute has been the most widely used packaging fibre for more than 100 years because of its strength and durability, low production costs, ease of manufacturing and availability in large and uniform quantities. Other traditional products include hessian cloth, carpet backing, yarn, twine, cordage, nonwoven felts and jute carpets.

More than 50% of the jute consumed annually is manufactured into sacks and bags for transportation and storage of agricultural and industrial commodities. Apart from this, jute is also used on a large scale for twines, carpet yarns, cloth backings for linoleum and carpets, etc., as well as for webbing, covering of cables and a host of other purposes of conventional use.

2.6.2 Unconventional/diversified applications

The steady decline in markets for traditional jute products during the 1970s–1980s forced governments and the jute industry to take up programmes for development of new jute products over the last few decades.

New technologies have been evolved for bulk use of jute as a raw material in the production of high value-added and price competitive intermediaries or final products. A host of innovative new products have been developed with high value addition such as home textiles, jute composites, jute geotextiles, paper pulp, technical textiles, chemical products, handicrafts, fashion accessories, etc. These products from new, alternative and nontraditional uses of jute are generally termed diversified jute products.

Growing consciousness about the use of renewable resources, and the need for environment friendly, energy-saving materials and processes in the industrial sector, has changed the focus and emphatically renewed world interest in jute and allied fibres. They are valued not only in the traditional fields of application but also in many other new and diverse application areas.

Jute is a versatile natural fibre finding a wide range of new applications in fabrics for furniture, upholstery, soft luggage, fancy bags, wall coverings, laminated sheets for packaging boxes and panelling, fibre-reinforced composites for automobile, construction and building materials, home textiles, technical textiles, geotextiles, floor coverings, blankets and semiapparels, nonwovens, insulation materials, asbestos replacing materials, particle boards, shopping bags, handicrafts and fashion accessories, etc.

Whole jute stems are suitable as raw material for paper pulp. The woody central core or 'jute stick' remaining after removal of the bast is used as a rural building material, for thatch, fences, fuel and for charcoal making. It can also be processed into paper, board and cellulose derivatives.

The leaves and tops of *C. olitorius* are eaten as a vegetable, for instance, in Malaysia, Indonesia and the Philippines. In Africa and the Middle East, *C. olitorius* is mainly grown and used for this purpose and not for its fibre. The leaves of *C. capsularis* are applied medicinally in Peninsular Malaysia to poultice sores, and in decoction, they are used to treat dysentery, phthisis and coughs and as a tonic for children. In the Philippines, the leaves of *C. capsularis* are used to treat headaches.

A great deal of research is presently going on in each of these fields; however, the largest potential markets are in composite products. These composites range from value-added speciality products to very large volume commercial materials. These markets are potentially larger than the past markets for jute and could lead to new dynamic uses. Technical textiles, namely geotextiles, especially for soil erosion control and rural road construction, are another area of potential use in huge volume.

2.7 Conclusions

A few success stories illustrate that jute, the golden fibre, could make a comeback. For example, a jute mill in Bangladesh recently reduced its losses by developing and introducing linoleum fabric, which is being used as an industrial material. Some manufacturing units of India are profitably making attractive bags and diversified jute products including shopping bags. It is perceived that demand for home textiles, particle board, jute-based composites, technical textiles, etc., is increasing. People around the world are becoming more conscious about the pollution caused by synthetics and are increasingly opting for natural fibre products.

It is evident that worldwide use of more traditional jute products and new, alternative and nontraditional items together with the diversified jute products would certainly rejuvenate the jute sector and would reduce pollution to a great extent. It is likely that development of the diversified sector would provide more employment opportunities and alleviate poverty.

2.7.1 *Environmental advantages*

Jute is a natural fibre and an annually renewable resource with high biomass production per unit land area. Moreover, leaf and crop waste, left out in the field, is transformed into organic materials, thereby reducing demand for supplementary chemical fertilizers for subsequent crops. In terms of global warming, a concern of great importance in the modern world, jute and jute products are proven to be absolutely harmless.

The following are among its numerous environmental advantages:

- In the 100 days of the jute-growing period, 1 hectare of jute plants can absorb about 15 MT of CO_2 from the atmosphere and liberate about 11 MT of O_2.
- Jute plants purify air and have carbon dioxide (CO_2) assimilation power several times higher than that of trees.
- The environmental impacts of jute production are much less harmful because in jute agriculture, the use of chemical fertilizers, pesticides and weed killers/fungicides is scanty.
- The biological efficiency of jute is much higher than that of wood as it is a fast growing annual crop that takes only 4–5 months to mature.
- Wastes from jute production are biodegradable and can always be used as manure.
- Jute products require only 7% of the energy required for the production of its synthetic counterparts.

- Total energy consumption during jute agriculture and jute sack production is 10.0 GJ per MT of jute fibre.
- Jute is biodegradable and therefore does not cause any environmental pollution during disposal.

2.7.2 Socioeconomic impact

It is worth mentioning that the subsistence farmers still grow jute mainly using family labour. So they can afford to continue with jute production. This crop is an important source of cash from the sale of jute fibres for the farming community. The farmers also need jute sticks for their domestic purposes.

There is a dilemma, however, for the producing countries in providing the farmers with adequate return while at the same time satisfying the requirements of the world market.

Farmers and processors also face technical and economical problems in cultivating and processing fibre crops. Information and research can solve technical problems, such as lack of expertise in cultivation of fibre crops, ignorance of new extraction methods, poor harvesting machinery and fibre quality inconsistency. Economic problems are the price level of plant fibres, inconsistency of prices, competition with other fibres, limited and small markets, uncertain financial returns, required capital investments and storage and transportation costs. The economic problems may be reduced when the costs of harvesting and processing become lower due to simpler methods and shorter production–trade chains or if productivity is increased along with fibre quality.

2.8 Sources of further information and advice

Further information may be available from the IJSG website—www.jute.org—and its publications uploaded in the site.

2.8.1 Key books

- Dempsey, J.M., 1975. Fiber Crops. University Press of Florida, Gainesville, FL.
- Frank, R.R. (Ed.), 2005. Bast and Other Plant Fibres. Woodhead Publishing in Textiles. CRC Press, London.
- Ghosh, P., 2004. Fibre Science and Technology. Tata McGraw-Hill Publishing, New Delhi.
- Ghosh, T., 1983. Handbook on Jute. FAO (Food and Agriculture Organization of the United Nations), Rome.
- Kirby, R.H., 1963. Vegetable Fibres. World Crops Books. London: Leonard Hill (Books).
- Kundu, B.C., Basak, K.C., Sarkar, K.C., 1959. Jute in India. Indian Central Jute Committee, Calcutta.
- Lewin, M., Pearce, E.M. (Eds.), 1998. Handbook of Fiber Chemistry. Marcel Dekker, New York.
- Morton, W.E., Hearle, J.W.S., 1986. Physical Properties of Textile Fibres. Textile Institute, Manchester.

- Nakamura, A., 2000. Fibre Science & Technology (Translated from Japanese). Oxford and IBH Publishing, New Delhi.
- NIIR Board of Consultants and Engineers, 2005. Natural Fibres: Handbook with Cultivation and Uses. National Institute of Industrial Research, Delhi, India.
- Roy, S., 2010. Jute Basics. International Jute Study Group (IJSG), Dhaka, Bangladesh.
- Sur, D., 2005. Understanding Jute Yarn. Anindita Sur, Kolkata, India.

2.8.2 Major trade and professional bodies

The major jute trade associations/bodies are provided below with their email addresses or other contact numbers.

- Bangladesh Jute Association (BJA), Dhaka, emails: bjute@bangla.net, beejay@bangla.net
- Bangladesh Jute Exporters Association (BJEA), Dhaka, Tel: ++880 2 955 2910, 955 1515
- Bangladesh Jute Goods Association (BJGA), Dhaka, Fax: ++880 2 955 0664
- Bangladesh Jute Mills Association (BJMA), Dhaka email: bjmajute-good@agnionline.com
- Bangladesh Jute Spinners Association (BJSA), Dhaka, email: bjsa_bd@yahoo.com
- Burlap and Jute Association (Association du Jute), Ohio, USA, email: sspiegel@daybag.com
- European Association for the Trade in Jute and Related Products (EUROJUTE), Germany, emails: info@wgc.de; corchorus@wgc.de
- Gunny Trades Association (GTA), Kolkata, Tel:++91 33 220 3233
- Hanoi Jute Company, Hanoi, Vietnam, email: hanoijute@fpt.vn
- Indian Jute Mills Association (IJMA), email: ijma@cal2.vsnl.net.in
- Japan Jute Products Import Council (JJPIC), Osaka, Japan, Tel: ++81 6 227 1841, Fax: ++ 91 6 202 5585
- Jute Baler's Association, Tel: ++91 33 2220 2805
- Myanma Jute Industries (MJI), Yangon, Myanmar, email: mji@myanmar.com.mm
- Nepal Jute Mills Association (NJMA), Biratnagar, Nepal, emails: rkg@golchha.com; swastik@cworld.com.np; brt@golchha.com; sanjay@ golchha.com
- The Pakistan Jute Mills Association (PJMA), Lahore, Pakistan, emails: secretary@pjma.com; info@pjma.com.pk; azhar@habib-jute.com; pjma2000@yahoo.com

2.8.3 Research and development organizations and groups

The important R&D organizations/interest groups for jute are listed below:

- Bangladesh Jute Mills Corporation (BJMC), Dhaka, email: bjmc@bttb.net.bd
- Bangladesh Jute Research Institute (BJRI), Dhaka, emails: info@bjri.gov.bd, infobjri@yahoo.com
- Central Research Institute for Jute & Allied Fibres (CRIJAF), Barrackpore, Kolkata, India, email: crijaf@wb.nic.in
- Indian Jute Industries' Research Association (IJIRA), Kolkata, India, emails: director@ijira.org; ijiraweb@ijira.org
- Institute of Bast Fiber Crops (IBFC), Hunan Province, China, emails: ibfczyx@163.com, jgsu@vip.163.com
- Institute of Jute Technology (IJT), Kolkata, India, email: ijt@cal2.vsnl.net.in
- Jute Corporation of India Limited (JCI), Kolkata, India, email: jutecorp@vsnl.net

- Jute Diversification Promotion Centre (JDPC), Dhaka, email: edjdpc@ yahoo.com
- Jute Manufactures Development Council (JMDC), Kolkata, India, emails: jmdc@jute.com; jmdcindia@vsnl.com
- National Centre for Jute Diversification (NCJD), Kolkata, India, email: ncjd@vsnl.com
- National Institute for Research on Jute and Allied Fiber Technology (NIRJAFT), Kolkata, India, email: nirjaft@wb.nic.in
- South India Textile Research Association (SITRA), Coimbatore, India, E-mails: sitra@vsnl.com; sitraindia@dataone.in

2.8.4 Websites

www.jute.org; www.motj.gov.bd; www.Juteenterprisebd.com; www.bjmc.gov.bd; www.bjri.gov.bd; www.bjsa.org; www.texmin.nic.in; www.jmdcindia.com; www.jute.com; www.juteworld.com; www.iuteenterprise.in; www.ijira.org; www.ijtindia.org; www.sitraindia.org; www.crijaf.org.in; www.nirjaft.res.in; www.jutecorp.com; www.worldjute.com; www.jute-industry.com/ www.pjma.com.pk; www.eurojute.com; www.iwn.inf.poznan.pl.

Bibliography

Abdullah, A.B.M., Lutfar, L.B., Matin, N., 1992. An Introduction to Jute/Allied Fibres Properties and Processing. IJO, Dhaka, Bangladesh.

Alam, A., 1993a. Improved retting and extraction of jute and kenaf. In: Proceedings of the International Workshop. IJO-FAO, Malang, Indonesia.

Alam, A., 1993b. Ecological attributes of cultivation and primary processing of jute and kenaf: retting and the environment. In: International Consultation on Jute and Environment. FAO, The Hague, Netherlands.

Braungrat, M., Englefried, J., Hansen, K., Mulhall, D., Neumann, M., 1992. Jute and polypropylene, environmentally intelligent products. In: Comparative Impact Assessment, EPEA. Umwelt Institute, Hamburg, Germany.

Central Pollution Control Board, Ministry of Environment & Forests, 1992. Study of Retting of Jute Fibre – Its Impact on the Environment. Government of India.

De, R.N., 1995. Jute and the Environment. International Jute Organisation, Dhaka, Bangladesh.

Dempsey, J.M., 1975. Fiber Crops. University Press of Florida, Gainesville, FL.

European Commission, 1994. Industrial Fibre Crops. Studies by Science Research Development, Agro-Industrial Research Division, EC.

Ghosh, T., 1983. Handbook on Jute. FAO (Food and Agriculture Organization of the United Nations), Rome.

Ghosh, P., 2004. Fibre Science and Technology. Tata McGraw-Hill Publishing, New Delhi.

Groenewegen, P., Van Overbeeke, G., 1994. Jute & Life Cycle Analysis – A Review on a Comparative Impact Assessment. EPEA, Department of Physics and Astronomy, Amsterdam, Netherlands.

Guha Roy, T.K., 1996. The Problems of Water Pollution Related to Jute Industry and its Control. Indian Textile Annual & Directory, p. 175.

IIT–Kharagpur, JMDC, 2000. Comparative Study of Jute & Polypropylene in Respect of Their Relative Costs and Advantages (Report). India).
James, A.D., 1983. Medicinal use of jute. In: A Handbook of Energy Crops. Purdue University. Center for New Crops & Plants Products. www.hort.purdue.edu/newcrop/duke_energy/dukeindex.html.
Jarman, C.G., 1985. The Retting of Jute. FAO Agricultural Services Bulletin No. 60. Food and Agriculture Organization of the United Nations, Rome, p. 53.
Karmakar, P.G., Hazra, S.K., Ramasubramaniam, T., Mandal, R.K., Sinha, M.K., Sen, H.S. (Eds.), 2008. Jute and Allied Fibre Updates: Production and Technology. CRIJAF, Kolkata, India.
Liu, A., 2000. 'Jute – an environmentally friendly product'. In: Chrispeels, E. (Ed.), International Commodity Organisations in Transition. UNCTAD, Geneva. Chapter 15.
Morton, W.E., Hearle, J.W.S., 1986. Physical Properties of Textile Fibres. Textile Institute, Manchester.
Nakamura, A., 2000. Fibre Science & Technology (Translated from Japanese). Oxford & IBH Publishing, New Delhi.
NIIR Board of Consultants & Engineers, 2005. Natural Fibres: Handbook with Cultivation and Uses. National Institute of Industrial Research, Delhi, India.
Pan, N.C., Day, A., Mahalanabis, K.K., 2000. Properties of jute. Indian Text. J. 110 (5), 16–23.
Rathindranath, D., 1995. Jute and the Environment. IJO, Dhaka.
Rowell, R.M., 1993. International Consultation on Jute and the Environment. Food and Agriculture Organization of the United Nations, The Hague.
Roy, S., 2010. Jute Basics. International Jute Study Group (IJSG), Dhaka, Bangladesh.
Stout, H.P., 1988. Fibre and Yarn Quality in Jute Spinning. Textile Institute, Manchester.
Sur, D. (Ed.), 2003. Indian Jute – A New Symphony. JMDC, Ministry of Textiles, Government of India, Kolkata.
Sur, D., 2005. Understanding Jute Yarn. Anindita Sur, Kolkata, India.

Bast fibres: ramie

3

S. Roy[†], Latifa Binte Lutfar[1]
[1]Former Operations Officer, Dhaka, Bangladesh

3.1 Introduction to ramie

Ramie is one of the oldest fibre crops, having been used for at least 6000 years. Ramie fibre is one of the strongest and longest natural fine textile fibres in the world. It is a bast fibre derived from the bast layer of the stem, that is, phloem of the vegetative stalks of the plants.

3.1.1 Origin

This dicot, angiosperm, semiperennial shrub of the Nettle family Urticaceae is a native of the Far East and probably originated in the mountain valleys of southwestern China.

3.1.2 History

Ramie is reported to have been used in mummy cloths in Egypt during 5000–3000 BC. Since prehistoric times, ramie has been used in China, India and Indonesia. It was used for Chinese burial shrouds over 2000 years ago. Ramie was mentioned and praised as grass cloth in the Sanskrit poems of *Kalidasa* and *Ramayana*. It was used in the south of Russia about 900 BC. It is said to have been grown in China for many centuries and was one of the principal fibres used in ancient China for making cloth previous to the introduction of cotton around AD 1300. Ramie was first introduced from the East Indies to Holland in 1733, France in 1844, Germany in 1850, England in 1851 and Belgium in 1860. In 1857, ramie plants were introduced into the United States from Java and planted in the Botanical Gardens in Washington.

Ramie fibre, also known as China grass, grass linen or Chinese silk, was exported by China to the Western world at the beginning of the 18th century. However, it was not until 1930 that ramie textile production was established on a commercial basis in Western Europe. Commercial ramie production in Brazil first began in the 1930s with production peaking in 1971. In Japan and the Philippines, concentrated efforts were made to produce ramie during the Second World War. With the establishment of sizeable ramie acreages in South Florida, commercial-scale processing equipment was developed and operated successfully from 1946 to 1955. Ramie's popularity actually increased in the mid-1980s with the fashion emphasis on natural fibres.

[†] Deceased author.

Handbook of Natural Fibres. https://doi.org/10.1016/B978-0-12-818398-4.00004-9

3.1.3 Adaptation/agroclimatic conditions

Ramie is adapted to a wide range of latitudes from almost equatorial conditions to about latitude 45°N in Russia. In the temperate areas, between latitudes 25° and 38°N or S, ramie may produce two to three crops annually. In subtropical areas, between 20° and 25°N or S, four to five crops may be harvested, usually with supplementary irrigation. At latitudes below 10°N or S, it may be possible to harvest six or more crops annually, but irrigation must be provided for 2–3 months when there is insufficient rainfall. Ramie has been found to grow well in humid climates under moderate temperature. It requires a uniformly well distributed rainfall of 1500–3000 mm annually. The optimum temperature for good harvest is around 20–31°C, and the relative humidity should be at least 25%. However, the crop is sensitive to waterlogging, frost and strong winds.

3.1.4 Areas of production

China, mainly the central and southern part, leads the world in the production of ramie. Other major producers of ramie fibre are Japan, Taiwan, Brazil, the Philippines, Korea, Indonesia and India.

3.1.5 Economic importance

Ramie, the longest and one of the strongest fine textile fibres, has been grown experimentally throughout the tropical, subtropical and temperate zones of the world.

Ramie has been proved quite remunerative when grown under favourable edapho-climatic conditions. The income generally starts from the second year and continues thereafter. Fibres up to 4%–5% by weight of total biomass may be obtained from ramie. Raw fibre yield of up to 1.6–2.2 ton per hectare may be harvested per year under ideal conditions.

However, it is of secondary importance in world trade despite its unique characteristics. This is mainly due to a lack of suitable large-scale fibre extraction equipment until recent years and costly methods of degumming, spinning and weaving of the fibre.

Only a small portion of the ramie produced is available in the international market to be imported mainly to Japan, Germany, France and the United Kingdom.

3.2 Types of ramie

Different types of ramie of different quality originate from *Boehmeria* species of different habitats, some being herbs, some others are trees or shrubs.

3.2.1 Taxonomy

Ramie is a perennial plant belonging to the genus *Boehmeria* under Urticaceae or Nettle family of the order Urticales and class Magnoliopsida. There are about 100 species under the genus *Boehmeria.*

3.2.2 Main species

Boehmeria nivea L. Gaud., commonly known as 'white ramie', and *Betula utilis* generally referred to as *B. nivea* var. *tenacissima* and also known as 'green ramie' are the two major species of ramie fibre. This green ramie is probably a hybrid of white ramie.

The true ramie or China grass also called 'white ramie' is the Chinese cultivated plant. The second type, the 'green ramie' or rhea, is believed to have originated in the Malay Peninsula.

The two varieties differ in their habitat: while *B. nivea* grows best in temperate and subtropical regions, green ramie occurs in tropical regions. Green ramie or *B. utilis* is said to give higher fibre yield and stronger fibre than *B. nivea,* although the fibre quality is not so good in regard with its fineness and colour.

3.2.3 Botanical description

Ramie is the only member of the family used commercially for fibre production and grows well in warm climates. It is a type of shrub. The plant is an erect, usually non-branching, tall, fast-growing herbaceous perennial reaching 1–2.5 m in height at maturity. The leaves of ramie are heart shaped, 7–15 cm long and 6–12 cm broad and white on the underside of white ramie and green on the underside of green ramie with dense small hairs – this gives a silvery appearance; unlike nettles, the hair strands do not sting.

3.2.4 Cultivation

Fertile, deep loam and well-drained sandy loam soils are suitable for ramie cultivation. Soil rich in organic matter with an optimum pH range from 6 to 7 is preferred (Fig. 3.1).

Because this is a perennial plant, plantings are mainly made from asexual plant materials. The seeds of ramie are used only for selection work, as the seedlings rarely develop into plants having desired qualities. Most ramie seedlings are almost unsuitable for fibre. Ramie, thus, may be propagated by the following four methods in order of importance: (1) rhizome cuttings, (2) division of parent rootstock, (3) laying and (4) stem cuttings.

Figure 3.1 A view of a ramie field.

3.2.5 Harvesting

Ramie, being a perennial plant, is normally harvested two to three times a year, but under favourable growing conditions, it can be harvested up to six times per year.

Ramie plants produce a large number of unbranched stems from underground rhizomes and have a crop life of 6–20 years. Thus, harvesting of ramie can be done intermittently for up to 20 years.

Ramie should be harvested just before flowering or soon after flowering starts. The time of harvesting is important for getting the best fibre and also to obtain maximum fibre yield from the plants. Stems are harvested by either cutting just above the lateral roots or bending the stem.

3.2.6 Fibre extraction

The extraction of ramie fibre occurs in three stages. First, the cortex or bark is removed just after harvesting while the plants are still fresh, by hand or machine. This process is called decortication. Present-day decortication is vastly mechanized. Second, the cortex is scraped to remove most of the outer bark, the parenchyma in the bast layer and some gums and pectins. Finally, the residual cortex material is washed, dried and degummed to obtain the spinnable fibre.

3.2.7 Degumming

Degumming, the next step, is the process of removal of gum from the decorticated fibre. About 20%–30% of natural adhered gums, holding the fibres in dense strands, are removed so as to make the fibres suitable for spinning and weaving. Degumming may be done by different methods, namely chemical method, microbial method, etc.

3.2.8 Fibre quality and grading

Ramie fibre generally is graded according to length, colour and cleanliness. Top grades usually are washed and sometimes brushed. There is no standard set of grades for ramie fibre, but several countries have set up their own grades.

Eight ramie grades are established in China, four in Japan, three in Brazil and four in the Philippines.

3.3 Fibre morphology

Ramie is a multicellular bast fibre, by and large cellulosic in nature, having very little lignin and hemicellulose (Table 3.1). The intercellular binding constituents present in significant amounts are natural gums and pectinous matters. The cells of ramie fibre may be as long as 40–45 cm, cylindrical in nature and characterized by thick walls and narrow, curved lumens. The surface of the cell is marked by distinct ridges.

3.4 Properties of ramie

The physical properties of ramie are given in Table 3.2, while the thermal properties are given in Table 3.3. The raw ramie fibre strand has an average length of 0.61–1 m. The longer fibres are sometimes more than 1.5–2 m in length. These are not single fibres, rather a bundle of shorter single fibres, as in other bast fibres, held together

Table 3.1 Chemical composition of ramie.

Constituent	Content (%)
Cellulose	68.6–76.2
Hemicellulose	13.1–16.7
Lignin	0.6–0.7
Pectin	1.9
Fats and wax	0.3
Other extractives	6.1

Table 3.2 Physical properties of ramie.

Property	Value
Length, L (mm)	60–250
Breadth, B (μm)	15–80
Width (μm)	50
L/B ratio	3500
Fineness (tex)	0.4–0.8
Microfibrillar angle (deg)	7.5
Density (mg/m^3)	1.50
Tenacity (gf/tex)	40–65
Breaking elongation (%)	3.0–4.0
Flexural rigidity (dynes.cm^2)	0.8–1.2
Diameter swelling (%)	12–15
Volume swelling %	32.0
Moisture regain (65%–100% RH)	6.5–17.5
Moisture content (wt%)	8.0

RH, relative humidity.

Table 3.3 Thermal properties of ramie.

Property	Values
Specific heat	1.36×10^3 J/kg/K
Thermal conductivity	427.3 mW/m/K
Heat of combustion	17.5 J/g
Ignition temperature	193°C
Heat of wetting	18.2 Cal

by gummy and pectinous matters. The elementary cells/single fibres of ramie are longer and thicker than all other bast fibres. Ramie is characterized by its exceptionally long ultimate fibre cells that average to about 150 mm in length and highest length/breadth ratio of ultimate cell (3000).

Properly degummed ramie fibre is the longest and the strongest of all vegetable fibres. It is lustrous, possesses high tensile strength, is extremely absorbent, gains strength appreciably when wet and is highly resistant to bacteria, mildew, insect attack and rot. Ramie absorbs moisture and gives it up quickly with almost no shrinking and stretching. Thus, ramie has very good comfort properties and is suitable for summer clothes. Very low elasticity, low abrasion resistance, lack of resiliency, stiffness and brittleness, necessary degumming process and high cost are some of the weak points of ramie.

Ramie resists action of chemicals better than most of the other natural fibres do. It is dye absorption efficient and fast and has considerable resistance to microbial attack.

3.5 Typical applications

Historically, ramie has been used as a precious material for very fine upper cloth including high-grade *kimono* cloth, especially in Japan. The Korean traditional costume, *hanbok* – made of ramie – is renowned for its fineness. Fabrics of ramie are lightweight and silky, similar in appearance to linen and suitable for a wide range of garments and home textiles.

Coarse ramie fibres are generally used for making twines and threads, for which its strength and lack of stretch make it most suitable. Because of its high wet strength, quick dryability and considerable resistance to bacterial action, it is very useful for making fishing nets.

Ramie is used in many diverse applications such as suiting, shirting, sheeting, dress materials, table cloths, napkins, towels, handkerchiefs, fine furniture upholstery, draperies, mosquito netting, gas mantles, industrial sewing thread, packing materials, fishing nets, fire hose, belting, canvas, marine shaft packing, knitting yarns, hat braids, filter cloths, etc.

Because ramie has low elasticity and resilience, it is usually blended with other textile fibres such as cotton, wool, etc. It increases the lustre and strength of cotton blends and reduces shrinkage in wool blends. It is also blended with silk.

Shorter fibres and waste, called noils, blended with cotton/stapled rayon are used for making low-grade fabrics such as dish towels. The noils are also used in the manufacture of high-grade specialty papers such as cigarette paper, bank note/currency paper, etc.

For the 2010 Prius, Toyota will begin using a new range of plant-derived eco-friendly bioplastics made from cellulose in wood or grass, instead of petroleum. One of the two principal crops to be used is ramie.

3.6 Conclusions

The use of ramie is limited by its price and spinning properties. Apart from any economic or technical reasons, the development of any new fibre needs market development. Despite its excellent properties, wide occurrence/cultivation/adaptation of the plant and reasonable publicity, ramie fibre has not been developed or used to the expected level especially outside China and Japan.

Reliable figures relating to production of ramie in China, as the leading ramie producer of the world, are still very difficult to obtain.

The economic factors that prevent farmers from cultivating fibre crops include price level, inconsistency of prices, competition with cheaper synthetic fibres, too few off-takers, uncertain financial returns and the required capital investment in machinery. In addition, the production costs of plant fibres are relatively high compared with man-made fibres.

3.7 Sources of further information and advice

Important literature and organizations supplying information on ramie and other fibre crops are listed below.

3.7.1 Key books

- Dempsey, J. M. (1975), *Fibre Crops.* Gainesville, FL: University Press of Florida.
- European Commission (1994), *Industrial Fibre Crops.* Studies by Science Research Development, Agro-Industrial Research Diivision, EC.
- Ghosh, T. (1983), *Handbook on Jute.* Rome: FAO (Food and Agriculture Organization of the United Nations).
- Nakamura, A. (2000), *Fibre Science & Technology* (Translated from Japanese). New Delhi: Oxford & IBH Publishing.
- NIIR Board of Consultants & Engineers (2005), *Natural Fibres: Handbook with Cultivation and Uses.* Delhi, India: National Institute of Industrial Research.

3.7.2 Research and development organizations and groups

- Central Research Institute for Jute and Allied Fibres (CRIJAF)
- Institute of Bast Fibres (IBFC), China
- Institute of Natural Fibres (INF), Poznan, Poland
- National Institute for Research on Jute and Allied Fiber Technology (NIRJAFT), India
- South Indian Textiles Research Institute (SITRA), India.

Further reading

Angelini, L.G., Lazzer, A., Levita, G., Fontanelli, D., Bozzi, C., 2000. 'Industrial crops and products: ramie *(Boehmeria nivea* (L.) Gaud.) and Spanish Broom (*Spartium junceum* L.) fibres for composite materials', *Agronomical aspects.* Morphol. Mech. Prop. 11 (2–3), 145–161.

Barbara, S., Smith, J., 2009. Ramie: Old Fiber – New Image. Ohio State University Extension Fact Sheet, Columbus, Ohio 43210-1295, USA. Available from: http://ohioline.osu.edu/hyg-fact/5000/5501.html.

Guo, Q.Q., Chen, J.R., Yang, R.F., Hu, R.S., 1998. Callus induction and shoot regeneration from leaves of Ramie *(Boehmeria nivea* Gaud). China's Fiber Crops 20, 1–4.

Kozlowski, R., Rawluk, M., Barriga-Bedoya, J., 2005. Ramie, chapter 5. In: Franck, R.R. (Ed.), Bast and other Plant Fibres. Woodhead Publishing, Cambridge, pp. 207–227.

Misra, S.P., 2000. A Textbook of Fibre Science and Technology. New Age International (P) Ltd, New Delhi, India.

Mohanty, A.K., Misra, M., Hinrichsen, G., 2000. Biofibres, biodegradable polymers and biocomposites: an overview. Macromol. Mater. Eng. 276–7 (1), 1–24.

Natural Fibre Ramie, 2009. International Year of Natural Fibres 2009 (IYNF 2009) (Rome, Italy).

Saha, M.N., Sen, H.S., 2007. Ramie – A Fibre of Prospect, Bulletin No. 16. CRIJAF (ICAR), West Bengal, India.

Susan, B.H., Mary, L.Y., 1989. Ramie: patterns of world production and trade. J. Text. Inst. 80 (4), 493–503.

Wood, I., 1999. Ramie: the different bast fibre crop. Aust. New Crops Newsl. (11) Queensland 4069.

Bast fibres: kenaf

4

Jiantang Xu [1], *Aifen Tao* [1], *Jianmin Qi* [1], *Yufu Wang* [2]
[1]Fujian Agriculture and Forestry University, Fuzhou city, Fujian Provence, China;
[2]Institute of Bast Fibres (IBFC), Changsha city, Hunan Province, China

4.1 Introduction to kenaf

Kenaf (*Hibiscus cannabinus* L.) is an annual herbaceous, short-photoperiod plant that contains high-quality cellulose. The genus Hibiscus has a cosmopolitan distribution and contains more than 300 species. Kenaf is classified in section Furcaria. The diversity of the chromosomes number and genomes found in this section is unusual in the plant kingdom. Morphological and anatomical characters are helpful for selection of suitable kenaf genotypes for fibre and paper production. Kenaf has a large, light yellow to purple, bell-shaped flower with a widely open corolla. The flowers are 8–10 cm in diameter with five petals (H'ng et al., 2009).

Kenaf is characterized as a fast growing, high-yield plant with strong stress resistance and adaptability, which stem has traditionally been grown as a source of soft fibre and used as a substitute for jute (Williams, 1994; Xu et al., 2013). It has been cultivated since around 4000 BC for food and fibre and was first introduced to Taiwan, China, in 1908. Currently there is considerable interest around the world in using kenaf stem as the raw material of paper pulp, which has been found to be an important source of fibre for composites and other industrial applications (Thiruchitrambalam et al., 2012).

4.1.1 *Origin and history*

Kenaf is native to east-central Africa, and a common wild plant of tropical and subtropical Africa and Asia. For over 6000 years, kenaf was primarily used as cordage crop and secondarily as a livestock feed. It has been a source of textile fibre for such products as rope, twine and rugs. Kenaf is a good source of raw material fibre for pulp, paper and other fibre products, and has been introduced since World War II in China, Russia, Thailand, South Africa, Egypt, Mexico and Cuba. The production and retting process for kenaf remained basically unchanged for thousands of years; it was first domesticated in northern Africa, including its introduction to India 200 years ago, to Russia in 1902, and to China in 1908. In the United States, kenaf research and production began in the 1940s during World War II to supply cordage material for the war effort.

More than 500 plant species were evaluated in the United States in the 1950s in order to cover the increasing future fibre demands in the USA. Kenaf was identified as a possible substitute for jute. In the late 1970s, USDA researchers were evaluating

Handbook of Natural Fibres. https://doi.org/10.1016/B978-0-12-818398-4.00005-0

various plant species to fulfil future fibre demands in the United States, it was determined that kenaf was an excellent cellulose fibre source for multiple paper products (newsprint, bond paper and corrugated liner board). In the early 1970s, kenaf was first introduced in Malaysia and it was highlighted in the late 1990s as an alternative and cheaper source of material for producing panel products such as fibreboard and particleboard, textiles and fuel (Ayadi et al., 2017).

4.1.2 Taxonomy

In English, the common name of kenaf in India is mesta, ambadi, palungi, deccan hemp, and Bimli jute; in Taiwan is ambari; in Egypt and Northern Africa is til, teel, or teal and Gambo or dad in West Africa, in South Africa is stokroos; in Indonesia is Java jute; in Brazil is papoula de Sao Francisco; in West African is dah, gambo and rama (Monti et al., 2013).

The *Hibiscus* genus is widespread with more than 400 species; it is divided into six different sections: *Furcaria, Alyogen, Abelmoschus, Ketmia, Calyphyllia*, and *Azanza*. Kenaf (*Hibiscus cannabinus* L.) and *roselle* (*Hibiscus sabdariffa* L.) are classified taxonomically in the *Furcaria* section. This group includes important fibre, food, and medicinal plants. In the *Furcaria* section, the chromosome number is a multiple of 18 in all the species, from $2n = 36$ to $2n = 180$ with high level of polyploidy, kenaf having 36 chromosomes and a genome diploid (Ayadi et al., 2017). *Hibiscus cannabinus* L. belongs to the Malvaceae family and the Hibiscus genus which is known for both its economic and horticultural importance and is closely related to cotton (*Gossypium hirsutum* L.) and okra (*Hibiscus esculentum* L.).

4.1.3 Botanical description

Kenaf is divided in two groups, the early and the late-maturity varieties according to their reaction to flowering. Most kenaf cultivars are photo-sensitive, but some do not flower until day length decreases to 12.5 h. Later-maturing varieties produced higher fibre yields compared with early-medium varieties.

4.1.3.1 Root

The plant has a long effective tap root system and a relatively deep, wide-ranging lateral root system making the plant drought tolerant (Stricker et al., 2001). Further to that, kenaf with its tap root system is considered to be an excellent user of residual nutrients from previous crops. Kenaf root has deep root exploration that reaches to more than 1 m depth. Kenaf roots are sensitive to plow-pan and compactions.

4.1.3.2 Stems

Kenaf stems are generally round with thorns ranging from tiny to large such as on a blackberry bush depending on variety. The stems colour are varying pure green to deep burgundy (Fig. 4.1). The stem breadth is often 1–2 cm and the thickness of the bast is often 1–2 mm. Kenaf plants tend to grow as a single unbranded stem

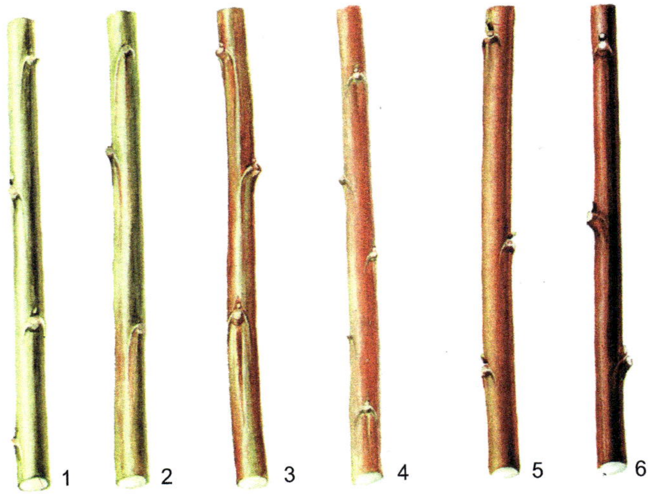

Figure 4.1 Different colours of stems. 1. Green, 2. yellow-green, 3. reddish, 4. light red, 5. red, 6. purple.

when planted at high production densities of 170,000–220,000 plants/ha. Kenaf has a high growth rate, rising to heights of 4–7 m in about 6–8 months. It can then yield a dry weight of 6000–10,000 kg/ha per year (Summerscales et al., 2010). The kenaf stem mainly consists of an outer part called the bark making up 35%–40% of total stem weight composed chiefly by long fibres (two to six mm) and an inner part called the core that has short fibres (0.6 mm) making up the remaining 60%–65% of total stem weight. The properties of the kenaf fibre are similar to that of jute fibres, and the structure of kenaf fibre is similar to that of other plant fibres (Table 4.1).

The bast contains average Cross and Bevan cellulose between 47% and 57%, alpha cellulose between 31% and 39%, lignin between 15% and 18%, pentosans between 21% and 23% and ash between 2% and 5%. The fibre has an average length of 2740 μm which is superior to wood (840 μm) and an average diameter of 20μ. As for the core, they contain average 34% alpha cellulose, 17.5% lignin, 19.3% pentosans and 2.5% ash. The average fibre length of the core is 600 μm (Table 4.1). However, this information varies related to the planting sites, planting time and how a kenaf is grown and harvested.

Table 4.1 The composition of kenaf fibre.

Name	Bast	Core
Cross and bevan cellulose	47%–57%	–
Alpha cellulose	31%–39%	34%
Lignin	15%–18%	17%
Pentosans	21%–23%	19.30%
Ash	2%–5%	2.50%
Fibre length	2740 μm	600 μm

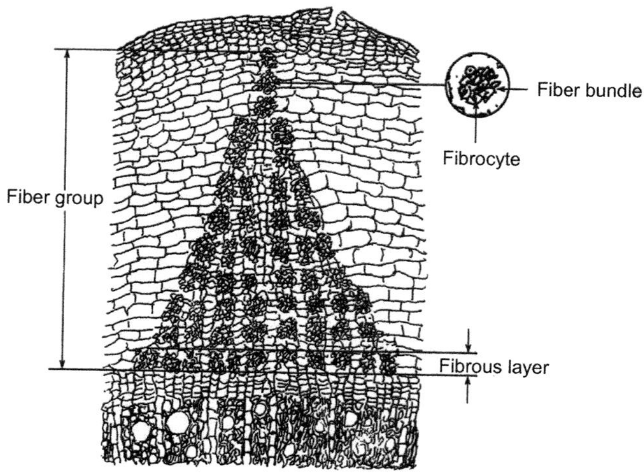

Figure 4.2 Strutcture of the position T.S of the dicot stem of kenaf.

Table 4.2 Chemical composition of some plant fibres.

Fibre type	Cellulose	Hemicelluloses	Lignin	Pectin
Kenaf	44–57	21	15–19	2
Jute	51–84	12–20	5–13	0.2
Ramie	68–76	13–15	0.6–1	1.9–2
Hemp	70–92	18–22	3–5	0.9
Flax	60–81	14–19	2–3	0.9
Cotton	82–96	2–6	0.5–1	5–7
Sisal	43–78	10–13	4–12	0.8–2
Wood	45–50	23–30	27	2–2.5

Source: Mwaikambo L. 2006. Review of the History, Properties and Application of Plant Fibres, vol. 7.

The fibres are formed by a combination of many cellulose cells (multicellular) which are bonded with non-cellulose materials. The vessels of the fibres are almost the oval and round shapes; with simple perforation, moderately large size, solitary and sometimes in pairs. The cell walls of the bast fibres appeared to be much thicker than that of core fibres (Fig. 4.2). Conversely, the lumens of the core fibres were found to be larger than that of bast fibres.

Table 4.2 shows the chemical composition of some plant fibres, and the physical properties of the plant fibres. It is noted that cellulose is the main constituent of plant fibres followed by hemicelluloses and lignin interchangeably and pectin, respectively. Cellulose is also the reinforcement for lignin, hemicelluloses and pectin. This makes plant fibres exhibit characteristics of a composites material (Mwaikambo, 2006).

These two fibres are so versatile because they can be used for many purposes such as to be converted to pulp, paper, cardboard, panels, traditional cordages, absorbent agent, packing materials, soil-less potting mixes, grass and flower mats and natural fuels.

4.1.3.3 Leaves

Kenaf leaf shape and petioles colour vary widely among varieties. It has two distinct leaf shapes, palmatifid and entire (Figs. 4.3 and 4.4). Petioles colour can be various shades of red, green, or purple.

Kenaf leaves can be consumed by human and animal because they contain high content of digestible protein. The average protein content in kenaf leaves range between 18% and 30%. Usually the leaves are consumed raw or boiled to greens. The leaves can be eaten as early as 10 days after planting and are also can be main ingredient for preparing delicious foods. The powder from dry leaves have been added in several kinds of food and it was found that the content of calcium and fibre was improved, while the taste remained the same. Its leaves can be also used as tea.

4.1.3.4 Flowers

The flowers of section Furcaria are characterized by having calyx with prominent central rib and two prominent marginal ribs. These rigid structures apparently are used for supporting the fragile and delicate petals. Also, the flowers of all species have more or fewer narrow bracts, which are borne below the calyx. The tip of these bracts may be unforked as in kenaf or forked, according to the sectional name Furcaria. Kenaf flowers are typical of *Hibiscus* genus, showing the characteristic fused statement column.

The flowers are large (7.5–10 cm), bell-shaped, and wide open with five petals. The flowers colour are range from light cream to dark purple, with a number of shades

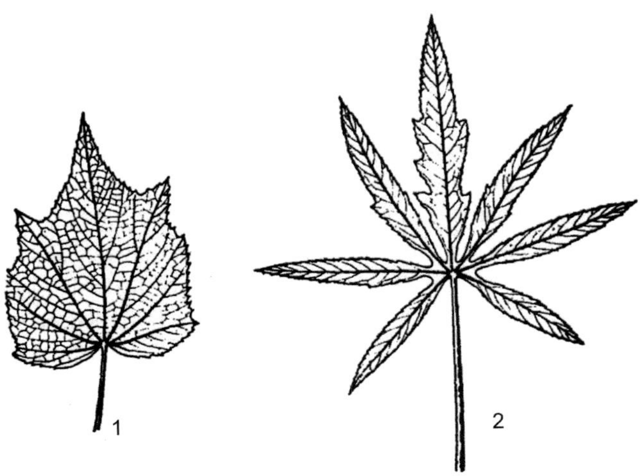

Figure 4.3 Leaf types of kenaf (1. whole leaf, 2. palmated leaf).

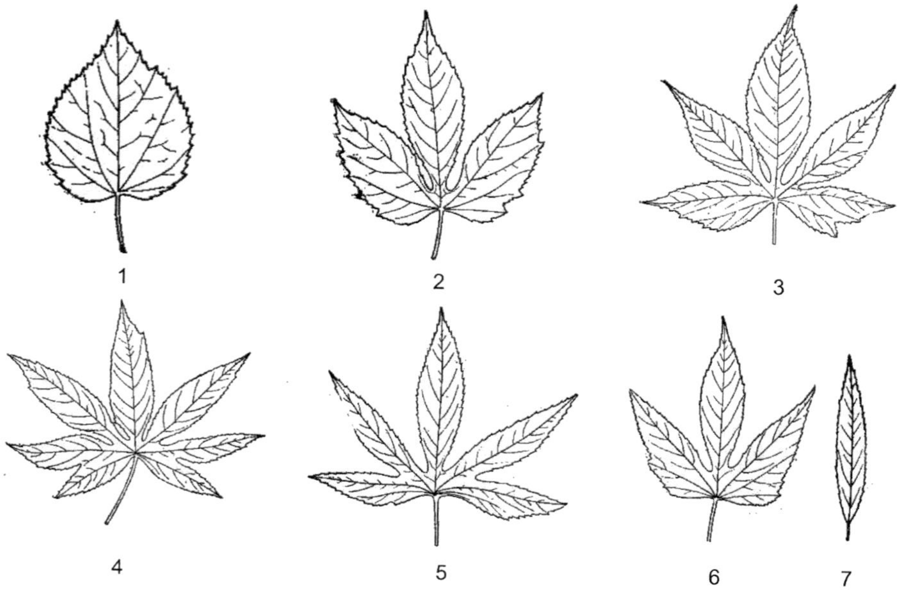

Figure 4.4 The different kinds of kenaf leaves during the growth. 1. Heart-shaped leaf, 2. three lobbed leaf, 3. five lobbed leaf, 4. seven lobbed leaf, 5. upside Five lobbed leaf, 6. upside three lobbed leaf, 7. lanceolate leaf.
Source: Bast fibre crops cultivation in China, 1993.

between them, but apparently never in bright yellow, pink, or red tones. Many cultivars have flowers with a deep red or maroon centre. They are borne on short stalks on the leaf axils on the upper portion of the stem and the main apical meristem retains its capacity for vegetative growth (Fig. 4.5).

While the flowering per plant can last 3–4 weeks or more, each individual flower blooms for only 1 day. The flowering early in the morning, begin to close about mid-day, and close in the mid-afternoon and never open again. The anthers release the pollen (which is bright orange or bronze colour) about the time when the flower opens, and the style emerges shortly thereafter. The five-part stigma expands; the lobes become turgid, but do not touch the anthers. The corolla closes spirally so that the anthers are pressed in contact with the stigma and self-pollination can be done unless cross-pollination has already been recorded.

Like its relative species such as cotton and okra, kenaf flowers and nectaries attract a number of insects, which pollinate (and cross-pollinate) the flowers. Kenaf has been classified on several occasions as a self-pollinating crop, but more recently, it has been classified as a cross-pollinate crop. Because of the moderate level of cross-pollination, the seed obtained is frequently not true to type. The cross-pollination crop in kenaf can be varied from 2% to 24%. Generally considered self-pollinated by the twisting closing movement of the petals, or cross-pollinated by domesticated honey bees (Ayadi et al., 2017).

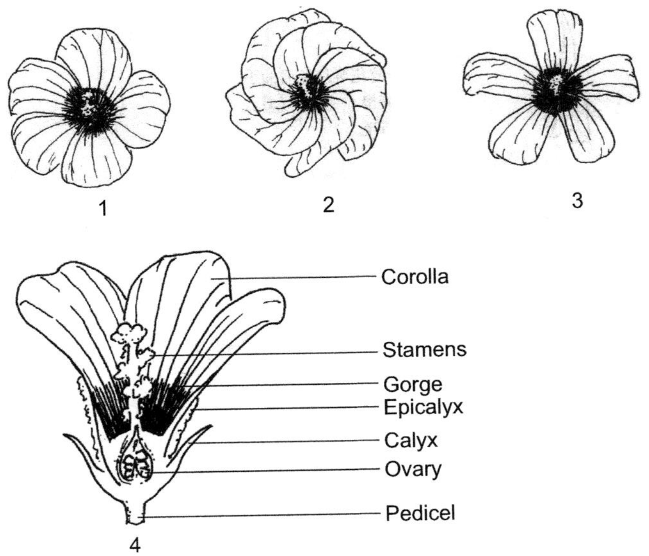

Figure 4.5 Morphological of kenaf flower. 1. Lapped, 2. spiral, 3. detached, 4. L.S of flower. *Source:* Descriptions and data standards for kenaf (*Hibiscus Cannabinus* L., 2006).

4.1.3.5 Fruits and seeds (Ayadi et al., 2017)

The fruit capsule that is about 1.9–2.5 cm long and 1.3–1.9 cm in wide (Fig. 4.6) is pubescent and contains five segments which are many-seeded (20–26 seeds). The seeds ovoid are brown, glabrous, wedge-shaped, 6 mm long and 4 mm wide, their weight is about 35,000 to 40,000 seeds/kg corresponding to 25–33 g 1000 grain weight.

The kenaf oil content in kenaf seed varies on average between 16% and 22% (Bakhtiari et al., 2011). Kenaf seed oil is considered edible for human because of the high quantities of monounsaturated and polyunsaturated fatty acids (Coetzee et al., 2008). The seed oil is reported to be cytotoxic towards leukaemia and cancer cells (Mohd et al., 2014).

Dempsey (1975) recommended kenaf as a profitable oil seed crop owing to its high oil content (20%) and seed yield. Kenaf seed oil is a great source of unique fatty acids (high oleic, linoleic and palmitic acid content) and it is considered as health promoting natural edible oil due to the presence of relatively high amount of monounsaturated and polyunsaturated fatty acids (Fig. 4.7) (Dhar et al., 2015).

Kenaf seed oil is a vegetable oil whose compositions are similar to cottonseed oil, soy bean oil, peanut oil, safflower oil, camellia oil and olive oil (Table 4.3), suggest that it is potential to be used as edible oil (Ruan et al., 2009). Depending on the processing temperature, kenaf seed oil producing from heated material is reddish brown with mild odour while it is odourless clear yellow oil if it is not heated. It is suitable for human consumption due to its unique fatty acid composition and plenty of function compounds (Cheng et al., 2016).

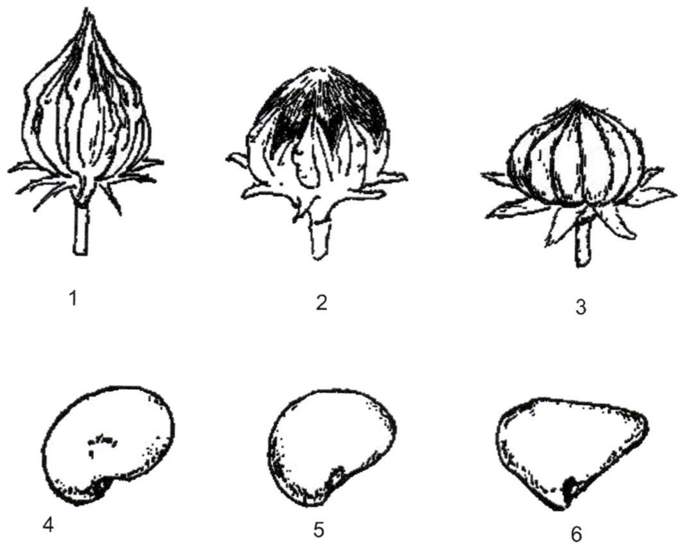

Figure 4.6 Kenaf fruits and seeds. 1. Peach-shaped, 2. suborbicular, 3. spheroidicity, 4. kidney-shaped, 5. sub-kidney-shaped, 6. triangle.
Source: Descriptions and data standard for kenaf (*Hibiscus Cannabinus* L.), 2006.

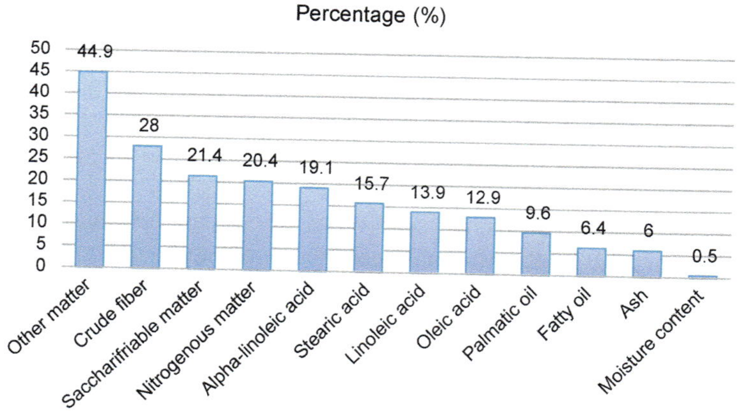

Figure 4.7 Kenaf seed fatty acids components.

Supply of high-quality seeds is needed for continuous and reliable kenaf production. However, seed production needs a particular climate and harvesting time. About 25% of kenaf seeds lose their viability because of suboptimal storage conditions. Kenaf seeds require drying and low temperature for storage. It is necessary to clean and dry the seed thoroughly for successful long-term storage. Artificial dryers can be used for this purpose. The seed can be stored at 8%−12% moisture level. Several studies have indicated that seeds should be stored at-20−5°C in a controlled humidity and temperature environment (Bakhtiari et al., 2011).

Table 4.3 The comparison of kenaf and other plants seed oil composition.

Cultivars	Palmitic acid C16:0(%)	Stearic acid C18:0(%)	Oleic acid C18:1(%)	Linoleic acid C18:2(%)	Linolenic acid C18:3(%)
Kenaf (Jinguang No.1)	17.31	1.67	28.3	51.81	1.09
Kenaf (FH952)	20.51	2.86	28.72	47.20	0.72
Kenaf (common)	15–19	2–6.8	28–51	25–43	0–1
Soy bean oil	10.8	3.6	20.8	50.2	7.7
Cotton seed oil	27	3.0	17.0	52	–
Peanut oil	10.7	2.64	46.1	29.0	1.1
Safflower oil	4–6	1–5	7–14	73–85	–
Camellia oil	8.8	1.1	82.3	7.4	0.2
Olive oil	9	1	80	8	–

4.1.3.6 Adaptation and agro-climatic conditions

Currently, many countries pay more attention to the collection of kenaf accessions and breeding of its high biological efficiency and wide ecological adaptability, for example, up to December 31, 2010, China had collected 1819 kenaf accessions (including roselle (*Hibiscus sabdariffa* L. var. *altissima* Wester)) (Zhigang Dai et al., 2012). The kenaf genetic resources could be classified according to the flowering time. It is argued that days of 50 % flowering and maturity may be the important traits for kenaf to differentiate the accessions according to the origins. The late to middle flowering accessions from different origins can be used to grow high fibre yield producing kenaf in the tropical environment (Golam et al., 2011). Kenaf genetic resources also could be divided into cultivars group, relatively wild species group, and wild species group by using molecular markers (ISSR and SRAP) (Xu et al., 2013). The results provide a useful basis for kenaf germplasm research and breeding. The main objectives of kenaf breeding were in developing the new varieties with higher yielding, resistance to pests and diseases, drought-resistant and ecological adaptability.

Kenaf is a tropical plant and adapt to a wide range of soils. Kenaf production fits well around the world. Planting and harvest activities will take place in the normally dry winter season while few if any field activities will be needed during the wet summer months. Kenaf needs a soil temperature of around 15°C for germination and growth. In China, kenaf could be planted beginning in early April (Stricker et al., 2001). Planting could continue through late April and harvest could begin in late October and extend through the end of February.

4.1.4 Cultivation

Like all the other cultivated crops, the determination of the appropriate crop management (varieties, sowing dates, plant populations, irrigation, fertilization, weed control and harvesting time) is a key factor for the successful insertion of the crop in the existing agricultural cropping systems with economic benefits. Kenaf sowing should take place in spring as soon as the soil temperature is higher than 15°C. When the crop is cultivated for its fibre stem, the plant population should fluctuate between 200,000 to 500,000 plants/ha and the row spacing from 35 to 50 cm (Fig. 4.8) (Alexopoulou et al., 2015).

In areas that the precipitation is limited, during the hot summer time, the irrigation is necessary for the crop to reach high biomass yields, the average yield of kenaf biomass can reach 23-25t/ha according to the conditions. Kenaf is also sensitive to nematodes, especially when it is cultivated in areas with sandy soil and this sensitivity should be taken under consideration in the rotation systems. Harvesting time and methods should be adjusted according to the purpose of the crop (e.g., for fibre, seeds, fibre and seeds, forage). The development of varieties or hybrids with higher yields (either for stem production or dual-purpose varieties for stems and seeds production) and improved resistance to drought as well as to nematodes and anthracnose is considered one of the main challenges for the future of the crop.

4.1.4.1 Areas of production

Kenaf originated from eastern Africa and through Egypt disseminated to India and China. During the World War II it was disseminated to USA and then to central and South America. Kenaf can be cultivated from 30°S to 45°N. Over the past decade, the total planting area of combined kenaf/jute in China ranks the third in the world next to India and Pakistan (together representing > 90% of the total kenaf area).

In China the mean cultivated area was 150,000 to 400,000 ha with an over 900,000 ha in 1985. However, kenaf production has been declined in the last years due to the strong competition by synthetic materials. In Far East countries

Figure 4.8 A view of kenaf field.

(2014−15), India contribute about 46% to world kenaf production(100 1000 tons), with China to contribute 26% to the total production, Pakistan with 0.46%, Indonesia with 1.6%, Africa with 6.8% (Fig. 4.9) (Alexopoulou et al., 2015). Kenaf was introduced to partially replace tobacco in Malaysia by 2010 (Ng et al., 2013).

Figs. 4.9 and 4.10 presented the world production for kenaf and other allied fibre crops during the year of 2010 and 2016. The far east countries contribute to 76.6%

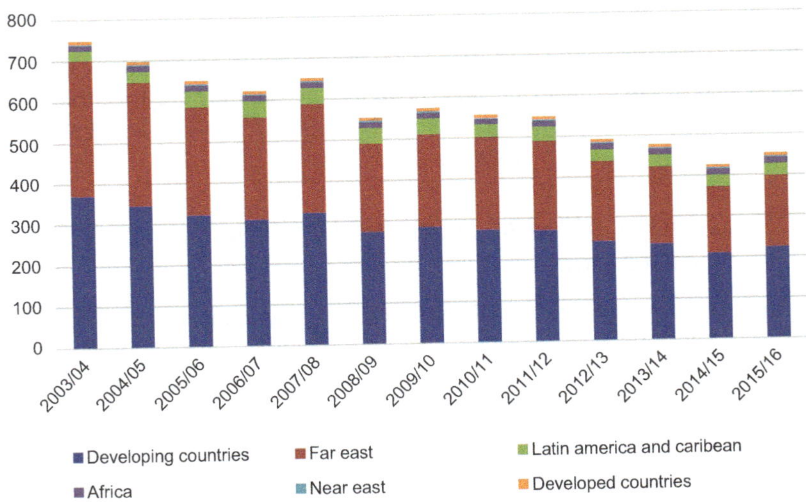

Figure 4.9 World production (1000 tons) for kenaf and other allied fibre crops (2003−2016). *Source:* Food and Agriculture Organization, FAO.

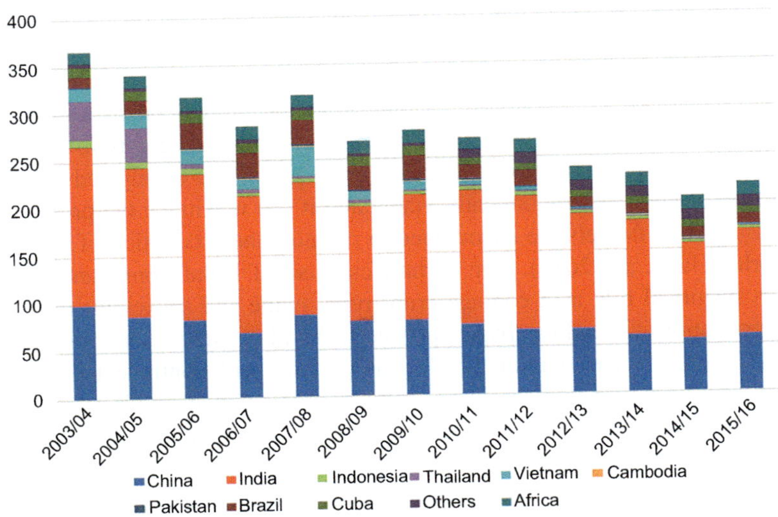

Figure 4.10 Kenaf and other allied crops production (1000 tons) in the main cultivation countries.
Source: Food and Agriculture Organization, FAO.

of world kenaf production in the years 2015–2016, with China and India contributing 74% of world production. But other countries only contribute about 26% of the total production.

4.1.4.2 Harvesting

Different harvest methods can be applied in the field of bast fibres. Some countries such as China, America and Malaysia had developed a whole stalk harvesting system that cuts the stalks and lays them in an orderly fashion at right angles to the row. Stalks are allowed to dry for around 2 weeks and are then gathered by a machine that picks up the stalks and arranges them in large bundles; the bundles are transferred to field trailers. The tractor-drawn field trailers haul the bundles to the field margin where they are stacked for shredding (Stricker et al., 2001).

4.1.4.3 Economy of kenaf production and their competitiveness to other crops

According to the statistic provided by FAO, the production of kenaf in the world tended to decrease from the year 2003 to 2016. Though jute and similar fibre crops play an important role in the social economy, the yield of jute and these have been stagnant in the past 10 years. While in recent years, there has been a downward trend, except for competitive fibres. In addition, the International Jute Organization believes that the main reasons for the decrease in fibre production are lack of high-yield and stable product species adapted to different cultivation conditions, and genetic resources are limited, due to the poor adaptability of existing varieties to agro-ecological conditions, which means that the existing varieties cannot be extended to a large area since the adaptability of fibre crops under the multiple cropping systems is poor and face many challenges such as resistance to certain pests and diseases in different environmental conditions.

Kenaf has important potentials economically as a commercial crop. According to different studies, it was determined that kenaf is an excellent fibre source for a large range of paper products (newsprint, bond paper and corrugated liner board). Production of kenaf is economically viable and it can maximize the profits of many farmers. As such farmers are encouraged to promote the production of kenaf to feed the world (Ha et al., 2016). The commercial use of kenaf continues to diversify from its historical role as a cordage crop (rope, twine and sackcloth) to its various new applications including paper products, building materials, absorbents and livestock feed. Choices within the decision matrix will continue to increase and involve issues ranging from basic agricultural production methods to marketing of kenaf products (Webber et al., 2002). In terms of competitiveness, most of the natural fibres, which kenaf is no exception, has been competing with synthetic products (Anandjiwala et al., 2007). Some studies argued that due to this competition along natural fibres and synthetic fibres, the prospect of kenaf in the future is unclear. As observed, the international market has been successfully dominated by synthetic products, such as plastics materials where most of the products are packaged in plastics containers because of their lower price, and ability to maintain the market. On the other hand, kenaf has to compete with other fibres such as jute since jute has a better quality compared to kenaf. However, in many countries, for instance, Malaysia, kenaf industry is a significant contributor to the Gross Domestic Product (GDP) (Ha et al., 2016). Kenaf has also been considered a substitute crop to replace the production of tobacco (Roslan et al., 2011). Nevertheless, emerging of substitute crops offer wider for international market competition, as well as achieve the progress of technology, which play a role to promote economic expansion.

4.1.4.4 Retting and extraction

Retting or so-called as degumming is a process for removing non-cellulosic material attached to the fibres to release the individual cellulosic fibres. Retting of kenaf typically involves stripping the bark from the harvested material, and then soaking the former in water (e.g., a pond or river) while endogenous microbes break down heteropolysaccharides binding the fibre bundles to facilitate their separation. Depending on the ambient air temperature, the retting process takes from 5 to 22 days, if the harvest time was too late, it will affect the process due to the temperature drop. Otherwise, it will spend less time on retting in the warm water (e.g. 32°C). During the retting process phloem-derived fibre bundles are loosened from other stem tissue composed of hemicelluloses, lignin and pectin (Amel et al., 2013).

Recently, some mechanical methods for separating the kenaf fibre have been described and practically performed. However, in some developing countries, plants are taken out of water after retting, and the fibre is traditionally extracted by using the hand rather than mechanically. Even though, some researchers have also reported that the mechanical decortications is a fast and simple process compared to water and chemical retting, and produces a big quantity of fibre with high quality. it was also reported that 6% NaOH was the optimum concentration for the chemical treatment to remove the impurities, high cellulose content can be obtained from kenaf bast fibre using water and NaOH retting processes (Amel et al., 2013).The resultant plant fibres find use in traditional areas such as cordage and fabrics. Mechanical separation of fibres and chemical retting using NaOH or other harsh treatments are easy to scale, but result in poor-quality fibres (Parikh et al., 2006; Ramaswamy et al., 1994). Investigations using enzymatic retting have reported high-quality fibres, but the process is likely cost-prohibitive for large-scale applications due to the high cost of purified pectinase enzymes (Evans et al., 2002; Akin et al., 2001). Microbial or 'water' retting therefore offers the best chance for large-scale production of high-quality fibres (Visi et al., 2013).

4.1.4.5 Grading systems in different countries

The principal criteria used are colour, lustre, strength, cleanliness, and freedom from retting defects, but the grading and classification of kenaf for commercial purposes is still done subjectively by hand and eye. Longer fibre and higher cellulose contents are required for good quality biocomposite materials (Withanage et al., 2015). Harvesting the plants at the correct time is most important and requires vast experience. For kenaf the optimum time for harvesting is when about 10 flowers are in bloom, and the older flowers have already set their seed.

4.1.4.6 Varieties and breeding

Kenaf breeding is the art and science of changing the genetics of plants in order to produce desired characteristics. It can be accomplished through many different techniques ranging from simply selecting plants with desirable characteristics for propagation, to more complex molecular techniques. The main purpose is breeding new varieties with high-yield, high-resistance to pests and diseases, drought-resistant or regionally adapted to different environments and growing conditions. Classical kenaf breeding uses crossing of closely or distantly related individuals to produce new kenaf varieties or lines with desirable properties. The development of varieties or hybrids with higher yields (either for stem production or dual-purpose varieties for stems and seeds production) and improved resistance to drought as well as to nematodes and anthracnose is considered one of the main challenges for the future of the crop. Over 30 kenaf varieties and hybrids were bred and approved in recent 30 years in China (Alexopoulou et al., 2015).

4.2 Application

4.2.1 Conventional applications

Historically, kenaf fibre was first used as cordage, rope, burlap cloth and fish net because of its rot and mildew resistance. It has also been identified as an excellent source of cellulosic fibre for the manufacturing of a large range of paper products in some countries as the substitute for wood. Because of environmental issues (artificial fibre produces long-time pollution) and increased paper consumption, the application of kenaf fibre has drawn tremendous attention in the world. In fact, pulping kenaf fibre can benefit the environment because it requires less energy and chemical inputs for processing with compare to standard wood pulping.

In addition, kenaf can be either pulped alone or blended with recycled paper. When used alone, one can produce high-quality kenaf fibre suitable for making speciality papers such as security paper, tea-bags, currency notes, etc. Kenaf paper is stronger, whiter, longer lasting, more resistant to yellowing, and has ink adherence better than wood paper. Moreover, the long bast fibre could be used to manufacture products such as high grade pulps for the pulp and paper industry, protective packing for fruits and vegetables, filters, composite board and textiles, and the short fibre or core fibre could be used to manufacture products such as animal bedding and horticultural mixes (Ayadi et al., 2017).

The traditional uses of the crop connected with its fibre stems (paper pulp, etc.) In Africa the leaves and the scions were used as food either raw or cooked. Its seeds were used for oil and various plant parts were used in medicines and in certain superstitious rites. Recent research and development efforts have further extended the diversity of uses for kenaf stems to include board or packaging materials, textiles, animal bedding and biocomposite materials mainly for the automotive and construction industries.

Kenaf competes in some applications with other bast fibre crops such as jute, hemp and flax and with wood or wood residues in other markets such as in wall panels and pulp and paper applications. Traditional gunnysack markets switched over to cheap synthetic man-made fibres based on fossil oil, resulting in a decline of demand and production of jute and allied fibres over the past decades. This declining trend may be reversed, only when the different new markets for fibre crops described in this chapter can be established on a viable scale. As well the policies for the transition from a petroleum-based economy to the bio-based economy are to be implemented, increased demand for these kinds of cellulose resources has to be anticipated for. This provides opportunities to develop kenaf-based industries and increased kenaf cultivation, especially in regions with limited supplies of wood (Monti et al., 2013).

4.2.1.1 Food use

The seeds of kenaf plant produce edible oil that is used for first class cooking oil and margarine production. They can also be used for cooking (flour) and lubrication, soap manufacture, linoleum, paints, and varnishes. Moreover, kenaf oil can be considered nutritionally healthy because of the relatively high amount of monounsaturated and polyunsaturated fatty acids (Coetzee et al., 2008; Peças et al., 2018). The seed production of kenaf is also economically important since it is a good source of oil (16%–22%) (Patane and Sortino, 2010). Additionally, the powder from dry kenaf leaves have been added in several types of food and it was found that the content of calcium and fibre was improved (Alexopoulou et al., 2015) (Ayadi et al., 2017).

4.2.1.2 Feed use

The nutrient content available for animals, particularly crude protein content, is known to vary depending on the region of inhabitation, climate, harvest time, and the portion of the plant used. It was reported that crude protein content in leaves could reach up to 20% crude protein during its immature stage. Dried kenaf leaves have 30% crude proteins and as a potential for livestock feed. It also has been eaten as vegetable in some countries (Zhang, 2013). Kenaf seeds have also been used as a protein supplement for ruminant livestock on pasture. The crude protein content of kenaf seeds ranges from 24% to 35%. The roughage is an additional nutrient, which is important in maintaining normal rumen function crucial to the health of the animal and better meat quality and dairy products (Oh et al., 2018).

4.2.2 Unconventional/diversified applications

4.2.2.1 Natural Fibre/plastic compounds

Increase of the awareness on the importance of protecting the environment has urged researchers around the globe to find ways to produce goods that can minimize harm to the environment. Kenaf is a natural plant which has began to gain attention as a material in the production of biocomposite. Kenaf is selected as an additional alternative material for producing biocomposite because of its fast-growing properties which makes it capable to deliver a large volume of raw material in a short period of time (Kamal et al., 2014). The plastic compounds of natural fibre based on kenaf can replace glass-reinforced plastics in many applications, such as automotive industry, packaging, and construction/housing. The compounds have the mechanical and strength characteristics of glass-filled plastics but are less expensive and completely recyclable which can be used in the automotive industry, construction, housing, and food package industry (Zhang, 2013).

4.2.2.2 Textile application

The idea of making fabrics from kenaf has been practiced since the early 1990s. But the quality of kenaf fabric is not yet good enough for making apparel. The bast fibres offer the advantage of renewability and biodegradability that is essential for making environmentally friendly textile products. Kenaf fibres have a long staple, meaning very fine and strong yarn can be spun. This makes them suitable for quality textiles. Kenaf textiles are also naturally very absorbent, and even fire-retardant, making it especially ideal for outerwear or shoes. Bast fibres are simple to process. Blended with cotton, kenaf fibres can be made into yarn and woven into fabrics. These textiles are aesthetically pleasing, lightweight, and have a soft feel and can provide a new texture for textiles to be used in apparel and home furnishing industry.But spinning kenaf fiber is still a big problem that needs to be solved. On the other hand, one of the major challenges in the textile industry is about environmental problem. Textile industries are facing great pressure to reduce pollutant emissions (Ayadi et al., 2017; Zhang, 2013).

4.2.2.3 Bioenergy production

Kenaf biomass would appear as a potential material for great sustainable energy, hence that is a promising lignocellulosic feedstock for bioenergy production (bioethanol, biohydrogen, bioenergy) provider in the coming future (Saba et al., 2015). Indeed, it is presently being used as a source of biofuel in several countries, including the USA and China (Meryemoglu et al., 2014; Singh, 2013).

4.2.2.4 Phytoremediation

Sarra Arbaoui et al. (2013) evaluated the potential of Kenaf for phytoremediation of contaminated soil. Kenaf can be used in a phytoremediation process of contaminated substrates in particular polluted sludge thus reducing the risks resulting from the presence of trace metals in these substrates. It can be cultivated in order to improve the ecological as well as economic values of marginal lands and for energy production, trace metals accumulated in shoots will remain on the ashes (Arbaoui et al., 2013). It was shown by Ramadan Mohamed Aishah et al. (2016) the excess amount of Zn and Cu could be removed from the soils by growing kenaf. Later, the plant biomass containing the metals has to be cut and subsequently discarded somewhere else. Both metals were largely retained in the shoots of the plants (Aishah et al., 2016).

4.2.2.5 Medicine

Kenaf has long been used as a folk medicine in India and Africa for the treatment of blood and throat disorders, bilious conditions, fever and puerperium. This plant composed of various active components including tannins, saponins, polyphenolics, alkaloids, essential oils and steroids, has long been prescribed in traditional folk medicine in Africa and India. For centuries, people have used this medicinal plant to treat its. Kenaf leaves were also applicable in treating dysentery, and blood and throat disorders. The seeds were also used externally to treat aches and bruises and they were also considered to be fattening. In addition, this plant has been reported to be an anodyne, aperitif, aphrodisiac, as well as fattening, purgative and stomachic (Lee et al., 2007).

4.3 Conclusion

Kenaf belongs to the Malvaceae family and it is widely grown in many countries, such as China and India. This plant is well-known for its fibre content, several industrial applications, medicinal usage, livestock feed, and used in the production of twine, rope, paper and others. Currently, many countries pay more attention to kenaf research and cultivation because of its high biological efficiency and wide ecological adaptability. The main characteristics of kenaf are high biomass, short life cycle, fast growing, wide growth area, strong adaptability to environment and low cost in cultivation. When the seed was sowed in mid-to-late April, the flowering for the early varieties starts in the beginning of August, while for the late ones starts in the middle of September. If kenaf for fibre use, the harvesting time should take place in early October. In agriculture practice, kenaf plant needs few treatments during the growth as, for example, low quantities of chemical fertilizers. In a sense, kenaf was identified as a possible substitute for jute. It is cheaper to produce and has less labour-intensive requirements. Furthermore, kenaf is a potentially suitable species for phytoremediation in the tropics with a combined purpose of biomass production especially to improve the ecological as well as economic valorization of degraded areas.

4.4 Future prospects

In the past, the importance of kenaf was mainly focused on paper pulp production, while the last two decades it has being developed as a multi-purpose crop.

The main new areas of kenaf applications would be in medicines, food additions, fodder, oil and chemical absorbents, medium for mushroom cultivation, natural fibre or plastic compounds, the automotive industry materials, building materials, biocomposites, environmental cleaning, animal bedding and poultry litter, sustainable energy production and as well as phytochemistry. Currently, the plant fibre reinforced composites are mostly used for indoor applications, and their potential in outdoor applications could be enhanced if the above issues related to plant fibres were addressed. In future, effort will be on the fibre surface modifications, their use in various thermoset and thermoplastic matrices, composite processing, hybrid composites, mechanics, modelling, applications and life-cycle assessment. Further research on fibres and their reinforcement potential has to be carried out. The moisture absorption of the fibre and fibre matrix has to be addressed to improve properties (Ramesh, 2018).

4.5 Sources of further information and advice

Important literature and organizations supplying information on kenaf and other fibre crops are listed as follow.

4.5.1 Key books

Jianguang Su, Zhigang Dai et al. (2006). Description and Data Standard for kenaf (*Hibiscus cannabinus* L.) (in Chinese). Beijing: China Agriculture Press.

Heping xiong (2009). The Breeding of Bast Fibre Crops (in Chinese). Beijing: China Agriculture Press.

Andrea Monti, Efthimia Alexopoulou (2013). Kenaf: A Multi-Purpose Crop for Several Industrial Applications. London: Springer.

Dempsey, J. M. (1975), Fibre Crops. Gainesville, FL: University Press of Florida.

Frank, R. R. (ed.) (2005), Bast and Other Plant Fibres. Woodhead Publishing in Textiles. London: CRC Press.

Ghosh, P. (2004), F ibre Science and Technology. New Delhi: Tata McGraw-Hill Publishing.

Ghosh, T. (1983), Handbook on Jute. Rome: FAO (Food and Agriculture Organization of the United Nations).

Kirby, R. H. (1963), Vegetable Fibres. World Crops Books. London: Leonard Hill (Books).

Lewin, M. and Pearce, E. M. (eds.) (1998), Handbook of Fibre Chemistry. New York: Marcel Dekker.

Morton, W. E. and Hearle, J. W. S. (1986), Physical Properties of Textile Fibres. Manchester: Textile Institute.

Nakamura, A. (2000), Fibre Science & Technology (Translated from Japanese). New Delhi: Oxford and IBH Publishing.

NIIR Board of Consultants and Engineers (2005), Natural Fibres: Handbook with Cultivation and Uses. Delhi, India: National Institute of Industrial Research.

Roy, S. (2010), Jute Basics. Dhaka, Bangladesh: International Jute Study Group (IJSG).

Sur, D. (2005), Understanding Jute Yarn. Kolkata, India: Anindita Sur.

4.5.2 Websites

FIBRA: http://www.fibrafp7.net/.
IJSG: www.jute.com.
IBFC: http://www.chinaibfc.com.
CRES: http://www.cres.gr/cres/index.html.
FAFU: www.fafu.edu.cn.
FAO: www.fao.org.
UPM: http://www.upm.edu.my/.

4.5.3 Research and development organizations and groups

Central Research Institute for Jute and Allied Fibres (CRIJAF).

Bangladesh Jute Research Institute(BJRI), Dhaka, E-mails: info@bjri.gov.bd; infobjri@yahoo.com.

Institute of Bast Fibres (IBFC), Changsha city, Hunan Province, China. wangyufu@caas.cn; chinaflax@126.com.

Fujian Agriculture and Forestry University (FAFU), Fuzhou City, Fujian Province China. xujiantang@fafu.edu.cn; qijm863@163.com.

Center for Renewable Energy Sources and Saving, Pikermi Attikis, Greece Agriculture of Athens. Athens, Greece.

Institute of Natural Fibres (INF), Poznan, Poland National Institute for Research on Jute and Allied Fiber Technology (NIRJAFT), India South Indian Textiles Research Institute (SITRA), India.

Institute of Tropical Forestry and Forest Products, Universiti Putra Malaysia(UPM). Selangor, Malaysia.

Acknowledgements

This work was supported by Scientific observing and experimental Station of Southeastern kenaf&Jute, Ministry of Agriculture and Rural Affairs of the People's Republic of China, Fuzhou city, China. Key Lab of Genetics, Breeding and Multiple Application of Crops (FAFU), Ministry of Education, Fuzhou city, China. Fujian Provincial Key Laboratory of Crop Breeding by Design, Fuzhou city, China. Fujian Agriculture and Forestry University, Fuzhou city, China. Institute of Bast Fibres (IBFC), Changsha city, Hunan Province, China.

This work was also supported by the foundation of China Agriculture Research System (nycytx-19-E06), Weng Hongwu Original Research Fund of Peking University and Weng Hongwu Academic Innovation Research Fund of Peking University. I would like to express my sincere appreciation to professor Jianguang Su and Zhigang Dai, for their academic guidance and personal support through every phase of this dissertation.

References

Aishah, R.M., Shamshuddin, J., Fauziah, C.I., Arifin, A., Panhwar, Q.A., 2016. Phytoremediation of copper and zinc in sewage sludge amended soils using jatropha curcas and *Hibiscus cannabinus*. J. Chem. Soc. Pakistan 38 (6), 1230–1243.

Akin, D.E., Foulk, J.A., Dodd, R.B., McAlister 3rd, D.D., 2001. Enzyme-retting of flax and characterization of processed fibers. J. Biotechnol. 89 (2–3), 193–203.

Alexopoulou, E., Li, D., Papatheohari, Y., Siqi, H., Scordia, D., Testa, G., 2015. How kenaf (*Hibiscus cannabinus* L.) can achieve high yields in Europe and China. Ind. Crops Prod. 68, 131–140.

Amel, B.A., Paridah, M.T., Sudin, R., Anwar, U.M.K., Hussein, A.S., 2013. Effect of fiber extraction methods on some properties of kenaf bast fiber. Ind. Crops Prod. 46, 117–123.

Anandjiwala, R.D., Blouw, S., 2007. Composites from bast fibres-prospects and potential in the changing market environment. J. Nat. Fibers 4 (2), 91–109.

Arbaoui, S., Evlard, A., Mhamdi Mel, W., Campanella, B., Paul, R., Bettaieb, T., 2013. Potential of kenaf (*Hibiscus cannabinus* L.) and corn (*Zea mays* L.) for phytoremediation of dredging sludge contaminated by trace metals. Biodegradation 24 (4), 563–567.

Ayadi, R., Hanana, M., Mzid, R., Hamrouni, L., Khouja, M.I., Hanachi, A.S., 2017. *Hibiscus cannabinus* L. – kenaf : a review paper. J. Nat. Fibers 14 (4), 466–484.

Bakhtiari, M.R., Ahmad, D., Othman, J., Ismail, N., 2011. Physical and mechanical properties of kenaf seed. Appl. Eng. Agric. 27 (2), 263–268.

Cheng, W.Y., Akanda, J.M.H., Nyam, K.L., 2016. Kenaf seed oil: a potential new source of edible oil. Trends Food Sci. Technol. 52, 57–65.

Coetzee, R., Labuschagne, M.T., Hugo, A., 2008. Fatty acid and oil variation in seed from kenaf (*Hibiscus cannabinus* L.). Ind. Crops Prod. 27 (1), 104–109.

Dhar, P., Kar, C.S., Ojha, D., Pandey, S.K., Mitra, J., 2015. Chemistry, phytotechnology, pharmacology and nutraceutical functions of kenaf (*Hibiscus cannabinus* L.) and roselle (*Hibiscus sabdariffa* L.) seed oil: an overview. Ind. Crops Prod. 77, 323–332.

Evans, J.D., Akin, D.E., Foulk, J.A., 2002. Flax-retting by polygalacturonase-containing enzyme mixtures and effects on fiber properties. J. Biotechnol. 97 (3), 223–231.

Golam, F., Alamgir, M.A., Rahman, M.M., Subha, B., Motior, M.R., 2011. Evaluation of genetic variability of kenaf *(Hibiscus cannabinus* L.) from different geographic origins using morpho-agronomic traits and multivariate analysis. Australian Journal of Crop Science 5 (13), 1882–1890.

Ha, A., M, S., Mt, P., Ara, S., Maaei, H., A, O., 2016. Financial and technical assessment of kenaf cultivation for producing fiber utilized in automotive components. Business and Economics Journal 7 (4).

H'ng, P.S., Khor, B.N., Tadashi, N., Aini, A.S.N., Paridah, M.T., 2009. Anatomical structure and morphology of new kenaf varieties. Asian J. Sci. Res. 2 (3), 161–166.

Kamal, I., Thirmizir, M.Z., Beyer, G., Saad, M.J., Rashid, N.A.A., Kadir, Y.A., 2014. Kenaf for biocomposite: an overview. J. Sci. Technol. 16 (2).

Lee, Y.G., Byeon, S.E., Kim, J.Y., Lee, J.Y., Rhee, M.H., Hong, S., Wu, J.C., Lee, H.S., Kim, M.J., Cho, D.H., et al., 2007. Immunomodulatory effect of *Hibiscus cannabinus* extract on macrophage functions. J. Ethnopharmacol. 113 (1), 62–71.

Meryemoglu, B., Hasanoglu, A., Irmak, S., Erbatur, O., 2014. Biofuel production by liquefaction of kenaf (*Hibiscus cannabinus* L.) biomass. Bioresour. Technol. 151, 278–283.

Mohd, H.A.B., Arifin, A., Nasima, J., Hazandy, A.H., Khalil, A., 2014. Journey of kenaf in Malaysia: a review. Sci. Res. Essays 9 (11), 458–470.

Monti, A., Alexopoulou, E., Kenaf, 2013. A multiPurpose crop for several industrial applications.

Mwaikambo, L., 2006. Review of the history, properties and application of plant fibres, vol. 7.

Ng, S.H., Tahir, P.M., Mohamad, R., Abdullah, L.C., Choo, A.C.Y., Liong, Y.Y., 2013. Effect of pretreatment process on bioconversion of kenaf (*Hibiscus cannabinus* L.) core to glucose. Bioresources 8 (2), 2010−2017.

Oh, S., Mbiriri, D.T., Ryu, C., Lee, K., Cho, S., Choi, N.J., 2018. In vitro and in vivo evaluation of kenaf (*Hibiscus cannabinus* L.) as a roughage source for beef cattle. Asian Austral. J. Anim. 31 (10), 1598−1603.

Parikh, D.V., Chen, Y., Sun, L., 2006. Reducing automotive interior noise with natural fiber nonwoven floor covering systems. Text. Res. J. 76 (11), 813−820.

Patane, C., Sortino, O., 2010. Seed yield in kenaf (*Hibiscus cannabinus* L.) as affected by sowing time in South Italy. Ind. Crops Prod. 32 (3), 381−388.

Peças, P., Carvalho, H., Salman, H., Leite, M., 2018. Natural fibre composites and their applications: a review. J. Compos. Sci. 2 (66).

Ramaswamy, G.N., Ruff, C.G., Boyd, C.R.J.T.R.J., 1994. Effect of bacterial and chemical retting on kenaf fiber quality. Text. Res. J. 67 (5), 305−308.

Ramesh, M., 2018. Hemp, jute, banana, kenaf, ramie, sisal fibers. In: Handbook of Properties of Textile and Technical Fibres. Bunsell AR: Woodhead Publishing, pp. 301−325.

Roslan, I., Shamshuddin, J., Fauziah, C.I., Anuar, A.R., 2011. Fertility and suitability of the spodosols formed on sandy beach ridges interspersed with swales in the kelantan. Malays. J. Soil Sci. 15.

Ruan, Q., Qi, J., Huang, L., Su, J., 2009. Fatty acid component analysis and enrichment process of Linoleic acid from kenaf seed oil (in Chinese with English abstract). J. Chin. Cereals Oils Assoc. 24 (9), 71−75.

Saba, N., Jawaid, M., Hakeem, K.R., Paridah, M.T., Khalina, A., Alothman, O.Y., 2015. Potential of bioenergy production from industrial kenaf (*Hibiscus cannabinus* L.) based on Malaysian perspective. Renew. Sustain. Energy Rev. (42), 446−459.

Singh, B., 2013. Biofuel Crops: Production, Physiology and Genetics. Fort Valley State University, Fort Valley, Georgia, USA.

Stricker, J.A., Prine, G., Riddle, T., 2001. Kenaf, a Possible New Crop for Central Florida. University of Florida Cooperative Extension Service, Institute of Food and Agriculture Sciences, EDIS.

Summerscales, J., Dissanayake, N.P.J., Virk, A.S., Hall, W., 2010. A review of bast fibres and their composites. Part 1-fibres as reinforcements. Compos. Part A-Appl S 41 (10), 1329−1335.

Thiruchitrambalam, M., Alavudeen, A., Venkateshwaran, N., 2012. Review on kenaf fiber composites. Rev. Adv. Mater. Sci. 32 (2), 106−111.

Visi, D.K., D'Souza, N., Ayre, B.G., Webber III, C.L., Allen, M.S., 2013. Investigation of the bacterial retting community of kenaf (*Hibiscus cannabinus*) under different conditions using next-generation semiconductor sequencing. J. Ind. Microbiol. Biotechnol. 40 (5), 465−475.

Webber III, C.L., Bhardwaj, H.L., Bledsoe, V.K., Janick, J., Whipkey, A., 2002. Kenaf production: fiber, feed, and seed.

Williams, R.L., 1994. Genetic variation of the control of flowering in kenaf with emphasis on the ambiphotoperiodic response of cultivar Guatemala 4. Ind. Crops Prod. 2 (3), 161−170.

Withanage, S.P., Hossain, M.A., Kumar, M.S., Roslan, H.A., Abdullah, M.P., Napis, S.B., Shukor, N.A., 2015. Overexpression of *Arabidopsis thaliana* gibberellic acid 20 oxidase

(AtGA20ox) gene enhance the vegetative growth and fiber quality in kenaf (*Hibiscus cannabinus* L.) plants. Breed Sci. 65 (3), 177–191.

Xu, J., Li, A., Wang, X., Qi, J., Zhang, L., Zhang, G., Su, J., Tao, A., 2013. Genetic diversity and phylogenetic relationship of kenaf (*Hibiscus cannabinus* L.) accessions evaluated by SRAP and ISSR. Biochem. Syst. Ecol. 49, 94–100.

Zhang, T., 2013. Improvement of kenaf yarn for apparel application. Louisiana State University.

Zhigang, D., Jian-guang, S., Ji-quan, C., You-cai, G., Ying, L., Xian-you, S., 2012. Research progress in conservation and utilization of bast fiber crops germplasm in China. J. Plant Genetic Res. 13 (5), 714–719.

Bast fibres: flax

5A

Ryszard M. Kozłowski[1], *Maria Mackiewicz-Talarczyk*[2], *Katarzyna Wielgusz*[3], *Marcin Praczyk*[3], *Anwar M. Allam*[a,†]
[1]Journal of Natural Fibers Editor-in-chief, coordinator of FAO/ESCORENA European Cooperative Research Network on Flax and other Bast Plants, former director of the Institute of Natural Fibres, Poznań, Poland; [2]Institute of Natural Fibres and Medicinal Plants (IWNiRZ), Associate Editor Journal of Natural Fibers, Secretary of FAO/ESCORENA European Cooperative Research Network on Flax and other Bast Plants, Poznań, Poland; [3]Institute of Natural Fibres and Medicinal Plants (IWNiRZ), Poznań, Poland

5A.1 Introduction

Various parts of the flax plant *(Linum usitatissimum* L.) – such as straw, seeds, leaves, bast, fibres, and the woody core – have potential as a source of valuable textile fibres and are also suitable for technical applications. Flax fibres, applied traditionally for textile utilization such as weaving, knitting and technical textiles (described in Chapter 4, vol. 2), have been used for many centuries.

Raw materials derived from flax can also be applied in biopolymers, in aerospace and automotive industries, as well as for insulation, building materials, production of agro fine chemicals, medicines, cosmetics and nutrition. Another application of short coarse fibres are also pulp, cigarette paper, packaging, laminates, coatings, particle-boards and nonwovens. The application of nonwovens is very wide ranging from furniture to geotextiles and chemotextiles (Kozłowski, 2010b).

In the last years, new achievements have been made in the field of research and development on bast plants (especially flax) such as emerging new varieties, new agrotechnology, new technology for processing, more competition with other fibres and, importantly, widening the area of application of flax. Generally, from an economic viewpoint, two types of flax are known in the agroindustry: fibrous flax and oleaginous flax (linseed). Oleaginous flax (called also linseed or oil flax) is cultivated on large areas in Canada, India, Argentina, USA, Kazakhstan, Ukraine, and the Czech Republic and has been successfully cultivated in Egypt. Its seed crop is much more rewarding than the textile flax crop – it does not need complicated processing operations, in addition to being highly demanded by international linseed markets at increasingly rewarding prices.

[a] Flax Expert/Advisor, Cairo, Egypt
[†] deceased

Handbook of Natural Fibres. https://doi.org/10.1016/B978-0-12-818398-4.00006-2

The different practical uses of the oleaginous flax crop (linseed) include the following:

- The oily seeds are crushed to produce linseed oil (edible, and for pharmaceutical and technical uses) and linseed cakes (see Chapter 18, vol. 2: The application of flax and hemp seeds in food, nutraceutical and cosmetics products).
- Broken capsules make an energizing tonic feed for animals, especially for race horses, sheep and calfs.
- Short broken straws can be used as a convenient raw material for the production of particleboards and composites, as a raw material for the production of particleboards glued by organic glue as well as cement or lime, and a convenient raw material for the production of other composites (Kozłowski et al., 1994, 1995, 1999, 2004).
- Seed capsules can be used, after being transformed into powder with the use of an appropriate hammer mill, as a good cellulose source in poultry feeding and according to recent research to obtain cyclolinopeptides (see Chapter 18, vol. 2: The application of flax and hemp seeds in food, nutraceutical and cosmetics products).

5A.1.1 Current world demand for flax fibres

Production of flax fibres is 0.4% of the total world production of fibres (Gherzi, 2017).

The capacity of the world market for flax fibres can currently be evaluated at about 400,000 tons per year (160,000 tons of scutched long linen fibres) and that for linseeds at not less than 2,500,000 tons per year. Both fibres and linseeds are bought at very high prices, as supply is much lower than demand.

Linen long and short fibres consumption is about 101,000 tons in 2016. Production of long linen fibres in 28 European Union countries (EU 28) is 86% of all world flax fibre production. Rest 14% is produced by Belarus, Russia, Ukraine, China and Egypt (in 2016). The biggest producers of good-quality linen fibres in EU are France, Belgium and the Netherland. World import trade of scutched linen fibres is dominated by China followed by India; in 2017, China concentrated more than 95% of scutched linen fibre import. Asia dominates world spindleage potential (all technologies included) and linen spinning; China 31%, India 21%, rest of Asia and Oceania 24%, Turkey 4%, West Europe 2%, East Europe 1%, North America 2% and South America 3%.

World estimate production of linen yarns is 100,000 tons in 2017 (Fiber Year, 2017), (Gherzi, 2017). Remark: Eighty percent of the world linen spinning capacities are concentrated in China and India. Long and short linen fibres represent approximately 0.4% of the world fibre mill consumption out of an estimated total of 101 million tons (staple and filament) in 2016. China dominates the export of linen yarn followed by Commonwealth of Independent States - CIS countries (Armenia, Azerbaijan, Belarus, Kazakhstan, Kyrgyzstan, Moldova, Russia, Tajikistan, Uzbekistan) and Egypt. China exported 30,371 tons of linen yarns in 2017; Egypt noted the sharp increase in exporting linen yarn with 7034 tons.

India and EU 28 dominate import of linen yarns: 60% of total import in 2017. Japan, South Korea and Turkey comprise 24% of world import of linen yarns in 2017.

Planning investments in linen production and processing globally: reorganization of the scutching units, modernization of 29 flax processing mills with 54% new lines, modernization of harvesting and scutching equipment, spending ca 60–80 mln Euro in the last 3 years. French original equipment manufacturer signed a 3-year contract in Russia to supply to the Smolensk Region modern machines for cultivation of flax and primary processing of linen.

It is connected with the governmental program of Russia of resurgence of the flax cultivation about 1 mln ha and linen processing in Russian Federation. China set up a new spinning plant in Heilongjiang province with a capacity of 4000 tons/a and 24,000 wet spindles; full production was planned for the second quarter of 2018. Leading Indian linen spinners plan to install 44,000 additional spindles by year 2020.

At present, there is an urgent demand for textile flax fibres of high quality, suitable for the production of fine yarns, without major spinning problems. They are, in fact, the natural potential characteristics of the pure flax fibres in general, which can be summarized in the following specifications:

- Homogeneous individually separated single filaments of the same thickness, strength, and suppleness.
- Thickness: diameter not more than 24 microns.
- Strength: tensile strength not less than 60 g per tex.
- Suppleness: smooth, glossy shiny velvety touch, free of rigidity.

5A.1.2 History

According to a very old legend, at one time late in the afternoon, a tired traveller while passing by a small lake considered that it was a good opportunity to refresh, renew his water supply and rest for a while under a nearby shady tree. When the traveller lay down on his back under the tree, he noticed a flying bird carrying pieces of straw that it took to the top of the tree. After several trips to and from the tree, the bird started carrying some fluffy shining fibres instead of the straw; the traveller deduced that the bird was finishing its nest by covering its interior with the fluffy fibres, to form a soft base for the future laid eggs. The curious traveller wanted to know the nature of these fluffy fibres and started to follow the bird from the tree to the source of the fibres. The bird landed down near the shallow water and then plunged its head into the lake and came out with a piece of straw and started to repeatedly beat the surface of water with it, until fibres appeared from the adhering pieces of straw. After the bird flew away, the traveller went to the place that the bird had just left, took off his shoes, waded in the shallow water and found the straw lying on the bottom. He took some pieces of the wetted straw and started repeatedly beating them on the surface of the water, and like the bird, the traveller obtained several glossy shining fluffy wet fibres. The straw was flax straw, which contained hidden fibres that only became visible and available after the straw remained under water for a sufficient time to expose them, after the separation from other plant tissues that surrounded them.

Thus, it is said that through the natural behaviour of a bird, mankind discovered valuable and useful knowledge about the existence of invisible fibres, hidden inside the flax plant and covered by other plant tissues. These fibres become visible and available for use only after the flax straw remains under water for a specified time, sufficient for their separation from the other plant tissues surrounding them (Allam, 2008).

This primitive information, enriched with additional personally acquired experience, was gradually transmitted from generation to generation and practically applied, in ponds, rivers and later in tanks, to obtain various flax textile fibres, without real knowledge of the exact details of how this separation takes place and how to successfully control it. A fragment of white woven cloth wrapped around a primitive tool was discovered by archaeologists in Cayonu (Turkey) near the headwaters of the Tigris river and examined by Gillian Vogelsang-Eastwood, a textile archaeologist at the National Museum of Ethnology in Leiden (The Netherlands). It was confirmed that the woven fragment is the oldest sample of textile ever found – about 10,000 years old – and it is a linen fabric. This fact helped to prove that our ancestors living in the early Neolithic period (10,000–8000 BC) changed their way of life after millennia of hunting and gathering and began to cultivate crops, establish villages and adapting to a more sedentary way of life (Ersserberger, 1993).

The textile samples found in 1962/1963 and examined by Ryder (1965) were provisionally dated to 6000 BC and were found to be made of flax (Taylor, 1995). The history of flax also goes back to the prehistoric era, and Egypt has been considered to be the historical cradle of dual-purpose flax cultivation and of the processing of its straw to produce textile fibres. Many thousands of years BC, the ancient Egyptians at the time of the pharaohs had already mastered flax cultivation (about 2000 ha) and excelled in processing its straw to obtain superior textile fibres that produced ultrafine yarns, which were used for the production of diaphanous linen fabrics for the higher members of the priesthood and the female members of pharaoh families.

The extrafine linen found in one of the pharaoh tombs, discovered at the beginning of the 20th century, showed that the flax yarn used for the production of these very light fabrics is difficult to be attained in modern times. Some sources say that fine and even transparent linen apparels were worn by Egyptian queens, and it was revealed that the fine yarn was obtained by enzymatic treatment of saliva in the human mouth.

Egypt's ancient flax technology was transmitted to other countries through Phoenician sailors' trading in Egyptian linen. Unfortunately, this flax technology faded and disappeared with time from Egypt. What remains of it are the numerous mural inscriptions on the walls of the pharaonic temples and other monuments referred to in thousands of related papyruses discovered in modern times but unfortunately still undeciphered.

When Spaniards brought cotton to Europe in the eighteenth century, cotton started to replace linen. Until the 18th century, flax dominated as a raw material for textiles in Europe. In the 19th and 20th centuries, especially after the First World War, cotton and artificial fibres (viscose) began to replace linen more and more intensively.

During the Second World War, the demand for flax textile fibres increased enormously because the United Kingdom, Germany and the Soviet Union had been almost completely deprived of their usual sources of flax fibre. The German military forces were widely using flax, for the first time in wars, in the manufacture of airplanes, parachutes and incendiary bombs. At that time, airplane wings were made of a solid wooden structure covered by a strong linen cloth. Parachutes' holding bands were made of very strong flax yarns specifically twisted and tightly woven. Linen fire hoses were essential to extinguish the fires caused by incendiary bombs. In response, the United Kingdom had to quickly produce more and more airplanes, parachute holding bands and linen fire hoses, over and above the usual military tarpaulins and other linen camping goods.

To replace and enlarge new sources of flax fibre supplies, the United Kingdom summoned its colonies and Commonwealth countries and other allies to grow as much flax as they could and to process the straw to produce flax textile fibres. The United Kingdom bought the crop and the production in advance to cover its military needs in the shortest possible time.

For this purpose, Egypt doubled its cultivation of flax, which necessitated the fast increase of its capability to process the additional flax crops to come. The annually cultivated area of flax in Egypt had always been positively proportional to the demand for fibres and did not exceed 14 000 ha, but during the Second World War, the cultivated area grew to 25,000 ha. The number of flax processing factories reached more than 60.

After the Second World War and since the 1970s, many changes occurred. After many centuries of worldwide practice, the conventional method of flax straw retting was found as polluting the environment, and all the European retting centres were officially asked to either neutralize the retting effluents before draining them into the river or close down within a 2-year period. The majority of those retting centres found that neutralization of retting effluents would cost them more than they could afford and decided to stop working altogether. Since the beginning of the 1980s, tank hot water retting, producing the highest textile grades of flax fibres, ceased in Europe. The reason was that for water retting, 1 ton of flax straw needs 30–40 m^3 of water (20 m^3 hot water for retting and 10–20 m^3 water for rinsing retted straw).

European flax straw crops had to shift to the ancestral method of dew retting, which is unable to currently produce selected high textile qualities of flax fibres, such as the ones that used to be produced by the more controllable method of tank hot water retting. The presence of high-grade flax fibre on the world markets became sporadic and insufficient to cover the demand for such raw material, which was essential to the production of fine counts of flax yarns that sold for high prices.

Dexterous, highly experienced flax spinners almost disappeared and were partly replaced by younger generations that, initially, preferred to discard flax altogether and shift to other easier-to-spin textile raw materials.

Nowadays, however, the many earlier unsolved problems of flax (and other bast fibres) have been satisfactorily solved. These ‘vegetal’ fibres are now considered normal textile raw materials, having naturally and potentially stable, favourable textile characteristics. They only have to be properly extracted and produced in a pure intrinsic form.

Initially, the produced flax fibres had inconsistent textile characteristics depending on the different circumstances of their cultivation and the skill of the individuals who performed producing them. Often the nature of the plant itself and the prevailing weather conditions were blamed for the unstable specifications of the fibres (Sultana, 1983).

With the advance of science, the first scientific explanation was given for the phenomenon of flax fibre separation when the straw was immersed in water for some time, resulting in the widely applied European flax extraction technology. According to this technology, the bacteria formed on the immersed flax straw develop and multiply and for their nutrition they secrete a specific enzyme that decomposes the pectin that ties the fibres together and to the other plant tissues. Once the pectin breaks down, the fibres are gradually liberated, starting from the root ends and up to the crop ends of the straw (Schilling and Muller, 1951; Sharma, 1992).

Information about common varieties of agricultural plant species, including flax and linseed, is presented in the Chapter 17, vol. 1: Developments in fibrous flax breeding and cultivation.

It is worth to put your attention to the chapters: Production of Flax, Retting of Flax, Processing and Testing of Fibre, Economic Aspect - in the book of Sharma, H. S. S. and Van Sumere, C. F. (eds) (1992), The Biology and Processing of Flax. Belfast: M Publications.

5A.2 Flax plant morphology

The vitally important nutritive system of flax is composed of the following.

5A.2.1 The root and rootlets (root hairs)

The root of the plant is composed of an absorbing tissue, allowing the storage of all the water (soil solution) collected by the fine rootlets projecting in all directions and at different levels. This structure also has the function of fixing the plant firmly into the soil. The collected water is channelled upwards through the wooden cylinder to each leaf of the plant for its transformation into nutritive sap. The actual capacity of the rootlets is a limiting factor to the development of all other plant tissues and thus determines the final amount of any plant crop.

To be most efficient, each root needs to be surrounded by an even, specific free space, to allow the full expansion of its root hairs without the interference of other rootlets belonging to other too closely positioned plants. Otherwise, the amount of available water will be shared between all closely positioned plants, which will only grow according to the percentage of the water each plant receives.

This is the reason why the maximization of the seed crop of closely positioned plants has been practically impossible. The vital free space needed to surround each plant has a maximum, which is equal to the furthest distance that the root hairs can reach, and an essential minimum without which the plant cannot grow normally.

The system of roots of plants, including flax and hemp, has got an important influence on resistance to drought (Jacoby, 2009).

The pictures of the root system of flax in comparison to roots of hemp are presented in Fig. 5A.1.

The recent discoveries in testing the resistance to drought show a modern way to make the plant more resistant to drought. Up to now in the United States, Monsanto and DuPont announced introduction in agriculture resistant to drought: rice, wheat, corn, soybean, rape (Canola), sunflower and cotton. Monsanto and BASF discovered that gene identified in bacteria, which is especially resistant to low temperature, help the plant to be tolerant to drought. This gene is called *cspB* - *Cold-shock proteins B*

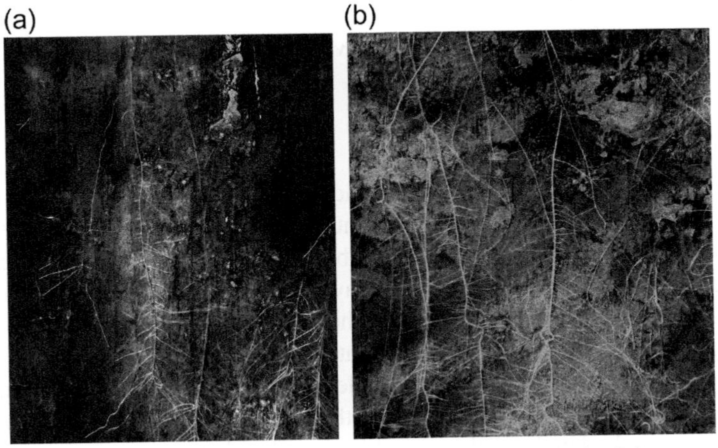

Figure 5A.1 (a) Photograph of the root system of flax; (b) Photograph of the root system of hemp.

coming from *Bacillus subtilis.* The first success was obtained when this bacterial gene was immersed (included) in corn. As a result, the breeders can obtain the corn resistant to drought 6%–15% depending on the intensity of drought in vegetation period. Another scientist put attention to sorghum, which is a plant tolerant to drought (Paterson et al., 2009).

It is known that genes that can help to feed the world are abscisic acid and pyrabactin.

When the pores of the plant open for the absorption of CO_2 and the emission of O_2, unfortunately, the water evaporates from the plant through the stomata. Plants from group C4 have closed stomata and minimally loose H_2O. This is a very important trick for the plant to grow well under the conditions of drought stress. The application of abscisic acid directly to plants is not a practical option - this agent is very expensive and sensitive to light.

The researchers found that pyrabactin, a stable synthetic sulfonamide, turns on the abscisic acid signalling pathway in the plant *Arabidopsis thaliana*, a widely studied model organism. They found that abscisic acid and its mimic, pyrabactin, trigger the signaling process by binding to a protein called PYR_1, which in turn binds to and inhibits the action of a phosphatase enzyme (Park et al., 2009).

Unfortunately, nobody has ever done such research in the field of bast plants, including flax. By the time when powerful and rich companies such as Monsanto, DuPont and BASF do not notice the important role of bast fibrous plant in agriculture, including flax, poorer and weaker centers are not able to solve the problem themselves (Jacoby, 2009).

5A.2.2 The wooden cylinder

The central wooden cylinder is composed of a porous wooden hard tissue acting as an upright standing structure for the plant. It also acts as a duct to bring up the soil solutions collected by the rootlets and stored in the root, to get them to the upper side of each leaf to be transformed into the nutritive sap of the plant.

5A.2.3 Flax fibers

Inside the plant, flax fibres are found as individually separated filaments of different lengths, which vary according to the height of the leaf they serve on the stem of the plant. Each filament is composed of a number of single taper-ended fibre cells evenly joined lengthwise, so that each cell underlays 50% of the preceding one and overlaps 50% of the following one, tightly tied together all along, to form single filaments of the same regular fine thickness, but of different lengths.

A set of 10 single filaments of the same length serves each plant leaf, running down from the height of the leaf to the surface of the soil. In each case, the 10 filaments have lengths equal to the height of the served leaf on the plant.

The 10 filaments do not stick together but are grouped together in bundles, circularly distributed around a supporting central wooden cylinder, and completely surrounded by a holding tissue, which fills the whole inner space between the wooden cylinder and the outer skin. These characteristics are responsible for the fine textile characteristics most favourable for the successful production of regular fine yarn, without major spinning problems.

In the plant, flax fibres are naturally and completely formed during the vegetative growth, as an essential part of the nutritive system. One of their functions is to receive and transport the already formed nutritive sap from the lower side of each leaf downwards to the holding tissue cells that store it and use it to fulfil the needs of all the other plant tissues during their development.

The scanning electron microscopy (SEM) micrographs of the cross-section and longitudinal section of flax fibre are presented in Fig. 5A.2.

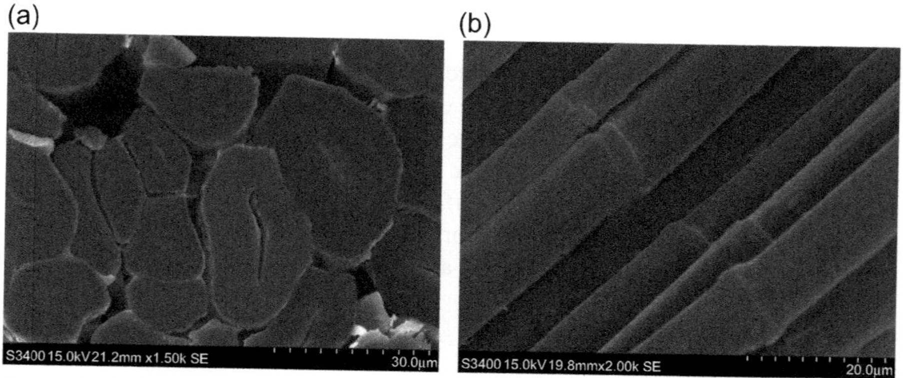

Figure 5A.2 (a) Cross section of flax fibre (SEM Micrograph); (b) Longitudinal view of flax fibre (SEM Micrograph).

5A.2.4 The holding tissue

The holding tissue is completely formed of dynamic cells that have semipermeable membranes to facilitate the easy exchange of nutritive sap; all the other plant tissues are formed of static cells that do not have semipermeable membranes. The holding tissue is essential to all future developments in the plant tissues or their replacement in case of damage. It also tightly surrounds the different bundles of fibres transporting the already prepared nutritive sap from each leaf to easily absorb and store all the upcoming sap.

The stem of fibrous flax has a maximum length in green maturity stage. According to the results of many research works (Department of Agriculture, Northern Ireland, 1985), the yield of flax straw and fibres is the highest in yellow maturity stage, but the highest quality fibre is obtained in green/yellow maturity stage of flax straw, when the fibre is fine and the level of lignin is lower than in yellow maturity stage.

Flax fibres are an important active part of the plant nutritive system and thus contribute to natural potentially excellent textile characteristics (quality), from the beginning of the vegetal growth phase, increasing in quantity with the growing plant's further development until the end of the vegetal growth phase.

During any stage of growth, flax fibres have definite stable natural potential characteristics in relation to their thickness, tensile strength and suppleness. These characteristics are only apparent when the fibres are properly and smoothly extracted and obtained in a single homogeneous filament form, although the various filaments differ in length, because in each case, they are equal to the height of the leaf they serve. It is worth adding that the 10 single filaments in each of the ten-bundle filaments are of equal length.

In the plant, flax fibres are individually separated but grouped in bundles completely and firmly surrounded by the holding tissue. They have smooth glossy surfaces that allow them to freely slide on each other when the stem bends in all directions.

The number of groups of flax filaments inside each stem is equal to the number of leaves appearing on this stem. Therefore, the number of single filaments equals the number of leaves multiplied by 10.

The single flax filament diameter (thinness, fineness) has been mathematically calculated and found to be 0.250 tex.

The development of the cellulosic fibres from flax occurs progressively during growth through two main steps:

1. The elongation of the mother fibre cells (from several mm to cm depending on its position in the stem) results in a cell wall of high plasticity because of a high level of neutral hydrosoluble polysaccharides, mainly β-1,4-galactan.

2. The progressive maturation of the fibres leads, on the one hand, to the synthesis of cellulose and, on the other hand, to the reinforcement of the linkages in the cellular junction and middle lamellae with specific components such as the acetylated polygalacturonic acids or rhamnoglacturonan I-like polysaccharides, resulting in a supraorganization of the fibres into bundles.

At the maturity stage, the cell wall in the cortex is enriched in polygalacturonic acids of various degree of methylation depending on their localization in the primary cell walls or in the middle lamellae.

Such polysaccharides are easily degraded by pectinolytic enzymes (see Chapter 7, vol. 2: Enzymatic treatment of natural fibres). Conversely, the cell walls of the fibres are enriched in β-1,4-galactan, rhamnogalacturonan I associated to acetylated polygalacturonic acids and rhamnogalacturonan II—like polymers.

The high heterogeneity of flax production is well known. There are a lot of heterogeneities in length, diameter, cellulose content in the fibre and polysaccharide cements of both the tissue cortex and fibre differences in pectin composition according to the tissue cortex and fibres (Morvan, 1991).

In Russia, 13 pure lines of flax genetic collection were evaluated at N. I. Vavilov Institute in St Petersburg, for their vegetative period, yield, quality of fibre, as well as structure and composition of the fibre. Correlation analyses showed that the strength of technical fibres was correlated to the plant height, straw yield and number of bundles per stem section with the diameter of elementary fibres. The fineness and flexibility were linked with the ratio between the areas of the fibre and xylem tissues and with the relative amount of encrusting components to cellulose in fibres (Brutch et al., 2008).

5A.3 Structure and chemical composition of flax

The group of fibrous plants including flax, hemp, jute, ramie, sisal and kenaf is the source of bast fibres. The main component of the fibres is cellulose, whereas secondary components are waxes, fats, hemicellulose, lignin and pectin (Kozlowski et al., 2006a, 2006b, Konczewicz and Kozłowski, 2007). The chemical composition of lignocellulosic fibres is presented in Table 5A.1.

Pectin plays an important role in the fibre as a component that binds the fibre into bundles and also determines the lustre and touch of the fibre. Pectins are macromolecular compounds of polygalacturonic acid. They are agglomerated in the middle blades, mainly in the ground tissue. In fibrous plants, two pectin fractions can be found, that is, fraction A which is soluble in water and fraction B which is water insoluble. Technical flax fibres consist of cells bonded with each other by a lamella, which in turn consists of mostly pectin B with a small amount of pectin A. Fibre bundles, located in the phloem, form rings around the stem, of more or less compact structure, bundles of which are attached to adjacent tissues with a layer of pectin A.

Proper removal of pectin substances during preliminary processing determines the divisibility and, as a result, the fineness of the fibre and its suitability for spinning (Kozłowski et al., 2010a,b).

The easiest to remove in the retting process is pectin A, which is decomposed by bacteria, fungi and enzymes, whereas pectin B remains within the fibre and is decisive for the compactness of the fibre. Excessive removal of pectin and natural waxes will cause the fibre to have an unpleasant dry and coarse touch. Complete removal of pectin

Table 5A.1 Chemical composition (%) of lignocellulosic fibres.

Fibre	Cellulose	Lignin	Pectins and hemicelluloses
Abaca	60–80	6–14	13
Bamboo	26–43	21–31	–
Cabuya	80	17	–
Coir	36–43	41–45	3–4
Cotton	83–99	6	5
Curaua	70–80	13	–
Flax	64–84	0.6–5	19
Hemp	67–78	3.5–5.5	17
Henequen	60–78	8–13	4–28
Isora	75	23	–
Jute	51–78	10–15	37
Kenaf	44–57	15–19	–
Nettle	53–82	0.5	0.9–4.8
Pineapple	80	13	–
Pita	80	17	–
Ramie	67–99	0.5–1	22
Sisal	60–80	6–14	13

Source: Kozłowski et al., 2010.

will result in disintegration of fibre bundles into elementary fibres (see Chapter 7, Enzymatic treatment of natural fibres).

Waxes and fats are also important in terms of technological parameters. They determine soft touch, low friction and thus ease of moving the fibre. In flax fibres, waxes are present mainly in the outer part of the stem, in the epidermis, and in smaller amounts in fibre cells.

Lignin functions as an inlaying component within amorphous areas of cellulose and causes the cellulose to be rigid. In the elementary fibre, lignin occurs in the primary wall and outer part of the secondary wall. For processing, the presence of lignin is undesirable as touch and flexibility of the fibre become worse. The presence of lignin means that the fibre is more easily breakable and mechanical parameters of tenacity and resiliency deteriorate. Moreover, the divisibility of the fibre lowers as a result of the presence of lignin.

The structure of elementary fibre and its microscopic, mechanical, physical and chemical properties also have an effect on the overall quality of the fibre.

Single fibres that constitute plant fibrous tissue are referred to as elementary fibres, either when they are a part of a compact bundle or are loosely located within the cortex tissue. The structure of elementary fibres is similar in all lignocellulosic fibres.

The quality of the extracted fibre depends on the extraction method and the amount of elementary fibres, their distribution and binding within the bundles and between the bundles. To separate the fibres from other tissues, various mechanical treatments are used — decortication, scutching, hackling, etc. Better results are achieved by preliminary processing such as water or dew retting, as well as enzymatic, chemical or physical processing of the fibre. (Kozłowski, 1970.)

The average chemical compositions of the flax stem and fibre are as follows:

Stem:

Cellulose 49%–60%
Hemicellulose 10%–25%
Lignin 17%–23%, pectin 3%–4%
Fats and waxes 1.5%–3%

Fibre:

Cellulose 85%–87%
Hemicellulose 7%–9%
Lignin 2.5%–4%, pectin 1.5%–2.5%

The differences in chemical composition depend on the kind of applied varieties, conditions of cultivation, date of harvesting, system of primary processing (decortication, dew retting, water retting, osmotic degumming), etc.

The structure and properties of flax in comparison to other bast fibres are presented in Tables 5A.2 and 5A.3.

A comparative study of the infrared spectrum has shown the presence of varying quantities of pectin, hemicelluloses and lignin in the fibres originating from various methods of flax straw degumming. The obtained data have revealed fluctuation of valence bands characteristic to flax and have shown the reliability of the Fourier Transform Infrared Analysis method for estimation of the degree of flax fibre fineness (Titok et al., 2010).

Modern tools for testing morphology and supermolecular structure of flax are

- light microscopy,
- scanning and transmission electron microscopy (SEM and TEM),
- X-ray diffraction,
- nuclear magnetic resonance spectroscopy.

More data on methods of testing and identifying natural fibres are given in Chapter 14, vol. 1: Identification of natural textile fibres.

There are some important rules for the production of high-quality fibrous flax, followed by degumming and extracting high-quality flax fibres, suitable for spinning.

- The choice of a proper soil with pH 6.5–8.0. Flax should not be grown on the same plot of land more frequently than once in 4 years, and the preferable interval is 7 years. Sowing should be carried out as soon as favourable weather conditions occur. In Central Europe, this can be done usually in the second week of March and preferably before the first week

Table 5A.2 Morphological characteristics of bast fibres.

Fibre type	Single bast fibre (cell)				Cellulose microfibril	
	Shape of cross-section	Length (mm)	Diameter (μm)	Wall thickness (μm)	Fibrillar bundle width (nm)	Microfibril width (nm)
Hemp	Flat/ polygonal	10–30	5–50	5–10	25–80	3–18
Flax	Round-oval/ polygonal	20–40	10–20	5–10	30–80	3–18
Jute	Round-oval/ polygonal	1–6	10–25	2–5	40–90	3–12
Ramie	Flat	50–250	10–60	3–5	40–100	3–12
Sisal	Round-oval/ polygonal	1–5	10–40	3–5	20–50	3-B

Source: Pinnov and Fink, 1999.

Table 5A.3 Degree of crystallinity and crystallite dimensions of native cellulosic fibres.

Fibre	Content of cellulose (%)	Degree of crystallinity (%)		Crystallite dimensions (nm)	
		Fibre	Cellulose	Lateral	Longitudinal
Flax	79	44	56	4–5	12–24
Hemp	75	44	59	3–5	12.5–29
Jute	61	33	54	3–5	12–21
Cotton linters (scoured and bleached)	100	56–63	56–63	5–6	10–20

of April. Later sowing is not recommended because the yield and quality are lower. Young plants survive to −3°C and for a short time even at −6°C.

- The use of a suitable compound fertilizer to provide approximately 20–22 kg N, 20–25 P_2O_5 and 80–90 kg of K_2O per 1 ha. It is also recommended to apply 4–6 kg of $ZnSO_4$ dissolved in 400 L of water per 1 ha, as well as 0.3–1 kg of borax.
- The choice of a suitable variety.

- Establishing the optimum sowing density (about 1800–2000 plants/m^2; 125–140 kg seeds/ha).
- Good crop of flax free from weeds.
- Freedom from lodging.
- Pulling at the stage of green (green/yellow) straw (straw yield between 5 and 7 tons/ha).
- In the case of dew retting – turning the straw in the field at least twice.
- Collecting the retted straw and storing at a uniform humidity below 16%.

In the case of appropriate dew retting, scutching gives a yield of 15%–20% of long fibre and tow 6%–3% (Sultana, 1991).

It is possible to maximize the production of either the fibres or the seeds by using two distinctive methods of flax cultivation. The maximization of the fibre crop is successfully obtained through optimizing the number of slender branching single stemmed plants, by giving each sowing seed only the minimum necessary vital free space around it aimed at preventing the plants from giving birth to tillers or many branches and sub-branches. The maximization of the seeds is successfully obtained through minimizing the number of plants and encouraging them to give birth to the greatest number of tillers, branches and sub-branches, to bear the greatest number of capsules containing seeds, by giving the plants the maximum necessary vital space around each plant (Allam, 2008).

5A.4 Flax agrotechnology (agrotechnics)

Prepared by Dr Marcin Praczyk and Dr Katarzyna Wielgusz, Institute of Natural Fibres and Medicinal Plants, Poznań, Poland.

5A.4.1 Climate and soil requirements

The best weather conditions for flax are not very high air temperature and high relative humidity (Agosti et al., 2005). The highest yields are promoted by temperatures 18–20°C and accompanying cloud cover (Xinwen, 1997). Due to the shallow root system, flax (especially fibre flax) plants are very sensitive to water deficit in the soil. Annual sum of precipitation in fibre flax cultivation should be least 600–650 mm, which in the vegetation period falls to 110–150 mm. Flax plants transpire very high amounts of water. The transpiration coefficient, determining the amount of water necessary for production of a dry matter unit for flax, is 400–600 (Heller et al., 2012).

Ensuring optimum conditions for flax yields requires cultivation on fertile soils kept in high culture, medium heavy, humus-rich sandy clays that produce no crust, with regulated air–water ratio. Besides fertile and high culture soils, also newly cultivated postforest, postmeadows, postpastures and postfallows are good for flax. Soil for fibre flax cultivation should be deep and have loose aggregate structure that allows for air access to the roots and soil microflora, as well as evacuation of excess water. High ability to absorb and hold water and nutrients is also important. Optimum degree of soil acidity should be close to neutral pH 6.5–6.9. However, fibre flax plants also tolerate slightly acidic soils and very good yields can be obtained at such fields (Rólski et al., 2000). This is an important positive property that allows cultivation of flax in countries with a high share of acidic soils.

5A.4.2 Crop rotation

Requirements of flax in terms of predeceasing crop are not high, and flax can be grown successively after any crop, providing it gave good yield and left the soil in good culture. Good forecrops in conditions of intensive farming are successful cereals, of which oat is the best. Oat intensifies the process of self-cleaning the soil from fungal spores of the genus *Fusarium,* which can cause very large losses in yield. In turn, flax is a very good forecrop for winter wheat and does not cause soil sterilization, leaving it clean (Wałkowski et al., 1998). Many experiments confirm obtaining higher yields of winter wheat cultivated after flax. On the light soils, the best forecrop is potatoes, which leave the soil clean, as well as, does not use up preferential soil nutrients and moisture (Sultana, 1983). Potatoes also reduce the risk of *Fusarium* wilt.

5A.4.3 Soil cultivation

Flax requires a clean and carefully cultivated field. After crops leaving more weeds, e.g., wheat, rye, barley, winter ploughing should be preceded by harrowing and cultivator treatment (Heller et al., 2006). Plants leaving the field late, e.g., potatoes usually leave the soil clean and loose; therefore, the harrow is applied to level the furrows followed by winter ploughing, no stubble-breaking in this case.

Deep ploughing should be eliminated in spring. The spring tillage should be limited to the most necessary treatments, e.g., dragging aiming at stopping transpiration from the soil and induction of weed seeds to germinate, as well as loosening soil structure with a harrow, a cultivator with stiff of semistiff teeth, before sowing, which also destroys germinating weeds, depending on the type and condition of the soil (Heller and Rolski, 2001). To reduce water transpiration and cover the planted seeds, the presowing and postsowing harrowing with light harrows should be applied, if it's necessary.

5A.4.4 Fertilization

The availability and use of ingredients provided to plants in fertilizers depends on the appropriate soil pH. Due to the poorly developed root system and the related limited possibilities of nutrient absorption from the soil, flax plants react strongly to their deficiency. Moreover, the majority of nutrients that are necessary for good yielding are accumulated in a short time by flax plants, i.e., from tying flower buds to full flowering. The most intensive absorption of nitrogen, phosphorus and potassium is in the flowering stage. Fibre flax requires very precise fertilization because particular nutrients have effect on qualitative features of fibre (El-Hariri et al., 1998). To ensure good yield of fibre of high quality, it is recommended to provide the macronutrients in the following ratio: N: P_2O_5: K_2O as 1:2:3, which should correspond to the following amounts: 30–40 kg N, 60–80 kg P_2O_5 90–120 kg K_2O per hectare. Sixty percent of nitrogen dose should be applied before sowing, and remaining amount should be applied after germination, depending on the condition of plants, soil and climate. Depending on the forecrop and weather pattern, a supplying dose of nitrogen can be applied after germination (Endres et al., 2002). Nevertheless, nitrogen has to be applied in the initial phase of growth because applied later can worsen the quality of the fibre (Mańkowski and Szukała, 1998). Beneficial effect of potassium is revealed when application of nitrogen is correct. Excessive doses of nitrogen, however, can reduce the length of stem and fibre content. Potassium has beneficial influence on fibre quality, strength and elasticity and also is crucial for proper dew retting process. The recommended doses of potassium found in literature vary from 50 kg ha^{-1} to 180 kg ha^{-1}.

Phosphorus is indispensable for straw to reach the proper length and form proper number of fibre bundles in the stem (Grant and Bailey, 1989). The excessive doses of phosphorus, however, lead to shortening and branching of the stem, which has a negative effect on fibre quality by reducing its tensile strength.

It has also been proved that boron fertilization has a positive effect on the growth and development of flax plants. Plants are then characterized by larger resistance to bacterial diseases. Flax plants' health condition is also stimulated by zinc. Many soils, e.g., in Europe show high or average content of zinc from 10 to 200 mg in 1 kg of soil; however, soil pH has a significant effect on zinc assimilation: the more acidic the soil reaction, the better the uptake of zinc. Because flax is grown on soil with pH close to neutral, it may be necessary to supplement zinc $ZnSO_4$ at 15 kg/ha (Marchenkov et al., 2003). The proper addition of Zn is very important because in case of fibrous flax, the plant will be slender and unbranched.

5A.4.5 Sowing

Flax is usually sown in the period from 15th March to 15th April in the mid and north Europe. Due to the specific climatic conditions and the different way of cultivation, there are deviations from this rule. For example, in Northern Ireland, flax is sown the in first decade of May, and if the flax is grown as a secondary crop, it is sown in October or November (El-Hariri et al., 2004a,b). Seed germination occurs already at a temperature of 2°C, but the optimal germination temperature is 5–7°C. Certified seeds of flax plants should be sown with proven high germination, purity and health. With proper agrotechnics and favourable weather conditions, stable and high yields are ensured. The European germination quality standard for flax seed is at least 92% for fibre flax and 85% for linseed. Seed purity should not be less than 99%. Optimum sowing amount of fibrous flax is 110–140 kg ha^{-1} (2000–2400 seeds per 1 m^2) for industrial plantations and 50–70 kg ha^{-1} (1000–1100 seeds per 1 m^2) for seed plantations (Fouilloux, 1988). Optimum sowing amount for linseed is 40–60 kg ha^{-1}. Row spacing depends on the way of plant protection on the field. In the case of chemical weed control, fibre flax is sown in rows of 8–12 cm and linseed 15–25 cm. For mechanical interrow care, the row spacing must not be less than 30 cm. Flax seeds should be sown shallow to a depth of 1–2 cm.

5A.4.6 Postemergent treatments

Postemergent treatments include the soil crust, weeds, diseases and pest control. For the soil crust control, using a spiked roller, ring roller or sometimes presowing light harrow is recommended, especially on heavy, crusty and confluent soils.

- Weed control

Weeds compete with flax plants for nutrients and water. They also shadow the crop (Easson, 1989). Weeds also often transmit diseases and pests and cause lodging, making pulling and dew retting of flax more difficult. Due to a delicate morphological structure (single, non branched stem, small leaves and poorly developed root system) flax as an annual spring crop poorly competes with weeds for its life space (Heller, 2001). Weed infestation accounts for 15%–20% reduction of yield. Efficiency of weed control has an indirect effect on fibre yield and its quality (Heller, 1992). The predominant weeds in flax cultivation in Europe are the typical species for the root crops and small grains. The most frequently used and most numerous weed species in fibre flax are *Chenopodium album* L., *Fallopia convolvulus* L. Löve, *Viola arvensis* Murr., *Stellaria media* Vill., *Lamium amplexicaule* L., *Thlaspi arvense* L., *Elytriga repens* L. Nevski, *Polygonum nodosum* Pers. and *Echinochloa cruss-galli* L.

For chemical weeding of flax, soil herbicides used immediately after sowing and on-leaf herbicides used after emergence of plants are recommended. There are two types of herbicides: for the dicotyledonous weeds and graminicides for the control of monocotyledonous weeds. To control grasses, the spraying should be conducted 5–6 days before or after the broad-leaf weed control treatments were applied (Šmirous, 1989). The regulations regarding recommended pesticides in flax cultivations are variable due to the emergence of new ones and the removing of existing active ingredients. It is absolutely necessary to use only currently authorized and recommended plant protection products.

Soil herbicides applied directly after sowing of flax are effective if the weather in spring is warm and wet and soil remains in good culture. On-leaf herbicides should be used when plants reach a height of 6–12 cm because flax plants are relatively sensitive to herbicides. In general, herbicides are better used in early plant growth. As the flax plants reach subsequent stages of development, they become more and more sensitive. It should also be remembered not to use herbicides in the absence of moisture in the soil when the flax plants show a weak turgor and their growth is inhibited.

- Disease management

Fusarium wilt and rust have been recognized as the main factors limiting the production of flax (Beaudoin, 1989). However, these two diseases have been successfully controlled by polygonal resistance contained in the majority of registered varieties of flax. Linseed varieties of flax are more susceptible to these diseases (Andruszewska et al., 2001). Recently, the real band and powdery mildew has spread, causing local epidemics of diseases (Rashid, 2003). Yellow-orange linseed varieties, valuable in terms of the taste values of the oil obtained, under favourable weather conditions, are affected by fusariosis. Maintaining resistance to rust and fusarium wilting and reproduction for resistance to powdery mildew are the goals when developing new varieties.

Fusarium wilt is caused by the seed fungus and transmitted through the soil *Fusarium oxysporum* f.sp. *lini*. The fungus infects plants with roots at any stage of growth during the growing season and continues to cause infection in the tissue carrying the root water. This interferes with water intake and warm weather and therefore worsens the symptoms of plants because of illness (Brayford, 1996).

Symptoms: Early infections can kill flax seedlings soon after sunrise, whereas late infections cause yellowing and wilting of leaves, followed by browning and death of plants. The roots of dead plants are grey. The tops of wilted plants often turn down and form a "shepherd crook". Affected plants are more often found in patches, but they can also be dispersed throughout the field. Mushroom remains in the soil because mycelium and spores will last for many years in the debris of flax and other organic matter in the soil. Release by wind and flowing water can spread mushrooms from one field to another. Control: The most important control measure is the use of available resistant or moderately resistant varieties. Low inoculum level in the soil can be achieved by stopping flax cultivation at the same locality for at least three years. Dressing seeds with recommended fungicides can protect the crop from early infection at the seedling stage and help maintain good tree stands and vigour.

Rust is potentially the most destructive disease affecting flax (Burdon and Jarosz, 1992). The last major rust epidemic occurred in the 1970s. Although it is effectively controlled by genetic resistance, it remains a potential threat to flax production as it can survive locally and complete its life cycle on flax, thus having the ability to produce new races that attack hitherto resistant varieties. The causal organism is *Melampsora lini*, a fungus that overwinters by means of teliospores on flax debris. Early infections produce the aecial stage with aeciospores on volunteer flax seedlings, which subsequently produce the uredial stage. Urediniospores can cycle through several generations during the growing season, resulting in completely defoliated flax plants and reduction of seed yield and fibre quality. Flax rust completes its life cycle on the flax plant, unlike many other rusts that require an alternate host.

Symptoms: Rust is readily recognized by the presence of bright orange and powdery pustules, also called uredia. Rust pustules develop mostly on leaves, but also on stems and bolls. The pustules produce numerous urediniospores that are airborne and cause new cycles of infections during the season. Spread and infections are favoured by high humidity during cool nights, warmer day temperatures and on plants growing vigorously. As the season progresses, the orange pustules turn black and produce overwintering telia and teliospores. The black pustules are most common on stems.

Control: Complete control is achieved by the use of rust-resistant varieties. Planting susceptible varieties may not only result in serious yield loss but also affords the fungus a chance to produce new races that may attack resistant varieties. Additional safeguards include destroying plant debris, using certified and disease-free seed of a recommended variety, crop rotation and planting the flax crop in a field distant from that of the previous year.

The next of the diseases occurring on flax is pasmo. This disease was once covered by quarantine. The causal organism of this disease is *Septoria linicola*, a fungus that attacks above-ground parts of flax and overwinters in the soil on infected flax stubble (Convey, 1962). Flax is susceptible to pasmo from the seedling stage to maturity. Epidemics can occur early in the season when favourable conditions of high humidity with frequent rain showers prevail. Pasmo can cause defoliation and premature ripening and can weaken the infected pedicels, resulting in heavy boll-drop under rain and wind conditions. Depending on the earliness and severity of the infection, pasmo reduces seed yield as well as seed and fibre quality. Commercial flax varieties lack resistance to this pathogen.

Symptoms: Pasmo is characterized by circular brown lesions on the leaves and brown to black infected bands that alternate with green and healthy bands on the stem. Infected flax tissue is characterized by tiny black pycnidia that are the fruiting bodies of the fungus. The debris carries numerous pycnidia that overwinter and produce masses of spores that cause the initial infections on leaves and stems. Spores are dispersed by rain and wind. High moisture and warm temperatures favour the disease. Lodging favours the development of pasmo because of increased humidity within the crop canopy and this may result in patches of dead plants completely covered with the fungus.

Control: Growers should follow a crop rotation of at least 3 years between flax crops to minimize the inoculum pressure of the pathogen.

The causal agent of powdery mildew is the fungus *Oidium lini*. Early infections may cause severe defoliation of the flax plant and reduce seed yield and quality (Saharan and Saharan, 1994). Some flax varieties are resistant/moderately resistant to this disease.

Symptoms: The symptoms are characterized by a white powdery mass of mycelia that start as small spots and rapidly spread to cover the entire leaf surface. Heavily infected leaves dry up, wither and die. Early infections may cause complete defoliation of flax plants.

Control: The most economical control is through the use of resistant varieties. Early seeding will reduce the impact of this disease on yield loss by avoiding early infections and buildup of epidemics.

Stem break and browning are phases of a disease caused by the seedborne and soilborne fungus, *Aureobasidium pullulan* var. *lini*, also called *Polyspora lini*.

Symptoms: Stem break is the first conspicuous disease symptom. Development of a canker at the stem base weakens the plant. The stem may break at this point when the plants are still young, or at a later stage. Plants may remain alive after stem breakage, but any seed produced may still be lost in harvesting because the seeds produced will be smaller and thin. Initial infections in spring may start from spores produced on diseased stubble and are spread by wind and rain. Infections may start during seedling emergence when seed coats of the diseased seed are lifted above the ground and the fungus produces the first cycle of spores of the season. The browning phase is initiated by infections on the upper part of the stem that appear as oval or elongated brown spots, often surrounded by narrow, purplish margins. The spots may coalesce, and the leaves and stem turn brown. Patches of heavily infected plants appear brown, giving the disease the name of 'browning'. The fungus may penetrate bolls as well as seeds or may produce spores on the seed surface. Affected seeds may remain viable.

Control: Use of disease-free seeds produced by healthy plants is the most important control measure.

Seedling blight and root rot may be due to soilborne fungi such as species of *Fusarium*, *Pythium* and *Rhizoctonia*. However, *Rhizoctonia solani* is the principal causal agent and can be particularly destructive in soils that are loose, warm and moist. *R. solani* survives as a composite of strains that differ in host range and pathogenicity. Seed coats of yellow-seeded flax varieties are more prone to cracking and splits, which renders them more susceptible to infections, causing seedling blight and root rot, than brown-seeded varieties.

Symptoms: Blighted seedlings turn yellow, wilt and die. Infected seedlings may occur singly or in patches. Seedling blight may be inconspicuous, and gaps in the row may be the principal sign of disease occurrence. Roots of recently affected plants show red to brown lesions and may later turn dark and shrivel. Diseased plants are often difficult to distinguish from those killed by the wilt fungus. Root rot symptoms usually appear in plants after the flowering stage. Plants may wilt on warm days and turn brown prematurely. Plants with root rot usually set little or no seed.

Control: Seedling blight and root rot can be controlled by a combination of farm practices: using certified seeds of a recommended variety and reducing cracking of seeds by adjusting combined settings during harvest.

The causal agent of Sclerotinia stem rot disease is *Sclerotinia sclerotiorum,* which is a widespread pathogen causing diseases on different other crops. This fungus survives for 3–4 years in the soil as compact masses of mycelia called sclerotia. The infection in flax is caused by mycelial infection on plants touching the infested soil in heavily lodged flax.

Symptoms: White coat appears on the stems, which causes the stem breaking. This results in earlier atrophy of plants. Mycelia grow on the stem surface, and cylindrical-shaped sclerotia are formed inside the stem.

Control: Producers should avoid fields with previous history of heavy *Sclerotinia* inoculum and water saturated or heavily irrigated fields.

Control: It is recommended to use lodging-resistant varieties, proper seeding rate and recommended fertilizer rate to avoid dense crop canopy. Producers should avoid fields with previous history of heavy *Sclerotinia* inoculum and water-saturated or heavily irrigated fields.

- Pest control

Plantations of fibre flax are infested with both pest characteristic for the crop such as flea beetles, flax thrips, cutworms of gamma moth and by polyphagous pests such as crane flies and nematodes (Haydock and Pooley, 1997; Jankauskiene, 2004). The biggest losses are caused mostly by flea beetles (*Longitarsus parvulus* Payk. and *Aphtona euphorbiae* Schrank.) and thrips (*Thrips linarius Ladureu* and *T. angusticeps* Uz.). The choice of insecticides recommended for flax also covers products such as lambda-cyhalothrin, acephate, alphamethrin, beta-cyfluthrin, deltamethrin, esfenvalerate, methomyl omethoate, vamidithion and acetamiprid. Early sowing reduces the risk of damages caused by flee beetles and thrips. Moreover, using seed dressing treatments with both fungicidal and insecticidal effect, e.g., those based on thiamethoxam + metalaxyl-M + fludioxonil, considerably limits the environmental costs of flax cultivation (Jankauskiene et al., 2004).

Harvesting is described in the paragraph below.

5A.4.7 Cultivars

The most important functional traits of flax and linseed, seed yield and quality and content of fibre (Decanniere, 1989, El-Hariri et al., 2004a,b), are characterized by the low inheritability and high effect of environmental factors which makes breeding difficult, especially in countries with less favourable climate (El-Shimy et al., 1997; Praczyk and Silska, 2013). Currently, the Register of European Union holds 182 cultivars of flax and linseed. The list excludes varieties deleted by the preservation institutions but still in the register until the end of the period during which the marketing of seed material of these varieties is allowed.

The list of the cultivars included officially in the Register of European Union in presented in vol. 1, chapter 17. Developments in fibrous flax breeding and cultivation by Pavelek et al.

The photographs of varieties of flax blossoming in various colours are presented in Fig. 5A.3.

Figure 5A.3 (a) Blue flowering flax; (b) White flowering flax; (c) Red flowering flax Rubrum.

5A.5 Flax harvesting

During harvesting, flax is pulled, not cut. In 1960, the first self-propelled flax pulling machine was developed by Depoortere in Belgium. Photographs of pulling, deseeding and turning machines used by Belgian company Depoortere are presented below (Fig. 5A.4).

The one raw machine has four pulling elements of 35 cm each, which means a total pulling width of 1.40 m. The two row machines are provided with a double pulling machine with total pulling width of 2.40 m. When used in this process, self-propelled pulling machines (e.g., those made by Depoortere, Belgium) achieve a pulling yield from 0.5 to 1.25 ha/h. Then, the flax straw next is laid in the field in rows, and the dew retting process is conducted.

After pulling is over, the turning operation is carried out, which can be done in different dew retting times. Most of the machinery used for turning is also self-propelled. In the case of dew retting, deseeding has to be performed 10 days after pulling. Generally, two types of deseeding machines are used: a deseeding machine can be mounted on the chassis of the self-propelled flax-pulling machine or a machine designed specifically for deseeding may be used. The deseeding has to be carried out when seed capsules are dry: 25% of moisture is allowed, but then the capsules have to be stored under cold ventilation to dry capsules to achieve the moisture content 12%−15% (Depoortere, 1989).

Figure 5A.4 (a) Depoortere_flax pulling machine, double row with cabin; (b) Deseeding machine for flax, double row, Depoortere company, Belgium; (c) Turning machine for flax, double row, Depoortere.

Flax harvesting from the field (unretted and dew retted) is performed by flax round balers, which are of two types: drawn round baler and self-propelled round baler. The drawn machine is cheap, but its work is not so precise as that of the self-propelled round baler, which bales flax perfectly. During round baling, two ropes are rolled around a bale and their consumption ranges between 25 and 30 kg/ha. The ropes have to be made from natural fibres, such as coarse flax, hemp, jute, kenaf or sisal fibres, never from polypropylene. Straw intended for water retting and for decortication is collected by a round baling system, whereas straw intended for dew retting is laid in rows in the field for 3–6 weeks and after that time, the retted straw with humidity below 16% is rolled and delivered to a scutching mill. Currently, as already mentioned, in most European countries, dew retting is widely used, but crop loss, associated with this process, occurs around once every 5 years. Water retting, which consumes a lot of water (see above) and results in the formation of a very large amount of sludge (about 30 m^3 of sludge/ton of retted straw), is not applied nowadays (Dambroth and Seehuber, 1988; Kozłowski, 1970).

5A.6 Degumming

To separate fibre from the woody core, several methods for fibrous plant degumming are used. In this process, fibres are separated from adhesive substances like hemicelluloses, pectin, and partially from lignin, waxes and fats. The methods of fibre plant degumming are presented in Fig. 5A.5. These methods of separation enable fibre to be obtained with the required length, fineness, purity, strength, optimal efficiency and homogeneity.

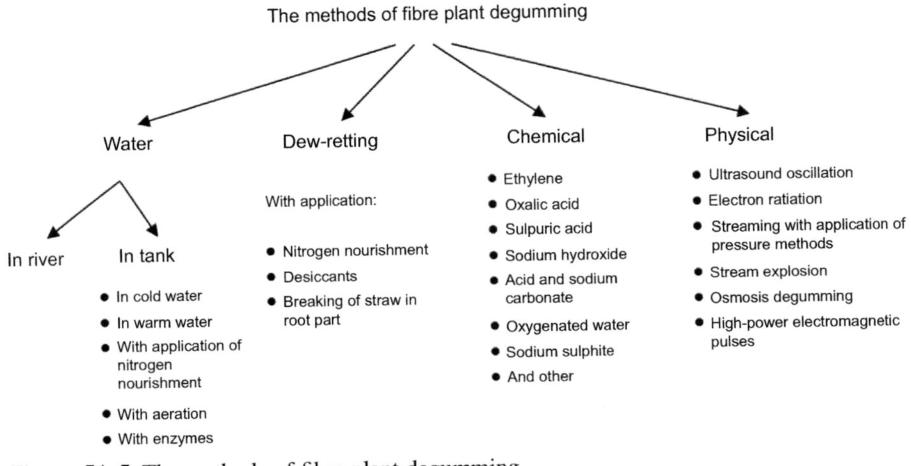

Figure 5A.5 The methods of fibre plant degumming.

5A.6.1 Retting of flax: water and dew retting

Retting is the subjection of crop or deseeded straw to chemical or biological treatment to make fibre bundles more easily separable from the woody part of the stem. Flax is described as water-retted, dew-retted, or chemically retted, etc., according to the process employed (McIntyre et al., 2002).

The biochemical process of retting can be performed under aerobic and anaerobic conditions.

5A.6.1.1 Water retting

In the process of water retting, depending on temperature, environmental conditions (the presence or absence of oxygen), and sometimes nutrients, different bacteria take part, namely *Clostridium* sp., *Clostridium butiricum, Granulobacter pectinovorum, Clostridium felsineum, Clostridium guerfelli* and *Bacillus amylobacter*. In the case of water retting under anaerobic conditions, pectin materials and hemicelluloses which bond the fibres decompose and form volatile fatty acids, mainly butyric (80%), responsible for the odour of retting liquor and artificially dried fibres. When sun-drying is performed, the volatile fatty acids undergo decarboxylation and decompose, thus after such drying, water-retted straw, as well as fibre dried in the air, does not have any odour (Jankauskiene et al., 2006; Kozłowski (1969), 1992; Schilling and Muller, 1951).

Once this fact became common knowledge, all who performed the retting of flax straws had to stimulate the multiplication of the bacteria by heating the treatment water to 30°C–34°C or by adding active bacteria cultures to the water, or even by adding pure enzymes to accelerate the decomposition, but in all these attempts, the inconsistent specifications of the produced fibres persisted.

Recently, this interpretation of the reason for the separation of flax fibres was found to be scientifically incorrect. Thanks to a newly discovered, more controllable method of fibre extraction, it was discovered that within the flax plant, flax fibres have naturally stable excellent textile characteristics; they exist in the form of single filaments grouped in bundles surrounded by a holding tissue. They are not glued together and to the other plant tissues; they are only grouped and held firmly by the holding tissue, consisting of dynamic cells having vulnerable semipermeable membranes.

By smooth and complete elimination of the holding tissue, the flax filaments are gradually liberated in the form of individually separated filaments without affecting their natural potential textile characteristics, which were recently found to be suitable for spinning the finest yarns without any major problems, in addition to being consistent for every variety of flax. Picture of water retted flax fibre bundles presents Fig. 5A.6.

The retting process may be carried out either without or with addition of a nutrient – urea. Research carried out on a commercial scale included the comparison of retting process without and with addition of 0.5% urea in relation to the dry mass of retted flax straw. This research included the mechanical processes: scutching, hackling, spinning, yarn bleaching, weaving, fabric dyeing and printing (Kozłowski, 1970). As a result of urea application, in addition to the reduction of the retting period (in this

Figure 5A.6 Watter retted flax fibres bundles.

case, above 34%), it appears that the most advantageous technological factors have been obtained at the optimum retting period of 52 h with the addition of 0.5% urea/weight of straw and 75–80 h for the conventional retting process. Both hackled fibre and grey and bleached yarns produced from these optimum retting conditions had the best parameters, but the parameters from 52-h retting with addition of urea are more advantageous than those obtained from the conventional retting.

The most advantageous factors for the finished fabrics have also been obtained from the optimum retting periods. The tensile strength of the fabrics produced from the longer retting periods is lower than that from the optimum retting periods, both with and without addition of urea. The fabrics produced from the fibre retted by the conventional process experienced higher losses during different finishing processes than those manufactured from fibre retted with addition of urea (Probulski et al., 1967; Kozłowski, 1970; Kozłowski, 1970).

There are two problems in the case of traditional use of fibre after water retting:

- pulling-binding, where the problem is in obtaining a good tossing in the shives for colour and the homogeneity of retting;
- pulling-spreading (mostly used for green flax harvesting) is associated with the same problem as that encountered in the case of field retting because swaths regular in thickness and alignment will give straw homogeneous in colour and make it easy to pick it up for baling (Kozłowski, 1992).

5A.6.1.2 Dew retting

During dew retting, various fungi take part in this process, such as *Alternaria alternata, Alternaria linicola, Cladosporium herbarum, Fusarium nivale, Cephalosporium* sp. and other, mainly pectinolytic and cellulolytic species. Flax straw lies on the field for about 2 months. Dew, sometimes rain, starts to develop the fungi activity as mentioned previously.

After dew retting, two factors have a great influence on fibre quality:

- the level of moisture for a good conservation, without risk of overretting in bales, which degrades the fibre strength;

- the alignment of swath to avoid picking up the neighbouring swaths and crushing the extremities of the stems in the baler.

The harvesting operations in field retting can preserve quality in several respects:

- homogeneity of dew retting and fibre alignment influence processing yield;
- degree of retting has an effect on divisibility of fibre and strength of the yarn;
- cleanness of the straws helps with avoiding irregularities after the dyeing of yarn or fabric.

Picture of flax fibre bundles after water retting Fig. 5A.6.

5A.6.2 Osmotic degumming

The discovery of a more controllable and nonpolluting method of vegetal fibre extraction, using regular water and taking advantage of the natural forces of fluid diffusion – osmosis and the effective force of the resulting osmotic pressure generated inside the cells that have semipermeable membranes – made it easy to gradually and smoothly extract the flax fibres without affecting their natural potential for excellent textile characteristics.

In Egypt, with the involvement of the Institute of Natural Fibres, Poznan, Poland, and the FAO/ESCORENA European Cooperative Research Network on Flax and other Bast Plants, a new, more controllable method of vegetal fibre extraction has been investigated, using osmotic pressure which enabled the avoidance of the pollution caused by bacteria and fungi. This method is capable of resulting in more homogeneous pure flax fibres in a single filament form which, once implemented, would fully satisfy the world demand for fine flax fibres.

Osmotic degumming consists in the extraction of fibres based on physical laws: water diffusion, osmosis and osmotic pressure. Using physical laws, especially those concerning osmosis, when the inner part of a fibrous plant is in contact with water, the extraction of fibres can occur with no effect on their natural properties. The fibre obtained by this method is delicate and thin and has colour adequate to the quality of raw material used (Konczewicz and Kozłowski, 2007; Konczewicz et al., 2006a, 2006b).

Osmotic degumming enables homogeneous fibre to be obtained and allows for objective evaluation of the amount and quality of fibre contained in the raw material. This is especially important when evaluating progress in breeding new cultivars of fibrous plants (Rolski et al., 2000). Osmotic degumming can be accelerated using ultrasound processing. From the point of view of industrial application of osmotic degumming method to bast fibrous plants, it is worth mentioning that fibres obtained by using the above method can be employed for producing better textiles (yarn, fabric, nonwoven) and composite products.

Osmotic degumming gives homogeneous long flax fibres with fineness of 0.91–1,00 (tex) and breaking tenacity of 62.79 (cN/tex) (Kozłowski et al., 2010a). This more controllable method of vegetal fibre extraction takes advantage of natural physical laws of fluid diffusion, osmosis and osmotic pressure, to consistently produce flax fibres of stable and more homogeneous textile characteristics, capable of successfully competing against all other textile raw materials. It is reliable, as it is based on the following scientific principles:

1. The tissue surrounding the fibres inside the plant is completely formed of dynamic cells, which have semipermeable membranes, whereas all other parts of the plant (the wooden cylinder, the fibres and the outer skin) are formed of static cells, which do not have semipermeable membranes.
2. According to the natural physical phenomena of fluid diffusion, osmosis and osmotic pressure, when two fluid media of different water molecule concentrations are separated by a semipermeable membrane, the water molecules in the medium of higher water molecule concentration will continuously move to the medium of lower concentration by crossing the semipermeable membrane, which only allows a one-way crossing, until the concentrations in the two media become equal.
3. When flax straw is submerged into fresh virgin water, each dynamic cell in the plant represents a medium of lower water molecule concentration, separated by a semipermeable membrane from the surrounding fresh water, which represents the medium of higher water molecule concentration.
4. Under these conditions, the water molecules in the surrounding treatment water continuously move and enter each dynamic cell, through its semipermeable membrane.
5. As a result, increasing outwardly osmotic pressure is generated on the inside of the semipermeable membranes of each dynamic cell, causing the gradual rupture of its semipermeable membrane when the osmotic pressure becomes more than these membranes can stand and the contents of the dynamic cells are then dispersed into the surrounding water.
6. This gradual destruction of the dynamic cells, forming the holding tissue firmly surrounding the fibres, continues until the complete release of the fibres, as long as the difference in water molecule concentration, inside and outside these dynamic cells, is kept at its maximum, by renewing the treatment water every time it becomes ineffective when saturated or polluted.
7. To follow up the course of the fibre separation, samples of the treated straw are frequently tested by trying to pull out the fibres sideways from a piece of straw from the root end to discover the degree of fibre separation.
8. Water molecules enter the flax straw through the central wooden cylinder and attack the dynamic cells layer after layer, starting with the first circular layer adjacent to this wooden cylinder up to the last circular layer adjacent to the outer skin.
9. During the extraction operation, all static cells forming the wooden central cylinder, the fibres and the outer skin are only wetted to saturation without being affected.
10. Once all the dynamic cells, forming the entire holding tissue, are eliminated, the fibres are totally liberated and appear individually separated; they are also in a pure form and clearly have natural potential for superior high textile qualities.

5A.6.3 Enzymatic degumming

Enzymatic treatment allows for faster and simultaneous degumming of the fibre from the woody part and decomposition of pectin in the fibre itself. This leads to shortening of the degumming time and improvement of fibre quality, mainly by higher divisibility and fineness of fibre strands (Kozłowski et al., 2005, 2006a, 2006b).

Enzymatic treatment (mostly with use of pectinolytic enzymes) enables the application of preparations that undergo complete biological degradation in sewage which leads to an eco-friendly technology as well as water and energy savings (Foulk et al., 2007, Foulk et al., 2009; Guebitz et al., 2006; Jayapriya et al., 2010; Kozłowski et al., 2009).

The commercial offer to improve the retting process by using enzymes has been put forward by Novo Industry, Rohm and Haas, Enzylin, Bioprogress Maceraze, Pektopol U and other companies.

Flaxzyme offered by Novo Industries for degumming of flax straw is used at a concentration of 3–6 mL per 1 L of water at temperature of 40°C. The enzyme solution can be recuperated many times. Flaxzyme consists of polygalacturonases, methylesterases, pectin lyases and, in smaller quantities, hemicellulase and cellulase. Enzymatic treatment can also be applied to green decorticated fibre. In the case of decorticated flax fibres, the following parameters were obtained: fineness 3.6–4.7 tex, strength 6.8–8.8 cN/tex and good softness and feel (Kozłowska, 1989). More about enzymatic treatment and about the mechanism of enzymatic action is presented in Chapter 7, vol. 2: Enzymatic treatment of natural fibres.

5A.6.3.1 New process of plant fibres degumming using electric resonance technology

A novel plant fibre degumming technology has been developed that consists in the use of electric resonance electrodes immersed into a water pond, possibly with the addition of specific pectolytic enzymes to the water, to accelerate the process.

The special electrodes used for degumming fibre crops in a pool develop an electric field, without producing electric current but propagating a particular resonance signal in the water.

The retting process occurs by combining two factors: (a) the utilization of a device called EL3 that generates signals at frequencies ranging from 50 Hz to 12.8 MHz and (b) the addition of a special mixture of selected enzymes.

The well-dried biomass bale should be fully submerged in the pool for a period of a few days. A continuous retting system is easily conducted; the electric resonance produced by the electrodes eliminates bacterial development.

The advantages of the technology are

- in the case of flax, the seeds can be harvested separately, before baling of the stems;
- the possibility of processing also well-dried biomass;
- the process produces neither bad odour nor contaminated water, in contradistinction to the traditional pond retting system, which at present is not accepted in most countries;
- the colour of fibres is not dark (as often occurs in the case of the dew retting process or full immersion retting); fibres are clear and brilliant, and they are characterized by higher quality.

To make the separation of fibres easier and faster, specific pectinolytic enzymes, dissolved in an appropriate amount of water, are used.

The optimal temperature for degumming is between 25°C and 30°C. Lower temperatures will slow down the time of processing. In optimal conditions of treatment, an increase of water temperature is observed because of molecular movements induced by the resonance. In a water volume of 5.5 cubic metres and with the power system used, two pairs of immersed electrodes are needed. Electrodes should be located under the water level, one in front of the other, but they should not be in contact with the walls and with the bottom of the container, nor with the bale of biomass. Each electrode used is of the size of 55 × 15 × 5 cm.

After processes involving the electrodes and enzymes are over, it is necessary to completely dry the biomass to preserve and, later on, to process the material for the extraction of fibres. This drying process could be energy consuming, if not properly done. Possibly, energy from solar panels and dry air forcing technologies should be used after the opening of wet bales to speed up the process and save electric and thermal energy.

If the stems could be properly cut and orderly baled by machine at harvest, all the process could be made easier and the final product quality could be better. Note: special electrodes were developed by Italian physicist Professor Mario Pandolfo; see the technical indications provided by Giovanni Mignoni, Pietro Cappelletto and other experts (Bozzini and Mignoni, 2010). According to our knowledge, this method is not largely implemented in the practice.

5A.7 Scutching

Scutching is the next operation after harvesting and retting, which is aimed at separating the woody part of deseeded or retted flax straw from the fibre (separating bast from woody core). For this operation, retted straw is prepared as parallel layer stalks. Next, the flax layer is gripped between two belts – U belt and T belt. During the transit, the flax tops are broken by 2 × 5 pairs of jagged cylinders: the first five pairs with rough teeth and the second five pairs with fine teeth. After this, broken parts of the stalks are taken over by two other belts, which continue the transit. The scutching removes broken stalks and remaining shives and also short fibres obtained.

The typical scutching machines consist of two drums 4 m long for scutching the foot of the straw and two drums 4 m long for scutching the top.

The shives and short fibre dropping under the drums are pneumatically sucked up and put on the shaker that removes the loose shives.

At the end of the shaking, the short fibre is transported to an automatic tow press or tow scutching machine. The scutched long flax fibre goes out of the end of these scutching machines and is laid down in handfuls, which are tied into bundles (16–20 kg). These bundles are pressed into 100 kg bales and sent to the spinning mill.

The picture of the scutching turbine (line) of the Depoortere company, Belgium, working at LENKON company in Steszew, Poland, is presented in the photograph, Fig. 5A.7.

The scutched flax is in the form of coarse fibre strands (bundles of many fibres joined together by hemicelluloses and residual pectin and also some woody impurities). The fibre strands consist of 15 fibre bundles held together in lattice formation

Figure 5A.7 Scutchine turbine (line) of the Depoortere company, Belgium.

and they are roughly parallel. The quality of the scutched fibre generally is estimated by the organoleptic method (features such as lustre, colour, handle, alignment of the strand, etc.), as well as by objective methods such as measuring the strength, fineness, length of fibres, level of impurities and also without naps and knots.

5A.7.1 Decortication of green straw

There are so many disadvantages of water retting and dew retting, which were mentioned before, that many research centres have attempted to develop direct extraction of the fibre from the straw by breaking or tearing off the raw straw by the process called decortication. Obviously, the quality of such fibres (bast) is very low, and such fibres are not useful for textile application without enzymatic or chemical treatment. Decortication is cheaper and faster in comparison to different degumming methods. One of the new concepts is the treatment of the decorticated bast by osmosis and enzymes and intensifying these treatments by exposure to ultrasound.

A photograph of the decorticator developed at the Institute of Natural Fibres (Polish patent, 199630, Kozłowski et al., 2008; Kubacki, 2010) is shown in Fig. 5A.8. This system is very promising from an economic viewpoint.

5A.8 Hackling (combing)

Hackling – a process in which stricks of scutched flax are combed from end to end to remove short fibres, naps (or neps) and nonfibrous material, and to subdivide and parallelise the fibre strands (McIntyre et al., 2002).

Hackling of scutched flax fibres is a very important process in the preparation of fibres to effective quality for spinning. In the linen industry, three types of hackling (combing) machines are used: traditional Russian system, Mackie system (Northern Ireland), Linificio system (Italy), N. schlumberger (France).

Figure 5A.8 Decorticator for bast plants, LENKON, Poland, developed at the Institute of Natural Fibres, Poznań.

Modern system of combing machine model Alfa lin 990452 GRIS AT, produced by the company N. schlumberger, Guebwiller, France, is presented in Fig. 5A.9.

Lummus Mackie developed a hackling machine that was mechanically simpler and more efficient than the Russian system. The patented system opens the combing sheets, moves the flax pieces into position and then closes for the hackling process to commence. The sheets then are opened again allowing the flax to be carried to the next stage. This system achieves a consistency in the rate of hackling, which is ensured by the inclusion of an inventor control motor to regulate the rate of hackling.

Figure 5A.9 Combing machine model Alfa lin_n.schlumberger_990452_GRIS_AT, France.

The corner motion is by a pneumatic mechanism, which ensures a quick and efficient method of transferring the flax pieces from one side of the frame to the other, while at the same time repositioning the flax piece to ensure a complete hackle. The holder facilitates the hackling of a larger amount of flax per lift.

Dust and dirt extraction is an important part of the hackling process. The above system comprises a doubler, four drawing frames and a roving frame.

The commonly used system of processing (hackling) of the scutched fibres (e.g., Linmack machine of Mackie International) is described in Table 5A.4.

The hackling machine consists of two sides or 'machines' to treat the top end of the fibre strand and the root end. The fibre bundles held in holders travel along the 'root side' of the machine, where the shorter root ends of the fibres are combed or hackled between the pinned sheets of the machine. At the other end of the machine, the fibre is turned around so that when travelling along the 'top side' of the machine, the tops of the fibres are combed, care being taken to see that there is no part of the fibre left untreated – this is achieved by an automatic pull-through at the change-over end.

The hackling machine therefore produces two types of fibres:

1. Long straight fibre bundles – 'line fibre' and
2. Short tangled fibre – 'tow fibre' (hackled noils).

The yield of the line fibre is defined as the following:

$$\text{Yield of line fibre} = \frac{\text{weight of line fibre produced}}{\text{weight of scutched fibre fed}} \times 100 \tag{6.1}$$

Table 5A.4 System of hackling of the scutched flax fibres.

Flax line	Flax tow (hackled noils)
Preparing	*Carding*
Doubler	
First drawing passage	
Second drawing passage	*Preparing*
Third drawing passage	
Roving	*Finisher*
Line roving machine	
Wet ring spinning	*Dry ring spinning*
	Wet ring spinning

Source: Mackie International, 1994.

The yield of hackled fibre is usually in the range of 55%–65% and depends on the degree of hackling, on fibre quality and purity. The production yield (capacity) of the hackling machine is normally between 40 and 65 kg/h (Ross, 1992).

Carding of linen – the disentanglement, cleaning and intermixing of fibres to produce a continuous web or sliver suitable for subsequent processing. This is achieved by passing the fibres between relatively moving surfaces covered with so-called card clothing. The process follows opening and blending (McIntyre et al., 2002).

The picture of the modern carding machine to process flax tow (Card for flax tow 030,801 GRIS AT) produced by the company N. schlumberger, Guebwiller, France, is presented in Fig. 5A.10.

5A.9 'Cottonization'

'Cottonization' is a process by which the flax fibre is made shorter and finer with the lowest possible level of impurities, similar in features to cotton fibre.

Due to overproduction of short fibres, the method of making this fibre cleaner (impurities of the fibre from retted straw make up less than 0.5%, with fibres finer and homogeneous) is very promising. Such a method was developed by using Rieter Elitex machines, which provide cleaner and homogeneous short fibres. This system can also be used for cleaning green decorticated fibres, which then can be used for the production of, e.g., cigarette paper or natural linen nonwovens and also coarse, thick technical yarns, e.g., strings.

Rieter Elitex line is presented in Fig. 5A.11.

There are several methods to make flax fibre 'cottonized'.

Figure 5A.10 Card for flax tow_030801_GRIS_AT_n.schlumberger, France.

Figure 5A.11 Rieter Elitex line for cotonization of bast fibres, including flax, LENKON, Poland.

5A.9.1 'Cotton' type flax fibre

'Cotton' type flax fibre has the following properties: staple length – unbleached 30–35 mm, bleached 20–30 mm, linear density unbleached 1.5–2.9 tex, bleached 1.0–2.0 tex, impurities below 0.5% (Kozłowski, 1996).

'Cottonized' flax can be processed in the same way as cotton (U.S. patent 666696 – a new mechanical treatment process for flax). The decortications can be used to separate the woody component, pectin and other impurities, which can be removed by chemical or cooking process of a fibre bundle that is then converted into a shorter staple fibre. Fibre bundle with a 10% of moisture content is carded and a fine fibre strip or sliver is created. The flax tow can be carded on Mackie card and the sliver has only 5% of woody impurities. The sliver is then drawn 5 m/min with 16 doubling. Next, it is padded and fed into an intermediate steamer followed by further wet processing, but finally drum dried and delivered into a card can coiler.

At the first stage, chemical treatment and subsequent steaming takes place at 60°C and a pressure of 2 psi (13.79 kPa), after which washing is carried out. Liquor pick up is 72%, and the sliver is then deposited in a steamer for 30 min and treated with saturated steam at 100% followed by drying at 140–150°C. This system is then used to bleach the sliver with intermediate steaming at 20°C and again 2 psi (13.8 kPa). The bleaching liquor composition is 0.15 g/L magnesium sulphate, 8 mL/L water glass, 38° Be 5 g/L sodium hydroxide sol, 50° Be 6 g/L cottoclarin OK, 2 g/L securon 540, 2 g/L Belsoft 200, plasticizer, 60 g/L H_2O_2 – 35% (Lennox, 1998).

5A.9.2 'Wool' type flax fibre

The thinness of 'wool' type flax fibre is a maximum of 30 nm (for the yarn 23 tex, length of the staple 40–120 mm, impurities below 1%, homogeneous colour).

'Wool' type flax fibre is used for ring, wrap, friction–DREF two spinning, mainly for the production of blended linen yarns, also for fabrics, knitting, apparels, socks, shoes, household goods and upholstery fabrics.

5A.9.3 Spinning

Flax fibres after extraction and hackling can be used for spinning. Both short and long fibres can be used, but different techniques are used for spinning short (tow and noils) and long fibres.

Generally, the flax fibres used in wool and cotton spinning come from classical scutching. Next, they are proceeded to adjust their length and purity according to the spinning system, which also provides with the possibility of blending with other fibres (Sultana, 1991).

5A.9.4 Problems with spinning linen

The established belief has been that flax fibres naturally have heterogeneous textile characteristics, making their spinning a complicated and difficult job, successfully achieved only by highly skilled spinners. For this reason, more than 75% of flax yarns usually produced worldwide have been of easier-to-spin, coarser counts whose production has stopped lately because of lack of demand. The more astute spinners who have been producing different specialized fine counts of flax yarns still exist but are loudly complaining about the scarcity and irregularity of the flax raw materials available.

The latest thorough scientific review of the plant's detailed morphology and fibre extraction operation, sustained by practical experience in flax cultivation, flax straw processing and flax fibre spinning, has successfully led to the firm confirmation that the stability of flax fibres has favourable natural potential textile characteristics.

Flax yarns have an irregular thickness that is often welcomed by weavers as an attractive and idiosyncratic feature. Their users have considered the characteristic irregularities in the textures of the linen fabrics produced as a welcome change to the monotonous regularity of other more uniform fabrics. These linen irregularities have even been industrially imitated with other homogeneous textile raw materials to get the same irregular effect.

5A.9.5 Developments in linen spinning

Fineness and regularity in length are two important factors for flax spinning. The good ripeness of the flax straw brings a lower heterogeneity between the fibrils of the top and bottom (Sultana, 1991).

The first flax spinning machines could produce only coarse yarns and could not compete with handmade 'workshop linen yarns'. The first step in splitting flax fibre bundles, not only in the hackling process, but also in further stages of spinning, was done in France. In particular, progress was made by a chemist Gay-Lussac at the Engineering College in Paris, who had adopted the chemical treatment to flax spinning. Further development in changing the length and thickness of flax fibres resulted in mechanical and chemical treatments in preparation stages of spinning including boiling or bleaching roving and in wet yarn formation on the spinning frame. In this way, despite difficulties for the spinners, these yarns up to tex 8 (Nm 125) could be produced on an industrial scale. Unfortunately, the cost of spinning fine yarns is relatively high, mainly because of the low speed of spinning and the low automation of the process (Kozlowski and Manys, 1994).

There are remarkable differences in the speed of spinning, and there is n further need for progress in the field of flax fibre processing. A similar conclusion can be drawn from the comparison of spinning costs. It is obvious that only high value end fibre products can be spun at low speed where there are high costs of the yarn production.

One hundred percent linen fine fabrics and apparel made of hackled flax fibre may compete with cotton and man-made fibres, despite low speed of spinning and high cost of the yarns. These linen fabrics and goods can compete because they provide high comfort, excellent physiological influence on human body, hygiene as well as elegance. Unfortunately, many excellent advantages are not enough to solve the competition problems of short flax fibre. By the use of traditional spinning technology only coarse yarn or yarns of medium density can be spun from short fibres (tow). Even worse, carded linen yarns are rigid, irregular in every aspect and have low elongation. This is why that for many years, short flax fibre lost its market in household and other uses. It should be mentioned that in the traditional flax fibre technology, short flax fibre represents about 50% of total flax fibre production. Information about the method of improving short flax fibre quality is presented in point 5A.9. 'Cottonization'.

Among traditional linen yarns, 100% linen yarns for knitting specially softened in or after traditional spinning can be produced. The features of these pure linen yarns are similar to these of traditional weaving yarns and in addition they have favourable qualities for rigidness and softness, which make them suitable for knitting. They suit perfectly the requirements for comfortable and hygienic apparel for hot, humid climatic conditions. Also, blended linen yarns (with cotton, wool and man-made fibres) are on the market. Tables 5A.5–5A.7 present the properties of blended linen yarns, spun by conventional cotton ring spinning systems (Krmela, 1993).

The features of blended linen yarns spun by the long staple ring spinning system are presented in Table 5A.8 and those spun by the rotor spinning system in Table 5A.9.

The rigidness of flax fibre and the low flax fibre elongation forced the development of wrap spinning technology, e.g., the application of flax-blended Parafil yarns in woven and knitted products.

The Parafil spinning system frames are characterized by the following:

- The length range of processed fibres, including short flax fibres after appropriate preparation.
- The linear mass range of manufactured yarns, including fundamentally the whole range of fineness for yarns made of flax and its blends.
- Lowering of energy consumption in the process of spinning.
- Significantly higher speed of spinning.
- Considerably lower labour consumption of the process – sliver spinning among others (Manys, 1989).

Table 5A.5 Blended linen yarns spun using conventional cotton ring spinning systems (flax tow).

	Flax 10% Cotton 90%	Flax 20% Cotton 80%	Flax 15% PAC 85%	Flax 35% PAC 85%	Flax 15% Viscose 85%	Flax 30% PES 70%	Flax 30% PES 70%	Flax 30% PES 70%
Linear mass (tex)	50	50	49.7	48.2	50	45.4	47.3	51.4
Linear mass irregularity (%)	–	–	1.2	1.1	–	2.7	4.1	4.4
Strength (cN/tex)	11.5	10.3	18.2	14.4	10.3	17.3	19.5	15.7
Strength irregularity V (%)	13.7	15.6	11.5	12.7	10.4	9.5	12.4	16.4
Elongation (%)	6.1	6.0	19.4	18.5	15.2	20.3	20.0	
Twists (n/m)	453	443	428	407	409	486	525	516
Twist irregularity V (%)	4.2	4.4	5.4	2.2	5.9	4.9	9.1	

PAC, polyacrylic fibre; *PES*, polyester fibre; *V*, variation.
Source: Krmela, 1993.

Table 5A.6 Blended yarns spun by conventional cotton ring spinning technology (30% flax, 50% viscose, 20% cotton) (Rieter).

Linear mass (tex)	**64**	**64**	**31**	**31**
Linear mass irregularity CV (%)	18.4	19.2	22.0	25.0
Breaking strength (cN/tex)	11.5	11.1	10.0	9.6
Elongation (%)	6.0	6.5	4.8	5.2
Faults (n/1000 m)				
Thin places	20	30	280	500
Thick places	600	980	2000	500
Naps	480	650	1900	2450

CV, coefficient of variation.
Source: Krmela, 1993.

Table 5A.7 Blended linen yarns spun by conventional wool spinning system (flax combings).

	Flax 11% Wool 22% PES 41% PAC 26%	**Flax 11% Wool 22% PES 41% PAC 26%**	**Flax 15% PAC 85%**	**Flax 35% PAC 47% Wool 18%**	**Flax 15% Viscose 17% PAC 68%**	**Flax 25% Viscose 17% PAC 58%**	**Flax 26% Viscose 14% PAC 60%**	**Flax 30% PAC 70%**	**Flax 30% Viscose 70%**
Linear mass (tex)	48.6 × 1	48.9 × 2	32.1 × 2	32 × 2	54	50	55	51	68
Linear mass irregularity (%)	1.8	1.3	2.1	2.7	2.48	2.6	4.4	2.3	5.4
Strength (cN/tex)	13.1	15.5	13.8	6.5	12.9	13.0	9.9	9.9	7.3
Strength irregularity V (%)	12.5	9.3	12.1	10.1	3.9	4.2	5.0	5.4	7.5
Elongation (%)					12.8	10.7	14.9	12.6	11.0
Twists (n/m)	434	180	–	–	360	340	326	330	334
Twist irregularity V (%)	6.5	7.2		6.7	3.3	3.5	1.9	3.6	4.3

PAC, polyacrylic fibre; *PES*, polyester fibre.
Source: Krmela, 1993.

Table 5A.8 Rotor blended linen yarns.

Blend	Linear mass		Breaking strength	
	Tex	CV%	cN/tex	CV%
10% Flax/90% cotton	30	3.5	8.5	13.5
20% Flax/80% cotton	42	3.5	8.0	13.5
30% Flax/70% cotton	50	3.5	7.5	14.0
40% Flax/60% cotton	64	3.5	7.0	14.0
10% Flax/90% cotton	30	3.5	6.5	13.5
20% Flax/80% cotton	42	3.5	6.5	13.5
30% Flax/70% cotton	50	3.5	6.5	13.5
10% Flax/90% PES	30	3.5	12.0	13.5
20% Flax/80% PES	42	3.5	12.0	13.5
30% Flax/70% PES	50	3.5	11.5	13.5
40% Flax/60% PES	64	3.5	11.5	13.5

Source: Krmela, 1993.

Table 5A.9 Wrapped yarns spun by spinning frame Parafil 2000.

Type of yarn	Linear mass (tex)	Breaking strength		Elongation (%)	Wrapping material (%)
		cN/tex	V%		
Flax-PL combed yarns					
Flax-PL combed yarn (fibre broken to 220 mm)	86.2	9.4	22.0	4.5	5.5
Flax-PL combed yarn (fibre broken to 200 mm)	75.8	7.5	18.0	1.9	5.5
Flax combed blended yarns					
Flax-PL combed blended yarn, 30% flax, 70% PES	56	20.2	18.0	17.9	6.2
PL combed blended yarn, 30% flax/70% viscose	100	11.5	13.9	9.7	2.3
PL combed blended yarn, 30% flax/70% viscose	50	11.5	19.0	17.9	4.6
PL combed blended yarn, 20% flax/30% PES/50% PAC	86	11.6	14.8	14.1	5.5

PL, Parafil yarns, PAC, polyacrylic fibre; *PES*, polyester fibre.
Source: Krmela, 1993.

Data on blended short and long staple Parafil linen yarns are presented in Tables 5A.9 and 5A.10.

Friction spinning parameters of linen-blended yarns spun by DREF 2 and DREF 3 spinning machines are presented in Tables 5A.11 and 5A.12. These yarns are mainly applied to the manufacture of upholstery yarns and knitted apparel (Kozlowski and Manys, 1995).

Summing up, the main features of the new short flax fibre types, characterized by shorter length, lower thickness, low content of impurities and sometimes improved surface properties, provide the possibility of producing the yarns mentioned above. Narrowing the staple and fineness diagram, reducing impurities and increasing softness are tasks for the future development in processing of short flax fibres. The design of apparel according to fashion expectations may change, but the flax fibre will remain untouched by fashion and market. Now, we have knowledge about producing new homogeneous and fine flax fibres, also functionalized for special application areas including composites, working in different extreme conditions. New linen woven and knitted textiles are in leading collections over the world, the main trend of which is to improve comfort and hygiene features, such as low electrostatic properties, natural blocking of ultraviolet (UV) radiation, obtaining noniron properties by liquid ammonia treatment, and giving the best sleeping conditions on 100% flax bedlinen (Kozlowski and Manys, 1994).

There are a number of advantages and disadvantages from the spinning point of view arising from the properties of flax fibres.

The well-known properties of flax fibres are high hygroscopicity, high heat absorption, low static resistance and nonallergic behaviour. These features provide comfort and hygiene of apparels and bedclothes, as well as bedlinen. However, technologically disadvantageous properties of flax fibres exist too:

- long and irregular fibres;
- linear mass of fibres – high and irregular;
- resistibility against splitting;
- rigidness;
- low elongation;
- high content of impurities;
- different colours;
- different mechanical strength.

These disadvantages of flax fibres make the competition against cotton and man-made fibres very difficult and sometimes impossible, especially in the spinning area (Kozlowski and Manys, 1994).

In countries with the highest level of textile technology, more than 70% of flax fibre processed is used in the manufacturing of clothing. The area of clothing manufacture is open to proposals for new types of flax yarns produced by means of new techniques, including decorative fancy yarns. Some examples of spinning techniques for flax are given below:

- Flax-adapted processing with the use of nonflax conventional techniques of spinning (with mechanical and physicomechanical processes at the pretreatment stage) is presented in Tables 5A.13 and 5A.14.

Table 5A.10 Wrapped yarns spun by spinning frame Parafil 1000.

Specification	Linear mass		Breaking strength		Elongation (%)	Wrapping material (%)	Number of wraps per 1 m	Spinning speed (nm/min)
	Tex	V%	cN/tex	V%				
40% Flax/30% PES/30% PAC	78.7	–	9.7	–	14.5	6.3	330	53
30% Flax/20% PES/50% PAC	98.0	3.5	10.2	14.7	14.1	4.9	130	80
30% Flax/70% PES	54.5	0.2	19.1	13.5	20.0	5.0	130	80
20% Flax/40% PES/40% viscose	57.7	2.1	14.4	13.0	14.7	5.0	130	80
20% Flax/80% PAC navy blue	51.4	2.9	9.6	20.2	12.7	5.0	130	80
20% Flax/80% PAC beige	55.4	2.5	10.5	12.8	14.6	5.0	130	80
An experiment on manufacturing crewel to be used in knitted top fabrics: 30% Flax/20% PES/50% PAC	450.6	—	14.6	8.5	14.7	3.7	150	120

PAC, polyacrylic fibre; *PES*, polyester fibre.
Source: Krmela, 1993.

Table 5A.11 Friction linen blended yarns (spun by spinning machine DREF 3).

Blend		Linear mass			Strength			Elongation (%)	Twist (n/m)
		Tex	Nm	V%	daN	cN/tex	V%		
DREF 3	Core: PAN 38 mm Coat: PES 50%, flax 50% 150 m/min	51.09	19.57	0.7	0.63	12.39	13.5	18.3	–
DREF 3	Core: PES 38 mm Coat: Cotton 60%, flax 40% 150 m/min	53.44	18.71	1.1	1.10	20.51	9.4	11.4	–
DREF 3	Core: flax 20%, PES 25%, PAN 55% Coat: PES 50%, flax 50% 150 m/min	82.57	12.11	1.8	0.71	8.65	13.7	9.9	–

PAN, polyacrylonitrile fibre; *PES*, polyester fibre.
Source: Krmela, 1993.

Table 5A.12 Friction linen blended yarns (spun by spinning machine DREF 2).

No.	Blend	Linear mass		Strength		Elongation		Twist	
		Tex	Nm	V%	daN	cN/ tex	V%	%	n/ m
4675	Core: fil. PES 167 dtex Coat: flax 20%, PES 25%, PAN 55%, 150 m/ min	99.70	10.03	1.6	0.94	9.43	4.0	21.1	–
4675	Core: fil. PES 167 dtex Coat: flax 20%, PES 25%, PAN 55%, 150 m/ min	161.45	6.19	1.1	0.78	4.83	20.9	13.3	–
4675	Core: fil. PES 167 dtex Coat: flax 20%, PES 25%, PAN 55%, 170 m/ min	152.00	6.58	1.0	1.28	8.42	6.3	20.2	–
4689	Core: fil. PES 167 dtex Coat: flax 20%, PES 25%, PAN 55%, 170 m/ min	105.26	9.50	1.0	0.95	9.03	5.6	19.7	–
4689	Core: fil. PES 167 dtex Coat: flax 20%, PES 25%, PAN 55%, 130 m/ min	159.06	6.29	1.1	0.71	4.44	16.8	12.5	–
	Core: fil. 167 dtex Coat: flax 100%, 130 m/min	108.34	9.23	0.4	1.04	9.60	8.0		–
	Core: fil. PES 167 dtex Coat: flax 100%, 130 m/min	149.48	6.69	1.0	0.85	5.68	21.3	6.3	–

fil. PES, filament polyester fibre; *PAC*, polyacrylic fibre.
Source: Krmela, 1993.

Table 5A.13 Spinning techniques – ring spinning for short and long staple fibres.

Ring spinning for short staple fibres	
Length of fibre (mm)	17–28
Fineness of fibre (tex)	0.6–1.6
Breaking strength (cN/tex)	34.0–18.0
Impurities (%)	0.0–0.7
Degree of polymerization, DP	2.600–2.200
Degree of whiteness	½ – ¾
Ring spinning for long staple fibres	
Length of fibre (mm)	52–110
Fineness of fibre (tex)	0.75–2.2
Breaking strength (cN/tex)	25.0–19.0
Impurities (%)	0.2–1.5
Degree of polymerization, DP	3.200–2.000
Degree of whiteness	Up to ¾

Source: Manys, 1991.

- Some new forms of spinning techniques for flax, presented above (excluding traditional ring spinning), have overcome barriers in its processing such as slow speed of ring yarn manufacturing from flax fibres, limited assortment of flax yarns and narrow range of using mixed systems of raw materials in yarn.

The data presented above show that the flax fibre has been accommodated for almost all the modern techniques of spinning of recent yarns. New forms of flax make it possible to manufacture new flax yarns of different characteristics in composition with all other natural and chemical fibres used in the production of yarns for household goods and clothing.

Some of the new forms of flax fibre have already been included in commercial offers, whereas some others, most likely, will be commercially offered, too.

While describing the new phenomenon in flax processing, one must stress the following: just like in the case of traditional flax technology, mechanical or physico-chemical processes used in obtaining new standards of flax quality must be accompanied by the use of highly effective spinning techniques to raise the spinning quality of flax fibre.

Nevertheless, the ultimate effect, i.e., fineness of the yarn obtained and quality of the final product (associated with fineness) depends on the anatomical structure of

Table 5A.14 Spinning techniques – rotor spinning, friction, friction/combined spinning.

Rotor spinning – open end (OE)	
Length of fibre (mm)	24–32
Fineness of fibre (tex)	0.4–0.8
Breaking strength (cN/tex)	18.0–14.0
Impurities (%)	0.0–0.1
Degree of polymerization, DP	2.000–3.000
Degree of whiteness	Up to $^3/_4$
Friction spinning – OE DREF 2	
Length of fibre (mm)	38–64.3
Fineness (tex/Nm)	1/1000–2.64/388
Impurities contents (%)	0.2–1.1
Degree of polymerization, DP	2.700–3.000
Degree of whiteness	–
Friction/combined spinning DREF	
Length of fibre (mm)	20/150
Fineness (tex/Nm)	0.8–1.5
Impurities (%)	0.3–4.0
Degree of polymerization, DP	2.200–3.200
Degree of whiteness	
Hollow spindle (wrapping)	
Short staple system, e.g., Parafil 1000	
Length of fibre (mm)	84–112
Fineness of fibre (tex/Nm)	0.8–1.7
Breaking strength (cN/tex)	15.0–27.0
Impurities (%)	0.6–1.2
Degree of polymerization, DP	2.000–2.600
Degree of whiteness	–
Long staple system, e.g., Parafil 2000	
Length of fibre (mm)	24–36
Fineness of fibre (tex/Nm)	0.7–1.8

Continued

Table 5A.14 Continued

Rotor spinning – open end (OE)	
Breaking strength (cN/tex)	19.0–23.0
Impurities (%)	
Degree of polymerization, DP	1.900–2.300
Degree of whiteness	Up to $^3/_4$

Source: Manys, 1991.

elementary flax fibre and on the structure of cellulose and its accompanying substances which in turn depend on vegetation conditions of flax and harvesting time.

New spinning techniques have been developed that are used to obtain new flax fibre types and 100% new fibre-containing yarn, as well as blended yarns.

Italian company Linificio e Canapificio Nazionale is the world leader in flax and hemp yarn production – produces 100% linen yarns from Nm 25 – Nm 105. Also produces extremely innovative and fine yarns, e.g., new-generation 100% linen yarn Capri, which allows getting a weaving performance 10%–20% higher in comparison to other top-quality linen yarns.

For instance, these fibres were adapted to the production of 'cottonlike' flax fibres (see 'cottonization') by spinning techniques such as open end, wrap, friction DREF 3, for production of 100% linen and blended linen yarns for fabrics, knitting, especially suitable for apparel, socks, shoes, household goods and upholstery fabrics (Manys, 1991).

5A.10 Bleaching and dyeing

Generally, raw linen fabrics and knitting are grey and coarse. They should be bleached, dyed, printed and finished to be of appropriate quality and performance for the needs of apparel and textiles for the home. After these processes, they become more attractive and softer, and their area of potential application is enlarged. These treatments are widely used in the linen textile industry. The following subsections present some data on bleaching, dyeing, printing as well as finishing of linen fabrics.

5A.10.1 Bleaching

Until the turn of the 19th and 20th centuries, our ancestors generally used open space and sun operation for bleaching flax fabrics. For this purpose, they put linen fabrics in supported positions above the field, towards the sun and exposed them to high humidity conditions by spraying water on the fabrics. The UV radiation from the sun (wavelength shorter than 400 nm) decomposed water on the fabric surface to create active oxygen 'in statu nascendi', part of which formed ozone O_3. Bleaching of linen fabrics by using this process took a long time, but it was not hazardous to the environment. To improve this bleaching process, alkaline potash (from burned wood) and organic acids coming from fermentation of starch and milk (buttermilk) were used. This traditional bleaching process was known as crafting or grassing. After the discovery of chlorine (1774) and the development of hypochlorites at the end of 19th century, calcium hypochlorite (bleaching powder) came into use. After 1920, chlorine gas and acidified hypochlorite were used. In the early 1930s, the modern procedure of boiling in peroxide was introduced.

The major bleaching agents for textiles are hydrogen peroxide (most widely applied), sodium hypochlorite, sodium chlorite and peracetic acid (oxidative bleaching agents). Silk and wool are bleached by using, for example, sodium dithionite, the reductive agent, belonging to a group called often 'hydros' (Holme, 2005).

The history of the achievements in the scope of bleaching of cellulosics are presented in Table 5A.15, whereas major organic stabilizers used in peroxide bleaching are given in Table 5A.16.

It is important in bleaching that there is no negative influence on the degree of depolymerization of cellulose, in parallel to achieving high whiteness and moderate weight loss. The principal operations in industrial modern linen bleaching are alkaline boiling or scouring, acid chlorination, sodium hypochlorite solution bleaching, hydrogen peroxide bleaching and sodium chlorite bleaching. Worth mentioning is the intensive development of the use of molybdenum catalyst for effective bleaching (without chlorine components) (Holme, 2005). There has been growing awareness that the use of chlorine compounds can result in the formation of hazardous dioxins, resulting in the elimination of chlorine bleaching and the utilization of mainly hydrogen peroxide with sodium silicate (Kernaghan and Kiekens, 1992a).

Table 5A.15 Nine major milestones in the bleaching of cellulosics.

Process	Approximate date of significant usage
Sulphuric acid scour	1756
Chlorine bleach	1790
Le Blanc process for soda ash	1791
High pressure kier	1815
Rope washer	1830
Enzyme desizing	1900
Peroxide bleaching	1925
Continuous peroxide bleaching	1939
Fluorescent brightening	1950

Source: Pratt, 1994.

Table 5A.16 Organic stabilizers (sequestrants) that have been used in peroxide bleaching.

Gluconic acid
Glucoheptonic acid
Aminocarboxylates
Nitrilotriacetic acid (NTA)
Ethylenediaminetetraacetic acid (EDTA)
Diethylenetriaminepentaacetic acid (DTPA)
Organophosphonates
Aminotris (methylenephosphonic acid) (ATMP)
1-Hydroxy ethylidene-1,1-diphosphonic acid (HEDP)
Ethylenediamine tetra(methylenephosphomic acid) (EDTMP)
Diethylenetriamine penta(methylenephosphonic acid) (DETPMP)

Source: Lewin (1984).

Single bath full bleaching of flax fibres using an activated sodium chlorite hexamethylene tetramine (HMTA) system for the bleaching of flax fibres is based on the activation of sodium chlorite by HMTA in the presence of an anionic wetting agent. The bleached flax fibres are assessed for critical properties, namely whiteness index, loss in fibre weight, tensile strength, content of carboxyl and carbonyl groups. Optimal bleaching conditions for the flax fibres are $NaClO_2$ 5 g/L, HMTA 0.25 g/L and wetting agent 1 g/L, temperature 90°C, time 3 h and liquor ratio 1:50 (Zahran, 2005).

Biobleaching of flax fibres by degradation of lignin with microorganisms such as *Phanerochaete chrysosporium* and *Trichosporon Cutaneum* R57 is also used. The enzyme complex secreted by the filamentous yeast *T. Cutaneum* R57 can be successfully applied to bleaching of flax fibres and fabrics. Also, *P. chrysosporium* produces an enzyme that shows a similar behaviour. After the treatment, the fibre increased its lightness with a stronger lignin degradation and removal during processing.

A sodium chlorite bath for scouring and bleaching of linen fabric in a one-step process activated by glyoxal as an organic reducing component with two aldehyde functional groups was investigated as well. This treatment was tested under a variety of conditions and the following properties were monitored: degree of whiteness, loss of fabric weight, wettability, carbonyl and carboxyl group content, percentage of sodium chlorite decomposition and tensile strength of fabrics. The optimum conditions were found at 60°C, pH of 10.35 with aqueous solution containing 5 g/L $NaClO_2$, 6 mmol/L of glyoxal, wetting agent 2 g/L for 120 min at the liquor ratio 1:30 (Zahran, 2010).

A new improvement of linen bleaching was made by the Technical and Natural Sciences Research Development Innovation Institute at the 'Aurel Vlaicu' University of Arad, Romania, according to which the acceleration of bleaching can be achieved by using a catalyst of the formula $Na_2 [Mo_2O_6 (C_2O_4)]\cdot 4H_2O$.

5A.10.2 Dyeing

All fibres and fabrics usually have to be dyed. The use of natural dyestuff for dyeing fibres and fabrics dates back to very early times and is associated with the human understanding of beauty. Primitive man, while watching nature, dreamed about capturing its colours. He found them in roots, berries, branches, barks, leaves, flowers, fruits, minerals, shells and insects. Skill mastering in plant usage and particulars of knowledge of the use of auxiliary chemical substances were a secret art in the past.

5A.10.2.1 Dyeing with natural dyestuffs (of plant and insect origin)

In ancient Egypt, dyeing was at an advanced level. The first information about dyeing of linen concerned early Dynastic Egypt and was reported by Plinius (AD 79). The earliest methods of dyeing used mordant and mineral dyes. This knowledge was adopted by the Israelites and Phoenicians from the Egyptians and then by the Greeks and Romans.

The discovery of America and its exploration showed the independent dyeing development on high cultural level in Mexico and Peru. Thanks to sea routes to East India, the following dyes became known: cochineal, indigo and dyes from trees such as *Pernambuco Caesalpinia brasiliensis* and *Sanders pterocarpus Santalinus*. Dye producers keep secret their knowledge and protected it from becoming commonly known.

Natural dyes can also be derived from plants, minerals, insects and shellfish, although they have many limitations in terms of fastness and brilliancy of shade.

The archaeological findings indicate that dyes from natural sources have been used to colour textiles for the last 6000 years. Dye research conducted within an ethnography and ecological framework has long been the focus of artists, conservationists, chemists, etc.

A dyestuff is a pigment or colouring agent dissolved in water or some other carrier so that it may be able to stain, more or less permanently fibre or fabric to which it is applied. Natural dyes are classified into two groups – substantive and adjective dyestuffs:

- *Substantive dyestuffs* dye the fibres directly. These kinds of dyes were probably the first type of dyes used by early people. Substantive dyes usually rich in tannins come from bark, leaves and fruit. This group includes also vat dyes – indigo *Indigofera tinctoria* and woad *Isatis tinctoria.* The colouring matter of vat dyes is not directly soluble in water. The colour develops on exposure to oxygen present in air. In indigo and woad, the dye matter is the substance indican.
- *Adjective dyestuffs* need a mordant to fix permanently to the textile fibres. Metallic salts of alumina, chromium, iron and tin help create affinity between the fibre and pigment. The

mordant often produces a much stronger colour on fibres. Examples of such dyes are madder, logwood, cochineal and fustic.

Some dyes do not have a solubilizing group. In such cases, a temporary solubilizing group is introduced in the dye at the time of application. A classic example of such a dye is indigo. Indigo is water-insoluble. After application, the reduced form of the indigo is converted back to its original form by air oxidation of the material. This process is known as 'vatting', and indigo is classified among vat dyes.

In scientific papers and books on colour, the primary colours are identified as red, blue, yellow and green (Schmidt-Przewozna, 2001; Schmidt-Przewoźna, 2016; Schmidt-Przewozna et al., 2016).

Blue

Indigo *(I. tinctoria)* is the king of dyes, originally closely related to life and culture. In Egypt, deep blue was reserved for the daughters of the pharaohs who painted their breasts with blue and gold. Indigo-coloured garments have been found in Thebes. In most traditional cultures, blue was the colour associated with social class status. Blue was also obtained from woad *(I. tinctoria).* This plant contains the same blue colouring matter as indigo, but in lower concentrations. Woad was probably the first blue dyestuff used in America. From the 13th century, many farmers and dyers in Germany, England, France and Italy grew rich on woad production. The woad vat was basically similar to the indigo vat requiring care in controlling the fermentation process by which the colouring agent was reduced to its soluble form. Nowadays, this dye is little used in England and France by artists and designers, although the designer Oliver Lapidus showed an indigo collection at a fashion show in Paris in 1999. The fabrics were dyed at the Lambert workshop in France.

Red

In European culture, red was reserved for noblemen. In ancient times, purple was obtained from shells. Historically, Poland was renowned over a long period for the source of one very important dyestuff, Polish cochineal *(Porphyrophora polonica* L.). From the Middle Ages until the end of the 16th century, this insect was used for dyeing silk with an alum mordant. It thrived on the perennial knawel *(Scleranthus perennis* L.), which can be found in parts of northern Europe including Poland. The precious dyestuff was obtained from the cysts of the insect. The insect is a root pathogen; therefore, it was necessary to dig the plant out to gather the cysts. Poland exported the dyestuff to Italy, the Netherlands, France and other parts of Europe, and also to Turkey. The high point of this trade was in the 15th century. In the Middle Ages, dyers used Polish cochineal and also kermes *(Kermes vermilio* Planchon); sometimes both dyestuffs were used together. There are sometimes mistakes in sources referring to these insects as the differences between the two dyestuffs are not always realized. After the discovery of America, the use of both of them declined as they were replaced by the richer dye insect, cochineal *(Dactylopius coccus* Costa), found in Mexico and Central America. Other important red dyes were the plant dyes madder *(Rubia tinctorum* L.) and safflower *(Carthamus tinctorius* L.), both of which were planted commercially. Other, more local, dyestuffs were also used in Polish workshops. These included the fugitive dyestuff obtained from bilberry or blueberry

(Vaccinium myrtillus L.) and the colourant extracted from the stems of the nettle *(Urtica dioica* L.), which were collected for use in the autumn when they are red. Dyers throughout Europe imported red dyestuffs from other parts of the world, and Polish dyers were no exception: brazilwood was imported from southern India and Malaysia *(Caesalpinia sappan* L., sappan wood) and, after the discovery of the Americas, from parts of Central and South America *(Caesalpinia echinata* Lam and other species). Red sandalwood *(Pterocarpus santalinus* L.), such as sappan wood, was imported from southern India (Schmidt-Przewozna, 2005).

Yellow

The colour of the sun symbolizes growth and possibility. The principal yellow of the ancients was reserved only for the emperors. The yellow dye for textile manufactures was obtained from the weld *Reseda luteola* L. It is possible to obtain a good yellow from turmeric *(Curcuma longa* L.). The main producer of this plant was India, but it was also cultivated in China, South East Asia and tropical countries. Yellow-dyeing plants are common in nature. It is possible to create a full palette of yellows by cultivated, wild and forest plants. Other sources of yellow with considerable tinctorial properties are onion *Allium cepa,* dyer's chamomile *Anthemis tinctoria,* barberry *Berberis* species, birch *Betula* species, heather *Calluna vulgaris,* safflower *C. tinctorius,* drug fumitory *Fumaria officinalis* L., common horsetail *Equisetum arvense* L., heart's ease *Polygonum persicaria* L., sorrel *Rumex acetosella* L., common yarrow *Achillea millefolium* L., dandelion *Taraxacum officinale* L., dyer's coreopsis *Coreopsis tinctoria* L., coreopsis *Coreopsis grandiflora* L., French marigold *Tagetes species* L., dyer's greenweed *Genista tinctoria* L. and many others.

In some parts of the world, natural dyes are still used to colour traditional textiles. In India, Eastern and Central Asia, South America and Africa, natural methods of dyeing are still alive in residual forms. The world tendencies aim towards their recovery (Schmidt-Przewozna, 2001, 2005, Schmidt-Przewozna and Zimniewska, 2007; Schmidt-Przewozna et al., 2009, 2016; Schmidt-Przewoźna, 2016, 2018).

5A.10.2.2 Dyeing with synthetic dyestuffs

Until the 1950s, only about 10% of the linen fabric produced was dyed. In the 20th century, natural dyestuff application in the textile industry almost disappeared and was replaced by chemical dyestuffs. After the 1980s, it was observed that especially fine apparel was dyed.

Direct dyestuffs were the first important synthetic class, developed in the 19th century. By the early 20th century, the vat, azoic and sulphur classes of dyestuffs as well as reactive dyestuffs were also applied to dyeing linen. Most commercial dyestuffs for linen are similar to those used for cotton, viscose and rayon – most of them are applicable to all lignocellulosic fibres. The simple method of dyeing involves exhaustion of dyestuff from electrolyte solution. These dyestuffs belong to a group of direct dyestuffs.

Vat dyestuffs represent the reference standard of fastness and resistance to most agents. In practice, linen fabrics are impregnated by pad mangle using vat dyestuffs.

Also, these vat dyestuffs are used for dyeing linen yarn, which is dyed in the hank form and also in the pack form.

The reactive class of dyestuffs achieve retention by formation of a covalent bond with the cellulose. This reaction fundamentally proceeds via hydroxyl group of cellulose, ionized under mild alkaline conditions to a form linkage between the dye molecule and substrate. There are two possibilities for this reaction mechanism: (a) nucleophilic substitution or (b) nucleophilic addition.

5A.10.2.3 Printing

For decoration purposes, especially for wall coverings, domestic furnishings, tea towels, upholstery, curtains and blinds made from linen, special dyestuffs are used. Very little printing of the expensive apparel weight fabrics is performed. Obviously, absorbency and penetration are potential problems with linen (Kernaghan and Kiekens, 1992b).

Nowadays, there is growing awareness regarding environmental pollution by textile effluents, e.g., from dyeing. There are many dyestuffs, especially those containing heavy metals that are forbidden. In Europe, the Registration, Evaluation and Authorization of Chemicals (REACH) system regulates the use of many chemicals and some dyestuffs as well. On the other hand, increasing awareness concerning environmental issues contributes to the return to natural dyeing and many research centres throughout the globe carry out research on making natural dyeing systems more effective and economically reasonable.

5A.11 Finishing

General definishing of finishing: descriptive of processes, physical or chemical, applied to a substrate to produce a desired effect.

There are many possibilities of finishing and functionalization of natural fibres and fabrics including linen.

Below are presented some examples of linen finishing.

Commercial modified glyoxal-based resin is used for the chemical end finishing of linen and ramie fabrics in wet fixation. This technology significantly reduces the tensile and tearing properties and decreases the resilience of the fabrics and results in a considerable increase in wet recovery angle (Kim, 2005).

Butanetetracarboxylic acid (BTCA) and citric acid are used to replace the formaldehyde-releasing agent in cellulosic fabric finishing. BTCA can esterify hydroxyl groups of cellulose to produce a graft copolymer of cellulose (Bertoniere, 1999).

Liquid ammonia treatment is applied in large-scale production for improving quality of cotton/linen warp/weft-blended fabrics. This treatment improves the dimensional stability, abrasion resistance, appearance after washing and creasing recovery (Dornyi, 2007).

Liquid ammonia treatment was first used by Norwegian company TEDECO in the 1970s to upgrade woven cellulosic fabrics. Then, it was used in Japan for cotton and also for linen. The treatment for linen was developed by the Institute of Natural Fibres (Poland) in cooperation with Nisshinbo Industries Inc. (Japan). The best results were obtained when liquid ammonia and low formaldehyde emission resins were used. The 'Super Soft Peach Phase' imparts the following properties to noniron products:

- low shrinking after washing,
- increased wrinkle resistance,
- increased elasticity,
- increased strength,
- softness to touch,

Liquid ammonia penetrates cellulose very rapidly and forms complexes with its hydroxyl groups after breaking hydrogen bonds in crystalline as well as in amorphous regions of the fibres. Ammonia molecules do not dissolve cellulose, but they are capable of increasing the distances between the cellulose chains through penetration into crystallites. The mechanism of the chemical reaction is not fully recognized; however, it should be mentioned that in finished linen fabrics, a small content of nitrogen was found. Probably ammonia was bonded to the carboxyl groups of cellulose chain. A significant increase in intensity of the absorption band at 895.78 cm^{-1} suggests the formation of $CONH_2$ species (Kozłowski and Helwig, 1996; Kozłowski and Manys, 1996).

Liquid ammonia treatment is similar to the mercerization process, which is very widely used for cotton. The mercerization process (with sodium hydroxide) improves some properties of linen, namely it increases strength, dyeability, water sorption and lustre. After the mercerization process, linen fibres are untwisted and become almost round in cross-section with almost complete disappearance of the convolution. This treatment involves a short exposure to aqueous solution of NaOH which results in the transformation of native cellulose I crystal structure into the regenerated cellulose II crystalline modification (Kozłowski and Helwig, 1996; Kozłowski and Manys, 1996).

5A.12 Recapitulation

The present situation in the flax market can be summarized as follows: the main demand for flax fibres is for the superior high grades suitable for the easy production of fine fibres for flax yarns with no major spinning problems. Since then, the great majority of the world producers of flax fibres have been facing the dilemma of either improving the specifications of their flax fibres to meet the high level required or totally stopping production because of the almost complete absence of demand for inferior-quality textile flax fibres.

Recent efforts aimed at finding a solution to the alarming situation of flax fibres resulted in establishing the following facts:

- In general, flax straws potentially contain very fine fibres, which naturally have superior high textile qualities; however, they become apparent only when these fibres are properly and smoothly extracted from the other plant tissues surrounding them and are obtained in pure, intrinsic, individually separated single filament form, which have stable more homogeneous textile characteristics, inherent to each variety.
- The only method of vegetal fibre extraction that steadily produces pure intrinsic flax fibres of obvious natural excellent textile characteristics is the newly discovered osmotic pressure method that liberates the fibres smoothly, without affecting their natural excellent textile characteristics and without polluting the environment.

To maximize the yield of the fibres, flax cultivation should be aimed at getting the greatest number of fine, slender, unbranched stems through increasing the density of population of the plants by giving each plant the minimum necessary free vital space around it and by evenly distributing the sowing seeds. In Egypt, the maximum potential number of fine stems per square metre has been practically increased to 2500.

To maximize the crop of seeds, flax cultivation should be performed in such a way that the greatest number of capsules is obtained, which is attained by giving the plants the maximum vital free space around each plant, through a regular distribution of the sowing seeds, to allow the plants to give birth to the greatest number of tillers and as much branching and subbranching, to bear the greatest number of flowers which would produce, after pollination, the greatest number of seed-containing capsules.

Specific free space around each plant is essentially needed to allow the full expansion of its root and rootlets, in all directions and at different levels, to collect all the water that touches them, without interfering with other nearby plant rootlets. Otherwise, the nearby plants will share the amount of available water and each one of the plants will only grow according to the amount of water it acquires. The crop maximization of any plants spaced too closely is practically impossible.

Plants grow according to a number of influential factors, some of which are natural factors that are out of our human control, whereas the other ones can be effectively controlled. If any of these factors is partially absent, the final overall effect will be limited to the level of the presence of this factor, in accordance with the physical law of the effective capacity of a "wooden barrel" being equal to its shortest stave.

In flax cultivation, if the required free space around each plant is not respected, the actual crop of each plant crop will be limited to the actual percentage of the required free space around each plant. If, e.g., the required free space around each plant is 10 cm and the actual free space around a plant is only 4 cm, the optimum growth and crop of this plant will not exceed 40% of its natural potential capacity, even if all the other influential factors are 100% present.

In equal weights, thinner flax straws contain relatively more fibres than the thicker ones and need less time to be properly extracted from other surrounding plant tissues. The best average diameter of flax straw varies between 0.7 and 1.0 mm. In this case, the total yield of fibre can reach the level of about 35%–37% and even to 40% of total straw mass. Currently, the best producers of fibrous flax – farmers in Normandy, France, are highly skilled in proper agrotechnology, including appropriate procedures of harvesting flax for fibre, when the straw colour is green-yellow.

5A.13 Conclusions and future trends

5A.13.1 Conclusions

Linen is the one of the oldest textile fibres. The properties and special physiological activity of flax fibres has many advantages in comparison to man-made fibres, and even other natural fibres. Generally, flax fibres have got the opinion as most healthy fibres, they never cause allergy and oxidative stress, and linen fabrics are the best for bedlinen and healthy apparel. By-products such as seeds and shives are also source of valuable agrochemicals for the production of oil (rich in omega 3 and omega 6 fatty acids), lignans and cyclolinopeptides. (see Chapter 18, vol. 2: The application of flax and hemp seeds in food, nutraceutical and cosmetics products).

Shives could be applied for the manufacturing of, e.g., composites.

Currently, there is ongoing discussion regarding bioengineering modification of flax, e.g., in creating 'in statu nascendi' polyhydroxyalkanoates, which are naturally created polyesters. The new techniques and technology of processing flax fibres give opportunities to obtain very fine yarn and thin noniron flax textiles.

The new improved aspects of flax are due to the innovations applied to its cultivation and to the extraction of its fibres. The maximization of flax crops to obtain either seeds or fibres, and the stabilization and improvement of fibre specifications, have been the results of first-hand scientific approaches, which updated and enriched the available knowledge about the morphology of this ancient plant and about the impact of the influential natural and human factors on its behaviour during growth and subsequent development.

From now on, flax manufacture will be much improved. During cultivation, its behaviour is under full control and can practically be directed to economically and beneficially maximize the production of excellent textile fibres or rich oily seeds through two completely different forms of flax sowing seed distributions: for the maximization of fibres, each plant should be surrounded by the minimum necessary free vital space; for the maximization of seeds, the plants should be surrounded by the maximum necessary free space.

For the first time, the extraction of flax fibres can be completely mastered and scientifically industrialized, in accordance with known natural forces acting according to physical laws, leading to the steady production of pure, more homogeneous flax fibres that show natural potential of superior high textile quality, suitable for the production of the finest possible yarns, without major spinning problems, at reasonable competitive production costs.

As a result, flax industries can look forward to a healthy revival and guard against future unexpected negative situations.

Useful flax knowledge, on the basis of Egyptian experience, in terms of flax cultivation, flax straw processing, linseed oil production, flax tow twine production, production of particleboard and flax hardboard production, is presented in Tables 5A.17–5A.21.

Table 5A.17 Useful knowledge – flax cultivation.

Item	Rate	Unit
Preparation of the field	4 labour hours	Per hectare
Sowing of seeds for fibres	2500 seeds	Per square metre
Sowing of seeds for seeds	1000 seeds	Per square metre
Mechanical sowing of seeds	$^3/_4$ labour hours	Per hectare
Hand spreading of seeds	4 labour hours	Per hectare
Artificial irrigation	600 cubic metres	Per hectare
Artificial irrigation	2 labour hours	Per hectare
Chemical weed control	$^1/_2$ labour hour	Per hectare
Manual weed control	100 labour hours	Per hectare
Mechanical spreading of fertilizer	$^1/_2$ labour hour	Per hectare
Manual spreading of fertilizer	$2^1/_2$ labour hours	Per hectare
Amount of fertilizer	Nitrogen 80 units Superphosphate 300 kg Potassium 350 kg	Per hectare
Mechanical harvest	2 labour hours	Per hectare
Manual harvest	140 labour hours	Per hectare
Mechanical deseeding and tying beets	6 labour hours	Per hectare
Manual deseeding and tying beets	160 labour hours	Per hectare
Weight of imported sowing seeds	5 g	Per 1000 seeds
Weight of local sowing seeds	9.5 g	Per 1000 seeds
Weight of deseeded straw of 1 mm in thickness	0.5 g	One stem
Weight of deseeded straw of 3 mm in thickness	1.5 g	One stem
Volume of flax straw with seeds	18 cubic metres (9.5–12.0 m^3 in Europe)	One metric tonne
Volume of deseeded flax straw	25 cubic metres (11.0–14 m^3 in Europe)	One metric tonne
Volume of flax seeds	1.5 cubic metres	One metric tonne
Even spaces around each stack	3 times its height	Stack height

Table 5A.17 Continued

Item	Rate	Unit
Flax straw stacking	2 labour hours	Per metric tonne
Stacking of flax seeds in bags	¾ labour hour	Per metric tonne
Permitted humidity for flax straw	10% (16%–18% in Europe)	
Permitted humidity for flax fibres	12%	
Permitted humidity for flax seeds	8%	
Straw loading on lorries	3 labour hours	Per metric tonne

Source: Allam (2008).

5A.13.2 Future trends

The growing fibrous flax in the European Community has increasing potential, although to a limited extent.

The capacity of the world market for flax fibres can currently be evaluated at about 400,000 tons per year including 160,000 tons of scutched long flax fibres.

The conditions to obtain the indicated growth are as follows:

(a) restructuring of the entire linen industry;
(b) research on the improvement of the quality of flax fibres (higher fineness, higher fibre content);
(c) enlarging the area of utilization (Coby, 1990).

The main reasons for fibrous flax expansion and flax technology modernization in the 21st century are the following:

- The exhaustion of resources of the major organic chemicals (coal, gas, oil), which dictates the trends towards organic renewable raw materials both for energy and products.
- In the long term, future polymers for textile and nontextile applications might be of plant origin (fibrous plants and wood) supplemented by algae and autotrophic bacteria, which can use carbon dioxide directly to synthesize polymers (e.g., bacterial cellulose).
- Fibrous flax and oil flax are perfect plant biopolymers.
- Rapidly growing ecological trends and ecological awareness create the demand for healthy clothing and environmentally friendly raw materials and technologies.
- The urgent need for sustainable agricultural development is well known, and flax could be cultivated in the areas extending from one arctic circle to the other, in contradiction to cotton, which can be cultivated only in hot climatic zones and requires huge amounts of water and plant protection chemicals.

Table 5A.18 Useful information – flax straw processing.

Item	Rate	Unit
Straw sorting	16 labour hours	Per ton
Filling of retting tanks	3/4 labour hour	Per ton
Emptying of retting tanks	2 1/2 labour hours	Per ton
Setting of wetted straw in cones for even dryness	10 labour hours	Per ton
Mechanical breakage and scutching using turbines	50 labour hours (20 h in Europe)	Per ton
Manual breakage and scutching	300 labour hours	Per ton
Sorting and classification	50 labour hours (20 h in Europe)	Per ton
Pressing of long filament flax fibres in bales of 250 kg	1 1/2 labour hours	Per ton
Pressing of tows in bales of 200 kg	2 labour hours	Per ton
Pressing of fibre waste in bales of 158 kg	2 1/2 labour hours	Per ton
Loading of pressed flax fibre bales	1 labour hour	Per ton
Unloading of pressed bales	1/4 labour hour	Per ton

Source: Allam (2008).

Table 5A.19 Linseed oil production.

Item	Rate	Unit
Linseed sieving	3 labour hours	Per ton cleaned
Crushing and extracting the oil	72 labour hours	Per ton in containers

Source: Allam (2008).

Table 5A.20 Useful information – flax tow twine production.

Item	Rate	Unit
Flax tow carding	20 labour hours (10 in Europe)	Per ton of sliver
Sliver drawings to needed weight	15 labour hours	Per ton of sliver
Spinning and twisting into twine	50 labour hours (30 h in Europe)	Per ton of twine

Source: Allam (2008).

Table 5A.21 Useful information – production of particleboard.

Item	Rate	Unit
Cleaning of flax shives	2 labour hours	Per ton cleaned
Shive mixing and material gluing	4 labour hours	Per ton of boards
Sanding and finishing the boards	5 labour hours	Per ton of boards

Source: Allam (2008).

- The other reasons for flax expansion in the current century are the valuable new products and by-products associated with flax, e.g., reinforced polymers, biocomponents based on flax (or other bast plants and derived fibres) instead of glass fibres.
- One hundred percent biodegradable packaging materials based on, e.g., waste short fibres, also special pulp and paper (e.g., cigarette paper).
- Brand new products based on linseed food and nutrition products rich in omega 3 and omega 6 fatty acids as well as in lignans and cyclopeptides, biooils for nutrition and cosmetics.
- Possibility of cultivation of flax and other bast fibrous plants on soils polluted with heavy metals as alternative crops for nonfood purposes and soil remediation (see chapter 11, volume 2. The role of hemp (*Cannabis sativa* L.) in remediation of degraded lands).
- New fibre processing technologies developed in the last decades are similar to those in the so-called yarn revolution that occurred some years ago in the area of cotton, wool and man-made fibre industries. This requires new types of flax fibres that are cleaner and highly homogeneous. The shortening, splitting and cleaning necessary to spin flax by modern techniques are far beyond the reach of traditional fibre assortments. These new types of flax fibres can be blended with all natural and man-made fibres. The newest 'dream concept' in the area of flax fibre production is to develop one integrated process which should include harvesting, deseeding and fibre extraction performed by one machine in the field at harvest time.
- The new research achievements and progress made in the field of flax production and processing have also resulted in the improvement of the competitiveness and coexistence of flax with other fibrous materials (both natural and man-made ones).

Summing up, new types of flax fibres have been developed that are characterized by a controlled level of lignin and pectin, adopted to conventional and unconventional spinning and other processing systems (e.g., nonwovens), and linen-blended yarns and new fields of the application of flax fibre and by-products emerged, namely healthy linen apparel, protective and anti-UV and antielectrostatic clothing (see Chapter 4, vol. 2 in the 1st edition), knits such as socks, pullovers, healthy nonwovens and upholstery, bedlinen, biocomposites, degradable packaging and linseed-based new products rich in omega 3 and omega 6 fatty acids as well as rich in lignans (like Bioflax – see Chapter 18, vol. 2).

These data give excellent confirmation that the Latin name *'Linum usitatissimum'* given to flax by Carl Linnaeus in the 1770s is well deserved (Kozłowski, 1996).

5A.14 Sources of further information and advice

For more information and knowledge development, the following literature on flax and linen is recommended:

Ageeva, M. V., Petrovska, B., Kieft, H., Salnikov, V. V., Snegireva, A. V., van Dam, J. E., van Veenendal, W. L., Emons, A. M., Goshkova, T. A. and van Lammeren, A. A. 2005. 'Intrusive growth of flax phloem fibers is of intercalary type', Planta, 222 (4) 565–574.

Allaby, R. G., Peterson, G. W., Merriwether, D. A. and Fu, Y. B. 2005. 'Evidence of the domestication history of flax (Linum usitatissimum L.) from genetic diversity of the sad2 locus'. Theoretical and Applied Genetics, 112 (1), 58–65.

Allam A. 2004. Flax Latest Diagnostic, Journal of Natural Fibers, 11,109–110.

Andème-Onzighi, C., Douchiche, O., Driouich, A. and Morvan, C. 2005. 'Composition of flax hypocotyl fibres', Journal of Natural Fibers, 2 (3), 1–15.

Baraniecki, P. 1997. 'Ecological friendly clothes based on flax and hemp fiber', in Proceedings of the 78th World Conference of the Textile Institute in association with the fifth Textile Symposium of SEVE and SEPVE, Thessaloniki, Greece, vol. 2, pp. 205–216.

Betcheva, R., Georgieva, N., Yotova, L., Valchev, I. and Chadjiska, C. 2007. 'Biobleaching of flax degradation of lignin with phanerochaete chrysosporium and trichosporon cutaneum R57', Journal of Natural Fibers, 4 (4), 31–40.

Blackburn, R. S. (ed.) 2005. Biodegradable and Sustainable Fibres. Cambridge: Woodhead Publishing.

Brutch, N. B., Soret-Morvan, O., Porokhovinova, E. A., Sharov, I. Y. and Morvan, C. 2008. 'Characters of fibre quality in lines of flax genetic collection', Journal of Natural Fibers, 5 (2), 95–126.

Buckley, S. A., Clerk, K. A. and Evershed, R. P. 2004. 'Complex organic chemical balm of Pharaonic animal mummies', Nature, 431, 294.

Crangle, A. A., Heaney, J. P., Hill, B. J., McIlhagger, R. M. and Lyttle, M. 2005. 'Improvements to the wet spinning of flax: Part I – The effect of two sets of feed rollers on the physical characteristics of wet spun linen yarns', Journal of the Textile Institute, 96 (1), 21–28.

Crangle, A. A., Heaney, J. P., Hill, B. J., McIlhagger, R. M. and Lyttle, M. 2005. 'Improvements to the wet spinning of flax: Part II – The effect of breast beam draft control measures on the physical characteristics of wet spun linen yarns', Journal of the Textile Institute, 96 (1), 29–36.

Dambroth, M. and Seehuber, R. 1988. Flachs Zuchtung, Anbau, Verarbeitung. Stuttgart: Eugen Umler GmbH & Co, Germany.

Day, A., Ruel, K., Neutelings, G., Cronier, D., David, H., Hawkins, S. and Chabbert, B. 2005. 'Lignification in the flax stem: Evidence for an unusual lignin in bast fibers', Planta, 222 (2), 234–245.

Dissanayake, N. P. J., Summerscales, J., Grove, S. M. and Singh, M. M. 2009. 'Energy use in the production of flax fiber for the reinforcement of composites', Journal of Natural Fibers, 6 (4), 331–346.

Dornyi, B., Csiszár, E. and Somlai, P. 2007. 'Improving quality of linen-cotton fabrics with liquid ammonia treatment', Journal of Natural Fibers, 4 (4), 41–57.

Dowgielewicz, S. 1954. Roślinne Surowce Włókiennicze (Plant fibrous resources). Warszawa. Państwowe Wydawnictwo Naukowe, 1-457

El-Beltagi, H. S., Salama, Z. A. and El Hariri, D. M. 2008. 'Some biochemical markers for evaluation of flax cultivars under salt stress conditions', Journal of Natural Fibers, 5 (4), 316–330.

Evans N, McRoberts N, Hitchcock D and Marshall G. 1995. Screening for resistance to Alternaria linicola Groves and Skolko in Linum usitatissimum L. using a detached cotyledon assay, Ann Appl Biol, 127, 263–271.

Evtimova, M., Vlahova, M. and Atanassov, A. 2005. 'Flax improvement by biotechnology means', Journal of Natural Fibers, 2 (2), 17–34.

Fakin, D., Golob, V. and Stana-Kleinschek, K. 2006. 'Influence of enzymatic pretreatment on the colours of bleached and dyed flax fibres', Journal of Natural Fibers, 3 (2/3), 69–81.

Fauque, C. 1997, Secrets de lin: Carnets du textile. Paris: Syros.

Flax as a Fibre and Oil Bearing Crop: Proceedings of the FAO European Workshop on Flax, Brno, Czechoslovakia, 18–20 June.

Food and Agriculture Organization 1994. Definitions and Classification of Commodities: Fibres of Vegetal and Animal Origin.

Foulk, J. A, Akin, D. E. and Dodd, R. B. 2009. 'Miniature Spinning Enzyme-Retted Flax Fibers', Journal of Natural Fibers, 6 (1), 1–13.

Franck, R. R. (ed.) 2005. Bast and Other Plant Fibres. Cambridge: Woodhead Publishing.

Fröier, K. and Zienkiewicz, H. 1979. Linboken. Hemodling och hemberedning. Stuttgart: LT.

Gorshkova, T. and Morvan, C. 2006. 'Secondary cell-wall assembly in flax phloem fibres: Role of galactans', Planta, 223 (2), 149–158.

Growing Flax: Production, Management & Diagnostic Guide, fourth edn. Saskatchewan: Flax Council of Canada and Saskatchewan Flax Development Commission.

Hann, M. A. 2005. 'Innovation in linen manufacture', Textile Progress, 37 (3), 1–48.

Harwood J, McComrick P, Waldron D and Bonadei R. 2008. Evaluation of flax accessions for high value textile end uses, Ind Crops and Products, 27, 22–28.

Herzog, R. O. 1930. Der Flachs, Abtl. 1: Botanik, Kultur, Aufbereitung Bleicherei und Wirtschaft des Flachses, Technologie der Textilfasern, Springer, Berlin.

Hilliers, D. R. (ed.) 1974. Discovered in the Wadi Ed Daliveh. The Annual of the American Schools of Oriental Research, 41, Cambridge, Massachusetts, American Schools of Oriental Research, Cambridge: MA.

Ilea, V. 2009. 'Genetic resources and breeding fiber flax for the next century', in Scientific Bulletin of ESCORENA, vol. 1, ed. Aurel Vlaicu University, Romania, Arad, 3–6.

Jackowski, T. and Frydrych, I. (eds.) 2009. Natural Fibres – Their Attractiveness in Multi-Directional Applications, Gdynia Cotton AssociationGdynia, Poland.

Jankauskiene, Z. 2009. 'Achievements of Lithuanian fibre flax breeding – new varieties: Dangiai, Snaigiai and Sartai', in Scientific Bulletin of ESCORENA, vol. 1, ed. Aurel Vlaicu University, Romania, Arad, seven to nine.
Jankauskiene, Z., Lugauskas, A. and Repeckiene, J. 2006. 'New methods for the improvement of flax dew retting', Journal of Natural Fibers, 3 (4), 59–68.
Kim, E. and Csiszár, E. 2005. 'Chemical finishing of linen and ramie fabrics', Journal of Natural Fibers, 2 (2), 39–52.
Kind, W., Koenig, P., Müller, W., Schilling E. and Steinbrinck C. 1930. 'Der Flachs. Abteilung 1: Botanik, Kultur, Aufbereitung, Bleicherei und Wirtschaft des Flachses: mit einer Einführung in den Feinbau der Zellulosefaserns', Berlin, Germany, Springer, 427.
Kozłowski R. 1969. 'Sposoby okreslania stopnia wyroszenia slomy lnianej [The methods of the assessment of the degree of flax straw retting]', in Biblioteczka dla praktykow, ed. Instytut Przemyslu Wlokien Lykowych, BOINTE Poznan, 1/69, 1–20.
Kozłowski, R. (ed.) 1989. Flax in Europe: Production and Processing, REUR Technical Series 9. Poznan: Institute of Natural Fibres.
Kozłowski, R. and Mackiewicz-Talarczyk, M. 2009. 'Natural fibres and their potential in diversified applications', Vom Agrarrohstoff zu neuen Produkten – Verfahrenstechnische Forschung im Nacherntebereich, Bornimer Agrartechnische Berichte, Heft 65, Potsdam-Bornim.
Kozłowski, R., Barriga-Bedoya, J., Batog, J., Konczewicz, W., Mackiewicz-Talarczyk, M., Walentowska, J., Wielgus, K. and Zimniewska, M. 2009. 'Outlook for the International Year of Natural Fibers (2009): The role of biotechnological research in the improvement of natural fibers', in Book of Abstracts of 10th International Conference on Frontiers of Polymers and Advanced Materials, Santiago, Chile, 28 September–2 October.
Kozłowski, R., Batog, J., Konczewicz, W., Muzyczek, M., Sedelnik, N. and Tanska, B. 2006. 'Survey and recent report on enzymatic processing of bast fibres', Journal of Natural Fibers, 3 (2/3), 113–129.
Kozłowski, R., Heller, K., Mankowski, J., Kolodziej, J., Kubacki, A., Grabowska, L., Mackiewicz-Talarczyk, M., Baraniecki, P., Praczyk, M., Burczyk, H. and Kolodziejczyk, P. 2009. 'Yielding potential of bast fibrous plants in Europe', in Scientific Bulletin of ESCORENA, vol. 1, ed. Aurel Vlaicu University, Romania, Arad, 27–44.
Kozłowski, R., Kozłowska, J. and Czarnecki, L., 'Skutecznosc wybranych preparatow grzybobojczych w zabezpieczeniu przed korozja podkladu lniano-jutowego wykladziny podlogowej "Lentex" [Effectiveness of selected fungicide agents for protection against biodecomposition of banking used in flax-jute floor covering "Lentex"]', Works of IKWN, 632.952:69.025.3, Poznań, Poland, 263–271.
Kozłowski, R., Kozłowska, J., Rawluk, M. and Barriga, J. 2004. 'Potential of lignocellulosic fibrous raw materials, their properties and diversified applications', Nonlinear Optics, Quantum Optics, 31 (1–4), 61–89.
Kozłowski, R., Mankowski, J. and Kubacki, A. 2004. 'Efficient technology for the production of decorticated hemp and flax fibres and linseed flax as a raw material for different industries', Journal of Natural Fibers, 1 (2), 107–108.
Kozłowski, R., Mankowski, J., Kolodziej, J., Mackiewicz-Talarczyk, M. and Baraniecki, P. 2009. 'Bast fibrous plants raw materials characteristic and their applications', in Scientific Bulletin of ESCORENA, vol. 1, ed. Aurel Vlaicu University, Romania, Arad, 53–63.
Kreže, T., Iskrač, S., Smole, M. S., Stana-Kleinschek, K., Strnad, S. and Fakin, D. 2005. 'Flax fibers sorption properties influenced by different pretreatment processes', Journal of Natural Fibers, 2 (2), 25–37.
La culture du lin graine. Paris: Cetiom, avec le concours de l'ITL.

'Le fil du Lin et du chanvre', *The Journal All about Linen and Hemp*, La Linen Community, **1** (2). France.
Lennox-Kerr, P. 1998. 'A new dawn for linen', Textile Month, October, 52–54.
Liu, Z., Erhan, S. Z., Akin, D. E. and Barton, F. E. 2006. '"Green" composites from renewable resources: Preparation of epoxidized soybean oil and flax fiber composites', Journal of Agricultural and Food Chemistry, 22 (6), 2134–2137.
Maggioni, L., Pavelek, M., van Soest, L. J. M. and Lipman, E. (compilers) 2002. Flax Genetic Resources in Europe, Ad hoc meeting, 7–8 December 2001, Prague, Czech Republic, International Plant Genetic Resources Institute, Rome, Italy.
Mapleston, P. 2009. 'The future for fibres', Compounding World, January, 11.
Marshal, G. (ed.) 1989. Flax: Breeding and Utilization. Dordrecht: Kluwer Academic Publishers for the Commission of the European Communities, 1–171.
McIntyre, J. E., Denton, M. J. and Daniels, P. N. 2002. Textile Terms and Definitions. Manchester: The Textile Institute.
Nilmini, P. J., Dissanayake, J., Summerscales, J., Grove, S. M. and Singh, M. M. 2009. 'Energy use in the production of flax fiber for the reinforcement of composites', Journal of Natural Fibers, 6 (2), 331–346.
Nilsson, D. 2006. 'Dynamic simulation of the harvest operations of flax straw for short fibre production – Part 1: Model description', Journal of Natural Fibers, 3 (1), 23–34.
Nilsson, D.2006. 'Dynamic simulation of the harvest operations of flax straw for short fibre production – Part 3: Simulation results', Journal of Natural Fibers, 3 (4), 43–57.
Nožinič, M., Markovič, M. and Trkulja, V. 2009. 'Experimental linseed oil production in Banja Luka region', in Scientific Bulletin of ESCORENA, vol. 1, ed. Aurel Vlaicu University, Romania, Arad, 45–47.
Nozkova, J., Brindza, J., Stehlikova, B. and Pavelek, M. 2006. 'Importance of collected flax germplasm (Linum usitatissimum L.): Characterization', Journal of Natural Fibers, 3 (1), 1–16.
Pengilly, N. L. 2005. The Essential Flax: A Compendium of Diet Reference, Information, Facts, Folklore, Recipes and Research. Saskatoon, Canada: Saskatchewan Flax Development Commission.
Plonka, F. and Anselme, C.1956. Les Varietes de Lin et Leurs Principales Maladie. Paris: Institut National de la Recherché Agronomique.
Poradnik plantatora lnu włóknistego (Guide to the grower of fibrous flax) ed. R. Kozłowski et al., 2006, Poznań, Institute of Natural Fibres, 1–84
Probulski, Z., Kozłowski, R., Tabisz, J. and Stanislawski, K. 1967. 'Porownanie wartosci technologicznej wlokna, przedzy i tkanin pochodzacych ze slomy lnianej roszonej z udzialem i bez udzialu mocznika [Comparison of technological value of fibre, yarn and fabrics manufactured from retted straw without and with addition of urea]', Works of IPWL, Poznan, 167–179.
Proceedings of the 58th Flax Institute of the United States, Fargo, North Dakota, 23–25 March 2000, 1–248.
Producing for the Market: Proceedings of the fourth European Regional Workshop on Flax, Rouen, France, 26–28 September.
Schilling, E. and Muller, W. 1951. Water retting. Len [Flax]. Warsaw: Panstwowe Wydawnictwa Techniczne, Poland, 226–276.
Schmidt-Przewozna, K. 1997. 'Application of flax and hemp fibres in 3D woven structures', in Proceedings of the 78th World Conference of the Textile Institute in association with the fifth Textile Symposium of SEVE and SEPVE, Thessaloniki, Greece, vol. 2, 287–293.
Scrive-Loyer, J. 1886. A propos du Lin, second edn. Lille: Imprimerie Verly, Dubar.

Sègalen, H.-A. 2005. Le Chanvre en France: Cannabis sativa L. vulgaris – culture, rècolte, applications. Rodez: Éditions du Rouergue.
Sharma, H. S. S. and Van Sumere, C. F. (eds) 1992. The Biology and Processing of Flax. Belfast: C.F.M Publications, Belfast, UK, 576 pp.
Simay I.E. 1994. Seedborne fungi of flax. Plant Genetic Resources Newsletter, (http://www.ipgri.cgiar. org/pgrnewsletter/article.) (viewed 10 08 2007).
Sohn, M., Himmelsbach, D. S., Morrison, W. H., Akin, D. E. and Barton F. E. 2006. 'Partial least squares regression calibration for determining wax content in processed flax fiber by rear-infrared spectroscopy', Applied Spectroscopy, 60 (4), 437–440.
Sorlino, D. 2005. 'Research applied to global knowledge of flax development', Journal of Natural Fibers, 2 (2), 111–116.
Titok, V. V., Lemesh, V. A., Jurenkova, S. I. and Khotyleva, L. V. 2010. Genetika, fiziologija i biochimija lna. Nacionalnaja Akademija Nauk Belarusi, Belaruskaja Navyka, Minsk, Belarus.
Titok, V., Leontiev, V., Shostak, L. and Khotyleva, L. 2006. 'Thermogravimetric analysis of the flax bast fibre bundle', Journal of Natural Fibers, 3 (1), 35–41.
Tiwari, M., Singh, A. and Singh, M. K. 2010. 'Enhancing the value of linen fabric with resist printing using vat dyes and different resisting agents', Colourage, 57 (7), 65–76.
Waldron, D. and Harwood, J. A. 2010. 'Study of the relationship between bending rigidity and the ease of decortication of flax (Linum usitatissimum) straw', Journal of Natural Fibers, 7 (1), 42–60.
Wałkowski T., Ladek A., Piotrowska A. 1998. Len oleisty (Linseed), Instytut Hodowli I Aklimatyzacji Roślin ed., Poznań, 36 pp.
Wang, H. M. and Wang, X. 2004. 'Evaluation of the fineness of degummed bast fibers', Fibers and Polymers, 5 (3), 171–176.
Wang, Y. F., Fu, Y., Hua, K. Q., Yan, L., Chen, L. X. and Jun, L. S. 2005. 'The development of the study on technique for introducing exogenous DNA into flax in China', Journal of Natural Fibers, 2 (2), 1–16.
Zahran, M. K. 2010. 'One stage ecological and economical pretreatment of linen fabrics', Journal of Natural Fibers, 7 (1), 1–16.
Zahran, M. K., Rehan, M. F. and El-Rafie, M. H. 2005. Single bath full bleaching of flax fibers using an activated sodium chlorite/hexamethylene tetramine system', Journal of Natural Fibers, 2 (2), 49–67.
Zhivetin, V. V., Ginzburg, L. N. and Olshanskaja, O. M. 2002. Len i jevo kompleksnoje ispolzovanije. Moscow: FLUP CNIILKA.

References

Agosti, M.B., Sorlino, D., Trapani, N., 2005. How does light intensity affect the elementary fiber length in flax? J. Nat. Fibers 2/1, 15–24.
Allam, M.A., 2008. Flax (Linum usitatissimum) remedial recent innovations & idealizing fibers transfiguration. Egypt 1–130 (unpublished material).
Andruszewska, A., Langner, K., Byczynska, M., 2001. The economical aspect of trace elements containing fungicides and fertilizers application in flax cultivation. J. Nat. Fibers 281–282, 2001/1.
Beaudoin, X., 1989. Disease and pest control. In: Proceedings of the EEC Flax Workshop, Held in Brussels, Belgium, 4–5 May, 1988, pp. 81–88.

Bertoniere, N.R., 1999. Chemical and biological modification of cotton cellulose. In: Proceedings of the 5th Asian Textile Conference on Diversity and Harmony, Asian Textile in the Twenty-First Century, Kyoto Research Park, Kyoto, Japan, 30 September−2 October, pp. 643−646.

Brayford, D., 1996. Fusarium oxysporum f. sp. Lini. IMI description of fungi and bacteria. Mycopath 133, 49−51.

Bozzini, A., 2010. A new technology of plant fibres retting using the electric resonance technology. In: Proceedings of Natural Fibres and Medicinal Plants Yesterday, Today and Tomorrow, Poznan, Poland, 25−26 October.

Brutch, N.B., Soret-Morvan, O., Porokhovinova, E.A., Sharov, I.Y., Morvan, C., 2008. Characters of fibre quality in lines of flax genetic collection. J. Nat. Fibers 5 (2), 95−126.

Burdon, J., Jarosz, A., 1992. Temporal variation in the racial structure of flax rust (Melampsora lini) populations growing on natural stands of wild flax (Linum marginale): local versus metapopulation dynamics. Plant Pathol. 41.

Coby, J., Rensema, R., Koster, A.C., Huften, T.J.H.M., 1990. Flax 2000: The Future of Flax in EC, vol. 429. Agriculture', Economics Research Institute LEI, Communication, Hague, Netherlands, ISBN 9052 42 0823.

Convey, R.P., 1962. Field resistance of flax to pasmo. Phytopathology 52, 1−34.

Dambroth, M., Seehuber, R., 1988. Flachs Zuchtung, Anbau, Verarbeitung. Stuttgart. Eugen Umler GmbH & Co, Germany.

Decanniere, R., 1989. Flax, an alternative crop in the EEC ?. In: Proceedings of the EEC Flax Workshop, Held in Brussels, Belgium, 4-5 May, 1988, pp. 149−154.

Department of Agriculture, Northern Ireland, 1985. Flax for Fibre: A Guide to Flax Production and Pre-harvested Retting. Short-Term Booklet.

Depoortere, M., 1989. Utilization and development in machines for flax harvesting and scutching. In: Kozłowski, R. (Ed.), Flax in Europe: Production and Processing, REUR Technical Series 9. Institute of Natural Fibres, Poznan, pp. 117−122.

Dornyi, B., 2007. Improving quality of linen-cotton fabrics with liquid ammonia treatment. J. Nat. Fibers 4 (4), 41−57.

Easson, D.L., 1989. The agronomy of flax and its effect on fibre yield and quality following glyphosate desiccation. In: Proceedings of the EEC Flax Workshop, Held in Brussels, Belgium, 4-5 May, 1988, pp. 61−70.

El-Hariri, D.M., Al-Kordy, M.A., Hassanein, M.S., Ahmed, M.A., 2004a. Partition of photosynthates and energy production in different flax cultivars. J. Nat. Fibers 1 (4), 1−15.

El-Hariri, D.M., Hassanein, M., Ahmed, M., 1998. Effect of different NPK rates productivity of flax. J. Nat. Fibers, Spec Ed 20−27. Proceedings of the Hemp, Flax and Other Bast Fibrous Plants − Production, Technology and Ecology Symposium, Poznań, Poland.

El-Hariri, D.M., Hassanein, M.S., El-Sweify, A.H.M., 2004b. Evaluation of some flax genotypes straw yield, yield components and technological characters. J. Nat. Fibers 1 (2), 1−12.

El-Shimy, G.H., Mostafa, S.H.A., Zedan, S.Z., 1997. Studies on yield and yield components, quality and variability in some flax genotypes. Egypt Agric Res 75, 697−715.

Endres, G., Hanson, B., Halvorson, M., Schatz, B., Henson, B., 2002. Flax response to nitrogen and seeding rates. Proc. 59th Flax Ins. United States 196−198. March 21−23.

Ersserberger, R., 1993. New York, with dissly M. In: Paris, 'A Remnant of Early Life. Anthropology: An Ancient Cloth Tells a Tale. Newsweek, p. 47, 26 July.

Fouilloux, G., 1988. Breeding flax methods. In: Proceedings of the EEC Flax Workshop, Held in Brussels, Belgium, 4-5 May, 1988, pp. 14−25.

Foulk, J.A., Dodd, R.B., McAlister, D., Chun, D., Akin, D.E., Morrison, H., 2007. Flax-cotton fiber blends: Miniature spinning, gin processing, and dust potential. Industrial Crops and Products 25, 8–16.

Foulk, J.A., Akin, D.E., Dodd, R.B., 2009. Miniature spinning enzyme-retted flax fibers. J. Nat. Fibers 6 (1), 1–13.

Gherzi Company materials, based on the Fiber Year 2017 (Figures 2016) and Gherzi interviews, presentations, 2017. Trade statistics.

Grant, C.A., Bailey, L.D., 1989. The influence of Zn and P fertilizer on the dry matter yield and nutrient content of flax. Linum usistatissimum L. on the soils varying in Ca and Mg level. Can. J. Soil Sci. 69, 461–472.

Guebitz, G.M., Cavaco-Paulo, A., Kozłowski, R. (Eds.), 2006. Biotechnology in Textile Processing. Haworth Press, New York, pp. 1–254.

Haydock, P.J., Pooley, R.J., 1997. Evaluation of insecticides for control of flax beetle in linseed. Tests Agrochem. Cultiv. 18, 4–5.

Heller, K., 1992. Concentration of segetal weeds on flax plantation and the possibilities of combating them by chemical methods. Yearbook J. Nat. Fibers 35/36, 23–28.

Heller, K., 2001. Monitoring and prognosis of weed infestation on fibre flax in Poland. Yearbook J. Nat. Fibers, Spec Ed 1, 286–288.

Heller, K., Rolski, S., 2001. Ecological aspects of utilization of herbicides in fibrous plant cultivation. Chem Agr 2, 96–99.

Heller, K., Andruszewska, A., Grabowska, L., Wielgusz, K., 2006. Fibre flax and hemp protection in Poland and in the world. Prog. Plant Prot. 46 (1), 88–98.

Heller, K., Baraniecki, P., Praczyk, M., 2012. Fibre flax cultivation in sustainable agriculture. In: Handbook of Natural Fibres, first ed., vol. 1. WoodHead Publishing Inc., pp. 508–527 508-527.

Holme, I., 2005. Textile bleaching: a modern perspective. Text. Mag. 2 8–12.

Jacoby, M., 2009. Getting by on little water. Advances in breeding and basic science confer drought tolerance to crops. C&EN Chicago 87 (39), 60–62. September 28.

Jankauskiene, Z., Gruzdeviene, E., Endriukaitis, A., 2004. Protection of fibre flax crop against flea beetles and seedling blight using compound seed-dressers. J. Nat. Fibers 14, 37–57.

Jankauskiene, Z., Lugauskas, A., Repeckiene, J., 2006. New methods for the improvement of flax dew retting. J. Nat. Fibers 3 (4), 59–68.

Jayapriya, J., Vigneswaran, C., 2010. Process optimization for biosoftening of cellulosic fiber with white rot fungi and specific enzymatic systems. J. Nat. Fibers 7 (1), 17–33.

Kernaghan, K., Kiekens, P., 1992a. Bleaching and dyeing of linen. In: Sharma, H.S.S., Van Sumere, C.F. (Eds.), The Biology and Processing of Flax. M Publications, Belfast, pp. 343–445.

Kernaghan, K., Kiekens, P., 1992b. Physical properties of linen and their influence on finishing. In: Sharma, H.S.S., Van Sumere, C.F. (Eds.), The Biology and Processing of Flax. M Publications, Belfast, pp. 475–500.

Kim, E., 2005. Chemical finishing of linen and ramie fabrics. J. Nat. Fibers 2 (3), 39–52.

Konczewicz, W., Kozłowski, R., 2007. Application of osmotic pressure for evaluation of quality and quantity of fibre in flax and hemp. In: Anandjiwala, R., Hunter, L., Kozłowski, R., Zaikov, G. (Eds.), Textiles for Sustainable Development. Nova Science Publishers, New York, pp. 95–102.

Konczewicz, W., Kozłowski, R., Heller, K., Byczynska, M., 2006a. Application of osmotic pressure for evaluation of fibre quality and quantity of flax cultivars. In: Proceedings of the 61st Flax Institute of the United States, Fargo, North Dakota, 22–24 March.

Konczewicz, W., Kozłowski, R., Heller, K., Byczynska, M., 2006b. Osmotic degumming for evaluation of fibre flax cultivation technologies. In: Proceedings of the International Conference on Natural Fibres: Vision 2020, New Delhi, India, 8–9 December.

Kozłowska, J., 1989. Possibilities of enzyme utilization in degumming of flax. In: Kozłowski, R. (Ed.), Flax in Europe: Production and Processing, REUR Technical Series 9, Proceedings of the European Regional Workshop on Flax, Poznan, 19–21 June. Institute of Natural Fibres, Poznan, pp. 137–144.

Kozłowski, R., 1969. Sposoby okreslania stopnia wyroszenia slomy lnianej [The methods of the assessment of the degree of flax straw retting]. In: Biblioteczka Dla Praktykow, Ed. Instytut Przemyslu Wlokien Lykowych, BOINTE Poznan, 1/69, pp. 1–20.

Kozłowski, P.D., 1970. Proces Roszenia Lnu Z Dodatkiem Mocznika Na Tle Klasycznego Sposobu, Ze Szczególnym Uwzglednieniem Lotnych Kwasów Tłuszczowych W Płynie Roszeniowym I W Paździerzach, Wydzielanych Gazów, Związków Azotowych Oraz Niektórych Własności Włókna, Przedzy I Tkanin. Retting Flax with Addition of Urea Compared with the Classical Method, with Particular Consideration to Volatile Fatty Acids in Retting Fluid and Shives, Produced Gases, Nitrogen Compounds and Some Properties of Fibre, Yarn and Fabrics. Poznań. PhD dissertation).

Kozłowski, R., 1992. Retting of flax in Poland. In: Sharma, H.S.S., Van Sumere, C.F. (Eds.), The Biology and Processing of Flax. M Publications, Belfast, pp. 251–259.

Kozłowski, R., 1996. Look on flax in the 21st century. In: Producing for the Market: Proceedings of the 4th European Regional Workshop on Flax, Rouen, France, 26–28 September, vol. 1, pp. 7–20, 13.

Kozłowski, R., Helwig, M., 1996. Critical look at the chemical modification of natural cellulosic fibres. In: Proceedings of the American Chemical Society Symposium. Division of Cellulose Modification, Hawaii, Honolulu, USA, December.

Kozłowski, R., Manys, S., 1996. The properties of liquid ammonia treated linen. In: Proceedings of the 212th ACSNM, Orlando, USA, pp. 25–29 (August).

Kozłowski, R., Barriga-Bedoya, J., Batog, J., Konczewicz, W., Mackiewicz-Talarczyk, M., Walentowska, J., Wielgus, K., Zimniewska, M., 2009. Outlook for the International Year of Natural Fibers 2009: the role of biotechnological research in the improvement of natural fibers. In: Book of Abstracts of 10th International Conference on Frontiers of Polymers and Advanced Materials, Santiago, Chile, 28 September–2 October.

Kozłowski, R., Batog, J., Konczewicz, W., Mackiewicz-Talarczyk, M., Muzyczek, M., Sedelnik, N., Tanska, B., 2005. Latest state of the art in bast fibres bioprocessing. In: Proceedings of the 11th International Conference for Renewable Resources and Plant Biotechnology, Poznan, Poland, 6–7 June.

Kozłowski, R., Batog, J., Konczewicz, W., Mackiewicz-Talarczyk, M., Muzyczek, M., Sedelnik, N., Tanska, B., 2006a. Enzymes in bast fibrous plant processing. Biotechnol. Lett. 28 (10), 761–765.

Kozłowski, R., Batog, J., Konczewicz, W., Muzyczek, M., Sedelnik, N., Tanska, B., 2006b. Survey and recent report on enzymatic processing of bast fibres. J. Nat. Fibers 3 (2/3), 113–129.

Kozłowski, R., Konczewicz, W., Sirghie, C., Wojtysiak, J., Podsiedlik, W., Kryszak, N., Nowaczkiewicz, E., 2010a. The methods of degumming lignocellulosic fibres. In: CD Proceedings of FIBRATEC 2010 – VI Symposium on Natural Fibers: Full Use and Applications, Havana, Cuba, 29 November–3 December.

Kozłowski, R., Kubacki, A., Mankowski, J., 2008. Patent PL199630 of 20.07.2000, Granted 22.12.2008, 'Sposob Mechnicznej Obrobki Włókien Naturalnych (Mechanical Method of Natural Fibres Processing)'.

Kozłowski, R., Mackiewicz-Talarczyk, M., Barriga-Bedoya, J., 2010b. Natural fibres production, processing, and application-inventory and future prospects. In: Contemporary Science of Polymeric Materials, ACS Symposium Series 061. American Chemical Society, Washington, DC, pp. 41–51.

Kozłowski, R., Mieleniak, B, B., Przepiera, A., 1995. Plant residues as raw material for particleboards. Zemědělská Technika 41 (3), 101–111.

Kozłowski, R., Wladyka-Przybylak, M., Helwig, M., 2004. Composites based on lignocellulosic raw materials. Mol. Cryst. Liq. Cryst. 418, 131/[859]–151/[879].

Kozlowski, R., Mieleniak, B., Przepiera, A., 1994. Plant residues as raw materials for particleboards. In: Proceedings of the 28th International Particleboard/Composite Materials Symposium. Washington State University, Pullman, WA, USA.

Kozłowski, R., Manys, S., 1994. A new outlook on flax fibre properties in textiles. In: Anderson, J.F., Schiavoni, M. (Eds.), Proceedings of World Fibre Flax Symposium, 41–62, New Haven, USA, 22–25 May.

Kozlowski, R., Manys, S., 1995. New outlook on Flax Fibres Features in Textiles (Nový pohled na vlastnosti lněných vláken v textile). Zemědělská Technika 41 (3), 89–99.

Kozłowski, R., Baraniecki, P., Mańkowski, J., Mieleniak, B., Helwig, M., 1999. New trends in use of natural fibres for technical textiles and compisites, Techtextil Symposium. In: 10th International Symposium for Technical Textiles, Nonwovens and Textile Reinforced Materials. April 12-15, 1999.

Krmela, S., 1993. Manufacture of blended yarns using cottonized flax and the properties of woven fabrics made thereof. In: Proceedings of the 3rd European Workshop on Flax, Bonn, Germany, 15–17 June.

Kubacki, A., 2010. 'Okreslenie Optymalnych Poziomow Czynnikow Agrotechnicznych W Uprawie Lnu Wloknistego Do Przerobu Na Wlokno Metoda Dekortykacji [Estimating the Optimal Agrotechnical Factors in Cultivation of Fibrous Flax Processed by Decortication]'. PhD thesis. University of Life Sciences, Poznan.

Lennox, K., 1998. 'A New Dawn for Linen', *Textile Month,* Chemical Henkel, vol. 5. Textile Technique Division, Düsseldorf.

Lewin, M., 1984. 'Bleaching of lignocellulosic and synthetic fabrics', in handbook of fiber science and technology. In: Lewin, M., Sello, S.B. (Eds.), Chemical Processing of Fibers and Fabrics, Part B, Fundamentals and Preparation, vol. 1. Marcel Dekker, New York, pp. 91–256.

McIntyre, J.E., Denton, M.J., Daniels, P.N., 2002. Textile terms and definitions. The Textile Institute, Manchester.

Mackie International, 1994. Linen and the restructuring of the polish textile industry. Materials of the Mackie Seminar 5–6.

Mańkowski, J., Szukała, J., 1998. The influence of agronomic factors stimulating obtaining of homomorphic flax fibre with refined utility features. Journal of Natural Fibers, Spec Ed. 1, 47–55.

Manys, S., 1989. Application of flax blended Parafil yarns in woven and knitted products. In: Kozłowski, R. (Ed.), Flax in Europe: Production and Processing, REUR Technical Series 9. Institute of Natural Fibres, Poznan, pp. 161–171.

Manys, S., 1991. Technological evaluation of flax fibres in terms of processing technology. In: Flax as a Fibre and Oil Bearing Crop: Proceedings of the FAO European Workshop on Flax, Brno, Czechoslovakia, 18–20 June, pp. 281–288.

Marchenkov, A., Rozhmina, T., Ushapovski, I., Muir, A.D., 2003. In: Cultivation of Flax in: Flax the Genus Linum, Saskatoon (Muir and Westcott).
Morvan, C., et al., 1991. The development of the cellulosic fibres from *Linum usitatissimum* L.: origin of different kinds of heterogeneity. In: Flax as a Fibre and Oil Bearing Crop: Proceedings of the FAO European Workshop on Flax, Brno, Czechoslovakia, 18–20 June, pp. 254–268.
Park, S.Y.1, Fung, P., Nishimura, N., Jensen, D.R., Fujii, H., Zhao, Y., Lumba, S., Santiago, J., Rodrigues, A., Chow, T.F., Alfred, S.E., Bonetta, D., Finkelstein, R., Provart, N.J., Desveaux, D., Rodriguez, P.L., McCourt, P., Zhu, J.K., Schroeder, J.I., Volkman, B.F., Cutler, S.R., 2009. Abscisic acid inhibits type 2C protein phosphatases via the PYR/PYL family of START proteins, 2009 May 22 Science 324 (5930), 1068–1071. https://doi.org/10.1126/science.1173041. Epub 2009 Apr 30.
Paterson, A.H., Bowers, J.E., Rokhsar, D.S., 2009. The Sorghum bicolor genome and the diversification of grasses. Nature 457, 551–556 of 29 January 2009.
Pinnow, M., Fink, H.-P., 1999. Structure and Properties of Bast Fibres. Fraunhofer-Institut Angewandte Polymerforschung (Fraunhofer Institute for Applied Polymer Research), Potsdam-Golm, Germany.
Praczyk, M., Silska, G., 2013. Analiza zmienności i sposobu dziedziczenia komponentów struktury plonu lnu włóknistego. Biuletyn Instytutu Hodowli i Aklimatyzacji Roślin 268, 173–182.
Pratt, H.T., 1994. Some milestones in the history of bleaching. Text. Chem. Colorist 26 (11), 21–27.
Probulski, Z., Kozłowski, R., Tabisz, J., Stanislawski, K., 1967. Porownanie Wartosci Technologicznej Wlokna, Przedzy I Tkanin Pochodzacych Ze Slomy Lnianej Roszonej Z Udzialem I Bez Udzialu Mocznika [Comparison of Technological Value of Fibre, Yarn and Fabrics Manufactured from Retted Straw without and with Addition of Urea]. Works of the Institute of Bast Fibre Industry, Poznan, Poland, pp. 167–179.
Rashid, Y.K., 2003. Principal diseases of flax in: flax, the genus Linum, Saskatoon. In: Muir and Westcott.
Rolski, S., Andruszewska, A., Grabowska, L., Heller, K., 2000. Breeding and Cultivation of Fibrous Crops. Natural Fibers, pp. 31–41. Spec jub ed.
Ross, T., 1992. Preparation and spinning of flax fibre. In: Sharma, H.S.S., Van Sumere, C.F. (Eds.), The Biology and Processing of Flax. M Publications, Belfast, pp. 275–296.
Ryder, M.L., 1965. Report of Textiles from Catal Hüyük, Anatolian Studies. J. Br. Inst. Archaeol. Ankara 15, 175–176.
Saharan, G.S., Saharan, M.S., 1994. Conidial size, germination and appressioral formation of Oidium lini Skoric, cause of powdery mildew of linseed. Indian J. Mycol. Plant Pathol. 24, 176–178.
Schilling, E., Muller, W., 1951. Water retting. Len [Flax]. Panstwowe Wydawnictwa Techniczne, Warsaw, Poland, pp. 226–276.
Schmidt-Przewozna, K., 2001. Application classes of natural dyes and their usage on animal and vegetable fibres. In: Proceedings of the 2nd Global Workshop of the FAO European Cooperative Research Network on Flax and Other Bast Plants, Bast Plants in the New Millennium. Borovets, Bulgaria.
Schmidt-Przewozna, K., 2005. Historical Dyes in Poland and Their Revival. Archetype Publications DHA. No. 20.
Schmidt-Przewozna, K., Zimniewska, M., 2007. The effect of natural dyes used for linen fabric on UV-blocking. In: Kozłowski, R., Zaikov, G.E., Pudel, F. (Eds.), Renewable Resources and Plant Biotechnology. Nova Science Publishers, New York, pp. 107–116.

Schmidt-Przewozna, K., Zimniewska, M., 2009. Natural dyeing plants as a source of compounds protecting against UV radiation. Herba Pol. 55 (3), 311–318.

Schmidt- Przewoźna, K., 2016. YOCOCU 2014. Professional's Experiences in Conservation of Cultural Heritage in America, Europe, Asia" A Short History of Natural Dye Materials Used in Dyeing and Printing of Polish Folk Fabrics. Scholars Publishing, Cambridge, pp. 106–120. Chapter in monography, ISBN-13: 978-1-4438-8519-5, ISBN (13): 978-1-4438-8519-5.

Schmidt –Przewozna, K., Brandys, A., 2016. In: Prasad, M.N.V. (Ed.), Utilization of Contaminated Lands for Cultivation of Dye Producing Plant - Bioremediation and Bio-economy. Copyright © 2016 Elsevier Inc, Netherlands, UK, USA, ISBN 978-0-12-802830-8, pp. 229–358. ISBN: 978-0-12-802830-8.

Schmidt-Przewoźna, K., 2018. The History of Red in Poland: Polish Cochineal Porypohora Polonica and Others Natural Dyes, Colour Culture and Science M. Godyń, B. Grobosz, A., Kwiatkowska-Lubanska, Kraków, ISBN 978-83-948545-2-2, pp. 33–43.

Sharma, H.S.S., Van Sumere, C.F. (Eds.), 1992. The Biology and Processing of Flax. Belfast. C.F.M Publications, Belfast, UK, p. 576.

Šmirous, P., 1989. Weed Control in Czechoslovakia, Flax in Europe – Proc European Regional Workshop on Flax. In: FAO Symposium Brno, pp. 71–77.

Sultana, C., 1983. The cultivation of fibre flax. Outlook Agric. 12, 104–110.

Sultana, C., 1991. The influence of flax harvesting techniques on fibre quality. In: Flax as a Fibre and Oil Bearing Crop: Proceedings of the FAO European Regional Workshop on Flax, Brno, Czechoslovakia, 18–20 June, pp. 234–237.

Taylor, J.H., 1995. Unwrapping a Mummy: The Life, Death and Embalming of Horemkenesi. University of Texas Press, Austin, TX.

Titok, V., Leontiev, V., Yurenkova, S., Nikitinskaya, T., Barannikova, T., Khotyleva, L., 2010. Infrared spectroscopy of fiber flax. J. Nat. Fibers 7 (1), 61–69.

Wałkowski, T., Ladek, A., Piotrowska, A., 1998. Len oleisty. (Linseed). IHAR Poznań, p. 36.

Xinwen, L., 1997. Analysis of ecological adaptation of flax in dry and cool areas in China, Natural Fibres Spec ed. In: Proceedings of the Flax and Other Bast Plants Symposium, pp. 43–48.

Zahran, M.K., 2010. One stage ecological and economical pretreatment of linen fabrics. J. Nat. Fibers 7 (1), 1–16.

Zahran, M.K., Rehan, M.F., El-Rafie, M.H., 2005. Single bath full bleaching of flax fibers using an activated sodium chlorite/hexamethylene tetramine system'. J. Nat. Fibers 2 (2), 49–67.

Bast fibres: hemp cultivation and production

5B

Matthew R.L. Horne
Shuttleworth College (Bedford College Group) Old Warden, Biggleswade, Bedfordshire, United Kingdom

5B.1 Introduction

5B.1.1 *What's in a name?*

The name 'hemp' is a term most commonly used in connection with the *Cannabis sativa* plant, its components (seeds, stems, leaves) and any products (foods, fibres or biomass) extracted and manufactured from them. The term is also used as a name applied to a large number of other fibre-bearing plants and their products, such as sisal hemp (*Agave rigida*) or Manila hemp (*Musa textilis*), but these plants and fibres are neither related nor associated with *Cannabis sativa*. The widely accepted common usage of the singular term 'hemp' in the English language as a description of a plant or product, for example, hemp crop or hemp fibre, is in reference to the plant source *Cannabis sativa*.

5B.1.1.1 *A brief history and background of hemp*

Hemp (*Cannabis sativa*) has been widely cultivated as a source of bast fibres and/or seeds. The plant is native to India and Persia, but over the last 6000 years, it has been cultivated in nearly all temperate and tropical countries of the world and is likely to be one of the oldest non-food crops known (Vavilov and Dorofeyev, 1926; Schultes, 1970). Historically, hemp fibres were used for the production of rope, cordage, fabrics and paper, and the debris of such industries is a significant source of archaeological evidence of hemp production. For example, in the British Isles, hemp fibres were widely used in the production of ship's rigging and sail cloth from 800 AD and retting sites used to identify the locations of production (Schofielda and Wallerb, 2005). During the 18th and 19th centuries hemp had a significant commercial role in Europe for the production of ropes and textiles.

Hemp seed has historically been harvested for the oil that can be crushed from it for use in culinaries and manufactures such as soaps, paints, lubricants and cosmetics. In addition, hemp has been grown as a medicinal plant, and investigations into the uses of hemp as a therapeutic and medicinal plant continue (Polezak et al., 2018; Zielonka et al., 2019; Makowiecka and Wielgus, 2014). More details are provided in the chapter 12. vol. 2.

Handbook of Natural Fibres. https://doi.org/10.1016/B978-0-12-818398-4.00007-4

During the 16th, 17th and 18th centuries the major fibre crops found growing in Europe, North America and Russia were hemp and flax (Pounds, 1979; Aber, 1980). However, during the 19th century the expanding cultivation of cotton and jute caused the world area of hemp to decline, while during the 20th century the widespread production and adoption of synthetic fibres as an alternative to natural fibres caused a further decline in hemp crop production (Atkinson, 1964). Additionally, during the 20th century the production of hemp was made illegal in many countries as part of controls on the recreational use of cannabinoids.

After the Second World War the major producer of hemp fibre was the Soviet Union, particularly in Russia and the Ukraine. In the latter hemp fibre production peaked at 3000 km^2 in the early 1950s (Holoborod'ko 1995). Large areas were also grown in Korea, China and France. In the late 20th century a number of countries, particularly the EU and Canada, encouraged the production of low-cannabinoid hemp for industrial end uses. As a result the area of hemp production expanded in the EU and Canada, and its cultivation developed and expanded in other countries such as Australia and Chile.

At present the greatest areas of hemp production are reported to be found in Canada, China and the European Union (EU), while production in North Korea (DPRK) is also reported to be significant (Table 5B.1). Within the EU, France, the Netherlands and Romania are the largest producers of hemp. However, the area of cultivated hemp in the EU has fluctuated significantly in individual member states, while the total EU cultivated area has generally remained at 15,000 to 20,000 ha per year over past decade (Table 5B.1).

Canada is now the world's largest grower of hemp but this is primarily seed hemp or dual-purpose hemp, rather than fibre hemp. The area of hemp grown in Canada has expanded significantly since 2010 and it now grows over 50% of the world area of hemp. China and the EU are likely to be the world's largest producers of fibre hemp (*Cannabis sativa*). However, the production of ambary hemp, which is otherwise known as Kenaf (*Hibiscus cannabinus*), is also widespread in China and Chinese fibre crop production statistics may also include the area of this crop in the hemp production statistics (Anon, 2018).

5B.2 The hemp plant

5B.2.1 Botany of fibre hemp

Hemp (*Cannabis Sativa*) is a tall, slender annual dicotyledonous plant, with a vigorous growth habit, of the Rosales order and sub-order Cannabaceae (Chase, 1998). The plant is most commonly recognized by the shape and pattern of its leaves. The first seedling leaves are sessile leaves, but all subsequent leaves have a petiole (Fig. 5B.1). The first pair of true leaves have a single elliptic blade with serrate margins, a leaf of the second pair of true leaves have three serrate leaflets emerging from the petiole, a leaf from the third pair of true leaves have five serrate leaflets, and so on. This pattern continues up to between 9 and 12 leaflets (Stearn, 1970). Phyllotaxis is

Table 5B.1 Hemp production area (ha) of major world hemp producers by country 2005 to 2017.

	2005	2006	2007	2008	2009	2010	2011	2012	2013	2014	2015	2016	2017
Austria	300	562	449	296	356	540	512	472	566	906	718	662	673
Canada[a]	9725	19,458	3237	1900	4047	10,968	16,188	23,472	25,090	43,707	33,590	29,900	55,853
China[a]	11,000	19,600	17,800	11,780	5920	5110	5710	5280	6500	7700	7757	5391	4841
DPRK[b]	19,000	19,000	19,000	20,000	21,000	20,000	20,000	20,000	20,000	20,460	20,961	21,070	21,247
France[a]	9300	8100	8000	8400	12,100	8500	7600	12,600	11,500	11,500	12,200	13,900	12,900
Germany	2000	1200	800	N/A	1200	1100	400	N/A	N/A	N/A	N/A	N/A	N/A
Hungary[a]	1192	1380	1470	1550	1550	1392	1366	1495	1572	1595	1622	1656	1659
Italy	250	250	250	200	175	209	100	227	276	223	530	731	459
Netherlands	100	0	100	300	900	1000	900	1274	1284	1633	2041	2262	1272
Poland	113	141	152	162	162	160	160	159	159	170	160	159	158
Romania	2957	2340	2512	2486	2841	3249	3213	3177	3139	3161	3193	3223	3255
Spain	735	0	0	135	126	0	0	0	22	56	85	94	150
UK	1300	1700	600	N/A	300	1100	1300	N/A	N/A	N/A	N/A	N/A	N/A

[a] = Total (Seed and Fibre) Hemp production, N/A = Not available.
[b] Democratic People's Republic of Korea.
Reproduced from FAO Stat (2018)

Figure 5B.1 Hemp seedlings.

opposing in young hemp plants, but as flowering begins phyllotaxis becomes alternate and the leaflet number declines (Heslop-Harrison and Heslop-Harrison, 1958).

Hemp plant stems are furrowed externally, hollow and can range from 5 to 25 mm in diameter. However, the diameter of a hemp stem is not uniform from the stem base to the stem top, with a stem base typically being ×5 to ×10 the width of the top. Consequently, there is a corresponding gradient of fibre fineness within stems: fine fibres are found at the top of the stem, while coarser fibres are found at the bottom (Amaducci et al., 2008a).

When grown at a high plant population they do not branch, and so when grown for fibre production the seeds are planted relatively close together (Schafer and Honermeier, 2006; Bennet et al., 2006). The height achieved by a hemp plant can range from 1 to 5 m at maturity, but this typically depends on growing conditions, husbandry and cultivar (Schafer and Honermeier, 2006).

Hemp plants have become adapted to a wide range of environments but are sensitive to photoperiod. Hemp is a short day plant; its flowering is delayed by long days and hastened by short days. In general hemp flowers in the autumn when the photoperiod falls to between 12 and 13 h per day and only flowers slowly at day lengths above 14 h per day, but it depends on the cultivar and geographical location (Borthwick, 1947). Some cultivars are adapted to maturity in short days in southern latitudes, while others are adapted to bearing seed in the comparatively longer days of northern latitudes (Bredemann et al., 1956).

Hemp is primarily dioecious; the pollen-bearing parts are found on one plant and the seed-bearing flowers on another. Hemp plants are wind pollinated. The male (staminate) and female (pistallate) plants are indistinguishable prior to flowering. The male inflorescence can be first identified by the development of round, pointed flower buds with five radial segments, while the female inflorescences can be identified by the presence of a calyx (Clarke, 1980). The proportion of male to female plants can vary widely from 40% to 60%. The male inflorescences go on to be strongly branched with few or no leaves, while the female inflorescences are leafy and unbranched. The life expectancy of the males is limited to a point soon after anthesis, while the female plants live 3 to 5 weeks after anthesis until the seed is fully ripe. As a result of the latter, where hemp has been grown for fibre production using manual labour, the male plants have been harvested, by hand pulling, approximately 3 weeks prior to the female plants (Kirby 1963).

Monoecious cultivars have been developed and they may be advantageous because of their uniform maturity in comparison to dioecious types, but they tend to exhibit lower fibre yields. Monecious types are perhaps suited to dual-purpose seed and fibre crops, rather than just fibre crops, where it may be possible to overcome the lower fibre quality and yield loss found when dioecious varieties are harvested at seed maturity (Berenji et al., 2013).

Hemp seeds are achene fruits; single seed contained within a shell and tightly encompassed in the thin wall of the ovary. The seed is typically an ellipsoid shape 2–7 mm in length and 2–4 mm wide at maximum diameter. The seed can appear in varying colours from a light brown to dark green and it contains about 35% oil and 25% crude protein (Dempsey, 1975).

5B.2.2 Morphology of fibre hemp stems

In cross-section, hemp stems can be thought to consist of two major parts (Fig. 5B.2). Firstly there is the tissue outside the vascular cambium later, often referred to as the 'bark' or bast, and secondly the tissues inside the vascular cambium which are often referred to as the 'core' and contain the xylem.

The outer bark contains two component parts: epidermis and cortex. Together these are often referred to as the bast later (McDougall et al., 1993). The epidermis is covered by a thin layer of wax called the cuticle, which is a barrier preventing excessive moisture loss and provides some protection for the plant. During the retting process bacteria

Figure 5B.2 Schematic diagram of a cross section of a hemp stem showing the position of primary and secondary fibres in the bast layer.

will access the stem via this layer. Hemp fibres are situated in the cortex in two distinct bundles: primary fibres (approx. 20 mm long and 10–40 μm diameter) and secondary fibres (approx. 2 mm long and 15 μm in diameter) van der Werf et al., (1994). The primary fibres are those used for textiles as the secondary fibres are too short.

The bast layer contains the bundles of fibres and runs the full length of the plant stem. The length of the ultimate fibres is reported to range from 2 to 60 mm, with a mean length ranging from 40 to 50 mm. The fibres tend to be thicker near the bottom of the stem in comparison to the top (Amaducci et al., 2008a; Kundu, 1942).

The inner core constitutes the cambium and the xylem. The cambium consists of tissue that separates the layer of fibres from the woody xylem tissue, sometimes referred to as 'hurd' or 'shiv'. The xylem consists of vessel members, ray and paratracheal cells and librform fibres (Easu, 1965); surrounding the pith cavity is a central hollow that runs the length of the stem. The libriform fibres are approximately 0.5–0.8 mm in length and about 25 microns wide. This inner core is separated from the outer bark layer during the primary fibre extraction process of decortication and is subsequently utilized as a by-product of fibre production. Typical end uses include animal bedding and construction materials.

5B.2.3 Chemical composition of hemp stems

The distribution of chemical constituents of hemp stems varies significantly between the outer bast and the woody core. Hemp fibres are 60%–70% cellulose, 15%–20% hemicelluloses, 2%–4% lignin, 2%–4% pectin and 1%–2% fat and wax (Kymäläinen et al., 2001). The chemical composition of hemp woody core is similar to that of hardwood with approximately 40% cellulose, 20% hemicelluloses and 20% lignin (McDougall et al., 1993; Thomsen et al., 2005).

5B.3 Hemp cultivation

5B.3.1 Hemp breeding and cultivars

The form taken by hemp plants and any resulting fibre is highly dependent on growing conditions (soil type, temperatures and water availability) and husbandry at a given site. However, the underlying pattern of growth and development of hemp plants is highly dependent on day length and the photoperiod sensitivity of the selected cultivar (genotype). In practice, cultivars and locations that provide a long vegetative growth phase achieve relatively high stem biomass yields in comparison to circumstances where vegetative growth phase is shortened. For example, when cultivars bred at northern latitudes are grown in southern latitudes total biomass yields are reduced due to shortened vegetative growth and early anthesis (Meijer et al., 1995), while those cultivars bred at low latitudes have increased biomass yields when moved from southern to northern latitudes due to extended vegetative growth periods and delayed flowering (Sankari, 2000). Selecting a cultivar that is suited to a specific end use (wovens, non-wovens, composites, etc) and is agronomically and economically viable at particular site is crucial for successful fibre hemp production.

Prior to the mid-1930s hemp crops were commonly cultivated from wild forms of the plant (Kirby 1963). Although there were very few efforts made to genetically breed and select the plant for cropping, selections were grouped into northern latitude (40–55° latitude) and southern latitude types (0–45° latitude) (Hoffmann, 1961). Types appropriate for a specific location and production objective could then be chosen.

Hemp breeding programs, particularly in Europe and China, have developed a number of hemp cultivars for cultivation according to end use: fibre hemp and/or seed hemp. Monecious cultivars are suited to the production of dual-purpose hemp crops (fibres and seeds), while dioecious cultivars are suggested as preferable for fibre crops (Berenji et al., 2013).

Modern dioecious fibre hemp cultivars have a tall, unbranched growth habit and a relatively low seed yield (De Meijer, 1995). A significant effort in fibre hemp breeding has developed the fibre content of fibre varieties from 12% to 15% up to 25%–33% (Hennink, 1994; De Meijer, 1995). Where the objective is to achieve a maximal stem biomass, such as when producing for pulp and paper production it is typically enough to select a late maturing cultivar with a long vegetative growth phase. However, where there is a requirement not just for stem biomass but also fibre yield and fibre quality the choice is less straightforward. This is in part as a result of the significant amount of variation in the characteristics of fibres extracted from individual cultivars in differing environmental conditions (Cromack, 1998; Bennet et al., 2006) and the large amount of variation in fibre characteristics within a single plant (intra-stem). In practice, cultivar selection may need to be based on pre-production trials at the intended locations.

Simultaneously, industrial hemp varietal development has also produced hemp varieties that contain a low level of the THC compound, a narcotic substance in marijuana (Small and Cronquist 1976). These authors report a concentration of 0.3% THC (dry weight basis) as the threshold to distinguish between industrial hemp and marijuana hemp. To prevent industrial hemp production (seed or fibre) being considered a narcotic substance a THC concentration upper limit of 0.5% (dry weight basis) was set in Europe in 1984 for the years to 1987 and further reduced to 0.3% THC for the subsequent years for the protection of public health. In 1999 the limit was further reduced to 0.2% THC to prevent the cultivation of drug cannabis in industrial hemp crops. Since 2013, each EU member state has been left to determine if a hemp variety is permitted to be grown or not. In many EU member countries only low THC cultivars may be grown and must be listed on the Hemp variety database (Table 5B.2). The permitted maximum THC content in Russia, United States, Canada and China is 0.3% THC. In practice, as a result of these differences certain varieties may be permitted for cultivation in some countries but not others.

Table 5B.2 List of 70 hemp (*Cannabis sativa* L.) cultivars approved for cultivation in the European Union.

Adzelvieši	Armanca	Asso	Austa SK	Balaton
Beniko	Bialobrzeskie	Białobrzeskie	Białobrzeskie	Cannakomp
Carma	Carmagnola	Carmaleonte	Chamaeleon	CS
Dacia Secuieni	Delta-405	Delta-llosa	Denise	Diana
Dioica 88	Earlina 8 FC	Eletta Campana	Epsilon 68	Fedora 17
Felina 32	Férimon	Ferimon	Ferimon	Fibranova
Fibrante	Fibrol	Fibror 79	Finola	Futura 75
Futura 83	Glecia	Gliana	Glyana	Henola
Ivory	KC Bonusz	KC Dora	KC Virtus	KC Zuzana
KCA Borana	Kompolti hibrid TC	Kompolti	Lipko	Lovrin 110
Marcello	Markant	Monoica	Orion 33	Rajan
Ratza	Santhica 23	Santhica 27	Santhica 70	Secuieni Jubileu
Silvana	Succesiv	Szarvasi	Teodora	Tiborszallasi
Tisza	Tygra	Uniko B	Uso-31	Villanova
Wielkopolskie	Wojko	Zenit		

Reproduced from: Common catalogue of varieties of agricultural species. A-85— HEMP - *Cannabis sativa* L., European Commission, Brussels (document accessed on Oct. 16, 2019 on the website: http://ec.europa.eu/food/plant/plant_propagation_material/plant_variety_catalogues_databases).

5B.3.2 Hemp cropping

The first stage in process of producing hemp fibre is the production of a hemp crop. As with all plant fibres, the environment and manner in which the crop is established and managed has a dominant influence on the quantity and quality of fibre to be found in the hemp stems.

Although it is not possible to give a prescription for hemp crop cultivation in all circumstances, because of the numerous types of hemp and the wide range of soils, climate types and agricultural systems in which the plant may be grown, there are certain aspects of hemp crop husbandry practice that can have a significant influence on the quality of hemp fibre produced and are critical to note if successful cultivation is to be achieved (van der Werf et al., 1996; Amaducci et al., 2015).

5B.3.2.1 Crop environment: climate and soil

Hemp is well adapted to be grown in many regions of the world, including most temperate, tropical and sub-tropical zones. Hemp crop growth has been found to peak at an average daily temperature of approximately 14°C, but growth occurs over a wide range of 5.6–27.5°C (Duke, 1982). Importantly, the crop is not very frost tolerant. Fibre hemp varieties require 4–5 frost free months from planting to produce a harvestable crop, while hemp seed types require 5–6 months to produce mature seed.

The availability of suitable moisture (rainfall, irrigation and humidity) throughout the growing season is vital to the production of fibre hemp crops. Hemp plants develop a significant root a system that will extend to extract water from approximately 2 m deep in well-structured, compaction free soils (soils that allow unrestricted rooting) (Amaducci et al., 2008b). It has been reported that hemp will achieve a peak yield at soil moisture levels of 80% of soil field capacity (Slovonov and Petinov 1980). Although hemp will tolerate dry conditions it will not 'thrive' in drought prone soil types (e.g. very light coarse sands) and climates, unless irrigation is employed. (Duke, 1982) suggests that hemp grows best in areas receiving 970 mm of rainfall per year, while during the growing season maximum fibre yields have been found to occur in crops receiving 535 mm consisting of precipitation (rainfall and irrigation) and soil moisture. In Europe studies suggest that hemp requires 500–700 mm of available moisture for optimum yield (Bosca and Karus, 2008). Hemp has been successfully cultivated with the use of irrigation (Dempsey, 1975; Lisson and Mendham, 1998).

Hemp can be grown on a very wide range of soil types, and it is well suited to be grown on soils of many textures and structures. Hemp grows best on well drained, loamy soils with high levels of organic matter (van der Werf, 1991). The crop is not well suited to drought prone soils (e.g. coarse sandy soils), nor is it suited to very heavy soils (e.g. predominantly clay soils) as these soils are often slow to warm up in the spring, prone to water logging and they are not always able to accommodate mechanical harvesting, especially in high rainfall years (Vessel and Black, 1947). Hemp can tolerate a soil pH as low as 5.0, but a pH around 6.0 is suggested (Duke, 1982, Bosca and Karus, 2008). However, where hemp is grown as one crop in a rotation the requirements for soil pH of an individual crop needs to balance with those of others from across the rotation.

5B.3.2.2 Crop environment: soil date and sowing density

The method and timing of crop establishment, at any one location, are significant factors determining the yield and quality of any fibre ultimately produced. The two factors most important to consider are the time of sowing and the density of sowing.

Hemp should be sown once the risk of very hard frosts has passed, making it a late sown spring crop. For example, in Europe this ranges from late March through to early June which would give a harvest date between mid-August and the of end September. Typically planting as soon as soil temperatures reach 8°C ensures rapid crop establishment. A number of studies have investigated the effect of sowing date on fibre yield and quality. Overall the major impact of sowing date is on fibre yield as a result of delayed sowing reducing crop biomass (Table 5B.3), rather than the size of the fibrous

Table 5B.3 The effect of date of sowing on hemp stem yield (t per ha) at 15% moisture content in two growing seasons in Ireland (1998 and 1999).

Sowing date	Season 1	Season 2
Late March	15.3	12.5
Mid-April	11.2	10.9
EarlyMay	10.5	8.3
Mid-May	8.4	6.7
Standard error	1.12	0.99

From Crowley, J.G., 2001. The Performance of Cannabis sativa (Hemp) as a Fibre Source for Medium Density Fibre Board (MDF). ADAS: Cambridge.

layers in the plants. However a delay in the sowing of a fibre hemp crop, without an appropriate delay in harvesting date to allow the crop to reach the required stage of maturity, could result in fibres being harvested at an earlier stage of maturity. This will then have a subsequent impact on fibre quality, particularly fibre fineness and fibre strength. Planting fibre hemp crops late may result in an uneven growth and development pattern in the crop that will increase the overall duration of flowering. This has the potential to increase the variation on the crop for subsequent harvesting and processing.

In practice, late maturing varieties of hemp should be sown earlier in a planting sequence, while early maturing varieties should be sown later to ensure the required stages of maturity are achieved in the available growing season.

The plant density of a hemp crop, as determined by the seed sowing density, has a significant impact on the properties of the fibres in the plants. Seeding rates for hemp crops can be found to vary widely from 50 to 500 seeds per m^2 in older recommendations (Dempsey, 1975), but in more recent studies seed rate range from 150 to 400 seeds per m^2 (van der Werf et al., 1995; Amaducci et al., 2008a,b). Crops planted at too lower seed density are likely to suffer from reduced biomass yield (t per ha) and may also suffer from weed competition in the early stages of growth. Individual plant stems from low-density stands have generally been found to have a wider diameter (Westerhuis et al., 2009a,b). At high seed densities self-thinning can occur as a result of the competition between plant stems in the latter stages of vegetative growth (van der Werf et al., 1995). Overall a high plant density produces a crop with increased early growth competition with weeds, and high plant densities will produce crops with a higher proportion of stems and lower proportions of leaves and seeds. However, higher densities do not increase crop biomass beyond a maximum. A higher plant density will also cause a relative decrease in the diameter of stems and increase the percentage of fibre and reduce the percentage woody core (van der Werf, 1991).

Seed sowing density can have a significant impact on fibre quality, specifically fibre fineness. Hemp plants grown at a higher plant density will correspond to high fibre yields of fibres that are finer and have less lignification of secondary fibres (van der Werf et al., 1995; Amaducci et al., 2008a,b). Given the influence of sowing density on fibre fineness it is important to target the end use of the crop prior to planting to ensure a suitable sowing density is used and the subsequent crop contains fibre of a suitable quality.

Seed should be drilled at a depth of 2–3 cm in rows spaced 12–50 cm apart depending on requirements (van der Werf et al., 1995). ITC (2007) recommends drilling in rows spaced at between 9 and 17 cm. However, this will depend on the coulter spacing on seed drills (Fig. 5B.3).

5B.3.2.3 Crop environment: husbandry

Hemp is a fast growing crop and the provision of adequate supplies of plant nutrients is very important. It has been estimated that to produce a tonne of hemp straw material 15–20 kg of nitrogen, 4–5 kg of phosphorus and 15–20 kg potash are required (Bócsa et al., 2000). The phosphorus and potassium may be available to the crop from adequate reserves in the soil or it may be applied at or prior to planting. The application of nitrogen fertilizer is an important input into hemp crops as it is needed to produce an economic crop yield and for stem uniformity, with the latter important for accommodating efficient crop post-harvest processing. However, high levels of nitrogen fertilizer (200 kg N per ha) can significantly increase stem variability in comparison to moderate levels (100 kg N per ha). It can also increase stem diameter above a desired range (Ranalli, 1999) or lead to self-thinning and a reduction in the number of live plants (van der Werf et al., 1995). In Europe nitrogen fertilizer rates typically range from 80 to 150 kg N per ha.

An attractive feature of hemp is that its fast growth habit means that it can compete with weeds with little or no pesticide use (Dempsey, 1975; van der Werf, 1991).

Figure 5B.3 Hemp seeds can sown with drilling machinery used for other small grain and oilseed crops.

In comparison to the impact of nitrogen on hemp biomass yield, the impact of nitrogen fertilization on hemp fibre qualities is less understood. Where high nitrogen doses have been found to increase yields they have also behind reported to reduce fibre strength (Jordan et al., 1946; Grabowska and Koziara, 2005). This has been linked to the cell fibres from plants grown at high nitrogen rates to have a much larger lumen (Grabowska and Koziara, 2005).

Hemp production does not at present incur many agriculturally significant pests or diseases, despite a number of diseases and over 300 insect pests being known to affect the crop (McPartland, 1996a; McPartland, 1996b; McPartland et al., 2000). The most important disease of hemp is grey mould (*Botrytis cinerea*), which attacks crops under conditions of cool temperatures and high humidity and has been observed in the United Kingdom (Cromack, 1998). In practice the most significant pest of hemp crops is birds feeding on seeds and grazing birds such as pigeons attacking seedlings. However, pest problems are not widespread. Although the crop is generally free of pests at present it may be possible that if the crop is more widely cultivated then pest and disease problems become more widespread. At present it is possible to cultivate fibre hemp crops without the use of any pesticides (Amaducci et al. 2015), This makes the cultivation of hemp attractive in comparison to other arable crops.

5B.3.2.4 Harvesting

When hemp is grown for the production of long fibres, similar to long flax fibre, the crop is harvested when the staminate plants have finished flowering but before the seed has ripened, also known as technical maturity (van der Werf, 1991). Keller et al. (2001) found that to delay harvesting until the end of flowering until the end of flowering made decortication of the stem more successful, while Amaducci et al. (2008a,b) found that fibre yield increased by 25% between the beginning and end of flowering and fibre homogeneity increased. Depending on the location, cultivar and planting date, the hemp crop harvest normally occurs in the late summer or early autumn, for example, in August and September in northern Europe. In Romania, the Ukraine, Czech Republic, Poland and Russia harvesting typically occurs in October. Once cut, the crop may need to be dew-retted and suitable weather conditions will need to occur to achieve the retting degree (Fig 5B.4) required while minimizing the chance of excessive precipitation that could lead to over retting and the subsequent degradation of fibres (Bennet et al., 2006).

After the stems have been cut they are laid on the ground in windrows for field retting, and to maximize the recovery of long fibres during stem breaking and fibre breaking the hemp stems are maintained in parallel alignment (Bruce, et al. 2005). Such a system is very similar to the process used to harvest flax for linen production, except that the plants are cut rather than pulled.

Alternatively, hemp can be cut with a disc or reciprocating knife type mower (Bednar, 2008) and then left on the field to ret in rows. The crop can then be turned to ensure the whole stems are retted and dry (<18% moisture content) evenly and then baled as a hay crop (Fig 5B.5). This method of harvesting has the advantage

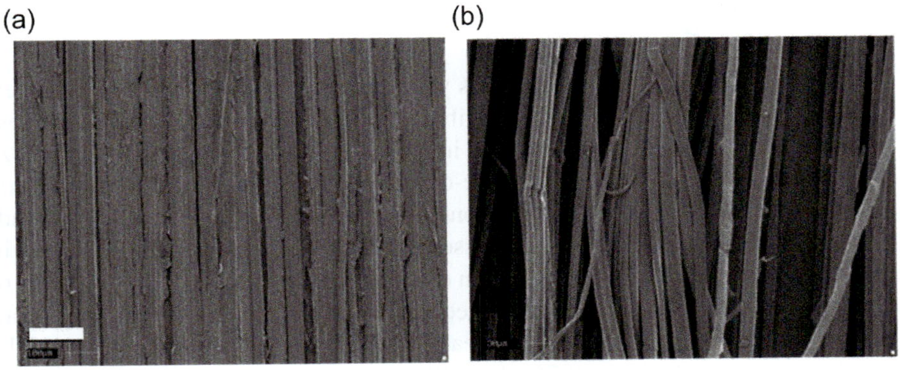

Figure 5B.4 Electronmicrographs of longitudinal section of hemp bast layer: (a) unretted hemp fibres and (b) retted hemp fibres.
Source: Used with permission from De Montfort University.

Figure 5B.5 Field retted fibre hemp being windrowed prior baling in Leicestershire UK.

of being very efficient and does not require specialist machinery (Chen et al., 2004). However, the method is not suited to crops grown for long fibres as the orientation of the stems is not maintained, but the system is suited to the production of short fibres for industrial end uses or textiles.

The fibre yield of a hemp crop will continue to increase until the plant reaches full maturity and senesces (van der Werf, 1991). However, as the plants proceed towards seed maturity the fibres in the stems become coarser and weaker, and greater levels of lignin are deposited. Increasing lignification has been suggested to potentially reduce the yield of fibre mechanically recoverable from the hemp straw (Struik et al. 2000). For end uses that do not require well retted and separated fibres these mature fibres may be adequate, but where fine fibres are required the crop may need cutting at an earlier development stage.

5B.4 Retting

The quality and quantity of hemp fibres yielded from a crop is very strongly determined by the degree of retting that the fibre has undergone. The dew retting process begins with the plants being cut and laid in the field. Fungi and bacteria then colonize the stems and produce a polysaccharide-degrading enzymes, especially pectinolytic enzymes, which will degrade the components in the middle lamella region surrounding the fibres (Akin et al., 2007; Henriksson et al., 1997; Kozłowski, 1992). This will have the effect of loosening the adhesion of the fibres to each other. The definition of retting is 'the subjection of crop or de-seeded straws to chemical or biological treatment to make the fibre bundles more easily separable from the woody part of the stem' (Farnfield and Alvey, 1975) (Fig. 5B.5).

The retting process can be undertaken chemically or biologically. In the case of the latter retting is achieved with the action of colonizing enzymes or the action of specialized enzymes secreted from colonizing fungi and bacteria.

Retting is of critical importance to the extraction and separation of fibres from the core of hemp stems, as it significantly affects the yield and quality of fibre achieved after the subsequent processing of the straw.

In practice hemp stems may be retted in several methods:

- Field, dew-retted
- Water-retted
- Enzyme-retted
- Chemical-retted

The chosen method will primarily be determined by location and the requirements of the target end use.

5B.4.1 Field (dew) retting

After the crop has been cut the stems are left spread over the surface of the field for 2–8 weeks (Fig. 5B.5), depending on the location, the degree of retting required and any consideration for subsequent land use such as the planting of the next crop. During this period of time the spread stems are normally turned a number of occasions to evenly expose the stems to light, weather and temperature conditions. Importantly with this method of retting, there can be a significant year to year variation in the duration of time required to achieve satisfactory retting as a result of varying weather conditions. The retting of hemp stems in the field is achieved by the application of moisture (rain and dew), warm temperatures, high relative humidity and still air in diurnal episodes. The stems are judged to be sufficiently well retted by examining the colour of the stems for an even dark grey colour throughout and/or by manual handling to examine the ease and cleanliness with which the woody core and fibres separate.

During the field retting process the plants are cut and laid in the field (Fig. 5B.5). Fungi and bacteria then colonize the stems and produce a polysaccharide-degrading enzymes, especially pectinolytic enzymes, which will degrade the components in the middle lamella region surrounding the ultimate fibres (Akin et al., 2007; Henriksson et al., 1997; Kozłowski, 1992). It is the most widely practised method of retting hemp and it is the source of most hemp fibre produced in Europe and the United Kingdom. Critical to the success of field retting is the presence of satisfactory temperatures and moisture, and so it can only be conducted in suitable geographical areas.

A major advantage of field retting is its agricultural efficiency and low cost (Tahir et al., 2011), as the technique does not require the crop to be moved from the field in which it was grown, and it can be fully mechanized to minimize labour input. Additionally, the technique is low input in comparison to water and chemical retting as it requires nothing more than the mechanical field operations. At present the field retting of hemp is the most commercially viable method of retting a fibre hemp crop.

The major disadvantages of field retting hemp arise from a comparative lack of control on the process by the hemp grower. The retting can be variable both within and between crops as a result of fluctuating weather and climate patterns, and crop variation. Subsequently the fibre extracted is often variable in fineness, length, strength and colour within a sample in comparison to chemical and water-retted hemp fibre. Other drawbacks of this method include the occupation of land until the retting is complete, which may impact on the planting of a subsequent crop, and that hemp fibre extracted from dew-retted crops tends to be dusty such that it can possess a hazardous health risk to operators during subsequent processing.

5B.4.2 Water retting

This method of retting is that which often produces the best separated and therefore finest hemp fibres, in a short period of time (Table 5B.4). The traditional method of water retting hemp involved steeping the stems in running water by submersion in ditches, streams or rivers but this has largely been replaced by the method of water retting in tanks through which water flows (Di Candilo et al. 2000, 2009). Water retting tanks can be sealed or open depending on the climate, and may be arranged in a series or cascade layout. In the cascade design a series of three to four interconnected tanks will have water passed through at a rate that will achieve one change of water in the system every 2 days. The use of cold water in tank-based water retting will result in the retting process taking only 7–14 days to complete, but this can be reduced to 4–5 days by retting in warm water (30–40°C). It is not uncommon in all methods of water retting to find the inclusion of chemical additives to enhance the process and offset some of the environmental impacts of the process. The latter water retting has been widely practised in Europe and can still be found conducted in some countries.

The primary advantage of this method of retting hemp is that it can be controlled and evenly applied to hemp stems, subsequently producing a superior quality of fibre in comparison to field retting. Additionally, it is not dependent on weather conditions, so enabling fibre hemp to be grown in locations that may not possess a suitable climate for field retting, and in its basic form it is a relatively simple technology that can be used in locations without access to chemicals. Furthermore the process is relatively

Table 5B.4 Chemical composition of water-retted hemp fibre at different times with and without the addition of bacterial inoculum.

	Treatment time (hours)	Cellulose (%)	Pentosans (%)	Soluable lignin (%)	Insoluable lignin (%)	Extractives (%)	Ash (%)
Untreated		82.0	8.4	2.2	0.7	4.4	2.3
Water retted without bacterial inoculum	1	85.4	6.3	2.2	0.7	3.2	2.2
	2	86.4	5.8	2.0	0.7	3.2	1.9
	3	86.4	5.2	3.0	0.6	3.2	1.6
	4	86.6	5.0	3.2	0.6	3.2	1.4
	5	87.4	4.7	3.2	0.6	3.1	1.0
	6	87.3	4.5	3.7	0.4	3.1	1.0
Water retted with bacterial inoculum	1	85.5	6.5	1.5	0.7	3.2	2.6
	2	87.0	5.5	0.6	0.6	3.2	2.0
	3	88.6	4.8	2.1	0.6	3.2	0.7
	4	88.6	4.6	2.4	0.6	3.1	0.7
	5	88.4	4.4	2.9	0.5	3.1	0.7
	6	88.9	4.1	3.1	0.5	2.8	0.6

From Di Candilo et al. (2009).

quick in comparison field retting (Table 5B.4) and ensures the field is available for subsequent cropping more rapidly after cutting.

On the other hand, the process suffers from a very high level of water consumption per tonne of straw and produces a significant amount effluent as a waste-product. Typically 20 m^3 is used per tonne of straw and a further 10 m^3 for rinsing. Water retting is also very malodorous. Additionally, it is necessary to dry the fibre after the water retting has been completed which can incur significant energy costs.

In order to combat some of the problems associated with high levels of water consumption and effluent aerobic bacilli are added to the retting liquor and the tanks are aerated. This gives slightly alkaline liquor, lowers the concentrations of volatile acids and reduces the odour that results from the presence of n-butyric acid. In practice it also enables the water to be recycled through the water retting system indefinitely and the bacilli do not attack or compromise the fibre. Despite water retting producing some of the best quality hemp fibre and the advances made with the technique to reduce the potential environmental impact of effluent, the use of water retting in Europe has largely been terminated as a result of environmental concerns and the high cost of artificially drying the fibre after retting (Kozłowski, 1992).

5B.4.3 Chemical retting

Chemical retting methods have been described and used for retting hemp stems (Wang et al., 2003; Obendorf and Song, 2006; Kostic et al., 2008) and are similar to those used for chemical treatment of flax. Chemical retting is commonly performed using a sodium hydroxide (NaOH) solution with or without the use of chelating agents such as EDTA (ethylenediaminetetraacetic acid). Other chemicals, such as sodium carbonate (Na_2CO_3) or sodium sulphite (Na_2SO_3), can also be used in retting solutions. Many of these chemical mixtures are similar to those used for degumming other plant fibres, such as ramie. The main factors determining the effectiveness of chemical retting solutions include

- Chemical concentration
- Solution acidity (pH)
- Stem to liquor ratio
- Temperature
- Duration of treatment
- Agitation degree

This method possess all the advantages of water retting over field retting, but the retting process can be further accelerated from the few days required for water retting to just a few hours.

Although the results achieved from chemical retting of hemp have generally been very good, in terms of fibre quality and fibre yield, the method is generally cost prohibitive for many end uses and is not widely commercially viable. The results of chemical retting are generally good (Table 5B.5).

Table 5B.5 Chemical composition and weight loss of hemp fibre treated with NaOH at room temperature (RT) or boil (B).

		Cellulose (%)	Hemicellulose (%)	Lignin (%)	Weight loss (%)
Untreated		76.12	12.28	5.65	
5% NaOH	RT	83.63	7.49	2.61	4.4
5% NaOH	B	89.16	1.93	2.07	25.5
18% NaOH	RT	86.18	3.83	2.44	14.1

Kostic, M., Pejic, B., Skundric, P., 2008. Quality of chemically modified hemp fibers. Bioresour. Technol. 99 94–99.

5B.4.4 Enzyme retting

A significant amount of research has been focused on retting hemp with enzymes by attempting to replace the bacteria which facilitate fermentation in water retting with enzymes in tanks. The objective of this technique is to ultimately produce a fibre quality equivalent to that achieved with water retting but without the problems of effluent waste and malodour.

A number of enzyme mixtures have been tested at the laboratory level and found successful at retting hemp fibres, although variations in performance have been noted (Kozlowski et al., 2006; Fischer et al., 2006; Nykter et al., 2006). With enzyme retting no field retting is required and the time for land occupation by the hemp crop is shortened to that of growing the crop without interference to subsequent cropping. Additionally the method is attractive because it allows retting to take place all year around and eliminates the risks associated with field retting.

The major disadvantage of this method is the potential high cost of retting, as a result of the cost of the enzymes and the equipment needed.

5B.5 Fibre extraction

The extraction of fibres from hemp stems is commonly achieved by the mechanical processing of the straw, although historically manual decortication has been widely practiced. The purpose of this processing stage is to completely extract the fibres by entirely separating them from the woody core of the hemp stem, which is often referred to as 'hurd'. Prior to processing, after retting, it is necessary that the straw be dried to a suitable moisture content to stop the retting process and to enable the straw to be stored without deterioration. This drying process enables the fibre bundles in the stem to contract and become partially loosened away from the woody core.

Although the objective of extracting, separating and cleaning the fibres from the other parts of the hemp plant is universal in hemp fibre production, the exact method of fibre extraction employed will ultimately depend on the necessary condition of fibre (length, fineness, etc) required for the end market and users. Conventionally, hemp fibres were extracted in a long fibre (50–60 cm) form, with a significant quantity of by-product short fibre, using breaking and scutching methods similar to those used for the production of long flax fibre (Fig. 5B.6). However as new markets have developed for hemp fibres, particularly in industrial products, decortication methods of fibre extraction that aim to produce a single quality of hemp fibre have been developed.

5B.5.1 Breaking and scutching

The first stage of mechanical processing is the breaking process whereby the straw is mechanically treated to break and loosen the woody core from the fibres. Typically the stems are fed into a series of fluted and smooth rollers, either horizontally or vertically arranged, which crush and break the straw along its length. The smooth rollers will crush and break the stems along their length, while the fluted rollers will break the woody core into small pieces which then drop out of the breaking unit. The fluted rollers are typically arranged in a series with declining pitches so as to break the woody core into increasingly smaller pieces.

Overall, the effectiveness of the breaking stage is almost entirely dependent on the degree of retting of the hemp straw. Well-retted hemp stems can be broken with ease and the content of hurd remaining at the end of the process is significantly reduced, to less than 40%, and the fibre is suited to further mechanical processing that will yield fibres suited for spinning for textiles. The processing of unretted material results in a comparatively less successful removal of hurd from the straw and the final fibre is only really suitable for use in paper manufacture.

The second stage of mechanical processing is the scutching stage, which has the objective of completely removing all the remaining hurd adhering to the plant fibres, which was loosened by previous retting and mechanical breaking. The process also has the effect of opening the fibre, making it finer and softened. The fibre is scutched by being beaten between turbine blades, bladed wheels operating against a curved plate. The broken fibre is held vertically in a continuous belt that passes the fibres between a pair of turbines, rotating in different directions. The turbines are operated in two pairs so that after passing through the first pair of turbine blades the fibres are reversed to pass through the second pair facing in the opposite direction. The primary output of this process is the long (line) fibres, but during the process some fibres are broken and contained in the scutching waste. These fibres are extracted from this waste and cleaned as short (tow) fibres for use in textile products, coarse yarns or ropes. The greater the degree of retting the straw the lower is the yield of tow fibre of total processed fibre.

After scutching, the tow fibres that have been removed during the process are separated from the hurd and other impurities (dust, leaf material, etc) by passing them through some form of cleaning unit. The tow fibres may then be further processed to open the fibres in preparation for the anticipated end use. They may also be dried to required moisture content if needed. The key factors of hemp tow fibre production are fibre processing efficiency, and fibre qualities such as fineness, colour and fibre length.

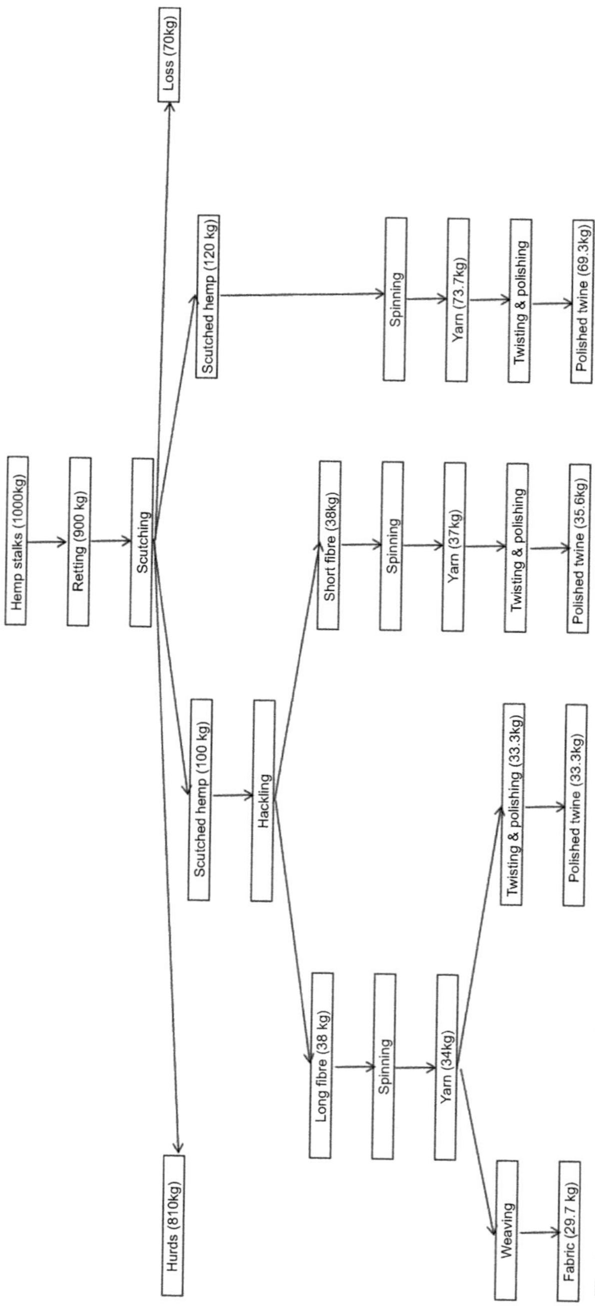

Figure 5B.6 Long fibre hemp production flow chart and typical yields.
Source: Sponner et al., 2004.

5B.5.2 Decortication and fibre cleaning

5B.5.2.1 Decortication

Although the use of breaking and scutching lines has been widely employed for the production of hemp fibre, particularly for the production of high quality textile fibres, alternative methods of extracting hemp fibres and subsequent cleaning have also been developed (Hobson et al., 2001; Horne, 2010; Wang et al., 2018). The objectives of decorticating the hemp straw, and subsequent opening and cleaning of the extracted fibre are the same with these alternative methods as for traditional breaking and scutching, but the actual mechanical methods employed are different. In particular, the fibres produced from these processing techniques are significantly different and make the fibres much less suited to the manufacture textiles but are well suited to the production of technical fibres. A significant difference between the scutching and hackling methods and the alternative decortication methods is that the harvested stems in the latter do not have to be kept in a single orientation, with all the stem bases kept together and all stems parallel, to ensure the effective extraction of the long fibres. Instead the stems can be harvested and the stems allowed to randomly orientate in the harvested bales.

The first stage in these processing methods is the opening of the bales of hemp straw and their feeding into the decortication machine. During this first stage of the processing, the hemp straw can be cut with a guillotine or bale cutter into shortened lengths (400–500 mm) to ensure efficient straw feeding and uniform subsequent processing. The opened and cut straw is then metered into the decortication equipment with either a box feeder or metering box device, or the straw can be transferred in an air flow to a decorticator.

As with breaking and scutching, the process of decortication has the objective of separating the outer layer of fibres of the hemp stems from the woody inner core, while also applying some opening actions to the bundles of fibres. The principle methods used to conduct a decortication operation are swing-hammer mills, with and without cleaning, and roller decortication. Hammer mills consist of a cylinder containing a rotor equipped with hinged beaters (hammers) that rotate at high speeds and decorticate the hemp straw, fed vertically into the top of the cylinder, by working it against the internal surface of the cylinder casing. Decorticated material passes down the cylinder pneumatically in an axial direction. Some hammer mill decorticators are equipped with a screen on the bottom surface of the cylinder to allow woody core extracted from the hemp straw to be separated from the decorticated fibre. The major advantage of these decorticators is that they operate at a very high throughput; 4–6 t flax straw per hour, and can decorticate both unretted and well-retted hemp straw (Horne et al., 2010). When cleaning is introduced into the hammer mill it is possible to achieve high levels of cleanliness and up to 60% of hurds can be discharged at this stage, making subsequent cleaning significantly easier. However, the major disadvantage of this processing technique is that fibre decortication is achieved by the impact of the hammers acting on the surface of the straw. Given that the fibres are contained in an outer layer on the straw this has the potential to cause mechanical damage to the fibres, which could result in inconsistency in the extracted fibre and limit its use to non-woven products and industrial textiles, rather than fine short-fibre hemp-based yarns suited produced.

An alternative to hammer mill-based decortication is the use of roller decorticators (Fig. 5B.7). Using this method opened hemp straw is fed through a series of fluted crimping rollers that will break and fracture the inner woody core and separate the fibre bundles. The rollers are operated in a horizontal series of 7–10 rollers, and the pitch of the rollers is reduced from one roller pair to the next to allow for the reduction in the width of the straw flow. Hurds fall out below the rollers as the straw passes through the series. The advantage of this system is that the fibres are not damaged by the decortication process and it is possible to extract long fibres after this method. Some work has been undertaken to develop decorticators for hemp fibre extraction and it is possible to remove the majority of hurd from the hemp straw (Hobson et al., 2001; Wang et al., 2018). The major commercial drawback of this system of decortication is that it generally only accommodates a very low throughput of straw, no more than 1t per hour. Also, roller decorticators are less able to decorticate unretted straw and they are less effective as the straw moisture content exceeds 15% (Hobson et al., 2001).

5B.5.2.2 Fibre cleaning

After the straw has been decorticated, the resulting mass will be a combination of fibre and broken hurds and the next stage(s) must separate these two component parts. The type and amount of cleaning required will depend on the cleanliness of the material after decortication and the cleanliness levels required by the end users of the hemp fibre. After hammer mill decortication, with a cleaning screen, in its cylinder the level

Figure 5B.7 Roller decorticator behind guard.

of hurd in the fibrous mass is likely to be lower than with roller decortications. Consequently, the output of the latter will require more cleaning than the former. Several types of cleaner can be employed in various combinations and sequences to achieve desired fibre cleanliness, but, in general, fibre cleaners operate on the principle of either a tambour turbine, step cleaner or comb shaker table.

5B.5.2.3 Picker

A picker is a hard faced cylinder covered in pins or beater blades that rotates to strike a downward motion against fibres fed into the beating cycle by feed rollers, typically via a feed belt (Fig. 5B.8). As the fibres are fed into the machine they first become arched by the rotating action of the cylinder and the fibres are then conveyed between the cylinder and a bar grid screen at the base of the machine. The hurd and non-fibrous particles drop through the grid, while the fibre adheres to the cylinder. The main advantage of these machines is that they are able to process quantities of fibre in excess of 1 tonne per hour. They can also be used as decorticators to process straw and extract fibre from bast fibre plants, including hemp (Horne, 2010).

5B.5.2.4 Step cleaner

A step cleaner is so because of the arrangement of its beater cleaner components. Up to six drums are positioned in an ascending manner like stairs, with bar or wire grids beneath each drum. These drums are hard faced but have beater arms which loosen the fibres fed into the lowest end of the machine (Fig. 5B.9). As the drum rotates the fibre is passed from one beater to the next by the air flow generated by the rotating drums, while the loosening action of the drum on the fibre results in an excellent separation of the fibre from the hurds, which fall through the grids beneath the drums. Commonly the bar or wire grids have adjustable gaps widths to accommodate different particle sizes.

Figure 5B.8 Picker decorticator feed table and clothed swift.

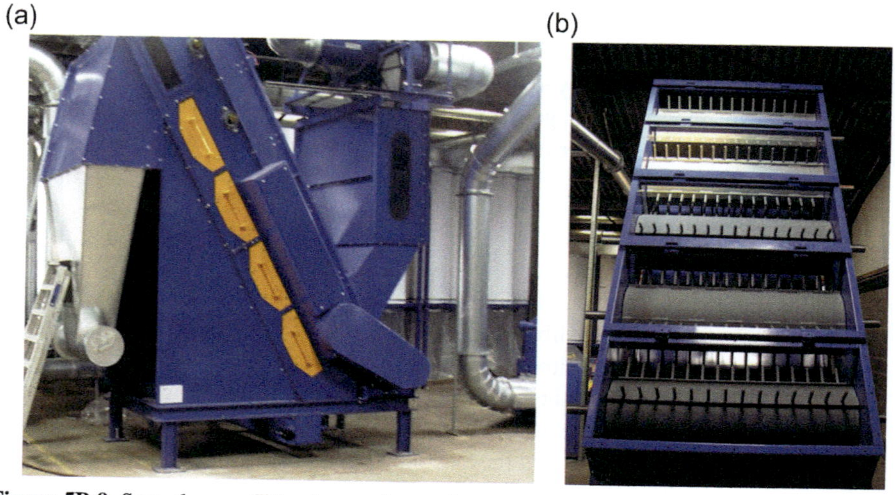

Figure 5B.9 Step cleaner. Side view and rear view showing rollers and beaters.

Step cleaners are highly efficient and are normally used for the production of fine fibres, and they can process fibres at a high throughput of up to 5 tonne per hour.

5B.5.2.5 Shaker table

A shaker table is a simple cleaner unit that is frequently used for the final cleaning stage of fibre. The comb shaker constitutes a grid positioned at a slight gradient with rows of oscillating teeth over the top of them, to operate as a comb reciprocating the fibre up the gradient. The hurd is loosened from the fibre and falls through the grid bars. Although relatively effective as a cleaner the major disadvantage of this method of cleaning is the slow throughput of the unit.

An example short fibre hemp decortication and cleaning line is shown in Fig. 5B.10.

5B.5.2.6 Fibre opening

The objective of fibre opening is to separate the bundles of fibre into smaller aggregations of fibre or into single fibres. The purpose is to increase the fineness of the fibre material, while simultaneously removing any non-fibrous material, particularly dust and small pieces of hurd.

5B.5.2.7 Opener

A fibre opener consists of one or more opening cylinders into which fibre is fed by feed rollers. The fibre opening cylinder is 'clothed' with needles, pins or garniture with different pitches. As the fibre is fed in the rotating cylinder the clothing pull the fibre over several carding plates to separate, parallelize and comb the fibres. Finally, the fibres are sucked off pneumatically. The process will primarily separate the fibres and increase the fineness of the fibre. A major advantage of this type of opener is

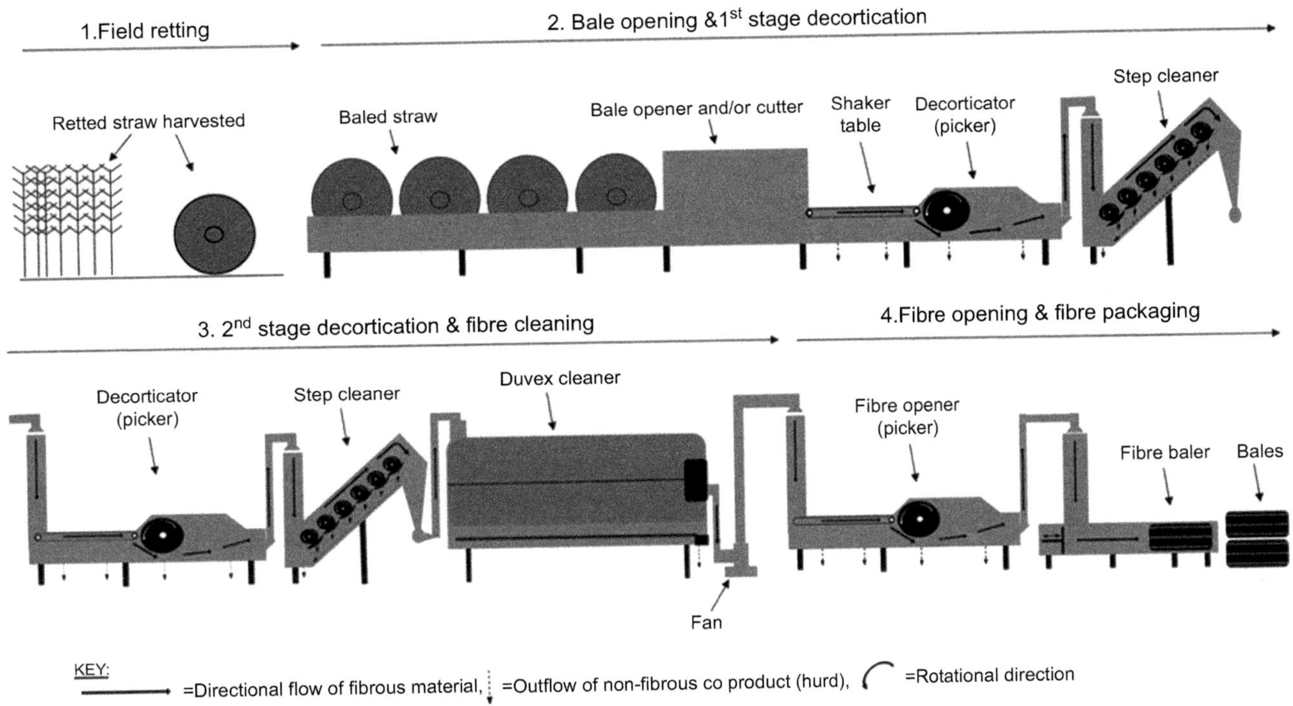

Figure 5B.10 Fibre processing sequence for cleaning and opening field retted fibre.

that it is relatively low cost in comparison to carding and it can be a flexible process that produces can utilize number of needle, pin or wire arrangements to produce fibre of a very specific fineness.

5B.5.2.8 Carding

Using a card to process hemp fibre extracted using decortication rather than breaking and scutching of straw is identical to that used open short fibre hemp tow. However, the production of hemp fibre for technical rather than textile end uses is a lower value market and the requirements for finer fibre less. As carding is an expensive process it is less suited to opening this hemp fibre.

5B.6 Hemp fibre spinning

Conventionally there have been two principle types of hemp fibre, long line fibres and shorter tow fibres, produced during the extraction of the fibres from the stems can be spun into yarns. The long line fibre is usually spun to produce a relatively coarse yarn, while tow fibre hemp may be used to produce finer yarns and it is often blended with other fibres prior to spinning.

5B.6.1 Long-fibre spinning

5B.6.1.1 Preparation of long-fibre hemp

Prior to the hackling process it is common to find that hemp fibres are processed to increase the fineness of the bundles of fibre produced during scutching. A softening of hemp fibres can be achieved with the use of ribbed rolls that will soften the hemp fibres as they move between the rolls. Further softening may be achieved during this stage with the application of a softening emulsion to the hemp fibres prior to treatment.

Long-fibre hemp rove weights are typically 2.5–5 g per metre.

Long fibre hemp spinning is a highly specialized operation and is now uncommon, although it is possible to undertake in areas where flax fibre is processed for long fibres for use in linen yarns.

5B.6.2 Short-fibre spinning

During the scutching and hackling operations a certain amount of short fibre is removed as part of the extraction and preparation of the long fibres. This fibre typically falls into one of two types: scutched tow or hackled tow. This fibre can be processed in preparation for spinning into a wide range of yarns, depending on the quality of the hemp being processed and uniformity of fineness, length and cleanliness. However, unlike long fibres hemp tow fibre tends to be processed and spun on more conventional fibre preparation and spinning equipment.

Alternatively, fibre hemp crops can be grown and processed only for short fibres for use in textile and woven fabrics. Increasingly this is an area of interest and developments to understand.

5B.6.2.1 Carding

The first processing stage for short-fibre hemp prior to spinning is carding, which may be completed in a single or multiple phases. Overall, the objective of the carding process is to breakdown the long fibres into ultimate or split compound fibres, untangle and open lumps of fibres, remove any remaining dirt and very short fibres and form a carded sliver. The process consists of the fundamental processes of 'working', 'stripping' and 'doffing' and a considerable degree of fibre opening and compound fibre breakdown can be achieved with retted hemp fibre on a breaker or finisher card. In addition the carding process will parallelize fibres, as a result of the directional bias of the carding machine, and it provides an opportunity for blending different qualities of hemp or hemp with other fibres.

The carding process of hemp can include a first carding stage that is conduct on a 'coarse' card, with large cylinders and rolls and stronger card wires than the main stage. This process will clean out the remaining foreign matter to be found in the fibre, cards the fibre and forms a sliver. A complimentary second stage card (finisher card) then unwinds the slivers prior to being fed into a card condenser and main cylinder that disentangles the fibres further. The fibre is then formed into a sliver at the drawhead prior to doffing. If hemp fibres are to be blended together or hemp fibres blended with other fibres, this can be done at entry into this final card where different coils can be fed into the card at the required ratio. The subsequent carding and drawing and doubling processes then blend the fibre further. Alternatively, hemp fibres may be carded in a single stage card and any required hemp fibres blended prior to and/or during carding without the formation of an intermediate sliver.

5B.6.2.2 Combing, drawing and doubling

Depending on the intended quality of yarn anticipated, the final sliver may undergo a combing process after carding but prior to subsequent preparation. Combing will further parallelize the fibres and remove any remaining impurities and short fibres. Combing is only necessary for the production of finer yarn counts, while coarse and medium count yarns can be spun directly from the finisher sliver on spinning frames. Any slivers intended to be combed after carding are normally passed through a draw frame first to limit the losses during the combing operation. After combing the slivers will undergo a further three draw frame passes prior to being doubled and then spun. Drawing attenuates the sliver to decrease the mass of fibre per unit length of the sliver. Drawing frames similar to those employed for the processing of line fibres can be used, but with a shorter drafting reach, although the use of gill drawing frames can be used as an alternative for producing slivers for coarser yarns. Intersecting gill frames less likely to be used with hemp processing than flax due to the cost involved.

After the drawing operation 'doubling' superimposes several slivers, processed together, to improve sliver regularity.

5B.6.2.3 Spinning

Dry spinning

Short-fibre hemp prepared for spinning as a sliver is typically spun using a dry spinning method, rather than wet spun, on frames similar to those used for spinning long fibre hemp but with a shorter drafting zone. As a result of the fibres still existing in groups, although sub-divided to some extent, and as they are drafted together there is a limit to the yarn fineness achievable of approximately 65 tex using this method. The final yarns tend to be soft and open structured.

Alternative spinning methods

Spinning methods used in the production of long-staple fibres, flax and hemp, are relatively specialized activities and have not received the attention of developments in innovation received by more widespread short-staple systems. The majority of recent developments in spinning systems have been directed towards the likes of warp spinning, ring spinning, open-end spinning and friction spinning and the use of cotton and short-staple man-made fibres. However there has been some development of the production of short-staple hemp fibre of a length suitable to be processed either alone or in blends with other short-staple fibres.

Over the last 10 years there has been an increasing interest in the use of hemp fibre in blends with cotton and man-made fibres in fabrics, woven and knitted, mainly driven by the attractive potentially low environmental impact of hemp fibre production. In dry spinning, hemp can be blended with cotton and most types of man-made systems as well as wool. In the cotton spinning system flax fibres should have length of 20–60 mm, while in wool spinning systems hemp fibres should be between 60 and 120 mm. In general, the potential number of options for spinning hemp fibres on short-staple systems with other fibres, and in what quantities, is very high. For the production of yarns for use in apparel fabrics the most important blend is hemp fibre with cotton and polyester, at percentages from 10% to 60%. Its blending with viscose is also important. Blends of hemp with wool are easier to achieve and can produce coarse yarns of a relatively high composition, up to 90%.

Winding

Immediately after being spun yarns are transferred to large containers, typically cones. This process is known as winding and is usually completed at high speed to produce a package of uniform density without disturbing the structure of the yarn. Although the process is not productive itself, winding provides an opportunity for quality control and the removal of faults. Fine hemp yarns, of a count between Nm 1 and Nm 6, spun from long hemp fibres and from short fibres of a count of Nm 1 to Nm 15 are normally wound onto comb winders. Thicker hemp yarns, of a count between 0.20 Nm and 1 Nm, are wound onto cross winders of a significant weight, up to 10 kg.

Figure 7.11 Furnishing fabric manufactured from wool and hemp blend fabric. Number.

Once packaged the yarn is then ready for the preparatory treatments of warping and warp sizing, assuming that the yarn is intended for weaving (Fig. 5B.11).

5B.7 Conclusions and future trends

Over the past 4 decades many efforts have been made to improve the cultivation, processing, extraction and properties of hemp fibres for textiles, which have received a wide variety of interest from a ranch of industries, including the textile sector. A key driver for this interest relates to the wide range of possibilities that are offered by the use of hemp fibres in textile products and. The aim of this work was to offer an overview of the production of hemp fibres with a description of the key stages in the cultivation of hemp crops and the processing of the crop to extract the fibre. The findings appear to show the cultivation of hemp plants and the processing of hemp crops is practically possible using methods and knowledge gained from the wide range of research efforts. However, retting methods are likely to produce varied results and mechanical processing to efficiently extract fibres from straw are likely to produce varied outputs.

Interest in natural fibres for use in textiles and composites is increasing and a very positive future is ahead for the application of hemp fibres in products (Angelini et. al., 2016). From this study it is clear that the production of hemp crops and fibres can have a direct impact on rural communities with the development of new markets for agricultural goods, while other jobs could be created in existing manufacturing sectors. A number of research projects have been undertaken to develop some of the required information needed to achieve this, but further work is needed on the mechanical processing of the harvested hemp straw to efficiently extract the fibre (decortication) from the gathered crop and also research is needed to improve the understanding and methods of retting the crop materials to achieve consistent fibre grades.

5B.7.1 Further information

European Commission (2019). Common catalogue of varieties of agricultural species. 59 *Cannabis sativa*, EC Commission: Brussels.

Horne. M.R.L. 2012. Hemp Cultivation and Production. In: Kozolowski, R.M (editor). *Handbook of Natural Fibres, Volume 1: Types, properties and factors affecting breeding and cultivation*. Woodhead Publishing: London.

McPartland J M, Clarke R C, Watson D P (2009) Hemp diseases and Pests. Wallingford: CABI.

5B.7.2 Research and development

5B.7.2.1 Institute of Natural Fibres

The Polish Institute of Natural Fibres and Medicinal Plants has played a crucial role in hemp research, growing and processing.

Address: Institute of Natural Fibres and Medicinal Plants, Wojska Polskiego 71 b, Poznań, Poland. Website: http://www.inf.poznan.pl.

5B.7.2.2 European Industrial Hemp Association

Address: nova-Institut GmbH Chemiepark Knapsack Industriestraβe 300 50354 Huerth, Germany. Website: http://eiha.org/about/

5B.7.2.3 Textile Engineering and Materials (TEAM) Research Group

Conducts research in Agriculture and Natural Fibre, including plant fibres from non-food crops, process engineering for plant fibre extraction and processing of protein fibres and textiles.

Address: De Montfort University, The Gateway, LE1 9BH. Website: https://www.dmu.ac.uk/team.

References

Aber, E.L., 1980. Marihuana, the First Twelve Thousand Years. Plenum Press, New York.

Akin, D.E., Condon, B., Sohn, M., Foulk, J.A., Dodd, R.B., Rigsby, L.L., 2007. Optimization for enzyme-retting of flax with pectate lyase. Ind. Crops Prod. 25, 136–146.

Amaducci, S., Zatta, A., Pelatti, F., Venturi, G., 2008a. Influence of agronomic factors on yield and quality of hemp (*Cannabis sativa*) fibre and implication for an innovative production system. Field Crop. Res. 107, 161–169.

Amaducci, S., Zatta, A., Raffanini, M., Venturi, G., 2008b. Characterisation of hemp (*Cannabis sativa*) roots under different growing conditions. Plant Soil 313, 227–235.

Amaducci, S., Scordia, D., Liu, F.H., Zhang, Q., Guo, H., Testa, G., Cosentino, S.L., 2015. Key cultivation techniques for hemp in Europe and China. Ind. Crops Prod. 68, 2–16.

Angelini, L.G., Tavarini, S., Di Candilo, M., 2016. Performance of new and traditional fiber hemp (*Cannabis sativa L.*) cultivars for novel applications: stem, bark, and core yield and chemical composition. J. Nat. Fib. 13 (2), 238–252. https://doi.org/10.1080/15440478.2015.1029193.

Anon, 2018. Agriculture. In: China Statistical Yearbook. China Statistics Press.

Atkinson, R.R., 1964. Jute-Fibre to Yarn. Temple Press, London.

Bednar, P., 2008. Harvesting technologies for industrial hemp. In: Proceedings of International Conference on Flax and Othe Bast Fibers, Saskatoon, Canada, pp. 308–310. July 21-23 2008.

Bennet, S.J., Snell, R., Wright, D., 2006. Effect of variety, seed rate and time of cutting on fibre yield of dew-retted hemp. Ind. Crops Prod. 24, 79–86.

Berenji, J., Sikora, V., Fournier, G., Beherec, O., 2013. Genetics and selection of hemp. In: Bouloc, P. (Ed.), Hemp Industrial Production and Uses. CABI, Wallingford, UK, pp. 48–71.

Bócsa, I., Karus, M., Lohmeyer, D., 2000. Der Hanfbau. Botanic, Sorten, Anbau undErnte, Märkte und Produktlinien. Landwirtschaftsverlag GmbH, Germany.

Borthwick, H.A., 1947. Day Length and Flowering-Hemp. The Yearbook of Agriculture, pp. 1943–1947.

Bosca, I., Karus, M., 2008. The Cultivation of Hemp. Hemptech, Sebatopol, CA.

Bredemann, G., Schawanitz, F., Sengbusch, V.R., 1956. Problems of modern hemp breeding with particular reference to the breeding of varieties of hemp containing little or no hasish. Bull. Narc. 31–35.

Bruce, D.M., Hobson, R.N., Hamer, P.J.C., White, R.P., 2005. Drying of Hemp for Long Fibre Production. Biosys. Eng. 91 (1), 45–59.

Chase, M., 1998. The Angiosperm phylogeny group: an ordinal classification for the families of flowering plants. Ann. Mo. Bot. Gard. 85, 531–553.

Chen, Y., Gratton, J.L., Liu, J., 2004. Power requirements of hemp cutting and conditioning. Biosyst. Eng. 87 (4), 417–424.

Clarke, R.C., 1980. Marijuana Botany. Ronin publishing, Berkeley, California.

Cromack, H.T.H., 1998. The effect of cultivar and seed density on the production and fibre content of *Cannabis sativa* in England. Ind. Crops Prod. 7, 205–210.

De Meijer, E.P.M., 1995. Fibre hemp cultivars: a survey of origin, ancestry, availability and brief agronomic characteristics. J. Int. Hemp Assoc. 2 (2), 66–72.

Dempsey, J.M., 1975. Fiber Crops. University of Florida press, Gainsville, Florida.

Di Candilo, M., Bonatti, P.M., Guidetti, C., Focher, B., Grippo, C., Tamburini, E., Mastromei, E., 2009. Effects of selected pectinolytic bacterial strains on water-retting of hemp and fibre properties. J. Appl. Microbiol. 108, 194–203.

Di Candilo, M., Ranalli, P., Bozzi, C., Focher, B., Mastromei, G., 2000. Preliminary results of tests facing with the controlled retting of hemp. Ind. Crops Prod. 11, 197–203.

Duke, J.A., 1982. Ecosystematic data on medicinal plants. In: Utilization of Medicinal Plants, by CK Aktal and Kapur KM, 13–23. United Printing Press, New Dehli.

Easu, K., 1965. Plant Anatomy. John Wiley and Sons, New York.

European Commission, 2019. Common catalogue of varieties of agricultural species. 59 *Cannabis sativa*. EC Comm.

FAO, 2018. FAOSTAT. Food and Agriculture Organization of the United Nations. Rome, Italy.

Farnfield, C.A., Alvey, P.J., 1975. Textile Terms and Definitions. The Textile Institute, Manchester.

Fischer, H., Müssig, J., Bluhm, C., 2006. Enzymatic modification of hemp fibres for sustainable production of high quality materials. J. Nat. Fibers 3 (2–3), 39–53.

Grabowska, L., Koziara, W., 2005. The effect of nitrogen dose, sowing density and time of harvest on development and yields of hemp cultivar Bialobrezskie. J. Nat. Fibers 2 (4), 1–17.

Hennink, S., 1994. Optimisation of breeding agronomic traits in fibre hemp (*Cannabis sativa*) by study of parent offspring relationships. Euphytica 78, 69–76.

Henriksson, G., Akin, D.E., Hanlin, R.T., Rodriguez, C., Archibald, D.D., Rigsby, L.L., Eriksson, K.L., Russell, R.B., Carolina, S., 1997. Identification and retting efficiencies of fungi isolated from dew-retted flax in the United States and Europe. Am. Chem. Soc. 63, 3950–3956.

Heslop-Harrison, J., Heslop-Harrison, Y., 1958. Studies on flowering plant growth and organogenesis III. Leaf Shape changes associated with flowering and sex differentiation in *Cannabis Sativa*. Proc. R. Ir. Acad. 57, 257–289.

Hobson, R.N., Hepworth, D.G., Bruce, D., 2001. Quality of fibre separated from unretted hemp stems by decortication. J. Agric. Eng. Res. 78, 153–158.

Hoffmann, W., 1961. Hanf *Cannabis sativa*. In: Kappert, H., Rudorf, W. (Eds.), Handbuch der Pflanzenzüchtung Band V. Paul Parey, Berlin-Hamburg, pp. 204–261.

Holoborod'ko, P., 1995. Hemp research and growing in the Ukraine. In: Proceedings of Bioresource Hemp Symposium, Frankfurt Germany, March 2nd 1995.

Horne, M.R.L., 2010. Yield and properties of fibre from linseed straw. Aspect Appl. Biol. 101, 8–9.

Horne, M.R.L., Waldron, D., Harwood, J.L., Harwood, R., 2010. The production and extraction of flax-fibre for textile fibres. J. Biobased Mater. Bioenergy 4, 98–105.

ITC, 2007. Le Chanvre Industrial. Guide Technique Institut Technique du Chanvre. Technopole de l'Aube en Champagne. Troyes Cedex 9, France.

Jordan, H.V., Lang, A.L., Enfield, G.H., 1946. Effects of fertilizers on yield and breaking strengths of American hemp, *Cannabis sativa*. J. Am. Soc. Agron. 38, 551–563.

Keller, A., Leupin, M., Mediavilla, V., Wintermantel, E., 2001. Influence of the growth stage of industrial hemp on chemical and physical properties of the fibres. Ind. Crops Prod. 13, 35–48.

Kirby, R.H., 1963. Vegetable Fibres Botany. Cultivation & Utilisation. London: Leonard.

Kostic, M., Pejic, B., Skundric, P., 2008. Quality of chemically modified hemp fibers. Bioresour. Technol. 99, 94–99.

Kozłowski, R., 1992. Retting of Flax in Poland, in The Biology and Processing of Flax. In: Sharma, H.S.S., Van Sumere, C.F. (Eds.). Belfast: M Publications, pp. 251–259.

Kozlowski, R., Guebitz, G.M., Cavaco-Paulo, A., 2006. Biotechnology in Textile Processing, first ed. CRC Press, London.

Kundu, B.C., 1942. The anatomy of two indian fibre plants, cannabis and corchorus with special reference to fibre distribution and development I. Indian Bot. Soc. J. 23, 93–129.

Kymäläinen, H., Hautala, M., Kuisma, R., Pasila, A., 2001. Capillarity of flax/linseed (*Linum usitatissimum*) and fibre hemp (*Cannabis sativa*) straw fractions. Ind. Crops Prod. 14, 41–50.

Lisson, S.N., Mendham, N.J., 1998. Response of fibre hemp (*Cannabis Sativa*) to varying irrigation regimes. J. Int. Hemp Assoc. 5 (1), 9–15.

Makowiecka, J., Wielgus, K., 2014. Therapeutic potential of cannabinoids —perspectives for the future. J. Nat. Fib. 11 (4), 283–311.

McDougall, G.J., Morrison, I.M., Stewart, D., Weyers, J.D.B., Hillman, J.R., 1993. Plant fibres: botany, chemistry and processing for industrial use. J. Sci. Food Agric. 62, 1–20.

McPartland, J.M., 1996a. A review of Cannabis diseases. J. Int. Hemp Assoc. 3 (1), 19–23.

McPartland, J.M., 1996b. Cannabis pests'. J. Int. Hemp Assoc. 3 (2), 52–55, 49.

McPartland, J.M., Clarke, R.C., Watson, D.P., 2000. Hemp Diseases and Pests. CABI Publishing, Wallingford, Oxon.

Meijer, W.J.M., van der Werf, H.M.G., Mathijssen, E., Vandenbrink, P.W.M., 1995. Constraints to dry matter production in fibre hemp (*Cannabis Sativa).* Eur. J. Agron. 4, 109–117.

Nykter, M., Kymäläinen, H.M., Thomsen, A.B., Lilholt, H., Koponen, H., Sjöberg, A., Thygesen, A., 2006. Effects of thermal and enzymatic treatments and harvesting time on the microbial quality and chemical composition of fibre hemp (*Cannabis sativa*). Biomass Bioenergy 32, 392–399.

Obendorf, S.K., Song, K.H., 2006. Chemical and biological retting of kenaf fibers. Text. Res. J. 76, 751–756.

Polezak, E., Wosko, S., Slawinska, K., Szopa, A., Wrobel, A., Serefko, A., 2018. Cannabinoids in depressive disorders. Life Sci. 218, 18–24.

Pounds, N.J.G., 1979. An Historical Geography of Europe 1500-1840. Cambridge University Press, Cambridge.

Ranalli, P., 1999. Agronomical and physiological advances in hemp crops. In: Ranelli, P. (Ed.), Advances in Hemp Research. Haworth Press, USA, pp. 61–84.

Sankari, H., 2000. Comparison of bast fibre yield and mechanical fibre properties of hemp (*Cannabis sativa*) cultivars. Ind. Crops Prod. 11, 73–84.

Schafer, T., Honermeier, B., 2006. Effect of Sowing date and plant density on the cell morphology of hemp. Ind. Crops Prod. 23, 88–98.

Schofielda, J.E., Wallerb, M.P., 2005. A pollen analytical record for hemp retting from Dungeness Foreland, UK'. J. Archaeol. Sci. 32, 715–736.

Schultes, R.E., 1970. Random thoughts and queries on the botany of Cannabis. In: Joyce, C.R.B., Curry, S.H. (Eds.), The Botany and Chemistry of Cannabis. J & A Churchill, London, pp. 11–38.

Slovonov, L., Petinov, N.S., 1980. The content of nucleotides and the ATPase activity in hemp leaves in relation to the water supply. Fiziol. Rastenii 27 (5), 1095–1110.

Small, E., Cronquist, A., 1976. A practical and natural taxonomy for Cannabis. Taxon 25 (4), 405–435.

Stearn, W.T., 1970. The cannabis plant: botanical chrarcteristics. In: Joyce, C.R.B., Curry, S.H. (Eds.), The Botany and Chemistry of Cannabis. J & A Churchill, London.

Struik, P.C., Amaducci, S., Bullard, M.J., Stutterheim, N.C., Venturi, G., Cromaci, H.T.H., 2000. Agronomy of fibre hemp (*Cannabis sativa* L.) in Europe. Ind. Crops Prod. 11, 107–118.

Tahir, P.M., Ahmed, A.B., SaifulAzry, S.O.A., Ahmed, Z., 2011. Retting process of some bast plant fibers and its effect on fibre quality: a review. Bioresources 6 (4), 5260–5281.

Thomsen, A.B., Rasmussen, S., Bohn, V., Nielsen, K.V., Thygesen, A., 2005. Hemp Raw Materials: The Effect of Cultivar, Growth Conditions and Pretreatment on the Chemical Composition of the Fibre. Risø-R Report 1507 (EN), Risø Nationa lLaboratory, Roskilde, Denmark, p. 30.

van der Werf, H.M.G., 1991. Agronomy and Physiology of Fibe Hemp' Center for Agrobiological Research Report 142, Wageningen, The Netherlands, Research Report 142, Wageningen, the Netherands.

van der Werf, H.M.G., Mathijssen, E., Haverkort, A.J., 1996. The potential of hemp (*Cannabis sativa*) for sustainable fibre production: a crop physiological appraisal. Ann. Appl. Biol. 129 (1), 109–123.

van der Werf, H.M.G., van der Veen, J.E.H., Bouma, A.T.M., ten Cate, M., 1994. Quality of hemp (*Cannabis Sativa*) stems as a raw material for paper. Ind. Crops Prod. 2, 219–227.

van der Werf, H.M.G., Wijlhuizen, M., Deschutter, J.M.M., 1995. Plant density and self thinning affect yield and quality of fibre hemp (*Cannabis Sativa*). Field Crop. Res. 40 (3), 153–164.

Vavilov, N.J., 1926. Centers of origin of cultivated plants. In: Dorofeyev, V.F. (Ed.), Origin and Geogrpahy of Cultivate Plants. Cambridg University Press, Cambridge.

Vessel, A.J., Black, C.A., 1947. 'Soil Type and Soil Management Factors in Hemp Production' Iowa Agricultural Experimental Station Bulletin, vol. 352. Iowa State College, Ames, Iowa, pp. 384–424.

Wang, S., Gusovius, H.J., Lühr, C., Musio, S., Uhrlaub, B., Amaducci, S., Mussig, J., 2018. Assessment system to characterise and compare different hemp varieties based on a developed lab-scaled decortication system. Ind. Crops Prod. 117, 159–168.

Wang, H.M., Postle, R., Kessler, R.W., Kessler, R., 2003. Removing pectin and lignin during chemical processing of hemp fibre for textile applications. Text. Res. J. 73 (8), 664–669.

Westerhuis, W., Amaducci, S., Struik, P.C., Zatta, A., van Dam, J.E.G., Stomph, T.J., 2009a. Sowing density and harvest time affect fibre content in hemp (*Cannabis sativa*) through their effects on stem weight. Ann. Appl. Biol. 155 (2), 225–244.

Westerhuis, W., Struik, P.C., van Dam, J.E.G., Stomph, T.J., 2009b. Postponed sowing does not alter the fibre/wood ratio or fibre extractability of fibre hemp (*Cannabis sativa*). Ann. Appl. Biol. 155 (2), 333–348.

Zielonka, D., Kiraga, Ł., Kozłowski, R., 2019. Medical potential of cannabis: an overview. In: Handbook of Natural Fibres in the press.

Further reading

Chaudry, N., 2008. Industrial Hemp Production in Canada. Government of Alberta, Canada.

Crowley, J.G., 2001. The Performance of *Cannabis Sativa* (Hemp) as a Fibre Source for Medium Density Fibre Board (MDF). ADAS, Cambridge.

Eurostat, 2008. Area, yield and production of fibre flax and hemp. EC Comm.

Schultes, R.E., 1973. Man and marijuana. Nat. Hist. 82 (7), 59–63.

Sponner, J., Toth, L., Cziger, S., Franck, R.R., 2004. Hemp. In: Franck, R.R. (Ed.), Bast and Other Fibre Plants. Woodhead, Cambridge, UK, pp. 176–206.

Technopole de l'Aube en Champagne. Troyes Cedex 9, France.

Abaca: cultivation, obtaining fibre and potential uses

6

Edwin A. Simbaña[1], Paola E. Ordóñez[1], Yadira F. Ordóñez[1], Víctor H. Guerrero[2], Moraima C. Mera[3], Elmer A. Carvajal[4]
[1]Pontificia Universidad Católica del Ecuador Sede Ibarra, Escuela de Ciencias Agrícolas y Ambientales, Productos Naturales Bioactivos, Ibarra, Imbabura, Ecuador; [2]Escuela Politécnica Nacional, Departamento de Ingeniería y Ciencias de los Materiales, Quito, Pichincha, Ecuador; [3]Pontificia Universidad Católica del Ecuador Sede Ibarra, Escuela de Ciencias Agrícolas y Ambientales, Ibarra, Imbabura, Ecuador; [4]Pontificia Universidad Católica del Ecuador Sede Ibarra, Escuela de Diseño, Ibarra, Imbabura, Ecuador

6.1 Cultivation and fibre extraction

6.1.1 Cultivation

Abaca is a name in Tagalog language (Palacios and Peña), mostly spoken in the Philippines, given to a plant that is similar to the banana tree, from which a very resistant fibre is extracted. This plant is also known as Manila hemp or Bandala in its language of origin. *Musa textilis* Nee is the scientific name of this herbaceous stolonic, monocotyledons and perennial plant of the family of the Musaceae (see Fig. 6.1), with similar aspect to *Musa paradisiaca* (banana plant). Abaca is mainly used in the production of fibrous filaments, concentrated in the vascular bundles – the most complex parts of the plant – where the fibre develops. They are the pods of the leaves that form a pseudostem of 15–20 cm in diameter and can reach up to 6 m in height of thin leaves (see Fig. 6.2). This plant can have more than 20 pseudostems in different stages of growth or development with different colours, ranging from green, and brown, to purple or black in the centre. The Musaceae family is very broad and has 54 species, of which *Musa textiles* are characterized by their members having the number of diploid chromosomes of 2n = 20 (Wong et al., 2002).

In the place of Abaca's origin, around 400 varieties have been identified, of which 20 have commercial value, whereas three of them are the most important for the region of America where they are grown, these varieties are briefly describe below:

1. Tangongon: A large and strong variety with approximately 4–5 m high. It does not demand for fertile soils and grows adequately in heavy clay soils; it is easily overturned because young shoots push through the soil surface; its pseudostems produce 2.5%–3% of a strong fibre, which is often difficult to extract (Wong et al., 2002).
2. Bongulanon: An early variety with medium-sized stems and narrow leaves. It has an early maturation; however, it shows a short productive life. Its yield decreases after 5 years; it requires wet, well-drained alluvial soils and does not tolerate heavy clay soils or dry sandy

Handbook of Natural Fibres. https://doi.org/10.1016/B978-0-12-818398-4.00008-6

Figure 6.1 Abaca plantation.

Figure 6.2 Abaca stem.

soils; its pseudostems yield around 2.3% of strong, white and good-quality fibre that is extracted easily (Wong et al., 2002).

3. Maguindanao: It possesses large stems. It is relatively resistant to tumbling; its first harvest takes place between 15 and 18 months after sowing. It has a long productive life of more than 15 years; it adapts to a variety of soils except for heavy clay ones; its root system is shallow and its pseudostems yields around 1.75% of strong, white and soft fibre, which is extracted easily (Wong et al., 2002).

Abaca's cultivation has spread from the Philippines, its centre of origin, to other countries with similar agroclimatic characteristics, such as Ecuador and Costa Rica, where it is grown commercially. According to FAO, there are also smaller-scale crops in Indonesia, Equatorial Guinea and Kenya. Abaca fibre is a very important raw material because of its high content of cellulose, which gives it a high resistance, for this reason, it is used in the cordage or spinning industry and mainly in paper, therefore known as the queen of natural fibres. The Philippines gives forth 90% of Abaca's world production and the crop employs 1.5 million workers (Anon, 2013).

The Abaca crop has a perennial production cycle. At the beginning it takes approximately 2 years to produce the first fibre, then it can be harvested every two or 3 months, depending on the climatic conditions since it is optimally developed in humid tropical climate with temperatures of 28–30°C and a good relative humidity. It requires 2000–2500 mm of water throughout the year. Humidity and sunlight are very important for the production of quality fibre. Under appropriate conditions and good agronomic management the crop can last up to 20 years (Wood et al., 2005).

The dry and clean fibre production is around 1.5%–to 2% of the weight of the stem. The optimum time for the obtaining of fibre is when the inflorescence starts to appear, since it influences in the quality of the fibre and the productivity of the crop.

6.1.2 Fibre extraction

The fibre extraction process is carried out in seven steps, described as follows:

(a) Stripping the leaves off – It consists of defoliating and blunting the stems that are ready to extract the fibre.

(b) Cutting the stems off – The selected stems are cut to 10 cm from the ground, with a bevel cut inclined towards the outside with the purpose of avoiding stem rot (see Fig. 6.3).

(c) Shelling – From the cut stem, the pods that surround it are separated. With a knife, it is cut into strips of 5–8 cm wide by 2–4 cm thick; the length depends on the size of the stem. The outer shells produce fibre of lower quality and are cinnamon-coloured or brown and the interiors have a better quality, whose colour is white (Fig. 6.4).

(d) Shredding – This process must be carried out before 8–12 h after cutting the stem. The opposite will result in a faded fibre of lower quality. This operation is carried out by means of a machine that consists of two parts: the diesel engine that generates movement to a roller and the shredder, with roller and blades, and a system of attachment and opening of the blades, operation that is carried out with the other half of the pod or tuxe, that is wrapped in the roller (Fig. 6.5).

Figure 6.3 Cut stems with leaves stripped off.

Figure 6.4 Division of pods.

Figure 6.5 Shredding.

(e) Drying – The fibre obtained has a high percentage of humidity; for this reason, it is necessary to dry it in the same farm. The drying time of the fibre can last for hours or days, depending on atmospheric conditions. Parallel to this activity, a preliminary classification is performed, according to the colour of the fibre. After drying, the fibre is placed in a dry and ventilated tank to guarantee its quality (see Fig. 6.6).

(f) Classifying – Once it has reached 8% moisture, the fibre is stored for classification. This classification is based on the colour and diameter of the fibre, a criterion that allows obtaining five quality grades. The white fibre is considered grade 1, whereas when until it reaches a dark brown colour, it is grade 5. The thinner fibre is considered of better quality.

(g) Packing – Once classified by colour and diameter, packages of 125 kg are organized, covering 0.93 m wide by 0.52 m, and ready to be presented to the international market (Fig. 6.7).

Figure 6.6 Abaca's fibre drying.

Figure 6.7 Fibre grade 5.

6.2 Characterization of fibre, physical and chemical properties

6.2.1 Structure and physical properties of Abaca fibre

The structure of Abaca fibre is formed by innumerable tubular structures called lumen; the characteristic diameter of the tubes are ~10 μm. These tubular structures are formed by secondary walls that are composed of nano-sized strands (fibrils) that have a typical cross-section diameter of ~10 nm. At the nanoscale, the fibrils are actually formed by many cellulose molecule chains. It is clear that mechanical properties of these fibres are closely related to the hydrogen bonding among the cellulose molecules along the direction of the fibrils with contribution of Van der Waals and hydrogen bond interactions that might exist in between the fibrils.

Abaca's fibre is obtained from the leaves of the plant and are among the strongest natural fibres because they exhibit exceptional mechanical strength, resistance to salt water damage and durability. Abaca has lustrous light beige colour and long fibre length – up to 3 m; Abaca fibres have higher tensile strength and lower elongation than synthetic ones like nylon. By contrast, glass fibres are around 10 times stronger than Abaca fibres (see Table 6.1). Moreover, Abaca fibre shows thermal stability up to 250°C; this makes Abaca suitable for fabrication of thermoplastic matrix composites (De Souza et al., 2018).

The mechanical properties of some natural fibres mainly applied in the plastic and paper industry, including jute, sisal, flax, Abaca, kenaf and softwood are shown in Table 6.1.

Table 6.1 Mechanical properties of some natural fibres.

Type of fibre	Density (g/cm^3)	Tensile strength (MPa)	Young's modulus (GPa)	Elongation %
Abaca	1.5	400–677	34.4–41	3.4–11.1
Flax	1.4–1.5	345–1500	27.6–80	2.7–3.2
Hemp	1.48	550–900	70	1.6
Softwood	1.5	1000	40	–
Cotton	1.5–1.6	287–597	5.1–12.6	7–8
Jute	1.3–1.5	393–800	10–30	1.16–1.5
Sisal	1.3–1.5	400–700	9–38	3–7
E-Glass	2.5	2000–3500	70	2.5
S-glass	2.5	4570	86	2.8
Nylon 6.6	1.14	950	4	18

Mohanty et al. (2001); Müssig (2010).

6.2.2 Chemical properties of Abaca

Most natural fibres have similar chemical composition, as observed in Table 6.2. They are composed of cellulose, hemicellulose, lignin, wax and water-soluble materials.

Cellulose constitutes the basic material of all natural lignocellulosic fibres. It is a linear chain consisting of several glucose units; hydroxyl groups that are present on a glucose molecule forms strong hydrogen bonds with the adjacent chains and is

Table 6.2 Chemical composition of some natural fibres.

Type of fibre	Cellulose (wt%)	Hemicellulose (wt%)	Lignin (wt%)	Lipids (wt%)	Wax (wt%)	Ash content (wt%)
Abaca	56–63	20–25	7–13	0.4	3.0	0.5
Flax	64–72	16	2–2.2	–	1.7	–
Hemp	70–74	18–22	3.7–6	–	0.8	0.8
Softwood	40–44	25–29	25–32	–	–	–
Kenaf	45–57	8–13	18–21.5	–	0.8	2–5
Cotton	83–95	3	–	–	0.6	–
Jute	61–71	18–22	12–13	–	0.5	0.5–2.0
Sisal	66–78	10–13	8–10	–	2.0	1

Sapuan et al. (2018); Heinze et al. (2018); Arthanarieswaran et al. (2014).

responsible for the high tensile strength of cellulose. Cellulose is crystalline, strong and resistant to hydrolysis (Heinze et al., 2018).

Hemicellulose is composed of several different sugar units; it has a random, amorphous structure with little strength. It can be easily hydrolysed by diluted acid or base to release products high in xylose.

Lignin is a type of complex biopolymers rich in both aliphatic and aromatic subunits. It exhibits high molecular weight – above 10,000 u; it is amorphous and hydrophobic in nature.

Waxes consist of different types of alcohols; they are lipophilic compounds, insoluble in water as well as in several acids (Kalia et al., 2011).

Abaca has been characterized by relative high lignin content, higher than other natural fibres used for textile and papermaking such as hemp or flax, but lower than wood. It has a content of lipids similar to other flax or kenaf. The low ash content of the fibre (0.5%) is justified by the low content of different metals (Del Río and Gutiérrez, 2006).

It has been found that the hemicellulose fraction is mainly constituted by xylose and that the most predominant lipids in Abaca fibres are fatty acids (24% of total extract) and free sterols (24% of total extract) (Del Río and Gutiérrez, 2006).

6.3 Cellulose and paper production

6.3.1 Pulp and paper

Abaca is one of the few plants, nonwood, which are industrially processed and they can be purchased as white pulp. This fibre is especially composed of cellulose, about 63%; hence, the greater use is in the manufacture of paper for various uses (Zambrano, 2017; PhilFIDA, 2017).

6.3.1.1 Processes for obtaining pulp

1. **Fibre cut** – The purpose is to reduce the size depending on the application.
2. **Cooking (delignification)** – The pulp is obtained by digestion with chemicals and the process can be divided into acid and alkali ones. Within said processes, the chemicals normally used are Kraft, polysulfide, sodium hydroxide, acid sulfite and bisulfite. The common goal is to sufficiently dissolve lignin to fully reduce the fibre. Aqueous systems of low pressure and temperature are used; the Kraft process contains sodium sulphide that accelerates significantly the rate of formation of the pulp. The Kraft cooking process period varies between 2 and 3 h and, in a solution of sodium hydroxide, between 4 and 6 h. The speed of reaction in the Kraft process is faster due to the presence of sodium sulphide. In both processes, the maximum cooking temperature is 180°C. Yields with paste in the sodium hydroxide operation usually reach 45%, whereas yields with Kraft activity reach the 45% or 50%, or even overcome it. These values depend on the conditions in which the manufacture has taken place and on the pulps' intended purpose. A process of hydrolysis is recommended for producing pulp before cooking. The reaction produced in this chemical operation is the hydrolysis of the ether phenolic bonds, making lignin fractions to be transformed into soluble compounds (Zambrano, 2017; Herrera, 2018; Villalba, 2014).

The delignification occurs in three stages: initial, main and residual. The initial stage is observed at low temperatures where 20%–25% of the lignin can be dissolved. The main stage is appreciated at temperatures above 140°C, when delignification speed is increased significantly by temperature and concentrations of hydroxide and sodium suphfide. The residual is seen when the concentration of lignin in the fibre is reduced. The chemical processes applied are the cause of the changes that occur in the physical and chemical properties of the resulting pulp; however, there are other important factors that affect these properties, such as the dimension of the fibre, lignin content, hemicelluloses and cellulose. The most representative properties of chemical pastes are high strength, easy bleaching, easily refined and medium–high performance (Herrera, 2018; Villalba, 2014).

3. **Washing** – Takes place with plenty of water to verify that there are no residues of the chemicals.
4. **Bleaching** – Chemical bleaching treatments eliminate or modify substances that give colour to the paste; usually those are remains of lignin and extracts. The bleaching process consist of adding oxidizing agents to the pastes that generate groups of acids in the residual lignin and others that remove water-insoluble degradation products. The main processes involve the application of the following chemical agents:
   - Chlorination (acid medium): It causes the oxidation and chlorination of lignin.
   - Chlorine dioxide (acid medium): It oxidizes, whitens and solubilizes lignin. Combined with chlorine, it protects the cellulose from degradation.
   - Hypochlorite (alkaline medium): It oxidizes, whitens and solubilizes lignin.
   - Oxygen (high pressure, alkaline medium): It oxidizes and solubilizes lignin.
   - Hydrogen peroxide (alkaline medium): It oxidizes and whitens the lignin in chemical and high-performance pastes.
   - Ozone (acid medium): It oxidizes, whitens and solubilizes lignin, among others (Zambrano, 2017; Herrera, 2018).
5. **Blending** – This procedure is carried out after the bleaching and consists of introducing the cooked substance into a container with two-thirds of water, to operate the blender and blend for 30 s. It is necessary to check if the fibres are reduced to a uniform smoothness with a proper consistency.
6. **Development of paper sheets** – The pulp (see Fig. 6.8) will be ready for the formation of paper sheet with the aid of a mechanical method through a mesh strainer type, where the relationship of the mass is directly proportional to the weight of the finished product. Once in the moulds, the pulp must be pressed. To start with the drying process, we must wait it to be drained (Zambrano, 2017; Gellerstedt and Monties, 1989).

6.3.1.2 Abaca paper applications

In the paper industry, Abaca is mixed with other fibres to increase resistance or to manufacture low weight papers; mainly it is used in the manufacture of special papers, such as bags of tea and coffee, paper for wrapping meat and sausages, paper money (as in the case of the yen), papers for cigarettes, wallpaper, for medication, for electrical plugs, for electrolytic capacitors, bags for vacuum cleaners and high-quality paper for

Figure 6.8 Abaca pulp.

writing, etc. Abaca allows the elaboration of heavy-duty paper, of translucent and silky touch. If an intense refining is performed, translucent and parchment-like paper can be obtained (García, 2007; Capellades, 2018).

6.4 Textile and nontextile products

Abaca fibre is also used in the textile industry especially in cordage or spinning because of its physical properties; however, the manufacture of cloths is not suitable because Abaca is a rough fibre. The use of softening techniques such as mercerization allows achieving good results for weaving processes; in addition, it can be mixed with other fibres such as cotton to make clothing (see Fig. 6.9).

Abaca fibre is long and resistant to traction. Textile use of this fibre is restricted because it is rough, hard and full of impurities (Chang and Montero, 2000); however, it is gaining importance as a good reinforcement for composite materials (Vijaya Ramnath et al., 2013). The processes of bleaching and elimination of oil, despite their

Figure 6.9 Mercerized Abaca fibre.

difference in their application and purpose, should be treated as a whole because the final characteristics of the fibre depend on the overall process (Chang and Montero, 2000). It is reported that Malays use Abaca fibre to make ropes, fishing nets, woven cloths, mats and bags (Mamun et al., 2015).

For the production of nontextile fibre, pulp is used to make manila envelopes, thermal and acoustic insulation is a fibre that absorbs water. It is found in products such as separators of electric storage batteries, cleaning cloths, geo textiles used in drainage networks for the collection of ashes, detection of coal and methane gas collection systems and other gases and nonwoven filter (Chang and Montero, 2000). It is currently used for pulping, cordage and yarns, and fabrics. Several surface modifications can minimize the drawbacks of these fibres and increase their applications in the automobile and packaging industries (Mamun et al., 2015). The interaction between Abaca fibres and gypsum matrix was studied to produce fibre-reinforced plasterboards with enhanced toughness performances (Iucolano et al., 2015) and 'soft'' applications in the automotive industry, as filler for cushions and seats. It is important to emphasize that, owing to the high resistance to stress, it can also be used for 'tougher' applications for external components as a substitute for fibreglass reinforced with plastic components (FAO, 2014).

The cars company, Mercedes Benz, has used a mixture of thermoplastic polypropylene (PP) and yarn of Abaca in various parts of the structure of automobiles to replace natural fibres. Glass fibres can reduce the weight of the parts of the car while facilitating a more environment-friendly production with the recycling of parts (FAO, 2014). Owing to the extremely high mechanical strength of the fibre, as well as to its length, the implementation of Abaca offers a great potential for different industrial applications (Chang and Montero, 2000); although the most valuable use of nonwoven products lies in the elaboration of disposable diapers, feminine hygiene, medical supplies, gas absorbers, clothes for surgery and disposable aseptic products to name some of the applications (Chang and Montero, 2000).

6.5 Crafts

The artisans of the countries that produce Abaca fibre have developed their creativity and ingenuity in different artistic expressions through handmade products together with their experience in improving their elaboration technique. Figs. 6.10–6.13 display samples of Ecuadorian handcrafts.

6.6 Prospective of potential uses of Abaca fibre

The Abaca fibre has several attractive features, such as high tensile properties; it is resistant to rotting, abrasion and UV rays; it is low cost, lightweight and it is environmentally safe. For these reasons, it constitutes an alternative as reinforcement for polymeric matrices, which could even replace glass fibres (Begum and Islam, 2013).

Figure 6.10 Hand-knitted bag.

Figure 6.11 Indigenous woman. Made with corn leaves and Abaca.

6.6.1 Abaca fibre–reinforced polymer composites

A number of researchers have proposed Abaca fibre in composite applications, not only to develop new sustainable resources, but also to open a potential field as engineering materials. The aim of adding Abaca fibre is the improvement of the mechanical properties of the polymers. The major concerns regarding the composite mechanical behaviour include impact strength, tensile strength and stiffness. Nevertheless, the effect of this natural reinforcement in the final composite depends, among others, on the Abaca fibre (i) dimension, (ii) amount, (iii) dispersion grade, (iv) orientation and (v) the interface between the Abaca fibres and the polymer matrix.

Figure 6.12 Indigenous woman. Made with different qualities of Abaca fibre.

Figure 6.13 Knitted rug with Crochet technique.

The dimension of Abaca fibre is variable, but usually the diameter is in the range of 100–300 microns, and its length (in composites) does not exceed the 5 mm. Moreover, the loading concentration is in general below 40% w/w. The distribution and orientation of the Abaca fibres will be determined by the processing conditions of the polymer matrix.

Regarding the interface, one of the difficulties related to the use of Abaca is the strong polarization associated to the lignocellulosic fibre. This condition impedes a good adhesiveness with the polymeric matrix which is hydrophobic. In this context, some surface treatment strategies such as using coupling agents, esterifications, alkali treatment and cyanoethylation have been proposed to locate compatibilizer groups on the Abaca surface and to develop polymer matrix affinity. On the other hand, the interactions in the interface are also influenced by the matrix nature. Therefore, the interface performance will depend of each particular arrangement. The Abaca fibre has been used as reinforcement in thermoset matrices as well as thermoplastics.

Several studies have been reported in the literature in which Abaca fibres were used to improve the mechanical properties of PP (Bledzki et al., 2010; Rahman et al., 2009; Vilaseca et al., 2010). One study has compared jute, flax and Abaca fibres as PP reinforcements, and the results exhibited better damping properties (notch Charpy and falling weight impact) with Abaca fibres than jute and flax, using the a same fibre loading of 30% w/w (Bledzki et al., 2007).

Abaca fibre was also incorporated in thermosetting matrices as epoxy (EP). The results obtained for this composite revealed a greater flexural and impact strength than the neat matrix and a jute/Abaca hybrid (Saba et al., 2016). On the other hand, in a urea formaldehyde matrix, the presence of Abaca fibres resulted in an improvement of tensile strength, flexural strength and Charpy impact strength (Suvarna et al., 2015).

6.6.2 Biodegradable composites

The worldwide environmental awareness has motivated the production of real green materials. For this reason, in the field of composite manufacturing, biodegradable polymers and reinforcements started to have utmost demand. For this purpose, aliphatic polyesters such as PCL (poly(e-caprolactone)), PHBV (poly(3-hydroxybutyrate-co-3-hydroxyvalerate)), PBS (poly(butylene succinate)), PEC (polyestercarbonate), and PLA (poly(lactic acid)) have been studied. In fact, in PHBV and PBS matrices, the presence of the Abaca fibre accelerates the weight loss of the composites, which is favourable when degradability is evaluated (Teramoto et al., 2004).

Regarding the mechanical properties, some authors reported that the addition of Abaca fibre increases the flexural moduli of PBS, PEC and PLA composites, in 193%, 90% and 67%, respectively, using 20% w/w of esterified Abaca (Shibata et al., 2003). Likewise, the tensile properties in PBS and PHBV matrices were improved. Actually, the results in PHBV were comparable with those obtained with glass fibre (Shibata et al., 2002). Another property, which is generally improved in comparison with the neat biodegradable matrix, is the stiffness (Bledzki et al., 2009).

6.6.3 Nanocomposites

Abaca fibres are also used for nanocomposite manufacturing, the latter being a multiphase solid material with at least one phase in the nanometric scale (<100 nm). Owing to its high cellulose composition of 56%–63% (Baptista et al., 2013), the Abaca fibres have been proposed as a source to acquire nanofibrillated cellulose (Ashraf et al., 2013) or nanocrystalline cellulose (Lapuz et al., 2017). These nanomaterials have potential applications in health composites, pharmaceutical, chemical and food additives. For its extraction from Abaca fibre pulp, procedures such as acid hydrolysis have been used. However, this technique has a yield of just around 15%.

6.6.4 Structural, automotive and industrial applications

In recent years, the usage of renewable and low-cost supplies has become a trend in many fields. For this reason, the Abaca fibre has a potential application in aeronautics, ship building, high-rise buildings construction and world-class furniture production (Shahri et al., 2014).

The Abaca fibres used as reinforcement to develop green composites has also been considered as an alternative to substitute the conventional fossil products (Faruk et al., 2014). For instance, in the automotive industry, big enterprises such as Mercedes Benz started to use Abaca fibre for manufacturing spare tire well covers and underfloor covering parts (along with PP) for their vehicles. The latter technique was patented by Daimler Chrysler's researchers, but Rieter Automotive industrialized the process (Bledzki et al., 2015; Koronis et al., 2013). These initiatives were stimulated by the 2015 European regulation that defined that 95% of a vehicle weight had to be recyclable.

6.7 International market

According to FAO (Food and Agriculture Organization of The United Nations, 2018), in 2016, the production and commercialization of Abaca fibre worldwide is led by the Philippines with more than 60% market share, followed by Ecuador with 35% and very distant from them, Costa Rica and Indonesia with 1%. The same source indicates that there is a small production by Equatorial Guinea and Kenya, as observed in Table 6.3 and Fig. 6.14.

The international fibre market is concentrated in a few countries. According to the FAO, the main buyer of Abaca is Japan, followed by the Philippines and Morocco.

Table 6.3 Production of Abaca in metric tonnes.

	Year						
Country	**2010**	**2011**	**2012**	**2013**	**2014**	**2015**	**2016**
Philippines	66,512	68,612	68,510	64,952	68,053	68,856	67,403
Ecuador	33,040	34,001	34,243	34,962	35,682	36,401	37,121
Costa Rica	1246	1247	1200	1225	1249	1246	1246
Indonesia	492	514	520	600	627	570	569
Equatorial Guinea	278	230	244	250	232	222	212
Kenya	43	44	45	50	47	48	47
Total	101,611	104,648	104,762	102,039	105,890	107,343	106,598

Source: Food and Agriculture Organization of the United Nations, 2018. FAOSTAT.

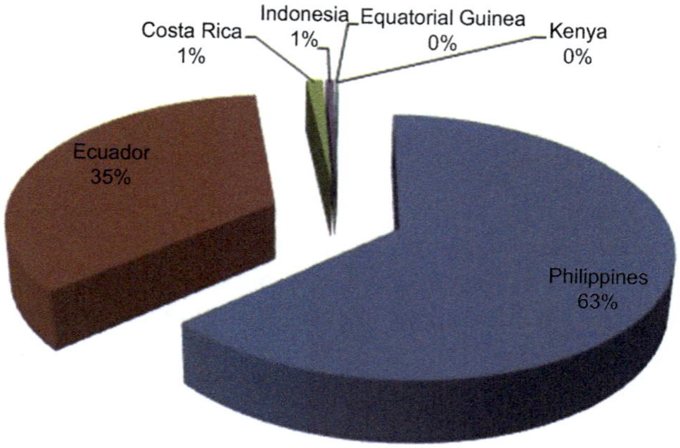

Figure 6.14 Abaca production in % by countries. Year 2016.

China and Mauritius have a significant trend, as seen in the following chart: The international fibre market is concentrated in a few countries. According to the FAO, the main buyer of Abaca is Japan, followed by the Philippines and Morocco. China and Mauritius have a significant trend, as can be seen in Table 6.4.

Table 6.4 Import of Abaca's fibre in metric tonnes.

Country	Year						
	2010	2011	2012	2013	2014	2015	2016
Argentina	0	0	0	0	0	0	3
Botswana	0	0	0	0	1	0	0
China	0	0	0	172	206	214	90
Congo	13	13	13	0	94	0	0
Cote d'Ivoire	212	0	0	0	291	445	0
Egypt	45	0	0	0	0	0	0
Philippines	311	631	103	725	931	165	2299
Gabon	0	0	0	0	0	8	0
Ghana	0	0	0	0	19	71	0
Japan	10,573	10,150	5928	6969	8714	10,341	9071
Liberia	0	0	0	0	124	0	0
Morocco	0	0	0	0	0	4972	5564
Mauritius	0	0	110	187	284	89	59
Nigeria	7	7	7	7	0	0	0

Source: Food and Agriculture Organization of The United Nations, 2018. FAOSTAT.

According to the foreign trade portal, NOSIS, the main exporters of Abaca fibre are the Philippines and Ecuador. Clients or buyers of Philippines are Spain, United Kingdom, Japan, China and the United States, countries that sold more than 44 million dollars in 2015 (FOB). In the case of Ecuador, the main buyer is the Philippines, followed by the United Kingdom, Japan, Spain and China, where more than 21 million dollars (FOB) have been negotiated at an average price of \$ 1000 per tonne. The largest buyer is the United Kingdom, since the world's largest pulp and paper industry is located there. See Table 6.5 and Fig. 6.15.

6.8 Conclusions

Abaca fibre is a raw material of multiple uses. It comes from the humid tropical ecosystems of Philippines, the main producer country. It manages to place in the international market around 67,000 tonnes a year of Abaca, which represents more than 60%. The second largest producer is Ecuador with an average of 35,000 tonnes per year, with a market share of 35%. The price of a tonne of fibre depends on the quality of the fibre. First quality fibre, the whitest one, can exceed 1000 dollars FOB per tonne, depending on market conditions. The most significant use of fibre is for the production of pulp to obtain paper, due to its high cellulose content (65% on average). Therefore, fibre demand is concentrated in the United Kingdom because of its paper industry. The plant is an excellent ally to fight deforestation of species used for this purpose.

Abaca fibre has been legendarily used for the elaboration of ropes and strings because it is resistant to salt water as it is to rupture, thus becoming useful for maritime transport. It is used in the textile field for clothing, curtains sewing and upholstery too.

Lately, nanomaterials, biodegradable and composite materials have been used by the automotive, aeronautical, shipping, construction, furniture and handicraft industries. The latter shows a growing market and represents a significant craftsmen's income increase in fibre producing countries. Owing to its multiple uses, it is known as the queen of natural fibres.

6.9 Future trends

Between the research, development and innovation trends of the use of Abaca, we highlight the improvement of good-quality fibre producing varieties; its resistance to pests and diseases attack; an increase of demand, as it is a sustainable raw material and therefore an increase of the sown surface; innovation of fibre softening techniques, mixed with other techniques for textile uses; development of nanomaterials for applications in compounds for health, pharmaceutical and food products; the development of composite materials to improve the mechanical properties of polymers and the development of biodegradable composites; new structural, automotive and industrial applications.

Table 6.5 Main importing and exporting countries of Abaca's fibre in FOB values.

	Importer							
Exporter	**China**	**U.S**	**Spain**	**U.K**	**Japan**	**Philippines**	**Portugal**	**Canada**
Philippines	6.926.875	158.367	13.923.930	13.900.155	9.602.908		12.469	1.917
Ecuador	147.437	1.283	1.049.321	6.821.147	3.811.051	9.924.077		
Total	7.074.312	159.650	14.973.251	20.721.302	13.413.959	9.924.077	12.469	1.917

Source: NOSIS. Available at: https://trade.nosis.com/es/Comex/Importacion-Exportacion/Mundial/coco-y- other-fibras-vegetales–coco-Abaca-canamo-de-manila-musa-textilis-nee- ramie-and-demas-fibra-te/WD/5305.

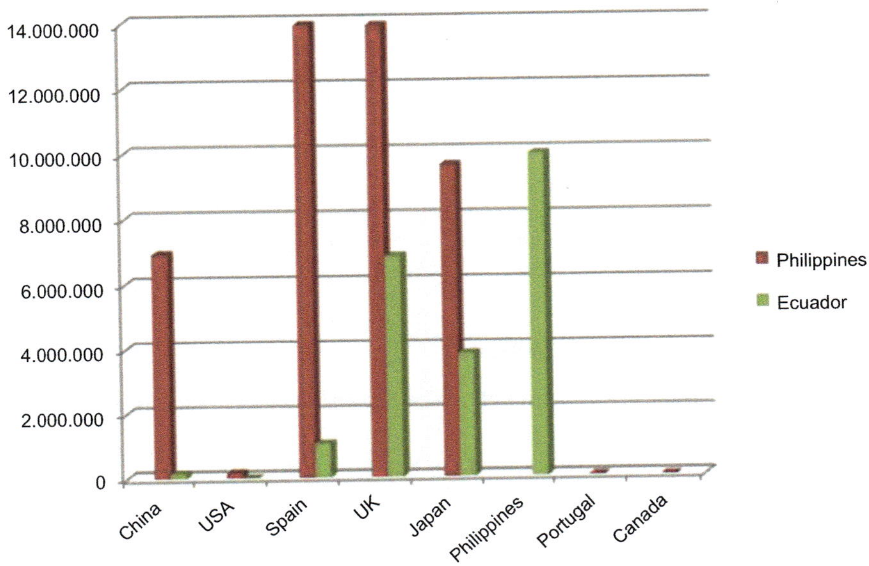

Figure 6.15 Main countries that buy Abaca's fibre. Millions of dollars (FOB). Year 2015.

The use of renewable and low-cost raw materials is becoming a trend in many industrial applications; thus Abaca fibre has a potential use in aeronautics, shipping, construction of buildings, production of furniture and crafts of international quality.

6.10 Source of further information and advice

For more information, consult the following websites:

6.10.1 *Agriculture*

Food and Agriculture Organization of the United Nations (FAO). http://www.fao.org.
Philippine Department of Agriculture. http://www.philfida.da.gov.ph/index.php.
Ministry of Agriculture and Livestock of Ecuador. https://www.agricultura.gob.ec/.
Ministry of Agriculture and Livestock of Costa Rica. http://www.mag.go.cr/.
Research Centers. University of Philippines. Los Baños. http://uplb.edu.ph/research/research-centers/.

6.10.2 *International trade*

FAO Food and Agriculture Dat. http://www.fao.org/faostat/es/#data.
World Trade Organization. https://www.wto.org/index.htm.
Market Research and Development. https://trade.nosis.com/en/comex.

6.10.3 Pulp and paper

About Glatfelter – Beyond Paper. http://www.glatfelter.com/products/default.aspxhttp://www.glatfelter.com/products/default.aspx.

6.10.4 Industrial textile

Philippine Textile Research Institute. http://ptri.dost.gov.ph/.

6.10.5 Industrial development

New materials laboratory. Escuela Politécnica Nacional. Ecuador. https://www.epn.edu.ec/laboratorio-de-nuevos-materiales/.

6.10.6 Handicrafts

Centro Interamericano de Artesanías y Artes Populares. CIDAP. http://www.cidap.gob.ec.http://www.cidap.gob.ec/.

References

Anon, 2013. (Research) Plant Molecular Biology and Plant Virology. National Institute of Molecular Biology and Biotechnology. Archived from the original on 4 June 2014. Retrieved 4 June 2014.

Arthanarieswaran, V.P., Kumaravel, A., Kathirselvam, M., 2014. Evaluation of mechanical properties of banana and sisal fiber reinforced epoxy composites: influence of glass fiber hybridization. Mater. Des. 64, 194–202.

Ashraf, C., Pere, M., Vilaseca Fabiola, B.S., 2013. Reinforcing potential of nanofibrillated cellulose from Nonwoody plants. Polym. Compos. 116 (1), 37–43. https://doi.org/10.1002/pc.

Baptista, A., Ferreira, I., Borges, J., 2013. Cellulose-based composite systems for biomedical applications. In: Biomass Based Biocomposites. UK Smithers Rapra Technology, pp. 47–60. Retrieved from:http://info.smithersrapra.com/downloads/chapters/Biomass-based_Biocomposites.pdf.

Begum, K., Islam, M.A., 2013. Natural fiber as a substitute to synthetic fiber in polymer composites: a review. Res. J. Eng. Sci. 2 (3), 46–53. https://doi.org/10.13140/RG.2.1.2685.6163.

Bledzki, A.K., Mamun, A.A., Faruk, O., 2007. Abaca fibre reinforced PP composites and comparison with jute and flax fibre PP composites. Express Polym. Lett. 1 (11), 755–762. https://doi.org/10.3144/expresspolymlett.2007.104.

Bledzki, A.K., Jaszkiewicz, A., Scherzer, D., 2009. Mechanical properties of PLA composites with man-made cellulose and Abaca fibres. Compos. Appl. Sci. Manuf. 40 (4), 404–412. https://doi.org/10.1016/j.compositesa.2009.01.002.

Bledzki, A.K., Mamun, A.A., Jaszkiewicz, A., Erdmann, K., 2010. Polypropylene composites with enzyme modified Abaca fibre. Compos. Sci. Technol. 70 (5), 854–860. https://doi.org/10.1016/j.compscitech.2010.02.003.

Bledzki, A.K., Franciszczak, P., Osman, Z., Elbadawi, M., 2015. Polypropylene biocomposites reinforced with softwood, Abaca, jute, and kenaf fibers. Ind. Crops Prod. 70, 91–99. https://doi.org/10.1016/j.indcrop.2015.03.013.

Capellades, 2018. Recuperado de. http://www.papercapellades.com/fibras/Abaca-musa-textilise Octubre 2018.

Chang, A., Montero, G., 2000. Analisis Del Sector Exportador De Abaca En El Ecuador. Retrieved from: http://repositorio.ucsg.edu.ec/bitstream/3317/3900/1/T-UCSG-PRE-ECO-CECO-66.pdf.

De Souza, N.C.R., d'Almeida, Tensile, J.R.M., 2018. Thermal, morphological and structural characteristics of Abaca (musa textiles) fibers. Polym. Renew. Resour. 5, 47–60.

Del Río, J.C., Gutiérrez, A., 2006. Chemical composition of Abaca (*Musa textilis*) leaf fibers used for manufacturing of high quality paper pulps. J. Agric. Food Chem. 54, 4600–4610.

FAO, 2014. Organización de las Naciones Unidas para la Alimentación y la Agricultura. Retrieved September 11, 2018, from: http://www.fao.org/economic/futurefibers/fibers/Abaca0/es/.

Faruk, O., Bledzki, A.K., Fink, H.P., Sain, M., 2014. Progress report on natural fiber reinforced composites. Macromol. Mater. Eng. 299 (1), 9–26. https://doi.org/10.1002/mame.201300008.

Food and Agriculture Organization of The United Nations, 2018. FAOSTAT. Retrieved September 15, 2018, from: http://www.fao.org/faostat/es/#data.

García, J.A., 2007. Fibras Papeleras. UPC, Barcelona.

Gellerstedt, L., Monties, 1989. Wood and Cellulosic Chemistry, Revised, and Expanded. Obtenido de: https://books.google.com.ec/Septiembre 2018.

Heinze, T., Seoud El, O.A., Koschella, A., 2018. Cellulose Derivatives. Cellulose Derivatives: Synthesis, Structure, and Properties. Springer International Publishing.

Herrera, M, 2018. Obtención de Nanocelulosa a partir de celulosa de puntas de Abacá. EPN. Quito- Ecuador.

Iucolano, F., Caputo, D., Leboffe, F., Liguori, B., 2015. Mechanical behavior of plaster reinforced with Abaca fibers. Constr. Build. Mater. 99, 184–191. https://doi.org/10.1016/J.CONBUILDMAT.2015.09.020.

Kalia, S., Kaith, B.S., Kaur, I. (Eds.), 2011. Cellulose Fibers: Bio- and Nano-Polymer Composites: Green Chemistry and Technology. Springer-Verlag Berlin Heidelberg.

Koronis, G., Silva, A., Fontul, M., 2013. Green composites: a review of adequate materials for automotive applications. Compos. B Eng. 44 (1), 120–127. https://doi.org/10.1016/j.compositesb.2012.07.004.

Lapuz, A., Torres, A., Lapuz, R., Mari, E., Basilia, B., 2017. Nanocrystalline cellulose from Abaca fibers. J. Mater. Sci. Eng. 06 (04), 4172. https://doi.org/10.4172/2169-0022-C1-067.

Mamun, A.A., Heim, H.P., Faruk, O., Bledzki, A.K., 2015. The use of banana and Abaca fibers as reinforcements in composites. Biofiber Reinf. Compos. Mater. 236–272. https://doi.org/10.1533/9781782421276.2.

Mohanty, A.K., Misra, M., Drzal, L.T., 2001. Surface modifications of natural fibers and performance of the resulting biocomposites: an overview. Compos. Interfac. 8, 313–343.

Müssig, J. (Ed.), 2010. Industrial Applications of Natural Fibers. John Wiley & Sons.

Palacios, M., Peña, S. (S/A). El Abacá. Available in: http://www.mag.go.cr/rev-histo/ra-09-04-119.pdf. [Consulted: 15/10/18].

PhilFIDA, 2017. Abaca Sustainability Manual. Recuperado de. Septiembre 2018. http://www.philfida.da.gov.ph/index.php/2017-09-22-05-23-59.

Rahman, M.R., Huque, M.M., Islam, M.N., Hasan, M., 2009. Mechanical properties of polypropylene composites reinforced with chemically treated Abaca. Compos. Appl. Sci. Manuf. 40 (4), 511–517.

Saba, N., Jawaid, M., Alothman, O.Y., Paridah, M.T., Hassan, A., 2016. Recent advances in epoxy resin, natural fiber-reinforced epoxy composites and their applications. J. Reinf. Plast. Compos. 35 (6), 447–470. https://doi.org/10.1177/0731684415618459.

Sapuan, S.M., Ishak, M.R., Sahari, J., Sanyang, M.L. (Eds.), 2018. Kenaf Fibers and Composites. CRC Press.

Shahri, W., Tahir, I., Ahad, B., 2014. Abaca fiber: a renewable bio-resource for industrial uses and other applications. In: Biomass and Bioenergy. Springer International Publishing, Cham, pp. 47–61. https://doi.org/10.1007/978-3-319-07641-6_3.

Shibata, M., Takachiyo, K.I., Ozawa, K., Yosomiya, R., Takeishi, H., 2002. Biodegradable polyester composites reinforced with short Abaca fiber. J. Appl. Polym. Sci. 85 (1), 129–138. https://doi.org/10.1002/app.10665.

Shibata, M., Ozawa, K., Teramoto, N., Yosomiya, R., Takeishi, H., 2003. Biocomposites made from short Abaca fiber and biodegradable polyesters. Macromol. Mater. Eng. 288 (1), 35–43. https://doi.org/10.1002/mame.200290031.

Suvarna, A.S., Katagi, A., Pasanna, J., Kumar, S., Badyankal, P.V., Vasudeva, S.K., 2015. Mechanical properties of Abaca fiber reinforced urea formaldehyde composites. Mater. Sci. Res. India 12 (1), 54–59.

Teramoto, N., Urata, K., Ozawa, K., Shibata, M., 2004. Biodegradation of aliphatic polyester composites reinforced by Abaca fiber. Polym. Degrad. Stab. 86 (3), 401–409. https://doi.org/10.1016/j.polymdegradstab.2004.04.026.

Vijaya Ramnath, B., Junaid Kokan, S., Niranjan Raja, R., Sathyanarayanan, R., Elanchezhian, C., Rajendra Prasad, A., Manickavasagam, V.M., 2013. Evaluation of mechanical properties of Abaca–jute–glass fiber reinforced epoxy composite. Mater. Des. 51, 357–366.

Wong, C., Kiew, R., Argent, G., Set, O., Lee, S.K., Gan, Y.Y., 2002. Assessment of the Validity of the Sections in Musa (Musaceae) using AFLP. Annals of Botany 90 (2), 231–238. https://doi.org/10.1093/aob/mcf170.

Wood, F., Roberts, G.A., 2005. In: Prance, G., Nesbitt, M. (Eds.), The Cultural History of Plants. Routledge, New York, NY.

Vilaseca, F., Valadez-Gonzalez, A., Herrera-Franco, P.J., Pèlach, M.À., López, J.P., Mutjé, P., 2010. Biocomposites from Abaca strands and polypropylene. Part I: evaluation of the tensile properties. Bioresour. Technol. 101 (1), 387–395. https://doi.org/10.1016/j.biortech.2009.07.066.

Villalba, V., 2014. Elaboración de pastas papeleras para uso artístico a partir de hoja caduca de árboles frutales. España.

Zambrano, M., 2017. Diseño de un digestor para la obtención de pulpa celulósica para papel a partir de fibra abacá. EPL. Guayaquil Ecuador 5–25.

Bamboo fibre

7

C. Prakash
Department of Fashion Technology, Sona College of Technology, Salem, Tamil Nadu, India

7.1 Introduction

The fastest growing woody plant on this planet is bamboo. It grows one-third times faster than the fastest growing tree. Some species can grow up to 1 m per day. Bamboo is just grass, but it varies in height from dwarf, one foot (30 cm) plants to giant timber bamboos that can grow to over 100 feet (30 m). It grows in many different climates, from jungles to high on mountainsides. Bamboos are further classified by the types of roots they have. Some, called runners, spread exuberantly, and others are classified as clampers (sympodial), which slowly expand from the original planting. There are also varieties of root systems that are a mixture of these types. Generally, the tropical bamboos tend to be clumpers and the temperate bamboos tend to be runners. There are many different types of bamboo. Countries as India, China and Brazil are large producers. Annual world bamboo fibre production is over 40,000 tons and is increasing. However, almost all of the bamboo fiber production takes place in Asia, throughout the Asia-Pacific region. It used to take place in Europe but it has moved overseas due to rising labour costs and stricter environmental rules. The biggest viscose fiber producing country nowadays is China. Jilin Chemical Fiber is the largest bamboo fiber production enterprise in China, with its capacity of bamboo staple fiber and bamboo filament yarn reaching 78 kt/a and 25 kt/a respectively in 2018.

Bamboo fibre and starchy pulp are made from bamboo that grows widely through Asian countries. Starchy pulp is a refined product of bamboo stems and leaves through a process of hydrolysis-alkalization and multiphase bleaching. Chemical fibre factories then process it into bamboo fibre. Bamboo is both decorative and useful. In many parts of the world, it is food, fodder, the primary construction material and is used for making great variety of useful objects from kitchen tools, to paper to dinnerware. Generally bamboos are commonly used for furniture, construction, musical instruments and many more things.

Bamboo is not only highly fashionable for decorative purposes but useful too. As it is a viable replacement for wood, in Far Eastern countries, it is the primary building material. Bamboo is in fact one of the strongest building materials available and even provided the first regreening in Hiroshima after the atomic blast in 1945. Bamboo's tensile strength is 28,000 per square inch versus 23,000 for steel.

Bamboo fibre is a natural, ecofriendly fibre. It is one kind of grass with hollow and woody stem. There are so many varieties of bamboo plant across the world. Botanically categorized as a grass and not a tree, bamboo just might be the world's most sustainable resource. It is the fastest growing grass and can shoot up a yard or more

Handbook of Natural Fibres. https://doi.org/10.1016/B978-0-12-818398-4.00009-8

in a day. Bamboo reaches maturity quickly and is ready for harvesting in about 4 years. Bamboo does not require replanting after harvesting because its vast root network continually sprouts new shoots which almost zoom up, pulling in sunlight and greenhouse gases and converting them to new green growth. Bamboo grows in a natural way without the need for petroleum-guzzling tractors and poisonous pesticides and fertilizers. Bamboos are a group of woody perennial evergreen plants in the true grass family Poaceae, subfamily Bambusoideae, tribe Bambuseae. Some of its members are giant, forming by far the largest members of the grass family. Although bamboo is a grass, many of the larger bamboos are very tree-like in appearance and they are often called "bamboo trees". The reason bamboos are so different from trees is they lack a vascular cambium layer and meristem cells at the top of the culm. The vascular cambium is the perpetually growing layer of a tree's trunk beneath the bark that makes it increase in diameter each year. The meristems make the tree grow taller (Sharma and Goel, 2011).

Few decades ago, bamboo was traditionally used for making variety of household goods such as furniture, sporting goods, handbags, flooring and cutting board etc. Nowadays, owing to developments in manufacturing processes, it is possible to produce fibre from bamboo stem, which has remarkable properties for its use in yarns and fabrics. Bamboo viscose is a new regenerated cellulosic fibre that has recently evinced keen interest primarily because the bamboo plant is a fast-growing renewable resource and therefore, the fibre can arguably be considered a 'sustainable' fibre (Sarkar and Appidi, 2009). Bamboo pulp is refined though a process of hydrolysis-alkalization and multiphase bleaching before converting in to the fibre. Physical and chemical properties of bamboo fibre are nearly close to viscose. It has good durability, stability, moderate tenacity and thus good spinnability. Bamboo products are further characterized by their good hydrophilic nature, excellent permeability, soft feel, excellent dyeing behaviour and colour vibrancy and overall its antimicrobial property.

It possesses unique properties such as anti-UV radiation, antibacterial, breathable, cool soft handle, etc. Owing to various microgapes and microholes in cross-section, it has better moisture absorption. Bamboo fabric absorbs and evaporates sweat very easily, and hence gives comfortable feel. It can be spun into 100% bamboo yarn and also be blended with natural and manmade fibres such as cotton, polyester, silk etc. Fabric made of bamboo fibre possesses better moisture transmission property, softer, handle, drapability and easy drying. It has many applications such as apparels, sweaters, bath suits, mats, towels, T-shirts and socks etc. Bamboo fibre has antibacterial function which makes it suitable for underwear, T-shirts and socks etc. It has also wide scope in the field of hygienic materials such as sanitary napkin, masks, mattress, bandages, surgical cloths, surgical gown, absorbent pads and food packing, etc. Owing to antiultraviolet radiation characteristics of bamboo fibre, it is suitable for making summer clothing for the protection of human skin against damages of UV radiation (UVR). Wallpapers, curtains and sofa covers are produced from bamboo fibre because it absorbs UVR of different wavelength from atmosphere, hence protecting human skin from UVR (Ajay and Avinash, 2012).

Exposure of human skin to ultraviolet radiation of different wavelengths causes some serious skin problems such as ageing of skin, roughening, wrinkles, blotches, sagging and also sometimes leads to skin cancer. The intensity of UVR depends on sun's altitude, geographical position, atmospheric conditions etc (Saravanan, 2007). With reference to above said skin problems, bamboo fabric can protect human skin strongly from damages of UVR because it naturally possesses anti-UV radiation characteristics.

Bamboo fibre constitutes a recent kind of natural material that has a huge application possibility in the textile field due to some of its unique properties (Liu and Hu, 2008). For example, it has a unique structure that makes them superior to other natural lingocellulosic (Ray et al., 2004). In recent years, bamboo fibres have attracted great attention as the most abundant renewable biomass materials that can be used in textiles (Xu et al., 2006) and composite reinforcement (Das et al., 2006). They possess many excellent properties when used as textile materials such as high tenacity, excellent thermal conductivity, resistant to bacteria, high water and perspiration adsorption. Natural bamboo fibres have excellent properties and therefore they have potential for use in textiles; however, they have not received the attention they deserve owing to their coarse and stiff quality. Regenerated bamboo fibre is 100% cellulose, biodegradable and is claimed to be 'green' and environmentally friendly (Erdumlu and Ozipek, 2008). Owing to microgaps in its structure, bamboo fibre has high air permeability and water absorption properties. Bamboo-based fabrics are antibacterial and very soft, with a low amount of pilling and creasing (Karahan et al., 2006). Fabrics made of bamboo fibre have very good physical properties. When compared with cotton fabrics, bamboo fabrics require a lower amount of dye for the same depth of shade. Moreover, the colourant is absorbed better and faster than in cotton fabrics and dyed bamboo fabric appears better than dyed cotton fabric (Wallace, 2005).

7.2 Morphology of bamboo fibres

The chemical constituents of bamboo are primarily cellulose, hemicellulose and lignin, with cellulose making up about 60% and lignin roughly 32% (Okubo et al., 2004; Ni et al., 2018). As a natural nanocomposite, bamboo possesses multinodes and functional gradient structures, macroscopically as well as microscopically. The fibres are spread compactly in the outer surface region and sparingly in the inner surface region (Rao and Rao, 2007). Its structure is intended to have uniform strength at all positions in both the radial direction on the transverse section and the lengthwise direction (Nogata and Takahashi, 1995). Parameswaran and Liese (1976) indicated that the thick-walled bamboo fibres exhibit a polylamellate structure with alternating broad and narrow lamellae. The concentration of lignin is higher in the narrow lamellae than in the broad ones. Xylan seems to occur in a higher concentration in the narrow lamellae (Parameswaran and Liese, 1976). Ikeda et al. (2007) make it in extracting pure fibres from bamboo culms and developed the method to differentiate bamboo fibres from other plant fibres using microscopy and infrared spectroscopy.

7.3 Chemical composition and natural durability

The choice of bamboo species for various applications is not only related to physical and mechanical properties but also to the chemical composition (Table 7.1) (Sun et al., 2015; Sowmya et al., 2017; Higuchi, 1957). Tomalang et al. (1980) bring into being that the main constituents of bamboo culms are holocellulose (60%–70%), pentosans (20%–25%), hemicelluloses and lignin (each amounted to about 20%–30%) and minor constituents such as resins, tannins, waxes and inorganic salts. The proximate chemical compositions of bamboo are similar to those of hardwoods, except for the higher alkaline extract, ash and silica contents. The carbohydrate content of bamboo plays a vital role in its durability and service life. Stability of bamboo against mould, fungal and borers attack is strongly associated with the chemical composition. In producing material such as cement-bonded particleboard, chemical content (starch and sugar) will hinder the absorption rate of H_2O^+ ion on the cement mineral surfaces and will slow down the setting reaction. The study by Chew et al. (1992) found out that *Bambusa vulgaris* contains glucose 2.37%, fructose 2.07% and sucrose 0.5%. The total sugar before and after soaking was 4.94% and 0.28%, respectively. This study showed that by the technique of soaking the sugar content could be reduced below 0.5%, a permitted level for the production of cement-bonded particleboard.

7.4 Manufacturing process of regenerated bamboo fibre

The manufacturing process for regenerated bamboo fibre using hydrolysis alkalisation with the multiphase bleaching principle is given below (Erdumlu and Ozipek, 2008; Gericke & Van Der Pol, 2010; Taraboanta, 2012; Chand et al., 2008):

Preparation: Bamboo leaves and the soft, inner pith from a hard bamboo trunk are extracted and crushed.

Steeping: The crushed bamboo cellulose is soaked in a solution of 15%–20% sodium hydroxide at a temperature of between 20°C and 25°C for one to 3 hours to form alkali cellulose.

Table 7.1 Chemical analysis of bamboo.

Species	Lignin (%)	Cellulose (%)	Pentosan (%)	Ash (%)	Ethanol toluene extractives (%)
Phyllostachys heterocycla	26.1	49.1	27.1	1.3	4.6
Phyllostachys nigra	23.8	42.3	24.1	2.0	3.4
Phyllostachys reticulata	25.3	25.3	26.5	1.9	3.4

Pressing: The bamboo alkali cellulose is pressed to remove excess sodium hydroxide solution.
Shredding: The alkali cellulose is shredded by a grinder to increase the surface area and make the cellulose easier to process.
Ageing: The shredded alkali cellulose is left to dry for 24 h to be in contact with the oxygen of the ambient air. During this process, the alkali cellulose is partially oxidized and degraded to a lower molecular weight because of high alkalinity. This degradation is controlled to produce chain lengths short enough to produce correct viscosities in the spinning solution.
Sulphurisation: In this stage, carbon disulphide is added to the bamboo alkali cellulose to sulphurise the compound, causing it to jell.
Xanthation: The remaining carbon disulphide from the sulphurisation is removed by evaporation because of decompression and cellulose sodium xanthogenate is the result.
Dissolving: A diluted solution of sodium hydroxide is added to the cellulose sodium xanthogenate, dissolving it to create a viscose solution consisting of about 5% sodium hydroxide and 7%–15% bamboo fibre cellulose.
Spinning: After subsequent ripening, filtering and degassing, the viscose bamboo cellulose is forced through spinneret nozzles into a large container of diluted sulphuric acid solution which hardens the viscose bamboo cellulose sodium xanthate and reconverts it into cellulose bamboo fibre threads which are spun into bamboo fibre yarns.

The above-mentioned manufacturing process for regenerated bamboo fibre resembles that of viscose rayon fibre; consequently, regenerated bamboo fibre is also called bamboo viscose.

7.5 Advantages of bamboo fibre

7.5.1 Smooth, soft and luxurious feel

It has a basic round surface which makes it very smooth and to sit perfectly next to the skin. Bamboo apparel is softer than the softest cotton, and it has a natural sheen such as silk or cashmere. Bamboo drapes such as silk or satin yet is less expensive and more durable. Bamboo/organic cotton blends are also extremely soft but heavier in weight.

7.5.2 Allergy reduced

Bamboo's organic and naturally smooth fibre properties are nonirritating to the skin, making it ideal for people with skin sensitivities or other allergies and dermatitis.

7.5.3 Good absorption ability

Bamboo fibre absorbs and evaporates sweat very quickly. Its ultimate breathability keeps the wearer comfortable and dry for a very longer period.

7.5.4 Temperature adaptability

Fabrics made from bamboo fibre are highly breathable in hot weather and also keep the wearer warmer in cold season. Bamboo is naturally cool to the touch. The cross-section of the bamboo fibre is filled with various microgaps and microholes leading to much better moisture absorption and ventilation. It is also very warm in cold weather because of the same microstructure as the warm air gets trapped next to the skin.

7.5.5 Natural antibacterial properties

Bamboo is naturally antibacterial, antifungal and antistatic. Bamboo has a unique antibacterial and bacteriostatic bioagent named 'bamboo kun', which bonds tightly with bamboo cellulose molecules during the normal process of bamboo fibre growth. This feature gets retained in bamboo fabrics too. It makes bamboo fabrics healthier, germ-free and odour-free.

7.5.6 Thermal regulating

Bamboo fabrics are warm in the winter and cool in the summer. Bamboo clothing's excellent wicking properties also make it ideal for warm summer days.

7.5.7 UV protection

Bamboo naturally provides added protection against the sun's harmful UV rays.

7.5.8 Antistatic

Owing to its high moisture absorption property, bamboo fabric results in the enhancement of antistatic property.

7.5.9 Green and biodegradable

As a regenerated cellulose fibre, bamboo fibre was 100% made from bamboo through high-tech process. They are all three- to four-year old new bamboo, of good character and ideal temper. The whole distilling and producing process in our plant is green process without any pollution. It produces natural and ecofriendly fibre without any chemical additive. As a natural cellulose fibre, it can be 100% biodegraded in soil by microorganism and sunshine. The decomposition process does not cause any pollution environment. Bamboo fibre is praised as "the natural, green, and ecofriendly new type textile material of 21st century".

7.5.10 Breathable and cool

Bamboo fibre gives human skin a chance to breathe free. What is notable of bamboo fibre is its unusual breathability and coolness. Because the cross-section of the bamboo fibre is filled with various microgaps and microholes, it has much better moisture absorption and ventilation (Karthikeyan et al., 2017). With this unparallelled micro-structure, bamboo fibre apparel can absorb and evaporate humans sweat in a split second. Just like breathing, such garments make people feel extremely cool and comfortable in the hot summer. It is never sticking to skin even in hot summer. According to authoritative testing figures, apparels made from bamboo fibres are 1–2° lower than normal apparels in hot summer. Apparel made from bamboo fibre is crowned as air conditioning dress.

7.6 Bamboo fibre fabric production flow

Bamboo fibre fabric production flow: bamboo - thick pulp - fine pulp - bamboo fibre - bamboo yarn - fabric.

Bamboo fibre is a kind of regenerated cellulose fibre, which is produced from raw materials of bamboo pulp. First, bamboo pulp is refined from bamboo through a process of hydrolysis-alkalization and multiphase bleaching. Then process bamboo pulp into bamboo fibre. Repeated test has proved that it has a strong durability, stability and tenacity. And the thinness and whiteness degree of bamboo fibre is similar to classic viscose.

Furthermore, it owns high abrasion-proof capacity. Bamboo fibre spins nicely. This fibre is a natural cellulose fibre, can achieve natural degradation in the soil and it will not cause any pollution to the environment. Bamboo can be spun purely or blended with other materials such as cotton, hemp, silk, Lyocell (Tencel), modal cotton chemical fibre and so no.

After hi-tech disposal, bamboo fibre is thinner than hair. It has a round and smooth surface. Thus it has no stimulation against human skin.

7.7 Dyeing and finishing of bamboo textile

Light sergeing, enzyme desizing, and moderate bleaching and semimercerizing should be applied to the bamboo fabric during its dyeing and finishing process, which also avoid drastic condition and use small mechanical tension.

Sergeing: Carried out in moderate condition.

Desizing: Should be consolidate, desizing rate should be over 80%.

Scouring: Pure bamboo normally need no scouring, sometimes wash it with a little alkaline soap. The scouring process should be made in terms of contents blended with cotton. When pure bamboo fabrics are under scouring, the alkali should not be over 10g/liter but be applied in accordance with the thickness of fabrics.

Bleaching: The processing should be made in terms of the specification and thickness of fabrics.

Mercerizing: The fabrics are normally free of mercerizing due to their sound lustre and bad antialkaline. However, some cases are found to increase their absorbance capacity to dyestuff.

Dyeing: It is better to use active dyestuff during dying process, and the alkali should not be over 20g/liter; the temperature should not be over 100°C. During drying process, low temperature and light tension are applied.

7.8 The end-use of bamboo fibre

Bamboo fabrics are made by pure bamboo fibre yarns, which have excellent wet permeability, moisture vapour transmission property, soft hand, better drapery, easy dying and splendid colours. It is a newly founded, great prospective green fabric.

The following are some of the end use of bamboo fibre: bathrobes and towels, foot mats, mats, bed clothes, underwear, bamboo T-shirts and bamboo socks

Bamboo intimate apparels include sweaters, bath suits, mats, blankets; towels have comfortable hand, special lustre and bright colours and good water absorbance. Bamboo fibre has such a sole function as antibacterial function, which is suitable to make underwear, tight T-shirt and socks. Its antiultraviolet nature is suitable to make summer clothing, especially for the protection of pregnant ladies and children from the hurt of ultraviolet radiation.

Bamboo nonwoven fabric is made by pure bamboo pulp, which has same property as viscose fibres. However, bamboo has wide prospects in the field of hygiene materials such as sanitary napkin, masks, mattress, and food-packing bags because of its antibacterial nature.

Bamboo sanitary materials include bandage, mask, surgical clothes, nurses wears and so on. The bamboo fibre has natural effects of sterilization and bacteriostasis; therefore, it has incomparably wide foreground on application in sanitary material such as sanitary tower, gauze mask, absorbent pads, food packing and so on. In the medical scope, it can be processed into the products of bamboo fibre gauze, operating coat, and nurse dress, etc. Because of the natural antibiosis function of the bamboo fibre, the finished products need not to be added with any artificial synthesized antimicrobial agent, so it will not cause the skin allergy phenomena, and at the same time, it also has competitive prices in the market.

Bamboo decorating series has the functions of antibiosis, bacteriostasis and ultraviolet proof. It is very advantageous for utilization in the decorating industry. Along with the badly deterioration of atmosphere pollution and the destruction to the ozonosphere, the ultraviolet radiation arrives the ground more and more. Long-time exposure to ultraviolet irradiation will cause skin cancer. But the wallpaper and curtains made from bamboo fibre can absorb ultraviolet radiation in various wavelength, thus to lessen the harm to human body farthest. What is more, bamboo decorating product will not go mouldy because of the damp. Curtain, television cover, wallpaper and sofa slipcover can all be made by bamboo fibre.

Bamboo bathroom series enjoy good moisture, soft feel and splendid colours as well as antibacterial property, which are well popular in home textiles. Bamboo towel and bath robe have soft and comfortable hand feeling and excellent moisture absorption function. Its nature antibiosis function keeps bacterium away so that it will not produce bad odour.

7.9 Fashion trend and bamboo clothing

Bamboo has attracted increasing attention from the designers around the world because of the growing environmental awareness among consumers. With the bamboo fibre having a number of advantages, one can expect the industry to develop new, value-added products from bamboo in future. In this context, there is an increasing demand for functional fibre products. With new spinning technologies achieving significant progress in recent years, it is now possible to make bamboo fibre tough enough to withstand abrasion and be able to satisfy relatively rigorous end user requirements. Bamboo fibre can instantly absorb and evaporate large amounts of water, which gives it excellent moisture properties. This can lead to some interesting textile end uses in future. Similarly, its natural antibacterial properties can also lend themselves to some interesting end-use applications/textile products in the near future. Bamboo clothing is naturally more wrinkle resistant than cotton. Bamboo clothing can be machine-washed in cool water with mild soap on the gentle cycle, although softeners or bleaches are good to be avoided for bamboo clothing. It is best if garments made of bamboo are hung to dry. Surprisingly, it dries faster than most other fabrics (Earth Easy, 2009). Bamboo socks, bed sheets, jackets are very common items among the web based retailers. Fashion retailer, Target and Sussan are already selling bamboo textile products in Australian market. Bamboo clothing may create a new dimension in the fashion trend of sunny region of the world like Australia for its multifunctionality.

7.10 Conclusion

The merits of bamboo for use in textile their high functionality and their beauty. The fashion world's fascination with all things and bamboo is attracting the interest of a number of designers. More and more of this material is expected to find way onto store shelves as time goes by. The available information suggests that bamboo fibres have unique properties such as excellent appearance and feel, natural antibacterial, UV-shielding and moisture-controlling characteristics. However, those unique properties may largely depend on the manufacturing processes, the details of which are usually close guarded by the manufacturing companies. However, bamboo based textiles are not yet achieved their full potential and cleaner production processes are appearing. At present, there are only a small number of manufacturing plants in China that manufacture natural bamboo fiber. Ecologically pioneer textile manufacturing companies like Litrax and Lenzing have already introduced greener manufacturing processes into bamboo textiles. Researches are going on full swings for the development of eco-friendly, natural bamboo fabric. With abundant sources of raw materials, relatively low cost, and unique performance of bamboo fiber it is only a matter of time to develop green and pure bamboo textiles. Further, bamboo textile industry has the potential to provide livelihood for millions of people worldwide.

References

Ajay, R., Avinash, K., 2012. Physical and UV protection properties of knitted bamboo fabrics. Text. Rev. 7 (10), 24–26.

Chand, N., Shukla, M., Sharma, M.K., 2008. Analysis of mechanical behaviour of bamboo (dendrocalamus strictus) by using FEM. J. Nat. Fibers 5 (2), 127–137.

Chew, L.T., Sudin, R., Kasim, J., 1992. Bambusa vulgaris for urea and cement-bonded particleboard manufacture. J. Trop. For. Sci. 3, 249–256.

Das, M., Pal, A., Chakraborty, D., 2006. Effects of mercerization of bamboo strips on mechanical properties of unidirectional bamboo–novolac composites. J. Appl. Polym. Sci. 100 (1), 238–244.

Erdumlu, N., Ozipek, B., 2008. Investigation of regenerated bamboo fibre and yarn characteristics. Fibres Text. East. Eur. 16 (4), 43–47.

Gericke, A., Van der Pol, J., 2010. A comparative study of regenerated bamboo, cotton and viscose rayon fabrics. Part 1: Selected comfort properties. J. Fam. Ecol. Consum. 38 (1), 63–73.

Higuchi, H., 1957. Biochemical studies of lignin formation, III. Physiol. Plant. 10, 633–648.

Ikeda, Y., Koshiba, T., Yoshida, Y., Miyamoto, K., Tominaga, M., 2007. Fiber Identification and Consumption Characteristics of Bamboo [Phyllostachys Heterocycla] Fiber, vol. 1. Bulletin of Tokyo Metropolitan Industrial Technology Research Institute, pp. 10–13.

Karahan, A., Öktem, T., Seventekin, N., 2006. Natural bamboo fibres. Tekstil Ve Konfeksiyon 16 (4), 236–240.

Karthikeyan, G., Nalakilli, G., Shanmugasundaram, O.L., Prakash, C., 2017. Moisture management properties of bamboo viscose/tencel single Jersey knitted fabrics. J. Nat. Fibers 14 (1), 143–152.

Liu, Y., Hu, H., 2008. X-ray diffraction study of bamboo fibers treated with NaOH. Fibers Polym. 9 (6), 735–739.

Ni, H., Li, Y., Fu, S., 2018. Morphological structure and properties of bamboo shell fiber. J. Nat. Fibers 15 (4), 586–595.

Nogata, F., Takahashi, H., 1995. Intelligent functionally graded material: bamboo. Compos. Eng. 5 (7), 743–751.

Okubo, K., Fujii, T., Yamamoto, Y., 2004. Development of bamboo-based polymer composites and their mechanical properties. Compos. Appl. Sci. Manuf. 35 (3), 377–383.

Parameswaran, N., Liese, W., 1976. On the fine structure of bamboo fibres. Wood Sci. Technol. 10 (4), 231–246.

Rao, K.M.M., Rao, K.M., 2007. Extraction and tensile properties of natural fibers: vakka, date and bamboo. Compos. Struct. 77 (3), 288–295.

Ray, A., Das, S., Mondal, S., Ramachandrarao, P., 2004. Microstructural characterization of bamboo. J. Mater. Sci. 39 (3), 1055–1060.

Saravanan, D., 2007. UV protection textile materials. Autex Res. J. 7 (1), 53–62.

Sarkar, A.K., Appidi, S., 2009. Single bath process for imparting antimicrobial activity and ultraviolet protective property to bamboo viscose fabric. Cellulose 16 (5), 923–928.

Sharma, V., Goel, A., 2011. Bamboo fiber versus cotton fiber: a comparative study. Man Made Text. India 24 (9), 313–318.

Sowmya, R., Vasugi Raaja, N., Prakash, C., 2017. Investigation of relationship between blend ratio and yarn twist on yarn properties of bamboo, cotton, polyester, and its blends. J. Nat. Fibers 14 (2), 228–238.

Sun, B., Huang, A., Wang, Y., Liu, J., 2015. Natural bamboo (Neosinocalamus affinis Keng) fiber identification using FT-IR and 2D-IR correlation spectroscopy. J. Nat. Fibers 12 (1), 1–11.

Taraboanta, I., 2012. Study on bamboo fiber process of creating properties-application. Ann. Univ. Oradea Fascicle of Textiles 13 (1), 242–245.

Tomalang, F.N., Lopez, A.R., Semara, J.A., Casin, R.F., Espiloy, Z.B., 1980. May. Properties and utilization of Philippine erect bamboo. In: International Seminar on Bamboo Research in Asia Held in Singapore, pp. 28–30.

Wallace, R., 2005. Commercial Availability of Apparel Inputs: Effect of Providing Preferential Treatment to Apparel of Woven Bamboo Cotton Fabric. US Patent 332465007.

Xu, X., Wang, Y., Zhang, X., Jing, G., Yu, D., Wang, S., 2006. Effects on surface properties of natural bamboo fibers treated with atmospheric pressure argon plasma. Surf. Interface Anal. 38 (8), 1211–1217.

Further reading

Alay, S., 2010. An investigation of knitted fabric performances obtained from different natural and regenerated fibres. Mühendislik Bilimleri ve Tasarım Dergisi 1 (2), 91–95.

Mohmod, A.L., Khoo, K.C., Ali, N.A.M., 1992. Carbohydrates in some natural stand bamboos. J. Trop. For. Sci. 4 (4), 310–316.

Coconut fibre: its structure, properties and applications

8

Leena Mishra, Gautam Basu
ICAR-National Institute of Natural Fibre Engineering and Technology, Kolkata, West Bengal, India

8.1 Introduction to coconut fibre

In the world of industrial fibres, coconut fibre (*Cocos nucifera* L.) is extracted from the husk of coconut, although a secondary crop occupies a prominent position as regards of its bulk production. Annual yield of coconut is 50–100 fruits per tree, which require a year to ripen. Mature fruits are ovoid or ellipsoid in shape, 30–45 cm in length, and 15–20 cm in diameter; have a thick, fibrous husk surrounding the single-seeded nut (Encyclopaedia Britannica., 1982). The fruit is around 1–2 kg in weight, which contains about 30% husk. Coconut fibre, production areas, are found mostly along the coastal regions in the wet tropical areas of India, Sri Lanka, Philippines, Indonesia, Malaysia, Fiji Islands, Vietnam, Hawaii Islands, Papua New Guinea, and Solomon Islands. For more than a century, India (largest producer) and Sri Lanka (Ceylon) have grown practically about 90% of world's production (Table 8.1) (FAO, 2013).

FAO report (FAO, 2013; FAO, 2015) shows that, there is a steady growth of production of coconut fibre (at the rate of 37%) and it's yarn (at the rate of 10%) in India (Fig. 8.1) due to growing consciousness for using eco-sustainable, biodegradable, natural materials in the world. It is interesting to observe, the trend of growth of world production for coconut fibre is greatly at par with the growth trend of production of coconut fibres in India (Table 8.1). It can be observed in Table 8.1 that the production of brown fibre (extracted through mechanical means from matured brown husk) is sharply enhancing from 335 to 459 thousand tonnes, whereas there is a steady reduction in production of white fibre (from 95 to 81 thousand tonnes) extracted through conventional backwater retting practicing in India.

The traditional method of fibre extraction i.e. 'retting' (dipping coconut husk for 6–12 months at backwater lagoon) to obtain whiter, softer and spinnable retted fibres is abolishing rapidly in India due to various technological and societal concerns. The major problem of the conventional water retting is pollution to open backwater bodies and its surrounding air. The conventional extraction (retting) causes occupational diseases especially to the women folk involved in this profession and subsequent manual yarn manufacture is also a reason of massive human drudgery, due to which production of yarn in India has been persistently declining (Fig. 8.1). Reduction in production of the yarn in India affects the global export of the yarn and the finished products (mats, matting and rugs) (Fig. 8.2) based on it (FAO, 2013; FAO, 2015). The raw

Handbook of Natural Fibres. https://doi.org/10.1016/B978-0-12-818398-4.00010-4

Table 8.1 Production of coconut fibre (in 1000 tonnes).

	2007	2008	2009	2010	2011	2012	2013	2014
Brown fibre								
India[a]	335.0	340.8	394.0	415.5	425.0	431.5	455.8	458.8
Sri Lanka[a]	100.1	101.6	169.6	149.8	135.0	131.2	123.7	150.4
Thailand	53.7	55.5	64.6	63.7	49.9	45.7	64.8	68.1
Vietnam	72.2	74.7	91.1	103.0	97.6	115.1	105.9	143.3
Other countries	11.0	11.0	57.7	66.5	81.9	84.8	79.3	97.6
White fibre								
India	95.0	97.0	97.0	100.0	100.5	100.0	80.4	81.0
Total	667.0	680.6	923.6	952.7	949.0	988.1	973.6	1064.0

[a]India and Sri Lanka production data does not include coir pith.

fibre extracted through mechanical means is bound to export (increase from 97 to 505 thousand tonnes) to global market without any value addition (Fig. 8.2). Thus, the present challenges of the coconut fibre sector is to find a suitable replacement of conventional fibre extraction method (backwater retting) to produce good quality of coconut fibre and mechanized processing of the fibre to manufacture finer and good quality yarns/products without any human drudgery and environmental pollution.

8.1.1 Husk of coconut

Coconut husk represents the entire fibrous material enveloping the fruit constituting both the inner endocarp (liquid and solid food part) and outer mesocarp (fibrous

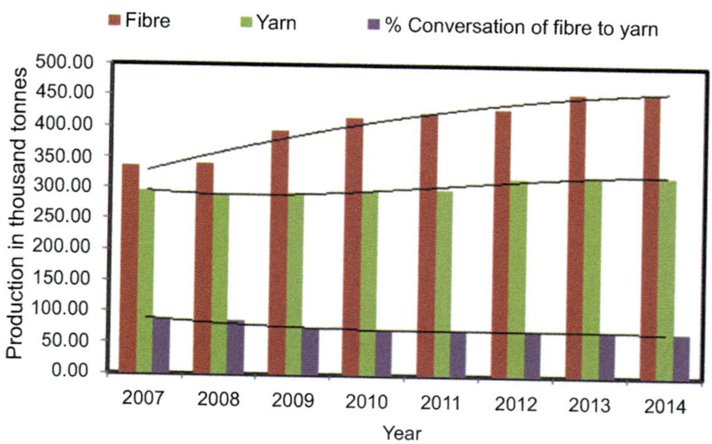

Figure 8.1 Production of coconut fibre and yarn in India.

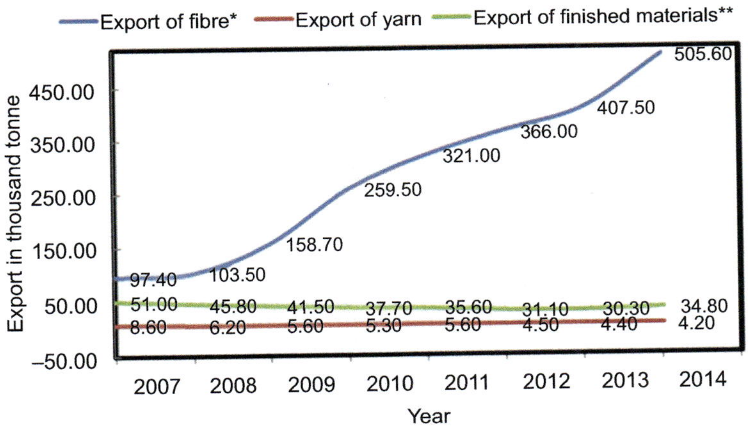

Figure 8.2 Export of coconut fibre yarn and finished material from India.

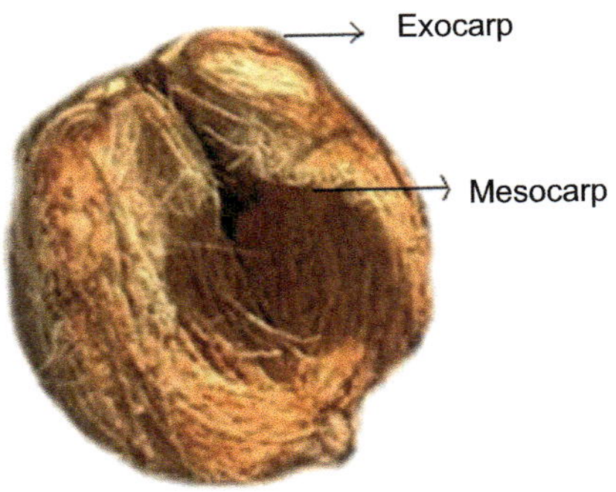

Figure 8.3 Coconut husk.

part) (Fig. 8.3). The mesocarp, which is an assemblage of fibres and elastic cellular cork like parenchymatous cells cementing the fibrous materials dispersed throughout the mass. Retting in water of this material causes separation of the leathery exocarp (thin outer slippery cover) from spongy fibrous mesocarp. The fibrous strands are composed of a highly lignified form of cellulose, hence are harsh and rigid. These non-fibrous parenchymatous cells of husk are referred as coconut pith and accounts for as much as 50%–70% of the total weight of husk. There are about 300 eco-types in coconut and there are variations in the quality and quantity of the fibre. Even in the same variety, the fibres vary in length and thickness. The husk is not collected qualitywise (eco-type) and hence the mixture of coconut fibre from different eco-types results in fibre of lower quality having high variation in length and diameter. Age of husk is an important factor for the selection of good quality fibre. Husks from near ripe to ripe nuts are reported to give best fibres (Meenatchisundarm, 1979).

8.2 Segregation and commercial grading of fibre

Fibre content of a husk distinctly shows three groups viz. coarse, medium and fine. Spinning of coconut fibre yarn is made from the mixture of all the three groups of fibres. Coconut fibre yarn is coarser and its products are also coarser and of low value. When we are aiming at production of value added diversified products of finer quality, segregation of fibres into three grades as coarse, medium and fine is needed. Thorough and good cleaning of fibres is also needed.

Depending upon fibre quality attributes viz. fineness and length; coconut fibres are graded into three varieties, viz. (1) finest and short length fibre, (2) moderately coarser and longer length fibre, and (3) hardest and very long fibre. The finest and shortest fibres are used as inner fillings of mattresses. The medium length and moderately softer fibres are the fibres, which have suitability in textile applications, whereas the hardest and longest length fibres are used in making of brushes and brooms. Again depending on colour, coconut fibres were segregated into white and brown. No other major fibre quality parameters (strength, moisture content, presence of non-fibrous material, rigidity), which are essential for commercial uses, are yet established. Also, no machine has so far been commercialized to segregate the fibres precisely according to the above grades.

8.2.1 Fibre extraction

The separation of tightly bound fibres i.e. extraction of fibres can be done by manual and/or mechanical means. Fibre is attached to the mesoderm with pectinous matter. So, for extraction of fibre, the husk is to be essentially softened. The conventional practice of harvesting coconut fruit is above 12 months to obtain high copra content (Encyclopaedia Britannica., 1982). The husk at this maturity stage is about to dried and thus the fibres were brown, coarse and rigid. Being a secondary crop, not much attention is given on the fibre quality.

8.2.1.1 Extraction of white fibres (manual) through conventional retting

Retting of coconut fibre is best suited using fresh husks separate from the nuts within a few days after the plucking. Husks exposed to the sun or longer storage become rigid and brown and thus do not yield good fibre. Conventional retting in saline backwater lagoons is comparatively cheap and manually manageable biological method based on fermentation. In the process of retting, generally green husks (10–11 month old nuts) are kept immersed in water (slow running or stagnant) for a very long period of about 6–10 months depending on the seasonal variation in temperature, humidity and other conditions. Each retting yards usually contains about 1000 husks either tied into a bundle or in a coconut fibre nets. These are allowed to remain in the water bodies until they get fully soaked. The soaked husks are either immersed down by piling their tops with mud and slimes or tied down to bamboos buried in the mud on the coast but within the reach of tidal waves. During high tides, the lagoons get filled with fresh saline water by exclusion of some muddy dark retting liquor, thus conventional back-water retting facilitate more whiter fibres and softer fibre than the artificially water retted coconut fibres.

Process of retting happens in two stages viz. physical and biological. In the first stage, the husk absorbs water and the plant tissues swell. The swelling helps in leaching of carbohydrates, glucosides, tannins, nitrogen compounds. During the second stage, a variety of micro-organisms develop at the expense of the extracted substances, thereby creating suitable conditions for the microbes decomposing the binding materials of the tissues. The microbes secrete various enzymes that act upon the carbohydrates and nitrogen compounds releasing different organic acids and gases, followed by a rise in temperature. The evolution of the gas results in the frothing of liquor. When the frothing subsides, a whitish film of fungi covers the surface of water (particularly seen in stagnant retting water bodies). At the stage, the anaerobic decomposers dissolve the pectin in the middle lamella of the parenchymatous tissues and separate the coconut fibres. After the retting, the soaked husks are subjected to manual beating with wooden mallets for removal of the outer epithelial layer (exocarp) and to loosen the retted white fibre (Fig. 8.4) and pith in the mesocarp. The thrashed husks are fed to a disintegrator first and then to a spiked opener for removal of pith from the fibres. The fibres possess good elasticity and colour. The fibres so extracted are of good quality and are suitable for spinning–weaving operations and are mainly used in export quality rope and woven mat (including tufted) manufacturing. Such retted fibres from green husks have also been found preferred for preparation of bleached and dyed coconut fibre products (Meenatchisundarm, 1979).

8.2.1.2 *Extraction of brown fibres (mechanical)*

Direct extraction of fibre from brown coconut 'husk' (12-month-old nuts) is mechanically done for production of 'Brown fibre' (Fig. 8.5). The husk either may be soaked in pits for a couple of days (2–3 weeks) or may not be soaked at all. Coconuts are dehusked with a double-ended sharp shovel. Then the husks are fed to a cylinder

Figure 8.4 Manual defibering of retted husk.

Figure 8.5 Opening of coconut fibre using spiked pin rotating roller.

with 6–8 bar lags that is driven by motor (30–50 HP) to disintegrate the fibre from the husk. The disintegrated fibres are soaked and washed with water and again fed to a defibering machine. This machine has a feeding passage and a cylinder with series of sharp nails projecting from its periphery. The nails help in further opening of the exposed disintegrated fibre coming from the feed passage and remove the pith and throw the cleaned fibres out. The fibres are sun dried for a day or two before further processing. All these operations are done sequentially involving excessive machining with unscientific utilization of space and power. Brown fibre is used mainly in making brushes, doormats, mattresses, sack of inferior quality, rubberized coir, automobile upholstery, and as geotextiles (nonwoven mat) for soil erosion control (Meenatchisundarm, 1979).

8.2.1.3 Yield of coconut fibre

The yield and quality of coconut fibre depends on the factors like age of husk, variety of husk, season local conditions and practices, and the location of retting yards. Husk weighs about 35% of the weight of a nut, containing 30%–50% of fibre. So apparently, the yield of fibre is 10%–17.5% of weight of a nut. Generally, in Sri Lanka 1000 husks yield about 140 kg of fibre as compared to 90 kg in India. Further a 1000 husks qualitywise produced 31.75 kg (13.9%) bristle fibre, 59.0 kg (25.9%) of mattress fibres with dust/pith of 137.0 kg (60.2%) (Meenatchisundarm, 1979).

8.2.2 Post-extraction processing

8.2.2.1 Yarn manufacturing

The households and co-operative sectors account for more than 75% of the coconut fibre yarn. For a good quality yarn there is a need for good quality fibre. In Southern India (Kerala), spinning is mostly done by using the simple traditional 'ratt' (spinning wheel), which is driven manually. To prepare a 2-ply coconut fibre yarn on the spinning wheel, a set of two wheels, one stationary and the other movable are needed. While one person rotates the stationery wheel (Fig. 8.6a), two people release the fibres in small quantity continuously, keeping the fibre bundle in their armpits. Simultaneously, they walk backward till the required length of strand is reached (Fig. 8.6b). The strands are then passed through a grooved wooden gadget (Fig. 8.7) and tied together into the notch of the spindle. While the grooved wooden gadget is being allowed to move forward, the movable wheel is turned in the opposite direction to insert the doubling twist. The people thus are to walk up and down in the spinning yard for drawing the individual strands and for the doubling operation. The yarn produced is uneven in cross-section and twist. This traditional wheel spinning requires at least three people per spinning head to produce about 12–15 kg of yarn per day (Senan, 2014).

As an improvement, the manually operated traditional ratt has been motorized. By rotating the spinning wheel with a motor, it has become possible to reduce one worker out of the three used earlier, thus increasing the labour productivity. However, there has not been much improvement in machine productivity and the quality of yarn problems exist in such methods of spinning. In a further development, an automatic version of spinning machine has been introduced with much higher productivity (50 kg/h) and which can be operated by a single operator. The coconut fibres in the form of either a loose stock or slivers are fed on the feeder of the machine before falling on to a 'W' shaped tray. The fibres are carried forward by a core high-density polyethylene/low-density polyethylene/nylon continuous filament thread before being twisted by a rotating grip. Two such strands are doubled to form 2-ply yarn and wound on to a bobbin. But the machine suffers from the following drawbacks:

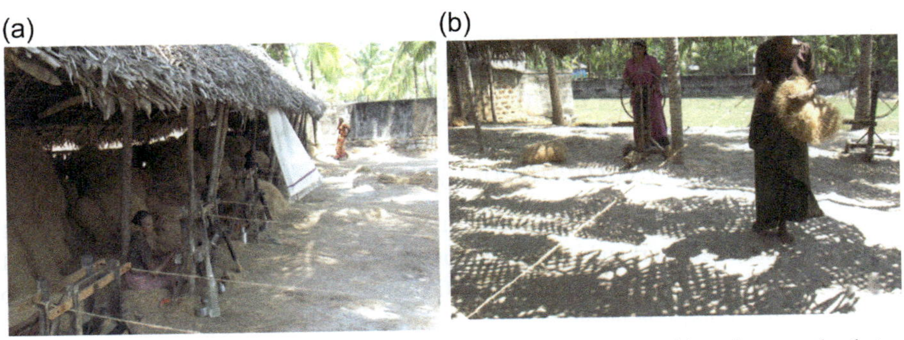

Figure 8.6 Traditional ratt/wheel spinning of coconut fibre showing (a) stationary wheels to form single strands (b) moving wheels to form yarns from twisting of two single strands.

Figure 8.7 Grooved wooden gadget used in traditional ratt/wheel spinning of coconut fibre.

(i) It is not 100% natural yarn, but a yarn with synthetic core covered with coconut fibres.
(ii) It lacks the soft feel which was prevalent in the yarn produced by the traditional ratt.
(iii) Yarn surface is more hairy and twist can be varied only in a narrow range.
(iv) Spinning limit is very low, i.e. produces much coarser yarn which is used for making rope.
(v) The machine is devoid of any safety mechanism including the covers for the rotating components.

8.2.2.2 Bleaching of coconut fibre yarn

Bleaching is a prior requirement to improve penetration of dyes resulting in the colouration with brighter shades. In coconut fibre sector, bleaching attempts at the yarn stage to impart ornamentation to coconut fibre prior to making value-added products, like mats, matting, rugs, carpets etc. Bleaching is mainly done using hydrogen peroxide and sodium silicate. Generally 100 kg of coconut fibre-based rope/yarn requires 12–20 times of water, respectively, using hydrogen peroxide. However, there are wide variations in the recipe of bleaching as well as the degree of whiteness (Anto et al., 1997; van Dam, 2002). Bleaching of coconut fibre yarn was also done using eco-friendly fungal *Trametes versicolor* and bacterial consortium *Coirret*. The bio-treatment could impart an improvement in softness of the yarn to an extent of 38% using bacterial treatment and 46% using fungal treatment. An increase of 10% in brightness index of the *Trametes versicolor* treated yarn was observed without any reduction of strength of fibre (Ravindranath and Sarma, 2010).

8.2.2.3 Dyeing of coconut fibre yarn

Dyeing of coconut fibre yarn is done to improve the aesthetic value of coir products leading to increased marketability. Traditionally, coconut fibre is dyed in large vessels heated by fuel wood. The required quantities of dyestuff (depending on the depth of shade required) and chemicals are added as a paste to the liquor in the vessel and dyeing is carried out at the required temperatures for the specific time period. For dyeing, 60 kg of coconut fibre yarn at fibre to liquor ratio of 1:12 is used, while smaller quantities of coconut fibre (30 kg) are dyed at 1:20 ratio. The dye bath is stirred manually and the material is turned frequently to obtain even distribution of the dye. After dyeing, the material is removed, washed with cold water and air dried in the shade. Some mechanical improvements include the use of special-purpose fixed stainless steel vats equipped with a valve for draining dye effluents, having heating facilities using electrical elements built inside the housing of the vat (van Dam, 2002). In the conventional process, dyeing is carried out in an indolium/copper/GI vat, which is heated from below using locally available wood. The dye bath is stirred manually and frequently for level dyeing. The material to liquor ratio is maintained at 1:12 and the batch size is 50–60 kg of yarn, which is boiled in the bath for about an hour. Thereafter, the yarn is taken out, washed in cold water and dried under shade (van Dam, 2002; Mohanan et al., 1997).

8.3 Properties of coconut fibre

Some discrete reports are available on investigation of physical, mechanical and other properties of coconut fibre (Bledzki et al., 1996; Stern and Stout, 1954; van Dam et al., 2006; Rahman and Khan, 2007; Coirboard, 2016; NIRJAFT, 1990; Wei and Gu, 2009; Satyanarayan et al., 2007; Brígida et al., 2010; Satyanarayana et al., 1982; Ray and Bandyopadhyay, 1966; Kaswell, 1953; Varma et al., 1984; Meredith, 1945; Banik and Ghosh, 2008), which have been compiled and given in Table 8.2. Similar properties were compared with jute (Satyanarayan et al., 2007; Kaswell, 1953; Mahato et al., 1993) (jute being the highest available lignocellulosic fibre). Coconut fibres' length is in range between 8 and 337 mm. The fibres amount with length range of 15–145 mm is 81.95%. Weight of fibres with the length range of 35–225 mm accounted for 88.34%. The average fineness of is 27.94 tex (Wei and Gu, 2009). Study of fibre diameter revealed a large variation between the fibres within a husk. Longer fibres (more than 10 cm) usually had higher diameters (van Dam et al., 2006; Wei and Gu, 2009) having average diameter, 337 ± 55 μm (Brígida et al., 2010). Density of unretted raw coconut fibre is reported to be 1.15 g/cm^3, considering the adhered pith particles of lower density (0.11 g/cm^3) on surface. The fibres after retting are found to have density (van Dam et al., 2006) of 1.32 ± 0.03 g/cm^3. Coconut fibre tenacity (10–15 g/tex) is low as compared to other popular rope making lignocellulosic fibres. Coconut fibres having short fibre length are stronger than the long fibres. The permanent increase in coconut fibre length (plastic deformation) was calculated by loading the fibres at 50 and 100 mm span length within 1 min. The permanent deformation was calculated to 11% of the span length (van Dam et al., 2006). Moisture regain of coconut fibre was reported to be 8%–12.5% at 65% relative humidity. Transverse swelling of coconut fibre in water has been observed to be about 15% in diameter and it increases appreciably up to 34% after 15 min. Longitudinal swelling was reported to be 0.9% after 15 min of wetting (van Dam et al., 2006). Untreated coconut fibre sample showed water uptake up to 61% (Rahman and Khan, 2007).

Table 8.2 Comparison of physical, mechanical and moisture properties of coconut fibre and jute fibre.

	Coconut fibre	Jute (C. Olitorius)
Dimension of ultimate cell		
Length (mm)	0.8−1.06 (van Dam et al., 2006; Rahman and Khan, 2007; Coirboard, 2016; NIRJAFT, 1990)	0.75−6.0 (Satyanarayan et al., 2007)
Width ($\times 10^{-3}$) (mm)	14−16 (Rahman and Khan, 2007; Wei and Gu, 2009)	5−25 (Satyanarayan et al., 2007)
Length/width	95−536 (Rahman and Khan, 2007; Wei and Gu, 2009)	110 (Satyanarayan et al., 2007)
Area of cross-section of cell(10^{-4} mm^2)	1.6 (van Dam et al., 2006)	–
Lumen (%)	38 (Rahman and Khan, 2007)	–
Dimension of fibre		
Length (mm)	8−337 (Brígida et al., 2010)	–
Fibre fineness (linear density) (tex)	25−50 (Satyanarayan et al., 2007)	2−5 (Satyanarayan et al., 2007)
Fibre diameter (μm)	69−870 (Coirboard, 2016; Satyanarayana et al., 1982; Ray and Bandyopadhyay, 1966; Kaswell, 1953)	–
Coefficient friction	–	0.45 (Satyanarayan et al., 2007)
Fibre density (g/cm^3)		
True density (g/cc)	1.40 (Satyanarayan et al., 2007)	1.45−1.52 (Satyanarayan et al., 2007; Kaswell, 1953)
Apparent density (g/cc)	1.15−1.32 (NIRJAFT, 1990; Satyanarayan et al., 2007; Brígida et al., 2010; Varma et al., 1984)	1.23 (Satyanarayan et al., 2007)
Tensile properties		
Single fibre tenacity(g/tex)	10.0−15 (Wei and Gu, 2009; Satyanarayan et al., 2007)	35−50 (Satyanarayan et al., 2007)
Bundle tenacity (g/tex)	10−15 (Satyanarayan et al., 2007)	16−35 (Satyanarayan et al., 2007)

Table 8.2 Continued

	Coconut fibre	Jute (C. Olitorius)
Single fibre breaking elongation(%)	15–37 (van Dam et al., 2006; Wei and Gu, 2009; Kaswell, 1953; Varma et al., 1984)	1–2 (Satyanarayan et al., 2007)
Torsional modulus ($x10^{10}$ dyn/ cm^2)	0.2–1.5 (Varma et al., 1984)	0.25–1.3 (Satyanarayan et al., 2007)
Flexural modulus (dynes-cm^2)	150–250 (Varma et al., 1984)	3.5–6 (Satyanarayan et al., 2007)
Initial modulus (g/denier)	38.3 (Varma et al., 1984; Banik and Ghosh, 2008)	130–220 (Satyanarayan et al., 2007)
Young's modulus (Gpa)	4–6 (van Dam et al., 2006; Coirboard, 2016; NIRJAFT, 1990)	0.86–1.94 (Satyanarayan et al., 2007)
Work of rupture	–	0.30 (Mahato et al., 1993)
Moisture relation		
Moisture regain(%) at 65% R.H. 27°C	8–12.5 (van Dam et al., 2006; NIRJAFT, 1990; Wei and Gu, 2009)	12.5 (Satyanarayan et al., 2007)
Transverse swelling in water (%)	5–15 (Satyanarayan et al., 2007)	20–22 (Satyanarayan et al., 2007)
Electrical resistance		
Volume resistivity, 100 V (Ω cm $\times$ 10^5)	9–14 (Satyanarayana et al., 1982)	–

8.4 Composition of coconut fibre

Main chemical constituents of coconut fibre are polysaccharides and lignin. Although a number of minor components, such as pectin, inorganic salts, nitrogenous substance, colouring mater, wax etc. are also found in them. Chemical composition of coconut fibre as compared to that of jute (popular rope making allied lignocelluloses fibre) is given in Table 8.3. Composition of coconut fibres contains slightly higher cellulose than its lignin content with varying amount of hemicelluloses. The total polysaccharide content (50.9%) of the fibre is substantially higher containing 32.8% glucose as the

Table 8.3 Comparison of chemical composition (in percent of bone dry weight) and fine structure parameters of coconut and jute fibre.

	Coconut fibre	**Jute (C. Olitorius)**
Cellulose	35–43 (Bledzki et al., 1996; van Dam et al., 2006; Rahman and Khan, 2007; Coirboard, 2016; Brígida et al., 2010; Satyanarayana et al., 1982; Varma et al., 1984; Meredith, 1945)	58–59 (Rahman and Khan, 2007; NIRJAFT, 1990; Meredith, 1945; Banik and Ghosh, 2008)
Lignin	40–45 (Bledzki et al., 1996; van Dam et al., 2006; Rahman and Khan, 2007; Coirboard, 2016; Brígida et al., 2010; Satyanarayana et al., 1982; Varma et al., 1984; Meredith, 1945)	12–14 (Rahman and Khan, 2007; NIRJAFT, 1990; Meredith, 1945; Banik and Ghosh, 2008)
Hemicellulose	0.15–24 (Bledzki et al., 1996; van Dam et al., 2006; Rahman and Khan, 2007; Coirboard, 2016; Brígida et al., 2010; Varma et al., 1984; Meredith, 1945)	14–25 (Rahman and Khan, 2007; Meredith, 1945; Banik and Ghosh, 2008)
Water solubles	5.25 (Coirboard, 2016; Meredith, 1945)	–
Pectins	3.3–4.0 (Bledzki et al., 1996; Rahman and Khan, 2007; Coirboard, 2016; Varma et al., 1984)	0.2–0.5 (Rahman and Khan, 2007; Banik and Ghosh, 2008)
Mineral matter (ash)	2.2 (Bledzki et al., 1996)	0.5–1.2 (Rahman and Khan, 2007; Banik and Ghosh, 2008)
Fats and waxes	–	0.4–0.9 (Banik and Ghosh, 2008)
Crystallinity (%)	25–33 (Ray and Bandyopadhyay, 1966; Varma et al., 1984)	52–60 (Treatise on Physical, 1988)
Herman's angle of orientation (degree)	41–45 degrees (Bledzki et al., 1996; Stern and Stout, 1954; Rahman and Khan, 2007; Satyanarayana et al., 1982; Ray and Bandyopadhyay, 1966; Varma et al., 1984)	7–9 degrees (NIRJAFT, 1990)

major constituent, largely derived from alpha cellulose. Major non-cellulosic polysaccharide contains of fibre is xylose and polyuronic acid present in the ratio of 3:1 present in hemicelluloses fraction of the fibre. Long and short fibres show similar composition even in different varieties. However, fibres at different maturity found to contain different chemical composition. Fibres harvested from 6 to 11 months showed a gradual enhancement in glucose may be associated in the formation of cellulose and lignin, while showed no change in other sugars and extractives (van Dam et al., 2006).

8.4.1 Chemical composition of coconut fibre

Coconut fibre consists of 49.6% carbon, 6.1% hydrogen. The elemental composition did not change much after the various treatments with water, alkalis and acids (Wei and Gu, 2009). The surface elemental distribution of coconut fibre studied by X-ray photoelectron spectroscopy (XPS) analysis revealed the major element of the fibres are carbon (69.1%) and oxygen (27.4%) along with other minor elements like nitrogen (2.0%), silicon (1.1%) and calcium (0.5%). The oxygen to carbon (O/C) of raw coconut fibre surface is much lower (0.40) than the O/C ratio of the cellulose, hemicelluloses and pectin (0.83). However, the O/C ratio of lignin is 0.35, which indicates the surface content of the fibre is majorly lignin along with other waxy materials (Mahato et al., 1993). The apparent hydrophobic nature of coconut fibre is also due to presence of high carbon to carbon bonds (C—C) as compared to lesser carbon to hydrogen (C—H) bonds (Yueping et al., 2010) and consequent lower oxygen to carbon (O/C) ratio.

8.5 Morphology and structure of coconut fibre

The scanning electron micrograph (Fig. 8.8) shows the micro-fibrils by longitudinal parallel ridges. Coconut fibres contain impurities; wax, fatty and globular protrusions making the fibre surface extremely heterogeneous (Fig. 8.8a) with prominent cracks, micro-pores and irregular wax-like deposition are detectable. The surface morphology of backwater retted fibre (Fig. 8.8b) has more irregularities and impurities than the raw fibre, which might be due to formation of an additional salt coating through its backwater retting followed by air drying without washing. Comparing to raw and backwater retted coconut fibres, treated fibres appeared to be clean, with a smother surface and it is possible to observe a reduction on fats and waxes. Washing with plain water could not remove these impurities; however, treatments of alkali can clean the surface exposing the surface pores called as pits (Fig. 8.8c) (van Dam et al., 2006; Rosa et al., 2009; Basu et al., 2015), which were not revealed on the surface of raw and backwater retted fibres.

The helical arrangements (Fig. 8.9 (left)) of micro-fibrils leads to much higher angle of orientation up to 45 degrees, which is reduced on extension (37 degrees) and cause enhancement in the crystallinity (by 3% on extension to rupture) of the fibre (Ray and

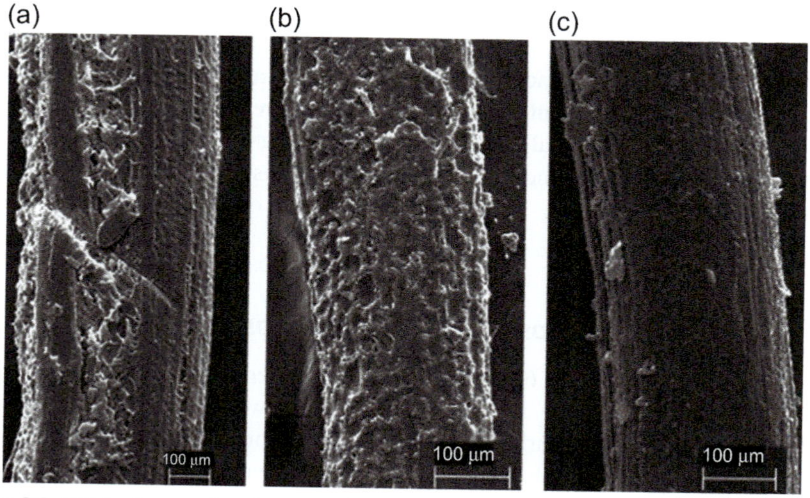

Figure 8.8 SEM micrographs of coconut fibre (a) raw (b) backwater retted (c) chemically retted.

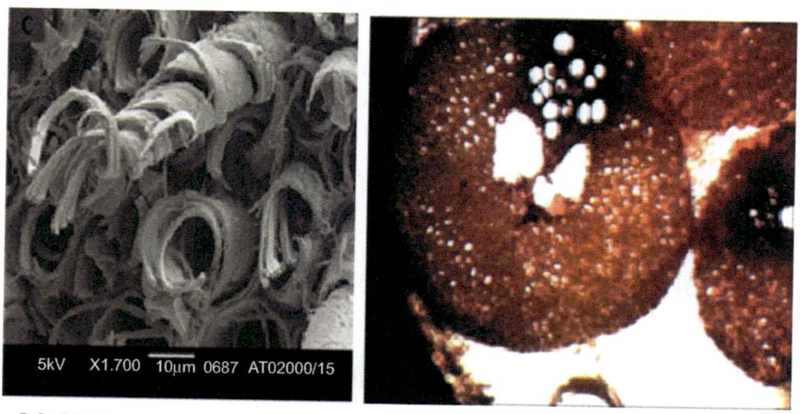

Figure 8.9 SEM of helical structure of elementary fibre cell wall and optical microscopic view of coconut fibre.

Bandyopadhyay, 1966). Optical microscopy (Fig. 8.9 (right)) illustrates that the cross-section of fibre is circular having lumen. Circular cross-section makes the raw coconut fibre lustrous in appearance. Presence of numerous voids around lumen indicates its multi-fibrillar structure.

Crystallinity of coconut fibre is reported to be considerably low due to presence high amount of non-cellulosic impurities like lignin and hemicelluloses. Diffraction pattern (Fig. 8.10) illustrates polycrystalline nature of the fibre. Coupling of peaks 101 and $10\bar{1}$ indicates amorphous contribution with non-uniform strain and texture. A broad peak at 002 is evident along with 004 peaks. Broadening of peak 002 shows bigger crystallite size and presence of large amount of non-cellulose matter

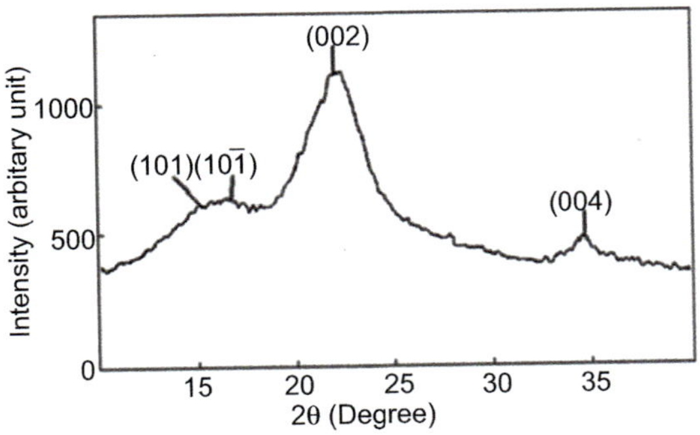

Figure 8.10 X-ray diffractogram of coconut fibre.

(Basu et al., 2015). Crystallinity of cellulose of the coconut fibre has been calculated 37%–45% (van Dam et al., 2006; Basu et al., 2015) having inter-planer distances of the lattice structure of the peaks of (10$\bar{1}$) and (002) of 5.49 and 4.29 nm at 15.7 and 22.1 degrees diffraction angles (2θ) (Basu et al., 2017).

Thermal analysis of coconut fibres (Fig. 8.11) showed that the initial decomposition of fibre takes place in case of evaporation of water and volatile substances (low molecular weight waxes and fats) and this occurs between room temperature and 150°C. The second weight loss corresponds to hemicelluloses degradation, starts at about 190°C. Weight loss occurs between 290 and 360°C, mainly corresponds to cellulose degradation. Lignin degradation starts at about 280°C and continues even above 500°C (Rosa et al., 2009). Thermal degradation for the temperature range 125–295°C was found to 40% as compared to the 24% in the later temperature range 300–600°C indicating the presence of higher methyl-aryl ether groups than aromatic groups in lignin of coconut fibre (Basu et al., 2015). Similar trend was also observed through NMR (Rencoret et al., 2013).

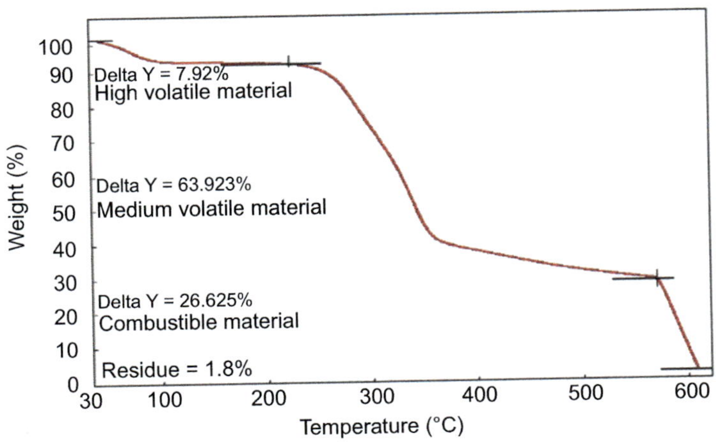

Figure 8.11 TG curve for coconut fibres.

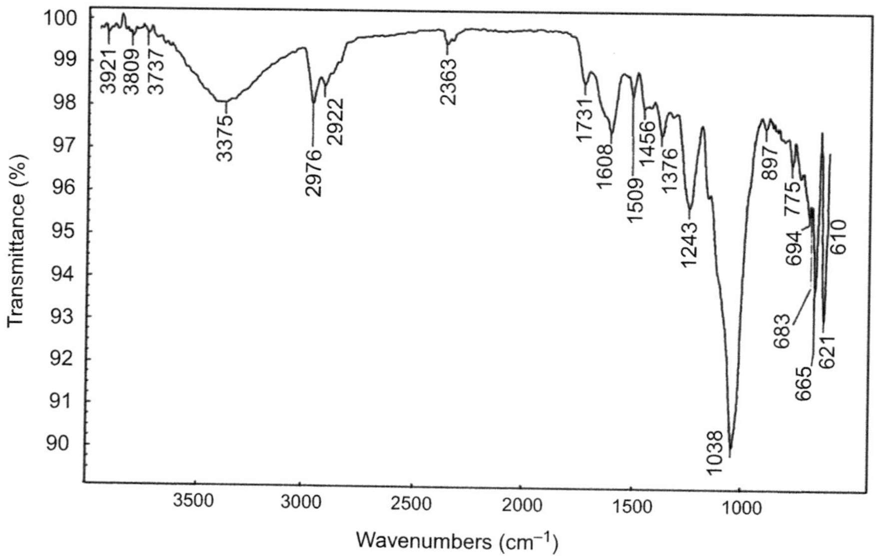

Figure 8.12 FTIR spectrum of coconut fibre.

FTIR spectroscopy of coconut fibre (Fig. 8.12) shows a broad and intense peak at 3375 cm^{-1} suggesting OH stretching vibrations from cellulose and lignin (Bilba et al., 2007). A peak at 2363 cm^{-1} is indicating the C–H stretching in polysaccharide chain units. The hemicelluloses, lignin and pectin are identified at the peak at 1731 cm^{-1} ascribed to conjugated C=O stretching of ester and aldehyde (Samanta et al., 2008). Peak 897cm^{-1} is characteristic for β-linkages, especially for hemicelluloses (Cao and Tan, 2002). IR spectrum of coconut fibre also shows a characteristic C–O–C stretching absorbency peak at 1038 cm^{-1} of cellulose (Vasi and Shah, 1981), absorption peaks corresponding to 1250 cm^{-1} (lignin aromatic C–O stretching), and 1509 cm^{-1} (lignin aromatic ring stretching) (Mahato et al., 1993; Samanta et al., 2008). The characteristic absorbency peak in the wave number of 1608 cm^{-1} is obtained for lignin (Yueping et al., 2010).

8.6 Treatments of coconut fibre affecting the structure and properties of coconut fibre

8.6.1 *Alkali treatments (NaOH)*

Alkali treatments caused maximum weight loss (9.3 ± 1.5%) and also changed the colour of the fibres from pale yellowish brown to dark brown. The helical arrangements of micro-fibrils at an angle of orientation of 45 degrees impart crimp on coconut fibre due to differential shrinkage on treatment with 10% alkali (Varma et al., 1984).

The weight loss of 12.65% is obtained by treatment with 20% NaOH at temperature 25°C (Rahman and Khan, 2007). The fibres also become finer due to reduction in diameter. The diametrical reduction of coconut fibre was found to be more for thicker fibres (43%) having thicker cell wall than thinner fibres (10%). The size of central lacuna also reduced considerably (van Dam et al., 2006; Varma et al., 1984). The coconut fibre bristle fibres got slightly lesser strength, due to possible rupture of alkali sensitive bonds existing between components of the fibre as a result of swelling and removal of hemicelluloses in alkali treatment (Varma et al., 1984). On the contrary, Rosa et al., 2009 reported a considerable improvement in tensile properties (10.2–13.6 Mpa) and initial modulus of the fibre on treatment with the NaOH (10%) (Rosa et al., 2009). The increase in tenacity in the fibre is resulting out of its reduction in non-cellulosic content enhancing the content and purity of cellulose, thus, enhanced the crystallinity index of the fibre (Varma et al., 1984). Also, the alkali treatment removes the cementing material between the fibrils causing slippage of molecular chains resulting in higher elongation at break. FTIR spectra of unretted and alkali-treated coconut fibre shows a disappearance of a small peak at 1740 cm^{-1}, due to removal of hemicelluloses (Brígida et al., 2010; Varma et al., 1984). The thermal analysis of alkali-treated fibre showed an inflection for mass loss in the temperature zone of 150–310°C possibly due to considerable loss in hemicelluloses and pectin (Rosa et al., 2009). Treatments of coconut fibre gradually from 5% to 30% NaOH convert the thermal curves having exothermic nature (cleavage of pectin, hemicelluloses and cellulose including a part of lignin) to endothermic nature. This may be interpreted as the initial treatment with lower concentration of alkali is reacting with the lignin and hemicelluloses of the fibre and degrades them only, while at higher concentrations of NaOH (above 15%) the cellulose also get degraded, which is reflected in its thermal instability. The activation energy, which also reveals the thermal stability shows initial enhancement of activation energy of coconut fibre on treatment with 5% NaOH followed by a drastic fall in the activation energy of the fibre up to treatment with 30% NaOH. Treatment of 5% NaOH, initially enhance the activation energy/thermal stability of coconut fibre due to removal of hemicellulosic mass only and give rise to lignin-cellulose complex. Hemicelluloses are less thermally stable than lignin and cellulose, hence, removal of it makes the fibre more thermally stable (Mahato et al., 1993). The coconut fibre treatments from 5% to 50% NaOH at different temperature ranges from 0 to 100°C indicates variable weight loss and fibre length shrinkage. The highest shrinkage (5.88%) was observed at 0°C with 20% alkali (Rahman and Khan, 2007). Such type of action is also seen in case of jute fibre when treated with 18% NaOH at low temperature (0–5°C), resulting in severe crimp generation in fibre due to differential shrinkage of the molecular chain (Saha et al., 1961; Lewin et al., 1959). Coconut fibre treated with NaOH, from 2% to 10% (w/v), showed enhancement in moisture regain (from 16.87% to 20.13%) (Samal et al., 1995). Treatment of the fibre with 2% NaOH increased tensile strength (from 130 to 161 kg/cm^2) but reduced the strength to 152 kg/cm^2 on treatment with 8% and 10% NaOH (Brígida et al., 2010; Samal et al., 1995).

8.6.2 Acid treatments (HCl, CH_3COOH, HNO_3)

Hydrochloric acid (10%) and acetic acid (10%) treatment caused about 7.5 ± 0.5% and 6.5 ± 1% weight loss in fibre, respectively, while no notable change in colour was observed. Moisture regains of coconut fibre increased considerably from 8% to 16% on hydrochloric acid treatment (Varma et al., 1984). Treatments of sulphuric acid (40% and above) at ambient temperature, resulted in surface carbonization with gradual solubilization (complete or partial) of the fibre. However, for concentration of 30% and below for H_2SO_4, the fibres resist the treatment for 1 h without any surface carbonization. The action of HCl was milder than H_2SO_4, which shows no surface carbonization under the treatment conditions. However, the treatments made the fibre brittle during storage even after alkali neutralization, probably due to hydrolysis of cellulose polymers by the acids (Meredith, 1945). The treatment of coconut fibre with concentrated nitric acid also reduced the fibre strength (Varma et al., 1984; Meredith, 1945).

8.6.3 Oxidizing treatments (NaOCl, H_2O_2)

Pre-treatment of NaOCl (4%–6%) prior to treatment caused higher removal of waxy materials, causing disappearance of the peak 1238 cm^{-1} for the CO– group attributed mainly to a presence of waxes and fats (esters, ethers and phenol). Hydrogen peroxide treatment oxidizes the OH groups in cellulose present on the surface and thus enhance the peak intensity of the existing C=O group at 1728 cm^{-1}. Treatments of NaOCl and H_2O_2 caused enhancement in oxygen to carbon (O/C) ratio due to reduction in hydrophobic surface impurities (fats and waxes), lignin and thus increasing cellulose exposition and making the fibre more hydrophilic. Although, the H_2O_2 treatment caused highest removal of fats and waxes but could not able to remove surface lignin hence not able to enhance the wettability of the fibre. The treatment of NaOCl (4%–6%) prior to NaOH (10%) caused considerable mass and volume loss (diameter reduced from 495 to 165 μm) enhancing the fibre density. The treatment also enhanced the percentage cellulose content (from 45.9% to 62.8%) probably due to much higher removal of hemicelluloses and alkali soluble lignin (reduced by 50%) and other impurities (Brígida et al., 2010).

8.6.4 Treatment with ethylene dimethacrylate (EMA) and UV radiation

Coconut fibre possesses high moisture absorption, poor wettability and insufficient adhesion between untreated fibre and polymer matrix that lead to debonding with age in the composites. To resolve these problems the fibres are pre-treated with UV radiation. In some cases, fibres are treated with NaOH (5%–50%) and then with ethylene dimethylacrylate and methanol solution followed by exposure of UV radiation for different time and doses. Notable change in mechanical properties of coconut fibre was observed after pre-treatment with UV radiation up to a certain doses (125 passes). Increase in UV radiation enhanced fibre tensile strength by 33%, elongation by 39% and modulus. The improved mechanical properties under UV exposure may be due to cross-linking between the molecular chains of cellulose. The EMA loading of 19.12% are highest at 50% in methanol at 125 passes of UV radiation. Grafting of alkali treated fibre showed an increase of polymer loading (about 56%) and tensile strength (about 27%) than 50% EMA grafted raw fibre. UV radiation exposure followed by NaOH treatment considerably improved the surface morphology and assists in better interfacial bonding in the modified fibre-matrix interfaces (Rahman and Khan, 2007).

8.6.5 Treatment involving dinitrophenylation, diazo coupling and combined diazo coupling-cyanoethylation

Treatment of coconut fibre (preswelled with alkali (NaOH) solution) with 5% 2, 4 – dinitro chlorobenzene in acetone at 60°C (dinitrophenylation) caused much higher improvement in tensile strength (from 130 to 185 kg/cm^2). The treatment also induces hydrophobicity to the fibre to a greater extent i.e. reduced the moisture regain from 16.9% to 4.1%. Treatment with diazotized aniline solution in ice bath (diazo coupling) followed by treatment with cyanoethylate (combined diazo coupling-cyanoethylation) for various time also showed enhancement in tensile strength (178 kg/cm^2) and reduction in moisture regain (11.78%). Reduction of moisture regain was confirmed through FTIR by reduction of the peak intensity at 3446 cm^{-1} for the –OH stretching in both dinitro-phenylated and diazo coupled cyanoethylation treated fibre (Samal et al., 1995).

8.6.6 Treatment with urea on mercerized and acid hydrolyzed coconut fibre (Cocos nucifera) fibres photocured 1,6-hexanediol diacylate

Coconut fibres were modified with 30% 1, 6-hexanediol diacrylate (HDDA) with photo-initiator Irgacure-500 (2%) in methanol by using ultraviolet radiation. The treatment leads to enhancement of polymer loading by 17% and tensile strength by 50%. Urea of different concentrations (0.5%–2%) was incorporated with 30% HDDA to monitor its effect on the properties and 1% urea produced the enhanced polymer loading (25%) and tensile strength (82%). Further improvement of the properties was attempted by treating the fibres with alkali (5% potassium hydroxide) at various mercerizing times in hot and normal conditions. Among all the mercerized fibres, the fifers treated with hot alkali for 6 h and cured under optimized condition demonstrated the maximum enhancement of polymer loading (35.5%) and tensile strength (130%). The fibres were also subjected to acid hydrolysis for different times with different concentrations of sulphuric acid (H_2SO_4). Again the effect of urea (1%) on the properties of the pre-treated fibres was scrutinized. Water uptake and degradable properties of the treated and virgin fibres were performed (Khan et al., 2005).

8.6.7 Accelerated chemical softening in combination of strong reducing agent and alkalis

It was observed that the optimum combination of 40% Na_2S, 15% Na_2CO_3 and 6% NaOH caused highest weight loss (about 28%) without affecting the tensile property. The treatment reduced flexural rigidity (from 1273 to 361 cN-mm (FAO, 2013)), linear density (from 53 to 34 tex), and diameter (from 345 to 225 μ) significantly resulting in enhancement of length to diameter ratio (from 335 to 991). It is assumed that, the combination of alkalis (NaOH, Na_2CO_3) might enhance the effect of Na_2S. In the chemical retting process, the presence of strong reducing agent Na_2S sustainably hydrolyzed to produce NaOH, making the bath highly alkaline. The rate of hydrolysis is increased with increase in temperature. In presence of high alkaline medium at elevated temperature, all components are reacting with alkali. Na_2S being a reducing agent, the incorporation of the same in the bath retards the alkaline hydrolysis and oxidation of cellulose reduce the possibility of cellulose degradation, while NaOH alone is used for softening process. Apart from this, NaSH reacts with lignin to introduce −SNa group which increases its acidity and consequently its solubility in alkaline medium (Basu et al., 2015).

8.7 Applications of coconut fibre

8.7.1 Yarns of coconut fibre

The most popular product of coconut fibre is its yarn which is made of two ply single thread spun from coconut fibre by hand with the help of traditional ratts, as well as by semi or fully automatic spinning machines, as discussed earlier. In India, the yarn is of 27 different qualities/grades based on the quality of fibre used, the nature of twist, presence of impurities etc., available in different forms like hydraulically pressed bales, spools bobbins, balls etc., cut length for various industrial and agricultural purposes. The price of Angengo 240 yarn (best quality yarn) started from $ 100−143 USD/100 kg and the price of inferior quality yarn available from Vaikom known as Vaikom −180, 160 yarn is available at the price of Rs. 3500 per 100 kg. In this regard, it is noteworthy to mention the price of yarn made from conventionally retted fibre ($ 97 USD/100 kg, Beypore yarn 90) is twice in price than that of yarn produced from their unretted fibre ($ 53 USD/100 kg) (Coirboard, 2015).

8.7.2 Coconut fibre mat and mattresses

Raw coconut fibre had been conventionally used to make mats and mattresses in Indian subcontinent. Presently, due to the technological advancements of natural as well as synthetic rubber latex, value added rubberized coconut fibre mattresses and mats are getting much popularity for its much better resiliency, lighter weight with adequate load bearing capacity. These are used as cushioning materials for beddings, settee, automobile seats, mats etc.

8.7.3 Geotextile and agrotextile mat

Geotextiles made out of coconut fibre made to give high tensile strengths by using high twisted with low runnage twines by weaving them closely. Depending upon the tightness of the weave (the grid openings) coconut fibre geotextiles can be made to prevent erosion of high gradient slopes and easily erodible loose or small particle soils. The coconut fibre mat as needle punched sheets present above soil has the ability to absorb and retain moisture and therefore invariably help the growth and development of vegetation.

8.7.4 Coconut fibre log

Coconut fibre log is a cylindrical tube made of coconut fibre netting or knotting filled with natural coconut fibre and thickly compacted. It is mainly used for shoreline protection to prevent scouring of soil from the shoreline. It can also be produced pre-seeded to allow establishment of vegetation when it comes to contact with water after installation.

8.7.5 Composites of coconut fibre

Efforts have also being made to manufacture synthetic matrix composites reinforced with coconut fibre (van Dam et al., 2006; Rahman and Khan, 2007; Rosa et al., 2009). Starch and polyethylene vinyl alcohol copolymer blended with the coconut fibre as fillers for reinforcement and decreasing stiffness has received increased attention for making of biocomposites, particularly for reduction in price and high volume application (Rosa et al., 2009). However, not much success story large-scale trials of this product are available. Coconut fibre in lightly twisted (hand-spun) yarns are cut in a specified length and flocked on an adhesive-coated (mostly polyvinyl-based) wide continuous length of rubber sheet to produce wider matting. Reports on effect of coconut fibre content on the physical and mechanical properties as well as fracture behaviour of composite cement reinforced material and also to partially remove sand are striking (Aggarwal, 1992; Asasutjarit et al., 2007).

8.7.6 Bioethanol production from coconut fibre

The alkaline pre-treated coconut fibres were hydrolyzed enzymatically and fermented with *Saccharomyces cerevisiae* yeasts for ethanol production. The alkaline pre-treatment promoted an efficient solubilization of lignin (80%) converted 87% of the sugars and the ethanol fermentation consumed 81% of the substrate, leading to a sugar to ethanol conversion efficiency of 59.6%. These results points out that green coconut husks are a promising alternative to the production of renewable energy (Silva et al., 2016).

8.7.7 Nano-whiskers from coconut fibre

Nanometre-sized single crystals of cellulose, commonly referred to as whiskers, nano-whiskers or nano-fibrils, can be obtained from delignified/bleached coconut fibres through sulphuric acid hydrolysis to obtain cellulose nano-whiskers with diameters as low as 5 nm and aspect ratio of up to 60. The presence of higher residual lignin content induces higher thermal stability to nano-whiskers (Rosa et al., 2010).

8.7.8 Coconut fibre absorption for dyes

Coconut fibre husk is a cheap and readily available agricultural by-product that can be turned into value-added product which effectively removed methyelene blue and malachite green dye to nearly 93.3%. Furthermore, this study also shown that granular form of coconut fibre husks absorbed methyelene blue and malachite green dye more effectively compared to filament form which may be due to the higher surface area which made readily contacted to the surrounding environment much better if compared to its filament form (Wong et al., 2013).

It was found that percentage adsorption varied linearly with the amount of adsorbent and concentration with time but varies non-linearly with pH. Adsorption equilibrium data were represented by isotherm, kinetics and thermodynamics models. FTIR analyses of the adsorbent suggest that adsorption of the dye was through a chemical interaction of the functional groups on the surface of the adsorbent (Etim et al., 2016; Ong et al., 2013).

8.7.9 Absorption of heavy metals by coconut fibre

The applications of coconut fibre for absorption of heavy metals like lead (Pb), zinc (Zn) (Kathrine, 2008) and chromium (Cr VI) (Tan et al., 1993) were studied extensively. It is found that metal ions are absorbed by the xylems and cell wall facing the lacuna across the secondary wall. Investigation also suggests the absorption of heavy metals like Pb and Zn are positively correlated with pectin of coconut fibre composition.

8.8 Conclusion and future trends

Coconut fibre possesses highly variable fibre length, and large diameter (320 μm) resulting in low length-diameter ratio. So segregation of long and finer fibre from the fibre bulk may give better spinnabilty and short fibres can be used for making composites. The fibre has low breaking tenacity (11.25 cN/tex), initial modulus (200 cN/tex), but higher breaking elongation (21.5%) property causing notably high specific work of rupture (13.4 mJ/tex.m). The fibre shows high flexural rigidity and lower inter-fibre frictional resistance for which it is difficult to spin finer yarn and indicates its suitability for coarse textile applications. Coconut fibre is circular and multifibrillar. Moisture regain is moderate (11.7%). Thermal and component analyses reveal the presence of similar amount of lignin and celluloses in the fibre. SEM image revealed that surface of the fibre was waxy, irregular, and having micro-imperfections. Microbial resistance makes the coconut fibre suitable for using it in different engineering and geotechnical applications in conjunction with soil and water.

The optimum value addition of coconut fibre can be realized, when a fully systematic supply chain of quality fibre with proper gradation system can be developed. Spinning of finer coconut fibre-based yarn using mechanized systems is the key area of the processing line for diversifying and improving the coconut fibre-based products line. The fibres can be highly suitable to form compressed mat for various products of geotextile, agro-textiles, and sound and heat insulators with improved softness and fineness property. Lastly, coconut fibre should not need to be treated as a secondary crop; rather some fibre crop varieties or dual varieties have to be developed, which will concentrate on the fibre quality than the food obtained from it. So, the future scope of coconut fibre sector is to establish the cash crop coconut from the status of food to the status of fibre.

References

Aggarwal, L.K., 1992. Studies on cement-bonded coir fibre boards. Cement Concr. Compos. 14 (1), 63–69.

Anto, I.R., Ravi, P.K., Sarma, U.S., 1997. Studies on imparting softness to coir yarns. In: Workshop on Wet Processing of Coir, Alleppey, December. 1–27/Coir News, 27, pp. 19–28 (3) (1998).

Asasutjarit, C., Hirunlabh, J., Khedari, J., Charoenvai, S., Zeghmati, B., Cheul, U.S., 2007. Development of coconut coir-based lightweight cement board. Constr. Build. Mater. 21 (2), 277–288.

A treatise on physical and chemical properties of jute. In: International Jute Council (International Jute Organisation, Dhaka, Bangladesh), IJC (X)/11, 1988, pp. 20–24, 30.

Banik, S., Ghosh, S.N., 2008. Pectinolytic effect of microorganisms in piling of jute. Indian J. Fiber Text. Res. 33, 151–156.

Basu, G., Mishra, L., Samanta, A.K., 2017. Investigation of structure and property of Indian *Cocos nucifera* L. fibre. J. Inst. Eng. (India) Ser. E (Springer) 98 (2), 135–140.

Basu, G., Mishra, L., Seiko, J., Samanta, A.K., 2015. Accelerated retting cum softening of coconut fibre. Ind. Crops Prod. 77, 66–73.

Bilba, K., Arsene, M.A., Ouensanga, A., 2007. Study of banana and coconut fibers, Botanical composition, thermal degradation and textural observations. Bioresour. Technol. 98 (1), 58–68.

Bledzki, A.K., Riehmane, S., Gassan, J., 1996. Properties and modification methods for vegetable fibers for natural fibers composites. J. Appl. Polym. Sci. 59 (8), 1329–1336.

Brígida, A.I.S., Calado, V.M.A., Gonçalves, L.R.B., Coelho, M.A.Z., 2010. Effect of chemical treatments on properties of green coconut fiber. Carbohydr. Polym. 79, 832–838.

Cao, Y., Tan, H., 2002. Effects of cellulase on the modification of cellulose. Carbohydr. Res. 337 (14), 1291–1296.

http://coirboard.gov.in/wp-content/uploads/2015/09/weeklyprice-of-coir-yarn-kerala.pdf.

www.coirboard.gov.in, 27.08.2016; 16:15 hrs.

Encyclopaedia Britannica, 1982. Micropaedia, Encyclopaedia Britannica, Inc, Chicago, USA, vol. 15, p. 1032 (2).

Etim, U.J., Umoren, S.A., Eduok, U.M., 2016. Coconut coir dust as a low cost adsorbent for the removal of cationic dye from aqueous solution. J.Saudi Chem. Soc. 20 (1), 67–76.

FAO, 2013. Jute, Kenaf, Sisal, Abaca, Coir and Allied Fibers Statistics. Food and Agriculture Organization of the United Nations, Rome. December, 2013.

FAO, 2015. Jute, Kenaf, Sisal, Abaca, Coir and Allied Fibers Statistics. Food and Agriculture Organization of the United Nations, Rome. December, 2015.

Kaswell, R.E., 1953. Textile Fibres, Yarns and Fabrics. Reinhold Publishing Corporation, New York, USA.

Kathrine, C., 2008. Correlation between the distribution of lignin and pectin and distribution of sorbed metal ions (lead and zinc) on coir (*Cocos nucifera* L). Bioresour. Technol. 99 (17), 8476–8484.

Khan, M.A., Taslima, R., Mustafa, A.I., 2005. Role of urea on the physico-mechanical and degradable properties of mercerized and acid hydrolyzed coir (*Cocos nucifera*) fibers photocured with 1,6-Hexanediol Diacylate. Polym.-Plast. Technol. Eng. 44 (5), 1003–1023.

Lewin, M., Shiloh, M., Banbaji, J., 1959. The crimp of alkali treated jute fibers. Text. Res. J. 29 (5), 373–385.

Mahato, D.N., Mathur, B.K., Bhattacharyay, S., 1993. Effects of alkali treatment on electrical and spectral properties of coir fiber. J. Mater. Sci. 12, 1350–1353.

Meenatchisundarm, R.I., 1979. Retting of coir—A review. Ceylon Coconut Plant. Rev. 7, 20–28, 1975–1980.

Meredith, R., 1945. The tensile behavior of raw cotton and other textile fibers. J. Text. Inst. 36, T107–T130.

Mohanan, K.G., Antonel Vaz, P.C., Sunilduth, J., Radhakrishnan, A., Sebastian, V.A., Sarma, U.S., 1997. In: Proceedings of the International workshop on Wet processing of Coir 8–9, pp. 58–81.

NIRJAFT, 1990. Fifty years of research 1939-89. In: Pandey, S.N., Anantha Krishnan, S.R. (Eds.), Jute Technological Research Laboratories. Indian Council of Agricultural Research, Kolkata, India.

Ong, S., Ho, L., Wong, Y., Zainuddin, A., 2013. Adsorption behavior of cationic and anionic dyes on to acid treated coconut coir. Separ. Sci. Technol. 48 (14), 2125–2131.

Rahman, M.M., Khan, A.M., 2007. Surface treatment of coir (*Cocos nucifera*) fibers and its influence on the fibers' physico-mechanical properties. Compos. Sci. Technol. 67 (11–12), 2369–2376.

Ravindranath, A.D., Sarma, U.S., 2010. Economical and ecofreindly wet processing of coir using *Trametes versicolor* and *Coirret*. Coir News 39, 11–16.

Ray, P.K., Bandyopadhyay, S.B., 1966. An X-ray study of coir fibre. Indian J. Phys. 39 (9), 421–427.

Rencoret, J., Ralph, J., Marques, G., Gutierrez, A., Martínez, T.A., Río, C.J., 2013. Structural characterization of lignin isolated from coconut (*Cocos nucifera*) coir fibers. J. Agric. Food Chem. 61, 2434–2445.

Rosa, M.F., Chiou, B., Medeiros, E.S., Wood, D.F., Williams, T.G., Mattoso, L.H.C., Orts, W.J., Imam, S.H., 2009. Effect of fiber treatments on tensile and thermal properties of starch/ethylene vinyl alcohol copolymers/coir biocomposites. Bioresour. Technol. 100, 5196–5202.

Rosa, M.F., Medeiros, E.S., Malmonge, J.A., Gregorski, K.S., Wood, D.F., Mattoso, L.H.C., Glenn, G., Orts, W.J., Imam, S.H., 2010. Cellulose nanowhiskers from coconut husk fibers: effect of preparation conditions on their thermal and morphological behaviour. Carbohydr. Polym. 81 (1), 83–92.

Saha, P.K., Chatterjee, K.K., Sarkar, P.B., 1961. Woolenised Jute as a Wool Substitute. Indian Central Jute Committee Monograph.

Samal, R.K., Rout, S.K., Panda, B.B., Mohanty, M., 1995. Effect of chemical modification on FTIR spectra; Physical and chemical behavior of coir. J. Appl. Polym. Sci. 12, 229–234.

Samanta, A.K., Basu, G., Ghosh, P., 2008. Enzyme and silicon treatments on jute fibre. Part I: effect on textile-related properties. Text. Inst. 99 (4), 295–306.

Satyanarayan, K.G., Guimaraes, J.L., Wypych, F., 2007. Studies on lignocellulosic fibres of Brazil. Part 1 – source, production, morphology, properties and applications. Compos. Part A 38, 1694–1709.

Satyanarayana, K.G., Pillai, C.K.S., Sukumaran, K., Pillai, S.G.K., Rohatgi, P.K., Vijayan, K., 1982. Structure property studies of fibres from various parts of the coconut tree. J. Mater. Sci. 17 (8), 2453–2462.

Senan, D.C., 2014. Occupational Diseases and Health Awareness through Multimedia: A Case Study Among Women at Risk Employed in Coir Retting in Kerela. Partridge Publishing, India.

Silva, C.E.F., Cabral, M.M.S., Abud, A.K.S., Almeida, R., 2016. Bioethanol production from coconut husk fiber. Ciência Rural. 46 (10).

Stern, F., Stout, H.P., 1954. Morphological relation in cellulose fibre cells. J. Text. Inst. Trans. 45 (12), T896–T911.

Tan, W.T., Ooi, S.T., Lee, C.K., 1993. Removal of chromium (VI) from solution by coconut husk and palm pressed fibre. Environ. Technol. 14, 277–282.

van Dam, J.E.G., 2002. COIR PROCESSING TECHNOLOGIES Improvement of drying, softening, bleaching and dyeing coir fibre/yarn and printing coir floor coverings Common Fund for Commodities. Netherland 35, 7–27.

van Dam, J.E.G., Van den Oever, M.J.A., Teunissen, W., Keijsers, E.R.P., Van der putten, J.C., Anayron, C., Josol, F., Peralta, A., 2006. Process for production of high density high performance binderless boards from whole coconut husk. Part 2: coconut husk morphology, composition and properties. Ind. Crops Prod. 19, 96–104.

Varma, D.S., Varma, M., Varma, I.K., 1984. Coir Fibres: Part I effect of physical and chemical treatments on properties. Text. Res. Inst. 54, 827–831.

Vasi, I.G., Shah, P.R., 1981. In Spectroscopic Data Tables. Mahajan Brothers, Ahemadabad, India), pp. 4–13.

Wei, W., Gu, H., 2009. Characterization and utilization of natural coconut fibers composites. Mater. Des. 30, 2741–2744.

Wong, Y.C., Senan, M.S.R., Atiqah, N.A., 2013. Removal of methylene blue and malachite green dye using different form of coconut fibre as absorbent. J. Basic Appl. Sci. 9, 172–177.

Yueping, W., Ge, W., Haitao, C., Genlin, T., Zheng, L., Feng, X.Q., Xiangqi, Z., Xiaojun, H., Xushan, G., 2010. Structures of bamboo fiber for textile. Text. Res. J. 80, 334–343.

Wool fibres

9

Umit Halis Erdogan [1], *Yasemin Seki* [1], *Figen Selli* [1,2]
[1]Dokuz Eylul University, Faculty of Engineering, Department of Textile Engineering, Tınaztepe Campus, Buca, Izmir, Turkey; [2]Advanced Fibers, Empa Swiss Federal Laboratories for Materials Science and Technology, St. Gallen, Switzerland

9.1 Introduction

Wool fibres have been used as a textile raw material for apparel and home textiles for years. Latterly, it is also increasingly being used in technical textiles applications. Wool has traditionally been a significant contributor to various producer and manufacturer countries' economy. High-value wool and wool blend products such as carpets, suits and sweaters have still both economic and social prominence in the market. However, man-made fibres dominated the traditional wool market and minimalized the importance of wool in the textile industry. The world clean wool fibre production has been around 1.2 million tons annually for the last 25 years. On the other hand, a great opportunity on wool's future is the growing consumer interest in natural, renewable and sustainable materials. Moreover, in the last decades, remarkable improvements in wool production in terms of yield and quality have been achieved. Utilization of poor and waste wool for technical use and application of nanotechnology to wool are other significant attempts in wool science and technology. Considering these intense scientific studies to enhance the properties of wool for producing new and high-value wool products, it can be said that wool still maintains its importance for the textile industry. Therefore, this chapter describes the current situation of wool production and industry, summarizes the unique physical and chemical properties of wool, and focuses on recent innovations in wool fibres and applications.

9.2 Wool production and wool industry

Fibres from various natural resources have been used as the raw materials of clothing since ancient times. These renewable and sustainable fibres are still raw materials of home textiles, technical textiles, and apparel. Each year, approximately 35 million tons of natural fibres are harvested by farmers from a wide range of plants and animals (http://dnfi.org/statistic). Among the natural fibres, wool from sheep is one of the oldest textile fibres used by mankind. Due to its exceptional characteristics and limited supply, wool is still known as the world's extremely versatile textile fibre.

Wool is produced in about 100 countries on half a million farms by approximately one billion sheep. Sheep were first domesticated about 10,000 years ago to provide a ready source of meat, milk, and wool in Central Asia. According to some estimates,

Handbook of Natural Fibres. https://doi.org/10.1016/B978-0-12-818398-4.00011-6

there are more than 1000 breeds of sheep worldwide, but only some of them have economic importance for the commercial industry (https://www.iwto.org/wool; http://www.naturalfibres2, 2009). Sheep come in all different sizes, shapes and colours, and there are many ways to classify sheep: according to their primary purpose (meat, milk, or wool), the type of fibres they grow (fine, medium, long or carpet wool; or hair), the colour of their faces (black, white, red, or muddled), and/or by specific physical or production characteristics (http://www.sheep101.info/).

Until the industrial revolution, wool fibres had responded to human's basic requirements such as protection, clothing, etc. The invention of man-made fibres dominated the wool market and minimalized the importance of wool in the textile industry. Even though the main aim of sheep husbandry is meat production, nowadays, various breeds of modern sheep produce a wide range of fine and coarse wool fibres for apparel fabrics and interior textiles, respectively. World sheep numbers for the last 10 years can be seen in Fig. 9.1 Sheep numbers are up slightly from 2010.

Over the last decades, remarkable improvements in wool production in terms of yield and quality have been achieved. In recent years, different attempts for utilization of poor quality wool fibres and wool wastes were also considered. Nevertheless, global wool production has remained fairly static since 2008. Besides, in the past 15 years, finer wool production typically used in clothing has declined steadily, while broader wool production for interior textiles has increased slightly and the gap between broader and finer wool production continues to widen. The world greasy wool and clean raw wool production has been around 2.1 and 1.2 million tons annually for the last 10 years, respectively (https://www.iwto.org/wool; http://www.fao.org/faosta; https://www.sheepusa.org/, 2017).

By country, Australia is the world's largest wool producer, with a share of 20%–25%, followed by China and New Zealand. Regarding fine merino wool for apparel, Australia's share is even higher. A large proportion of the fine wool is harvested from Australia and the value of Australian wool exports in 2015–16 is

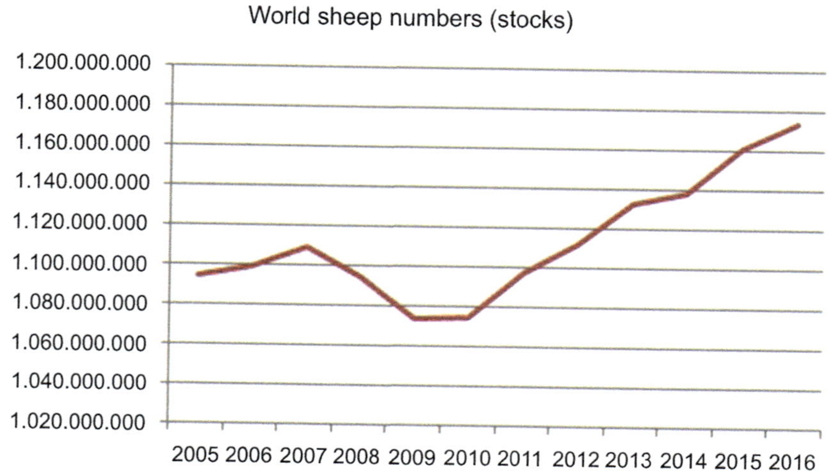

Figure 9.1 World sheep numbers for the last 10 years (http://www.fao.org/faosta).

estimated to be around $2.6 billion (https://www.iwto.org/wool; http://www.naturalfibres2, 2009; Pattinson et al., 2015; Wilcox, 2018). Fig. 9.2 summarizes world clean raw wool production by country in the year 2017.

The industry employs millions of people in wool production, harvesting, and throughout its many textile processing stages. There are number of factors which will fundamentally influence the conditions of global wool industry such as economy (demand, consumption and prices), population, seasonal conditions, technology, and fashion. Wool accounted for only 1%–2% of world textile consumption and production since the 2000s, but holds a much higher level of importance in terms of unit price and status (https://www.iwto.org/wool; Man-made fibre yearbook, 2017). The value of wool and its use category are largely determined by the diameter of the fibre. The classifications of wool are very broad and general, while various wool growing countries may grade their wool into more distinct categories. In the industry, the classifications of wool according to fibre diameters are generally as follows:

- *Fine wool:* Wool fibre of <25 μm from merino breed for the manufacture of apparel, grown in throughout Australia, South Africa, South America, New Zealand and the western United States.
- *Long stapled wool:* Wool fibre of 25–35 μm from merino crossbreeds and other sheep breeds for the manufacture of interior products such as upholstery, grown in England, Scotland, New Zealand, etc.
- *Coarse wool:* Wool fibre of >35 μm from merino crossbreeds and another sheep breed for the manufacture of interior products such as carpets, grown in throughout Asia especially in the Middle East, United States, etc (https://www.iwto.org/wool; http://www.sheep101.info/; Kuffner et al., 2012).

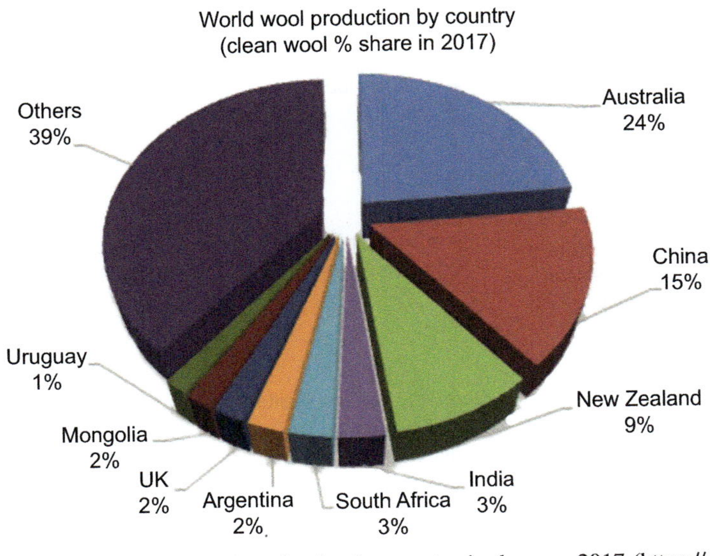

Figure 9.2 World clean raw wool production by country in the year 2017 (https://www.sheepusa.org/, 2017; Wilcox, 2018).

The wool supply chain starts with sheep farms in different regions of world. Thereafter, world wool processing and manufacturing industry is very complex since the trade of wool can be made in different forms such as raw wool, semi-processed and finished wool products. Retailer and consumer demands also play an important role in the wool industry supply chain. Simplified diagram of the wool industry can be seen in Fig. 9.3.

Processing and manufacturing of wool are economically important industries in several countries. Australia, China and New Zealand are the leading producers of raw wool. China is also a major raw wool importer and the leading manufacturer of semi and finished wool products. Therefore, China is the world's leading exporter of semi and finished wool products and dominates the world wool textile industry, particularly for wool clothing. Other semi and final wool product manufacturer countries are Italy (yarn and garment), India (carpet and rug), United Kingdom (garment), Germany (yarn and garment), Turkey (garment and carpet), Belgium (carpet), United States (carpet), and Bangladesh (garment). Major wool consuming countries are European countries, Japan and the United States (Pattinson et al., 2015; Erdem and Erdogan, 2005).

9.3 Chemistry and morphology of wool fibres

Keratin fibres are very complex in nature due to the consistency of a large number of histological components made of various proteinous agents (Bradbury and Peters,

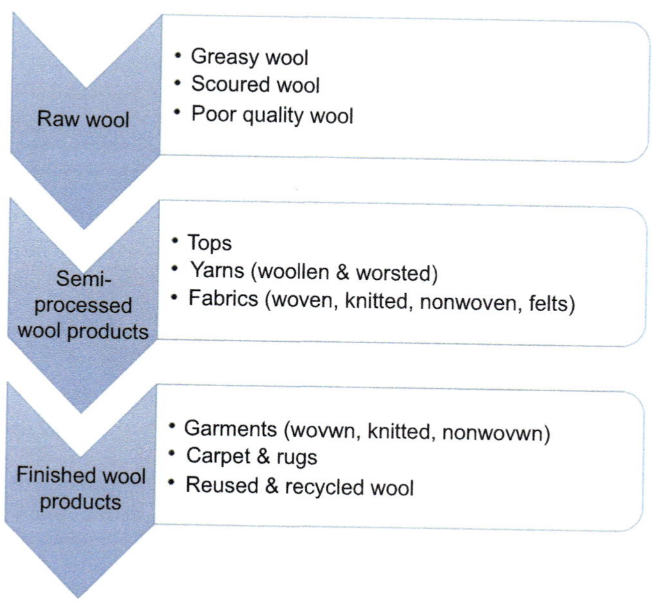

Figure 9.3 Simplified diagram of the wool industry.

1972). Wool is a complex multi-cell system composed of inanimate cells which vary in composition, shape and properties (Wojciechowska et al., 1999). The chemical material of the wool fibres is keratin – a fibrous protein (Wojciechowska et al., 1999). Wool and other keratin fibres are protein-based fibres which are mainly composed of carbon, hydrogen, oxygen, nitrogen and sulphur (Höcker et al., 2002). The primary constituent of the wool fibre is proteins. In addition to many types of amino acids, wool fibres also largely consist of sulphur-containing groups, lipids, mineral salts, nucleic acid residues and carbohydrates. Sulphur creates disulphide bond which is formed two sulphur atoms and principal cross-linking between wool elements (Höcker et al., 2002). The large quantities of lipids and carbohydrates play important role on fibre function and properties (Jones et al., 2006).

Wool fibres consist of more than 20 types of amino acids which are characterized by their side chain. The amino acids can be characterized by acidic and basic structure (e.g. aspartic acid and histidine), hydroxyl groups as the side chain (e.g. Serine), the presence of sulphur (e.g. cysteine), and non-existence of reactive groups as the side chain (e.g. glycine) (Höcker et al., 2002). They impart different features such as hydrophobic, hydrophilic, acidic or basic. The sequence of amino acids defines the possibility of intermolecular links, the molecular cohesion of keratin and access of amino acids to chemical reaction (Wojciechowska et al., 1999). The fibre elements having oppositely charged side chains are linked through salt bridges which can be stated as the second kind of cross-linking (Höcker et al., 2002). There is also one more type of linking, isopeptide bond, that is recognized as third of cross-linking and formed between a glutamic or an aspartic acid and a lysine residue (Höcker et al., 2002). Wool fibres also include hydrogen bonds which are formed between amide and hydrogen-related groups.

Wool and other mammalian fibres also contain lipids, namely wool grease, partially blocks application of textile finishing processes to these fibres. The lipids can be removed by souring in pre-treatment processes of the wool fibres. Lanolin can be classified as the most commonly known lipid for the wool fibres. There are many types of lipids such as cholesterol, fatty acids and polar lipids such as ceramides, cerebrosides and cholesterol sulphate (Höcker et al., 2002). In addition, wool fibres contain mineral salts, nucleic acid residues and carbohydrates.

The basic components of a wool fibre can be noted as medulla (for coarse wool fibres), cortex and cuticle (Fig. 9.4). The wool fibre morphology is composed of cuticle layers, cortex, macro-fibrils and micro-fibrils from the outer side to inner side in details. Cuticle surrounds fibre elements and consists of three major components labelled; epicuticle, exocuticle (two segments, A Layer and B Layer) and endocuticle (Fig. 9.5). The surface structure of wool fibre is the cuticule (Wojciechowska et al., 1999). The epicuticle layer is the exterior resistant membrane and hydrophobic in nature that causes of low-degree surface wettability of the fibre. This membrane layer mainly consists of lysine and also the low content of lipid and protein. The exocuticle is the key part of the cuticle layer containing a great amount of cystine. The existence of one cross-link per five amino acid residues and also few polar groups can be the reason of the resistant structure of wool to dyes and acids (Bradbury, 1976). The endocuticle is the inner part of the cuticle membrane that is produced from cuticle cell

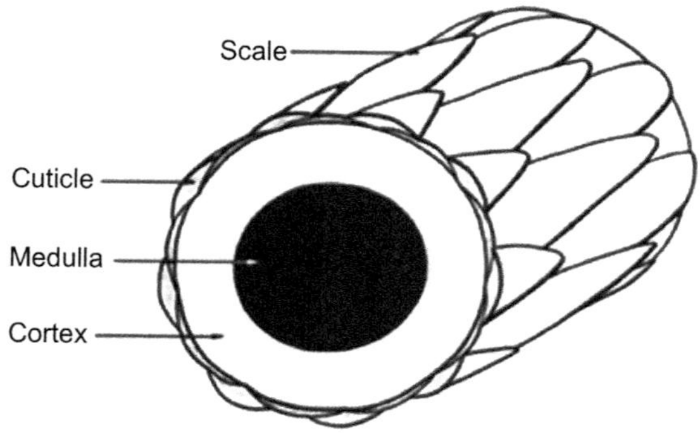

Figure 9.4 Morphology and components of wool fibre (Tridico, 2009).

cytoplasm, nucleus, and cell organelles. The endocuticle is possibly the most vulnerable part of the keratin fibres. The main difference of the endocuticle apart from the exocuticle is high tendency to absorb dyes and other reagents (Jones et al., 2006).

Wool has a complex structure that can cause processability problems such as wettability and accordingly dyeability. Additively, the diffusion and absorption of the liquids onto wool fibres are also affected by surface morphology. Therefore, understanding the surface structure of the wool is essential. The unique properties of wool fibres like other keratin fibres are their overlapping cuticular scales which present on the outermost surface of the fibre. The position of these scales looks like overlapping tiles on a roof. These scales determined the felting tendency of the keratin fibres. Felting is a physical transformation which is interlocking of the scales and matting together that is caused by heat, moisture and mechanical effects. As a result of felting, the fabric shrinks and gets thicker and undergoes characteristic changes such as losing its elasticity. Although felting is an undesired physical formation for keratin fibres, felt padding and felted fabrics are used in apparel, furniture and hat industry.

Cortex is a significant element of wool and other mammalian fibres that includes two types of cortex cells; para-cortex and ortho-cortex cells. Cortical cells play a determinative role in physical, chemical and mechanical properties of the fibres. The proportions of ortho-cortex and para-cortex cells change in the range of 60%–90% and 40%–10% of a wool fibre, respectively (Jones et al., 2006). These two cortical cells arrange in a specific helical structure along the fibre axis which ends up with bilateral manner. Spindle-shaped 'cortical cells' orients parallel to the fibre axis (Rajabinejad et al., 2018). In the curvature structure of cortical cells, para-cortex places in inner side, whereas ortho-cortex locates in the outer part. This bilateral arrangement possibly results in the crimp structure of the fine wool fibres (Jones et al., 2006). There are basic differences between ortho-cortex and para-cortex cells which are linked to the differences in amino acid composition. Para-cortex contains more (about 25%) cystine, whereas tyrosine, phenylalanine and glycine contents are higher than those of para-cortex by 42%, 23% and 18%, respectively (Bradbury, 1976). Ortho-cortex is more capable to absorb dyes and treat with other reagents in comparison with para-cortex. The third type of cortex cell has been lately identified in Merino wool

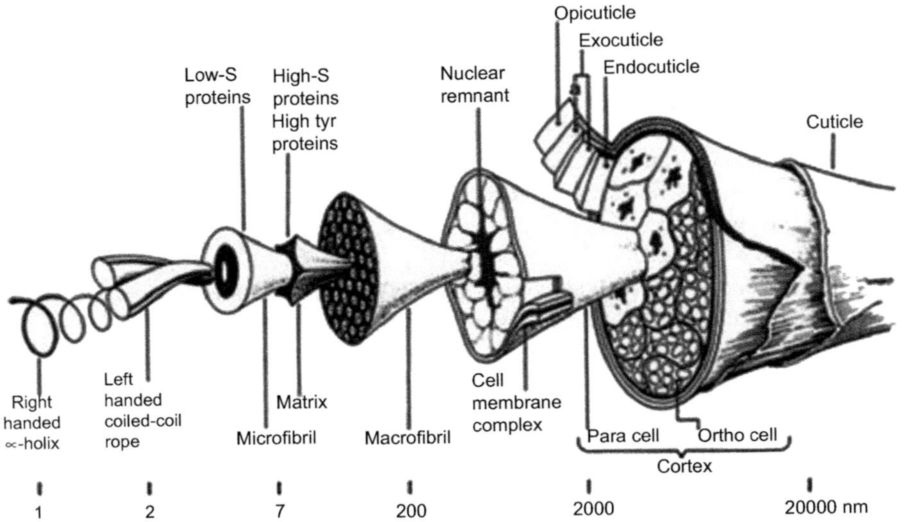

Figure 9.5 Schematic illustration of a broken cross-section of a fine wool fibre (Kim et al., 2015).

sections as called meso-cortex. Meso-cortex is hardly differentiated from para-cortex cells. It is noted that the fibrils run parallel to the fibre axis and are tightly and regularly packed in a hexagonal array (Morton and Hearle, 2008).

Cortical cells consist of 5–20 micro-fibrils which have a diameter of 100–300 nm in the inter-macro-fibril matrix. The matrix is composed of cytoplasmatic and nuclear remnants of the keratinocytes (Höcker et al., 2002). macro-fibrils comprise 500–800 micro-fibrils having a diameter of nearly 72°A. micro-fibrils are arranged in the matrix in a low degree of order, which are also called as hard keratin intermediate filaments.

Wool is a high-quality protein fibre and widely used as a high-quality textile material (Xu et al., 2006). Like other natural fibres, wool fibres are vulnerable to acids, alkalis, bleaching agents and even water. The effects of chemicals on wool are explained in detail under the following subtitles.

9.3.1 *Effect of alkali*

Wool is vulnerable to alkali even in mild conditions. Alkalis degrade proteins at low concentrations and temperatures, at a significant rate. As a result of the alkali attack, two new cross-links are created, namely lanthionine and lysinoalanine (Ziegler, 1964). The newly formed cross-links are strong and alkali-modified wool has a less solubility in urea-bisulphite solution. In mild alkali conditions, the wool fibres achieve more hydrophilic nature as a result of partial cleavage of linked fatty acids. Moreover, the wetting and dyeing time can be also enhanced by a combination of alkali and a cationic surfactant. It must be emphasized that alkali can deteriorate mechanical properties and cause yellowing of the wool fibre.

9.3.2 Effect of acid

Wool can hydrolyze to peptide bonds when treated with acids. Peptide bonds on both sides of amino acid residues with acidic side-chains (aspartic and glutamic acid residues) are hydrolyzed faster than other peptide bonds. Peptide bonds having cysteic acid residues on both sides can also easily hydrolyze. In finishing processes of wool such as shrink-proofing, the fibres exposed to oxidation become more vulnerable to acids. The dying of wool fibres is generally performed in acid solutions (Sawada and Ueda, 2004; Ferreo and Peiolatto, 2012). The dying process of wool takes place 100°C at pH 2–7 in general (Jones et al., 2006). The pH values should be controllable, thus in acidic conditions the wool fibres can be damaged and fibre strength can decrease.

9.3.3 Effect of water

Wool fibres contain −CONH and other polar groups which can react with water. When wool fibres are exposed to water above the temperature of 50°C, lanthionine and lysinoalanine cross-links are formed and the solubility of the wool decreases. The temperature should be controllable when treating the wool fibres with water. The wool garments can contract at temperatures between 128°C and 140°C, depending on the rate of heating (Jones et al., 2006; Lewin, 2006).

9.3.4 Effect of oxidation

Oxidative agents are utilized in wool fibre treatment process. Oxidized wool is soluble in alkali solutions. When treating wool with hydrogen peroxide, it should be paid attention because alkaline solutions can damage the wool fibres. Further, performic acid and peracetic acid are also used in the oxidation of the keratin fibres. Some hydrolysis of peptide bonds occurs during oxidation with peracetic acid, less with performic acid. Oxidation is also performed to develop shrink-proof wool fibres for mild conditions particularly before the application of a polymer on the surface of the wool fibres. In the concept of commonly used procedure of producing shrink-proof wool, chlorination is carried out. Chlorination targets the surface structure of wool fibres. Besides, the treatment also decreases the damage to bulk properties of the wool fibres. Wool is attacked by chlorine bleaches and progressively yellowed by the short ultraviolet wavelengths in sunlight (Moody and Needles, 2004). Hypochlorite can react and damage wool fibres irreversibly. Even dilute solutions can change the colour of the wool fibres (Moody and Needles, 2004). In addition to these chemical effects, wool fibres are vulnerable to moths and microorganisms. Being exposed to UV light having short wavelengths, wool fibres can degrade and yellow. Additively, hydrogen peroxide has been widely applied in the finishing process of textile, especially for bleaching of wool fabrics. Hydrogen peroxide treatment is effective on enhancing moisture transport, water absorbing and shrink-proof properties of the wool fibres (Wang et al., 2012). Ozone treatment can be categorized in oxidative treatments of the wool. Ozone is ecologically friendly due to effectiveness in reducing water consumption. Ozone treatment can increase whiteness degree of the wool that depends on procedure time and temperature (Gülümser et al., 2009).

9.4 Physical properties of wool fibres

Based on microscopic observation of wool, for both fine and coarse fibres, the surface of the fibres is covered with scales (Fig. 9.6). The outer surface consists of flattened overlapping cuticle cells that provide a protective sheath around cortical cells (Barani and Calvimontes, 2014). In the case of fine wool, one scale can enfold one fibre, whereas more than one scale is required to enfold the coarser fibres. Medulla, inside layer of wool, varies with the fineness of the fibre (Figs. 9.6 and 9.7). Medulla exists and becomes more distinct and coarser when the fibre gets coarser. Medulla that can also be expressed as air pocket in the fibre can vary with the origin and the size of the mammalian fibre. In finer fibre, the medulla layer cannot be seen because finer fibres may not have this layer.

As mentioned before, wool is a natural protein fibre composed of keratin-type protein (Barani and Calvimontes, 2014). Some typical physical properties of wool fibres are given in Table 9.1. Moreover, wool fibres have a highly crimped structure in nature. The fineness, physical, mechanical and chemical properties can vary with the variety of sheep (age, cultivated area, etc.) and the part of the sheep where the fibres were obtained from. In addition to given physical parameters of wool fibres in Table 9.1, wool fibres are thermally insulative materials in nature. Wool fibres have more flame retardant tendency particularly more than cellulosic fibres and comparable with synthetic fibres. In case of touch with fire, wool fibres can char easily. Upon removing the char, the fibres will seem undamaged.

The physical quality of wool affects the processing route and processing efficiencies, the quality and potential uses of end products and thus the commercial price of both raw wool and wool textiles. Various subjective and objective methods available for assessment of wool quality were presented and discussed in the literature for years (Cottle and Baxter, 2015). Methods of measuring wool fibre parameters

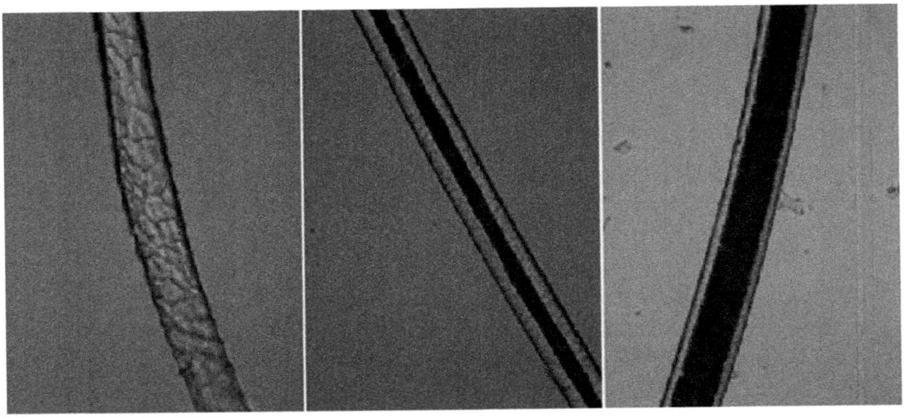

Figure 9.6 Optical images of unmedullated, medullated, and kemp wool (from left to right).

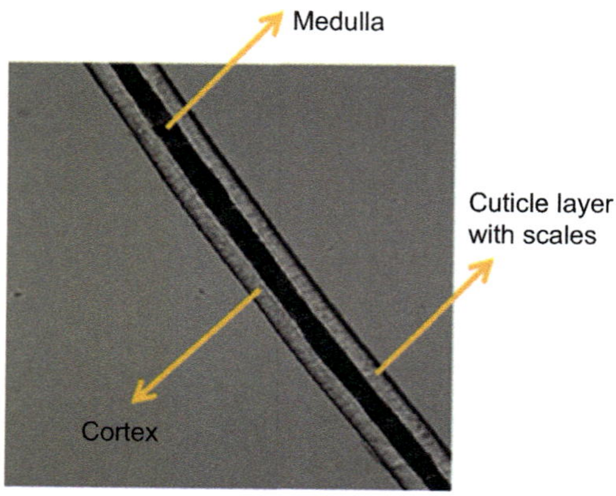

Figure 9.7 Basic components of the optically viewed coarse wool fibre.

Table 9.1 Some typical parameters of the wool fibres (Barani and Calvimontes, 2014; Hearle et al., 2002).

Parameter		Value
Crystallinity (%)		47.8
Tenacity (g/tex)	Dry	9–18
	Wet	7–16
Elongation at break (%)	Dry	25–40
	Wet	25–60
Initial modulus (g/denier)	Dry	25
	Wet	10
Moisture regain (%)		13–18
Specific gravity (g/cm^3)		1.28–1.32
The temperature that wool fibres are thermally stable up to (°C)		150
Limit oxygen index (LOI) (%)		25
Thermal conductivity (mWm^{-1}K^{-1}) at 0% moisture regain		194
Thermal conductivity (mWm^{-1}K^{-1}) at 30% moisture regain		290

developed in various research laboratories have been published as test method specifications by the international organizations, and then adopted for industry after extensive evaluation and verification. Several instruments have been also developed for testing wool. The wool industry now has a number of objective measurements defining a wide range of characteristics of raw and processed wool, which are supported by internationally recognized Standard Test Methods (Cottle and Baxter, 2015; Sommerville, 1998). These include

- *Fibre diameter* (IWTO projection microscope, airflow, laser scan and optical fibre diameter analyzer-OFDA test methods)
- *Fibre diameter CV* (IWTO laser scan and OFDA test methods)
- *Yield and vegetable matter* (IWTO Test method 19–2012)
- *Staple length and strength* (IWTO Test method 30–2007)
- *Colour* (IWTO Test method 56–2013)

These methods are now well established and commonly use in mill specifications, performance monitoring and commercial transactions.

9.5 Area of application

Wool is the oldest textile raw material and it has been used as a clothing raw material since ancient times. After civilization of human being, wool also has become the major raw material of interior textiles such as floor coverings. In the early 20th century, humankind met with the synthetic polymers that are currently being used as raw materials for several industrial products. The invention of man-made fibres dominated the wool market and minimalized the importance of wool in the textile industry. However, due to the concerns about increased environmental pollution and the influence of widely used synthetic polymer-based products on human health, new raw material sources and improvement on production process have become compulsory. There are two main approaches to dealing with the problems derived from these issues; recycling of synthetic polymers for sustainability and production of renewable industrial goods from bio-based polymers such as wool. On the other hand, in recent times, textile raw materials find an alternative and a new area of applications in technical textile products. Wool, with its exceptional properties, can be a potential raw material for some technical applications such as biodegradable composites, insulation and filtration materials. Wool fibres having enhanced-functional properties and technical performances with finishing technologies have also the potential for usability in technical textiles. Wool as the renewable and sustainable fibre is still an important raw material of apparel, home textiles, and technical textiles. It can be used in every field of textiles, see Fig. 9.8. Traditional applications of wool for clothing and interior textiles were reviewed and discussed in various publications (Kuffner et al., 2012; Simpson and Crawshaw, 2002; Von Bergen, 1972; Johnson and Russell, 2009; Muthu and Gardetti, 2016).

Wool is increasingly being used in technical applications in which its unique properties and the opportunities for specific enhancements can be profitably utilized. Contrary to common belief, wool is an extremely technical fibre and its presence in technical textiles cannot be neglected. For instance, due to its natural low flammability characteristics, wool has traditionally been the fibre of choice in many technical applications. Moreover, the availability of alternative flame-resist treatments and the ability to blend wool with flame-resist fibres ensures a good future for wool in technical end

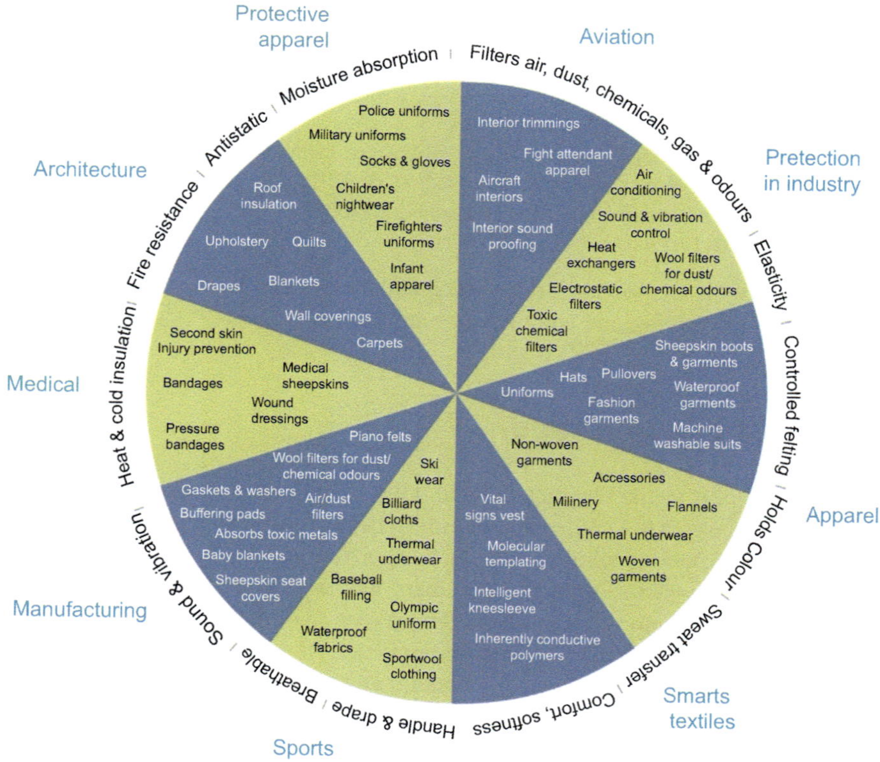

Figure 9.8 Wool usage in textile world (https://www.iwto.org/wool; Kuffner et al., 2012).

products such as protective garments, transportation and military textiles. Wool has also well-known medical applications such as bedsore and pressure sore prevention. Further its range of applications can be widened by means of chemical modification of the wool (Sawant et al., 2014; Naylor). Technical textile applications of wool were also considered in various paper and research. Johnson, N.G.A et al. examined wool's attributes for technical textiles and introduced a wide range of new and future applications. Advanced of wool in various technical textile application was discussed considering the structure and properties of wool (Johnson et al., 2003). Evidence of the role of wool in human health and well-being was reviewed by Laing R. and Swan P., considering 240 documents and principal parameters such as physical contact between textiles/garments and human skin, thermal and moisture properties (Laing et al., 2016). A research on innovative geotextile design with poor quality and recycled wool was also presented by Broda et al. Biodegradation properties of wool geotextiles, which were designed for erosion control, were examined and discussed (Broda et al., 2016). How wool is currently being used in the field of intelligent textiles and how it may be exploited in this field in the future was described by Collie, S.R (Collie et al., 2009). Applications of nanotechnology to wool at both research and commercial levels and performance of wool in sport suits are also summarized in the same book by Denning R. and Holcombe, B., respectively (Johnson and Russell, 2009). Various technical applications of wool were also discussed in *Handbook of Natural Fibres Volume* 2 edited by Kozlowski R.M (Kozlowski, 2012). Consequently, wool has still a valuable raw material for various textile applications and new innovations also expand the field of wool applications beyond traditional high-value products.

9.6 Recycling of wool

Humankind dominates the life on earth and shapes the future of all species. The current world population of 7.6 billion is expected to reach 8.6 billion in 2030 and 9.8 billion in 2050 (https://www.un.org/develo, 2017) which increased the demand for higher production and consumption leading to water, air and soil pollution. World Health Organization (WHO) reported that 9 out of 10 people breathe polluted air and environmental pollution annually causes 1.7 million deaths in children under 5 years old (http://www.who.int/news-r, 2017). The world economy has been 'linear' for decades in which the raw materials are being used to make goods and at the end or even in some cases before their end of lifetime eventually thrown away. Nonetheless, a promising alternative 'circular economy' has received a significant worldwide attention with the aim of improved sustainable production. In a circular economy, products are manufactured, used, re-used and recycled which is also called 'closing the loop' (https://www.ellenmacarthu). Recycling of materials can help to grow economies, conserve natural resources, reduce pollution, saves energy and creates jobs (http://europa.eu/rapid/pr). Textile industry being the second most polluting industry with high environmental costs have to take the responsibility and move towards a circular economy to close the loop (Snoek, 2017). Recently, sustainable production, recycling, bio-based and biodegradable materials gained importance along with research on textile and wool recycling (Fig. 9.9) (Sandin and Peters, 2018).

Wool is the leader among animal fibres with 1.160 million kg annual production (clean raw wool) in 2015 (https://www.iwto.org/wool). This fibre has high added value and unique properties such as heat insulation and low flammability thanks to its high LOI (limiting oxygen index) value (Kuffner et al., 2012). Such unique properties make it a preferred candidate not only for conventional textiles but also for products with demanding applications such as insulators, separators, filters and fillers (Russell and Blackburn, 2009). Conventionally, wool-based products were re-used or sent to landfills. Since landfills become less available in the major consuming countries and also in order to prevent pollution, present promising solutions of reusing and recycling of wool fibre should be considered as a privilege. Additionally, decomposition of wool in landfills produces methane which contributes to greenhouse gas (http://www.bir.-org/indust). In such a situation, wool recycling remains an essential choice to reduce environmental damage with sustainable production and promote the circular economy. Recycling also offers the most convenient way towards valuable and top quality wool-based products.

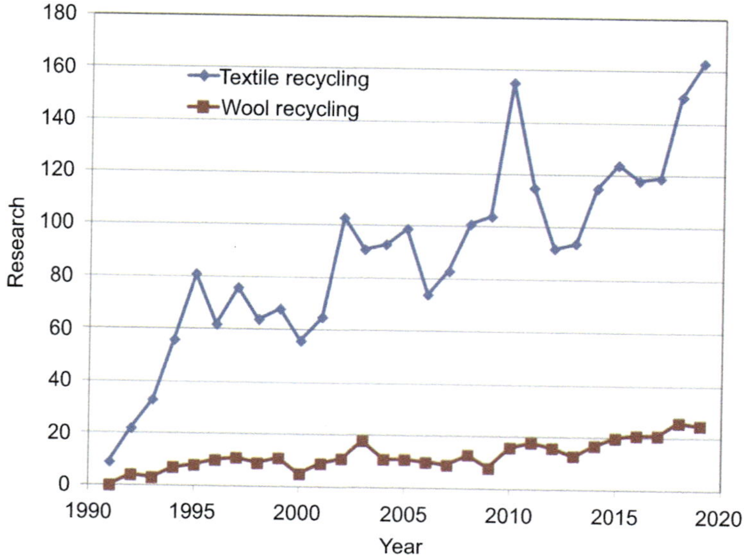

Figure 9.9 Published research on the recycling of textiles and wool with respect to years, x-axis refers the years (1990–2018) and y-axis the number of articles published (Data taken from Scopus, Keywords: textile recycling*, wool recycling* 14.03.19).

Wool fibre has been recycled for many years to produce a material called shoddy, where recycled woollen rag was blended with virgin wool (Shell, 2015). Wool can be pre-consumer or post-consumer waste in different forms; fibre, yarn, knitted, woven and non-woven fabrics. Pre-consumer products are by-products (waste products) from yarn spinning and fabric manufacture, while post-consumer waste comes out after disposal/use. Post-consumer wastes which are not suitable for yarn spinning and fabric production can be fed into the making of products such as mattresses, reinforcement of composites, fillings or non-woven products of various types. The classification of wool fibre waste process is given in Fig. 9.10.

Collection of post-consumer waste is supplied by bins and curbsides, recycling companies and donations to charities and also in the stores of big fashion retailers. According to the results of textile waste project SIFO (Consumption Research Norway), 68% of garments mainly with wool or hair are donated to charities, 18% directly send to the bin, 11% given to family or friends and 4% stored for the museal purpose (Laitala and Klepp, 2018). World's leading fashion retailers have successful attempts to encourage their clients for recycling, one of these retailers is collecting unwanted and worn garments of any brand to be re-worn, re-used and re-cycled, with the project 'The Garment Collection Program' in its stores all over the world (http://www2.hm.com/en_gb/). Besides, another company focuses on up-cycling of wastes instead of down-cycling refers to lower quality materials that are the most prevailing type of recycling in the textile industry. In the year 2016, the company have collected 7.102 tonnes of garment, footwear and accessories in 392 stores in six countries (http://static.inditex.com, 2016).

Figure 9.10 The simplified classification of the wool fibre waste process. Open-loop recycling refers to manufacturing of lower quality products from waste wool whereas closed-loop recycling refers to similar quality products.

Following the collection, wastes are processed either with open/close-loop recycling or incineration. Incineration is the method to recover energy by burning wastes and turning into heat, electricity and steam. This recycling method is mostly preferred by developed countries but existing concerns about pollution are still debatable (Wang, 2010).

Closed-loop recycling is a process through which a product is recycled back into the similar or an alike material without a significant value lost. High-quality woollens are generally closed-loop recycled for instance; they are shredded and re-used as yarns in new garment production. Additionally, recycling of woollen garments can be classified as up-cycling and down-cycling refers to a higher and lower quality material, respectively. As an example of up-cycling, high-value protein keratin which is used as a raw material for cosmetics and pharmaceuticals can be extracted from wool fibre waste with green technology (Moore et al., 2016; Jiang et al., 2018). Wang et al. (2016) extracted keratin from wool using L-cysteine as a reducing agent for the dissolution of wool (Wang et al., 2016). Besides keratin extraction, wool-based nonwovens have decent oil sorption properties which can be used to reduce the removal of diesel fuel, crude, base, vegetable and motor oil from water (Radetic et al., 2008). Radetic et al. (2003) proposed that recycled wool-based nonwovens (78% wool and 22% polyester) have decent oil absorption properties for several such applications (Radetic et al., 2003). Moreover, wool-rich carpets (ground-up wool carpet) can be used as a fertilizer to improve the yield of soil (Zoccola et al., 2015) and wool-waste may be used as nutrient source and growth medium constituent for container grown plants (Zheljazkov et al., 2009).

In the open-loop recycling, the wool fibre has a lower quality but anyhow gets a second life chance such as insulation materials in the automotive industry, pads for mattresses and filling materials. Wool fibre waste is a potential source for insulation and eco-friendly alternative of glass fibre based materials in the construction industry (Bosia et al., 2015). Semi-rigid insulation panels for the eco-building sector which is made from recycled sheep wool combined with hemp have promising acoustic absorption and thermal conductivity properties with low environmental impact (Pennacchioa et al., 2016). Patnaik et al. (2015) produced nonwoven mats made from waste wool and recycled polyester with good thermal insulation, acoustic absorption, moisture absorption and fire properties that can contribute to green building applications (Patnaik et al., 2015). Whether it is open or closed-loop recycling, it is possible to increase the life-span of the wool fibres by up to 10 more years and to have a continuous cycle after every expiration (Russell et al., 2016).

There are limitations for collection, separation (wool blended with natural and synthetic fibres) and characterization of wool fibres and blends which confines the attempt for closing the loop. Likewise, during mechanical recycling of wastes (closed-loop recycling), the length of fibres are shortened which limits the yarn spinning. According to the Waste and Resources Action Programme (WRAP), branding of garments (labelling) also makes the re-use of wastes challenging (i.e. separation of label fibre material, procedures to protect the brand) which may be solved with de-labelling of garments with various kind of methods such as; manual unpicking, microwave treatment, over-branding and quick release adhesives (Philpot et al., 2013). In order to decrease the amount of waste that is sent to landfill, these limitations need to be addressed (Peterson, 2015).

In conclusion, wool fibre is a valuable and expensive raw material which has to be considered as a resource, not a waste. It has a wide range of application with the advantages of its unique structure such as; its natural ability to breathe, multi-climate properties, insulation properties, sound and oil absorption and low flammability. Reusing and recycling of woollen garments are viable alternatives and a step towards circular economy and sustainable production. Besides the reusing and recycling of wool fibre waste, individuals can help to close the loop by; reducing consumption, using second-hand clothing, buying new products made from recycled materials or new products that can be easily recycled and donating worn garments.

9.7 Future trends

Wool is an environment-friendly, high-value fibre and its production consumes significantly less energy than popular synthetic fibres (Henry et al., 2015). It is found in many sectors including ready wear, furnishing, architecture, medical applications and protective clothing. Contrary to common belief, wool clothing is not an irritant rather superfine merino wool may assist the management of skin disease for children (Su et al., 2017). Considering the advantages of this unique fibre, there will be sustained esteem for wool in long-term even it is non-competitive for the price but competitive on value and fibre quality. The assistance and support of governments for sheep husbandry and wool production in developing countries may help to lift the wool market. In addition to this, recent developments in nanotechnology, composites and additive chemistry are continually pushing the wool fibre potential further.

Demands for sustainable and environment-friendly products will increase 'certified organic wool' production. Prospects for the development of animal welfare will demark the sheep husbandry regulations for an enhanced level production (https://www.iwto.org/anim). Wool fibre quality will be improved by selective sheep breeding (Gong et al., 2016) and transgenic sheep (Powell et al., 1994) which meet the expectations for raw wool requirements. Moreover, ongoing researches on the genetics of wool can widen the properties.

In near future, attention will be growing towards the development of smart textiles via nanotechnology (Denning, 2009). Self-cleaning wool textiles, antimicrobial treatments on wool fabrics, stain blocking, water repellency, colouration of fabrics with nanoparticles and novel techniques such as wool dyeing by microencapsulation will be developed (Ki et al., 2007; Pakdel et al., 2015; Li and Sun, 2007; El-Zawahry et al., 2007).

The role of wool in chemistry is on the rise and expected to expand further. Wool itself can act as a catalyst to achieve Henry reaction (Haring et al., 2016). Wool metal complexes and nanoparticles supported on wool can also be used effectively as a catalyst for several reaction/transformations (Wu et al., 2011; Xue et al., 2004; Shaabani et al., 2015, 2016). Wool as catalyst support helps to retain good activity during reaction without any additional activation treatment and also can be used for several times. Their easy availability and reusability make them a cheaper alternative to their counterparts. Such new and innovative applications of wool ensure that it will bring several profits to products, humankind and the planet as a whole.

9.8 Conclusion

Wool has been used as traditional raw material of textiles since ancient times. The fleece of sheep was most probably the first clothes of mankind. As a garment, wool had lots of function at those times such as temperature regulation of the human body, protection against cold and fire. It may be used as an armour against wild animals and plants and its antibacterial properties cured to injured bodies. It was also a floor covering for caves. Moreover, colouring of wool with natural dyes is another historical fact. From past to present as a result of civilization and technological developments life-style of consumers changes continuously, however the role of wool as a textile raw material is still similar with ancient times. Currently, wool garments are the best choice for us to feel warm, since the moisture transportation properties of wool are superior. Wool carpets and interior textiles also have several application areas now such as houses, cinemas, theatres, and automotive in which natural flame resistance property of wool is important. Wool is still being used as a medical textile to prevent bedsore. Besides, wool maintains its importance for the textile industry with new applications especially in technical fields. Considering the advantages of this unique fibre, there will be sustained esteem for wool in long-term even it is non-competitive for the price but competitive on fibre quality and function. In conclusion; parallel to advances in social life and technology, researches on wool focus on new areas such as smart materials, nanotechnology, genetics, sustainability, recycling, and chemical modifications. This chapter describes the current situation of wool production and industry, summarizes the unique physical and chemical properties of wool, and focuses on the recent developments in wool functions and applications considering the above mentioned future trends.

9.9 Sources of further information and advice

- American Sheep Industry Wool Journal.
- Journal of Natural Fibers.
- The Fibre Year, https://www.thefiberyear.com/home/.
- IWTO Market Information Review and Outlook.
- IWTO Guidelines for Wool LCA.
- Kuffner, H., Popescu, C., 2012. Wool Fibres in Handbook of Natural Fibres Volume I edited by Kozlowski, R.M, Cambridge. Woodhead Publishing Limited, 171–195.
- Tao, X. (Ed.), 2001. Smart Fibres Fabrics and Clothing. Woodhead Publishing, Cambridge.
- Koncar, V. (Ed.), 2016. Smart Textiles and Their Applications. Woodhead Publishing, Cambridge.
- Shishoo, R. (Ed.), 2005. Textiles in Sport. Woodhead Publishing, Cambridge.
- Breniquet, C., Michel, C. (Eds.), 2014. Wool Economy in the Ancient Near East and the Aegean, Oxford, Oxbow Books.
- Cottle, D.J., Baxter, B.P., 2015. Wool metrology research and development to date. Textile Progress, 47:3, 163–315.
- https://www.woolmark.com/innovation/developments/.
- Muthu, S.S. (Ed.), 2015. Handbook of Life Cycle Assesment (LCA) of Textiles and Clothing. Woodhead Publishing, Cambridge.
- Blackburn, R.S. (Ed.), 2009. Sustainable Textiles Life Cycle and Environmental Impact, Woodhead Publishing, Cambridge.
- IWTO Re-use and Recycling. https://www.iwto.org/re-use-and-recycling.

References

Barani, H., Calvimontes, A., 2014. Effects of oxygen plasma treatment on the physical and chemical properties of wool fibre surface. Plasma Chem. Plasma Process. 34, 1291–1302.

Bosia, D., Savio, L., Thiebat, F., Patrucco, A., Fantucci, S., Piccablotto, G., Marino, D., 2015. Sheep wool for sustainable architecture. Energy Procedia 78, 315–320.

Bradbury, J.H., Peters, D.E., 1972. The chemical composition of wool. Text. Res. J. 42, 471–474.

Bradbury, J.H., 1976. The morphology and chemical structure of wool. Pure Appl. Chem. 46, 2–4.

Broda, J., Kobiela-Mendrek, K., Rom, M., Grzybowska-Pietras, J., Przybylo, S., Laszczak, R., 2016. Biodegradation of wool used for the production of innovative geotextiles designed to erosion control. In: Fangueiro, R., Rana, S. (Eds.), Natural Fibres: Advances in Science and Technology towards Industrial Applications from Science to Market. RILEM Book Series 12.

Collie, S.R., 2009. Intelligent Wool Apparel. In: Johnson, N.A.G., Russell, I.M. (Eds.), Advances in Wool Technology. Woodhead Publishing Limited, Cambridge.

Cottle, D.J., Baxter, B.P., 2015. Wool metrology research and development to date. Text. Prog. 47 (3), 163–315.

Denning, R., 2009. Enhancing Wool Products Using Nanotechnology. In: Advances in Wool Technology, pp. 248–264.

El-Zawahry, M.M., El-Shami, S., El-Mallah, M.H., 2007. Optimizing a wool dyeing process with reactive dye by liposome microencapsulation. Dyes Pigments 74 (3), 684–691.

Erdem, N., Erdogan, U.H., 2005. In: An Overview of Textile Raw Material Production in Turkey, the 1st Euro-Mediterranean Textile & Clothing Conference, (Cairo, Egypt).

Ferreo, F., Peiolatto, M., 2012. Ultrasound for low-temperature dyeing of wool with acid dye. Ultrason. Sonochem. 19 (3), 601–606.

Gong, H., Zhou, H., Forrest, R.H., Li, S., Wang, J., Dyer, J.M., Hickford, J.G., 2016. Wool keratin-associated protein genes in sheep-A Review. Genes 7 (6).

Gülümser, T., Akça, C., Bahtiyari, M.İ., 2009. Investigation of the effects of ozone on whiteness degree in wool finishing. Tekstil ve Konfeksiyon 1, 52–55.

Haring, M., Pettignano, A., Quignard, F., Tanchoux, N., Diaz Diaz, D., 2016. Keratin protein-catalyzed nitroaldol (Henry) reaction and comparison with other biopolymers. Molecules 21 (9).

Hearle, J.W.S., 2002. Physical properties of wool. In: Simpson, W.S., Crawshaw, G.H. (Eds.), Wool Science and Technology. CRC Press.

Henry, B.K., Russell, S.J., Ledgard, S.F., Gollnow, S., Wiedemann, S.G., Nebel, B., Swan, P., 2015. LCA of wool textiles and clothing. In: Muthu, S.S. (Ed.), Handbook of Life Cycle Assesment of Textiles and Clothing, vol. 172.

Höcker, H., 2002. Fibre morphology. In: Simpson, W.S., Crawshaw, G.H. (Eds.), Wool Science and Technology. CRC Press.

http://dnfi.org/statistics/.

http://europa.eu/rapid/press-release_IP-15-6203_en.htm.

http://static.inditex.com/annual_report_2016/en/our-priorities/commitment-to-the-excellence-of-our-products/focus-on-a-circular-economy.php.

http://www.bir.org/industry/textiles/?locale=en_US.

http://www.fao.org/faostat/en/#data/QA.

http://www.naturalfibres2009.org/en/fibres/wool.html.

http://www.sheep101.info/sheeptypes.html.

http://www.who.int/news-room/detail/06-03-2017-the-cost-of-a-polluted-environment-1-7-million-child-deaths-a-year-says-who.

http://www2.hm.com/en_gb/ladies/shop-by-feature/16r-garment-collecting.html.

https://www.ellenmacarthurfoundation.org/circular-economy.

https://www.iwto.org/animal-welfare.

https://www.iwto.org/wool-production.

https://www.sheepusa.org/Resources_Publications (ASI Wool Journal May 2017).

https://www.un.org/development/desa/en/news/population/world-population-prospects-2017.html.

Jiang, Z., Yuan, J., Wang, P., Fan, X., Xu, J., Wang, Q., Zhang, L., 2018. Dissolution and regeneration of wool keratin in the deep eutectic solvent of choline chloride-urea. Int. J. Biol. Macromol. 119, 423–430.

Johnson, N.A.G., Russell, I.M., 2009. Advances in Wool Technology. Woodhead Publishing Limited, Cambridge.

Johnson, N.A.G., Wood, E.J., Ingham, P.E., McNeil, S.J., McFarlane, I.D., 2003. Wool as a technical fibre. J. Text. Inst. 94 (3–4), 26–41.

Jones, L.N., Rivett, D.E., Tucker, D.J., 2006. Wool and related mammalian fibres. In: Lewin, M. (Ed.), Handbook of Fiber Chemistry. CRC Press.

Ki, H.Y., Kim, J.H., Kwon, S.C., Jeong, S.H., 2007. A study on multifunctional wool textiles treated with nano-sized silver. J. Mater. Sci. 42 (19), 8020–8024.

Kim, N.K., Lin, R.J.T., Bhattacharyya, D., 2015. Effects of wool fibres, ammonium polyphosphate and polymer viscosity on the flammability and mechanical performance of PP/wool composites. Polym. Degrad. Stab. 119, 167–177.

Kozlowski, R.M., 2012. Handbook of Natural Fibres Volume 2: Processing and Applications. Woodhead Publishing Limited, Cambridge.

Kuffner, H., Popescu, C., 2012. Wool Fibres. In: Kozlowski, R.M. (Ed.), Handbook of Natural Fibres Volume I. Woodhead Publishing Limited, Cambridge, pp. 171–195.

Laing, R., Swan, P., 2016. Wool in human health and well-being. In: Fangueiro, R., Rana, S. (Eds.), Natural Fibres: Advances in Science and Technology towards Industrial Applications from Science to Market. RILEM Book Series 12.

Laitala, K., Klepp, I.G., 2018. Use Phase of Apparel: A Literature Review for Life Cycle Assessment with Focus on Wool, Technical Report, SIFO Consumption Research Norway.

Lewin, M., 2006. Handbook of Fiber Chemistry, third ed. CRC Press.

Li, D., Sun, G., 2007. Coloration of textiles with self-dispersible carbon black nanoparticles. Dyes Pigments 72 (2), 144–149.

Man-made fiber yearbook, 2017. In: Published by Chemical Fibers International, October 2017 (Germany).

Moody, V., Needles, H.L., 2004. Tufted Carpet: Textile Fibers, Dyes, Finishes and Processes, first ed. William Andrew Publishing, New York.

Moore, K.E., Mangos, D.N., Slattery, A.D., Raston, C.L., Boulos, R.A., 2016. Wool deconstruction using a benign eutectic melt. RSC Adv. 6 (24), 20095–20101.

Morton, W.E., Hearle, J.W.S., 2008. Physical Properties of Textile Fibres, fourth ed. Wood head Publishing Series in Textiles.

Muthu, S.S., Gardetti, M.A., 2016. Sustainable Fibres for Fashion Industry, vol. 2. Springer.

Naylor, G. The wool fibre and its applications, CSIRO Textile and Fibre Technology. www.woolwise.com.

Pakdel, E., Daoud, W.A., Afrin, T., Sun, L., Wang, X., 2015. Self-cleaning wool: Effect of noble Metals and Silica on visible-light-induced Functionalities of nano TiO_2 colloid. J. Text. Inst. 106 (12), 1348–1361.

Patnaik, A., Mvubu, M., Muniyasamy, S., Botha, A., Anandjiwala, R.D., 2015. Thermal and sound insulation materials from waste wool and recycled polyester fibers and their biodegradation studies. Energy Build. 92, 161–169.

Pattinson, R., Wilcox, C., Williams, S., Curtis, K., 2015. NSW Wool Industry and Future Opportunities: A Report to the NSW Department of Primary Industries Paper 1: Trends and Drivers for the Global and Australian Wool Industry (Australia).

Pennacchioa, R., Savioa, L., D. B., Thiebata, F., Piccablottoa, G., Patruccob, A., Fantuccic, S., 2016. Fitness: sheep-wool and hemp sustainable insulation panels. In: Paper Presented at the 8th International Conference on Sustainability in Energy and Buildings (Turin/Italy).

Peterson, A., 2015. Towards Recycling of Textile Fibers. Chalmers University of Technology. MsC Thesis.

Philpot, B., Pye, A., Stevens, G., 2013. De-labelling Branded Corporate-Wear for Re-use. WRAP, pp. 1–70.

Powell, B.C., Walker, S.K., Bawden, C.S., Rogers, G., 1994. Transgenic sheep and wool growth: possibilities and current status. Reprod. Fertil. Dev. 6 (5), 615–623.

Radetic, M., Jocica, D., Petar, M.J., Petrovica, Z., Thomas, H.F., 2003. Recycled wool-based nonwoven material as an oil sorbent. Environ. Sci. Technol. 37 (5), 1008–1012.

Radetic, M., Ilic, V., Radojevic, D., Miladinovic, R., Jocic, D., Jovancic, P., 2008. Efficiency of recycled wool-based nonwoven material for the removal of oils from water. Chemosphere 70 (3), 525–530.

Rajabinejad, H., Zoccola, M., Patrucco, A., Montarsolo, A., Rovero, G., Tonin, C., 2018. Physicochemical properties of keratin extracted from wool by various methods. Text. Res. J. 88 (21), 2415–2424.

Russell, I.M., 2009. Sustainable wool production and processing. In: Blackburn, R.S. (Ed.), Sustainable Textiles Lifecycle and Environmental Impact. Woodhead Publishing, pp. 63–85.

Russell, S., Swan, P., Trebowicz, M., Ireland, A., 2016. Review of wool recycling and reuse. In: Fangueiro, R., Rana, S. (Eds.), Natural Fibres Advances in Science. RILEM Book Series 12.

Sandin, G., Peters, G.M., 2018. Environmental impact of textile reuse and recycling. J. Clean. Prod. 184, 353–365.

Sawada, K., Ueda, M., 2004. Evaluation of the dyeing mechanism of an acid dye on protein fibres in supercritical CO_2. Dyes Pigments 63 (1), 77–81.

Sawant, R., Bhute, A., Basuk, M., 2014. Wool in technical textile. https://www.textiletoday.com.bd/wool-in-technical-textiles/.

Shaabani, A., Hezarkhani, Z., Badali, E., 2015. Wool supported manganese dioxide nano-scale dispersion: a biopolymer based catalyst for the aerobic oxidation of organic compounds. RSC Adv. 5 (76), 61759–61767.

Shaabani, A., Hezarkhani, Z., Faroghi, M.T., 2016. Wool-SO_3H and nano-Fe_3O_4@Wool as two green and natural-based renewable Catalysts in one-pot isocyanide-based multicomponent reactions'. Monatshefte für Chemie - Chemical Monthly 147 (11), 1963–1973.

Shell, H.R., 2015. Shoddy Heap: A Material History Between Waste and Manufacture. History Technol. 30 (4), 374–394.

Simpson, W.S., Crawshaw, G.H., 2002. Wool Science and Technology. CRC Press.

Snoek, S., 2017. Circular Economy in the Textile Industry. Wageningen UR University. MsC Thesis, Supervised by Kris Van Koppen.

Sommerville, P., 1998. In: Objective Measurements – More than Pretty Numbers, Working with Wool Seminar, 15th September 1998. Canberra Institute of Technology.

Su, J.C., Dailey, R., Zallmann, M., Leins, E., Taresch, L., Donath, S., Lowe, A.J., 2017. Determining effects of superfine sheep wool in INfantile eczema (DESSINE): a randomized paediatric crossover study. Br. J. Dermatol. 177 (1), 125–133.

Tridico, S.R., 2009. Natural Animal Textile Fibres: Structure, Characteristics and Identification. Identification of Textile Fibers, pp. 27–67.

Von Bergen, 1972. Wool Handbook, third ed. John Wiley & Sons Inc, England.

Wang, X., Shen, X., Xu, W., 2012. Effect of hydrogen peroxide treatment on the properties of wool fabric. Appl. Surf. Sci. 258, 10012–10016.

Wang, K., Li, R., Ma, J.H., Jiana, Y.K., Cheb, J.N., 2016. Extracting keratin from wool by using L-cysteine. Green Chem. 18, 476–481.

Wang, Y., 2010. Fiber and textile waste utilization. Waste Biomass Valori. 1 (1), 135–143.

Wilcox, C., 2018. The Global Wool Market: The Good, the Bad and the Prospects, the 87th IWTO Congress, 14-16 May 2018 Hong Kong.

Wojciechowska, E., Włochowicz, A., Wesełucha-Birczyńska, A., 1999. Application of Fourier-transform infrared and Raman spectroscopy to study degradation of the wool fibre keratin. J. Mol. Struct. 511–512, 307–318.

Wu, S., Ma, H., Jia, X., Zhong, Y., Lei, Z., 2011. Biopolymer-metal complex wool–Pd as a highly active heterogeneous catalyst for heck reaction in aqueous media. Tetrahedron 67 (1), 250–256.

Xu, W., Ke, G., Wu, J., Wang, X., 2006. Modification of wool fibre using steam explosion. Eur. Polym. J. 42, 2168–2173.

Xue, L., Jia, B., Tang, L., Ji, X.F., Huang, M.Y., Jiang, Y.Y., 2004. Asymmetric hydration of alkenes catalyzed by wool-palladium complex. Polym. Adv. Technol. 15 (6), 346–349.

Zheljazkov, V.D., Stratton, G.W., Pincock, J., Butler, S., Jeliazkova, E.A., Nedkov, N.K., Gerard, P.D., 2009. Wool-waste as organic nutrient source for container-grown plants. Waste Manag. 29 (7), 2160–2164.

Ziegler, K., 1964. New crosslinks in the alkali-treated wool. J. Biol. Chem. 239 (8), 2713–2714.

Zoccola, M., Montarsolo, A., Mossotti, R., Patrucco, A., Tonin, C., 2015. Green hydrolysis as an emerging technology to turn wool waste into organic nitrogen fertilizer. Waste Biomass Valori. 6 (5), 891–897.

Mohair, cashmere and other animal hair fibres

10

L. Hunter
CSIR and Nelson Mandela University (NMU), Port Elizabeth, Eastern Cape, South Africa

10.1 Introduction

This chapter covers the various animal hair fibres that are generally referred to as luxury or rare animal fibres, excluding the wool from sheep and also silk (which are described in Chapter 13, volume 1: Silk fibres, and Chapter 11, volume 1: Wool fibres). These luxury fibres have been discussed in two previous books (Hunter, 1993; Franck, 2001) to which the reader is referred for more information and a more detailed list of references. The fibres considered here represent considerably less than 0.1% of global fibre production. Therefore, in terms of production volumes, they are insignificant. Nevertheless, they are highly sought after and play a significant role in the luxury, high-value-added end of the market, notably the apparel market. Most of these fibres are produced by animals that inhabit inhospitable mountainous regions, covering a range of altitudes and extreme climates, where a highly insulating and protective fibrous coat (mostly a double coat) is essential for survival. The hair and fibres often tend to be medullated, which combines good insulation with lightness. Owing to the extremes of temperature they encounter, most of these animals have developed two distinctly different coats: (i) an outer coat consisting of coarse-medullated guard hairs, produced by the primary follicles, which offers protection from the sun, rain and dust and (ii) a finer and shorter down hair or fibre (undercoat or inner coat), produced by the secondary follicles, which provides outstanding insulation against extremes in temperature. The only exceptions are the Angora goat, alpaca and vicuña which, like sheep, are essentially single coated, although still containing a combination of secondary and primary follicle fibres. In some cases, e.g., llama, there is also a group of intermediate fibres. For a given species, the younger the animal, the higher the altitude, and the sparser the food, the finer the fibre generally. The nature of their environment also largely determines the natural colour of animal fur (Stoves, 1976), for camouflage and other purposes; for example, the white winter fur of arctic animals, the grey-brown fur of forest and marshland animals and the yellowish fur of animals living in sandy areas (Stoves, 1976; Bereck, 1994).

The fine undercoat (down) fibres of two-coated animals, which are generally shed during spring, are most valuable from a textile perspective, because of their combination of fineness, softness, lightness and good thermal insulation, and need to be separated from the undesirable coarse guard hair, either by hand or mechanically, a process called dehairing. The finer the fibre and the lower the percentage of coarse fibres (guard hair) remaining in the fine component after dehairing, the better the textile

Handbook of Natural Fibres. https://doi.org/10.1016/B978-0-12-818398-4.00012-8

quality and value of the fibre. In this respect, successful dehairing is largely dependent upon the differences between the two components in terms of fibre diameter/rigidity/linear density, friction and length and interfibre cohesion. Townend et al. (1989) and Algaa and Mägel (1992a) have dealt with the dehairing of luxury animal hair fibres.

This chapter focuses on the down fibres (fine undercoat) of those animals with two coats, the coarser guard hair (outer coat/hair) not being dealt with, except in a few isolated cases, and then only briefly. The ASTM D 2816 defines coarse hair as hair or fibres coarser than 30 μm, the criterion often used to differentiate between guard hair (outer coat) and down hair (undercoat).

Essentially, the main speciality animal hair fibres covered can be grouped into the following three main families or groups (see Fig. 10.1):

- Goat *(Capra hircus)*
  - Cashmere (*Capra hircus laniger*)
  - Angora (*Capra hircus aegagrus*)
  - Cashgora
- Camel *(Camelid/Camelus)*
  - Alpaca (*Lama pacos*)
  - Bactrian camel (*Camelus bactrianus*)
  - Guanaco (*Lama hunchus* or *Lama guanicoe*)
  - Llama (*Lama glama*)
  - Vicuña (*Vicugna vicugna*)
- Bovine *(Bos)*
  - Musk-ox (*Ovibos moschatus*)
  - Yak (*Bos (poephagus) grunniens*)

Angora rabbits *(Oryctolagus cuniculus)* do not belong to any of the aforementioned family groupings.

The South American Camelid (Camelidae) family consists of four species, three, namely llama *(Lama glama)*, and alpaca *(Lama pacos)*, both domesticated, guanaco *(Lama hunchus* or *Lama guanicoe)* a wild species (Bustamante et al., 2002), representing the genus *Lama* and the fourth, vicuña *(Vicugna vicugna)*, being a separate and also wild species (Greaves and Rainsford, 2005). They are now largely found in the Southern Andes, centred mainly on Peru's Altiplano region, known as the 'puna', a flat and rather barren terrain some 4600 m above sea level. There are also hybrids of the llama and alpaca. According to Greaves and Rainsford, 2005, the origins of the Camelid family (alpaca, guanaco, llama and vicuña) are not very clear, it being generally agreed that the two domesticated breeds, namely alpaca and llama, descended from the guanaco. It is also fairly widely believed that the alpaca originated as a hybrid between the llama and guanaco (Greaves and Rainsford, 2005), occurring after the Spanish Conquest of Peru in the early 1530s. According to Peruvian Andes archaeological sites, evidence suggests that the domestication of Camelids occurred between 7500 and 12,000 years ago (Greaves and Rainsford, 2005).

McGregor et al. (2018) recently gave a concise resumé of the various techniques of identifying and differentiating between the various animal fibres.

Table 10.1 compares the fibre characteristics and production figures for the different fibre types. It should be noted, however, that figures quoted in the literature often do

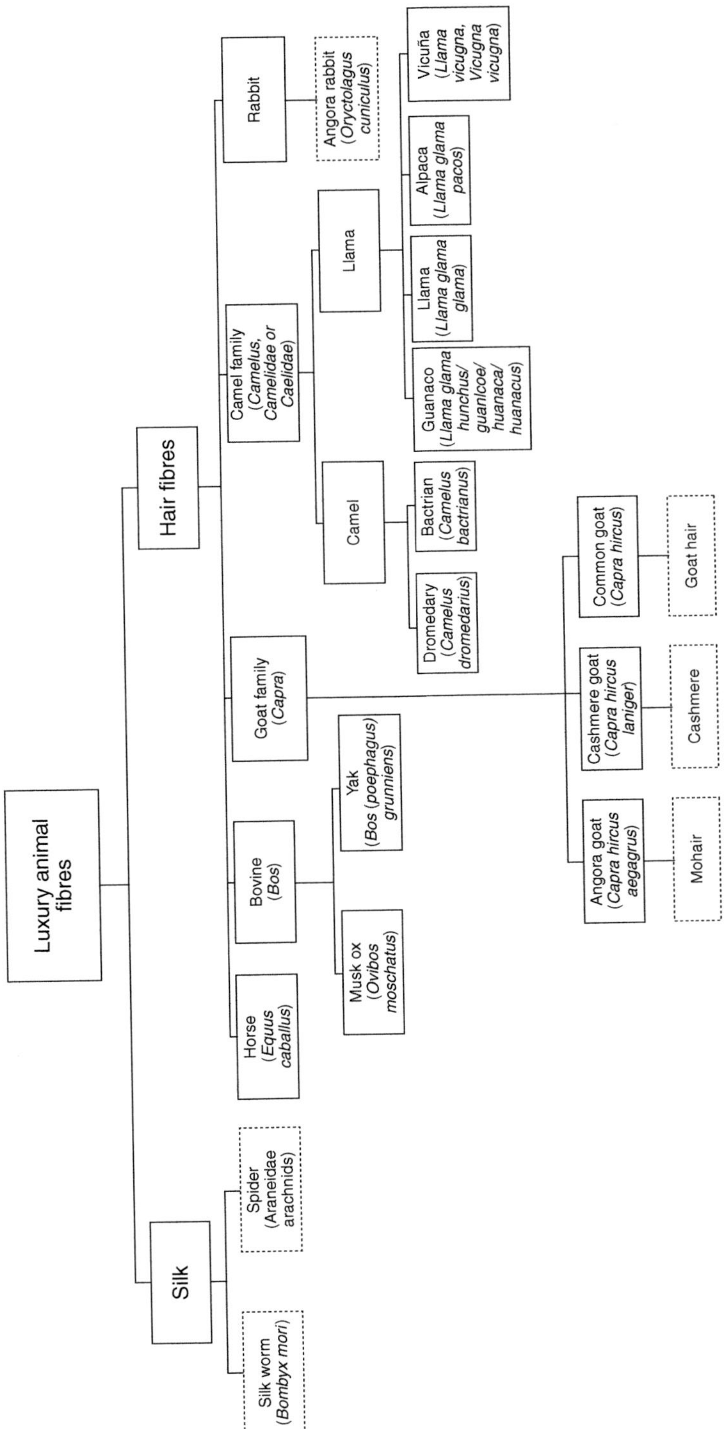

Figure 10.1 Main luxury animal fibre groups.

Table 10.1 Approximate commercial fine textile fibre properties and production[a].

Fibre	Mean fibre diameter range (μm)	Typical (average) mean fibre diameter (μm)	Processed mean fibre length range (mm)	Typical processed mean fibre length (mm)	Fibre curvature (°/mm)	Raw fibre production per animal (kg/year)	Global fibre production (mkg/year)
Alpaca	18–35	27	60–80	75	30	2–4	7
Angora rabbit	10–18	13.5	30–60	45	60	0.5–1	4
Camel	15–25	18	30–60	50	80	2–5	1
Cashgora	18–23	20	30–60	55	35	2–4	0.05
Cashmere	13–19	15.5	20–50	35	60	0.1–0.4	10
Guanaco	14–20	15	20–50	35	-	0.3–0.9	0.01
Llama	20–32	27	50–80	65	35	1.1–3.5	1
Mohair	22–40	30	50–110	80	12	2–6	5
Musk-ox	13–20	17	40–80	60	-	2–3	0.003
Vicuña	11–15	13	20–40	30	80	0.2	0.05
Yak	15–25	20	25–50	35	75	0.2–1.3	7

[a]Compiled from various sources, with inputs also from Alpha Tops SA, Incalpaca TPX (Textiles Peruanos de Exportación), Seal International, Dr B. McGregor: Deakin University.

not specify whether they refer to greasy, raw, whole fleece or down fibres, or to staple or fibre length, etc. Animal fibres are composed mainly of keratin protein that comprises 19 amino acids (McGregor et al., 2018). Their chemical composition, on a mass (weight) basis, is typically ≈16%–17% nitrogen, ≈3.2%–3.7% sulphur, ≈0.38%–0.42% ash (which includes ≈0.09%–0.12% calcium, ≈0.017%–0.023% phosphorus and some small amounts of sodium), ≈27% oxygen, ≈47% carbon and ≈6% hydrogen (McGregor et al., 2018). Wuliji et al. (2000) reported on the production performance and repeatability in lightweight, fleece weight and fibre characteristics of alpacas in New Zealand, highlighting the fibre length and diameter variability over the body of the animal.

Tucker et al. (1990a) presented the composition of the whole greasy (raw) fleeces of various speciality fibres (see Table 10.2). McGregor and Quispe Peña (2018) found that species, fibre diameter and animal age in many cases played a role in the cuticle and cortical cell morphology of rare animal fibre.

Most luxury or speciality animal fibres tend to be finer, less crimped and smoother than wool, their cuticular scales also being less pronounced (flatter or thinner), typically 0.4 μm thick or even thinner (generally 0.3%–0.4 μm, with mohair sometimes up to 0.5 μm) compared to wool which is generally 0.6 μm and thicker (typically about 0.8 μm) (Wortmann, 1991). The scales are also more widely spaced, average scale frequency per 100 μm varying from about 4 for mohair to about 12 for vicuna (McGregor and Quispe Peña, 2018). The ellipticity of rare animal fibres varies from about 1 (i.e., perfectly circular) to about 2, being on average between about 1.2 and 1.3 (McGregor and Quispe Peña, 2018). Where crimp is present, it is generally not as pronounced, or of as high a frequency, as that of fine wools and in some cases is better described as a curling or waviness. Fibre cohesion and friction are consequently lower than for wool, requiring special conditions, or blending with other fibres, such as wool, for acceptable mechanical processing performance and yarn quality. After dehairing, the down fibres are generally processed into yarn following either the worsted route

Table 10.2 The composition of raw whole fleeces (Tucker et al., 1990a).

Fibre	Moisture (%)	Grease (%)	Water solubles (%)
Wool	11.0–11.7	9.5–27.0	3.9–7.1
Mohair	12.0–14.4	1.2–8.0	1.8–4.2
Australian cashmere	10.7–13.9	0.7–2.5	1.2–3.5
Chinese cashmere	11.1–12.9	5.0–7.2	2.3–3.0
Cashgora	13.2	1.2–2.8	0.6
Llama	12.0	2.8	–
Alpaca	10.9–14.4	2.8–3.9	0.6–2.4
Camel	9.9	0.5–1.1	–
Yak	10.4	12.3	–

for the longer fibres or the woollen route for the shorter fibres. Mechanical processing, although mostly done on machinery similar to that used for wool, needs to be adapted and optimized to suit the specific requirements of each of these fibres. Most of the expert knowledge to do so is held as a closely guarded secret by the various companies which process these speciality fibres. This also applies to the dehairing process, which depends, amongst other things, on the differences in length and diameter of the respective fibre populations, as well as the percentage of down fibres present.

Chemically, these fibres belong to the same protein (keratin) family of fibres as wool, although their fine structures (morphological) and surface structures do differ, they also often being medullated. Table 10.3 (Tucker et al., 1988) gives the chemical (amino acid) compositions of some of these fibres. Because of their similarities in chemical composition, it is not easy to differentiate between them chemically. Ways in which they can be differentiated are covered in Chapter 8. Mengüç et al. (2015) found Angora rabbit hair, cashmere and silk fabrics to be smoother, softer and less prickly than alpaca and mohair fabrics.

McGregor and Schlink (2014) investigated the felting propensity of a wide range of rare animal fibres, finding that resistance to compression, fabric curvature (possibly also crimp form) and mean fibre diameter affected felting (felt ball density).

These hair fibres are mostly processed, dyed and finished using similar machinery to that for wool, but the settings and the conditions (e.g., dyeing and finishing recipes, etc.) are adjusted to suit the specific characteristics, notably length and fineness, of each fibre. Because of the smooth and medullated nature of many of these speciality fibres, which differs from that of most apparel wools, dyeing recipes need to be adjusted from those used for wool, if a specific dye shade is to be achieved. Great care is also taken in practice to select the dyeing and finishing conditions, such as treatment time, temperature and pH, so that the desirable characteristics of the fibres, for example, lustre and softness, are not deleteriously affected. Danforth (2017) has dealt with cashmere production, beneficiation and export in Mongolia. Aspects concerning animal fibre identification per sé are covered in Chapter 8, and not here.

10.2 Alpaca

10.2.1 Fibre production, harvesting and properties

The ancient Incas already treasured the alpacas for their fine and soft fleeces which were reserved for royalty. After the Spanish Conquest of the Incas in the 16th century, alpacas were banished from lowland pastures to the hostile highland terraces of Peru (Liddle, 2002), their numbers decreasing greatly. Alpaca (*Lama pacos* or *Lama glama pacos)* from the genus *Lama,* descended from the Guanaca, is a member of the Camelidae family and mainly inhabits the 'Altiplano', a vast high, arid plateau of southern Peru, Bolivia, Chile and Argentina (South American Andes) at altitudes around 4000–5000 m (Briggs, 1973), most (close on 90%) being found in Peru. Alpacas (Fig. 10.2) are also found in significant numbers in other countries, such as the USA, Australia and Europe.

Table 10.3 Diameter and amino acid composition (mol.%) of speciality animal fibres (Tucker et al., 1988).

Amino acid	Aust. cashmere[a]	Chinese cashmere	Aust. cashgora[b]	Mongolian yak	Camel	Wool	Guanaco	White alpaca	Black alpaca	Llama	Vicuña
Diameter (μm)	**12.7–17.9**	**17.1 ± 2.4**	**16.2 ± 3.3**	**18.4 ± 1.9**	**18.7 ± 2.6**	**17.1**	**13.9 ± 1.7**	**26.4 ± 6.0**	**40.7 ± 9.8**	**19.5 ± 2.6**	**12.3 ± 1.7**
Amino acid											
Cysteic acid	0.1–0.2	0.1	0.1	0.4	0.3	–	0.5	0.2	0.6	0.4	0.5
Aspartic acid	6.6–7.1	6.7	7.1	6.6	8.1	6.9	7.2	7.3	6.9	7.2	7.1
Threonine	6.6–7.3	7.0	6.9	6.5	6.6	6.8	6.5	6.3	6.2	7.0	6.4
Serine	10.7–12.7	10.9	11.5	10.3	10.4	12.0	11.1	9.6	10.3	11.3	10.6
Glutamic acid	11.2–13.0	13.0	13.5	12.5	13.6	12.8	13.7	14.6	14.0	16.0	14.3
Proline	8.1–9.0	7.7	7.5	7.5	7.2	8.1	7.9	7.6	7.8	8.4	7.9
Glycine	9.0–10.2	8.8	8.4	9.3	7.8	9.5	8.1	7.9	7.9	5.9	8.1
Alanine	5.8–6.2	5.5	5.7	5.8	5.9	5.8	5.5	5.6	5.4	6.8	5.5
Cystine	4.2–5.6	5.5	4.8	5.4	4.6	4.6	6.0	6.0	7.6	6.3	5.9
Valine	5.0–5.7	5.7	6.0	5.9	5.9	5.2	5.8	6.0	5.9	6.3	6.1
Methionine	0.3–0.5	0.4	0.4	0.5	0.7	0.5	0.5	0.4	0.5	0.5	0.4
Isoleucine	2.6–3.0	3.1	3.2	3.4	3.3	2.8	3.0	3.2	3.0	3.3	3.0
Leucine	7.4–8.4	7.4	7.7	8.0	7.7	7.9	7.2	7.8	7.2	8.3	7.5
Tyrosine	3.4–4.1	4.1	3.5	3.5	3.3	4.0	2.9	2.8	2.6	2.8	2.6

Continued

Table 10.3 Continued

Amino acid	Aust. cashmere[a]	Chinese cashmere	Aust. cashgora[b]	Mongolian yak	Camel	Wool	Guanaco	White alpaca	Black alpaca	Llama	Vicuña
Diameter (μm)	**12.7–17.9**	**17.1 ± 2.4**	**16.2 ± 3.3**	**18.4 ± 1.9**	**18.7 ± 2.6**	**17.1**	**13.9 ± 1.7**	**26.4 ± 6.0**	**40.7 ± 9.8**	**19.5 ± 2.6**	**12.3 ± 1.7**
Phenylalanine	2.6–3.0	2.9	2.8	3.2	3.0	2.7	3.1	3.0	2.5	3.2	2.6
Lysine	2.5–3.0	2.8	2.8	3.0	2.7	2.9	2.5	2.8	2.6	2.9	2.7
Histidine	0.6–0.8	0.8	0.6	1.0	0.8	0.9	0.8	0.9	1.0	1.0	1.0
Arginine	6.4–7.2	7.5	7.4	7.5	8.0	6.7	7.7	7.9	8.2	8.7	7.7

[a]10 samples from individual goats.
[b]Sample from one goat.

Figure 10.2 Group of alpacas.
Source: Inca Tops, SA, Arequipa, Peru.

Moore et al. (2011) found that at relatively high wind speeds, Huacaya alpaca pelts with finer fibres resulted in a greater heat loss than coarser fibres.

Alpacas are ruminants (or pseudoruminants), the only member of the Camelid family bred specifically for their fibres, being smaller than llamas but larger than vicuñas. They thrive on ichu (ychu) grass and at high altitudes. Originally shearing (mainly hand shearing) took place biennially, but now annually during the summer rainy season, from November to April, which is also their breeding period. In the case of alpaca, there is a less distinct difference between the fibre diameter of the outer coat and the undercoat (down). There are, however, essentially two types of alpaca, namely the relatively rare Suri which has long (some 140–170 mm) (Siles, 1998), almost straight (or slightly wavy) (Baretta et al., 2012) and silky (lustrous) hair (somewhat similar to lustre long wool sheep fleeces) (Sponenberg, 2010). It tends to be finer, with fewer coarse fibres (coarser than 35 or 30 μm) (Valbonesi et al., 2010), and more valuable than that of the second type, namely the bigger and heavier Huacaya (Wakaya/Wayko), which is relatively short and crimpy and has some 90–110 hair/mm (Franck, 2001), and a diameter ranging from ≈25 to 30 μm (Siles, 1998). The coat of the latter consists of compact and crimped fibres (similar to Corriedale wool) (Sponenberg, 2010), the Huacaya accounting for close to 90% of the alpaca population and fibre production. Lupton and McColl (2011) reported a low correlation between the subjective and instrument measured lustre of Suri alpaca.

The average length of alpaca fibre ranges broadly from about 125 to over 200 mm (it grows about 100–125 mm per year), and the individual fibre diameters range from about 10 to 75 μm. Alpaca fibres, notable Suri, tend to be lustrous and are mostly medullated, elliptical in cross-section and pigmented. The alpaca fleece average diameter varies from about 17 to 35 μm (more typically 25–35 μm), with an overall average of around 26/27 μm (Greaves and Rainsford, 2005) (coefficient of variation [CV] from 25% to 35%). Alpaca fibres have become coarser over time, recently less than 10% being around 22.5 μm (20%, 20–22.5 μm) (Townend et al., 1989), 40% around 26.5 μm (35%, 24.5–26 μm) (Rainsford, 2005), 20% about 31 μm and more than 30% around 34 μm and coarser (Rainsford, 2007a), 45%–50% coarser than 31 μm (Rainsford, 2005). Some findings (Greaves and Rainsford, 2005) on mummified remains indicate that 1000 years ago, alpaca fibres may have been some 10 μm finer, and more even and less hairy than today, ranging from 14 to 21 μm, and mostly around 18 μm (Rainsford, 2007a). Nevertheless, according to Greaves and Rainsford, 2005, the fineness and related quality of alpaca fibres have not changed significantly from 1970 to 2005. Cervantes et al. (2010) obtained the following values for Huacaya alpaca: mean fibre diameter 23 μm (23.3% CV) and 88% for comfort factor, the corresponding values for Suri alpaca being 24.7 μm (26% CV) and 81%.

Alpaca has a variety of natural colours, including white, cream, light and dark fawns, piebald, greys and blacks; the lighter the colour, the more valuable the fibre, grey being particularly popular. Most alpaca is either pure white or flecked with a few brown or black hairs (Patthey, 2002), the balance being sorted into more than 20 shades, ranging from light fawn to black (Patthey, 2002), sorting being by hand, also according to fineness. Valbonesi et al. (2011) investigated the inheritance of white, brown and black coat colours in alpaca.

Today, alpaca fibre is mainly produced in Peru and also in significant quantities in Bolivia, Chile and Europe. It was estimated that in 2007, there were some 4.5 million alpacas in Peru, mainly being farmed by alpaca farmers (Alpagueros) in the Altiplano region bordering the city of Puno, 3871 m above sea level in the south of Peru (Liddle, 2002), Arequipa, being the centre of Peru's alpaca trade and processing centre (Liddle, 2002). Covered shelters built in Peru for alpacas significantly reduced the mortality rate of baby alpacas, which can be around 40% due to respiratory and digestive illnesses under normal cold conditions, increasing to 80 or even 100% under severe cold conditions (Anon.a). Alpacas have a productive life of up to 20 years, adults weighing between about 55 and 70 kg, and measuring about 1–1.1 m in height (up to the shoulder) (Coetsee, 2001). Peru, in around 2007, embarked on an important alpaca genetic improvement project (including DNA markers, artificial insemination and embryo transplants), recognizing that 95% of alpaca farmers are small breeders, without an awareness of, or access to, modern technology and genetic improvement technologies (Rainsford, 2007a).

Lupton et al. (2006) generated, and reported, comprehensive data on the fibre properties of the Huacaya alpaca produced in the USA at that time. In brief, the mean fibre diameter was 27.9 μm (CV 23.5%), mean staple length was 116 mm (CV 7%), mean curvature was 33% (CV 94%), mean staple strength was 50N/ktex (CV 39%), mean medullated fibre was 1720/10,000 and objectionable medullated fibre was 356/10,000. They concluded that the alpaca had a much higher staple strength and degree of medullation than wool, but a much lower crimp and resistance to compression, males producing stronger staples than females.

Staple strength, mean fibre diameter medullation and laboratory scoured clean yield all increased with age (Lupton et al., 2006).

Huarizo, which resembles llama and produces coarser hair than the alpaca, is the term used in Peru for the hybrid produced by crossing a male llama with a female alpaca, the reverse crossing (i.e., female llama with a male alpaca) being referred to as Misti (Greaves and Rainsford, 2005), the fleeces of which are not of the same quality as those from alpaca. When referring to the fibre per se, the term Huarizo is applied to alpaca fibre ranging from 29.1 to 31.5 μm, this quality tending to have a relatively broad medulla (Greaves and Rainsford, 2005). Crossing an alpaca male with a vicuña female produces the paco-vicuña, and the reverse cross, the vicuña-paco.

Braga et al. (2007) reported on the effect of altitude on alpaca fibre production. Indications are that alpaca grown at high altitudes suffer considerable weathering damage, particularly to their tips, partly due to higher UV irradiation at such levels (Villarroel-Leon, 1959), and the damage is more severe when the animals are only shorn every second year. The UV range from 290 to 310 nm causes the most damage by far to the fibre, the intensity of UV radiation increasing with increasing altitude.

Alpacas are now shorn, usually by hand, rarely by machine, annually (less often only every 18–24 months for combing, producing a staple length of over 200 mm), producing some 3 kg of fibre. Australian alpacas reportedly produce 6–8 kg of greasy fleece annually (Wang et al., 2008). The lack of grease (lanolin) on the fleece necessitates oiling of the shears. The fibre yield of a fleece is around 85%–90%, the raw 'white' alpaca fleece tending to be greyish in colour due to contaminants, such as dust, suint and grease (Stapleton, 1992); some typical values are given in Table 10.4 (Villarroel-Leon, 1959; Stapleton, 1992).

According to Valbonesi et al. (2010) (quoting various references), both alpaca and llama fibre diameter (fineness) is affected by the age, sex, fleece colour and position on the body of the animal, nutritional conditions, period of the year, herd and/or origin. The diameter of alpaca fibres increases with age (up to an age of about 4 years (Pumayalla, Barnett, Osorio)), for example, from 21 μm at 10 months of age to 25 μm at 4 years of age (Pumayalla and Levva, 1988), the CV being around 28%. Finer fleeces have a lower CV (e.g., 18% for 22 μm) than coarser fleeces (e.g., 26% for 36 μm), the fibres from the upper part of the body tending to be finer than those from the lower part of the body (Villarroel-Leon, 1959), with those from the rear

Table 10.4 Typical values for greasy alpaca fleeces (Villarroel-Leon, 1959; Wang et al., 2008; Stapleton, 1992).

Fibre	Grease (%)	Dust/vegetable matter (%)	Suint (%)	Yield (%)
Alpaca (Huacayo)	1–3	3–10	≈1	87–94
Wool (Med. Cross-bred)	15–30	5–20	4–7	43–57
Mohair (Cape)	1–8	5–10	2–4	78–92

part of the animal (britch area) invariably containing highly medullated, coarse and stiff fibres, similar to kemp fibres in cross-bred wool. Fibre length variation within a fleece is considerable (CV >40%). McGregor and Butler (2004) found that, for Australian alpacas, the Suri produced fleeces (midside saddle fleece samples) 2 μm coarser than the Huacayas, with mean fibre diameter increasing with both the age (up to 7 years) and liveweight of the animals, the CV of diameter ranging from 15% to 37% (average 24%), dropping from around 30% at 1 year of age to around 22% at 4 years of age. Mean fibre diameter of Huacaya alpaca (Chile) tends to increase with age, and follicle density tends to decrease (Crossley et al., 2014). Fibre curvature and resistance to compression decrease with an increase in mean fibre diameter, as is the case with most animal fibres (McGregor, 2014). On average, alpaca staple strength is some 40% higher than that of wool, and fibre strength is some 7% (Wang et al., 2003), greasy alpaca containing considerable less grease (≈1%-3%) and suint (≈1%) than wool (Wang et al., 2003).

McGregor et al. (2012a) found large variations in fibre attributes between 24 sites across a Huacaya fleece and that fibre curvature was highly negatively correlated with mean fibre diameter, decreasing from roughly 45°/mm at 20 μm to about 10°/mm at 55 μm. McGregor (2006) gave the following average values: clean washing yield 92%, greasy fleece weight 2.44 kg, fibre diameter 28 μm (CV 23.6%), staple length 112 mm, staple strength 76N/ktex, fibre curvature 28°/mm, Huacaya fibres having a greater curvature than Suri fibres.

Pinares et al. (2018) identified the genetic relationship between fibre diameter and medullation for white Huacaya alpaca, as a means to reduce problems related to prickliness.

Alpaca tends to have relatively more coarse fibres, e.g., coarser than 33 μm, than wool of the same mean diameter (Stapleton, 1992). Alpaca fibres are generally medullated, the degree of medullation and noncirculatory (ellipticity ≈ 1.2) increasing with increasing fibre diameter (Stapleton, 1992), the fibres becoming increasingly ovoid or kidney shaped as they become coarser. Fine alpaca has virtually no kemp, extrafine fibres (≈15−20 μm) being largely unmedullated, or have a fragmented medulla (≈20−30 μm), relatively fine fibres (≈30 μm) having an interrupted medulla, medium fibres (≈40 μm) a continuous medulla and coarse fibres (≥60 μm) a broad medulla. Coarse alpaca, with a mean fibre diameter around 34 μm, is strongly medullated, the medulla being slightly more than 50% of the fibre diameter (Greaves and Rainsford, 2005). Because of their medullae, the airflow method of fibre diameter measurement will underestimate the fibre diameter, the degree to which this happens depending upon the degree of medullation, the apparent density being about 1.24g/cc due to medullation (Vineis et al., 2011). More accurate diameter results can be obtained by projection microscope, Laserscan and Optical Fibre Diameter Analyser (OFDA)-based methods.

The ASTM standard specification (D2252-85, approved in 1991) for the fineness of different types of alpaca is given in Table 10.5.

Baby alpaca hair (≈22.5 μm) represents the finest hair from the first shearing, superfine alpaca (25.5/26 μm) forming a larger proportion of the production, also spanning the full range of more than 24 natural colours, coming from adult animals

Table 10.5 Fineness specifications for types of alpaca.

Type	Description	Average diameter (μm)
T extra	–	<22.0
T	Tui, 12 months of age	22.00–24.99
X	Extrafine adult	22.00–24.99
AA	Medium adult	25.00–29.99
A	Coarse	30.00–35.99
SK	Skirtings	>30.00
LP	Locks and pieces	>30.00

Source: Standard specification for fineness of Types of Alpaca (ASTM-D2252–85, approved 1991).

(Michell, 2002), almost half of Peru's production being >31 μm. The mean fibre diameter can range from as fine as 21 μm for Bolivian llama thampuls (Iñiguez et al., 1998a) to over 37 μm for Huacaya alpaca from Australia (McGregor, 2002), extra fine (<23.5 μm), fine (23.1–26.5 μm), coarse (>31.5 μm) and inferior (>30 μm) being some categories used.

In Peru, the Peruvian Technical Norm NTP 231:301 (Classified Alpaca Fibre), published in 2004 (Greaves and Rainsford, 2005) defines the qualities of alpaca fibre as follows:
Baby alpaca: ≈ 23 μm
Superfine alpaca: 23.1 ≈ 26.5 μm
Huarizo alpaca: 29.1 ≈ 31 μm
Coarse alpaca: ≈ 31.5 μm
In Australia, alpaca fibre is classed into five diameter categories, five length categories and six colour categories (see Table 10.6 (Wang et al., 2008)).
Pepper (Pepper, 1997) has given the alpaca fibre classification shown in Table 10.7.
Other alpaca grades encountered include:
Inferior: > 39 μm
Superfine: ≈ 26 μm
Baby: ≈ 22 μm

Table 10.6 Australian alpaca diameter, length and colour classing lines (Wang et al., 2008).

Fibre diameter	Length	Colour
Superfine (SF) < 20 μm	A: 120–150 mm	White (W)
Fine (F) 20–23 μm	B: 80–120 mm	Fawn (F)
Medium (M) 23–26 μm	C: 60–80 mm	Brown [light brown (BR), dark brown (DKBR)]
Strong (S) 26–30 μm	D: < 60 mm	Black (BLK)
Extra strong (XS) > 30 μm	O: (overgrown) > 150 mm	Rose grey/roan (RG), grey (G)

Table 10.7 Alpaca fibre classification (Pepper, 1997).

Type	Name	Average fineness (μm)	Minimum average length (mm)
Micheli alpaca fibre classification			
BL	Baby	22.5	55
SU	Suri	26.5	68
FS	Fleece	26.5	66
AG	Coarse	32.0	68
LL	Llama	28.0	60
Inca Tops' classification			
BA	Baby	22–23	55
SU	Suri	25.5–26	68
SF	Super fine	25.5–28	66
AD	Adults	28–30	68–60

and
Coarse: > 31.5 μm
Fine: 23.1%–26.5 μm
Extrafine: ≤ 23.5 μm

Alpaca tops and yarn are classified as follows:

Baby (22.5 μm)
Fine spinning
Adult (34 μm)
Suri (26/27 μm)
Superfine (26 μm)
Coarse alpaca (31–34 μm)

Alpaca fibres are smoother and have a lower resistance to compression (Liu et al., 2005) than wool. McGregor (2014) found that the resistance to compression of alpaca and other rare animal fibres was largely a function of fibre curvature and diameter, which is also the case for wool. Wang et al. (2003) found alpaca produced in Australia to be much softer, both in fibre and fabric form, than wool of equivalent diameter, attributing this, at least in part, to the lower resistance to compression/curvature and scale prominence (thickness) and greater smoothness of alpaca.

The mean scale height of alpaca fibres coarser than 19 μm is about 0.4 μm compared to that of 0.8 μm for wool fibres of similar diameter (Phan et al., 1988). It is therefore hardly surprising that the directional frictional effect (DFE) is much lower (Villarroel-Leon, 1959; Lipson and Howard, 1946) for alpaca (around 0.20 for Huacaya alpaca and 0.16 for Suri alpaca (Villarroel-Leon, 1959)) than for wool which is around 0.40 (Lipson and Howard, 1946). The with-scale and against-scale coefficients of friction (μ) of Suri fibres were found to be 0.29 and 0.40, respectively, the corresponding values for Huacaya being 0.28 and 0.42, respectively (Villarroel-Leon, 1959), and wool 0.15 and 0.32, respectively. The directional frictional effect (DFE) is 0.16, 0.20 and 0.36 for the fibres in the aforementioned order. The initial modulus of alpaca is higher than that of wool (Zhang et al., 2008). According to Liu and Wang (2007), Alpaca shrinks more in the Aachen felting test than wool of similar diameter, possibly due to the higher cuticular scale frequencies of alpaca, although the higher scale thickness and DFE of wool should be expected to cause a higher felting shrinkage.

Cuticular scale frequency for alpaca varies from around 10 to 11 per 100 μm (Liu et al., 2005; Liu and Wang, 2007; Antonini et al., 2004; Langley and Kennedy, 1981; Liu, 2003; Sich, 1990), which is similar to that for llama. The difference between the scale lengths of wool and alpaca fibres decreases as the fibres becomes finer (Villarroel-Leon, 1959). Because of the similarities of the cuticular and other surface structures of llama and alpaca (Camelid) fibres, it was considered well-nigh impossible to differentiate between them, particularly between Suri, Huacaya and llama fibres (Liu et al., 2005; Vinella et al., 1994). Nevertheless, Valbonesi et al. (2010) found that cuticular scale frequency, together with fibre diameter, enabled Suri fleeces (fibres) to be accurately and reliably differentiated from Huacaya and llama fleeces, and also to differentiate between llama and Huacaya fleeces, but less reliably so. Medium to coarse alpaca fibres (23–35 μm) have no distinct ortho-para-cortex differentiation (Villarroel-Leon, 1959).

The breaking strength of Suri alpaca is lower than that of Huacaya alpaca, both wet and dry (Villarroel-Leon, 1959), the dry tenacity of Suri being similar to that of wool but significantly lower than that of Huacaya, while its wet tenacity is lower than those of Huacaya and wool which are similar. The dry elastic moduli of the three types of fibre (Suri, Huacaya and wool) are similar. In one study (Halboth and Heidemann, 1971), the wet breaking tenacity and stress at 30% extension of Huacaya fibres were found to be double that of Suri fibres, the extension at break being similar for the two fibres. The greater strength of the Huacaya fibres was attributed to its more ordered structure (higher crystallinity) in the alpaca helices making up the microfibrils as revealed by X-ray diffraction. Australian alpaca was found to have much lower fibre curvatures (crimp frequencies) than wool (or cashmere (Wang et al., 2008)) of similar diameter, curvature decreasing with increasing fibre diameter (Wang et al., 2004) from about 50°/mm at 15 μm to about 15°/mm at 40 μm, the corresponding values for wool being 125°/mm and 58°/mm, respectively (Wang et al., 2004). Fibre crimp (curvature) decreases gradually from the scoured state to the top (Wang et al., 2004).

The complete amino acid composition of alpaca fibres has been determined. It varies widely between animals, because of factors such as nutrition, climate-induced stress and genetics (Stapleton, 1992). Cystine plays an important role in the molecular architecture of a fibrous protein, the level affecting fibre and yarn properties (Stapleton, 1992). Table 10.8 compares the levels of cystine for wool and alpaca (Stapleton, 1992).

The lipid component of the cell membrane complex (CMC) of different animal species differed, for example, the CMC lipids of alpaca and llama fibres do not contain

Table 10.8 A comparison of the cystine levels for wool and alpaca (Stapleton, 1992).

Fibre	Cystine content (mmol/kg)
Huacaya (Villarroel-Leon, 1959)	950
Suri (Villarroel-Leon, 1959)	1250
Wool (McLaren and Milligan, 1981)	900

any sterols (e.g., cholesterol) (Stapleton, 1992; Logan et al., 1989). The amount of extractable CMC lipid in alpaca fibre was found to be 2.6% by weight (Logan et al., 1989), which was significantly higher than that of the other animal fibres examined.

10.2.2 Fibre processing

Alpaca hair is hand-sorted according to fineness and colour and then scoured on a scouring line similar to that used for wool (e.g., five bowls), the clean yield being around 85%–90% (Franck, 2001). Czaplicki and Ruszkowski (2014) optimized the ultrasonic scouring of alpaca, which resulted in little, if any, fibre entanglement and damage.

Alpaca fleece contains up to 10% coarse fibres that need to be removed by dehairing (Harizi et al., 2007), which takes place after scouring. Little information is available, i.e., public knowledge, concerning the processing, including dehairing, of alpaca, since this is kept secret by the companies processing the fibre. The less distinct difference between the diameter of the outer coat and the undercoat makes effective dehairing more difficult (Ryder, 1996). Wang et al. (2008) investigated the technical feasibility of dehairing alpaca using a prototype cashmere dehairing machine, the dehairing efficiency being assessed by the OFDA-measured fibre diameter distribution, with changes in fibre length used to assess fibre damage during dehairing. Dehairing of alpaca is not as effective as that of cashmere, because of the smaller difference between the fine and coarse components in terms of diameter, stiffness and crimp (Wang et al., 2008).

After dehairing, mechanical processing, generally on the worsted system (rectilinear combing), sometimes on the woollen system, is similar to that for wool. Because of its smoothness and low fibre crimp, alpaca is difficult to process, as is the case with other speciality animal fibres, such as mohair. Blending with wool is common practice, which improves cohesion and processing performance on modern wool processing machinery (Wang et al., 2005). Wang et al. (2005) found that blending alpaca with low crimp wool produced knitted fabrics that shrunk less than those produced from alpaca blended with high crimp wool, but the latter was softer. They concluded that high crimp wool may be preferable for blending with alpaca, particularly in terms of processing performance. Knitting yarns from 100% alpaca are typically around 16 Nm ($\approx$60 tex) and weaving yarns around 28 Nm ($\approx$35 tex) (Rainsford, 2002), or even 2/16 Nm (i.e., $\approx$ R120/2 tex resultant).

Wang et al. (2003) dealt in detail with alpaca processing, with specific reference to Australia.

10.2.3 End-uses

Alpaca fabrics, particularly those containing the fine fibres, are generally considered to be soft, warm, light and luxurious. The medullated nature of the fibres (up to 70% medullation) suggests that these fibres will, on a weight for weight basis, provide better thermal insulation than unmedullated wool, although the relative amount of air trapped within the fabric structure will also play a significant role. Alpaca is popularly used in knitted and woven apparel, primarily in knitwear, often brushed or raised, finding application in products such as ladies coats and skirts, suits, sports jackets, upholstery, pile fabrics for rugs and simulated furs, hand-knitting yarn, blankets, pullovers, cardigans, stoles and scarves (Briggs, 1973). White alpaca tends to be used in the knitting sector, dyeing to fashion shades, while the natural colours tend to be used in woven fabrics, particularly coating fabrics, which are often raised (teaselled). Baby alpaca is particularly used in ladies' accessories, such as scarves, stoles, throws and capes (Anon., 2000a). Alpaca is frequently blended with other natural fibres, notably wool and silk, even cotton, since price is an important factor. In blends with wool and mohair, it is often used in overcoats. Pelin Gürkan Ünal et al. (Ünal et al., 2018) reported on the properties of woven fabrics containing wool, silk and alpaca. Alpaca is considered an ideal fibre for fancy yarns, such as bouclé and frisé variations (Michell, 2002), bouclé, in pure alpaca, or alpaca blended with silk, mohair or Angora rabbit hair, being popular for knitted garments. Poorer quality alpaca is also often used as filling in quilts, duvets and pillows (Wang et al., 2008), often in blends with wool, polyester or cotton because it is desirable to remove coarse fibres and contaminants to avoid discomfort (Wang et al., 2008). Alpaca and camel hair reportedly felt at a slower rate than cashmere (Fröhlich, 1969a).

The International Alpaca Association (IAA), based in Peru, runs Alpaca Marks (Chávez, 2002), such as gold, silver and white alpaca, which relate to fibre fineness and blend percentages. The Gold Mark applies to 100% alpaca, not coarser than 28 μm (Chávez, 2002). There is a Huarizo Mark for alpaca coarser than 28 μm, 31–34 μm representing coarse qualities. Peru's Ministry of Agriculture's National Council for South American Camelids the Consejo Nacional de Camélidos Sudamericanos (CONACS) launched a registered 'Alpaca Peru' Mark for all alpaca products produced by artisan knitters and weavers (Rainsford, 2007b).

Chemical bleaching of pigmented alpaca typically involves treatment with an iron salt (mordant) under acidic conditions, followed by the removal of the mordant not bound to the pigment and then bleaching the pigment with hydrogen peroxide, now that improvements in the second step have been researched (Cegarra et al., 1990). Alpaca is used either in its natural shade (colour) or dyed to the required shade. Alpaca generally requires more dye than wool to achieve the same depth of shade, probably because of the medullated nature of the fibre. For more information, the reader is referred to the references for more information (Franck, 2001; Wang et al., 2008; Stapleton, 1992; McGregor and Butler, 2004; León-Velarde and Guerrero, 2001), particularly Wang et al. (2003) who deal in great detail with alpaca production, fibre properties, processing and application as well as industry in Australia.

10.3 Angora rabbit hair

10.3.1 Fibre production, harvesting and properties

Angora fibre, from the Angora rabbit (*Oryctolagus cuniculus*) (see Fig. 10.3), is one of the lightest natural fibres because of its highly medullated (largely hollow) nature, making it ideal for thermal insulation and for maintaining a steady skin temperature, while still allowing the skin to breath (Anon.b).

The good qualities of the hair of the Angora rabbit have been appreciated for many centuries, and it is the only rabbit bred solely for its hair. The Angora rabbit essentially has two coats of hair, namely guard hair which is coarse (up to 60 μm), long (20–100 mm) and spike like (spiky), offering protection against the rain and elements, and a shorter (10–40 mm), fine (12–14 μm) down, offering thermal insulation. Angora rabbit hair occurs in various colours, mainly white, but also grey, black-and-white and camel brown, although the main strain (albino) bred for its fibres produces white fibres only (Franck, 2001). Raising Angora rabbits successfully is very labour intensive and a highly skilled work. They are fairly resistant to most illnesses, provided their metabolism is not disturbed and is 100% functional (Brand, 1989), in which case their resistance to diseases is reduced, and it is important to keep them under highly hygienic conditions.

The Angora rabbit reportedly originated from Ankara in Turkey, having been raised for over 200 years in Europe, reportedly being sighted in England as far back as 1708 (Reddy and Mudaliar, 1985). In France, the numbers increased to such an extent that it was estimated that around 1860 some 10,000 kg of 'Angora Silk' was produced annually (Brand, 1989). The rabbits were referred to as 'Silk Rabbits' or sometimes 'Combing Rabbits' (Brand, 1989). Intensive farming with Angora rabbits took place in

Figure 10.3 Angora rabbit.
Reproduced from Pier Giuseppe Alvigini, The Fibres Nearest the Sky, Mondadori Editore, Verona, by kind permission of Mr Pier Alvigini at Alvigini S.A.S., 13,900 Biella Via Dante, 12 Casella Postale 430, Italy.

Western and Central France, and in 1935, some 90,000 kg of rabbit hair was being produced (Brand, 1989). French Angora rabbits were introduced in Germany during the second half of the 20th century where breeding expanded rapidly, as well as selective breeding of improved animals (Brand, 1989). The French bred larger animals with more hair (better suited for fashion articles), while the Germans bred for finer and softer hair, the German rabbit mostly being shorn and the French rabbit 'plucked' (epilated).

There are essentially four types (strains) of Angora rabbits, namely 'French', 'English', 'German' and 'Chinese'. The 'French' type generally contains more guard hair (up to 14%) and is 'spikier', longer and more difficult to dye than the 'German' type (maximum of 3% of guard hair) and is more suitable for the production of a brushed appearance and fashion type garments. The 'German' type produces a softer product, being very suitable for, and popularly used in, various types of underwear and medical wear. German breeders wanted as fine and a low level of coarse hair (spike hair) as possible because it facilitates the spinning of fine yarn used in underwear. The English Angora rabbit is smaller than the German and French types and has hair of different colours that are very popular in home industries, where the natural colours are desirable. English Angora rabbits are normally grey, with black heads. In the early 1900s, China imported Angora rabbits, breeding rabbits for both their hair and meat, and in the mid-1960s became a major producer of Angora rabbit hair, mainly obtained from young animals (8–12 weeks old). The Chinese strain, developed in the early 1950s, accounts for the bulk of the Chinese and South American production of hair, the fibres produced generally being shorter and finer than those of the French and German Angora rabbits (Brand, 1989), the Tanghang representing an improved strain (Chowdhry, 1983).Over time, many other countries introduced Angora rabbit farming. Today, most Angora rabbit hair is produced in China; China has some 50 million angora rabbits (Yigit, 2014), with an annual production of 8–10 million kilograms of fibre, which represents approximately 90% of world production (Yigit, 2014). In China, the rabbits are farmed on a highly intensive small-scale factory farm system by individual farmers. Significant quantities of hair are also produced in South America (e.g., Chile and Argentina), Europe (e.g., France) and India.

The hair from Angora rabbits can be harvested by pulling/plucking, or combing by hand, or by shearing using either hand shears or electrical clippers. The animal is generally brushed before this so as to facilitate fibre harvesting and classing and the removal of any contaminants. When shearing in winter, it is necessary to leave a short length of fibre to protect the rabbit from severe cold. Animals are shorn (clipped) for the first time around 6–8 weeks of age to remove their 'nest' hair, and plucked for the first time around 100 days and then every 100 days thereafter. The fibre from the first shearing is not as valuable as that from the subsequent shearing (Brand, 1989), but shearing at an early age encourages subsequent hair growth. In China, the hair is mostly removed by hand plucking; the ripe or mature fibres (after 3 months of growth) that easily come out are mainly removed by this method. The French mainly use a method called 'epilation' (plucking or 'combing'), which involves the use of a type of 'comb' with a serrated edge to pluck the fibres four times annually (every 90 or so days). This produces a greater variation in fibre length than the hand plucking method. Plucking generally takes place when the hair is most easily removed (i.e., is 'ripe' or 'mature') and tends to produce more 'spiky' hair than shearing. Both methods of plucking encourage the growth of coarse hair but slowly reduce the total hair production (Brand, 1989). The Angora rabbit has around 180–200 hairs per mm^2 of the skin, which can drop to between 100 and 120 after four to five pluckings. Shearing is popular in Germany. This method tends to produce 'double cut' waste and bits of cut skin which can cause problems during spinning and dyeing but has advantages in that it subjects the rabbits to less pain and stress, only taking 10–20 min when using an electric clipper. This method produces hair with a greater proportion of fine hair, with very little seasonal difference, and only a difference of less than 10% between the summer and winter clip, compared to 20% with the combing method. The rabbits are generally shorn every 3 months, before the hair starts falling (shedding) which can cause felting (matting). Hand-plucked hair, which contains only the 'ripe mature fibres', is considered a better quality than shorn hair; it also leaves the animal with a coat 25–35 mm long as a protection against the cold. The different methods of harvesting Angora rabbit hair and the comparative qualities have been described in detail (Brand, 1989).

A healthy, good-quality Angora rabbit produces about 1 kg of first-grade hair per year. Shorn hair production can be as high as 1.5 kg per year (i.e., ≈ 380 g four times per year). Commercial yields of down hair vary between about 420 and 820 g a year in China and up to 1000 g in France and 1200 g in Germany (Franck, 2001), the yield being affected by nutrition and the quality of the rabbit.

The Angora rabbit can be considered to produce essentially three kinds of hair (Franck, 2001), the lengths referring to staple length:

Guide hairs: 100–110 mm long; they guide and cover the growth of the other hair.

Guard hairs: 80 mm long; these have rough points that lock together, lie over the down and seal it off.

Down: 60 mm long, with a diameter of about 14 μm, very smooth, with few cuticle scales.

10.3.1.1 Hair classification schemes

Angora rabbit hair is essentially graded according to its staple length and cleanliness, its appearance being of paramount importance. In certain cases, the quality of the hair is specified quite independent of the source of the hair (e.g., whether from German or French Angora rabbits).

In Germany, the official classification and trading scheme, based upon DIN-60,407, is as follows (Brand, 1989):

First grade: pure white, very clean, at least 6 cm long and not entangled (matted), without any double cuts.

Second grade: as for first grade, but shorter than 6 cm, and longer than 3 cm.

Third grade: as for the above two grades but shorter than 3 cm.

Felted 1*:* pure white, very clean but felted and matted.

Felted 2: white, felted and matted and dirty or with plant residues.

In France, the third grade is omitted, all hair with a staple length greater than 4 cm being included in the first grade. This allows for shorter intervals between harvesting (Brand, 1989).

The following grades, into which Angora rabbit hair is sorted after grooming and where fleeces or part fleeces with lower quality fibre have been removed, have been given: (Franck, 2001).

Grade 1: clean, free of felting, over 6 cm long (70% of the coat).

Grade 2: clean, free of felting, under 6 cm but over 3 cm (15% of the coat).

Grade 3: clean, felted, second cut.

Grade 4: all dirty, discoloured fibres.

The guard hair ranges from between about 20 and 100 μm, while the mean fibre diameter of the fine down ranges from about 9 to 14 μm, typically 12–14 μm, with the diameter of single fibres ranging from about 5 to 30 μm, 60% being in the range of 10–15 μm (Krishnan et al., 1991). The fibre diameter distribution of French Angora rabbit hair is stated to be better described by a log-normal rather than a normal distribution (Wortmann et al., 1988), the geometric mean, as opposed to the arithmetic mean, therefore being a more satisfactory representation of the 'true' mean fibre diameter, as determined by the evaluation of fibre cross-sections (Wortmann et al., 1988). Stephani and Wortmann (1986a) concluded that nutrition and climatic conditions, rather than differences in origin and quality, determined differences between the specimens they tested.

Mean fibre length of good-grade Angora rabbit hair is around 45 mm, with a CV of around 50% and a mean fibre diameter of around 13 μm (CV = 25%). Number-based mean fibre length, based on the Zweigle Staple Sorting instrument, varies from around 16 mm for the poorer quality fibres to almost 53 mm for the top quality (super, pure white, unfelted 6 cm), the CV ranging from about 65% to 90%, depending on the source and quality of the hair. The number-based short-fibre content ($\leq$10 mm) ranged from about 15% to 45%, and for fibres $\leq$ 20 mm from about 25% to 80%, the corresponding weight-based values being 2%–15% and 6%–48%, respectively (Fröhlich, 1969b). In some countries, such as South Africa, the percentage of hair with an average staple length shorter than 30 mm cannot exceed 10% in the better grades.

The cross-section of Angora rabbit hair is generally not circular (Blankenburg and Philippen, 1988), occurring mostly as elliptical (ellipticity $\approx$ 4) (Wortmann et al., 1988), rectangular, square or bilobal. Only a small percentage of the hair has a broken medulla or no medulla, the vast majority having one, or some even two or more medullae (channels) (Blankenburg and Philippen, 1988). The density of Angora rabbit hair can vary fairly widely, from about 0.95 to 1.3 (average between about 1 and 1.25 g/cm^3, often taken as 1.14g/cm^3), depending on the type and degree of medullation (Hohls, 1951). Blankenburg and Philippen (1988) reported that the density of the hair they studied varied from about 1.1 to 1.2 g/cm^3, with an overall average density of 1.2 g/cm^3, with between 5% and 17% of the fibres they studied having no medulla, 83%–89% having one medulla and 1%–7% having two or more medullae (kemp), the corresponding values for fibre diameter being 9–11 μm, 12–14 μm, 13–37 μm, respectively, when measured on a projection microscope. In general, the hair from the female rabbit (doe) is stronger (30 kgf/mm^2) than that from the ram (buck), with that from the young rabbit being weaker than that from the adult (Brand, 1989), winter hair tending to be weaker than summer hair. The extension at break of the hair lies between about 30% and 40% (Brand, 1989). Angora rabbit hair scale height is of the order of 0.4 μm on average (Wortmann, Wortmann, Arns) as is the case for other luxury animal fibres, such as cashmere and mohair. Angora hair felts less than wool in the felt ball density test.

Stephani and Wortmann (1986b) gave values for the three different breeds of Angora rabbit hair which they tested (Tables 10.9 and 10.10).

Fröhlich (1969b) listed the quality parameters for Angora rabbit hair (Table 10.11). He also gave average values for the chemical properties of the aforementioned fibres (Table 10.12) (Fröhlich, 1969b).

Gupta et al. (1992) gave the properties of three Angora rabbit genotypes, German, British and Russian, as well as their crosses (Table 10.13). They also gave a comparative table for Angora rabbit hair and other fibres (Table 10.14) (Gupta et al., 1992). Gupta et al. (1989) gave the average values for the six genetic groups of rabbit hair which they tested (Table 10.15).

Table 10.16 compares the properties of Angora rabbit fibre (hair) with those of wool (Anon., 1997). The comparative values (Table 10.17) for Angora rabbit hair and wool (Zhaogeng and Bo, 1987) have also been given for a selection of physical properties.

10.3.2 Fibre processing

Angora rabbit hair needs not be scoured before processing, generally containing less than 1% of natural grease (generally 0.7%–1.0%, sometimes even exceeding 1.2%, of fatty matter) (Anon., 1997). Dust and vegetable matter are often removed from the fleeces before sorting, by a process called grooming (Franck, 2001). The fibre yield is some 97%, the hair containing very little contamination – only around 1%–2% of hair scales and skin flakes. When hair is obtained from the skin, blowing of the hair is the traditional way of cleaning it, i.e., removing skin pieces, etc. Because of its smooth fibre surface, relatively short length, lack of crimp and cohesion and greater propensity to generate static electricity, Angora rabbit hair presents more difficulty and waste during mechanical processing than wool; an antistatic lubricant (e.g., 0.3%), applied before carding, is essential if it is to be processed successfully (Anon., 1997).

Table 10.9 The percentage of medullated fibres in Angora rabbit hair from Germany, France and China (Stephani and Wortmann, 1986b).

	Percentage of fibres		
Fibre type	**Germany**	**France**	**China**
Unmedullated	33.1	22.3	12.2
Single medulla	66	74.7	85.1
Multiple medullae	0.9	2.9	2.9

Table 10.10 Average diameter (μm) of different types of Angora rabbit hair from Germany, France and China (Stephani and Wortmann, 1986b).

Fibre type	Average fibre diameter (μm)		
	Germany	France	China
All fibre types	13.4	18.7	12.2
Single medulla	14.1	19.0	12.2
Multiple medullae	31.3	42.5	31.5
Unmedullated	11.4	14.5	8.8

Table 10.11 Quality parameters for Angora rabbit hair from various origins (Fröhlich, 1969b).

Sample no.	Quality/grade	Source/origin	Fat/grease content (%)[a]	Vegetable matter content (%)[b]
1.	First grade, pure white, unfelted 6 cm	Czechoslovakia	0.8–1.0	<0.1
2.	Second grade pure white, unfelted 3–6 cm	Czechoslovakia	0.8–1.0	<0.1
3.	Third grade, pure white, unfelted, 3 cm	Czechoslovakia	0.8–1.0	<0.1
4.	Third grade, white, slightly felted and contaminated	Czechoslovakia	0.9–1.2	0.1–0.2
5.	Fourth grade, white, heavily felted and contaminated	Czechoslovakia	1.2–1.5	0.3–0.5
6.	Super choice	France	0.9–1.0	<0.1
7.	Premium choice	France	0.9–1.0	<0.1
8.	'Tout venant'	France	1.2–1.5	0.2–0.3
9.	Quality I	Argentina	0.6–0.7	<0.1
10.	Quality II	Argentina	0.8–1.0	<0.1
11.	Quality AB	Japan	1.1–1.2	<0.1
12.	Quality B	Japan	1.1–1.2	<0.1
13.	Quality 95%	China	1.2–1.3	0.4–0.6
14.	Quality I	Germany	0.6–0.7	<0.1

[a]Methylene chloride extractable matter (IWTO-8–61).
[b]Boiled in 0.2N NaOH – solution for 15 min.

Table 10.12 Average values for chemical and physical properties of Angora rabbit hair from various countries (Fröhlich, 1969b).

Property	Average value	CV (%)
Grease content (%)	1.0	–
Vegetable matter content (%)	<0.1	–
Cystine content (%)	13.6	1.5
Alkali solubility (%)	9.0	16
Acid solubility (%)	7.5	25
Urea bisulphite solubility (%)	65.3	4.2
Regain (65% RH)	13.7	1.2
Water retention (%)	52.9	8.4
Mean fibre diameter (μm)	12.8	5.7
Tenacity (kgf/mm^2)		
Dry	19.9	12.4
Wet	17.8	10.8
Extension at break (%)		
Dry	33.1	5.0
Wet	45.9	2.0

CV, coefficient of variation; *RH*, relative humidity.

The properties and processing of Angora rabbit hair have been discussed (Anon., 1997), being stated that it was possible to process the hair on the worsted and woollen systems, economically and to the desired quality, provided special precautions and care were taken in terms of machinery fittings and adjustments. Spinning problems can arise due to the fibre surface characteristics, fineness, length and length distribution (Anon., 1997). Because of the difficulties often experienced in spinning pure Angora hair, it is generally not spun in 100% form, being blended with wool (e.g., 50%–90% of wool), and also nylon, with which it is highly compatible, a blend level of 15% rabbit hair, 75% wool and 10% nylon being popular. It is processed on either the woollen or worsted systems, depending on fibre length and blend. Blending with wool has a highly beneficial effect in terms of processing performance. Nevertheless, even then, special precautions, care and skill are required to successfully process the blends. Before mechanical processing, the fibre length variation (CV %) of Angora rabbit hair can vary from about 60% to 90% (Anon., 1997), accompanied by a high short-fibre content. The moisture content tends to be lower ($\leq$12%) and more variable than that of wool. A blend of Angora rabbit hair and wool generally produces more hairy yarns than pure wool.

Table 10.13 Properties of Angora rabbit hair (Gupta et al., 1992)[a].

Genotype	Fibre fineness (μm)	Fibre length (mm)	Grease content (%)	Suint content (%)	UB[b] solubility (%)	Tenacity at break (gf/tex)	Elongation at break (%)
German Angora	13.5 ± 0.38 (50.6)	41.3 ± 0.45 (18.8)	0.58	2.65	59.0	14.8	39.7
British Angora	11.8 ± 0.33 (48.4)	45.5 ± 0.60 (25.2)	1.30	2.36	77.7	13.1	40.1
Russian Angora	11.8 ± 0.31 (45.3)	48.9 ± 0.74 (23.3)	0.79	2.57	64.1	14.9	39.2
German × British	11.6 ± 0.30 (44.9)	40.2 ± 0.51 (21.9)	0.62	2.18	66.0	14.7	40.6
German × Russian	12.5 ± 0.39 (54.1)	52.5 ± 0.85 (28.0)	0.61	–	61.1	13.2	38.8
German × (Russian × British)	11.6 ± 0.29	55.7 ± 0.82	0.79	2.15	68.5	14.0	36.2

[a]The CV values are given in brackets.
[b]UB—urea bisulphite.

Table 10.14 Physical and mechanical properties of various fibres (Gupta et al., 1992).

				Dry			Wet		
Fibre	**Fibre diameter (μm)**	**Linear density (tex)**	**Fibre length (mm)**	**Breaking load (gf)**	**Ext. (%)**	**Tenacity (gf/tex)**	**Breaking load (gf)**	**Ext. (%)**	**Tenacity (gf/tex)**
Rabbit									
Fine hair	10.4	0.16	32.3	2.0	34.5	12.5	1.4	65.0	8.7
Guard hair	44.0	0.92	43.8	10.7	34.2	11.2	7.2	54.0	7.8
Cashmere									
Fine hair	10.9	0.26	30.5	3.2	32.5	12.3	2.7	59.6	10.4
Guard hair	48.4	1.67	50.7	19.4	34.0	11.6	13.6	58.0	8.1
Mohair	30.4	1.35	–	19.5	39.5	14.4	16.1	87.8	11.9
Rambouillet sheep wool	16.8	0.50	25.0	6.0	35.0	12.0	3.2	68.0	6.4
Avivastra sheep wool	21.5	0.87	45.1	9.0	29.7	10.3	5.2	54.0	6.0
Cotton (Coimbatore H-4)	12.0	0.17	28.0	5.0	5.2	29.0	6.6	6.3	37.0
Polyester	–	0.17	51.0	8.1	35.5	48.0	–	–	–
Acrylic	–	0.17	51.0	4.7	25.0	28.0	–	–	–

Table 10.15 Average values for six genetic groups of rabbit hair (Gupta et al., 1989).

Property	Average value	Range
Fineness (tex)	0.18	0.14–0.23
Tenacity (gf/tex)	12.1	13.1–14.9
Extension at break (%)	39	36–40

Table 10.16 Mechanical and physical properties of Angora fibre and sheep's wool (Anon., 1997).

	Angora fibre	Sheep's wool
Fibre length (mm)	Approx. 15–60	20–250
Fibre fineness (μm)	12–17	16–40
Fibre cross-section (shape)	Rectangular	Circular
Fibre surface structure	Relatively smooth scales	Distinct scale sheath
	Very low crimp	Good crimp
Relative density (g/cm^3)	1.1	1.32
Fibre tensile strength (kg/mm^2)	17–26	15–25
Wet fibre strength as % of dry	70–90	70–90
Breaking extension (%)	30–35	25–60
Wet fibre extension as % of dry	120–140	110–130
Regain at 21°C/6% RH (%)	13–15	15–17
Moisture retention (%)	45–55	40–45

On the woollen system, it is advisable for the fibre to have a moisture content (regain) of at least 14%–15%, and a fatty matter content of around 0.6%–0.8% (Anon., 1997) before carding, adequate time for conditioning being important. It is advisable to use, for blending, relatively fine wool (≈21 μm) with good crimp and not too long when processing on the woollen system.(McGregor and Butler, 2004) Nm12 woollen-spun yarn is fairly common, and it is advised (Anon., 1997) that the yarn should preferably have between 140 and 160 fibres in their cross-section. The use of carbonized wool in the blend has some advantages because of the lack of contaminants in such fibres. The relatively high Angora fibre waste produced necessitates that at least 10% more Angora fibre be incorporated in the blend than that which has been specified. Excessive drafts during ring spinning need to be avoided, with a relatively high twist factor ($\alpha > 80$) being advisable. For the worsted system, it is advisable to use Angora with a staple length of at least 45–50 mm, with fibre fineness and length even more critical than for the woollen system. Fine pinning during gilling

Table 10.17 Comparative physical characteristics of rabbit hair and wool (Zhaogeng and Bo, 1987).

Property	First-grade rabbit hair	Wool
Crimp/cm	2.7*	5.5
Breaking strength (gf)	2.8	10
Breaking elongation (%)	37	40
Coefficient of friction (μ)	*With scale*	*Against scale*
Rabbit hair on rabbit hair	0.21	0.49
Rabbit hair on wool	0.21	0.62
Wool on wool	0.5	0.80

or drafting, precise drafting zone (nip) settings, low drafts and delivery speeds are important (Anon., 1997). Mixing (blending) with wool generally takes place in sliver form. It is maintained (Anon., 1997) that provided all the necessary precautions are taken and processing settings and conditions optimized, it is possible to produce Angora/wool blend yarns on both the woollen and worsted systems which are of the same quality as the corresponding pure wool yarns. It is important to inspect the Angora blend yarns using an inspection (wrap) board because this gives an indication of the visual appearance of the ultimate knitted garment. Onal and Korkmaz (2006) have reported on the effect of fibre blend, yarn twist and fabric relaxation on the abrasion and pilling resistance of Angora rabbit hair blended knitted fabrics.

Until the early 1990s, it was not possible to spin Angora rabbit hair on the conventional cotton spinning system (Krishnan et al., 1991), but in more recent years, it has also been processed on the short staple (i.e., cotton) system, mostly in blends with cotton, and then either ring-spun or rotor-spun (OE). Because of its length, processing on the cotton system has its benefits. Krishnan et al. (1991) describe the conditions required to process Angora rabbit hair (38 mm) in blends with other fibres, such as cotton, viscose, wool, silk and acrylic, on the cotton system, ending in rotor (OE) spinning of hosiery yarn destined for knitting.

The same dyes used for wool can be used for Angora rabbit hair. Acid milling and 1:2 metal complex dyes are popularly used and the Chinese varieties are found (McGregor, 2002) to have a greater dye affinity than other varieties. Wortmann et al. (1987) found that blank dyeing of Angora rabbit hair yarns caused changes in the morphological structure of the medullated fibres, a significant proportion of the medulla cells starting to collapse at temperatures around 70°C, where the extent of the collapse depends on the stress on the fibre in the yarn that is released due to a softening process in the keratin. Fibres with a 'ladder'-type medulla passed through an intermediate stage where their medullae appear to be continuous before collapsing completely.

10.3.3 End-uses

Angora rabbit hair is very popular in knitwear and is used in both knitted and woven outerwear, ladies underwear, hosiery, gloves and knitted millinery and felt hats, although fibre shedding can sometimes present problems. Oglakcioglu et al. (2009) reported on the comfort-related properties of angora rabbit/cotton fibre blend knitted fabrics. It is very popular in medical and thermal underwear, under-blankets in hospitals, nightwear and blankets, etc., particularly the hair from the German Angora rabbit. The hair from the French type is very popular in fashion wear (e.g., shawls).

The German textile industry was the greatest user of Angora rabbit hair during the latter part of the 20th century, with popular end-uses being medical and other types of underwear. In medical underwear (Brand, 1989), it is considered to reduce muscular and other pains, attributed to its electrostatic, thermal and moisture absorption properties, the medullated and fine structures of the fibre resulted in having excellent thermal insulation properties.

Angora rabbit hair knitted products can present serious fibre shedding problems, because of their low fibre friction, length and strength, together with the fact that low yarn twists are generally employed to maximize the softness of the garments. Zhaogeng and Bo (1987) concluded that Angora rabbit hair shedding during wear was largely due to the fibres breaking rather than slipping, resulting from the low breaking strength of the hair. The low strength also results in serious fibre damage and breaking during carding, which can reduce the average fibre length by as much as 10 mm.

Additional information in relation to Angora fibre can be found in the references (Franck, 2001; Brand, 1989; Anon., 1997; Moses and Amayappan, 2006; Wortmann and Wortmann, 1988).

10.4 Camel

10.4.1 Fibre production, harvesting and properties

10.4.1.1 General

Camels belong to the Camelidae family, which has two genera (Harizi et al., 2007), *Camelus* and Lama. The genus *Camelus,* which is part of the Camelidae grey family (Franck, 2001), is made up of two species: the one-humped Dromedary camel *(Camelus dromedarius)* (see Fig. 10.4), also referred to as the Syrian, European or 'Arabian camel', mainly from Arabia, Northern India and the Mediterranean, and the two-humped Bactrian camel *(Camelus bactrianus)* (see Fig. 10.5), also referred to as the Asiatic or Central Asian camel, mainly found in Northern China and Mongolia, in areas bordering the Gobi desert, and to some extent in other parts of Asia (Harizi et al., 2007). The latter produces the best and softest fibres. The finest quality comes

Figure 10.4 One-humped Dromedary camel.

Figure 10.5 Bactrian camel.
Reproduced from Pier Giuseppe Alvigini, The Fibres Nearest the Sky (second ed. 1984), Mondadori Editore, Verona, by kind permission of Mr Pier Alvigini at Alvigini S.A.S., 13,900 Biella Via Dante, 12 Casella Postale 430, Italy.

from China, from young animals. It is thought to have descended from the cross-breeding of the ancient pure Bactrian (Bactriana) camel and the Dromedary camel, which produced the Bokhara type of camel, also known as the Boghdi (Moses and Amayappan, 2006).

According to Harizi et al. (2007), there were about 14 million Dromedaries and 4 million Bactrian camels worldwide in 2002, with the greasy hair production at the time being estimated (Franck, 2001) at 3–3.5 million kg, the bulk being from China and Mongolia. Khatami (2016) estimated 25 million camels in total worldwide. In the 1980s, India reportedly (Chowdhry, 1983) had 1.5 million camels (half Bikaneri breed), representing some 10% of the world camel population.

Camel hair used in textile applications almost solely comes from the Central Asian or Bactrian camel, the hair of the one-humped Dromedary generally being not important for textile applications because of its relative coarseness and inadequate length and strength (Sawbridge and Ford, 1987). The Bactrian camel yields a light fawn (tan) lustrous fleece mixed with coarser brown hairs. It has a double coat of hair, with the coarse outer hair or guard hair, 30–120 μm in diameter and from 60 to 375 mm in length, and the fine down or undercoat, 10–30 μm, or even 40 μm (average about 18 μm) (Pulling, 1973) in diameter (CV from about 25% to 35% (Phan, 1991)) and about 25–125 mm (typically 50 mm, CV ≈ 24%), in length (Sawbridge and Ford, 1987), and characteristically reddish brown (tan) in colour. Most camel hair is produced in China (including Tibet) and Mongolia, some also being produced in Afghanistan, Iran and Russia.

10.4.1.2 Bactrian camel

As already stated, the Bactrian camel is the main source of camel hair used in textiles, the predominant colour being brown. When the camel moults, it sheds its hair (fleece) in large tufts over a period of time (≈6–8 weeks). Starting during late spring, the neck hair falls off first, then the mane, and finally the body and belly covering (Shlomm, 1990), the hair on the humps remaining. When the camels moult, the hair can obtained by combing, shearing (done more commonly today) and collecting of shed hair, resulting in a combination of coarse guard hair and fine down, referred to as raw (i.e., undehaired) camel hair. Raw camel hair contains approximately 75%–85% fibre, 4%–5% fat (grease) and 15%–25% sand and dust (Ghoshal et al., 1993a). Iñiguez et al. (2014) obtained a clean fibre yield of 37%, mean fibre diameter of 18.0 (CV 31%), 88°/mm curvature and 46 mm staple length for the fine fibres of the Bactrian Camels.

Today the camels are increasingly being shorn (clipped) (Ghoshal et al., 1993a) once a year, the hair on the back generally not being shorn, serving as a cushion for the pack or saddle. The hair is also not shorn too short, so as to leave sufficient fibre to protect the animal from excessive heat and cold (Ghoshal et al., 1993a). The whole fleece weighs typically between 2.5 and 3.5 kg, while in cold regions, the hair yield can be as high as 5.5 kg, with the average annual yield being about 4 kg (Ghoshal et al., 1993a). The Bikaneri camels (in India) produce some 700 g of hair per annum, with an average fibre diameter around 25 μm, some 15% of the fibres being completely medullated and 45% partially medullated.

Msahli et al. (2008) (quoting Franck et al.) stated that the characteristics that determine the fibre quality of South American domestic camelids are mean fibre diameter, colour, type of fleece, fibre length and uniformity of diameter and length. The finest quality camel hair is a light, bright fawn while the lower qualities are deeper in colour

(Nagal, Naik). There is a relationship between fibre diameter and fibre length (Algaa and Mägel, 1992b). Algaa and Mägel (1992b) review the properties of camel hair.

The surface of camel fibre is covered by a layer of fine and irregularly shaped scales, having diagonal edges that are not very prominent (Mahapatra, 2009). The fibres are circular to oval, the cortical layer exhibiting regular striations due to strings of pigment granules (Mahapatra, 2009), which give the fibre its characteristic pale red-brown colour, and some fibres have a medulla which is often fragmented (Mahapatra, 2009). Camel hair has a tensile strength of about 16 cN/tex and a moisture regain of 13% (Sawbridge and Ford, 1987). The finer fibres do not have a medulla, while the coarser fibres have a medulla that varies in character and size (Sawbridge and Ford, 1987). Table 10.18 (Mahapatra, 2009) compares the properties of camel hair with those of certain other fibres. Weng et al. (2004) gave a table comparing camel hair fibre properties with those of 70s Australian wool (Table 10.19).

10.4.1.3 Dromedary camel

In Tunisia, there were some 100,000 one-humped Dromedaries in 2003 (Stapleton, 1992) (quoting Sghair) producing more than 100,000 kg of hair annually, only a little of which is harvested and used by Bedouin people to make traditional clothing, such as 'bernous' and 'wazra' worn by men.

The Dromedary camel also has a two-layer coat of hair, a coarse outer coat or guard hair (outer hair) and a fine undercoat or down (Pepper, 1997), brown being the

Table 10.18 Comparison of properties of camel hair and other fibres (Mahapatra, 2009).

Properties	Camel hair	Silk	Wool	Cotton
Tenacity (gf/den)	2–2.5	1–1.5	1.5–2.0	2–5.5
Elongation (%)	39–40	25–40	25–45	6–10
Density (g/cm^3)	1.32	1.34–1.38	1.33	1.50–1.54
Moisture regain (%)	13.0	11.0	14–16	9
Acid resistance	Excellent	Excellent	Excellent	Bad
Alkali resistance	Bad	Good	Bad	Excellent
Resistance to moth/fungus	Resistance to fungus but not to moth	Resistance to fungus but not to moth	Resistance to fungus but not to moth	Resistance to moth but not to fungus
UV resistance	Bad	Bad	Bad	Good

Table 10.19 The properties of camel hair and 70s Australian wool (Weng et al., 2004) [a].

Property	Camel hair	70s Australian wool
Average length (mm)	43	55
Average fineness (μm)	17.8	19.1
Fineness dispersion (%)	28.6	21.2
Absolute strength (cN)	4.9	5.7
Relative strength (cN/dtex)	1.47	1.53
Elongation at break	39.0	38.5
Initial modulus (cN/dtex)	20.4	15.3
Work of rupture (cNmm)	13.8	16.2
Crimp factor (times/cm)	3–4	4–7
Crimp ratio (%)	5.7	7.8
Residual crimp ratio (%)	3.6	5.3
Crimp-elastic recovery percentage (%)	64.0	68.5
Compression elasticity (%)	60.0	57.4
Fast-elasticity (%)	33	33
Slow-elasticity (%)	27	23
Elasticity (%)	60	56
Milling ball density (g/cm^3)	0.13	0.18
Mass resistivity (Ω)	2.3×10^{13}	3.3×10^{13}
Moisture content (%)	13.9	13.3
Oil (grease) content (%)	1.3	13.8
Handle	Good	Fine
Lustre	Good	Good

[a]Values rounded off. 70s refer to the wool count, a measure of quality.

dominant colour (Iñiguez et al., 2014). The raw fibre obtained from the camel has to be dehaired to separate and remove the more valuable undercoat from the coarse guard hair. In general, the efficiency with which the two components can be separated during the dehairing process depends on how much the two components differ, particularly in terms of fibre length, fineness (diameter) and rigidity, as well as in their surface characteristics and interfibre cohesion (Msahli et al., 2008): the greater the differences, the greater the dehairing efficiency generally.

Iñiguez et al. (2014) obtained the following average values for the fine fibres of the Dromedary camels they studied; 29% clean fibre yield, 21 μm mean fibre diameter (CV 30%), 78°/mm curvature and 53 mm staple length. Harizi et al. (2007), (2014) evaluated and reported on the production, physical and mechanical properties and yield of the Tunisian dromedaries, the down hair yield varying from 34% to 61% (average ≈ 45%), this being affected by the age of the camel. The greasy fleece weight varied from 0.3 to 2.42 kg (average ≈ 0.75 kg). They found the scale frequencies for the underhair and guard hair to be between 6 and 8 and 13 and 15 per 100 μm, respectively, compared to 6 and 7 for cashmere (Harizi et al., 2007) (quoting Phan et al.). The scales of the Dromedary fibres are, on average, extremely long and quite visible but do not protrude much from the fibre surface, appearing almost convex (Harizi et al., 2007). The colour of the hair varies from beige to brown to almost black (Harizi et al., 2007). The average fibre diameter is around 17.7 μm (CV = 27%) for the undercoat and 90 μm (CV = 28.5%) for the guard hair (Harizi et al., 2007), with an average regain of 15% (Harizi et al., 2007), which is slightly less than that for wool (Harizi et al., 2007). The average Almeter (AL100) fibre length (weight based) of the dehaired underhair of a one-year-old Dromedary was found to be about 53 mm (CV = 41%), with 7.7% short fibres (<15 mm), the hair being shorter than that of Mongolian camel (>70 mm) (Harizi et al., 2007), but longer than cashmere (≈44 mm) (Harizi et al., 2007). Harizi et al. obtained a tensile tenacity of 212 MPa, an elongation at break of 37% and an initial modulus of 3.87 GPa for the underhair (≈20 μm), the corresponding values for the guard hair (≈105 μm) being 121 MPa, 49.7% and 2.02 GPa, respectively. The average values for the underhair were as follows:

Mean fibre diameter: 21.2 μm (35.9% CV)
Mean fibre length (Hauteur): 18.1 mm (CV = 53.3%)
Tenacity: 11 cN/tex (CV = 12%)
Strain: 37% (CV = 21.6%)

The average stress and strain were higher, and the modulus was lower, than those of Merino wool, camel hair and cashmere.

Harizi et al. (2007) found that dehaired fibre taken from the neck, chest, shoulders, nape of the neck, throat, tail and back (area 1) had, on average, similar mechanical properties to those of fibre taken from the belly, side and kidney areas (area 2) but generally showed different trends with the age of the camel in terms of diameter and length. The average diameter of the fibres from area 1 was greater, also increasing more with age, than that of fibres from area 2. Fibre length increased with age for area 1 and decreased with age for area 2. Fibre bundle tenacity first decreased, and then increased with age, showing a minimum at about 8 years of age (Harizi et al., 2007).

10.4.2 Fibre processing

10.4.2.1 Bactrian camel hair

The processing of camel hair essentially consists of sorting, willowing (to remove dirt, dust and plant material), washing (scouring), dehairing, spinning, fabric formation (knitting and weaving), dyeing and finishing. Sorting involves grouping the hair according to the fibre colour and the age of the animal, and visually separating the coarse hair from the fine, soft hair, after which the fibres can be willowed and then washed to remove any dirt or debris, which is followed by the dehairing process (Mahapatra, 2009).

Raw (i.e., un-dehaired) camel hair from Mongolia contains between 20% and 35% guard hair and that from Iran between 50% and 65% (Algaa and Mägel, 1992b), and as much of this guard hair as possible has to be removed before textile processing. Although camel hair was originally dehaired fairly satisfactorily by carding and combing operations (even by hand in the 19th century) a mechanical dehairing process, involving specially developed machinery, is more commonly applied today (Pulling, 1973), which is more efficient, slower, gentler and more sophisticated. Details of the dehairing process are generally a closely guarded secret, with relatively few firms able to do it. Dehairing removes not only the coarse hair but also extraneous matter (contamination), such as vegetable matter and bits of skin (dandruff) (Mahapatra, 2009). When combing is involved, the fine short fibre (i.e., noil) removed can often be of a higher commercial value than the tops.

Dehaired down fibre diameter ranges from 16 to 20 μm, intermediate hair from 20 to 29 μm and guard hair from 30 to 120 μm (Franck, 2001). Baby camel hair, which is the softest, has a diameter of about 16–17 μm. The mean fibre length of the down typically ranges from about 30 to 40 mm, with the guard hair being up to 375 mm long (Franck, 2001), while the length of baby camel hair is similar to that of the adults (Franck, 2001). Fine camel hair tops typically have a length of 50 mm and a CV $\approx$ 25%.

After dehairing, camel hair is processed on both the woollen and worsted systems, often in blends with wool, sometimes also with nylon for hosiery and knitted products (Weng et al., 2004). Weng et al. (2004) concluded that the low cohesion of camel hair relative to that of 70s Australian wool was mainly responsible for its relatively poor spinning performance, its low cohesion being mainly attributed to its relatively low crimp and friction and relatively high initial modulus. Its shorter fibre length and greater fibre diameter variability may also have contributed to its inferior spinning performance vis-à-vis that of the 70s Australian wool. The long and coarse guard or outer hair, such as from camel manes (Mongolia) and third-grade Chinese camel, is carded, gilled and Noble combed (sometimes twice), the very long fibres often being reduced in length (broken) before combing.

Since, like most speciality animal fibres, camel hair occurs in nature in various shades of grey or brown, due to the presence of a natural pigment (melanin), a bleaching process is often used to remove its natural colour. The camel hair undercoat is characteristically reddish brown, reacting to chemicals very much like mohair and cashmere (Pulling, 1973). For white or pastel shades, such fibres need to be bleached, ideally with minimal damage to the fibre. Bereck (1994), Khishigsuren et al. (2001a) and Kang et al. (1994/95) have reviewed and reported on the bleaching of naturally melanin-pigmented animal fibres. In practice, pigmented fibres are bleached using either oxidative or reductive processes or their combination and commonly involve a treatment with ferrous salts (mordanting) followed by rinsing and bleaching with hydrogen peroxide (Khishigsuren et al., 2001a). Khishigsuren et al. (2001a),

(2001b) described an improved ferrous mordanting process for bleaching camel hair, which resolved serious problems of discolouration and excessive damage of bleached fibres associated with iron deposition during mordant bleaching, and there was no need for an after-treatment with a reducing agent. The process used a mixture of thiourea and hydrogen peroxide to produce thiourea dioxide. This is an effective reducing agent when applied during mordanting (Khishigsuren et al., 2001c), and lower mordanting temperatures are possible.

10.4.2.2 Dromedary camel hair

Msahli et al. (2008) reported on the effect of various parameters on the laboratory dehairing of Dromedary camel hair using a Shirley Analyser. They found that four passages through the Shirley Analyser were optimum for dehairing, the hair being scoured before dehairing, with an opening process before the Shirley Analyser being of benefit, particularly in terms of fibre length. After the fourth Shirley Analyser passage, the down still contained about 10% of coarse hairs (>30 μm) (Msahli et al., 2008). If an opening process was first carried out, the Shirley Analyser component had a mean fibre length of 35 mm (CV = 47%), that of the original raw fibre having been 48.3 mm (CV = 60.4%), and the diameter 22 μm (CV = 39%) compared to that of the raw hair which was 26.5 μm (CV = 64.4%). The Stelometer bundle tenacity for the down was 15 cN/tex and that of the raw hair 10 cN/tex.

10.4.3 End-uses

Camel hair (down) fabrics, both woven and knitted, are well known for their warmth, comfort and hard-wearing properties. They also drape well (Sawbridge and Ford, 1987; Pulling, 1973) and are used in both ladies' and men's high-quality clothing (apparel). Good-quality pure camel hair products are highly sought after, but expensive, resulting in the fibre often being blended with wool. It is used, particularly in blends with wool, for men's and women's coats, jackets and blazers, skirts, hosiery, sweaters, gloves, scarves, mufflers and caps, dressing gowns and robes (Mahapatra, 2009) as well as in high-quality blankets (Pulling, 1973). As an example, camel hair and noils are blended with fine wool and spun on the woollen system for use in fabrics for overcoats, knitwear and rugs (Nagal, Naik). Camel hair is also blended with nylon to produce hosiery and other knitted products (Mahapatra, 2009). Products containing camel hair should be dry-cleaned or hand-washed (Mahapatra, 2009).

The long and coarse guard hair, after being combed into tops, is spun into yarn on the worsted system and used in good-quality interlinings, ropes, industrial belting (Sawbridge and Ford, 1987), tent fabrics (Nagal, Naik) in carpets and warm waterproof coats (Mahapatra, 2009) and even paint brushes (Pulling, 1973). The hair from the North African camel is used in carpets, being harsh and unsuitable for apparel (Pulling, 1973).

Further information on camel hair can be found in the references (Franck, 2001; Algaa and Mägel, 1992b).

10.5 Cashgora

10.5.1 Fibre production, harvesting, properties, processing and end-uses

Although Cashgora initially attracted much interest, this was not maintained and the production of dehaired fibre dropped from about 200,000 kg in 1990 to about 60,000 kg in 2000 (Franck, 2001). Today, little Cashgora fibre appears to be produced commercially, particularly in New Zealand and Australia.

The name Cashgora was coined in Victoria, Australia, in the late 1970s (Browne, 1987). It comes from the Cashgora goat (Fig. 10.6), and it has been labelled the first new natural textile fibre of the last 100 years. The name 'Cashgora' was accepted as a generic term by the International Wool Textile Organisation (IWTO) in 1988.

The double-coated Cashgora goat is the progeny of a cross between a male (ram or buck) Angora goat *(Capra hircus aegagrus)* and a female (doe) down-bearing (cashmere-bearing) feral or cashmere goat (*Capra hircus laniger*), the first crossing producing the finest fibres. These goats were originally predominantly reared in New Zealand (Crawshaw, 1992) and Australia and shorn twice a year, as is the case for Angora goats. It has been stated that Cashgora is normally produced in the first and second cross and can be regarded as fine mohair (Ryder, 1987) or coarse cashmere (Rainsford, 2008a). In Australia, the first cross between a female feral goat and a male Angora goat is called a Cashgora, the fibre having some of the characteristics of both

Figure 10.6 Cashgora goat.
Source: http://flyinggoatranch.blogspot.com.

cashmere and mohair. Nevertheless, crosses of the Angora goat with cashmere goats, the *Anglo-Nubian*, and dairy goats have also been recorded (Dalton and Franck, 2001)—goats similar to the Cashgora goat and involving cross-breeding native Kirghis from the former USSR and Angora goats from Turkey were already being produced in 1820 (Friedlin and Petit, 1988). It has been stated (Rainsford, 2008a) that Cashgora is a classification for coarse cashmere: the Cashmere and Camel Hair Manufacturers Institute classifies only fibre not exceeding 19 μm as cashmere and everything exceeding this as Cashgora and not entitled to be classified as cashmere (Rainsford, 2008a). Friedlin (Friedlin and Petit, 1988; Friedlin, 1987) reported on the production and characteristics of New Zealand Cashgora, which was defined (Anon., 1989) as the down component from a two-coated fleece (down and guard hair) having a mean fibre diameter between 17.5 (sometimes it can be as fine as 17 μm) and 22 or 23 μm, a standard deviation below 6 μm, a CV of fibre diameter below 28% and less than 6% of fibres coarser than 30 μm. Three types of Cashgora have been defined, ranging from the top end (18.5 μm), marketed as 'Ligne Or', the medium range (20 μm) marketed as 'Ligne Emerande' and the lower range (just below 22 μm) marketed as 'Ligne Saphir' (Friedlin, 1990). At René Friedlin, the dehaired Cashgora was classified in three classes according to diameter, namely 17−18.5 μm, 19.5−21 μm and 22−23 μm (Friedlin and Petit, 1988).

Cashgora fibre (or hair) can therefore be taken as the 'down' (fine dehaired secondary fibre) component of the fleece of the two-coated Cashgora goat, under 22 μm (in some cases as 18−22 or 23 μm) in mean fibre diameter, with a length generally between about 30 and 90 mm (usually between 40 and 60 mm, individual fibres even exceeding 100 mm). The mean fibre diameter of Cashgora can in fact range from about 19 to 24 μm and the CV from about 25% to 35%. Finer fibres are down to 12 μm, with the coarser end running up to 45 μm. The fibres are medullated in some cases, and the fleece has almost the same lustrous appearance of mohair, none of the fibres being crimped. It has to be dehaired (i.e., the down fibre has to be mechanically separated from the coarse, rather kempy hair or primary fibres). The down fibre has a low to medium (gentle) lustre and is generally white, soft and delicate to the touch. Cashgora is dehaired using the same criteria as for cashmere, namely fibres coarser than 30 μm are classified as guard hair, the fine inner down fibre representing approximately 50% of the mass of the fleece.

Phan et al. (1990) discussed the morphological features of Cashgora, showing that they differed sufficiently from those of cashmere to allow the two types of fibres to be distinguished. Phan et al. (1991a) stated that the scale structure of Cashgora is more similar to that of mohair than to that of cashmere. Nevertheless, Cashgora fibres are considered to possess either cashmerelike features (i.e., cylindrical and semicylindrical scales) or the characteristics of mohair, with 'splits', lance-shaped scales and subscales (Phan et al., 1988). In the kid and young goat stage (up to 2 years of age), the fleece of the Cashgora contains fibres that are similar to cashmere and also fibres which are of the mohair type, both with lustre (Smith, 1987; Knott, 1990). As the animal ages, the fine cashmere type fibres disappear and the fleece reverts to superfine mohair in characteristics. A guard hair is always present.

Cashgora fibres range from a bilateral to nonbilateral structure, some resembling the bilateral structure of the cashmere, others resembling the nonbilateral structure of mohair, with the majority being intermediate (Tucker et al., 1988; Ryder, 1990a). It contains both orthocortex and paracortex, with fewer fibres exhibiting a bilateral structure than is the case for cashmere. After dehairing, Cashgora generally follows a similar processing route as wool and mohair. Cashgora can generally be woollen-spun and used in most articles of clothing (e.g., jackets, coats, scarves and stoles, with the exception of underwear) as well as in blankets. It is considered more suitable for the weaving trade for high-grade lightweight suiting fabrics (Friedlin and Petit, 1988). Albertin et al. (1990) have compared the behaviour and properties of Cashgora during finishing operations.

Knitted Cashgora garments do felt and shrink during machine washing (Cowan, 1989), although they can safely be hand-washed. The pilling of the knitted garments was low (good). A chlorination shrink-resist treatment, involving sodium hypochlorite, reduced felting shrinkage during machine washing to acceptable levels but caused some yellowing and scale modification, the latter assessed by means of the methylene blue test (Cowan, 1989).

10.6 Cashmere

10.6.1 Fibre production, harvesting and properties

Cashmere goats *(Capra hircus lasinger)* (see Fig. 10.7), also known as 'Shawl goats' or 'goats of Tibet' (Singh et al., 2003; Mauersberger, 1948), have two fibrous coats (usually white): a coarse guard hair (outer coat) and fine down hair (undercoat).

Figure 10.7 Cashmere goat.
Reproduced from Pier Giuseppe Alvigini, The Fibres Nearest the Sky, Mondadore Editore, Verona, by kind permission of Mr Pier Alvigini at Alvigini S.A.S., 13,900 Biella Via Dante, 12 Casella Postale 430, Italy.

The thick 'guard hair' provides physical protection while the fine undercoat provides thermal insulation (Ryder, 1990b), an ideal combination for extreme climatic conditions, particularly extremely cold and wet conditions. Generally, the fine down hair grows during summer, preparing the animal to withstand the winter, and is then shed in spring when it can be harvested, the animal not requiring this highly insulating layer under warm, even hot, summer conditions. The outer hair of the ordinary goat is mostly too coarse for textile use, although raw goat hair (i.e., a mixture of the down and guard hair) was used in earlier times in fabric (e.g., tenting) and carpets (Ryder, 1990b). Goat hair is still used to make brushes and interlinings, with the down removed to make felts (Ryder, 1990b). Cashmere goats, found in extreme climatic mountainous regions, around the Himalayas and Central Asia, particularly in the regions around the Gobi desert, (Anon.b) are domesticated in various countries, such as China, Inner Mongolia, Iran (from the Passang or Iranian goat), and Afghanistan, and also bred in countries such as Australia and New Zealand. There is, however, no specific or distinct breed of cashmere goat/[07] although cashmere goats are generally white with spiral horns (Ryder, 1990b), ranging between about 40 and 70 kg in weight and 0.6 and 0.8 m in height, with an average life span of about 7 years (Franck, 2001).

The fine down hair (undercoat) grown by the two-coated 'cashmere' goats is referred to as cashmere, being finest in the first year of the goat's life. The term cashmere, which originated from the Kashmir region (where in the 1700s Europeans first saw the fibre woven into shawls) and goats of Tibetan origin, is sometimes used interchangeably with the words Kashmir, pash and Pashmina (Ling, 1993) (the latter is the Persian term for cashmere (Ryder, 2001), and the local term for the cashmere or Pashmina goat and cashmere from the southern Himalayan region in India and Nepal; see also Section 12.13.8). Cashmere, also referred to as Tibet hair (Phan et al., 1995), 'soft gold' and 'fibrous diamond' (Gui-Fang and Yong, 2008), can be defined (Spilhaus, 2007), according to the US Wool Products Labelling Act, as the 'fine (dehaired) undercoat fibres produced by a cashmere goat, with an average fibre diameter not exceeding 19 μm, and with no more than 3% of the fibres (by weight) having an average diameter that exceeds 30 μm, and a CV of diameter not exceeding 24%'. According to this definition, therefore, the coarse guard hair from the cashmere goat is not regarded as cashmere. Cashmere is generally taken as the fine (dehaired) undercoat (down) fibre from a two-coated goat, which is generally finer than 19 μm according to Zhong et al. (2016) (quoting Hy et al., Report, Erdos Group, China, July 2015); the diameter and length of cashmere appeared to have increased over a 10-year period. The diameter of the down from cashmere goats is similar to that ($\approx$ 14 μm) from 'ordinary' or common goats (*Capra hircus*), although the former normally produces more down fibre than the latter. The AATCC definition of cashmere was coarsened in 2001 (Anon., 2007), with the ASTM not regarding fibres coarser than 30 μm as being cashmere.

Phan and Wortmann (1995) proposed the following definition of cashmere:

- The dehaired, fine undercoat fibres produced by a double-coated species of goat.
- An upper limit of 19 μm for mean fibre diameter and 24% for the CV of 'diameter', with the level of coarse hairs (>25 μm) less than 3% by weight.

- Goat fibre samples with a mean fibre diameter (MFD) between 19 and 23 μm are classified as Cashgora.
- First class cashmere should have an MFD below 15.5 μm and a soft handle.
- Cashmere with an MFD higher than 15.5 μm is second class.

The bulk of the global cashmere production comes from China and Mongolia (combined over 90%), with the rest coming from Iran, Afghanistan, Russia, India, Pakistan, Turkey, Australia, New Zealand, Britain and the USA (Singh et al., 2003). China's production of raw cashmere fibre (i.e., guard hair plus cashmere) was estimated (Anon., 2009) at some 15 million kg; in 2005 (9–10 mkg estimated in 2011) (Vineis et al., 2011), some 6.5 million kg being produced in Inner Mongolia (Qi, 2006).

The guard hair (outer coat) of the cashmere goat is long (around 15 cm), coarse (from 30 to 150 μm, average about 60 μm) and medullated and is produced by the primary follicles. The fine, short, unmedullated down or undercoat (with a prominent surface scale structure), termed cashmere, ranging in single fibre diameter from 4 to 30 μm (Badmaanyambuu et al., 2003), is produced by the secondary follicles. Between 100 and 300g of fine down is produced per goat (maximum 500 g from the Chinese Liaoning breed) (Vineis et al., 2011), the doe (female goat) producing less than the buck (male goat), with the average diameter ranging from about 12 to 19 μm, and average length from about 20 to 50 mm. The finest cashmere generally comes from China and Mongolia, although the average diameter of Mongolian cashmere increased from about 16 μm to between 17 and 19 μm after the mandatory culling of older male goats was revoked, the fineness of cashmere from male goats increasing by up to 1.5 μm at the age of 2 years. Nevertheless, the introduction of the Liaoning breed in Inner Mongolia is bringing the diameter down to about 15.5 μm. The Arbas breed of cashmere goat in Inner Mongolia is considered to produce excellent-quality cashmere, with a high yield of up to 400 g of fine down.

The cashmere goats moult over several weeks during spring, when the cashmere is harvested either by combing (e.g., in China), or shearing (e.g., Iran, Afghanistan and Australia) (McGregor and Butler, 2008a), the long, outer or guard hair often being shortened (clipped) before combing. The method of harvesting affects the fleece characteristics, such as the ratio of downfibre (cashmere) to guard hair, and the level of contaminants, also being influenced by nutrition and goat quality/(Anon.a). It also affects dehairing efficiencies, for example, it is more difficult to dehair shorn hair than combed hair.

The total hair removed from a cashmere goat can be about 800–900 g. Of this, some 500 g is clipped guard hair. The balance of 300 g (raw cashmere) is hand-sorted which removes ±20% of residual guard hair. Willowing removes about 10% of sand and dirt and the remaining 200 g is then scoured, leading to a further 20% loss. Dehairing then decreases the yield by another 30% (due to guard hair removal, known as cashmere waste), leaving 50% fine dehaired cashmere down (i.e., ≈ 100 g) (Blackburn, 1990). The fleece of the cashmere goat typically contains between 50% and 75% of fine down (i.e., undercoat).

Australia's cashmere industry, based on their feral goats (Smith et al., 1973), was launched in 1972 (Smith, 1988). Australian feral goats have a fine, 15 μm or less

(Buxton, 1986) (even down to 13 μm), and ≈60 mm long undercoat (down) and ≈60 μm hair fibres (≈55 mm). They are probably the progeny of the original domestic goats from various European and Asiatic areas, during the latter 1800s (Phan et al., 1991b). Only about one in 10 of these feral goats produce cashmere type fibres, which are often pigmented, the average downproduction per feral goat being between 75 and 125 g per year in 1981 (Evans, 1983). To produce white Australian cashmere, white top-quality Angora goat males (bucks) are crossed with female feral goats carrying coloured down fibres, producing mostly white soft down fibres, > 18 μm, which is called Cashgora in New Zealand (Phan et al., 1991b). Cashgora, 18–23 μm, has a CV 25%–30%, whereas well dehaired cashmere usually has a CV less than 20% (Phan et al., 1991b) (see Section 10.4). Australian cashmere, which is produced in both Australia and New Zealand, is shorn, is predominantly white in colour, and considered to be longer, stronger, cleaner, softer, less crimped (i.e., lower curvature) and more lustrous than that produced in other countries (Singh et al., 2003). It is therefore considered suitable for processing on the worsted system. The average fine fibre yield is some 200 g per animal, with an objective of 500 g per animal ultimately. Singh et al. (2003) gave a table (Table 10.20) containing comparative data of cashmere production in various regions. See also section 10.13.3.

Cashmere is not as lustrous as mohair and ranges in colour from white to brown, even black. According to the surface structure as well as handle, cashmere can be differentiated into 'cross-bred cashmere' and 'classical Asian cashmere'.

Various 'cashmere fibre types', classified according to average fibre diameter, presence of guard hair and colour, are produced commercially (white is the most valuable; even 5 fibres/g of coloured fibres in white cashmere impact negatively on quality). International markets generally evaluate (grade) raw cashmere quality according to fineness (fibre diameter), colour, length and level of contamination (e.g., bits of skin, vegetable matter, etc.). Fibre diameter greatly influences the market price of cashmere: the finer the fibre, the better the perceived quality and price. The colour of the fibre also has a great impact on its price, with white being the most expensive. Black cashmere generally has to be bleached, particularly if it is to be dyed to pastel shades. Price is also affected by the fibre length, level of guard (coarse) hair and level of contamination (e.g., bits of skin, vegetable matter, etc.).

Table 10.20 Cashmere produced in different countries (Singh et al., 2003).

Country	Average scouring yield (%)	Average dehairing yield (%)	Average diameter (μm)
China[a]	75	51	14–16
Mongolia[a]	75	51	16–17
Iran[b]	80	35	17–19
Australia[c]	92	30	15.5–18.5

[a]Blackburn, 1990.
[b]Ekhtiyari 2000.
[c]Australian Cashmere Marketing Corporation.

Fibre diameter can be measured by projection microscope (IWTO-8-97), Airflow (IWTO-6-98), Sirolan-Laserscan (IWTO-12-98) and OFDA (IWTO-47-95), as well as scanning electron microscopes and cross-sectional methods (Phan et al., 2000a). Some differences between the different techniques have been observed, however (Badmaanyambuu et al., 2003). Fibre length, for example, can be determined by the time-consuming single fibre length test (IWTO-16-67), or using an Almeter (e.g., AL-100) according to IWTO-17, values of 40 mm being obtained on Mongolian cashmere, when using the latter method (Badmaanyambuu et al., 2003).

Cashmere fibres generally have a very little crimp. Work by McGregor (2003a) and McGregor and Postle (2004) shows that cashmere from newer producing regions, such as Australia, New Zealand and the USA has an even lower fibre curvature (i.e., crimp) than cashmere from the traditional sources, such as China, Mongolia, Iran and Afghanistan, with cashmere fleeces not exhibiting the well-defined staple crimp structure (formation) found in Merino fleeces. Fibre curvature was, as in the case of wool, negatively related to fibre diameter and nutrition affected the curvature of Australian cashmere. For the Chinese cashmere, the age and sex of the goats affected the fibre curvature, as follows: bucks 52°/mm, does 65°/mm and kid bucks 78°/mm (McGregor, 2003a). For Mongolian goats, crimp curvature varied from 55 to 66°/mm, decreasing with mean fibre diameter, and also varying between certain regions (Badmaanyambuu et al., 2003). McGregor and Postle (2004) gave two tables representing results obtained on commercial dehaired cashmere and cashmere tops from various origins (Tables 10.21 and 10.22). McGregor (1998) reviewed various aspects related to the quality and production of mohair and cashmere, with specific reference to Mediterranean and temperate climate zones. McGregor (2003b) also reported on commercial lots of Australian cashmere, giving average values of 17 μm (CV~21%) and 52°/mm curvature.

Chinese cashmere is considered to be the best quality, with an MFD between about 13.0 and 16.5 μm. Iranian/Afghan cashmere is 1–2 μm coarser (around 17–19.5 μm) and that from Australia and New Zealand is around 16–18 μm. According to tests conducted in Germany on commercial cashmere samples from different origins, the MFD ranges between about 18 and 19 μm (Phan et al., 1991b), with less than 1% (even 0.2% for knitwear and 1% for wovens (Gray, 1990)) of coarse fibres (i.e., coarser than 30 μm). For good-quality Chinese cashmere, the fineness of the individual fibres ranges from about 8 to 24/25 μm (Phan et al., 2000a), with fibres coarser than 22 or 23 μm rare in well dehaired cashmere; it is stated that fibres between 25 and 30 μm are morphologically conspicuous and in appearance more like the coarse guard hair (Phan et al., 1991b). The upper threshold of the mean fibre diameter of commercial cashmere is around 19 μm, being set at 16 (+0.5 μm) by the Chinese National Standard, at 18.5 to ±0.5 μm by the CCMI (Cashmere and Camel Hair Manufacturers Institute) in the USA, and at 19 μm by the ASTM (Phan et al., 2000a). McGregor et al. (2009) reported on the quality of Kyrgyz cashmere from the Osh and Naryn provinces of Kyrgyzstan, the mean fibre diameter being 15.7 and 16.7 μm, respectively, and fibre curvature 59 and 56°/mm, respectively.

Well dehaired cashmere samples typically have a CV of diameter of 20% (Phan and Wortmann, 1995) (even up to 25%) (Phan, 1991), compared to that of 25%–30% for

Table 10.21 Median, standard deviation (SD), and range of pooled data for attributes of dehaired cashmere (McGregor and Postle, 2004).

Top attribute	Median	SD	Maximum	Minimum
Mean fibre diameter (MFD) (μm)	16.4	1.4	19.3	13.5
CV of MFD (%)	21.7	1.6	29.0	19.8
% Fibres >30 μm	0.5	0.5	2.1	0.1
Fibre curvature degree/mm	60.0	9.3	79.7	40.1
Resistance to compression (kPa)	5.6	0.6	7.7	4.5
Incidence of medullated fibre (%w/w)	0.43	2.4	8.8	0.0
Mean medullated fibre diameter (μm)	34.8	12.3	62.9	23.6
LAC[a] (mm)	22.8	7.8	36	15
CV (LAC)[a] (%)	67.0	5.7	75.6	52.1
LAC fibres <25 mm (%)	63.1	11.3	86.7	38.0
LAC longest 5% (mm)	53.6	10.6	77	32
Inferred 'Barbe' (mm)	33.6	7.1	50	19
Ratio LAC:MFD (mm/μm)	1.41	0.30	2.21	0.93
Bundle tenacity (cN/tex)	9.9	0.9	12.0	8.2
Bundle extension (%)	40.8	4.5	50.0	31.3
Lightness	61.1	3.6	65.4	44.8
Yellowness	−0.3	1.8	3.7	−4.3

[a]LAC: length after carding.

Cashgora. The mean fibre length of the fine undercoat fibre of Chinese raw cashmere typically varies from about 21 to 40 mm (super grade), the lengths of the individual fibres ranging from about 5 to 80/90 mm (Phan et al., 2000a), depending on the quality of the sample. Cashmere from China and Mongolia, as well as from Iran and Afghanistan, has shades of white, light grey, dark grey and brown; cashmere from Iran and Afghanistan is typically cream, fawn and dark brown. Chinese cashmere (dehaired) average fibre length ranges from about 24 mm for the poorer qualities (e.g., Brown, Second Grade) to 36 mm for the top qualities (e.g., Super Grade) (Phan et al., 1991b, 1995; Phan and Wortmann, 1995). That of Asian and Australian cashmere is measured between 21 and 40 mm (Phan et al., 1995; Phan and Wortmann, 1995), the individual fibre lengths ranging from 10 to 90 mm. Super Class cashmere has a fibre length of 38 mm (CV ≈ 33%) and First Class 36 mm. On the basis of the IWTO 5−66 test method, values for mean fibre length between 21 and 40 mm were found for Asian as well as Australian cashmere (Phan et al., 1991b). Coarse fibre (>30 μm) contents for Chinese Super Grade, First Grade and Second Grade are less

Table 10.22 Median, SD and range of pooled data for attributes of cashmere tops (McGregor and Postle, 2004).

Top attribute	Median	SD	Maximum	Minimum
Mean fibre diameter (MFD) (pm)	17.5	1.2	19.3	15.2
CV of MFD (%)	21.2	1.2	23.8	19.8
Fibres >30 μm (%)	0.6	0.4	1.6	0.1
Fibre curvature (degree/mm)	58.0	5.5	68.6	48.9
Resistance to compression (kPa)	5.9	0.4	8.3	3.6
Incidence of medullated fibre[a] (% w/w)	0.3	0.4	1.5	0.1
Mean medullated fibre diameter[a] (μm)	29.8	9.2	51.7	23.6
Hauteur (mm)	41.1	4.9	50	28
CV of hauteur (%)	41.2	6.9	57.4	31.8
Hauteur fibres <25 mm (%)	16.8	11.6	51.1	6.9
Hauteur longest 5% (mm)	69.7	6.6	82	59
Barbe (mm)	46.7	4.9	57	37
CV of Barbe (%)	36.7	5.0	47.8	31.2
Ratio hauteur:MFD (mm/μm)	2.37	0.36	2.91	1.52
Bundle tenacity (cN/tex)	10.4	1.0	12.0	8.3
Bundle extension (%)	39.1	6.6	50.0	19.5

[a]Adapted from McGregor and Postle (2004).

than 0.1%, 0.2% and 0. 5%, respectively (National Criteria of China for Cashmere Analysis, practised by CCIB offices in China).

McGregor (2016) obtained the following average values for processed (commercial) white cashmere from seven countries. Mean fibre diameter ≈17 μm (CV˜22%), fibre curvature ≈58 degrees/mm, Y (reflectance) ≈66 (T units), yellowness (Y-Z) ≈8.2 (T units), Hauteur ≈30 mm (coefficient of variation of Hauteur [CVH]≈60%), Bundle tenacity ≈10.2 (cN/tex) and Bundle extension ≈40%.

Mongolian cashmere mean fibre diameter tends to increase with goat age, on average, by roughly 1 μm from the age of 1 year to the age of 5 years (Badmaanyambuu et al., 2003; Phan et al., 2005). It was found that the mean fibre diameters of cashmere from different provinces of Mongolia ranged from about 15.5 to 18 μm (CV ≈ 20%), increasing with age (Badmaanyambuu et al., 2003), and mean fibre curvature from about 62 to 70 degrees/mm (Phan et al., 2005) (55 to 66,(Badmaanyambuu et al., 2003)), with a mean of 68.4°/mm. The mean bundle tenacity was ≈12.5 cN/tex and the bundle extension 45%.(Badmaanyambuu et al., 2003) In all, 94% of the samples had a down yield of over 60%, the average down yield being 80.4%, ranging from about 65% to about 85%. Mean fibre length of the two best provinces exceeded 38 mm in most cases, with 50% exceeding 40 mm. Yak and camel hair are about 1–3 μm coarser than cashmere hair, yak having some two scales per 100 μm more than Mongolian cashmere (Phan et al., 2005).

McGregor (2016) found that weathering reduced bundle tenacity and extension and reflectance and increased the yellowness of white cashmere, there being differences according to the origin of the cashmere.

Scale height of speciality fibres, such as cashmere, is about 0.4 μm and that of wool about 0.8 μm (Phan and Wortmann, 1995). Different strains of classical Asian cashmere exhibit similar fibre surface characteristics (Phan et al., 2000a), with typically regular cylindrical and semicylindrical scale shapes, i.e., each cuticle cell envelopes the entire or half of the fibre shaft, with a mean scale frequency of 6–8 per 100 μm fibre length, e.g., 6–7 per 100 μm and a scale height <0.5 μm for Mongolian cashmere (Badmaanyambuu et al., 2003). Cashmere from the relatively new sources exhibits quite different scale shapes than the Asian types, usually having a higher scale frequency (>8 per 100 μm). It is visually more complex and less regular, with many scales arrowhead-shaped like mohair (Phan and Wortmann, 1995; Phan et al., 2000a), and tends to be more lustrous and smooth (slippery), with a harsher handle than its Asian counterparts. The cortex of cashmere mainly consists of orthocortical and mesocortical cells (Tester, 1987). Mc Gregor and Liu (2017) found that cuticle scale frequency and thickness and cortical cell dimensions and ellipticity were affected by nutrition, also varying with fibre diameter and growth rate and the size of the animals.

10.6.2 Fibre processing

Raw cashmere can have between 25% and 30% of sand and dust, 4%–5% (8%–10%) of grease and 65%–70% of fibre (clean fibre yield). Typically, cashmere processing commences with sorting, willowing (to remove dirt, grass, etc.) and aqueous scouring, to remove natural oils (grease), etc. Scouring yield is about 70% (Gray, 1990), followed by dehairing (McGregor and Butler, 2008a). The most important step in cashmere processing is dehairing, the effectiveness of which has a major impact on the quality, price (value) and processing of the fibre. The residual grease after scouring should be around 0.5%. Dehaired cashmere (down yield), based on the raw fibre weight, can vary from around 30%–70%, this being largely dependent on the method of fibre harvesting, i.e., combing or shearing, and the type and quality of the goat. Originally, dehairing was done by hand, when it took some 2 hours to dehair less than 60 g of fibre (Ryder, 1990b). A cashmere dehairing process was developed by Joseph Dawson in the 1870s, with the first successful commercial dehairing machine being invented by Joseph Dawson and Sons in the UK some 100 years ago (McGrattan, 1996) with many different patents and designs of dehairing machines appearing subsequently. Mechanical dehairing essentially consists of three phases: opening up (individualizing) of the fibres, separation and removal of the guard hair and fibre mixing (Algaa and Mägel, 1992c). Commercial dehairing production rates are around 3 kg/h/m width.

The mechanical dehairing of various animal fibres, such as cashmere, camel and yak, has been discussed (Algaa and Mägel, 1992c). Fibre rigidity/stiffness (which is largely determined by the fibre diameter) as well as fibre length play an important role in efficient mechanical dehairing. A diameter threshold of around 30 μm is often used to distinguish and separate the fine down from the coarse guard hair. The separation criterion used for dehairing by the combing principle is fibre length (Algaa and Mägel, 1992c); it merely removes relatively short fibres, and hence it is only effective if the fibre lengths of the fine down and the coarse guard hair do not overlap. A roller carding principle of dehairing has also been developed (Liu et al., 2005). The carding stage is particularly important in producing good-quality cashmere yarns, requiring appropriately fine card wire and sharp points, as well as a varying level surface. The world standard for Super A grade cashmere is 38 mm, modified cotton combing machines providing a length of 40 mm (Anon, 1994).

It is possible to use both a modified Shirley Analyser (Couchman, 1986, 1989; Couchman and Holt, 1990) and the Shirley Trash Separator (Couchman and Holt, 1990) for the laboratory dehairing of cashmere, and for determining the yield, although some caution is necessary in the interpretation and application of the results, particularly those pertaining to fibre length. Aspects relating to the objective measurement of cashmere have been dealt with by Stubbs and Marler (1990). The Australian Wool Testing Authority Ltd (AWTA Ltd) developed (Stubbs and Marler, 1990) (see also www.awta.com.au) standard procedures for cashmere fleece testing, as well as for yield and mean fibre diameter measurement of sale lots. Glasbey et al. (1994) reported on the use of digital image processing of microscopically measured diameters of raw cashmere (i.e., un-dehaired) to determine the mean fibre diameter of the down and guard hair, respectively, as well as the cashmere yield, without the need for first dehairing. Lupton et al. (1995) reported on a method they developed for estimating cashmere down yield and average fibre diameter of raw (i.e., un-dehaired cashmere) using an OFDA. The fibre diameter results, but not the yield, correlated well with Shirley Analyser dehaired sample results. The density of guard hair decreased with increasing fibre diameter, due to concomitant increases in fibre medullation (Lupton et al., 1995,

Phan and Wortmann, 1987) have dealt with morphological aspects of cashmere). Singh (Wang et al., 2008) and coworkers (Singh et al., 2005) have reported on a prototype cashmere dehairing machine purpose built for 'Australian' shorn cashmere, which contains 60%–70% coarse hair.

McGregor and Butler (2008a) found that the following factors were associated with more efficient dehairing and/or longer dehaired cashmere: white colour, longer raw cashmere (staple length), greater fibre curvature, lower vegetable matter, normal length guard hair, and the absence of visible coating, with coarser diameters also dehairing more efficiently. Generally, commercial cashmere samples have a coarse hair content below 1% (Phan and Wortmann, 1995). The level of coarse (guard) hair (usually taken as the fibres > 30 μm) present in good-quality dehaired cashmere should be below 1%, or less than 0.5% by weight, according to Wang et al. (2008).

Because cashmere is so expensive, and for improved processing, performance and yarn and fabric quality, it is often blended with silk and fine wool, such as lambswool, sometimes with a small percentage of nylon to provide added stability and strength. Nevertheless, unless it is blended with an extremely fine fibre, its handle and softness can suffer.

Because of its length and other factors, dehaired cashmere has been mainly processed on the woollen system, often mule spun. If it is long enough, it is processed on the worsted system, but only very rarely on the semiworsted system. Similar machinery to that used for fine wool is used, but with settings, conditions, etc., which are most suitable for cashmere and generally kept secret. Until the mid-1970s, most cashmere was processed in Europe, mainly Scotland and Italy, but since the late 1970s, China has increasingly processed cashmere locally (Singh et al., 2003); this also applies to Mongolia. It has been stated (Ryder, 1991) that in Scotland, a down (i.e., dehaired) yield of some 120 g of fibre can reduce to about 66 g at the spinning (i.e., yarn) stage. Some work has even been done on processing cashmere (35 mm, 18.7 μm) on the cotton system, followed by rotor (open-end) spinning (Suadipradja and Wang, 1996). A good-quality cashmere top typically has a mean fibre length of about 38 mm and a CV of 33%.

Cashmere is generally fairly sensitive to dyeing and finishing conditions and the severity of the conditions (e.g., temperature, time, pH, agitation, etc.) must be minimized. Soft water is recommended for producing the lather which is so important in the optimum washing (scouring) processing of cashmere knitwear. It is also more sensitive to alkali than wool.

Pigmented keratin fibres contain black melanin pigment granules (Knott, 1990) (0.9 μm × 0.3) which are inert to all solvents. Oxidative and reducing agents, except hydrogen peroxide (Volooj et al., 1999), are used in what is essentially a two-stage bleaching process, involving an initial fast solubilization of the melanin, followed by a slow decolouration. Bleaching is often associated with severe fibre damage: iron mordanting improves whiteness but with the penalty of increased keratin damage (Volooj et al., 1999). The binding of the ferrous ions to melanin was far stronger than that to the keratin. Khishigsuren et al. (2002) showed that an after-bleaching rinsing process, involving sodium bisulphite on cashmere, produced a better whiteness and less fibre damage than the conventional bleaching process. Pigmented fibre bleaching generally involves three stages: mordanting with ferrous salts, rinsing, and treatment with hydrogen peroxide in an alkaline medium (Khishigsuren et al., 2002). 18-MEA is the major lipid bound to the surface of a cashmere fibre (Volooj et al., 1999), which is removed by the hydrogen peroxide bleaching (Volooj et al., 1999). Afterchrome dyeing of cashmere with SCA-Cr mordant decreased the level of hexavalent Cr (VI) in the residual dye bath, and damage to the fibre (Xing and Pailthorpe, 2000) was notably reduced in terms of surface oxidation damage to the fibres.

10.6.3 End-uses

The quality of cashmere, notably fineness and length, plays an important role in determining its specific end-use (i.e., product), with the higher quality fine cashmere preferred for knitwear, where softness is of primary importance, while the coarser (lower) qualities often go into weaving. It is considered desirable for very high-quality knitting and weaving yarns that the cashmere should be 40 mm or even longer. Good-quality cashmere, notably the finer qualities, is popularly used in knitwear (particularly fully fashioned), scarves, shawls, ladies' underwears, etc. Scottish manufacturers set an 'unofficial' standard for the upper limit of cashmere to be used in such knitwear as 15.5 μm (Phan and Wortmann, 1995). Fully fashioned knitting of cashmere has always been popular because it reduces the amount of costly fibre waste and produces superbly shaped quality garments. Because pure cashmere knitting yarns tend to be weak, optimum lubrication is important to ensure relatively trouble-free knitting, so too are proper machine settings, low yarn tensions, straight as possible yarn paths, low take-down tensions and relatively slow knitting speeds.

McGregor and Postle reported on the consequences in knitwear, of blending cashmere with superfine wool, the differences in the crimp/curvature of the wool and cashmere playing a major role (McGregor and Postle, 2008, 2009). Naebe and McGregor (2013) and McGregor and Naebe (2016) found that, for blends of superfine wool and cashmere, both the blend ratio and the wool fibre curvature affected the yarn comfort-related properties. Li et al. (2014) found that longer cashmere fibres in knitted sweaters tended to shrink less, while dyeing and decolourization increased yarn and fabric shrinkage.

Because cashmere is such a sought after fibre and so expensive, together with the rather limited supplies, it is in some cases even reused (recycled or reprocessed). Reused (recycled) cashmere should, however, be clearly labelled as 'recycled cashmere'. Fisher's discriminant analyses have been used to distinguish between new and recycled cashmere (Langley and Coskuntuna, 2000).

Raja et al. (2013) reported on a novel finishing process for cashmere/PVA-blended fabrics.

Although machine washable cashmere garments have been produced, they are generally cleaned by dry-cleaning or gentle handwashing, using water which is not very hot and a pure soap powder or liquid or detergent specifically recommended for cashmere. The garment must not be rubbed or wrung during washing and should be rinsed thoroughly in cool water without softener and dried flat, out of direct sunlight. Cashmere is said to become more quickly saturated with water than wool

(Hughes and Nelson, 2001). The measurement of pilling in cashmere knitted garments has been investigated (Zhang et al., 2005). For further information on cashmere, see the references (Franck, 2001; Knott, 1990; McGregor, 1998) and www.cashmere.org.

10.7 Guanaco

10.7.1 Fibre production, harvesting and properties

Some 2 million years ago, the guanaco (genus: *Lama hunchus* or more commonly *Lama guanicoe),* a wild (now semicaptive) member of the South American Camelid family, first appeared in the palaeontological record, evolving in South America from *Hemiauchenia* (Rainsford, 2006a). The guanaco (see Fig. 10.8) is larger than the alpaca and vicuña, but smaller than the llama, and is a very timid animal and was originally killed for its meat and fleece (Briggs, 1973), although today it is farmed in semicaptivity and shorn for its fleece.

The *lama guanicoe* is the guanaco subspecies located in the Argentinian Patagonia and represents some 70% of the total world population of guanacos (Bacchi et al., 2010). Bustamante et al. (2002) investigated the genetic diversity of guanaco and llama populations in Argentina and found a high genetic variability.

The guanaco represented an important source of meat and fibre, the latter used in textiles for thousands of years by pre-Columbian cultures (Rainsford, 2006a), and

Figure 10.8 A guanaco.
(Source: Inca Tops SA, Arequipa, Peru.)

reserved for the clothing of ruling families during the Inca Empire (Rainsford, 2006a). According to Rainsford (2006a), there were an estimated 30–35 million guanacos roaming the South American continent before the Spanish conquest of eastern South America in the early 1530s, but their numbers decreased drastically because of excessive hunting, competition for pasture from European livestock and the introduction of fencing (Rainsford, 2006a). It is indigenous to Southern Patagonia, both in Argentina and Chile. It is estimated (Rainsford, 2006a; Allolio, 2008) that, in more recent times, there were well over 600,000 animals in the world (including some 500,000 in Argentina, 90,000 in Peru and 70,000 in Chile), mainly in small herds in Patagonia and the island Tierra del Fuego (Allolio, 2008), about 1% being bred in semicaptivity. Guanaco, also known as Huanaco in Peru, can survive at almost any altitude, from sea level at Tierra del Fuego to 4500 m above sea level along the Andes of Northern Chile, Peru and Argentina (also referred to as luan), and almost every climate from semitropical to extremely cold snow-capped glaciers (Rainsford, 2006a; Anon., 2000b). They are both grazers and browsers, with areas of lush bog, known as 'bofadels', ideal for breeding males (Rainsford, 2006a), adult males weigh between 100 and 150 kg (Rainsford, 2006a) and adult females between 100 and 120 kg and their height ranges from about 1 to 1.2 m (Rainsford, 2006a). Very young guanacos (chulengos) are hunted for their meat and valuable skins, and fibre is also shorn from their skins (Allolio, 2001).

In 1975, the guanaco was listed on the Convention on Trade in Endangered Species of Wild Flora and Fauna (CITES) list (regulations in Appendix II), meaning that legally guanaco fibre can only be shorn and processed into luxury goods in accordance with CITES regulations covering protected animal species (Rainsford, 2006a). Guanacos are now bred in semicaptivity, whereas wild adults are captured, shorn and released (Allolio, 2007). A challenge to guanaco fibre breeders is to achieve the critical fibre production of between 4000 and 4500 kg of raw fibre required for industrial processing (Allolio, 2007). In Argentina, growers were allowed (in 2000) to capture a maximum of 10% of their total wild guanaco population with a view of adapting the young 'chulengos' to a new life which requires high fencing, special baby feeding and separation of families to avoid adult male fights (Anon., 2000c). To raise guanacos in semicaptivity involves rounding up pregnant wild females and keeping then in corrals until they give birth (Rainsford, 2006a), after which the mothers are kept with their youngs for 3 months to familiarize them with human contact, and then they are released back into the wild, making them easier to round up and recapture for shearing (Rainsford, 2006a). Attempts have also been made to raise the guanaco entirely in captivity, by capturing the very young, preferably week-old 'chulengos', bottle feeding them for 120 days, taking great care to avoid mortality (Allolio, 2008), and then releasing them into specially prepared paddocks with fences up to 2.1 m to graze on nutritional pastures, supplemented during the first winter (Allolio, 2008).

The guanaco fleece is shorter, coarser ($\approx$ 18–24 μm) and lighter in colour than that of the vicuña (Briggs, 1973), weighing some 150 g (Rainsford, 2006a) to 1 kg, and consists of two coats: a protective relatively coarse outer coat ($\geq$50%) and a much finer ($\approx$ 15 μm) undercoat ($\leq$50%). The animal is honey (fawn) coloured with a white underbelly and legs, and greyish-colour head and ears and the colour of the fleece ranges from golden/beige for the younger animals (chulengos) to dark beige and reddish brown for the adult animals (Allolio, 2001). The guard hair is up to 45 μm (usually 23–35 μm) in diameter and up to 140 mm in length. The diameter of the undercoat ranges from about 13 to 19 μm, the fibre diameter varying according to feeding conditions. Good animals achieve a mean fibre diameter of 14 μm (Desmond, 1993) while a young guanaco fleece can even be as fine as 13.3 μm, and 34 mm long (Allolio) with a weight of 450 g (Allolio, 2008). Commercial dehaired guanaco fibre ranges in diameter from about 13 to 11 μm (Greaves and Rainsford, 2005), with the bulk being 15–16 μm.

In an attempt to standardize the quality of fibre offered by the farmers (estancias), the Argentinian government has implemented regulations requiring its representatives to consult with the farmers concerning their fibre-harvesting systems (both from wild or semicaptive animals) (Rainsford, 2006a). The price paid for guanaco fibres mainly depends on fineness, length and clip preparation (Allolio).

10.7.2 Fibre processing

Hair harvested (shorn) from the guanaco needs to be dehaired to separate the coarse guard hair from the fine undercoat fibres. Guanaco fibre is mainly processed on the worsted system, although some fibre is also processed on the woollen system, scouring yields varying from about 85% to 95% (Allolio, 2001), with greasy to clean yield quoted (Franck, 2001) as 65%–70%. In one case, some 400 high-quality guanaco fleeces were processed into 14.91 μm tops, having a CV of 20.22%, Hauteur of 33 mm, a Barbé of 39.1 mm and a relatively low CVH of 43%, having a fawn shade similar to camel (Allolio).

10.7.3 End-uses

Guanaco fibre end-uses are similar to those of vicuña, being mainly used in luxury woven fabrics, supplied to exclusive tailoring companies, the suits of which can retail for US$10 000 (Rainsford, 2006a). It is less commonly used in knitwear, although up to 20% guanaco fibre blended with Merino wool is used in fine fashionable sweaters and other knitted garments (Allolio, 2008). Fine-count worsted yarns, in various blends with Merino wool and/or cashmere, are also supplied to exclusive tailoring businesses in London, Milan and Tokyo, where blazer cloth is popular, and where dress overcoats are made from a fabric containing guanaco, silk and cashmere (Rainsford, 2006a), each garment being individually certified. Woollen-spun guanaco yarns are used in suiting and overcoating fabrics for exclusive top-quality fashion brands (Rainsford, 2006a). Guanaco is also making inroads into markets for scarves and shawls. Pure (100%) guanaco is more common in accessories.

10.8 Llama

10.8.1 *Fibre production, harvesting, properties, processing and end-uses*

The llama *(Lama glama glama),* a ruminant or pseudoruminant living in the high altitudes of the Andes, has traditionally, and mainly, been a beast of burden of the Andes (Briggs, 1973), from the time of the Incas. It is the largest of the South American branch of the camelids. It is smaller than the camel (see Fig. 10.9) measuring about 1.2 m at the shoulder and weighs around 100–120 kg (Rainsford, 2006b). It has been domesticated since time immemorial.

The guanaco or huanaco *(Llama glama hunaca),* a ruminant (or pseudoruminant), is the common ancestor of the domesticated llama and alpaca (Wildmann and Wildmann, 1954). The natural habitat of the llama is the 'Antiplano', the vast, high and arid plateau of Peru, Bolivia, Chile and Argentina, at altitudes between 4000 and 5000 m (Briggs, 1973), feeding mainly on ichu grass. It has been estimated (Rainsford, 2006b) that in 2006, there were some 3.4 million domesticated llamas in the world, with some 2.0 million in Bolivia. Peru has some 1.1 million and there are also significant numbers in the USA.

Figure 10.9 Llamas.
Source: Inca tops SA, Arequipa, Peru.

Frank et al. (2006) gave the following average values for the fibre quality of llamas in Argentina: greasy fleece weight 1615 g, staple length ≈190 mm, mean fine fibre diameter ≈22.9 μm (26% CV), 28% total medullation, 20% continuous medullation, 52% nonmedullated fibres. Fleece mass and mean fibre diameter increased with age, whereas staple length decreased, medullation increased with fibre diameter. Brodtmann et al. (2017) obtained a value between 40 and 50 μm for the unclassed/undehaired fleeces, and between 23 and 25 μm for the fine (dehaired) fibres, the crimp frequency ranging from 0.5 to 7 per cm, with a mean 2.8 crimps/cm.

Martinez et al. (1997b) reported the following values for the fibre properties of some 143 llamas from Bolivia:

Average fibre diameter
- Fleece: 31.6 μm
- Coarse fibres: 40.8 μm
- Fine fibres: 25.5 μm
- Unmedullated fibres: 22.7 μm

Average fibre length**
- Fleece: 67 mm
- Coarse (guard) hair: 83 mm
- Fine (undercoat): 50 mm

Average medullation: 43%

** Excluding 2-year-old llamas.

Medullation and diameter increased and fibre length and crimp decreased with increasing age, while medullation increased with fibre diameter.

The llama fleece, typically shorn every 2 years (Iñiguez et al., 1998b), has a similar mean fibre diameter as the alpaca but is a double coat. There are, however, essentially two main types of llama. One type, the Kara (K'cara, Kh'aras, Kjaras, qara/carguera or light fleece, compact guard hair/down fleece (Iñiguez et al., 1998b)), is mostly used as a beast of burden and is the typical double-coated animal (short and coarse staple length ≈120 mm, single fibre diameter of the fleece ranging from about 10 to 150 μm, average ≈ 25 μm), which has predominantly coarse guard hair and relatively fewer fine undercoat fibres or down (Valbonesi et al., 2010; Antonini et al., 2004). The second type, 'C'haku/chachu' or Th'ampullis (woolly, finer or heavy fleece, staple length ≈155 mm, average diameter ≈21μm) (Iñiguez et al., 1998b), which is the main source of llama hair used for textiles, has essentially a single coat of relatively fine and soft fibres, but with a relatively high level of medullated (kemp) fibres amongst the secondary unmedullated fibres (Antonini et al., 2004). Interbreeding between llamas and alpacas has resulted in many and varied types (hybrids) of animals and fleeces, the cross of llama male with an alpaca female producing the Huarizo, known in India as Huaro, the reverse cross producing the Paco-llama or Misti.

Wurzinger et al. (2006) reported on genetic and nongenetic factors influencing the fibre quality (midside samples) of Bolivian llamas, which represent some 60% of the total world population of llamas. They obtained the following average values for the midside samples which they tested: K'hara mean fibre diameter 22.7 μm (36%CV), mean diameter of fibres finer than 30 μm ≈ 20.7 μm, 37% medullated fibres and 0.75% kemp. The corresponding vales for Th'ampulli were as follows: 21.7 μm (30%CV), 19.9 μm, 14.1% and 0.10%, respectively.

The fleece of the llama (typically 15–60 μm, mean ≈ 22 μm and CV˜30%) (Frank et al., 2011a) is highly medullated (≈80%) (Martinez et al., 1997a) and not as attractive as that of the alpaca. It contains both coarse, long, kempy hairs and fine undercoat hairs and it is not easy to separate (Briggs, 1973) the coarse guard hair from the fine down, also due to the presence of 30–40 μm intermediate fibres. The coarse guard hair is typically 30–50 μm in diameter (these are also referred to as intermediate fibres) but can even be up to 150 μm and ranges from about 50 to 200 mm in staple length (typically over 100 mm) (Martinez et al., 1997a) and is kempy. The fine undercoat or down ranges from about 10 to 35 μm in diameter (average around 22–25 μm; and CV from 25% to 35% (Phan, 1991)) (Martinez et al., 1997a) and 50–65 mm in length (staple length ≈130 mm) (Frank et al., 2011a) with baby llama fibre diameter being 20–21.5 μm, which is slightly finer than baby alpaca. Wildmann and Wildmann (1954) suggested 35 μm as the limit (or criterion) for distinguishing between the woolly (undercoat) fibres and the coarse (outer coat) hair. According to tests conducted by Greaves and Rainsford (2005), the fineness and related quality of llama hair have not changed significantly from 1970 to 2005. The fleece of the llama contains less than 4% grease and less than 25% total impurities, and average fibre yield is about 80% (85%–90% greasy to clean (Franck, 2001), quoting Rainsford).

The staple strength of K'jaras fibre is˜35N/ktex and that of Thampulli ≈52N/ktex, decreasing with age (Iñiguez et al., 1998b), CV of strength ≈37%.

The natural colours of the fleeces are similar to those of the alpaca, and there are often different colours within a fleece. It is mostly brown (fawn) in colour, although greys and even white, used alone, are used in blends in knitwear and outerwear chiefly used by the locals in clothing (e.g., ponchos), carpets, rugs, etc., while the coarse llama guard hair is used in ropes, sacking, braids, carpets, etc. The llamas breed, and are also shorn (mainly by hand), during the rainy season, from November to March. A fleece grown over a 2-year period weighs approximately 2.5 kg and only about 1 million kg of normal quality llama fibre is produced annually.

Frank et al. (2007) have objectively described the different fleece types, subjectively defined as styles, in terms of different fibre types, differentiated according to length, fineness, type of waves and the presence or absence of lustre. They also reported (Frank et al., 2011b) on the dehairing and processing of different types of llama fleeces from Argentina, the mean fibre diameter ranging from about 21 to 24 μm and the CV of diameter from about 23% to 41%, depending on the fleece type.

Llama fibres are nearly all medullated, only a few of the finest fibres being unmedullated. Cuticular scale frequency is around 10–11 per 100 μm (Antonini et al., 2004; Langley and Kennedy, 1981; Liu, 2003; Sich, 1990).

Llama is dehaired in Bolivia and other countries. Townend et al. (1989) review the various dehairing techniques that have been developed over time and report on experimental work on the variables affecting the dehairing of llama when using a single swift woollen card. They refer to the problems created by the wide range of fibre diameters (10–80 μm) present in un-dehaired llama hair. They conclude that modifying and optimizing the carding arrangement, conditions and settings could lead to significant dehairing, although probably not to the extent that is commercially desirable.

For more information on llama fibre, see the references Franck (2001), http://www.llama.org//llama-fiber.htm and Brodtmann et al (2017).

10.9 Mohair

10.9.1 Fibre production, harvesting and properties

Mohair, derived from the Arabic word Makhayar (Mukhayar or Mukhaya), is the fleece of the single-coated Angora goat *(Capra hircus aegagrus)*, named after the Turkish province of Ankara (Angora or Ancyra) (see Fig. 10.10). The Angora goat is thought to have originated in the Asian Himalayas (Asia Minor) or highlands of Tibet.

The Angora goat is regarded as unique amongst goats in that it is essentially single coated, with the fibres from the primary and secondary follicles not differing very widely, and it does not moult, its fibres growing continuously throughout the year. For centuries, mohair has been regarded as one of the most luxurious and best quality fibres available to man. It is generally a long, straight (uncrimped but often wavy), smooth and naturally lustrous fibre, and predominantly white in colour which can be dyed to deep, brilliant and fast colours. Mohair is characterized by, and renowned for, its high lustre, durability (hard wearing), elasticity, resilience, resistance to soiling, soil shedding, setting, strength, abrasion resistance, comfort (including moisture absorption) and pleasing handle, and by relatively low flammability, felting and

Figure 10.10 Angora goats.
Source: Mohair South Africa, Port Elizabeth, South Africa.

pilling. Although mohair has proved extremely popular in many applications, it has some limitations in certain 'close to the skin' apparel applications because of its coarseness relative to certain other types of luxury animal fibres. Its outstanding properties, such as resilience and durability, also make it particularly suitable for household textiles, such as upholstery fabrics, curtains and carpets. The low flammability of mohair renders it useful in several applications. The Limiting Oxygen Index of untreated mohair is about 25, with 27 generally regarded as the minimum required to pass the vertical flame test.

The mohair industry was first developed in Ankara, Turkey, which was also the first country to supply mohair as a raw material (Haigh, 1949). The first Angora goats to leave Turkey went to South Africa in 1838 (Anon., 1981; Kennedy-Sloane, 1977), and Angora goats arrived in the USA around 1849 (Ryder, 1989).

Angoras were also introduced into Australia during the 1850 and 1860s (Ryder, 1981), but attracted little interest until around 1970 (Ryder, 1981). Angora goats were introduced to Britain in 1881 (Ryder, 1981). Today, mohair is largely produced in South Africa (which presently accounts for over 50% of global production), and the USA (Texas), but also in Turkey, Argentina, Lesotho, Australia and New Zealand (Table 10.23).

McGregor (1998) has reviewed various aspects related to the quality and production of mohair and cashmere in Mediterranean and temperate climatic zones, also reporting (McGregor, 2010) on the effect of stocking rate on mohair production and quality in Southern Australia.

Angora goats can survive extreme temperatures, but they are very sensitive to cold after shearing, particularly a combination of cold, wind and/or rain. Angora goats can thrive on widely different types of pasture, grazing from about 30 cm to 1.6 m above ground level. It appears that the Angora goat is very efficient in converting feed into fibre (Gallagher, 1971) and more effective than woolled sheep (Van Der Westhuysen et al., 1985; Shelton et al., 1981); the latter are more effective in converting feed into body mass. Mohair grows about 20–25 mm in length per month (i.e., 240–300 mm per year), irrespective of age, and Angora goats are generally shorn six-monthly in South Africa and the USA and annually in Turkey and Lesotho. Mohair fibre is generally classified according to the age of the goats and when they are shorn.

McGregor and Butler (2011) found that 30% wear of the permanent first incisors reduced fleece production very significantly. McGregor et al. (2012b) found mohair mean fibre diameter to be allometrically related to the cube root of "fleece-free liveweight" over the lifetime of angora goats. Mueller et al. (2016) found that the introduction of Texan angora bucks improved Tajik goats in terms of fleece weight, and mohair staple length (longer) mean fibre diameter (finer), and both kemp and medullated fibre levels (lower).

According to Van Der Westhuysen et al. (1985), the age of the goat is probably the most important factor determining the quantity and quality of mohair produced, with fineness (i.e., diameter), independent of age, also very important. Kids have a birth coat of fibres that grow mainly from the primary follicles, those being the follicles which produce kemp and medullated fibres (Kingsbury, 1985). From about 3 to 6 months, the goats shed their birth coat ('mother hair') as the fibres grow

Table 10.23 World mohair production (million kg greasy).

Year	South Africa	Turkey	USA	Argentina	Australia	New Zealand	Lesotho	Other	Total
1972	3.7	4.1	4.6	1.0	–	–	0.8	0.0	14.2
1973	3.4	4.1	4.5	1.0	–	–	0.6	0.0	13.6
1974	3.7	4.1	3.8	1.0	–	–	0.6	0.0	13.2
1975	3.8	3.9	3.9	1.0	–	–	0.6	0.0	13.2
1976	4.1	4.0	3.6	1.0	–	–	0.6	0.0	13.3
1977	4.6	4.1	3.6	1.0	–	–	0.4	0.0	13.7
1978	4.9	4.5	3.7	1.0	–	–	0.5	0.0	14.6
1979	5.4	4.5	4.2	1.0	–	–	0.5	0.0	15.6
1980	6.1	4.5	4.0	1.0	–	–	0.6	0.0	16.2
1981	6.9	4.5	4.5	1.0	–	–	0.6	0.0	17.5
1982	7.6	4.5	4.5	1.0	–	–	0.6	0.0	18.2
1983	7.2	3.8	4.8	1.1	–	–	0.7	0.0	17.6
1984	8.1	3.5	5.0	1.0	–	–	0.7	0.0	18.3
1985	9.2	3.5	6.0	1.0	–	–	0.8	0.0	20.5
1986	11.0	3.0	7.2	1.0	–	–	0.8	0.0	23.0
1987	11.5	3.0	7.3	1.0	1.0	–	0.8	0.0	24.6
1988	12.2	2.9	7.8	1.0	1.0	0.4	0.7	0.0	26.0
1989	11.7	2.0	7.8	1.0	1.2	0.6	0.6	0.0	24.9

1990	10.1	1.8	7.3	1.0	0.6	0.4	0.6	0.0	21.8
1991	7.6	1.2	7.4	0.9	0.5	0.3	0.5	0.0	18.4
1992	6.7	1.2	7.1	0.6	0.5	0.3	0.4	0.0	16.8
1993	6.0	0.8	6.5	0.6	0.4	0.3	0.4	0.0	15.0
1994	5.7	0.8	5.4	0.4	0.4	0.2	0.4	0.0	13.3
1995	5.4	0.6	4.8	0.5	0.4	0.2	0.5	0.0	12.4
1996	5.6	0.4	3.5	0.4	0.4	0.2	0.5	0.0	11.0
1997	5.2	0.4	2.5	0.4	0.3	0.2	0.4	0.0	9.4
1998	5.0	0.4	1.5	0.4	0.3	0.2	0.4	0.0	8.2
1999	4.5	0.4	1.2	0.25	0.25	0.2	0.4	0.0	7.2
2000	4.3	0.4	1.0	0.3	0.3	0.2	0.5	0.0	6.9
2001	4.2	0.3	0.8	0.3	0.3	0.2	0.5	0.3	6.8
2002	4.2	0.3	0.8	0.3	0.2	0.1	0.5	0.3	6.6
2003	4.0	0.3	0.9	0.3	0.3	0.2	0.5	0.3	6.6
2004	3.7	0.2	0.85	0.3	0.3	0.2	0.5	0.2	6.1
2005	3.6	0.3	0.8	0.3	0.2	0.2	0.6	0.3	6.2
2006	3.4	0.3	0.8	0.4	0.2	0.1	0.75	0.2	6.1
2007	3.0	0.35	0.55	0.45	0.2	0.1	0.75	0.2	5.6
2008	2.9	0.35	0.5	0.45	0.2	0.05	0.75	0.1	5.3
2009	2.6	0.3	0.5	0.7	0.2	0.1	0.75	0.2	5.3

Continued

Table 10.23 Continued

Year	South Africa	Turkey	USA	Argentina	Australia	New Zealand	Lesotho	Other	Total
2010	2.3	0.170	0.48	0.7	0.180	0.05	0.75	0.20	4.8
2011	2.23	0.150	0.35	0.7	0.155	0.045	0.75	0.20	4.6
2012	2.32	0.190	0.21	0.6	0.160	0.050	0.77	0.30	4.6
2013	2.4	0.260	0.15	0.5	0.170	0.031	0.80	0.20	4.5
2014	2.45	0.2	0.15	0.6	0.17	0.02	0.8	0.32	4.71
2015	2.48	0.3	0.15	0.6	0.12	0.02	0.73	0.35	4.75
2016	2.48	0.22	0.15	0.6	0.12	0.02	0.76	0.35	4.7
2017	2.4	0.30	0.15	0.65	0.12	0.02	0.80	0.35	4.79
2018	2.24	0.34	0.23	0.71	0.06	0.03	0.80	0.35	4.76

Source: Cape Mohair South Africa.

increasingly from the secondary follicles which produce the finer and unmedullated hairs (Kingsbury, 1985; Harmsworth and Day, 1990). Fibre production increases from birth, reaching a maximum fleece weight at an age of between approximately 3 and 4 years (see Fig. 10.11).

With age, the fibre diameter increases, reaching a maximum at approximately 5 years (Van Der Westhuysen et al., 1985) (see Fig. 10.11). The mohair fibres are finer towards their tips, because of the fact that the fibres become coarser as the goat ages. Fibres from the neck and britch areas of the goat tend to be coarser than those from the other parts of the body. Taddeo et al. (2000) found that for angora goats in Argentina, the midrib sample was generally representative of the fleece.

In South Africa, the first shearing takes place around 6 months after birth and the second shearing 6 months later, the goats producing more fibre in the summer than in the winter. In South Africa, mohair obtained from the first two shearings (i.e., at 6 and 12 months) is generally classified as Kids. That obtained from the third (and also sometimes from the fourth) shearing (i.e., at 18 months and sometimes at 24 months) is classified as Young Goats and after that (i.e., from the fourth or fifth shearing, i.e., from the age of 24 or 30 months), the hair is classified as Adults. Generally, mohair from Kids is finer than 30 μm (varying between about 20 and 30 μm), that from Young Goats finer than 34 μm (varying from about 27 to 34 μm) and that from Adults generally coarser than 34 μm (but ranging from about 30 to 40 μm). Goats are classed as Young Goats up to the age of 3 years in Turkey. The grades of mohair vary in different countries. In general, the best grades of mohair are from Kids (e.g., Super Summer Kids) under 6 months old (i.e., first shearing). In South Africa, mohair to be classified as Kids must be 30 μm or finer, and Young Goats 34 μm and finer. Young goats and Adult goats produce about 2–2.5 kg of greasy mohair every 6 months, rams generally producing considerably more and coarser hair than ewes (Van Der Westhuysen et al., 1985). In the case of Kids, the fleece barely weighs 1 kg at the first shearing and is generally less than 2 kg at the age of 1 year (i.e., at the second shearing).

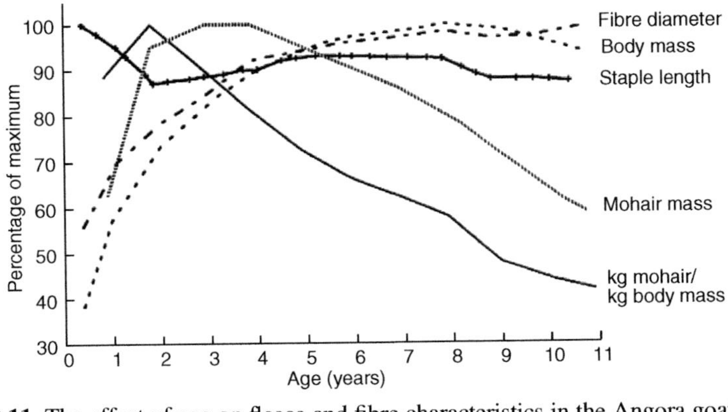

Figure 10.11 The effect of age on fleece and fibre characteristics in the Angora goat (Van Der Westhuysen et al., 1985, Van Der Westhuysen et al., 1981).

Snyman (2002) found that breeding could reduce the fibre diameter of Angora goats, while maintaining, or even increasing, body weight.

Visser et al. (2009) derived genetic parameters for the physical and quality traits of South African Angora goats. (Visser et al., 2011 identified Quantitative Trait Loci for mohair traits in South African Angora goats. Ghituleasa et al. (2004) compared the fibre properties of Romanian mohair with those of their wool of similar diameter and length, the mohair being almost twice as strong, with a much lower fibre friction and crimp, but a higher elastic modulus (407 vs 290 cN/tex).

Mohair does not have crimp in the true sense of the word but exhibits waviness or curl. Curvature values for mohair and some other animal fibres are given in Table 10.24 (Brown and Onions, 1961; Gupta and George, 1978). Table 10.25 shows some average values and ranges of various mohair properties (Turpie, 1985).

The fleece of the Angora goat, when shorn, contains natural and applied impurities; typically a total of 10%–15% of nonfibre is present. The sweat or suint, the water-soluble component, and grease (wax) combined are termed yolk. The grease (wax) is secreted by the sebaceous glands and the sweat (suint) by the sudoriferous glands. Other natural impurities contained in mohair include sand and dust (i.e., inorganic matter), vegetable matter (e.g., burr, grass seed) and moisture. Applied impurities include branding fluids and dipping compounds. Generally, mohair contains considerably less grease than wool (4%–6% on average, compared with an average of about 15% for wool). Because the yolk content of mohair is lower than that of wool, shearers are said to have to change combs and cutters more often than with wool.

Mohair, by virtue of its open fleece structure on the goat, is more exposed to weathering than is wool, the tips of the mohair fibres covering the back of the animal being damaged by sunlight or weathering, especially during the summer months (Von Bergen, 1969). This damage has an influence on the dyeing properties of the affected fibre part. Its wax is also more oxidized than that of wool (Kriel, 1964a), making it more difficult to remove during scouring (Van Der Westhuysen et al., 1981). Ilse (1958)

Table 10.24 Curvature and diameter values of several different animal fibres (Brown and Onions, 1961; Gupta and George, 1978).

Fibre	Mean curvature (cm^{-1})		Mean diameter (μm)
	Wet	Dry	
White alpaca	2.0	8.4	30.2
Fawn alpaca	1.2	6.0	40.0
Lincoln	2.4	5.0	36.0
Mohair	1.2	1.4	43.6
Cashmere	6.8	12.7	13.8
Southdown	18.8	32.0	23.8
Corriedale	10.0	16.4	29.7

Table 10.25 Some average or typical values and ranges of various mohair properties (Turpie, 1985).

Property	Range	Average/typical value
Diameter (μm)	22–42	32
CV (%)	20–33	26
Staple length (mm)	80–140	110
Medullation (%)	0.3–2.8	1.0
Curls per 10 cm	2.5–6.5	4.5
Vegetable matter content (%)	0.1–1.7	0.3
Grease (%)	2.0–8.0	4.5
Suint (%)	1.8–4.2	2.8
pH of suint	3.3–6.2	5.2
Scoured yield (%)	75–95	85
Compressibility (mm)[a]	10–13	11

[a]SAWTRI compressibility test.

compared the composition of mohair, Karakul and Merino wool waxes as shown in Table 10.26 and concluded that the mohair and Karakul waxes had the usual Merino wax components in surprisingly similar proportions.

McGregor and Stapleton (2016) investigated factors which affected the colour (particularly whiteness and brightness) of commercial mohair sale lots in Australia, finding that increasing cotting, kemp fault and mean fibre diameter (up to about 30 μm) were associated with a decrease in brightness and an increase in yellowness.

Mohair from Kids and Young Goats contains more grease than that from Adults, with the grease content higher in winter than in summer (Uys, 1963; Kriel, 1965)

Table 10.26 Characteristics of the waxes (Ilse, 1958).

Characteristics	Merino wax	Mohair wax	Karakul wax
Wax content of the fleece (%)	14–16	5	3
Saponification value (mg KOH/g)	92–102	128	110
Acid value	4	14	9
Hydroxyl value	54	57	58
Iodine value	15–30	36	56
Acids (%)	49	55	50
Unsaponifiable material (%)	51	45	50

and also higher towards the root (e.g., tip = 2.0%, middle = 4.6% and root = 6.0%). Uys, quoted by Kriel (Kriel, 1965), found an average grease content of 4.5% for summer hair and 5.8% for the winter hair, with a melting point of 39°C. He found the acid value to be 14.6 compared with a published value of 14. The unsaponifiable fraction was 46%. Kriel (1965) published values (given in Table 10.27) for the chemical constants for mohair grease.

10.9.1.1 Fibre dimensional and tensile properties

Fibre diameter (fineness) is generally the most important textile quality aspect and price-determining factor by far of animal fibres, such as mohair. The finer the fibre generally, the more sought after and expensive it is. Fibre fineness determines processing behaviour and performance and product type and quality, having a major influence on the finest yarn and lightest fabric which can be produced, as well as the handle and against body comfort of the fabric.

Hunter et al. (1985) studied the diameter and variation in diameter, as measured by projection microscope, of some 852 samples of raw and scoured mohair and 380 mohair tops. They found that, although standard deviation tends to increase with increasing mean fibre diameter, the relationship was a tenuous one and the scatter was large. There was a tendency for CV to decrease as mean fibre diameter increased up to a mean fibre diameter of somewhere around 35 μm, after which the reverse occurred. For most practical purposes, however, the CV of diameter could be regarded as largely independent of mean fibre diameter, with an average value of approximately 27%. Table 10.28 gives average (typical) values for CV of fibre diameter.

Wang et al. (1999) showed that there was a relationship between CV of mohair fibre diameter and CV of single fibre strength as predicted theoretically. Fibre length is of secondary importance to fineness, having an important effect on the processing route and behaviour and on yarn quality. Fibre tensile properties are also important from a textile point of view, fibre strength playing an important role in fibre breakage during mechanical processing, including spinning, yarn strength, fabric manufacturing, and in

Table 10.27 Chemical constants for mohair grease (Kriel, 1965).

Characteristics	Value	Literature
Saponification value	126–135	128
Acid value	14.6	14.0
Iodine value	14.8	36
Percentage acids	54	55
Percentage unsaponifiable fraction	46	45
Ester value	117	114

The experimental values of Kriel (1965) are shown in the column ‘Value’.

Table 10.28 Average values of coefficient of variation (CV) of fibre diameter corresponding to different mean fibre diameters (Hunter et al., 1985).

Mean fibre diameter (μm)	CV of fibre diameter (%)
25	30
30	27
35	26
40	27
45	29

the ultimate strength of the fabric. Generally, in the case of animal fibres, fibre strength increases almost linearly with the fibre cross-sectional area, more particularly the cross-sectional area of the thinnest (generally the weakest) place along the fibre. Therefore, fibre diameter, more specifically, at the point of break, often the thinnest place along the fibre, has the main effect on the absolute strength of the fibre (i.e., uncorrected for the fibre cross-sectional area). The fibre strength divided by the fibre cross-sectional area at the thinnest place or point of break (i.e., the intrinsic strength or intrinsic tenacity) is fairly constant for mohair.

Smuts et al. (1981) found that mohair generally had a higher single fibre tenacity, initial modulus and extension at break than wool of the same diameter (Table 10.29). The mohair cross-section corrected tensile characteristics (i.e., tenacity) were fairly constant over the whole range of diameters, probably because of the absence of crimp and variations in crimp and associated fibre characteristics. Lustre wools (e.g., Lincoln and Buenos Aires) had tenacities and initial moduli close to those of mohair (Smuts et al., 1981).

Hunter and Smuts (1981) found both bundle and single tenacity to be independent of mohair fineness, although the initial modulus increased slightly with an increase in fibre diameter (Table 10.30).

10.9.1.2 Fibre stiffness

The static bending and extension moduli of mohair fibres are similar and of the order of 308 cN/tex (King, 1967), with the optical cell densities in the medullae of kemp fibres affecting the bending but not the extension moduli. For kemp, having a virtually empty medulla, the bending and extension moduli were similar at about 77 cN/tex, whereas for the kemp with a virtually filled medullae, the bending modulus was about 365 cN/tex, which was higher than that of mohair (King, 1967). The extension moduli of the two types of kemp fibres were similar, indicating that any material in the medullae did not contribute significantly to the tensile properties of the fibre, confirming the results of the studies by Hunter and Kruger (1966), (1967).

Table 10.29 Average values for some tensile properties of wool and mohair[a] (Smuts et al., 1981).

Property	Mean	SD	CV %	Range	*n*
Wool[b]					
Fibre diameter (μm)	22.7	3.3	15	18.1–33.1	56
Linear density (dtex)	6.6	2.0	30	3.5–12.8	56
Staple crimp (cm^{-1})	4.2	1.2	27	1.9–6.5	56
Resistance to compression (mm)	17.5	2.8	16	13.6–24.7	56
Bulk/diameter ratio (μm/mm)	0.79	0.19	24	0.41–1.29	56
Tenacity (cN/tex)	12.7	0.9	7	10.9–15.0	56
Initial modulus (cN/tex)	290	27	9	230–392	56
Extension at break (%)	37.0	2.6	7	31.5–41.2	56
Mohair					
Fibre diameter (μm)	32.1	5.8	18	20.7–44.3	29
Linear density (dtex)	11.9	3.3	28	5.8–20.1	29
Tenacity (cN/tex)	16.7	0.7	4	14.6–18.1	29
Initial modulus (cN/tex)	407	13	3	384–430	29
Extension at break (%)	42.7	2.1	5	38.0–45.8	29

[a]20 mm test length and rate of extension 20 mm/min.
[b]Low crimp wool excluded.

Table 10.30 Typical tensile properties of mohair (Hunter and Smuts, 1981).

Property	Bundle test[a]	Single fibre test
Tenacity (cN/tex)	14.0	16.7
Extension (%)	14. 6[b]	43.0
Initial modulus (cN/tex)	–	407

[a]Leather linings were used and the tenacity values were multiplied by a correction factor of 1.16.
[b]The bundle test is not considered to give reliable extension values.

10.9.1.3 Fibre surface and frictional properties

Mohair, wool and hair are covered by a layer of sheet-like hardened cuticle cells (epidermal scales) which overlap each other, with their exposed edges towards the tip of the fibre. The cuticle plays an important role for the whole fibre because it is, on the one hand, exposed to environmental influences and, on the other hand, responsible for the surface properties of the fibre. Although, under a microscope, mohair is similar in appearance to wool, the epidermal scales (cuticle scales) of mohair are generally less pronounced and only faintly visible. The scale structure described previously is responsible for mohair's smooth handle, high lustre, low against-scale friction and low felting propensity. The cuticle scales are quite thin and flat (Appleyard, 1960), generally being less than about 0.6 μm (typically 0.4 μm) in thickness, and hardly overlap (Von Bergen, 1969). They are anchored much more closely to the body of the fibre (Von Bergen, 1969; Von Bergen and Mauersberger, 1954; Cook, 1968; Zahn, 1990), i.e., they lie near to the stem or are piled more tightly upon one another (Tagawa et al., 1972), giving the fibre its characteristic lustrous and smooth appearance.

In general, mohair has a relatively low-scale frequency, with a wide distance between the cuticle scale margins. The number of scales per 100 μm is generally in the order of 5 compared with between 9 and 11 in fine wools, with the scale lengths ranging from 18 to 22 μm. In the case of kemp, the number of scales per 100 μm is 10 or more, which is twice that for mohair; and they are arranged in a coronal or ring pattern, with smooth margins (Von Bergen, 1969). The width to length ratio of mohair fibre scales is of the order 2 (Ryder, 1990c). Ryder and Gabra-Sanders (1988) found that the width to length (W/L) ratios of scales from various goat fibres showed a clear sequence from the wild ancestor *(Capra aegagrus)* to mohair. They defined the scale width as equal to the fibre diameter. Indications were that the W/L ratio was independent of fibre diameter.

As in the case of wool, mohair fibres have a lower friction when rubbed from the root to the tip (i.e., with the scales) than when rubbed in the opposite direction (i.e., from tip to root, termed against scale). The low against-scale friction of mohair, relative to wool (Table 10.31) (Harris, 1954), which is one of its distinguishing features, can be largely attributed to its relatively smooth (unpronounced) scale structure. Mohair has a very small directional friction effect (DFE), owing to the extremely easy deformation of the thin distal edges in mohair and also to the absence of tilted outer surfaces and other high asperities. The against-scale (μ_2) to with-scale (μ_1) friction ratio of mohair is about 1:1 compared to about 1:8 for Merino wool (Onions, 1962). The 'scaliness' ($(\mu_2 - \mu_1) \times 100\%/\mu_1$) of mohair, measured dry, is about 5 compared to about 60 for a fine Merino wool (Speakman and Stott, quoted by Onions, 1962). When measured wet, the respective values are about 16 for mohair and 120 for Merino wool. These characteristics give mohair its low felting propensity.

Table 10.31 Fibre frictional properties (Harris, 1954).

Fibre	μ_2	μ_1	$\mu_2 - \mu_1$	$\mu_2 + \mu_1$
Wool	0.40	0.22	0.18	0.66
Mohair	0.23	0.15	0.08	0.38
Human hair	0.19	0.09	0.10	0.28

Note: All measured in distilled water against felt.
μ_1 with-scale, μ_2 against-scale.
Source: Frishman et al., quoted by Harris (1954).

10.9.1.4 Moisture-related properties

Although mohair, as in the case of wool, can absorb large quantities of moisture (up to about 30%) without feeling wet or damp, its surface is naturally water repellent, largely due to the presence of a strongly bound thin surface layer of waxy or lipid material which requires strong chemical action to remove it. The moisture-related properties of textile fibres are extremely important as they play a crucial role in the comfort of the fibre and in its behaviour during wet treatment and drying. Temperature and moisture also play an important role in the viscoelastic properties of wool and mohair, which in turn is related to wear properties, such as wrinkling.

Speakman (1930) published a table (Table 10.32) illustrating the absorption and desorption of moisture by wool and mohair at different relative humidities. Watt presented a comparative table (Table 10.33) of equilibrium water content (regain) for seven keratins, including mohair.

10.9.1.5 Medullation and kemp

Medullated fibres in mohair can be a source of problems in many end-uses when they differ in appearance from the rest of the fibres that are not medullated. They are characterized by having a central canal (medulla) containing cell residues and air pockets, running in either a continuous or fragmented form along their length (Fig. 10.12 (Franck, 2001)). The term 'kemp' is probably more familiar, but this traditionally refers to the more problematic and extreme form of medullated fibres (more recently referred to as 'objectionable medullated' fibres), which are clearly distinguishable by the naked eye, due to their chalky white appearance. Kemp is usually straight, and oval in cross-section. Of all the types of medullated fibres that occur in both wool and mohair, those collectively called kemp, which tend to have a relatively large medulla and to be relatively coarse, are the most visible and unwanted in the final product. Kemp occurs as short kemp, long kemp and heterotype fibres. The 'short kemp' is generally the most common, being short, chalky white, medullated, and pointed at each end when it has fallen out and has not been shorn off. Kemp or 'objectionable medullated fibres' tends to be much coarser than the parent population (on average 1.8 times coarser than the mean fibre diameter of the parent population (Hunter, 1987a).

McGregor et al. (2013a) found medullation to increase with increasing mean fibre diameter and increasing animal size (McGregor et al., 2013b).

Good-quality mohair, such as Cape Mohair, contains very few objectionable (kemp type) fibres, and the presence of even a small amount of such fibres in a high-quality mohair can have a pronounced adverse effect on its value and price. The main problems associated with the presence of kemp (objectionable medullated fibres) are their chalky white appearance and their lighter appearance after dyeing. The chalky white appearance is largely caused by the decreased length of the light path through the dyed fibre material and light refraction at the fibre/medulla interface and within the hollow network of cells (aerian vesicles). This, and not a difference in the dyeability of the solid fibre material (i.e., fibre wall), is considered to be the main cause of the different (paler) appearance of kemp fibres after dyeing.

Table 10.32 The absorption and desorption of moisture by wool and mohair at different relative humidities (Speakman, 1930).

Relative humidity (%)	Percentage increase in weight of wool					
	Geelong 80s Merino	Southdown	Oxford down	Leicester	Wensleydale	Mohair
Absorption						
7.0	3.40	3.37	3.17	3.40	3.46	3.41
25.0	6.96	6.90	7.03	6.96	7.01	6.93
34.2	8.41	8.62	8.79	8.54	8.67	8.64
49.8	11.22	11.48	11.68	11.44	11.59	11.51
63.3	13.97	14.19	14.41	14.46	14.51	14.41
75.0	16.69	17.03	17.30	17.43	17.44	17.33
92.5	23.81	24.17	24.49	24.59	24.90	24.24
100.0	33.3	32.9	35.3	32.9	33.9	31.8
Desorption						
92.5	24.70	25.70	26.33	25.98	26.13	25.82
75.0	18.69	18.79	19.05	19.02	19.16	18.91
63.3	16.12	16.16	16.43	16.28	16.46	16.26
48.7	13.36	13.38	13.47	13.39	13.46	13.46
34.2	10.57	10.55	10.64	10.58	10.63	10.68
7.0	4.77	4.73	4.83	4.79	4.76	4.87

Table 10.33 Equilibrium water contents for seven keratins at 35°C (in percentages).

Relative humidity	Merino wool	Corriedale wool	Lincoln wool	Mohair	Monkey hair	Horse hair	Rhino horn
5	2.6	2.5	2.5	2.5	2.2	2.3	2.5
10	3.9	4.0	4.0	3.7	3.3	3.5	3.8
20	5.9	6.1	6.1	5.7	5.1	5.5	5.6
35	8.6	9.0	9.0	8.3	7.5	7.9	8.4
50	11.3	11.8	11.5	10.7	10.0	10.7	11.4
65	14.4	15.0	14.5	13.7	12.4	13.8	14.8
80	18.6	19.6	19.2	17.5	16.3	18.2	20.1
90	23.6	25.0	25.4	22.2	21.4	22.7	28.0
95	27.7	28.2	29.7	26.1	24.9	26.9	35.5
100	34.2	33.5	36.0	32.3	30.0	32.8	49.0

Source: Watt.

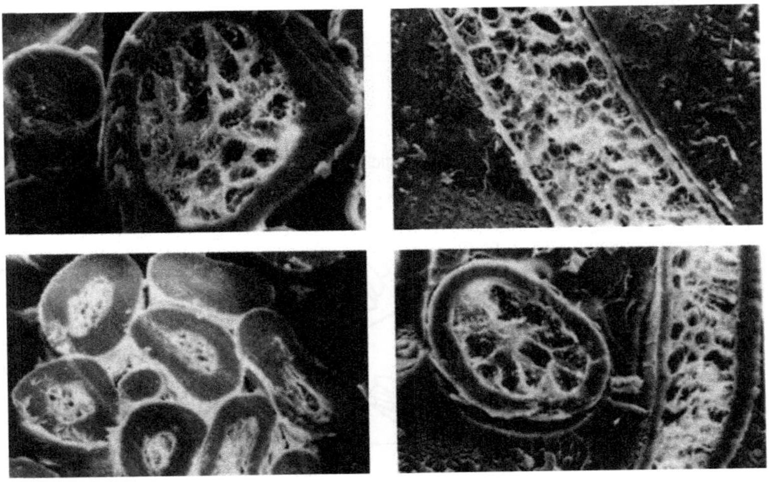

Figure 10.12 Cross-sections and longitudinal sections of medullated fibres illustrating the cellular nature of the medullae (Franck, 2001).

10.9.1.6 Fibre chemical, morphological and related structure and properties

The reader is referred to excellent and detailed reviews of this subject by Zahn (1990), (1991), Spei and Holzem (1991) and Tucker et al. (1990a), (1990b). Zahn (1995) and Zahn et al. (1980) reviewed the chemical and biological composite structure of wool, including mohair.

Mohair falls into the class of protein materials known as keratins characterized by its long filament-like molecules and insolubility in dilute acids and alkalis. Keratin can be regarded as a long fibrous composite, comprising crystalline, relatively water-impenetrable microfibrils, lying parallel to the fibre axis and embedded in an amorphous, water-penetrable matrix (Xiaoming and Postle, 1989). They generally have a high-sulphur content when compared with other proteins (Bamford, 1958). All mammalian keratin fibres contain three main protein fractions (Gillespie and Frankel, 1974), termed low-sulphur, high-sulphur and high-tyrosine proteins, with the low-sulphur proteins generally representing the largest proportion. All animal fibres contain approximately 3%–4% sulphur, largely as cystine. The mohair fibre generally consists of a cortex (cortical cells), the solid and main part or bulk of the fibre, which is predominantly orthocortex (cortical cells), and epidermis (cuticle cells) of numerous overlapping scales. The cuticle scales form a protective covering for the cortex and consist of three layers: epicuticle, exocuticle and endocuticle (see Fig. 10.13) (Smith, 1988). Each cuticle scale is enveloped by a thin semipermeable (Wortmann and Wortmann, 1988; Harizi et al., 2007) membrane called the epicuticle, which comprises protein and lipid.

For further details and discussion of the physical and chemical composition of mohair, see the references (Hunter, 1993) and (Franck, 2001).

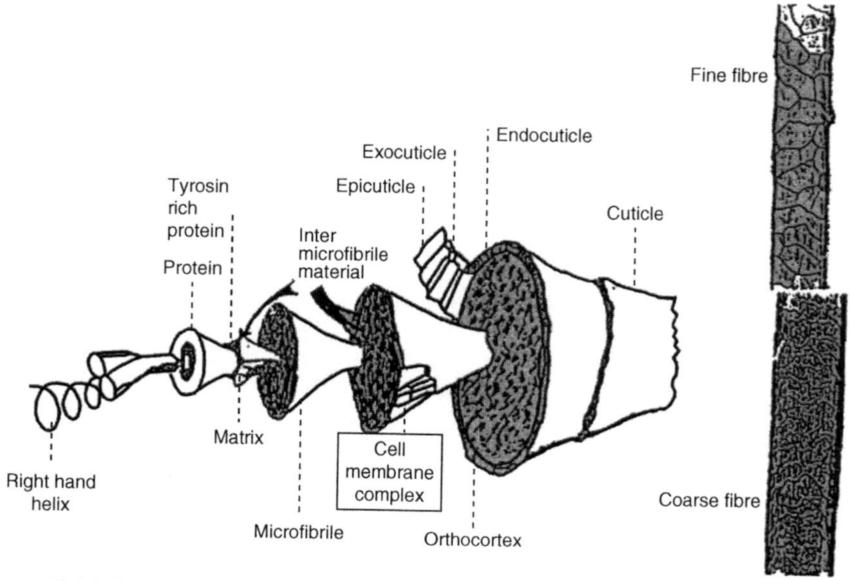

Figure 10.13 Structure of adult mohair fibre (Smith, 1988).

10.9.1.7 Objective measurement and trading

In practice, the quality of mohair is described as a combination of style and character, freedom from kemp, lustre, handle, yolk and uniformity of length and fineness (Terblanche, 1990). The presence of kemp is often the most undesirable quality characteristic of mohair. Handle is largely determined by fineness, although a soft natural yolk and oleaginous dips also improve softness of handle. Mohair characteristics of economic importance are fineness (fibre diameter), length, style and character, contamination (kemp, coloured fibres and vegetable matter), and clean yield, lustre and uniformity in general. Fibre diameter is particularly important, as is the presence and level of kemp, with length having a smaller, though still important, effect on price and processing.

Style and character are judged subjectively, high-quality style being described as solid-twisted ringlets (staples or locks), while character is described as the waviness or crimp shown in the staple (Van Der Westhuysen et al., 1981, 1985). Style without character or vice versa is undesirable, and a good balance between these two characteristics is considered to be of paramount importance (Van Der Westhuysen et al., 1981, 1985). Table 10.34 is an attempt to consolidate and rationalize some of the different systems of quality, fineness and grades encountered in the literature.

The textile processing performance, applications and general quality, and therefore value and price, of mohair are largely determined by the characteristics of the raw (greasy) mohair. It is therefore hardly surprising that considerable effort has been directed over the years towards the objective (i.e., instrumental) measurement of these characteristics to replace the subjective techniques traditionally used. Today,

Table 10.34 Some approximate quality types (Hunter, 1993; Franck, 2001).

Spinning count									
Tex	Worsted	English grades	Fineness/quality bradford count	Age group	Crimp[a] per 10 cm	Maximum mean diameter (μm)	Mean fibre diameter (μm)	Description	Age (years)
14.5–15.5	58–60s	–	8	Kids	6.5–8.0	25	<26	SSK	
16	56s	Kid	7	Kids	5.5–6.5	28	26–28	SWK	1
16.7–17.5	50–54s	30	6	Kids	5.5	30	29–30	WSK	–
–	–	–	6/5	–	–	32	–	–	–
18.5–19.5	46 48s	32	5	Young Goat	5.0–5.5	34	31–34	SYG	$1\frac{1}{2}$[a]
20	44s	34	4	Adult	4.0–5.0	36	35–36	SWH	2
22–24.5	36–40s	36	3	Adult	3.0–4.0	39	37–39	SSF	$2\frac{1}{2}$
24.5–27.5	32–36s	38	2	Adult	2.5–3.0	–	>40	SFO	2
31.5	28s	40	1	–	1.5–2.5	–	–	WHO	2
								ARH	–
								CBH	–

ARH, Adult (Ram's Hair); *CBH*, Cross-bred Hair (Adult); *SFO*, Summer First and Older; *SSF*, Super Summer Ferals; *SSK*, Super Summer Kids; *SWH*, Super Winter Hair; *SWK*, Super Winter Kids; *SYG*, Summer Young Goats; *WHO*, Winter Hair and Older; *WSK*, Winter/Summer Kids.

[a]Preliminary.

characteristics such as fibre diameter and yield can be, and often are, measured objectively with high accuracy.

Properties that need to be measured ultimately to characterize greasy mohair completely include the following:

1. Fibre diameter and its distribution (variability)
2. Yield (i.e., amount of clean fibre)
3. Staple (or fibre) length and strength, and their variability
4. Vegetable matter content and type
5. Inorganic matter content (e.g., sand, dirt, etc.)
6. Colour
7. Lustre
8. Medullation/kemp
9. Style/character

Douglas (1988) discussed the advantages of objective measurement of mohair. He stated that the mohair top must achieve strict specifications to satisfy the spinning requirements such as

- Quantity of top
- Mean fibre diameter
- Mean fibre length
- Maximum percentage of:
  - Short fibres (shorter than 30 mm)
  - Dark fibres
  - Vegetable matter specks
  - Entanglement (Neps)
  - Fatty matter
- Moisture regain
- Maximum percentage of kemp

In addition to the above, some spinners may have specifications which include (Douglas, 1988)

- Colour
- Distribution (CV%) of fibre diameter
- Bundle strength

Mohair base (i.e., the amount of clean dry fibre, free from all impurities, expressed as a percentage of the greasy fibre mass) is converted into the IWTO scoured yield (Douglas, 1988). This relates the tested yield to normal commercial yields for scoured greasy mohair. This yield is calculated from the mohair base to include all vegetable matter, standard residuals of grease and dirt, which would normally be retained in commercial scouring, and allows for a moisture regain of 17%, which means that yields of over 100% are possible.

There can be little doubt that mohair fineness (diameter) is one of its most important trading characteristics from the point of view of price and textile application and performance, with even a 1 μm change in diameter having a significant effect on price. It is therefore not surprising that mean fibre diameter, which can be measured by airflow, projection microscope, OFDA or Laserscan, is generally the main objectively measured and reported mohair characteristic, although the distribution of fibre diameter, in terms of CV, also has some textile significance. A major step forward in improving and standardizing the interlaboratory measurement of mohair fibre fineness occurred on the introduction of Mohairlabs International Round Trials and associated issuing of Mohairlabs stamps in the early 1970s. Unfortunately, this was terminated in 2003 with the dissolution of the International Mohair Association. In 2009, a similar body to Mohairlabs, called International Mohair Laboratories (IMLABS), was established in South Africa and became functional in 2011.

Turpie and coworkers (Turpie and Steenkamp, 1989, 1990; Turpie et al., 1989a, 1989b) as well as others (Blankenburg et al., 1992) reported on the calibration and application of the Fibre Diameter Analyser (FDA)200 for the rapid measurement of mohair fibre diameter and its distribution. It was concluded that, within the ranges covered, kemp level had little effect on the relationship between FDA, projection microscope and airflow diameter values. It was found that different calibrations are required for mohair and wool on both the FDA (Zahn et al., 1980) and the OFDA.

Turpie and coworkers (Turpie, 1985; Cizek and Turpie, 1985; Turpie and Cizek, 1985; Turpie et al., 1986; Turpie and Cizek, 1986; Turpie and Cizek, 1987) showed that the staple profile and length distribution could be used to predict the fibre length distribution of the staple and the top. The mohair staple has a very pronounced taper, indicating a fairly wide variation in fibre length within the staple. They found a reasonably good correlation between mohair staple length measured manually and that measured by the automatic staple length/strength tester.

Mohair burr and grass seed contaminants of mohair result in serious price penalties. Coloured (e.g., black or red) fibres, if present, could affect the finished cloth, particularly if light shades are dyed, and thereby the value of the mohair. Burrs or excessive vegetable matter in the fleece also have to be removed (Van Der Westhuysen et al., 1985). Urine and certain types of soil and vegetable matter contain substances that stain mohair permanently (Van Der Westhuysen et al., 1985). These affect the dyeing and the value of the mohair and the quality of the final product. Precautions must be taken to limit such stains, particularly urine stains (Van Der Westhuysen et al., 1985). Clean yield (i.e., the percentage of actual fibre plus commercially allowed moisture content in raw mohair) generally varies between about 80% and 90% in most fleece classes but may be as low as 60% in some outsorts, such as lox (locks), the remaining portion being made up of grease, dirt, dust and sweat.

10.9.2 Fibre processing

10.9.2.1 Scouring and carbonizing

Scouring is a critical process in mohair production and often it is at this stage that the ultimate state of the finished article is decided. As previously mentioned, mohair generally contains far fewer impurities than does wool (e.g., 4%–6% of grease compared to about 15% for Merino wool) and scouring generally causes a loss in mass of between 15% and 20%. Mohair is generally regarded as more sensitive to alkali than wool. Therefore, less, or even no, soda-ash (alkali) should be used during scouring, nonionic detergents being preferred today.

Before scouring, individual mohair bales are often sorted on screens for style and quality, efficient sorting and blending playing an important role in the eventual quality of the yarn. The fibre can then be willeyed (opening/cleaning) before it is scoured, and this is advisable. Scouring conditions for mohair are generally gentler than they are for wool, and scouring rates were lower; alkali not being necessary, the pH must be strictly controlled. Excess alkali in the fibre can lead to discolouration in dyeing. Care must be taken during scouring not to impair the lustre of mohair. Kriel (Kriel, 1964a, 1964b), quoting unpublished work by Veldsman, stated that a higher consumption of detergent was required to remove 1 g of grease from mohair than from wool. The relevant factors were the generally lower level of grease in mohair as well as its more oxidized nature, because of greater weathering than in the case of wool.

For the continental worsted system (French or rectilinear comb) of processing, which is very popular today, scouring to a residual grease content of 0.2%–0.3% is advisable, with a total fatty matter level between 0.7% and 0.9% (up to 1.2% for flexible card clothing) before carding.

Very little mohair (±2%) is normally classified as carbonizing type, although in high rainfall areas and seasons it can rise to as high as 15%; mohair with vegetable matter exceeding 3% is normally carbonized.

10.9.2.2 Mechanical processing

Mohair is considered to be difficult to mechanically process because of its smoothness and lack of cohesion. Nevertheless, provided the correct processing additives (lubricants and antistatics) and conditions and raw materials are used, very high-quality mohair yarn can be spun with acceptable efficiencies. In converting mohair into yarn, similar machinery is used as in the case of wool. Considerable secrecy exists even today concerning the precise processing conditions used; firms that have built up this specialized knowledge and skills do not share it because it provides them with a competitive edge.

It is generally easier to disentangle mohair than wool during carding, with less fibre breakage in this process, although problems with static and fly generation often necessitate lower carding speeds. Mohair's low cohesion often necessitates that the fibres (slivers) be supported, for example by aprons, during processing. Mohair blends well with wool, the wool facilitating its processing, by increasing interfibre friction and cohesion. The application of the correct types and levels of processing lubricants and additives (such as antistatics) and the selection of the most appropriate processing machinery and conditions (including atmospheric) are crucial in the efficient processing of mohair into a quality product.

Traditionally, mohair was processed on the Bradford worsted (oil-combed) system (drafting against twist) followed by flyer spinning (Srivastava, 1984). Today, the bulk of mohair is processed on the continental or dry-combed (French/rectilinear combing), as opposed to the oil-combed system. The French (continental or dry-combed) system of drafting and spinning involves French (rectilinear) combing, intersecting gilling and double apron drafting (drawing). It is possible to use either flyer (twisted) roving or rubbed (twistless) roving for subsequent yarn spinning. Most of the shorter mohair and also a significant amount of longer hair as well as mohair waste, such as carbonized noils, are processed on the woollen system. For the woollen system, a minimum amount of vegetable matter is essential. In woollen spinning, mohair shorter than about 75 mm staple length is generally used, while for the worsted system, the staple length is generally 90 mm and longer, with a staple length of some 120 mm often required.

The finest yarn that can be spun largely depends upon the mohair fibre diameter or fineness, traditionally expressed in terms of 'quality or quality counts', and these are related to the minimum number of fibres in the yarn cross-section. Wang and Khan (2000) have reported on the use of pinned apron and bottom roller for the improved drafting of mohair during ring spinning.

10.9.2.3 Fancy (novelty) yarns

Mohair is used for particular advantage in fancy or novelty yarns, such as loop, knop, brushed, bouclé, flame, snarl, slub and gimp, where its properties provide outstanding aesthetic appeal and comfort. Such yarns are used in blankets, stoles, shawls, scarves, knitwear (sweaters, cardigans, jerseys, etc.), travel rugs, curtaining, table coverings, upholstery, furnishings, pram covers, women's dress-wear, suitings and coatings. Adult hair is often used to form the loops of bouclé yarn properly (Anon. et al., 1983).

It reduced yarns imperfections, notably neps, hairiness and unevenness, and increased strength (Khan and Wang, 2000).

10.9.2.4 Fabric production and machinery

Generally, mohair yarn is converted into knitted and woven fabrics using similar equipment as for wool, though sometimes in a modified or adapted form and under special conditions, which allow for the more hairy, and often weaker, nature of the yarns.

10.9.2.5 Dyeing and finishing

Dyeing and finishing represent crucial stages in the manufacture of mohair products of the outstanding quality and appearance associated with items bearing the label 'mohair', acid and metal complex dyes being popular. It is generally the case that firms which dye and finish mohair also dye and finish wool and hence similar machinery is used for the two fibres. Furthermore, it is very rare to find pure mohair in yarns and fabrics. It is mostly present in blends with wool, often yarn blends, which means that the dyeing and finishing machinery and conditions used must be suited to both fibres.

There is a vast literature on the dyeing and finishing of wool, much of which is to a large extent also applicable to mohair. There is far smaller literature available on the specialized knowledge of conditions and procedures required for the dyeing and finishing of mohair products because most of such knowledge is a well-kept secret. In general, milder conditions (temperature and time) are used for the dyeing and finishing of mohair than for wool, partly because of the need to conserve the lustre of mohair and partly because mohair is more sensitive to wet treatments than wool. It is common practice to dye at temperatures below the boil, preferably below 90°C, and to limit the time of dyeing at high temperatures, so as to curtail any adverse effects on lustre and other desirable properties. It is also possible to limit damage to the fibre by using fibre protective agents.

10.9.3 End-uses

The textile application of mohair goes back many thousands of years and the fibre has found application in almost every conceivable textile end-use. Today, up to 80%–90% of mohair consumption, especially of the Adult hair, can be affected by fashion.

Mohair, often in blends with other natural fibres, notably wool, with which it blends well, is used to great advantage in knitwear (ladies sweaters) and hand knitting, mostly in brushed, loop or some other fancy form where brushing imparts softness and warmth without weight. Knitwear traditionally represented some 80% of mohair's outlets, but this sector is fairly sensitive to cyclical fashion changes.

Mohair also finds significant application in woven suiting and coating-type fabrics, particularly in men's lightweight summer (tropical) suits where it provides the wearer with considerable comfort and good wrinkle resistance. Mohair is widely recognized as having very good wrinkle resistance and recovery, which, together with its stiffness, make it an ideal fibre for use in comfortable lightweight tropical type fabrics.

Table 10.35 (Buxton, 1986) lists some of the end-uses of mohair, and a detailed list of mohair applications is given by Hunter (Hunter, 1993; Franck, 2001). In lean worsted-type lightweight tropical suits, mohair is regarded as a cool fibre, whereas in brushed articles, such as shawls, stoles, rugs, sweaters and blankets, mohair provides warmth without weight. Mohair's characteristics of hard-wearing durability, resilience or springiness, crease resistance, moisture absorption, comfort, lustre and smoothness make it ideally suited to many applications in apparel and interior textiles, such as upholstery and any pile fabric (e.g., plush, velour, velvet and moquette, etc.), furnishings, rugs and curtains and it is virtually unsurpassed for general durability, recovering very quickly after being crushed. Because of its general smoothness and low static propensity, except under dry conditions, the smooth fibres do not allow dirt to collect readily, and stains are generally fairly easily removed.

Table 10.35 Mohair consumption by end-uses (Buxton, 1986).

End-uses	Share (%)
Hand-knitting yarns	65
Men's suiting fabrics	15
Women's woven accessories and rugs	12
Woven furnishings and velours	8

Mohair, in blends with wool, bamboo and other fibres, has found significant application in comfort socks, particularly for active sports wear (e.g., cricket, mountaineering, etc.) and medical applications (e.g., for diabetics).

Further information on mohair is given in the references (Hunter, 1993; Franck, 2001; McGregor, 1998).

10.10 Musk-ox

10.10.1 *Fibre production, harvesting, properties, processing and end-uses*

The musk-ox *(Ovibos moschatus)*, a hollow horned ungulate (ruminant), native to the Arctic, is in fact not an ox, and has no musk, its nearest relative being thought to be the goat (Anon., 1987a). It reportedly has the largest and thickest hair of any mammal on earth (Schweiss, 2009). The prehistoric animal, also called 'omingmak' by the Eskimo people, is huge, and looks like a Bison (Schweiss, 2009) (see Fig. 10.14). Adult male animals have an average weight of some 360 kg (up to 450 kg), and a shoulder height up to 1.8 m (Anon., 1987a). It originally bred wild and undomesticated in the Northern territories (Arctic lands) of Canada, and wild animals are still found in Canada's North West Territory and in Greenland (Anon., 1987a). The musk-ox essentially has a two-layered fibrous coat, namely a relatively coarse, dense and long (up to 600 mm) guard hair (outer coat, dark brown in colour) and a relatively fine ($\approx 13-17$ μm) and short (40–80 mm) undercoat (brown-grey) of down fibres, known as qiviuk; the latter has been gathered and used since time immemorial by the Eskimo people (Schweiss, 2009).

Figure 10.14 Musk-ox. (By kind permission of Ms Nancy Bender, The Musk Ox Company, 633 Hatchery Road, Hamilton, Montana 59,840, USA.)

The musk-ox skin has an exceptionally high secondary primary follicle ratio between 40:1 (Ryder, 1997) and 37:1 (Rowell et al., 2001), the qiviut being primarily produced by the secondary follicles and the guard hair by the primary follicles. Rowell et al. (2001) reported, however, that there appears to be an intermediate (medium) group of fibres, coarser than qiviut but finer than the guard hair, variable in diameter and produced by the primary follicles, these being easily visible in combed fleeces. They are fine, like qiviut, near their roots and coarsen considerably towards the tips (Rowell et al., 2001). They are shed annually and can be medullated (Rowell et al., 2001). This group of fibres is believed to account for most of the fibres between 30 and 50 μm (Rowell et al., 2001) and is important in terms of the dehairing process.

The musk-ox became extinct in Alaska around 1850, (Ryder, 1997) but was reintroduced around 1953. The Canadian government placed the animals, which are easily hunted, under some protection in 1926, and in the 1930s, the US Department of Interior obtained animals from Greenland and released them on the uninhabited Nuaivak Island in Alaska (Anon., 1987a). First attempts to domesticate the musk-ox for its fibre occurred in the 1950s. The calves were captured in game sanctuaries in the North West Territories and the first successful musk-ox farm was established in Alaska during the 1960s (Rowell et al., 2001), when fibre production also started. In the 1980s, there were two domesticated herds, one at College, Alaska near Fairbanks, and the other in Northern Quebec (Anon., 1987a). In a joint venture between the United States and the USSR, around the 1980s, musk-ox calves were captured at Nuaivik Island and sent to the Soviet Union, the latter attempting to reestablish the musk-ox in Siberia (Anon., 1987a). Originally mainly hunted for their fat-rich meat, the musk-ox is now mainly valued and bred for its fibre (Anon., 2006a; Wilkinson, 1975). Selective breeding improved the fibre colour and yield and animal domestication and reduced the size of the animal, thereby increasing its cost-effectiveness (Laycock, 1973). In recent times, commercial harvesting of fibre from wild musk-oxen on Banks Island and elsewhere in the North West Territories of Canada has made available relatively large quantities of dehaired qiviut fibre, yarn and finished products (Rowell et al., 2001), fibre harvesting taking place semiregularly since around 1982.

The musk-ox sheds its soft 'silvery grey/white' undercoat (qiviut) in a tightly synchronized moult each spring (Rowell et al., 2001), from around April to June (i.e., around late spring, with temperature playing a role) (Laycock, 1973). The animals are placed in pens and the fibre is readily and easily combed out or plucked (pulled off) by hand, over a number of days, the undercoat being easily removed in sheets, which continues until early June (Laycock, 1973). The total collection (plucking) time per animal is slightly more than 2 h (Laycock, 1973). Fibre is also collected from objects against which the animals brush and which retain the fibre, the undercoat being found in great tufts on the bushes of the boreal tundra (Schweiss, 2009) where the animals graze. The guard hair is separated and removed by hand from the fine down.

The fibre removed from the musk-ox is referred to as raw fibre (i.e., containing a mixture of qiviut and hair) (Rowell et al., 2001). Adult animals annually yield between about 2.5 and 3.5 kg of hair (raw fibre) (Anon., 1987a; Rowell et al., 2001), yielding about 2−3 kg of qiviut (Anon., 2006a). Females produce about 2 kg of qiviut, while mature bulls yield more fibre than females and younger animals (Laycock, 1973).

For the purposes of their study, Rowell et al., 2001 defined qiviut to be all fibres ≤ 30 μm. The fine undercoat (qiviut) of the musk-ox has an average fibre diameter of about 20 μm (Phan et al., 1988), numerically, more than 90% of fibres in a 'raw' sample being ≤ 30 μm (Rowell et al., 2001). The scale frequency is ~7 per 100 μm and thickness (height) ≈0.4 μm (McGregor and Quispe Peña, 2018).

Rowell et al. (2001) reported that the average diameter for raw fibre shorn from the shoulder region of musk-oxen was 21.5 μm for adult males and 19.5 μm for 2-year-old females and that of the qiviut (taken to be fibres ≤ 30 μm) 16.5 μm for yearling males and 18.2 μm for adult males. The fibre diameter for both the raw fibre and qiviut increased slightly with age. For example, both the raw fibre and qiviut increased by about 1 μm from 1 year to 4+ years of age. The CV of diameter of the raw fibre was around 65%, and that of the qiviut around 27% (Rowell et al., 2001), the CV of diameter decreasing with age, for both the raw fibre and qiviut. The female animals tended to have slightly lower mean fibre diameters than the male animals, except at the age of 1 year (Rowell et al., 2001). The root section of the fibre is generally 1–2 μm finer than the middle section of the fibre (Rowell et al., 2001). Guard hair and qiviut staple length increases slightly with age, averaging around 120 and 52 mm, respectively, for the shoulder region of the musk-ox. The relatively low qiviut staple length (±50 mm) reported by Rowell et al. (2001) was attributed to both their sampling site (shoulder) and the fact that the fibres were shorn (i.e., the fleeces were cut from hides) and not shed.

Qiviut fibre quality is judged technically in terms of length and diameter, the fibres being similar in diameter, on average, to cashmere, and considerably longer (Anon., 1987a; Laycock, 1973). Musk-ox hair can be dehaired on the same machines as cashmere (Laycock, 1973).

Musk-ox fleeces are virtually free from contamination, such as vegetable matter and dirt, and contain very little grease (<7%) and suint (Rowell et al., 2001), it being almost possible to spin it directly from the animal's back (Laycock, 1973). Rowell et al. (2001) reported an average scoured yield of about 93% for 3-year-old and adult musk-ox, increasing only slightly with age.

Alaskan Inuit (Eskimo) women hand knit caps, scarves, stoles, sweaters, shawls and tunics from qiviut (Anon., 1987a; Elsner, 1998), about 115 g of qiviut fibre being sufficient for a sweater for a large man.

The fibre bleaches and dyes well, with garments made from musk-ox fibre being very light and comfortable, and highly suitable for temperatures below freezing. Garments made from musk-ox fibre reportedly do not shrink when washed (Anon., 2006a).

10.11 Vicuña

10.11.1 Fibre production, harvesting and properties

Vicuña (Lama genus: *Vicugna vicugna*), a ruminant (pseudoruminant) descended from the guanaco, is the smallest (0.7–0.9 m in height and weighing about 50 kg) and most

Figure 10.15 Vicuña.
Reproduced from Pier Giuseppe Alvigini, The Fibres Nearest the Sky, Mondadori Editore, Verona, by kind permission of Mr Pier Alvigini at Alvigini S.A.S., 13,900 Biella Via Dante, 12 Casella Postale 430, Italy.

timid member (see Fig. 10.15) of the South American Camelid family (Nagal, Naik) producing the finest and softest underhair of any animal. Its fibre, sought after for around 10,000 years (Rainsford, 2000), has been termed (Anon, 2008) 'Fibre of the Gods' and 'Silk of the New World' (Gwyther, 1996). The individual fibres range from about 6 to 35 μm in diameter and (CV~20%) 12–65 mm in length, the chest hairs being longer and lighter (almost white) in colour (Siles, 1998). Only some 5000 kg of 'raw' (harvested) fibre is available annually (Ferrero, 2007), which is reduced to almost half after scouring and dehairing (Ferrero, 2007). The vicuña originally had to be killed to obtain its fine coat of fibres (Ryder, 1980), although, reportedly (Anon., 1990a), the Incas some 500 years ago had captured, shorn and released thousands of vicuñas. Today, the fibre is obtained by shearing live animals (Chamut et al., 2016). During the time of the Inca empire, around 1400 AD, only the royal family and the handful of youthful ladies, chosen by the sovereign as ladies-in-waiting were allowed to wear vicuña garments (Gwyther, 1996). Quispe et al. (2010) distinguished between two geographic subspecies, the Peruvian (*vicugna vicugna mensalis*) with long white hair on the bib, and the Argentinian (*V. vicugna vicugna)*, with shorter hair on the bib.

Vicuña is native to the central Andean countries of Peru and Bolivia (also Chile and Argentina), living in the Cordelleros of Peru and surrounding country, at altitudes around 3500–6000 m, and grazing on ichu grass (Briggs, 1973; Gwyther, 1996). They can exist at higher altitudes than other members of the Camelid family (Rainsford, 2000). Millions of vicuña once roamed Peru's 4000 m high Altiplano. During the pre-Columbian Inca civilization, vicuñas were hunted only once every

4 years (Anon., 1990b). After the Spanish invasion in the early 1530s, the vicuña population dropped dramatically to only some 5000 in Peru in the mid-1960s, (Anon., 1990b) due to uncontrolled hunting for their fibre and meat. Simon Bolivar passed the first law protecting vicuñas in 1825 (Anon., 1990b). Peru passed a similar law in 1921, while a South American Agreement was introduced in 1969 which prohibited the commercial use of vicuña. The number of animals increased from some 5000 in Peru in the mid-1960s to between 10,000 and 15,000 by the end of the 1960s (Anon., 2000d), these being kept in the area Pampas Galeras (the 15,000 acre Pampas Galeras Naturel Reserve established in 1967), started as a programme of conservation and managed by the Peruvian government (Ryder, 1980), resulting in the vicuña numbers increasing to 38,000 by 1978, and the world population to 60,000 (Ryder, 1980).

The vicuña was considered endangered between 1960 and 1990 (Maycumber, 2002). According to one report (Anon., 1991), the vicuña had been hunted to virtual extinction for its fleece (Anon., 1991), and by the 1970s, the species had virtually been wiped out in certain areas, the fabric fetching up to $1500 per square metre at the time. CONAF, Chile's forestry commission, in 1970 created the Lauca National Park in Northern Chile to protect the vicuña's natural habitat and increase their numbers. As a result of this, there were some 28,000 vicuñas in the park around 1990 (Anon., 1991). Attempts also started at around that time to determine whether the animals could be safely trapped and shorn in the wild, leaving enough fleece to protect them from the cold (Anon., 1991).

Because of the danger of extinction, export bans on fibre in Peru and Bolivia were supported by import bans in manufacturing countries (Briggs, 1973). In the USA, vicuña was listed as endangered on 2 June 1970 (Anon., 2002), reclassifying it to 'threatened' in 2002, other markets having sanctioned imports in 1995. The Peruvian government, and afterwards in 1976, the UN Convention on International Trade in Endangered Species (CITES), stepped into make trading in vicuña an offence (Gwyther, 1996), with a resultant dramatic increase in their numbers. In 1987, CITES changed the designation of the vicuña to Appendix II, allowing fleeces to be harvested and converted into fabric (Anon.b). By 2001, it was estimated that they numbered over 200,000 in Peru (Anon., 2002) alone. In 1994, there were an estimated 66,500 vicuñas in Peru (Rainsford, 2000), of the purest bloodline, the number thereafter increasing as shown in Table 10.36. There were estimated to be some 55,000 vicuñas in Northern Chile, Bolivia and Argentina (Rainsford, 2000) in 2000.

Table 10.36 Vicuña population in Peru (Rainsford, 2000).

Year	Number
1994	66,500
1997	103,000
1998	120,000
1999	140,000
2000 (Est.)	165,000

The project for Rational Utilisation of Vicuña, under which thousands of vicuña were culled for their fibre and meat in Peru, was terminated early in the 1980s (Colman, 1982). The vicuña has never been domesticated, though bred in semicaptivity and herded. The rounding up in Peru is referred to as a Vicuña Chaccu/Chakku (also done by the Incas [Rainsford, 2000]) and the animals are shorn and released. This is called a 'Chakku' (Allolio, 2007) in Bolivia, while the 'Compesinos' herd vicuñas for shearing in Peru (Gwyther, 1996). In Peru, vicuña are shorn between 15 May and 15 November (Rainsford, 2007c), the total production amounting to a little over 6000 kg in 2007. Electric shearing is used where feasible, although clippers are considered preferable (Rainsford, 2007d) for shearing.

The fleece of the vicuña ranges in colour from golden to typically a rich, deep fawn (cinnamon) colour, with that of the underbelly and extremities being off-white. Hair from six vicuñas is required to produce one sweater and from 35 to produce one coat (Ferrero, 2008). Vicuñas have a double coat, a relatively coarse guard hair and a very fine undercoat or down, the most prized fibre being from the 'apron' of long whitish hair that falls between the front legs (Gwyther, 1996).

The fleece from the dorsal part of the neck, back, sides and upper part of the legs (the vellon) is finer than that from the rest of the body (Chamut et al., 2016). Fabric shorn from the chest, lower and inner parts of the legs and belly is coarser and is known as "belly-legs" (barriga-garra) (Chamut et al., 2016). The secondary to primary follicle ratio (S:P) is 42 in the fleece (vellon) part and 12 in the belly-leg samples, which explains the differences in fibre diameter, both the follicle and fibre diameters differing in the two regions of the body (Chamut et al., 2016).

Vicuña fleece ranges from 12 (11) to 15 μm in mean fibre diameter, and from about 15 to 50 mm in mean fibre length (Rainsford, 2008a), more typically 20–25 mm. The fine (13 μm) vicuña hair is found on the lower chest just behind the front legs and is a short ($\approx$50 mm) downy undercoat, the outer hair being much coarser and longer (up to 45 μm in diameter and 65 mm in length (Anon., 2000d). Commercial vicuña typically ranges from 13 to 15 μm (Greaves and Rainsford, 2005). Greaves and Rainsford (2005) reported little change in vicuña fibre fineness and quality from 1970 to 2005. An adult vicuña only produces about 200–250 g of fine hair every 2 years (Gwyther, 1996; Ferrero, 2008).

Average scale frequency is~12 per 100 μm (McGregor and Quispe Peña, 2018), scale thickness (height) $\approx$0.3 μm and ellipticity $\approx$1.3 (McGregor and Quispe Peña, 2018).

First-quality fibre averages about 37 mm in length and 12 μm in diameter (Anon., 1987b) and contains a small quantity of coarse and dead (kemp) hair. Second-quality fibre comes from skirtings, is 3–4 μm coarser (Anon., 1987b) and contains a relatively large proportion of coarse and dead hair (Anon., 1987b).

CONACS (Consejo Nacional de Camélidos Sudamericanos), the National Council for South American Camelids, is the government authority of the Ministry of Agriculture in Peru charged with commercialization of vicuñas and the certification of fibre shorn from live animals (Anon. et al., 2001). Hair from slaughtered animals is illegal under Peruvian law.

10.11.2 Processing

Vicuña is sorted by hand to remove coarse fibres, a worker taking a week to separate (sort) by hand 1 kg of mixed fibre and down (Gwyther, 1996). This is followed by dehairing. Vicuña is traditionally dehaired by hand, but more recently also by machine. By hand, it can easily take seven 8-hour shifts for a sorter to obtain 700 g of fine fibre (Patthey, 2002), of which only a small percentage is as fine as 13 μm. Its length (typically 20–25 mm) largely limits processing to the woollen system, as opposed to the worsted system (Patthey, 2002). Specially collected vicuña fibres, 35 mm in length, have, however, also been processed, in 100% form, on the worsted system into blazer cloth. This was claimed to be a 'first' in textile history (Stewart, 2008) (fibres from 5 vicuñas being required for one blazer length fabric). It was also claimed that the world's first 100% vicuña suiting from worsted spun yarn was produced in 2008 (Dyson, 2008).

Woollen yarn as fine as Nm25 (40 tex) and worsted yarn as fine as Nm46 (≈26 tex) have been produced from pure vicuna fibre (Rainsford, 2008b), manual removal of kemp and impurities taking place at the fabric stage.

10.11.3 End-uses

Vicuña reappeared on the market in the mid-1990s after an absence of 20 years. In 1994, the International vicuña Consortium won the rights to exclusively process and market vicuña globally (Patthey, 2002). According to the agreement in 1995 to commercialize vicuña hair, only 100% vicuña could be used in fabrics, this being amended around 2000 to include blends with wool (Anon., 2000d), 80% vicuña/20% fine wool (20.5 μm) blend fabric (580–600 g per linear metre), for example, being used for skirts, dresses, scarves, jackets and overcoats (Anon., 1987b; Middlebrook, 1987). Legally exported vicuña fabric is identified by a CITES-designated trademark 'Vicun–Andes' (Anon., 1990b).

Vicuña is popularly used in cardigans and sweaters and is highly sought after because of its very fine and soft nature. Scarves, parkas, throws, shawls, dresses and blankets, in 100% vicuña, are also popular.

Vicuña fabric can cost US$4000 per metre (Gwyther, 1996), and a full-length 100% vicuña overcoat costs as much as US$12 000.

In the case of woven fabrics, mainly plain weave fabrics are produced, with some common weights and end-uses as follows (Rainsford, 2000; Anon., 2000d)

- Average fibre length: 20–25 mm
- Yarn count: Nm 18 to Nm 40 (55–25 tex)
- Fabrics
  - 430 g/m for suitis
  - 550 g/m for jackets
  - 690 g/m for overcoats
  - 100 g/m (150 × 30 cm) for men's scarves
  - 190 g (180 × 30 cm) for ladies' scarves

10.12 Yak

10.12.1 Production, harvesting and properties

The yak (*Bos grunniens*), a member of the hoofed Bovidae family, is chiefly found in Tibet, living above the snowline on the Qinghai–Tibet Plateau, called 'The Roof of the World'. It is used in the mountainous region as a beast of burden (Xinhe, 1989), and for subsistence through its milk and meat (Anon.c). It is bred primarily in the Himalayan region of South Central Asia and the Tibetan plateau, as far north as Mongolia (Vineis et al., 2014), herdsmen in more recent times earning additional income from selling its hair. It was estimated that towards the end of the previous century, there were some 13 million domesticated yaks on the plateau flanking the Himalayas, of which some 12 million were in China (Xinhe, 1989), one-third of which were in the Qinghai province bordering on Tibet, making up the single biggest herd in the world (Xinhe, 1989). Yaks were also dispersed on the vast pastures in the adjoining provinces of Sichuan and Gansu and in Tibet (Xinhe, 1989), and there were also over 100,000 in India (Ghoshal et al., 1993b). It was estimated that there were some 4 million wild yaks in the deep valleys and remote mountain areas. In the Himalayas, the domestic YAK is usually found at altitudes above 2000 m.

The yak weighs about 300–350 kg and its thick coat of long hair almost reaches to the ground (see Fig. 10.16). Wild yaks tend to have black hair but domestic yaks, kept by most herdsmen, have hair which is often a piebald colour, predominantly black and brown, with white markings (Xinhe, 1989; Anon.c), that of down being brown, grey and very rarely white (Lu et al., 2013), it being very difficult to distinguish between it and cashmere (Lu et al., 2013). The yak has a double-coated fleece, consisting of a long coarse guard hair (outer coat), and a soft down (undercoat) which sheds annually

Figure 10.16 Yak.
Reproduced from Pier Giuseppe Alvigini, The Fibres Nearest the Sky, Mondadori Editore Verona, by kind permission of Mr Pier Alvigini at Alvigini S.A.S., 13,900 Biella Via Dante, 12 Casella Postale 430, Italy.

in spring and which is either combed out by hand or shorn (Phan et al., 2008). Based on a relatively low yield of 0.5 kg of fine fibre (down) per animal, some 7 million kg per annum of fine fibre should be produced according to the number of yaks from which hair is harvested. In fact, in 2006, the production of yak hair was estimated (Phan et al., 2008) at 7 million kg, mainly coming from China and Mongolia. A specially selected 'fibre line' of Jiulong yak can yield up to 10–12 kg of fine fibre (without coarse hair) per head which is some 10 times the normal yield (1–1.2 kg) (Phan et al., 2008) per animal.

The yak hair varies in fineness according to the part of the body from which it comes (Ghoshal et al., 1993b). The underside of the animal is covered with long coarse hairs, which form a fringe, the outer sides of the legs being covered with the same type of hair (Ghoshal et al., 1993b). The back none on the neck and spine as well as the occipital and frontal parts of the head are also covered with coarse, but much shorter, hair. The yak's tail is similar to that of a horse (Phan et al., 2008), but much thicker. A steel comb is used in Mongolia to harvest the downfibre from yaks. The lower sides and belly of the animals are combed soon after the spring moult (Anon.c) to remove the soft fine undercoat, whereas the neck, back and hind quarters are not combed (Anon.c). The hair clip varies from 300 to 900 g, at 2 years being from 400 to 500 g, and under 1 year 500–1300 g. For female yaks aged 3 years or more, the figures are 200–600 g, and for 2 years, it is 300–500 g (Ghoshal et al., 1993b).

Yak fine hair (down) varies in diameter from roughly 15–30 μm (typically 18–20 μm) (Lu et al., 2013; Anon.d) and that of the coarse guard hair from about 35 to 80 μm (Algaa and Mägel, 1992c, 1992d), the corresponding fibre length values being about 10–60 mm, and 45–160 mm, respectively. The underhair (down) of a 1-year-old calf has a diameter of 15–17 μm and is 40–50 mm in length. The corresponding figures for the adult are 18–20 μm and 30–35 mm, respectively (Anon.c).

It is laborious to remove the hair from the yak, first involving thinning the hair as summer approaches and before the yaks are put out to pasture, the soft hair being pulled from the animal by hand (Xinhe, 1989). Shears tend to be forbidden because they can expose the flanks of the animal as winter approaches (Xinhe, 1989). The white hair is the most precious. Fine yak hair is mainly produced in China (Yan, Höcker, Schäfer) where the hair is pulled or combed during the spring (Anon.c) moult, and the coarse outer 'guard' hair has to be separated from the down hair thereafter. Well dehaired yak fibre can have a mean fibre diameter between 19 and 21 μm (Phan, 1991), with a CV between 20% and 25%.

In India, *Clipped yak* fibre refers to fibre from the body of the yak obtained by shearing with hand shears or a shearing machine (Ghoshal et al., 1993b), while *Pulled yak* fibre refers to fibre removed from the yak by pulling out the fibre, with or without the aid of a soap or depilatory solution (Ghoshal et al., 1993b). *Mixed fibre* refers to a combination of clipped and pulled yak fibre (Ghoshal et al., 1993b), with 'Lot' meaning the total quality of yak fibre offered for grading, of one colour and type only (Ghoshal et al., 1993b). Laboratory yield percentage of yak fibres varies from 65% to 80%. According to one study (Ghoshal et al., 1993b), the percentage of foreign matter varies from about 1% to 6%, with length varying from 28 to 37 mm. A maximum of 3%

vegetable matter for different grades of yak fibre has been suggested (Ghoshal et al., 1993b)

Tentative specifications for Indian yak fibres have been proposed as follows (Ghoshal et al., 1993b):
Colour of fibres: (a) white, (b) brown, (c) black
Texture: (a) extrafine, (b) soft, (c) coarse
Foreign matter: 2%
Moisture percentage: 3%
Length: 10 cm

Yak fibres tend to be oval to circular in shape and generally do not contain a medulla. Yak fibres have a fine-scaled structure, the scales not being very prominent, and are typically deeply pigmented, with an average diameter around 20 μm. In one study (Ma et al., 2008), yak fibre scale height was found to be 0.36 μm on average, varying from about 0.21 to 0.65 μm, for an average fibre diameter of 20.7 μm, the scale frequency averaging 6.9 scales per 100 μm, varying from about 4 to 10 per 100 μm, the figures for scale height and frequency being very similar to those found for cashmere fibres (15.6 μm mean fibre diameter) in the same study (Ma et al., 2008).

10.12.2 Fibre processing

Yak hair does not appear to be sorted at the time of harvesting, being packed into large sacks (bags) that are sent to the warehouses or sorting factories where it is hand-sorted to reduce the quantity of guard hair present in the down and also to separate the natural colours (Anon.c). Sorters are only able to handle some 10 kg per day. After sorting, the fibre is packed into bales and sent to the processor for converting into yarn (Anon.c).

The approximate proportions of the different natural colours are as follows (Anon.c):

- White ≈ 10%, the most valued due to its dyeing potential
- Fawn ≈ 20%
- Dark grey (blue) ≈ 10%
- Dark brown ≈ 60%

Because fine yak hair is much cheaper than cashmere, it is economical to bleach it (Yan, Höcker, Schäfer) although bleaching tends to impair its handle. Reasons for the deterioration in handle and ways of improving it by, for example, adding a chelating agent have been discussed (Yan, Höcker, Schäfer). The deterioration in handle was ascribed to cleavage of the disulphide bonds. Work has also been carried out (Rainsford, 2009) to reduce the diameter of yak fibres by a stretching (i.e., slenderizing) treatment.

10.12.3 End-uses

The coarse yak hair (outer coat) is used locally for tents, ropes, huts, blankets and mats (Anon.c), while the fine downfibre is used in apparel fabrics (Anon.c). Dehaired fine downfibre can be spun into yarn which is comparable to cashmere (Anon.c) and, blended with other animal hair and polyamide (nylon), is used in knitwear (Xinhe, 1989). Yak hair is used in the felt industry and that of yak calves in the textile industry for the manufacture of high-quality, generally thick, fabrics (Ghoshal et al., 1993b).

10.13 Other animal hair fibres

10.13.1 Bison hair

The properties of hair shed by the North American bison *(Bison americanus, Bison bison)* have been measured (Williams and Braaten, 1998) and it was found that the guard hairs were hollow and ranged in diameter from 21 to 110 μm, while the fine down hair was solid (i.e., unmedullated), ranging in diameter from 12 to 29 μm, with a scale structure similar to that of wool ($\approx$9 scales/100 μm and $\approx$0.4 μm scale height) (McGregor and Quispe Peña, 2018). The moisture regain ranged from 13% to 20% (Williams and Braaten, 1998).

10.13.2 Cervelt (soft deer hair)

Cervelt is the soft undercoat (down) from the red deer (McGhee, 2008), having a mean fibre diameter of about 13 μm (CV = 18%). Each animal produces only about 20 g of fine fibre and the total world production is some 1000 kg (McGhee, 2008). The downfibre from about 14 deer is required to produce a sweater and from about 40 deer to produce a man's overcoat (McGhee, 2008). The fibre has a curl rather than crimp. It is used in menswear, women's fashion, knitwear, accessories and furnishings (McGhee, 2008). When the name cervelt is used, it is in its pure form, i.e., unblended with other fibres (McGhee, 2008).

10.13.3 Common or domestic goat hair (boer, Feral, etc.)

Common goats *(Capra aegagrus hircus)* or milk goats generally have a double coat containing very little, but rather fine ($\approx$15 μm), down fibre. See also Section 12.6.1.

Mitchell et al. (1991) reported on "cashmere-bearing" does in New Zealand, obtaining down (cashmere) staple lengths of around 50 mm, and mean fibre diameter of around 18 μm, concluding that optimal cashmere returns should be achieved by shearing in June/July.

In his review, McGregor (1998) quotes values of around 30–50 g per goat for downproduction of Australian feral goats, with mean fibre diameter around 14–15 μm.

Couchman (1988) reported on the "cashmere down" properties of South African Boer Goats, and their progeny, as well as those mated with Australian bucks, mean fibre diameter of the down generally falling within the range of 15–17 μm, and length between 15 and 45 mm. Down yield tended to be rather low, from around 12 to 40 g per goat.

Debnath et al. (1987) gave the following properties for the goat hair waste available from tanneries in India:
Average fineness (tex): 3.6–5.1
Average length (mm): 20–21
Average tenacity (gf/tex): 13–19.4
Average extension at break (%): 49–88

Paul et al. (2003) obtained the following results on the hair from Gaddi goats:

Average fibre diameter (μm): 76.7 (range = 46–88)
Average fibre fineness (tex): 4.6 (range = 2.1–7.0)
Average medullation (%): 75
Average fibre length (mm): 101 (range 60–121)
Average fibre tenacity (gf/tex): 13.8 (range = 7.8–18.7)
Average fibre elongation (%): 26.0 (range = 2.4–39.0)

10.13.4 Horse hair

Horse *(Equus caballus)* hair has been used since prehistoric times to make fishing lines and nets, but over the past 200 years, it was also commonly used for upholstery (soft furnishing), the fabric being prized for its durability (up to 100 years) (Ryder, 1999). It is also used in linings and handbags. Horse hair fabrics represented one of the most traditional coverings for fine furniture, widely used by the 18th-century masters, like Chippendale and Hepplewhite (Roberts and Wilkey, 1993), but then coming down market in the 19th century to become the favourite covering of the parlour sofas of the rising middle class (Roberts and Wilkey, 1993). It can be combined with other natural fibres, such as cotton and silk.

The horse has short (≈10–30 mm) and coarse (≈80–100 μm) body hair and much longer (≥300 mm) mane and tail hair. Horse tail hair generally has an average fibre diameter greater than 140 μm (Von Bergen, 1961), ranging from about 75 to 280 μm. The mane hair is finer (average ≈ 110 μm) and ranges from about 50 to 150 μm. Bolormaa et al. (2007) reported that the Mongolian horse tail hair they tested were medullated and had a mean fibre diameter of around 180 μm, with the diameter increasing from about 130 μm at the tip to about 230 μm at the root. The fat content (according to MNS 379:200 test method) was 2.4% and the moisture content (according to MNS 380:2001 test method) was 10.5% for the greasy hair and 16.8% for the scoured hair. The breaking extension ranged between about 43% and 52%, increasing with increasing diameter. Initial breaking modulus ranged from about 2.6 to 4.9 GPa, decreasing with increasing fibre diameter. The cuticle scales of the hair were smooth, with the height (thickness) of the cuticle scale edge lower than that of speciality fibres (Bolormaa et al., 2007).

10.13.5 Shatoosh, Shahtoosh, Shah-tus, Shah-tush, Tosh

There is some confusion concerning the precise definitions, origins and sources of fibres or products variously referred to as Shatoosh, Shahtoosh, Shah-tus, Shah-tush and Tosh. Nevertheless, it appears that they are variously applied to fibre and fibre products from wild animals, notably the ibex wild goat and the Tibetan antelope.

The official taxonomy currently lists four wild *Capra* species *(aegagrus, ibex, falconeri and cylindricornis),* although there may be three or more distinct species of Ibex (Tonin et al., 2002), the only CITES-listed species being the markhor *(Capra falconeri),* with the *Capra ibex* the typical ibex (Anon., 1990b).

Tosh fibres are obtained from an animal (antelope?) belonging to the ibex family (Ahmad and Kiramani, 1993), which is found in Tibet and the adjoining areas of the Ladakh region of India. Its fibre is collected from the bark of trees. It is mainly used to manufacture shawls in the Kashmir Valley, which are known as 'ring' or 'shah-tosh' (Ahmad and Kiramani, 1993). The following results were obtained on a sample of raw Tosh fibres:

Fine down yield (%): 71
Mean fibre diameter (μm): 11.9 (CV = 3.9%)
Mean fibre length (mm): 42.7 (CV = 6.2%)
Scouring yield (%): 86
The percentage of fine fibres (71%) was reportedly similar to that found for a Pashmina fleece.

The Asiatic ibex or wild goat, Siberian ibex (*Capra ibex sibirica*), also called the Yangir (wild cashmere) mountain goat, is considered to be the wild species most commonly exploited for textile fibres (Tonin et al., 2002). It lives in the mountainous areas (Himalayas) from India to Mongolia. The fibres are obtained from the skins of dead animals and dehaired manually, fibres from the throat areas being regarded (Burns et al., 1962) as the most desirable. It is thought that 15,000 animals have to be killed to produce 1000 kg of dehaired fibre (Tonin et al., 2002; Vineis et al., 2011), i.e., each animal produces around 70 g of dehaired fibre. Tonin et al., 2002 found a value of 13.6 μm (range 13−15 μm (Vineis et al., 2008)) for the mean fibre diameter of the dehaired Yangir fibre sample they tested, the CV of fibre diameter being 22.6% and coarse fibre (>30 μm) content 0.1%. They also found the scale density to range from 5/6 to 18/20 scales per 100 μm, and the scale height >0.8 μm (Vineis et al., 2008).

According to tradition, and certain reports, Shatoosh (Shah-tus) is the hair from the Tibetan antelope *(Pantholops hodgsoni),* called 'Chiru' in Tibetan (Tonin et al., 2002), which is present mainly in the Tibetan region of China. Shahtoosh/shahtus, which means 'king of wools' in Persian or Urdu (Ryder, 2003) (reference (Anon.b) and CNN Italy, www.cnnitalia.it/1999/STILE/11/04/sciarpa.tibetana, No. 15, 1999, quoted in reference (Tonin et al., 2002)), is one of the finest fibres and used to manufacture 'ring shawls' (they can pass through a wedding ring), each weighing ≈ 120−150 g. The animal is mostly killed to harvest its hair. Hair from at least five animals is necessary to produce a shawl. It was first thought that shah-tus was the shed under wool of the wild ibex collected during the spring moult, such fibre being used (Ryder, 2003) and the shawls referred to as *Shahtoosh* (Cizek and Turpie, 1985). As much as 30% of ibex guard hair is present in Tosh shawls (Rollins and Hall, 1999). According to Ryder (Ryder, 2000), the finest fibre in India was named Shah-tush and collected from wild animals, which could include goats, and that it had become illegal to sell shawls made from this fibre because antelopes were being killed to obtain the fibre.

Tibetan antelope is classified by CITES in Appendix I, its highest level of protection, banning all international trade in the species (Tonin et al., 2002). It was estimated that there were fewer than 75,000 of the antelopes at the beginning of the 21st century.

Tonini et al. (Debnath et al., 1987) obtained a value of 11.5 μm for the mean fibre diameter of the dehaired Shatoosh they tested, the CV of diameter being 20.1% and coarse fibre (>30 μm) 0.1%.

The fibres contain 5–6 scales per 100 μm (Langley, 1997), their natural colour being grey-brown. It takes between three and 5 Chiru to produce the 300–600 g of raw fibre required to produce a Shahtoosh shawl (Anon.b), each animal producing about 150 g of fibre. Shamina shawls, made from the fibre of domestic cashmere goats, are sold in India to protect the Chiru (Ryder, 2003).In South Africa, a fringed wrap called a 'shu-shu', consisting of blend of Cape summer kid mohair and superfine Cape Merino wool has been produced (Ryder, 2003).

Shah-tush also refers to very fine fibre shed by wild animals, including wild goats (Ryder, 2000) (e.g., wild ibex) and the Tibetan antelope, and the name Shahtoosh (Shatoosh) is also used to describe shawls from the ibex goat (Rollins and Hall, 1999). Rollins and Hall (Rollins and Hall, 1999) found that the fibres from the Tibetan antelope rarely exceeded 10 μm, generally ranging from 7 to 9 μm, while those from the ibex goat were generally 10 μm or coarser, the guard hair from the ibex having a thicker cell wall (i.e., relatively narrower medulla) than that from the Tibetan antelope (i.e., the latter had a relatively wide medulla). The fibre diameters of the guard hair of the Tibetan antelope varied from 50 to 100 μm or even greater, and those of the ibex goat were rarely over 50–60 μm (Rollins and Hall, 1999). The finer fibres generally did not have distinct medulla.

For further reading, see the references (Tonin et al., 2002; Burns et al., 1962; Ryder, 2003) and (Langley, 1997; Phan et al., 2000b; Schaller, 1977).

10.13.6 Opossum (possum) fur

Opossum *(Didelphis* genus, also known as possum) fur has received renewed attention in countries such as New Zealand, where it is blended with superfine Merino wool (e.g., 70%) to produce luxury knitwear. It is also used as duvet filling; it takes 15 bush-tailed possums to produce 1 kg of useable fur (Liddle, 1999), the fur being removed from dead animal skins. It was estimated (in 1999) that there were some 70–80 million possums in New Zealand. The fur ranges in colour from almost black to grey, in mean fibre diameter from about 12 to 16 μm and in mean fibre length from about 15 to 25 mm (Liddle, 1999), the latter making it suitable for spinning on the woollen system, although processing on the worsted system also takes place for the long fibres (Liddle, 1999). The fibre need not be scoured and is blended with wool before carding.

Gore and Laing, 2002 obtained the following results on the fibres (excluding regrowth) from New Zealand possum *(Trichosurus vulpecula)* fur:

Mean fibre diameter (um): 19.9 (CV = 49%)
Mean fibre length (mm): 26 (CV = 46%)

For regrowth fibres, they found a typical fibre length of 10 mm and diameter of 18.9 μm (CV = 61%).

10.13.7 Pashmina

Pashmina is a Persian word (Ryder, 2003) used to describe cashmere in India, Kashmir and Nepal (Spilhaus, 1999) but is often also used to describe products containing a blend of cashmere and silk (e.g., 70% cashmere/30% silk) (Ryder, 2003). The 'Pashmina' goat *(Capra hircus)* is bred by nomadic herdsman (Tchang Tang shepherds) in the Ladakh region of Northern India, at altitudes of some 4800 m (Schweiss, 2009) and in Nepal (Pelant et al., 1999). There are two breeds of pashmina goats in India, Chegu and Changthangi (Misra et al., 1998) quoting Acharya, R.M. 1982 in "Sheep and goat breeds of India". Anim. Prod. Health Pap. 30, FAO, Rome, pp151-153).

Pashmina (Pashm, Pashim, Pushima) can be taken as the fine, downy fibre (undercoat), white or light brown in colour (Misra et al., 1998), obtained from, or shed by, the Himalayan Pashmina goat and several other species of goats (ordinary and mountain) in Kashmir and other parts of India (Ryder, 2003). It is generally combed out or gathered from the ground or bushes when the warmer season (May/June) arrives. Coarse fibre production is ~0.5–1.1 kg per goat per year and that of the fine fibre is 120–170g per goat per year (Pelant et al., 1999), the yield (production of the Changthangi goat) increasing from about 80 g at 1 year of age, to approximately 200 g after the second year (shearing) (Misra et al., 1998), the average fibre diameter being 12 μm.

The fibre properties, including diameter (12–14 μm) (Anon.b) (or 14–16 μm) (Ryder, 2003) and length (≈50 mm) (Misra et al., 1998), processing and applications are generally the same as for cashmere.

Pashmina shawls and scarves were originally produced in Kashmir from hand-spun and woven fine cashmere fibre gathered from the ground and bushes where the goats had been feeding (Franck, 2001).

10.13.8 Karakul

The karakul *(Ovis)* breed of sheep, also known as Persian lamb, originally came from fat-tail sheep in the Bokhara district of Russia, which resembled the karakul sheep in their zoological characteristics, and were found in this region as early as 1600 BC (Anon, 1987). The word 'Karakkul' originated from the ancient Assyrian word 'Kara-Gjull' which means a flower (black rose) (Strydom, 1980) or something else of beauty (Anon, 1987). For many centuries, this breed was found only in Russia, and it was only at the end of the 19th century that some of these sheep were exported to Austria, the first karakul show in the world being held in Vienna in 1898 (Anon, 1987). In 1907, the German Governor of South West Africa (now Namibia) imported 10 ewes and two rams from Austria and the country became the largest supplier of karakul pelts in the Western world (Anon, 1987). The karakul pelts were the main purpose of farming with the breed, the wool being considered as only a byproduct. The adult animals were normally shorn twice a year producing about 2–3 kg of wool annually (Nicholas, 1998), the wool having a clean yield of some 65%. The colour of the natural pigmented karakul fibre varies from white to grey, brown and black, the pigment being fast to water if initially scoured under neutral or acid conditions (Anon, 1987).

In South Africa, karakul wool was sorted into 54 different types before sale. Production was some 7.1 million kg in the 1978/79 season (Strydom, 1980), and in 1996, some 278,000 kg of karakul wool was sold in South Africa and Namibia (Nicholas, 1998). The karakul wool staples consist of a short fine undergrowth and coarse long fibres, the latter having a diameter up to 70 μm, whereas the undergrowth can be as fine as 20 μm. The mean fibre diameter of karakul wool ranges from about 30 to 40 μm, with an overall average (typical) value of about 35 μm (CV ≈ 35%).

The length of karakul wool varies from about 20 to 150 mm (typically 50–100 mm), the length being important in determining the type into which it is sorted. The shorter, free from vegetable matter, karakul wool is processed on the woollen system and the longer types on the worsted system. Karakul processed on the woollen system, often in blends with other wool, were mainly used in carpets, blankets, upholstery and curtaining, and also in protective clothing (Anon, 1987) and much fabrics.

Karakul types, 65 mm and longer, lend themselves to combing, with the noils produced being used in the woollen system, needle punching (e.g., protective clothing, insulation, carpeting) or in industrial felts. The good felting properties of karakul have been ascribed to the presence of the fine undercoat fibres with the coarser outer coat fibres. Yarns produced on the worsted system were mainly (some 80%) used in interlinings. The processing of karakul on the woollen, worsted, and semiworsted systems as well as the manufacturing of products have been investigated by various researchers (Van Der Merwe, 1985; Van Der Merwe and Brydon, 1986; Strydom and Van Der Merwe, 1983; Hunter et al., 1982; Robinson and Slinger, 1970; Robinson et al., 1971; Anon., 2006b; Hunter, 1987b).

The natural pigmentation (colours) of karakul limited its application, although the lighter types of karakul wool (e.g., white, steel, silver and fawn) could be dyed to almost any colour, the most suitable dyes being reactives and chrome. Because of the limitations resulting from its natural colours, considerable work was carried out on the cost-effective bleaching of Karakul (Chen et al., 2001; Trollip et al., 1985; Van Rensburg and Heinrich, 1983) with the minimum amount of damage. A number of processes were developed, mordant bleaching in particular (Trollip et al., 1985), and ferrous mordanting before peroxide bleaching.

China is the largest producer of karakul today. The virtual demise of the karakul pelt industry, largely resulting from pressures by animal lovers and environmentalists, drastically reduced the production of karakul wool in South Africa and Namibia. Current production is only some 250,000 kg clean wool in Namibia.

10.13.9 Chiengora (dog hair)

Chiengora fibre, obtained from dog hair (Ramamoorthy et al., 2018), got its name from French ("chien" referring to dog and "gora" to "Angora" rabbit fur) (Greer. et al., 2007). Greer et al. (2007) and Ramamoorthy et al. (2018) reported on the fibre (Chiengora) properties of various breeds of dogs, it being used from ancient times as a textile fibre in garments and other uses. The fibre properties of Chiengora fibres vary according to the breed of dog, the mean fibre diameter, for example, ranging from 16 μm for the Schnauzer to over 60 μm for the Golden Retriever and Australian Shepard (Greer. et al., 2007) (most fall within the range of 25−35 μm), while fibre length varied from 40 mm for the Schnauzer to over 90 mm for the Yorkie, most fibre lengths falling within 50−60 mm range. The fibre tenacity of most dog breeds fall within the range of about 2−2.5 cN/tex (Greer. et al., 2007), elastic modulus between about 150 and 100 cN/tex, and elongation at break between about 60% and 70% (Greer. et al., 2007) (also 40%−50%) (Ramamoorthy et al., 2018). The keratin content of Chiengora fibres range from about 45% to 55%, the wax content from about 10% to 30%, moisture content from about 10% to 12%, suint from about 2% to 6% and alcohol extractable matter from about 2.5% to 4.5% (Ramamoorthy et al., 2018). Chiengora fibres are some 25% stronger than wool, cashmere, mohair and camel hair (Ramamoorthy et al., 2018). Chiengora fibre density ranges between about 1.31 and 1.34 g/cc (Ramamoorthy et al., 2018).

10.14 Future work

Sound scientific and technical information on the properties (physical, morphological, dimensional and chemical) and processing of all the rare animal fibres is incomplete and gaps need to be addressed and to be published in peer-reviewed journals. This also applies to aspects concerning the comfort and antimicrobial-related properties of these fibres because claims are often made in this regard, without sound scientific proof. Attention can also be given to the optimized harvesting, dehairing (where applicable), processing, blending and application of rare animal fibres.

Acknowledgements

The author would like to thank his wife, Mrs Edna Hunter and Mrs Kelly Matthews of the CSIR, for doing the word processing, as well as Mrs Jenny Wooldridge and Mrs Kelly Matthews of the CSIR for assistance in sourcing material.

References

Ahmad, S., Kiramani, M.A., July Sept., 1993. Wool woollens India 30 (3), 19.

Albertin, J., Souren, I., Rouette, H.-K., 1990. Textil. Praxis 45, 11 and 719.

Algaa, S., Mägel, M., 1992. Melliand Textilberichte 73 (2), 860.
Algaa, S., Mägel, M., 1992. Melliand Textilberichte 73, 860.
Algaa, S., Mägel, M., 1992. Mell. English 11, 392.
Algaa, S., Mägel, M., 1992. Melliand Textilberichte 73, 60 and 860.
Allolio, J., April 2001. Wool. Rec. 160 (3678), 40.
Allolio, J., April 2007. Wool. Rec. 166 (3750), 34.
Allolio, J., April 2008. Wool. Rec. 167 (3762), 44.
Allolio, J., Wool. Rec., 44.
Anon, 1981. SA Text 29 (10), 18.
Anon, June 1987. Karakul, South Africa wool board, department of technical services, Port Elizabeth, South Africa, and Anon. Wool. Rec. 146 (3572), 65.
Anon, July 1994. Text. Asia 25, 80.
Anon, 2008. Textiles 2, 6.
Anon., Wool. Rec. 142 (3469), 1983, 14.
Anon., Wool. Rec. 160 (3678), April 2001, 35.
Anon, April 1987. Wool. Rec. 146 (3510), 23.
Anon, February 1987. Wool. Rec. 146 (3508), 15.
Anon, 1989. Wool. Rec. 148 (3535), 1.
Anon, September 1990. L'Industrie Text. 1213, 96.
Anon, November 1990. Wool. Rec. 149 (3553), 43.
Anon, November 1991. Knitt. Int. 98, 74.
Anon, 1997. Melliand Textilberichte 78 (11−12), E 173.
Anon, July 2000. Wool. Rec. 159 (3669), 48.
Anon, October 2000. Wool. Rec. 159 (3672), 37.
Anon, October, 2000. Wool. Rec. 159 (3672), 43.
Anon, October 2000. Wool. Rec. 159 (3672), 25.
Anon, September 2002. Wool. Rec. 161 (3695), 9.
Anon, September 2006. New Sci. 23, 54. www.newscientist.com.
Anon, 2006. Melliand Textilberichte 87 (4), 207.
Anon, December 2007. AATCC rev. 7 (12), 22. www.AATCC.org.
Anon, Jan., 2009. Text. Asia 40, 34.
Anon, Wool. Rec. 24.
Anon, BTTG (Shirley technologies), www.bttg.co.uk.
Anon, Harvesting of textile animal fibres, http://www.Fao.org/docrep/v9384e/v9384e03.htm.
Anon, Natural plant and animal fibres − a return to renewable resources, yak, (http://www.binhaitimes.com/yak.html).
Antonini, M., Gonzales, M., Valbonesi, A., 2004. Livest. Prod. Sci. 90, 241.
Appleyard, H.M., 1960. Wool Ind. Res. Assoc. (WIRA). Leeds, England.
Bacchi, C.S., Lanari, M.R., von Thüngen, J., 2010. Small Rumin. Res. 88, 54.
Badmaanyambuu, R., Alimaa, D., Phan, K.-H., Augustin, P., Wortmann, F.-J., May 2003. Text. Asia 34, 24.
Bamford, G.R.E., 1958. The Sorption of Hydrochloric Acid and Potassium Hydroxide by Mohair and Wool. MSc Thesis. Rhodes University, Grahamstown.
Baretta, J., Iñiguez, V., Saavedra, V., Romero, F., Callisaya, A.M., Echalar, J., Gutiérrez-Gil, B., Arranz, J.-J., 2012. Small Rumin. Res. 105, 97.
Bereck, A., 1994. Rev. Prog. Col. 24, 17.
Blackburn, D.L., 1990. Journal 49.
Blankenburg, G., Philippen, H., 1988. Proc. 1st Int. Symp. Spec. Anim. Fibres, Aachen, DWI 103, 242.

Blankenburg, G., Philippen, H., Spiegelmacher, P., Hahnen, J., 1992. IWTO Rep., vol. 3. Nice, France.
Bolormaa, B., Drean, J.Y., Enkhtuya, D., 2007. J. Nat. Fibers 4 (4), 1.
Braga, W., Leyva, V., Cochran, R., 2007. Small Rumin. Res. 68, 323.
Brand, J.T., 1989. Die Volledige Angorakonyn-Handleiding Vir Suid-Afrika. Bridal Printers.
Briggs, D., June 29, 1973. Wool. Rec. 123, 25.
Brodtmann, L.I., Hick, M.H., Castillo, M.F., Frank, E.N., 2017. Text. Res. J. 00 https://doi.org/10.1177/0040517517700199.
Brown, T.D., Onions, W.J., 1961. J. Text. Inst. 52, T101.
Browne, J., Dec., 1987. Australasian Angora Mohair J 54.
Burns, R.H., Von Bergen, W., Young, S.S., 1962. J. Text. Inst. 53, T45.
Bustamante, A.V., Zambelli, A., De Lamo, D.A., von Thüngen, J., Vidal-Rioja, L., 2002. Small Rumin. Res. 44, 97.
Buxton, A., November 1986. Text. Outlook Int. 67.
Cegarra, J., Gacén, J., Cayuela, D., Caro, M., 1990. Melliand Textilberichte 71, 883.
Cervantes, I., Pérez-Cabal, M.A., Morante, R., Burgos, A., Salgado, C., Nieto, B., Goyache, F., Gutiérrez, J.P., 2010. Small Rumin. Res. 88, 6.
Chamut, S., Cancino, A.K., Black-Decima, P., 2016. Small Rumin. Res. 137, 124.
Chávez, S.C., October 2002. Wool. Rec. 161 (3696), 55.
Chen, W., Chen, D., Wang, X., 2001. Text. Res. J. 71, 441.
Chowdhry, B.R., January March, 1983. In: Wool woollens India, 20, p. 36.
Cizek, J., Turpie, D.W.F., 1985. In: Proc. 7th Int. Wool Text. Res. Conf., vol. 2. The Society of Fiber Science and Technology, Tokyo, Japan, p. 137, 1985.
Coetsee, J., March 23, 2001. Landbouweekblad 1191, 24.
Colman, B., June 1982. New Sci., vol. 24.
Cook, G.J., 1968. Handbook of Textile Fibres: I – Natural Fibres. Watford, Merrow.
Couchman, R.C., Holt, C.M., 1990. J. Text. Inst. 81, 142.
Couchman, R.C., 1986. J. Text. Inst. 77, 255.
Couchman, R.C., 1988. Small Rumin. Res. 1, 123.
Couchman, R.C., 1989. J. Text. Inst. 80, 129.
Cowan, N., March 1989. Knitt. Int. 96, 40.
Crawshaw, G., June 1992. Text. Horizons 12 (6), 37.
Crossley, J.C., Borroni, C.G., Raggi, A.S., 2014. J. Appl. Anim. Res. 42, 27. https://doi.org/10.1080/09712119.2013.795899.
Czaplicki, Z., Ruszkowski, K., 2014. J. Nat. Fibers 11, 169. https://doi.org/10.1080/15440478.2013.864577.
Dalton, J., Franck, R., 2001. Cashmere, camelhair and other hair fibres. In: Franck, R. (Ed.), Silk, Mohair, Cashmere and Other Luxury Fibres. Woodhead Publishing, Cambridge, pp. 133–174.
Danforth, W., February 2017. MICC, Cashmere Industry Report. https://frontcap.com/wp-content/uploads/2017/02/Mongolias-Cashmere-Report-February-2017.pdf.
Debnath, C.R., Bhowmick, B.B., Das, P.K., Ghosh, S.K., June, 1987. Text. Trends 35.
Desmond, S., November 1993. Wool. Rec. 152 (3589), 35.
Douglas, S.A.S., 1988. Aust. Angora Mohair J. 5 (4), 19.
Dyson, J., April 2008. Wool. Rec. 167 (3762), 14.
Elsner, W., 1998. The Spinning Wheel.
Evans, J.V., March 1983. Wool. Rec. 142 (3461), 51.
Ferrero, D., October 2007. Wool. Rec. 166 (3756), 56.
Ferrero, D., August 2008. Wool. Rec. 167 (3766), 20.

Franck, R.R. (Ed.), 2001. Silk, Mohair, Cashmere and Other Luxury Fibres. Woodhead Publishing, Cambridge.
Frank, E.N., Hick, M.V.H., Lamas, H.E., Gauna, C.D., Molina, M.G., 2006. Small Rumin. Res. 61, 141.
Frank, E.N., Hick, M.V.H., Adot, O., 2007. J. Text. Inst. 98 (3), 251.
Frank, E.N., Hick, M.V.H., Molina, M.G., Caruso, L.M., 2011. Small Rumin. Res. 99, 54.
Frank, E.N., Hick, M.V.H., Adot, O.G., 2011. J. Text. Inst. 102, 41.
Friedlin, R., Petit, M., 1988. Proc. 1st Int. Symp. Animal Fibers, Aachen, Germany, DWI 103, 221.
Friedlin, R., 1987. Wool. Rec. 146 (3517), 53.
Friedlin, R., 1990. Int. Text. 716, 18.
Fröhlich, H.G., 1969. Ges. Text. Ind. 71 (9), 588.
Fröhlich, H.G., 1969. Z. Ges. Text. Ind. 71 (1), 39.
Gallagher, J.R., 1971. Sheep and goat, wool and mohair. Text. Agric. Exp. Sta. 63. PR-2932.
Ghituleasa, C., Visileanu, E., Ciocoiu, M., 2004. J. Nat. Fibers 1, 13.
Ghoshal, S.P., Raut, S.D., Jaiswal, P.K., July–September 1993. Wool Woollens India 30, 34.
Ghoshal, S.P., Raut, S.D., Jaiswal, P.K., October-December. Wool Woollens India, vol. 30, p. 15, 4.
Gillespie, J.M., Frankel, M.J., 1974. B. Comp. Biochem. 47, 339.
Glasbey, C.A., Hitchcock, D., Russel, A.J.F., Redden, H., 1994. J. Text. Inst. 85, 301.
Gore, S.E., Laing, R.M., 2002. Text. Res. J. 72, 201.
Gray, R.F., 1990. Text. Technol. Int. 45.
Greaves, P., Rainsford, F.E.B., 2005. Text. 32 (3/4), 46.
Greer, S., Banks-Lee, P., Jones, M., 2007. AATCC Rev. 7 (5), 42.
Gui-Fang, W.U., Yong, H.E., 2008. In: Proc. 4th Int. Cashmere Identification Technique Symp., Erdos City, China, p. 79.
Gupta, B.S., George, W.T., 1978. J. Appl. Polym. Sci. 33 (Symp.), 225.
Gupta, N.P., Bapna, D.L., Patni, P.C., Mathur, J.P., September 1989. Indian J. Text. Res. 14, 141.
Gupta, N.P., Arora, R.K., Patni, P.C., April 1992. Indian Text. J. 103, 66.
Gwyther, M., June 30, 1996. Sunday Life. the Sunday Times, p. 14.
Haigh, H.S., 1949. J. Text. Inst. 40, 794.
Halboth, H., Heidemann, G., 1971. Text. Res. J. 41, 860.
Harizi, T., Msahli, S., Sakli, F., Khorchani, T., 2007. J. Text. Inst. 98, 15.
Harizi, T., Mekki, M., Msehli, S., Sakli, F., 2014. J. Nat. Fibers 11, 113.
Harmsworth, T., Day, G., 1990. Wool and Mohair – Producing Better Natural Fibres, second ed. Inkata Press, Melbourne, Australia.
Harris, M., 1954. In: Handbook of Textile Fibers, 75. Harris Research Laboratories, Washington, DC.
Hohls, H.W., February, 1951. Mell. Textilber. 32, 99.
http://www.llama.org//llama-fiber.htm.
Hughes, V.L., Nelson, G., March 2001. AATCC Rev. 1 (3), 39. www.AATCC.org.
Hunter, I.M., Kruger, P.J., 1966. SAWTRI Techn. Rep. 84.
Hunter, I.M., Kruger, P.J., 1967. Text. Res. J. 37, 220.
Hunter, L., Smuts, S., 1981. SAWTRI Bull 15 (2), 18.
Hunter, L., Taylor, H., Erdursun, H.H., December 1982. SAWTRI Bull 16 (4), 34.
Hunter, L., Braun, A., Gee, E., 1985. J. Text. Inst. 76, 289.
Hunter, L., 1987. Angora Goat Mohair J 29 (2), 31.
Hunter, L., November 1987. SAWTRI Special Publication, WOL78.

Hunter, L.'M., 1993. A Review of its Properties, Processing and Applications'. CSIR, Division of Textile Technology. International Mohair Association and the Textile Institute, Port Elizabeth.
Ilse, D., 1958. SA Ind. Chem. 12, 18.
Iñiguez, L.C., Alem, R., Wauer, A., Mueller, J., 1998. Small Rumin. Res. 30, 57.
Iñiguez, L., Mueller, J.P., Ombayev, A., Aryngaziyev, S., Yusupov, S., Ibragimov, A., Suleimenov, M., El-Dine Hilali, M., 2014. Small Rumin. Res. 117, 58.
Kang, T.J., Park, M., Oh, K., 1994/95. In: Proc. AATCC Conf. p. 135.
Kennedy-Sloane, B.A., 1977. In: 'Dimensional and Design Aspects of V-Bed Knitwear Using Wool and Mohair Yarns'. Index to Thesis N.4356, MPh, vol. 27. Leeds University, Leeds, ASLIB.
Khan, Z.A., Wang, X., 2000. J. Text. Inst. 91, 21.
Khatami, A.K., 2016. African J. Agric. Econ. Rural Develop. 4, 393.
Khishigsuren, A., Nakajima, M., Takahashi, M., 2001. Text. Res. J. 71, 487.
Khishigsuren, A., Nakajima, M., Takahashi, M., 2001. J. Text. Eng. 47 (1), 9.
Khishigsuren, A., Nakajima, M., Takahashi, M., July 2001. J. Text. Mach. Soc. Jpn. 53 (7), 51.
Khishigsuren, A., Nakajima, M., Takahashi, M., 2002. Text. Res. J. 72, 51.
King, N.E., 1967. Text. Res. J. 37, 204.
Kingsbury, E., 1985. In: Wool – 1985 Annual. Massey Wool Assoc. New Zealand, p. 6.
Knott, J., 1990. 'Fine animal fibres and their depigmentation process', comett. Eurotex 1.
Kriel, W.J., 1964. SAWTRI Techn. Rep. 33.
Kriel, W.J., 1964. SAWTRI Techn., Rep 29.
Kriel, W.J., 1965. Composition of Wool Wax. MSc Thesis. Potchefstroom University.
Krishnan, K.B., Krishna, T.B., Irudayasamy, R., September 1991. Text. Asia 22, 157.
Langley, K.D., Coskuntuna, E., 2000. Text. Res. J. 70, 181.
Langley, K.D., Kennedy, T.A., 1981. Text. Res. J. 51, 703.
Langley, K.D., 1997. EFFN News (2).
Laycock, H.W., June 29 , 1973. Wool. Rec. 123, 10.
León-Velarde, C.U. and Guerrero, J., Improving quantity and quality of Alpaca fiber; using a simulation model for breeding strategies, 2001. https://www.researchgate.net/publication/254416124_Improving_quantity_and_quality_of_Alpaca_fiber_using_a_simulation_model_for_breeding_strategies.
Li, M.-Z., Ma, W.-X., Chen, X.-Y., 2014. Text. Res. J. 84, 1200.
Liddle, J., December 1999. Wool. Rec. 158 (3662), 39.
Liddle, J., October, 2002. Wool Rec 161 (3696), 30.
Ling, M.Y., 1993. In: Proc. Text. Inst., Ann. Conf. Asia and World Textiles, Hong Kong, p. 585.
Lipson, M., Howard, P., 1946. J. Soc. Dye. Colour. 62, 29.
Liu, X., Wang, X., 2007. Text. Res. J. 77, 957.
Liu, X., Hurren, C.J., Wang, X., 2005. In: Proc. 11th Int Wool Text. Res. Conf., Leeds (CD-Rom).
Liu, X., 2003. A Study of Australian Alpaca Fibres. Deakin University, Geelong, Australia.
Logan, R., Rivett, D., Tucker, D., Hudson, A., 1989. Text. Res. J. 59, 109.
Lu, W., Fei, J., Yang, J., Tang, M., Dong, Z., Zhou, Z., Ye, J., Wu, H., 2013. Text. Res. J. 83, 773.
Lupton, C.J., McColl, A., 2011. Small Rumin. Res. 99, 178.
Lupton, C.J., Minikhiem, D.L., Pfeiffer, F.A., Marschall, J.R., 1995. Proc. 9th Int. Wool Text. Res. Conf., Biella, Italy 2, 545.
Lupton, C.J., McColl, A., Stobart, R.H., 2006. Small Rumin. Res. 64, 211.

Ma, H., Hong, X., Gao, A., Liu, T., November 2008. Proc 4th Int. Cashmere Identification Technique Symp, p. 174. Erdos City, China.
Mahapatra, C., 2009. May 56 (5), 75.
Martinez, Z., Iñiguez, L.C., Rodríguez, T., 1997. Small Rumin. Res. 24, 203.
Martinez, Z., Iniguez, L.C., Rodríguez, T., 1997. Small Rumin. Res. 24, 203.
Mauersberger, H.R. (Ed.), 1948. Mathews' Textile Fibers, fifth ed. John Wiley, New York.
Maycumber, S.G., July 8, 2002. Daily News Rec. 32 (27), 22.
McGhee, B., November 2008. Twist 30.
McGrattan, I., February 1996. Wool. Rec. 155 (3616), 15.
McGregor, B.A., Butler, K.L., 2004. Aust. J. Agric. Res. 55, 433.
McGregor, B.A., Butler, K.L., 2008. Text. Res. J. 78, 486.
McGregor, B.A., Butler, K.L., 2011. Small Rumin. Res. 100, 37.
McGregor, B.A., Liu, X., 2017. J. Text. Inst. 108, 1754.
McGregor, B.A., Naebe, M., 2016. J. Text. Inst. 107, 562.
McGregor, B.A., Postle, R., May 2004. In: Proc. Text. Inst. 83rd World Conf, p. 372. Shanghai, China.
McGregor, B.A., Postle, R., 2008. Text. Res. J. 78, 399.
McGregor, B.A., Postle, R., 2009. Text. Res. J. 79, 876.
McGregor, B.A., Quispe Peña, E.C., 2018. J. Text. Inst. 109, 767.
McGregor, B.A., Schlink, A.C., 2014. J. Text. Inst. 105, 927.
McGregor, B.A., Stapleton, D.L., 2016. J. Text. Inst. https://doi.org/10.1080/00405000.2015.1046290.
McGregor, B.A., Kerven, C., Toigonbaev, S., 2009. Small Rumin. Res. 84, 89.
McGregor, B.A., Ramos, H.E., Quispe Pena, E.C., 2012. Small Rumin. Res. 102, 191.
McGregor, B.A., Butler, K.L., Ferguson, M.B., 2012. Anim. Prod. Sci. 52, 35.
McGregor, B.A., Butler, K.L., Ferguson, M.B., 2013. Small Rumin. Res. 115, 40.
McGregor, B.A., Butler, K.L., Ferguson, M.B., 2013. Small Rumin. Res. 113, 90.
McGregor, B.A., Liu, X., Wang, X.G., 2018. J. Text. Inst. 109, 813.
McGregor, B.A., 1998. Small Rumin. Res. 28, 199.
McGregor, B.A., 2002. Small Rumin. Res. 44, 219.
McGregor, B.A., 2003. Aust. J. Exp. Agric. 43, 1199.
McGregor, B.A., 2003. Wool Technol. Sheep Breed. 51, 224.
McGregor, B.A., 2006. Small Rumin. Res. 61, 93.
McGregor, B.A., 2010. Anim. Prod. Sci. 50, 168.
McGregor, B.A., 2014. J. Text. Inst. 105, 597. https://doi.org/10.1080/00405000.2013.828448.
McGregor, B.A., 2016. J. Text. Inst. 107, 1193.
McLaren, J., Milligan, B., 1981. The Chemical Reactivity of the Wool Fibre: Wool Science. Science Press, Merrickville.
Mengüç, G.S., Özdil, N., Hes, L., 2015. Text. Res. J. 85, 2155.
Michell, D., October 2002. Wool. Rec. 161 (3696), 43.
Middlebrook, R., April 1987. Wool. Rec. 146 (3508), 15.
Misra, R.K., Singh, B., Jain, V.K., 1998. Small Rumin. Res. 27, 97.
Mitchell, R.J., Betteridge, K., Gurnsey, M.P., Welch, R.A.S., Nixon, A.J., 1991. N.Z. J. Agric. Res. 34, 287.
Moore, K.E., Blache, D., Maloney, S.K., 2011. Small Rumin. Res. 96, 165.
Moses, J.J., Amayappan, L., May 2006. Colourage 53, 53.
Msahli, S., Harizi, T., Sakli, F., Khorchani, T., 2008. J. Text. Inst. 99 (5), 393.
Mueller, J.P., Kosimov, F.F., Brent, L., Nishanov, N., Rischkowsky, B., 2016. Small Rumin. Res. 134, 74.

Naebe, M., McGregor, B.A., 2013. J. Text. Inst. 104, 634.
Nagal, K. and Naik, D., Int. Dyer, 18.
Nicholas, G., April 1998. Farmers Wkly. 3, 49.
Oglakcioglu, N., Celik, P., Ute, T.B., Marmarali, A., Kadoglu, H., 2009. Text. Res. J. 79, 888.
Onal, L., Korkmaz, M., December 2006. Indian J. Fibre Text. Res. 31, 507.
Onions, W.J., 1962. Wool: An Introduction to its Properties, Uses and Production. Ernest Benn, London.
Patthey, J.F., October 2002. Wool. Rec. 161 (3696), 52.
Paul, S., Garg, S., Grover, E., April/June 2003. Syn. Fib. 32 (2), 10.
Pelant, R.K., Chandre, B., Pu, J.B., Lohani, M., Suknaphasawat, N., Xu, G., 1999. Small Rumin. Res. 34, 249.
Pepper, J., November 1997. Wool. Rec. 156 (3637), 71.
Phan, K.-H., Wortmann, F.-J., 1987. DWI. 101, 137.
Phan, K.-H., Wortmann, F.-J., May 1995. In: Proc. European Symp. Metrology and Identification of Spec. Anim. Fibres, (SAF '95). European Fine Fibre Network Occasional Publication No. 4, Aachen, Germany, p. 45.
Phan, K.-H., Wortmann, F.-K., Wortmann, G., Arns, W., 1988. Proc. 1st. Int. Symp. Spec. Anim. Fibers, p. 137. DWI Report No. 103, Aachen.
Phan, K.-H., Wortmann, F.-J., Arns, W., 1990. Proc. Text. Conf., Aachen, Germany, DWI 105, 135.
Phan, K.-H., Wortmann, F.-J., Arns, W., 1991. Proc. Text. Conf., Aachen, Germany, DWI 108, 235.
Phan, K.-H., Arns, W., Wortmann, F.-J., Höcker, H., June 1991. IWTO Rep. No. 4 (Lisbon, Portugal).
Phan, K.-H., Wortmann, F.-J., Arns, W., 1995. In: Proc. 9th Int. Wool Text. Conf., vol. II, p. 571. Biella, Italy.
Phan, K.-H. and Wortmann, F.-J., DWI, A., Germany, May 2000.
Phan, K.-H., Wortmann, G., Wortmann, F.-J., 2000. In: Proc. 10th Int. Wool Text. Res. Conf., Aachen, Germany. SF-2, 1.
Phan, K.-H., Augustin, P., Wortmann, F.-J., Enkhjargal, D., Badsuren, S., Badmaanyambuu, R., Alimaa, D., 2005. In: Proc. 11th Int. Wool Text. Res. Conf. CDRom, Leeds, UK.
Phan, K.-H., Rütten, S., Popescu, C., 2008. In: Proc. 4th Int. Cashmere Identification Technique Symp, p. 49. Erdos City, China.
Phan, K.-H., 1991. Electron Microscopy and the Characterization of Keratin Fibres. Comett Eurotex, Portugal.
Pinares, R., Gutiérrez, G.A., Cruz, A., Morante, R., Cervantes, I., Burgos, A., Gutiérrez, J.P., 2018. Small Rumin. Res. 165, 93.
Pulling, J.A., June 29, 1973. Wool. Rec. 123, 6.
Pumayalla, A., Levva, C., 1988. Proc. 1st Int. Symp. Spec. Anim. Fibers 235. DWI Report No. 103, Aachen.
Qi, W., April 2006. Wool. Rec. 165 (3738), 27.
Quispe, E.C., Ramos, H., Mayhua, P., Alfonso, L., 2010. Small Rumin. Res. 93, 64.
Rainsford, F.E.B., 2000. In: Text. Mag., 2000, Issue 4, 18 and Proc. Text. Inst. Annual World Conf. p. 4.
Rainsford, F.E.B., October 2002. Wool. Rec. 161 (3696), 61.
Rainsford, F.E.B., 2005. Text. Mag. No. 2, 18.
Rainsford, F.E.B., Oct., 2006. Wool. Rec. 165 (3744), 27.
Rainsford, F.E.B., April 2006. Wool. Rec. 165 (3738), 41.
Rainsford, F.E.B., April 2007. Wool. Rec. 166 (3750), 30.

Rainsford, F.E.B., April 2007. Wool. Rec. 166 (3750), 32.
Rainsford, F.E.B., August 2007. Wool. Rec. 166 (3754), 55.
Rainsford, F.E.B., November 2007. Wool. Rec. 166 (3757), 57.
Rainsford, F.E.B., October 2008. Twist, p. 46.
Rainsford, F.E.B., July 2008. Wool. Rec. 167 (3765), 7.
Rainsford, F.E.B., 2009. Alpacas Aust. 58, 20.
Raja, A.S.M., Shakyawar, D.B., Pareek, P.K., Temani, P., Sofi, A.H., 2013. J. Nat. Fibers 10, 381.
Ramamoorthy, S., Ramadors, M., Ramasamy, R., Thangavel, K., 2018. J. Nat. Fibers. https://doi.org/10.1080/15440478.2018.1479996.
Reddy, V.S., Mudaliar, A.S.R., October–December, 1985. Wool Woollens India 22, 29.
Roberts, S., Wilkey, R., November, 1993. Wool. Rec. 152 (3590), 37.
Robinson, G.A., Slinger, R.I., October 1970. SAWTRI Tech. Rep. No. 142.
Robinson, G.A., Ellis, R., Van Der Merwe, J.P., September 1971. SAWTRI Tech. Rep. No. 150.
Rollins, C.K., Hall, D.M., 1999. Text. Res. J. 69, 856.
Rowell, J.E., Lupton, C.J., Robertson, M.A., Pfeiffer, F.A., Nagy, J.A., White, R.G., 2001. J. Anim. Sci. 79, 1670. http://jas.Fass.org.
Ryder, M.L., Gabra-Sanders, T., 1988. J. Text. Inst. 79, 330.
Ryder, M.L., April 1980. Wool. Rec. 137 (3426), 44.
Ryder, M.L., 1981. Wool. Rec. 140 (3445), 33.
Ryder, M.J., 1987. Cashmere Mohair and Other Luxury Animal Fibres for the Breeder and Spinner. White Rose II (Booklet), Southampton.
Ryder, M., December 1989. New Sci. 16, 34.
Ryder, M.L., 1990. In: Proc. 2nd Int. Symp. Spec. Anim. Fibers, vol. 106. Aachen, Germany, DWI, p. 175.
Ryder, M.L., 1990. Text. Mag. 19 (Issue 1), 9.
Ryder, M.L., 1990. In: Proc. 8th Int. Wool Text. Res. Conf., II, p. 241. Christchurch, New Zealand.
Ryder, M.L., September 1991. Wool. Rec 150 (3563), 73.
Ryder, M., November 1996. Wool. Rec 155 (3625), 4.
Ryder, M.L., November 1997. Wool. Rec 156 (3637), 67.
Ryder, M.L., September 1999. Wool. Rec 158 (3559), 5.
Ryder, M.L., December 2000. Wool. Rec. 159 (3674), 5.
Ryder, M.L., 2001. Text. Mag. 30 (1), 24.
Ryder, M.L., 2003. Text. Mag. 30 (1), 24.
Sawbridge, M., Ford, J.E., 1987. Textile Fibres under the Microscope, vol. 50. Shirley Institute, Manchester. Shirley Institute Publication S.
Schaller, G.B., 1977. Mountain Monarchs: Wild Sheep and Goats of the Himalaya. University of Chigago Press, Chicago.
Schweiss, J., February 2009. Twist, vol. 26.
Shelton, J.M., Bassett, J.W., 1981. Biology and efficiency of animal fiber production. In: Bowker, E. (Ed.), Sheep and Goat Handbook, pp. **2** 17–22.
Shlomm, B., November 1990. Wool. Rec. 149 (3553), 37.
Sich, J., 1990. Proc 2nd Int. Symp. Spec. Anim. Fibers, DWI, Aachen 91.
Siles, F., January 1998. Wool. Rec. 157 (3639), 21.
Singh, A., Wang, X., Wang, L., February 2003. Text. Asia 34, 41.
Singh, A., Wang, X., Wang, L., 2005. Proc. 3rd Int. Cashmere Determination Technique Sem. Erdos Group, Inner Mongolia, p. 375.
Smith, I.D., Clarke, W.H., Turner, H.N., June 1973. J. Aust. Inst. Agric. Sci. 128.

Smith, G. J., 1987.
Smith, G.A., 1988. In: Text. Technol. Int., 1988, 22 and Proc. 1st Int. Symp. Speciality Animal Fibers, Aachen, p. 8. Germany, DWI 103.
Smuts, S., Hunter, L., Van Rensburg, H.L., 1981. SAWTRI Techn. Rep., No. 482.
Snyman, M.A., 2002. Small Rumin. Res. 43, 105.
Speakman, J.B., 1930. J. Soc. Chem. Ind. 49, T209.
Spei, M., Holzem, R., 1991. Melliand Textilberichte 72. E 174 and 431.
Spilhaus, K., 1999. Wool. Rec., Dec. 158 (3662), 29.
Spilhaus, K., January 2007. AATCC Rev. 7 (1), 21. www.AATCC.org.
Sponenberg, D.P., 2010. Small Rumin. Res. 93, 210.
Srivastava, T.V.K., 1984. Text. Mach. Accessories Stores 20 (6), 25.
Stapleton, I.W., July 1992. 'Alpaca as a Textile Fibre: Fact or Fiction', Rep. Australian Association, Textile and Fibre Research Institute (TFRI), p. 1.
Stephani, G., Wortmann, F.-J., Jan., 1986. IWTO Rep. No. 9. Paris, France.
Stephani, G., Wortmann, F.-J., 1986. DWI No 99, 605.
Stewart, C., 2008. Wool. Rec. 167 (3766), 26.
Stoves, J.L., 1976. J. Soc. Dye. Colour. 92, 213–226.
Strydom, M.A., Van Der Merwe, J.P., Sept., 1983. SAWTRI Tech. Rep. No. 529.
Strydom, M.A., May 1980. SAWTRI Tech. Rep. 458.
Stubbs, C., Marler, J.W., 1990. In: Proc. 8th Int. Wool Text. Res. Conf, 2. Christchurch, New Zealand, p. 312.
Suadipradja, A., Wang, X., 1996. Proc. 77th Text. Inst. World Conf. 1, 221.
Taddeo, H.R., Duga, L., Almeida, D., Willems, P., Somlo, R., 2000. Small Rumin. Res. 36, 285.
Tagawa, T., Mori, J., Kondo, T., 1972. Sen-I-Gakkaishi, 28(14/570), and world text. Abstr. 4, 170, 1424.
Terblanche, E. le F., 1990. (Part 3), Angora Goat Mohair J 32 (1), 25.
Tester, D.H., 1987. Text. Res. J. 57, 213.
Tonin, C., Bianchetto, M., Vineis, C., Bianchet, M.F., 2002. Text. Res. J. 72, 701.
Townend, P.P., Smith, P.A., Lam, C.H., 1989. Hollings App. Indus. Rev. 6 (3), 15.
Trollip, N.G., Maasdorp, A.P.B., Van Rensburg, N.J.J., 1985. In: Proc. 7th Int. Wool Text. Res. Conf., Tokyo, Japan, vol. IV, p. 141.
Tucker, D.J., Hudson, A.H.F., Ozolins, G.V., Rivett, D.E., Jones, L.N., 1988. Proc. 1st Int. Symp. Animal Fibers, Aachen, Germany, DWI 103, 71.
Tucker, D.J., Hudson, A.H.F., Rivett, D.E., Logan, R.I., 1990. In: Proceedings of the 2nd International Symposium on Speciality Animal Fibers. DWI, 106, Aachen, Deutsches Wollforschungsinstitut, p. 1.
Tucker, D.J., Hudson, A.H.F., Logan, R.I., Rivett, D.E., 1990. In: Proc. 8th Int. Wool Text. Res. Conf., vol. 2. Wool Research Organisation of New Zealand, Christchurch, p. 364.
Turpie, D.W.F., Cizek, J., 1985. SAWTRI Techn. Rep., No. 572 IWTO Rep, vol. 9. Barcelona, Spain.
Turpie, D.W.F., Cizek, J., 1986. IWTO Rep. Oostende.
Turpie, D.W., Cizek, J., 1987. SAWTRI Techn. Rep. 596.
Turpie, D.W.F., Steenkamp, C., 1989. IWTO Rep. No., vol. 8. Perth, Australia.
Turpie, D.W.F., Steenkamp, C.H., 1990. IWTO Rep, vol. 1. Nice, France.
Turpie, D.W.F., Strydom, M.A., Cizek, J., 1986. In: Proc. 2nd World Merino Conf, Madrid, Spain, 3, 234.
Turpie, D.W.F., Steenkamp, C., Lüpke, E.E., Kritzinger, N.M., Lupton, C., 1989. IWTO Rep, vol. 9. Perth, Australia.

Turpie, D.W.F., Steenkamp, C., Lüpke, E.E., Kritzinger, N.M., Lupton, C., 1989. IWTO Rep, vol. 10. Perth, Australia.
Turpie, D.W.F., 1985. SAWTRI Spec. Publ., WOL 69.
Ünal, P.G., Atav, R., Ilarslan, S., Kastaci, B.B., 2018. J. Nat. Fibers. https://doi.org/10.1080/15440478.2018.1448319.
Uys, D.S., 1963. Angora Goat and Mohair J 6 (1), 31.
Valbonesi, A., Crisofanelli, S., Pierdominici, F., Gonzales, M., Antonini, M., 2010. Text. Res. J. 80, 344.
Valbonesi, A., Apaza, N., LaManna, V., Gonzales, M.L., Huanca, T., Renieri, C., 2011. Small Rumin. Res. 99, 16.
Van Der Merwe, J.P., Brydon, A.G., February 1986. SAWTRI Tech. Rep. No. 577.
Van Der Merwe, J.P., August 1985. SAWTRI Tech. Rep. No. 569.
Van Der Westhuysen, J.M., Wentzel, D., Grobler, M.C., 1981. Angora Goats and Mohair in South Africa. Nasionale Koerante, Port Elizabeth.
Van Der Westhuysen, J.M., Wentzel, D., Grobler, M.C., 1985. Wool. Rec. 144 (3493), 35.
Van Rensburg, N.J.J., Heinrich, A., November 1983. SAWTRI Tech. Rep. No. 536.
Villarroel-Leon, J., 1959. A Study of Alpaca Fibre. MSc Thesis. University of New South Wales, Australia.
Vineis, C., Aluigi, A., Tonin, C., 2008. Autex Res. J. 8 (3), 68.
Vineis, C., Aluigi, A., Tonin, C., 2011. Text. Res. J. 81, 264.
Vineis, C., Tonetti, C., Paolella, S., Pozzo, P.D., Sforza, S., 2014. Text. Res. J. 84, 953.
Vinella, S., 1994. In: Gerken, M., Camerino, C.R. (Eds.), Proc. 1st European Symp. South American Camelids. University of Camerino Press, Italy, 155−000.
Visser, C., Snyman, M.A., Van Marle-Köster, E., Bovenhuis, H., 2009. Small Rumin. Res. 87, 27.
Visser, C., van Marle- Köster, E., Bovenhuis, H., Crooijmans, R.P.M.A., 2011. Small Rumin. Res. 100, 80.
Volooj, S., Carr, C.M., Mitchell, R., Vickerman, J.C., 1999. J. Text. Inst. 90 (3), 60.
Von Bergen, W., 1954. Speciality hair fibers. In: Mauersberger, M.R. (Ed.), Matthews' Textile Fibers, sixth ed. John Wiley, New York. 615−000.
Von Bergen, W., 1961. Wool Handbook, vol. 1. Textile Book Publishers, New York.
Von Bergen, W., 1969. Speciality Hair Fibres: Von Bergen's Wool Handbook, vol. 1. Interscience Publishers, New York.
Wang, X., Khan, Z.A., 2000. J. Text. Inst. 91 (1), 16.
Wang, X., Chang, L., Wang, L.J., 1999. J. Text. Inst. 90, 456 (Part 1, No. 3).
Wang, X., Wang, L., Liu, X., 2003. Rural in. Res. Develop. Corp., RIRDC publication No 03/128, 1. https://www.agrifutures.com.au/wp-content/uploads/publications/03-128.pdf.
Wang, L., Liu, X., Wang, X., May 2004. In: Proc. 83rd Text. Inst. World Conf., p. 449. Shanghai, China.
Wang, L., Wang, X., Liu, X., 2005. In: Proc. 11th Int. Wool Text. Res. Conf., Leeds (CD-Rom).
Wang, L., Singh, A., Wang, X., 2008. J. Text. Inst. 99 (6), 539.
Weng, X., Oyang, X., Zhang, X., May 2004. In: Proc. 83rd Text. Inst. World Conf., Shanghai, China, p. 471.
Wildmann, A.B., 1954. In: Wildmann, A.B. (Ed.), The Microscopy of Animal Textile Fibres. Wool Industries Research Association, Leeds, UK.
Wilkinson, P.F., 1975. J. Zool. 177, 363.
Williams, R., Braaten, A., 1998. ITAA Proceedings 54, 1. English ITT Cat. No. TS 767.172, 1998.
Wortmann, G., Wortmann, F.-J., June 1988. IWTO Rep. No., vol. 3. Avignon, France.

Wortmann, G., Wortmann, F.-J., Knott, J., 1987. DWI, vol. 73. VII.
Wortmann, G., Wortmann, F.-J., Roes, J., 1988. International Wool Textile Organisation Report No. 7. Palm Beach, California, USA, Jan.
Wortmann, F.J., 1991. Text. Res. J. 61, 371.
Wortmann, F.-J., Wortmann, G. and Arns, W., AIF-project No. 6435. Deutsches Wollforschun gsinstitut, Aachen, Germany.
Wuliji, T., Davis, G.H., Dodds, K.G., Turner, P.R., Andrews, R.N., Bruce, G.D., 2000. Small Rumin. Res. 37, 189.
Wurzinger, M., Delgado, J., Nürnberg, M., Zárate, A.V., Stemmer, A., Ugarte, G., Sőlkner, J., 2006. Small Rumin. Res. 61, 131.
Xiaoming, T., Postle, R., 1989. J. China Text. Univ. 2, 1.
Xing, J., Pailthorpe, M.T., March 2000. J. Soc. Dyers Col. 116, 91.
Xinhe, Y., April 1989. Wool. Rec. 148 (3534), 27.
Yigit, G.K., 2014. American J. Mat. Eng. Techol. 2, 8.
Zahn, H., Föhles, J., Nienhaus, M., Schwan, A., Spei, M., 1980. Ind. Eng. Chem. Prod. Res. Dev. 19, 496.
Zahn, H., 1990. Proc. 2nd Int. Symp., Speciality Animal Fibers, DWI 106, Aachen, Germany. Deutsches Wollforschungsinstitut, p. 195.
Zahn, H., 1991. Melliand Textilberichte 72. E 371 and 926.
Zahn, H., 1995. In: Proc. 9th Int. Wool Text. Res. Conf., vol. 1, p. 1. Biella, Italy.
Zhang, Z., Yang, G., Meng, L., September 2005. In: Proc. 3rd Int. Cashmere Determination Technique Symp. Berduoso City, Dongsheng, China, p. 68.
Zhang, P.-F., Li, Y., Zhang, Y., 2008. Wool Text. J. 19, 26.
Zhaogeng, H., Bo, L., June 1987. Text. Asia 18, 68.
Zhong, Y., Lu, K., Tian, J., Zhu, H., 2016. Text. Res. J. 86, 1.

Further Reading

Atav, R., Türkmen, F., 2015. Text. Res. J. 85, 1331.
McGregor, B.A., Butler, K.L., 2008. Text. Res. J. 78, 486.
Proc. 1st Int. Symp. Speciality Animal Fibers 103, 1988. Aachen, Germany, DWI.
Proc. 2nd Int. Symp. Speciality Animal Fibers 106, 1990. Aachen, Germany, DWI.
Proc. 4th Int. Cashmere Identification Technique Symp, November 2008. Erdos City, China.
Pumayalla, A., Barnett, J. and Osorio, S. (quoted in Ref. 24).
Quispe, E.C., Rodríguez, T.C., Iñiguez, L.R., Mueller, J.P., 2009. Amim. Gen. Resources Inf. 45, 1.
Sakli, F., Dubois, R., Van Parys, M., Knott, J., 1988. Melliand Textilberichte 69, 191 and E 101.
Ukhnaa, S., Drean, J.Y., June 2001. Etude des Proprietes Physiques et Mecaniques de la Fibre de Cachemire, Universite De Haute Alsace (ENS).
Yan, K., Höcker, H. and Schäfer, K., Text. Res. J., 200, 70, 734.

Silk fibres – structure, properties and applications

11

K. Murugesh Babu
Bapuji Institute of Engineering and Technology, Davangere, Karnataka, India

11.1 Introduction to silks from silk worms

Silk is one of the oldest fibres known to man. Silk is an animal fibre produced by certain insects to build their cocoons and webs. Although many insects produce silk, only the filament produced by the mulberry silkmoth, *Bombyx mori*, and a few others in the same genus, is used by the commercial silk industry (Jolly et al., 1979). The silk produced by other insects, mainly spiders, is used in a small number of other commercial capacities, for example, weapon and telescope cross hairs and other optical instruments (Spring and Hudson, 2002)

Silk fibres from silkworms have been used in textiles for nearly 5000 years. The primary reasons for this long-time use have been the unique lustre, tactile properties, durability, and dyeability of silks. Silk fibres are remarkable materials displaying unusual mechanical properties: strong, extensible, and mechanically compressible (Akira et al., 2006). Silk is rightly called the queen of textiles for its lustre, sensuousness and glamour (Manohar Reddy, 2009). Silk's natural beauty and properties of comfort in warm weather and warmth during colder months have made it use in high-fashion clothing. Silk fibres have outstanding natural properties which rival the most advanced synthetic polymers, yet the production of silk does not require harsh processing conditions, and hence, widespread investigations are on, even for artificial synthesis of silk fibres (Chen et al., 2003).

11.1.1 Introduction to sericulture

Sericulture is the rearing of silkworms for the production of raw silk. The major activities of sericulture comprises food-plant cultivation to feed the silkworms which spin silk cocoons and reeling the cocoons for unwinding the silk filament for value added benefits such as processing and weaving. Although there are several commercial species of silkworms, *Bombyx mori* is the most widely used. Sericulture is ideally suited for improving the rural economy of the country, as it is practiced as a subsidiary industry to agriculture. Recent research has also shown that sericulture can be developed as a highly rewarding agro-industry.

Handbook of Natural Fibres. https://doi.org/10.1016/B978-0-12-818398-4.00013-X

11.1.2 Types of silk and their importance

11.1.2.1 Types of silk

There are five major types of silk of commercial importance, obtained from different species of silkworms which in turn feed on a number of food plants: Except mulberry, other varieties of silks are generally termed as non-mulberry silks. India has the unique distinction of producing all these commercial varieties of silk.

The bulk of the commercial silk produced in the world comes from this variety and often silk generally refers to mulberry silk. Mulberry silk (Fig. 11.1) comes from the silkworm, *Bombyx mori* L. which solely feeds on the leaves of mulberry plant. These silkworms are completely domesticated and reared indoors. In India, the major mulberry silk producing states are Karnataka, Andhra Pradesh, West Bengal, Tamil Nadu, and Jammu and Kashmir which together accounts for 92% of country's total mulberry raw silk production.

Tasar (Tussah) is copperish colour, coarse silk mainly used for furnishings and interiors. It is less lustrous than mulberry silk, but has its own feel and appeal. Tasar silk (Fig. 11.2) is generated by the silkworm, *Antheraea mylitta* which mainly thrives on the food plants asan and arjun. The rearings are conducted in nature on the trees in the open. In India, tasar silk is mainly produced in the states of Jharkhand, Chhattisgarh and Orissa, besides Maharashtra, West Bengal and Andhra Pradesh. Tasar culture is the main stay for many a tribal community in India.

It is a finer variety of tasar generated by the silkworm, *Antheraea proyeli* J (Fig. 11.3) in India which feed on natural food plants of oak, found in abundance in the sub-Himalayan belt of India covering the states of Manipur, Himachal Pradesh, Uttar Pradesh, Assam, Meghalaya, and Jammu and Kashmir. China is the major producer of oak tasar in the world and this comes from another silkworm which is known as *Antheraea pernyi*.

Also known as endi or errandi, eri is a multivoltine silk spun from open-ended cocoons, unlike other varieties of silk. Eri silk (Fig. 11.4) is the product of the domesticated silkworm, *Philosamia ricini* that feeds mainly on castor leaves. Ericulture is a household activity practiced mainly for protein rich pupae, a delicacy for the tribal. Resultantly, the eri cocoons are open-mouthed and are spun. The silk is used indigenously for preparation of *chaddars* (wraps) for own use by these tribals. In India, this

Figure 11.1 Mulberry Silk (a) Worm (b) Moth (c) Cocoons.

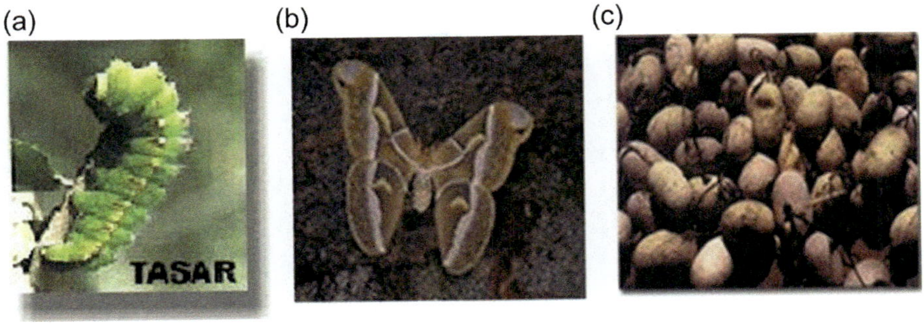

Figure 11.2 Tasar Silk (a) Worm (b) Moth (c) Cocoons.

Figure 11.3 Oak Tasar Silk (a) Worm (b) Moth (c) Cocoons.

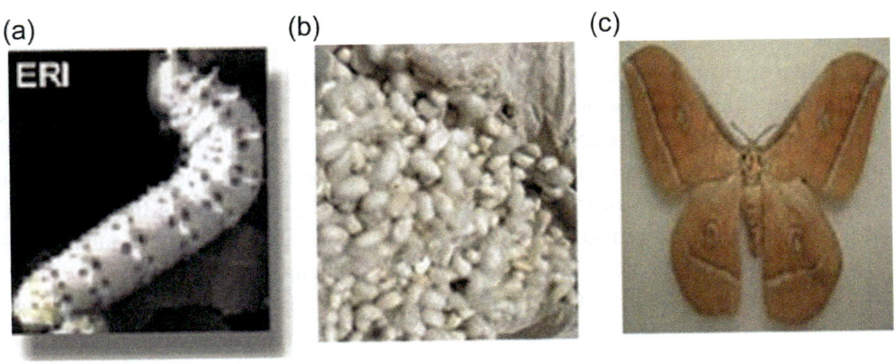

Figure 11.4 Eri Silk (a) Worm (b) Caterpillar (c) Cocoons.

culture is practiced mainly in the North Eastern states and Assam. It is also found in Bihar, West Bengal and Orissa.

This golden yellow colour silk is prerogative of India and the pride of Assam state. It is obtained from semi-domesticated multivoltine silkworm, *Antheraea assamensis*. These silkworms (Fig. 11.5) feed on the aromatic leaves of Som and Soalu plants and are reared on trees similar to that of tasar. Muga culture is specific to the state of Assam and an integral part of the tradition and culture of that state. The muga silk, a high value product, is used in products like sarees, mekhalas, chaddars, etc.

Figure 11.5 Muga Silk (a) Worm (b) Caterpillar (c) Cocoons.

11.2 Silk as a high-performance fibre

11.2.1 Requirements for composites

The silk fibre produced by spiders, silkworms, scorpions, mites and flies may have different composition, structure and material properties depending upon the specific source. These flaws can be avoided by spinning under controlled conditions to produce uniform cross-sectional area of silk fibre. Many of the properties of native silks (as opposed to regenerated silks) also make themsustainable for reinforcement materials for engineering (i.e. non-biomedical) composites besides plant fibres. Two key issues that require specific attention in the development of SFRPs with useful properties are (i) selection of a complementary matrix and (ii) understanding the role of the interface.

In order to produce silk fibre composites, the essential requirements are (i) using a high-failure strain and low-processing thermoset matrix, (ii) maximizing the reinforcing effect of low-stiffness ductile silk, (iii) facilitating impregnation and avoiding fibre degradation and (iv) employing high fibre volume fractions to ensure that the fibres carried a larger fractions of the load. The fundamental philosophy in fibre reinforced polymers is the development of a brittle fibre—ductile matrix system (Harris, 1999) where the fibres, compared to the matrix, have a much lower failure strain coupled with a much higher stiffness. Commonly used polymer matrices (including thermosets like epoxy, and thermoplastics like polypropylene) have a relatively low stiffness (<4 GPa) and high failure strain (>5%, even up to 1000%). The strong and stiff fibres are required to carry most of the load, while the ductile matrix provides crack blunting and bridging mechanisms (Harris, 1999). Silk fibres are different from traditional reinforcing fibres (including E-glass and flax) in that mulberry silk fibres have a relatively low stiffness (5—15 GPa) and high failure strain (15%—25%), while traditional reinforcing fibres have a high stiffness (>50 GPa) and low failure strain (1%—5%).

11.2.2 Silk medical products

The silk fibre has a long history of use in medicine since decades because of its non-septicity and degradability. Further, the silkworms efficiently converts the plant nutrients consumed by its larvae in to silk protein, a major advantage of producing such high-value protein at cheaper cost in shorter span of time. Hence, the silk can be used as a biomaterial in various forms, such as films, membranes, gels, sponges, powders, artificial ligaments and scaffolds. The silk applications also include the burn-wound dressings for faster healing, enzyme immobilization matrices, nets, vascular prostheses and structural implants (Cao and Wang, 2009). The silk fibroin, a natural protein of domestic silkworm, *Bombyx mori,* provides an important set of options for biomaterials and scaffolds because of its high tensile strength, controllable biodegradability, haemostatic properties, non-cyto toxicity, low anti-genicity and non-inflammatory characteristics (Li et al., 2003; Jin et al., 2004; Mauney et al., 2007). Importantly, the biocompatibility and bioresorbable properties of silks, their amenability to aqueous or organic solvent processing into various 'regenerated' forms (including aqueous solutions, films, hydrogels, porous sponges, regenerated fibres and cords, and nonwoven mats), alongside their unique combination of high strength and toughness, make them ideal for a wide range of clinical applications: from braided suture threads for surgical options, to porous, reinforced-composite scaffolds for cartilage and bone repair (Vepari and Kaplan, 2007; Hardy and Scheibel, 2010; Altman et al., 2003; Hakimi et al., 2007). Naturally, considerable research has focussed on biocomposites based on regenerated silks for such biomedical applications (Vepari and Kaplan, 2007; Hardy and Scheibel, 2010, Altman et al., 2003, Hakimi et al., 2007).

The prime requirement of biomaterials is biocompatibility; as like silk fibres, which can perform phenomenal mechanical properties as biopolymers suitable for medical uses. The properties of high oxygen and vapour permeability makes silk fibre ideal for soft tissue applications and in spite of its higher adhering ability to tissue cells, still be biodegradable. The silk fibre shows no toxicity to living bodies during degradation because it is composed of amino acids similar to those found in humans (Sehnal, 2008) as like silk sutures have lesser incidence of swelling and infection than any other material. The additional applications of silk biomaterials include new generation soft contact lenses, artificial corneas, skin grafts and epilepsy drug permeable devices (Dalpra et al., 2005).

11.3 Synthesis of silk fibres from silkworms

11.3.1 Introduction to synthesis of silk from silkworms

Sericulture is the cultivation of cocoons for their filaments. The best raw silk is obtained from the species of moth called *Bombyx mori.* Breeding of silkworm occurs once in a year but under scientific conditions, they may be hatched three times a year. The female moth lays around 350 to 400 eggs and the moths die soon after. As they are subject to hereditary infection, the eggs from infected moths are destroyed which results into production of fine silk. Larvae of about 3 mm are hatched from the eggs. For about 20–30 days, they are carefully nurtured and are fed five times a day on chopped mulberry leaves. In the meantime, the larvae change their skin for four times and are formed into caterpillar of about 9 cm long. Now they are ready to spin cocoon for which racks, clusters of twigs or straw are provided.

The caterpillar has small openings under their jaws called spinnerets through which they secret a protein-like substance. This substance solidifies when it comes in contact with air and the filament thus formed is spun around the silkworm in the figure resembling the digit 8. In 3 days the cocoon gets completed which is about a peanut shell's size. The filament is held together by sericin or silk gum. The life of the worm is ended by the process of 'stoving' or 'stifling' in which the cocoons are heated. Some of the cocoons are preserved so that the pupa or chrysalis inside them develops into moths for further breeding.

11.3.2 Silk reeling and silk fibre manufacture

Silk reeling is the process of unwinding the silk filaments from the cocoons and a process by which a number of cocoon baves are reeled together to produce a single thread. This is achieved by unwinding filaments collectively from a group of cooked cocoons at one end in a warm water bath and winding the resultant thread onto a fast moving reel. Raw silk reeling may be classified by direct reeling method on a standard-sized reel, indirect method of reeling on small reels, and the transfer of reeled silk from small reels onto standard-sized reels on a re-reeling machine. The last technique is primarily applied in modern silk reeling processes (Mahadevappa et al., 2001).

The reeled silk is formed into silk yarn or silk thread through the process called 'throwing'. The spinning process of other natural fibres corresponds to it. The raw silk skeins are sorted according to their colour, size, and quantity and washed in warm water with soap or oil for softening the sericin. After drying the skeins, they are placed on reels from where the silk is wound on bobbins. During winding, the silk strands are given desired amount of twist. The strands may be doubled and then given twist in similar or opposite directions. To get equal diameter throughout the length, the yarn is run through rollers. Many kinds of silk yarns are manufactured by giving different amount of twists. Still remaining sericin is removed from silk yarn by the process of 'degumming' in which the yarn is washed with soap and water for bringing out its natural shine and the soft feel.

11.3.3 Stages in silk filament production

11.3.3.1 Cocoon sorting

Cocoon sorting is a process of sorting out defective cocoons from a lot of good cocoons. The process also includes segregating cocoons according to their size. The defective cocoons can be classified as (a) Double cocoons (b) Pierced cocoons (c) Urinated cocoons (d) Flimsy cocoons (e) Pointed or constricted cocoons (f) Mould attacked cocoons (g) Immature cocoons.

The process of sorting according to the size is carried with a special equipment known as riddling machine. The segregation of uniform-sized cocoons is extremely important since size has influence on cooking and reeling properties.

11.3.3.2 Cocoon stifling

The main purpose of stifling is to kill the pupa inside the cocoon so as to avoid its emergence as a moth and thereby preserving the continuity of the filament. In addition, this operation enables the cocoons to get dried so that the cocoon can be stored for a long period. Cocoon stifling is generally done by the following methods such as sun drying, steam stifling, hot air drying. Other methods of killing the pupa include the use of infrared rays, cold air killing, use of poisonous gases, etc.

11.3.3.3 Cocoon cooking

The object of cocoon cooking is to soften the sericin so that the cocoon shell is loosened enabling the filament unwind smoothly during reeling. Different methods of cooking are adapted such as open pan cooking, three-pan cooking, pressurized cooking, conveyor cooking, etc.

11.3.4 Types of silk reeling machines

11.3.4.1 Country charka

This is a traditional type of reeling machine with crude way of unwinding the filaments from the cocoons. The silk produced from this machine mainly is utilized in the handloom sectors wherein the cost factor of the raw material has to be kept low to enable the rate of the finished product saleable. In addition, the establishment of charka unit does not need any big investment or special skill. This enables charka silk to be sold at a minimum price which is very much wanted by the handloom industry. Also, inferior quality multivoltine and defective cocoons available at lower prices can be reeled more economically on charka than on cottage basin or multi-end basins.

Charka silk is generally coarse and suffers from many defects since no improved devices such as button/slub catcher and standard croissure system in reeling are used. Charka silk is not re-reeled. Production of silk per charka per day is about a kilogram with coarse denier.

11.3.4.2 Cottage basin

This reeling device is an improved version over charka and it is designed on the principle of Japanese multi-end reeling machine. Here, cocoon cooking is done separately in a boiling water basin and reeling is done in a hot water basin. Each basin has six ends and each thread is first passed through a button to clean the slubs, waste, etc. Later on, it is independently passed through a travellette-type croissure, which is more efficient than that in charka. After the croissure, the thread passes through a traverse guide and finally on to a small reel.

Since re-reeling is done to prepare standard sized hanks from small reels, the quality of silk produced on this machine is superior to charka silk. Production of silk per day per basin is about 800 g. Superior quality cocoons like bivoltine can be reeled on this device.

11.3.4.3 Multi-end reeling machine

This reeling device is a further improved version over the cottage basin and it is power driven. In multi-end basins, there are some additional attachments such as Jetteboute which picks up the filaments to increase the efficiency of feeding. The distribution system is further improved and individual brake motion for each reel is also provided so that the overall working efficiency of the basin is enhanced. Normally each basin consists of 10 ends.

In principle, the multi-end reeling machine is supposed to be a modern reeling device and it is possible to use superior quality cocoons like bivoltines on these machines with better performance. Raw silk production per day per basin is varying from 600 to 800 g in the above machines and quality of silk is superior compared to that of cottage basin/domestic basin.

11.3.4.4 Automatic reeling machine

The automatic reeling technique has as its components pressurized cocoon boiling machine, automatic cocoon feeder and a mechanical brushing unit. In addition, it is equipped with automatic denier control device, wherein human error in maintaining the uniformity of denier is avoided to a large extent and thereby ensuring minimum size deviation. Automatic reeling is the improved version over the multi-end reeling method designed and extensively used in Japan for reeling univoltine and bivoltine cocoons. It is a Japanese version or sunken system, which is suitable for better quality cocoons in producing quality raw silk.

11.4 Structure and properties of silk from silkworms

11.4.1 Introduction

Silks belong to a group of high molecular weight organic polymers characterized by repetitive hydrophobic and hydrophilic peptide sequences (Altman et al., 2003). Silks are fibrous protein polymers that are spun into fibres by some arthropods such as silkworms, spiders, scorpions, mites and fleas (Altman et al., 2003, Craig, 1997). There are thousands of silk-spinning insects and spiders, yet only a few have been investigated in detail. Silks differ in composition, structure and properties depending on their specific source and function (Altman et al., 2003; Craig et al., 1999; Sheu et al., 2004). They are naturally produced by spiders or insects, such as *Nephila clavipes* and *Bombyx mori*, respectively (Becker et al., 2003, Bell et al., 2002). Silkworm fibres are classified as domestic silk and wild silk. Wild silks are produced by caterpillars other than the mulberry silkworm and they differ from the domesticated varieties in colour, size and texture. The cocoons gathered in the wild usually have been damaged by the emerging moth before the cocoons are gathered, so the silk thread that makes up the cocoon has been torn into shorter lengths. Domestic silkworm like '*Bombyx mori*' are commercially reared and the pupae are killed by dipping them in boiling water before the adult moths emerge allowing the whole cocoon to be unravelled as one continuous thread. There are other commercially exploited silk-worms other than '*Bombyx mori*' which can be used to rear the silk which are known as non-mulberry silkworms like *Antheraea mylitta* (tasar), *Philosamia ricini* (eri) and *Antheraea assama* (muga). Silks are also produced by spiders and insects which secret glycine-rich silks characterized by their unique synthesis and processing features, as well as strength and extensibility.

11.4.2 Composition of silk

Silk in contrast to all the other natural fibres does not have a cellular structure. In this respect and in the way it is formed, it closely resembles man-made fibre. The double thread or bave separated from the mulberry silkworm cocoon consists of two filaments or brins of fibroin enveloped and bound together by silk glue, sericin. Morphologically silk is a very simple fibre. It consists of two single compact, endless threads that are extruded by the silkworm as it spins it's cocoon, which are surrounded and covered by the silk gum or sericin as shown in Fig. 11.6. The denier of the filament varies within the cocoon.

Besides fibroin and sericin (which are proteins), raw silk contains small quantities of either or alcohol soluble organic matter. Silk which is naturally coloured yellow, green, etc. contains a small amount of colouring matter. Some ash remains after

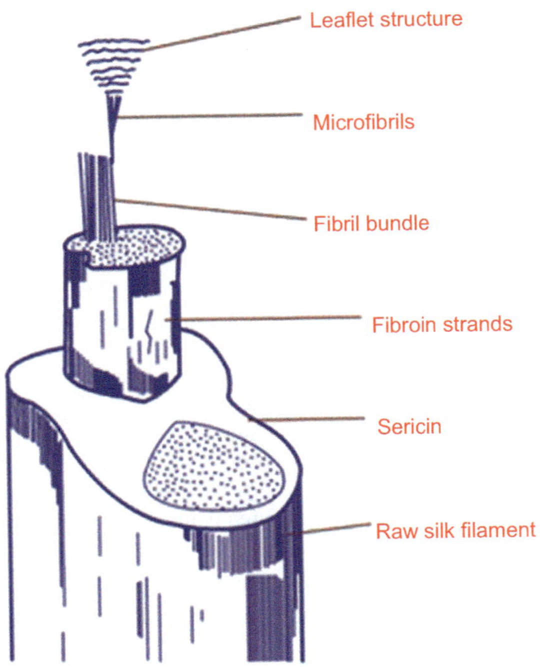

Figure 11.6 Composition of silk.

burning silk. The content of all these substances is not constant and varies within wide limits depending on the species of the silkworm and on the place and conditions of rearing.

The silk filament contains 72%–81% fibroin, 19%–28% sericin, 0.8%–1.0% fat and wax and 1.0%–1.4% colouring matter and ash of the total weight. The silkworm extrudes the liquid fibre from the two excretory canals of sericteries which unite in the spinneret in its head, each of them termed as brin. The two brins cemented together in the spinneret by sericin become a single continuous fibre called the bave or filament. The seric bave is thus made by the union of two brins held together by sericin (Carboni, 1952). Fibroin is a valuable protein along with sericin, sericin acts as a glue to fix fibroin fibre together in a cocoon. Sericin and fibroin protein are useful because of its properties and has been found to possess various biological functions.

11.4.3 Microstructure and appearance

Silk fibres (*Bombyx mori*) spun out from silkworm cocoons consists of fibroin in the inner layer and sericin in the outer layer. Each raw silk thread has a lengthwise striation, consisting of two fibroin filaments of 10–14 lm each embedded in sericin. The chemical compositions are, in general, silk fibroin of 75%–83%, sericin of 17%–25%, waxes of about 1.5%, and others of about 1.0% by weight. Silk fibres are biodegradable and highly crystalline with well-aligned structure. It has been known that they also have higher tensile strength than glass fibre or synthetic organic fibre, good elasticity, and excellent resilience. Silk fibre is normally stable up to 140°C and the thermal decomposition temperature is greater than 1500°C. The densities of silk fibres are in the range of 1320–1400 kg/m^3 with sericin and 1300–1380 kg/m^3 without sericin. Silk fibres are also commercially available in a continuous fibre type.

11.4.4 Longitudinal view

Scanning electron micrographs of longitudinal views of undegummed and degummed silk fibres are presented in Fig. 11.7a, respectively.

It may be observed that, mulberry shows a more or less smooth surface (Fig. 11.7a) whereas the non-mulberry silks such as Tasar, muga, and eri (Fig. 11.7b–d) all have striations on their surface compared to mulberry.

11.4.5 Cross-sectional view

The scanning electron micrographs of cross-section of silk fibres are presented in Fig. 11.8. The cross-section of silk fibre which is made of two types of protein namely, sericin and fibroin is shown in figure. It is found that two strands of fibroin filaments are enveloped by non-fibrous sericin. When a strand of fibroin filament is enlarged for its inner structure, it appears like a bundle of fibrils in which a large number of fibrils are accumulated (Minagawa, 2000).

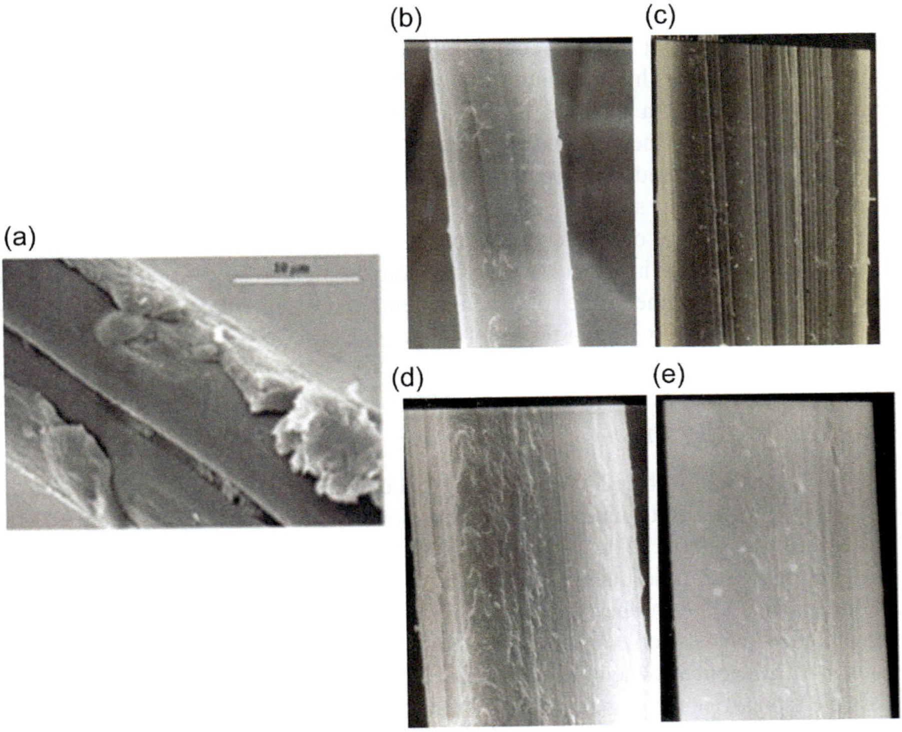

Figure 11.7 (a) The longitudinal view of silk fibres (degummed) (b) Mulberry (c) Tasar (d) Muga (e) Eri.

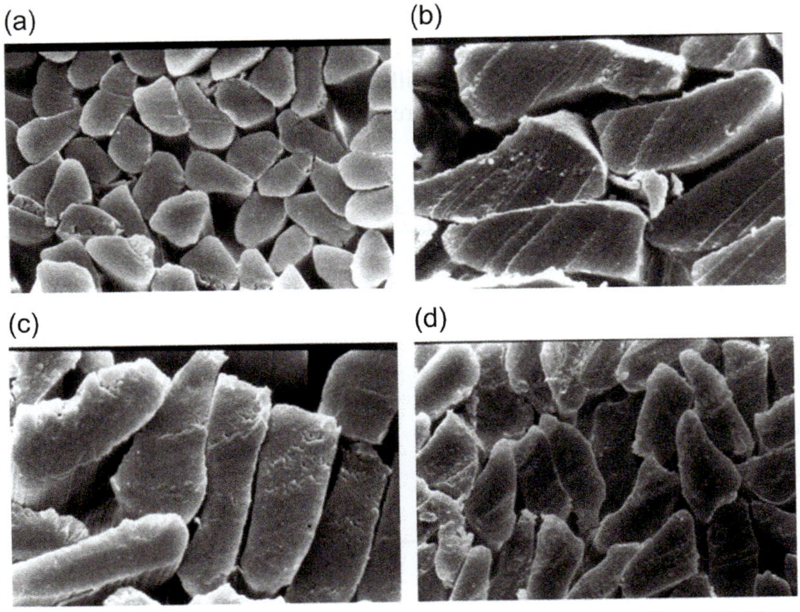

Figure 11.8 The cross-sectional view of silk fibres (degummed) (a) Mulberry (b) Tasar (c) Muga (d) Eri.

There are variations depending upon the variety of silkworms and also among the individual cocoons. It may be observed that, in this respect, the mulberry and non-mulberry silks exhibit an altogether different cross-sectional morphology. The mulberry silks show a more or less triangular cross-section and a smooth surface (Fig. 11.8a). Among the non-mulberry varieties, tasar and muga exhibit an elongated rectangular or a wedge-shape cross-section and a large cross-sectional area (Fig. 11.8b,c). The eri silk has a more or less triangular shape (Fig. 11.8d).

Usually in the case of the cocoon fibres of domestic silkworms like mulberry and eri, the cross-section is irregular ranging from triangular shape to circular shape. Moreover, even in the same fibroin filament, there are variations in the cross-sectional area depending upon the level of the cocoon layer.

11.4.6 Density of silk

Sen and Murugesh Babu (2004) have reported density values of different varieties of silks (Table 11.1). Both mulberry varieties show higher density values compared to those of non-mulberry silks. The higher density values of mulberry silks also indicate a higher degree of order and compact molecular packing compared to those of non-mulberry silks. The mulberry (bivoltine) variety shows higher density than that of the mulberry (crossbreed) variety. It is interesting to note an increasing trend in density values from the outer to the inner layers for all the varieties, which definitely suggests a possible increase in the degree of crystallinity and crystallite orientation as one moves from the outer to the innermost layers. Among the three non-mulberry silks, muga exhibits the highest density, followed by tasar and eri.

11.4.7 Moisture regain

Moisture regain of different varieties of silk has also been reported (Sen and Murugesh Babu, 2004). The results of the moisture regain of different varieties of silk fibres

Table 11.1 Density ρ Values for Different Varieties of Silk.

	ρ (g/cm^3)		
Type of silk	**Outer layer**	**Middle layer**	**Inner layer**
Mulberry (bivoltine)	1.350	1.361	1.365
Mulberry (crossbreed)	1.342	1.35	1.356
Tasar	1.300	1.33	1.340
Muga	1.332	1.34	1.348
Eri	1.28	1.29	1.295

Table 11.2 Moisture Regain (%) of silk fibres.

Type of silk	Moisture regain (%)		
	Outer layer	**Inner layer**	**% Change**
Mulberry (bivoltine)	8.52	8.14	4.46
Mulberry (crossbreed)	8.63	8.28	4.05
Tasar	10.76	10.27	4.55
Muga	9.82	9.47	3.56
Eri	10.21	9.79	4.11

determined under standard conditions are presented in Table 11.2. All three non-mulberry silk fibres show higher moisture regain values compared to those of the mulberry varieties. Among these, tasar shows the highest value (10.76%), followed by eri (10.21%) and muga (9.82%) for the outer layers. On the other hand, mulberry bivoltine and crossbreed varieties show lower values of 8.52% and 8.63%, respectively. The higher moisture regain of non-mulberry silks suggests that all three non-mulberry silk varieties may consist of a higher ratio of hydrophilic to hydrophobic amino acid residues in their chemical architecture compared to that of the mulberry varieties. Interestingly, moisture regain of the inner layers is about 4.0%–4.5% less compared to that of the outer layers, suggesting compactness of the inner layers.

11.4.8 Amino acid composition

The amino acid composition varies in different varieties of silk. Three major amino acids such as serine, glycine, and alanine may be found in mulberry and non-mulberry varieties. Among the other major amino acids present are tyrosine and valine. In general in mulberry silks glycine, alanine, and serine together constitute about 82%, of which about 10% is serine. Tyrosine and valine may be considered next to these at about 5.5% and 2.5%, respectively. The overall composition of acidic amino groups (i.e. aspartic and glutamic acids) in the mulberry variety is greater than that of the basic amino acids. The other important aspect is the composition of amino acids with bulkier side groups. The presence of bulky side groups can hamper close packing of molecules and hinder crystallization process. In general, a large portion of the mulberry fibroin is made up of simple amino acids such as glycine and alanine, suggesting a favourable condition for crystallization (Sen and Murugesh Babu, 2004)

Compared to the mulberry silks, the total amount of glycine, alanine, and serine constitute about 73% in the non-mulberry variety, less by about 10%. All the non-mulberry silks exhibit a high proportion of alanine compared to that in the mulberry variety. The proportion of alanine is about 34% in tasar, 36% in eri, and 35% in muga. This value is consistent but is lower than that particularly for muga (~44%).

On the other hand, the glycine content in these varieties is about 27%–29%, which is lower than that found in the mulberry varieties (~43%).

In addition, the non-mulberry varieties have a substantial proportion of amino acids with bulky side groups, especially aspartic acid (4%–6%) and arginine (4%–5%), which means that not only the acidic but also basic amino acid levels are greater. It is interesting to note the presence of sulphur-containing amino acids (i.e. cystine and methionine) in all the varieties of silk. Methionine content in non-mulberry silks is slightly higher (0.28%–0.34%) compared to that found in mulberry varieties (0.11%–0.19%), whereas the cystine content is comparable (Sen and Murugesh Babu, 2004). Amino acid composition of different varieties of silk is presented in Table 11.3.

Table 11.3 Amino acid Composition of Silk Fibres.

	Amino acid composition (mol %)			
Amino acid	***Bombyx mori* (mulberry)**	***Antheraea mylitta* (Tasar)**	***Antheraea assama* (Muga)**	***Phylisomia ricini* (Eri)**
Aspartic acid	1.64	6.12	4.97	3.89
Glutamic acid	1.77	1.27	1.36	1.31
Serine	10.38	9.87	6.11	8.89
Glycine	43.45	27.65	28.41	29.35
Hystidine	0.13	0.78	0.72	0.75
Arginine	1.13	4.99	4.72	4.12
Threonine	0.92	0.26	0.21	0.18
Alanine	27.56	34.12	34.72	36.33
Proline	0.79	2.21	2.18	2.07
Tyrosine	5.58	6.82	5.12	5.84
Valine	2.37	1.72	1.5	1.32
Metheonine	0.19	0.28	0.32	0.34
Cystine	0.13	0.15	0.12	0.11
Isoleusine	0.75	0.61	0.51	0.45
Leusine	0.73	0.78	0.71	0.69
Phenylalanine	0.14	0.34	0.28	0.23
Tryptophan	0.73	1.26	2.18	1.68
Lysine	0.23	0.17	0.24	0.23

11.4.9 X-ray diffraction studies

Silk fibroin, fibrous protein of silk cocoon, has been studied extensively by a number of investigators from various points of view. Many researchers have reported extensive studies on X-ray diffraction of different varieties of silk.

11.4.10 Crystalline structure

The silkworm cocoon and spider dragline silks are characterized as an antiparallel β-pleated sheet wherein the polymer chain axis is parallel to the fibre axis. Other silks are known to form α-helical (bees, wasps, ants) or cross-β-sheet (many insects) structures. The cross β-sheets are characterized by a polymer chain axis perpendicular to the fibre axis and a higher serine content. Most silks assume a range of different secondary structures during processing from soluble protein in the glands to insoluble spun fibres (Kaplan, 2004). The crystalline structure of silk was first described in the 1950s as an antiparallel, hydrogen-bonded β-sheet based on the characterization of *B. mori* fibroin (Marsh et al., 1955), and further modifications to this early model have been made over the years (Fraser and MacRae, 1973, Colonna-Cesari et al., 1975).

Most silkworm cocoon and spider dragline silk fibres contain assembled antiparallel β-pleated sheet crystalline structures (Marsh et al., 1955; Lucas et al., 1960; Fraser and MacRae, 1973). Silks are considered semi-crystalline materials with 30%–50% crystallinity in spider silks, 62%–65% in cocoon silk fibroin from the silkworm *B. mori,* and 50%–63% in wild-type silkworm cocoons. In the β-sheet crystals, the polymer chain axis is parallel to the fibre axis. Furthermore, the polyalanine repeats or the glycine–alanine repeats are the major primary structure sequences responsible for b-sheet formation. The β-sheets consisting of the glycine–alanine crystalline repeats in the silkworm fibre are asymmetric, with one surface primarily projecting alanyl methyl groups and the other surface of the same sheet containing hydrogen atoms from the glycine residues (Akira et al., 2006).

Molecules of silk fibroin are the polypeptide chain with formula (–CHR–CO–NH–)n with n = 1100 and R being one of the various amino acid residues. The polypeptide chain has not only a backbone but also side chains consisting of radicals R of amino acids. Naturally, the degree of branching of the polypeptide chain depends on the amino acids contained in the protein. Thus side chains form 19% of the weight in silk fibroin.

The side chains may be non-polar, as for instance, hydrocarbon residues

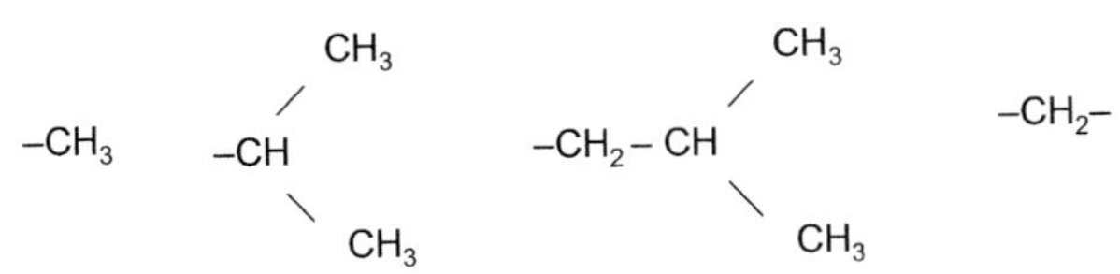

and may contain polar groups such as

$$-COOH,\ -CH_2OH,\ -C_6H_4OH,\ -SH,\ -S{-}S-,\ NH,\quad NH_2,\ -CO{-}NH_2$$

$$-NH{-}\underset{\underset{NH}{\|}}{C}{-}NH_2,$$

11.4.10.1 Crystallinity

Silk as mentioned earlier is a semicrystalline material. Earlier studies by X-ray diffraction analysis indicated 62%–65% crystallinity in cocoon silk fibroin from the silkworm, 50%–63% in wild-type silkworm cocoons, and lesser amounts in spider silk (Warwicker, 1956).

Physical parameters such as crystallinity, crystallite size and crystallite orientation have been determined from X-ray diffraction. In a study on mulberry, tasar, muga and eri (Bhat and Nadiger, 1980), it was observed that the mulberry and non-mulberry silks exhibit different WAX diffraction patterns (Fig. 11.9). Mulberry shows a broad 28 peak at 20 degrees corresponding to (201) reflection. On the other hand, all the non-mulberry silks exhibit similar X-ray diffraction patterns showing two major peaks at 28 equal to 17.1 and 20.2 degrees corresponding to (002) and (201) planes. Using X-ray diffraction Warwicker (1960), observed that the different silk fibroins could be classified into five groups. Mulberry *(Bombyx mori)* has been placed in group I and tasar, muga and eri (Antheraea silks) in group 3a. Group 3a includes most Antheraea species, together with other silks belonging to the family Saturniidae. The anti-parallel p-sheet structure is the molecular conformation assumed by the

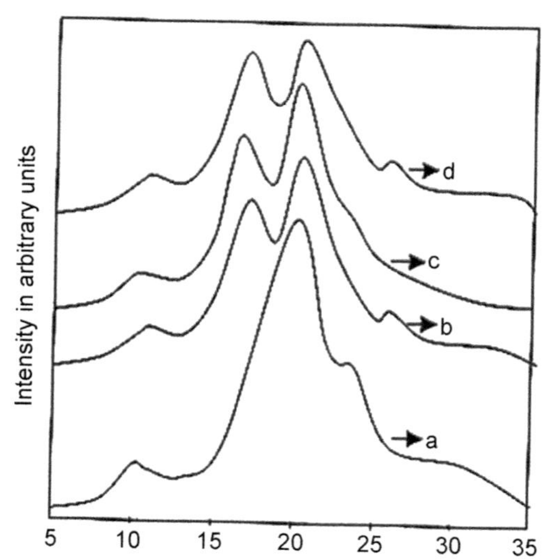

Figure 11.9 WAXD patterns of (a) Mulberry (b) Tasar (c) Muga (d) Bhat and Nadiger (1980).

fibroin chains in the crystalline regions, with unit-cell dimensions similar to that of $-(ala)_n-$ in β-form.

The density and the degree of crystallinity values for male and female varieties as obtained by X-ray diffraction analysis for mulberry, tasar, and mug a have been reported in Table 11.4. The data have been compiled from the work of Iizuka et al. (1996, 1994, 1993a,b,c, and 1997). No definite trend may be observed between the male and female varieties. It may be observed from the data that no significant differences exist between the mulberry and the non-mulberry varieties. However, the data obtained for the temperate tasar, muga and mulberry (extrafine) may be compared with those for unhydrolyzed samples obtained by Bhat and Nadiger (1980). No literature is available which gives any idea about any differences in crystallinity or lateral order along the length of the silk filament. All the values seem to be the average values. Considering the fact that some workers have reported gradual reduction of denier, elongation-at-break, degumming loss, etc. in the inner layers of cocoons, it is reasonable to expect the change in the above parameters along the length of the filament.

11.4.10.2 Crystallite size

Bhat and Nadiger (1980) investigated the crystallite size of hydrolysates (silk treated with HCl for different duration) of Indian silk varieties by wide angle X-ray diffraction (Table 11.5). A quick glance at the results reveals that the non-mulberry silk varieties seem to have larger crystals as compared to mulberry silk. The higher alanine content as also the predominance of Ala—Ala links in crystalline region may be responsible for higher size. From their results, they concluded that after the HCl treatment, the crystallite size calculated from 002 plane increased from 27 to 60 Å for tasar, 30 to 56 Å for eri and 32 to 59 Å for muga. Interestingly, the crystallite size for mulberry control

Table 11.4 Density and Crystallinity values for various silk fibres.

Variety	Sex	Density (g/cm^3)	Percentage crystallinity (X-ray)
A.Mylitta	M	1.329	39.5
(Tropical tasar)	F	1.334	39.6
A.Proylei	M	1.345	44.6
(Temp tasar)	F	1.344	41.8
A.Assama	M	1.314	42.7
(Muga)	F	1.327	43.7
Ariake	M	1.354	37.2
(Mulberry extracoarse)	F	1.356	35.2
Shunreix shongetsu	M	1.333	43.1
(Mulberry extrafine)	F	1.320	41.9

Table 11.5 Average crystallite size for silk and their hydrolysates.

Variety of silk	Planes	Control sample (Å)	48-hr hydrofibroin (Å)	(48 + 48) hr Hydrofibroin (Å)
Mulberry	002	10	15	20
	201	19	20	25
Tasar	002	27	68	60
	201	47	56	52
Eri	002	30	56	56
	201	47	47	60
Muga	002	32	47	59
	201	39	43	47

fibres was 10 Å from 002 plane and the size increased to 20 Å after 96 h treatment. Further, there was no significant increase in the crystallite size of mulberry silk calculated from 002 plane, whereas all the three non-mulberry silks showed a considerable increase in their crystallite size for 002 plane after the acid treatment. Similar trends were seen from the data obtained for 201 plane too.

The authors opined that, in the case of wild silks, the process of acid hydrolysis leads to the Ala-Ala segments aligning parallel to the existing microcrystal and thus causing the crystal to grow. However, the higher crystallite size that may also be related to the crystal geometry of non-mulberry silks compared to mulberry silks and needed to be further investigated.

11.4.10.3 Orientation (birefringence and sonic modulus)

Birefringence of different varieties of silk fibres has been reported by Sen and Murugesh Babu (2004). The Becke line method was used for determination of the birefringence. A series of mixtures using two liquids [i.e. liquid paraffin (refractive index, $\eta = 1.465$) and 1-chloronaphthalene ($\eta = 1.633$)] were prepared for determination of the refractive index of single filaments in both the parallel and the perpendicular directions. The refractive index of the mixture, in which the Becke line vanished under a polarizing microscope, was taken as the refractive index of fibre. The birefringence (Δ_n) was calculated using the relation:

$$\Delta_{\mathrm{n}} = \eta \parallel - \eta \perp$$

where $\eta\parallel$ is the refractive index of the fibre in the parallel direction and $\eta \perp$ is the refractive index of the fibre in the perpendicular direction to the plane of the polarizer light. An average of five readings was reported.

The birefringence (Δ_n) value gives an idea about the overall molecular orientation, whereas the sonic modulus (S_ε) gives a combined effect of order and orientation. The values of Δ_n and S_ε are listed in Table 11.6. Both birefringence and sonic modulus were found to increase from the outer to the inner layers within the same variety. This trend was observed in all varieties.

11.4.11 IR-spectroscopy

The IR-studies on silk fibroin (Bhat and Ahirrao, 1983; Bhat and Ahirrao, 1985; Nadiger and Bhat, 1985; Baruah et al., 1991; Magoshi et al., 1979) suggest that the appearance of absorption bands at 1660, 1540, 1235 and 650 cm^{-1} corresponds to amide one, amide two amide three and amide four, respectively, characteristic of the random coil conformation (amorphous) and those appearing at 1630, 1535, 1265 and 700 cm^{-1} indicate the p-conformation (crystalline).

The calculation of crystallinity index from IR-spectroscopy has been done by the previous workers by using the ratios of intensity bands corresponding to crystalline region to that of amorphous region. Drukker et al. (1953) have reported crystallinity value of 63% for *Bombyx mori* silk using bands at 1528 cm^{-1} and 1560 cm^{-1}. However, the authors emphasized that the value obtained was not exact, as the resolution of these bands was not exceptionally good. This value certainly appears high compared to those obtained from WAXD. Similar studies on mulberry and wild varieties of Indian silk were done (Bhat and Nadiger, 1980) using the ratio of absorbance of bands at 1265 cm^{-1} (β-form) and 1235 cm^{-1} [α- (random coil conformation) form]. It may be observed from their data (Table 11.7) that, mulberry and muga silks showed higher crystallinity indices (0.66 and 0.60, respectively) compared to tasar (0.50) and eri (0.50).

Table 11.6 Birefringence (Δ_n) and sonic modulus (S_ε) values for different varieties of silks.

	Δ_n			S_ε (g/d)		
Type of silk	**Outer layer**	**Middle layer**	**Inner layer**	**Outer layer**	**Middle layer**	**Inner layer**
Mulberry (bivoltine)	0.054	0.055	0.056	165	170	206
Mulberry (crossbreed)	0.051	0.052	0.052	153	167	192
Tasar	0.041	0.042	0.042	106	107	115
Muga	0.040	0.041	0.042	103	108	114
Eri	0.034	0.035	0.035	101	106	109

Table 11.7 Crystallinity indices for various silk fibres.

Variety	X-ray index		IR-crystallinity index		Electron diffraction index	
	Control	Hydrolyzed (96 h)	Control	Hydrolyzed (96 h)	Control	Hydrolyzed (96 h)
Mulberry	0.42	0.48	0.66	0.74	0.62	–
Tasar	0.43	0.77	0.50	0.66	0.60	–
Muga	0.44	0.72	0.60	0.64	0.62	–
Eri	0.43	0.65	0.50	0.82	0.63	–

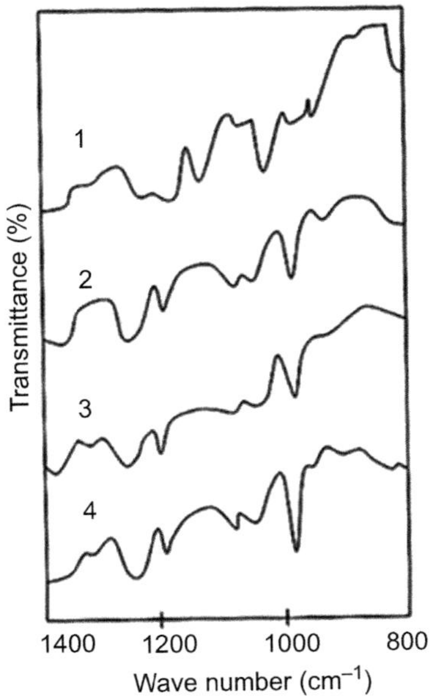

Figure 11.10 Infrared Spectra of different varieties of silk (a) Mulberry (b) Tasar (c) Muga (d) Eri.

In a study on crystallinity of silk hydrolysates, Bhat et al. (1980) reported the IR-spectra of different varieties of silk in the range of 1400–800 cm^{-1} (Fig. 11.10). They assigned the band at 1015 cm^{-1} to Gly–Gly linkage, the band at 970 cm^{-1} to Ala–Ala linkage and those at 998 and 975 cm^{-1} to Ala–Gly linkages. In mulberry, the bands at 1015 cm^{-1} and 970 cm^{-1} were totally absent while the bands at 975 and 998 cm^{-1} were present. The non-mulberry silks showed a strong band at 970 cm^{-1}. The intensity of these bands at respective wave numbers further enhanced in hydrolyzed silk samples of both the mulberry and the non-mulberry varieties. The band at 1015 cm^{-1}, which was corresponding to Gly–Gly linkage, totally disappeared in the hydrolyzed samples of wild varieties. On the basis of the above results, they concluded that, the crystalline structure of mulberry must have been formed mostly by the sequence of glycine and alanine linkages and that of wild silks by non-labile Ala–Ala linkages.

11.4.12 Properties of silk

11.4.12.1 Tensile properties

The tensile properties of different varieties of silks in terms of tenacity, elongation-at-break and initial modulus have been determined by a number of workers (Tsukada and Freddi, 1994; Iizuka et al., 1994, 1993; Iijuka and Itoh, 1997).

Studies conducted on some mulberry and non-mulberry varieties by E. Iizuka et al. reveal that the tenacity, elongation and modulus are all dependent on the linear density of the filament, and the linear density or the mean size in turn depends on the silkworm race. The tenacity is found to be linearly related to the linear density of the filament (Fig. 11.11). The correlation is negative, i.e. as the linear density increases, tenacity decreases. Similar trend has been observed for modulus too. Elongation, on the other hand, increases with an increase in linear density.

The tenacity ranges between 2.5 and 4.82 g/d, for Japanese and Chinese mulberry varieties, 2.4–4.32 g/d for Indian mulberry varieties, 3.74–4.6 g/d for Indian tasar varieties (Iizuka, 1995). In a study on chemical structure and physical properties of *Antheraea assama* (muga) silk, it has been reported that, the tenacity of muga varies between 3.2 and 4.95 g/d (Tsukada and Freddi, 1994; Iizuka et al., 1993). Another important non-mulberry variety, eri, showed lowest tenacity value, ranging between 2.3 and 4.0 g/d (Iijuka and Itoh, 1997).

Elongation-at-break on the other hand, showed a higher value for all the non-mulberry silks compared to mulberry varieties. The values range between 31% and 35% for tasar, 34%–35% for muga and 29%–34% for eri silks, respectively. The elongation values for mulberry varieties ranged between 19% and 24%. Some of the mechanical properties of different varieties of silk are summarized in Table 11.8.

11.4.12.2 Optical properties

Silk fibroin extracted from silkworm cocoons is a unique biopolymer that combines biocompatibility, implantability, along with excellent optical properties. Silk may be used as an optical material for applications in biomedical engineering, photonics and nanophotonics. Silk can be nanopatterned with features smaller than 20 nm. This allows manufacturing of structures such as (among others) holographic gratings,

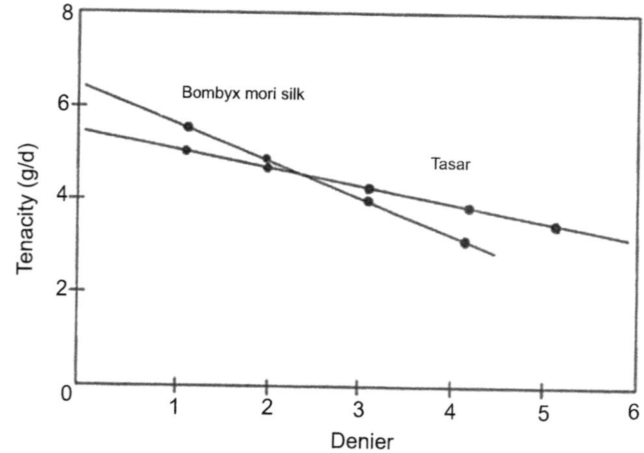

Figure 11.11 Tenacity vs denier of thread relationship.

Table 11.8 Mechanical properties of different varieties of silk.

Variety	Sex	Dynamic modulus (10^{10} dyn/cm^2)	Tan δ	Tenacity (g/d)	Elongation (%)
Shunreix shougetsu	M	1.847	–	5.265	20.36
(Mulberry)	F	1.808	–	5.207	21.48
A.Mylitta	M	1.132	0.030	3.412	31.36
(Tropical tasar)	F	1.087	0.035	3.256	31.12
A.Proylei	M	1.305	0.023	4.123	31.45
(Temp.tasar)	F	1.087	0.025	4.128	31.48
A.Assama	M	1.205	0.020	3.170	34.83
(muga)	F	1.230	0.023	3.823	34.10

phase masks, beam diffusers and photonic crystals out of a pure protein film. The properties of silk allow these devices to be 'biologically activated' offering new opportunities for sensing and biophotonic components.

Many interesting bio-optical devices can be fabricated by doping silk films with fluorescent materials (such as quantum dots shown above). Ways can be explored to enhance the light emission by patterning the silk film surfaces, as well as making tuneable wavelength devices and printing specific patterns on silk film surfaces.

Lustre associated with silk is due partly to the influence on the pattern of light — reflection of its triangular shape. In an attempt to understand the optical properties of silk many researchers have determined the refractive index and birefringence of fibres. The refractive index of silk generally varies through its cross-section. The birefringence (n) value for vary between 0.051 and 0.0539 mulberry silk and 0.030 to 0.034 for non-mulberry silks (Tsukada et al., 1994)

11.4.12.3 Viscoelastic behaviour

Silk fibre exhibits viscoelastic behaviour. Time dependent mechanical properties of silk fibres such as stress relaxation, creep, creep recovery, etc have also been the subject of interest. Creep is a phenomenon associated with time dependent extension under an applied load. The complimentary effect is stress relaxation under a constant extension.

Creep and stress relaxation behaviour of silk has been reported (Das, 1996). The instantaneous extension and secondary creep are both higher for tasar silk compared to those for mulberry silk. The stress relaxation was also found to be more in non-mulberry silks than in mulberry silk.

Silk has also been shown to exhibit inverse stress relaxation phenomenon (Das, 1996). The inverse relaxation could be observed for both mulberry and tasar silks when the level of strain was maintained below a certain value. Inverse relaxation becomes higher with the increase in peak tension. Cyclic loading has been found to reduce the extent of inverse relaxation.

11.4.3 Applications of silk

Silk is one of the most beautiful fabrics available, with a long and colourful history and changing applications in the world today. Be it for gowns, medical use, home decor and more, the uses of silk is a wide and varied topic.

11.4.3.1 Textile and apparels

Silk's good absorbency makes it comfortable to wear in warm weather and while active. Its low conductivity keeps warm air close to the skin during cold weather. It is often used for clothing such as shirts, blouses, formal dresses, high fashion clothes, negligees, pyjamas, robes, skirt suits, sun dresses and kimonos. Silk's elegant, soft lustre and beautiful drape makes it perfect for many furnishing applications. It is used for upholstery, wall coverings, window treatments (if blended with another fibre), rugs, bedding and wall hangings.

Silk has several uses in daily life. It is a common cloth used for high-end garments, including wedding gowns and blouses. It appears frequently in accessories like handbags and headbands and scarves. Because of its lustre and texture, it also is commonly used for home decoration, especially as sheets, bedding, curtains and cushions. It is also used for warmth and skiing garments because of its remarkable ability to retain heat; paradoxically, it also is a popular summer cloth because it keeps the wearer cool in warmer weather.

Silk continues to be used as a material to produce fine dresses. Chinese women continue to have their wedding cheongsam dresses made out of silk. They choose this material because it is one of the finest materials known in ancient Chinese culture. Delicately woven dragons, flowers and butterflies are sewn into the silk dresses. The material is thick so it allows a woman to look leaner and have cleaner lines. Additionally, the shininess of the material is very flattering and alluring. Women's evening gowns are also made of silk. It also falls well and is a material that is slightly warmer allowing the women's body to feel a bit of warmth, even as she wears a sleeveless gown in the winter.

Silk shirts remain popular amongst women. Especially for women's blouses, silk is a great material. It is seen as more classic and flattering since it drapes a woman's body. Additionally, the thinness of the material allows women to easily tuck the shirt into their skirts or wear a jacket on top without having the jacket look bulky. For men, silk shirts are becoming more and more popular. Men have started wearing silk short sleeve shirts as well as long sleeve shirts. Men used to only wear silk shirts for the evening, but the lightness of the material has made it a popular alternative to cotton.

Silk is also used very frequently for pajamas. Silk pajamas feel soft and smooth and very comfortable while sleeping. Both men's and women's pajamas are now made of this fine thread. For women, silk is often used to make lingerie as well. Using this fine thread has become a great option since it is smooth, tends not to catch on your hair, and is very breathable for the body.

Silk scarves have also been around for a long time. The number of gorgeous patterns and designs on the scarves has made these items very expensive. Women tend to tie the scarf around their necks for warmth and for style. Men tend to carry one in their pocket as an alternative to the cotton handkerchief. Be prepared to pay more for a silk scarf with a fancy design.

More and more, manufacturers of bedding have started to make silk sheets and silk pillowcases. The health benefits of silk are starting to be more widely known. Sleeping on them helps prevent coughing and sneezing, especially for those allergic to dust mites which do not like silk. Additionally, sleeping on silk sheets is sometimes beneficial to women's hair, helping it to have less tangles and less split ends. Beyond the solid colour sheets, many makers of luxurious bedding are now offering their own version of sheets with lots of different patterns and designs in stores.

Silk has come a long way since the Silk Road. It is still a highly sought after material. Having pieces of fine clothing or a scarf or even a silk sheet is often seen as the ultimate extravagance.

11.4.3.2 Bio-medical field

Silk, and especially *Bombyx mori* silk, has a very long history in biomedical applications. In recent years, the reported exceptional nature of silk lead to increased interest in silk for biomedical applications. Silk fibroin has been increasingly studied for new biomedical applications due to the biocompatibility, slow degradability and remarkable mechanical properties of the material. In addition, the ability to now control molecular structure and morphology through versatile processability and surface modification options have expanded the utility for this protein in a range of biomaterial and tissue-engineering applications. Silk fibroin in various formats (films, fibres, nets, meshes, membranes, yarns, and sponges) has been shown to support stem cell adhesion, proliferation, and differentiation in vitro and promote tissue repair in vivo. In particular, stem cell-based tissue engineering using 3D silk fibroin scaffolds has expanded the use of silk-based biomaterials as promising scaffolds for engineering a range of skeletal tissues like bone, ligament, and cartilage, as well as connective tissues like skin. More recent studies with well-defined silkworm silk fibres and films suggest that the core silk fibroin fibres exhibit comparable biocompatibility in vitro and in vivo with other commonly used biomaterials such as polylactic acid and collagen.

Silk from the silkworm, *B. mori*, has been used as biomedical suture material for centuries. The unique mechanical properties of these fibres provide important clinical repair options. Silk has been used in native fibre form as sutures for wound ligation and became the most common natural suture, surpassing collagen over the past 100 years.

Silk sutures are used in ocular, neural, and cardiovascular surgeries, as well as a variety of other tissues in the body. Silk's knot strength, handling characteristics, and the ability to lie low to the tissue surface make it a popular suture in cardiovascular applications. The above features of silk fibroin have led to the recent emergence of silk-based biomaterials for a wide range of cell and tissue studies such as scaffolds, films, sponges, hydrogels, bone materials, cardiovascular devices, etc.

11.4.3.3 Fibre-reinforced composites

Silk is a fibre with remarkable mechanical properties. This unique characteristic of silk has led its use in fibre reinforced composites for various applications. Silk yarn is easily available as the waste product of textile industry, so the composite is cost effective and perfect utilization of waste product. Though the silk is extensively used as a valuable material for textile purposes, in recent years it is being widely considered as a reinforcing material for composites made from epoxy and other bio-degradable bio polymeric resins. The organization of the silk fibres can contribute significantly to the impact resistance by ensuring either or both a sufficient strength of the composite and a good deformability of the composite.

11.4.3.4 Silk non-wovens

Silk non-woven fabrics can be developed from silk reeling waste and hard waste generated during twisting and weaving on shuttle-less looms. In the process of conversion of cocoon to fabric, about 4000 MT of silk waste of different forms is being generated, annually (in India alone). At present, this waste is used for manufacture of spun silk yarn, noil yarn, throwster yarn and carpet yarn, besides hand spun yarn. This waste can be more effectively utilized for development of silk non-woven fabrics for diversified end uses. In addition, the hard waste generated to the extent of 300 MT in silk twisting and weaving by 100% export oriented units (EOUs) is not being used for any value added purpose except in manufacture of coarser yarn for carpets. The same may effectively be utilized for development of non-woven silk fabrics. The other wastes like pierced and cut cocoons may also be tried. The web formation by air laid method and bonding by chemical/needle punching may be attempted *for production of no-woven webs.*

Based on the end-uses, non-woven fabric of specific weight can be produced for various applications including technical and medical textiles. The possible end-use applications are rather unlimited and non-woven silk fabrics could be a potential input in the areas of inner lining for warm garments, head wears, ties, garments and blankets, carpets, furnishing and home apparels, automotive carpeting and insulation, wall coverings, handicrafts like wall paintings, wall hangings, wall coverings, gift tags, purse/wallets, table mats, matting for pictures, journal/book/album covers, greeting/invitation/business cards, lamp shades and many more.

11.5 Future trends

With improved analytical techniques, together with the tools of biotechnology, a new generation of silk-related materials is envisioned that depart from traditional textiles and medical sutures, which are currently the focus for these materials. These proteins are already finding broadened applications in medical fields as biomaterials. The unique and remarkable mechanical properties of silks have already prompted studies for their application in high performance materials, as well as modes to mimic these features through more traditional synthetic organic polymer chemistry routes. Since the specifics of the silks can be modulated through genetic manipulation, through processing and through choice of starting protein, a great deal of control of structure, morphology, and functional attributes in materials derived from these proteins can be obtained. This control is already exploited in the biomaterials arena and suggests that this family of novel proteins can serve as an important blueprint for understanding structure–function relationships as well as for new and novel materials. A number of consumer products are already promulgated based on silks, including cosmetics, hair replacements and shampoos, among others. Sutures, biomaterials for tissue repairs, wound coatings, artificial tendons, bone repair, and related needs may be possible applications since silks are biocompatible and slowly degrade in the human body. Genetically engineered silks have also been commercialized as cell culture plate coatings to improve cell adhesion. Genetic variants of silks are also actively pursued as controlled release systems to deliver pharmaceuticals in a variety of systems.

New trends in production and applications of silk have to be deployed for maintaining a sustainable market for silk. As a result researchers in Japan have developed fluorescent silks. The world's first silks exhibiting fluorescence and other pioneering properties have been successfully developed as a result of transgenic silkworm research conducted by Japanese researchers. **Fluorescent Silk** thread is made by pulling threads from cocoons made by silkmoth larvae (silkworms). Researchers have now developed three lines of transgenic silkworms. The first line produces silk threads that emit green, red, or orange fluorescent light. These threads were created by introducing into silkworm eggs genes that promote the generation of fluorescent proteins. It has been possible to achieve green fluorescence using genes extracted from jellyfish, a technique developed by Nobel Prize-winning chemist Osamu Shimomura, and red and orange fluorescence with genes extracted from coral, a technique that has already been used in commercial applications. The fluorescent silk threads have great potential for use in the fashion industry, and there is expected to be considerable demand for them from producers of high-end apparel.

Though, the traditional practices make silk to construct only textiles, the new approach extend its application towards nutritional, cosmetic, pharmaceutical, biomaterial, biomedical and bioengineering, automobile, house building and art craft applications. The inclination rightly suits to silk due to faster production rate and increasing global demand for its variable eco-friendly composites and viable contributing impact on value, employment and environmental safety. This move, however, requires more of awareness among stakeholders; trainings and idea exchange between entrepreneurs, besides service accessibility to consumers. Increased applications of natural fibres for varied applications will create awareness on insect based natural fibre, silk for its continued use and innovative research in future.

11.6 Summary

Silk fibres are remarkable materials displaying unusual mechanical properties: strong, extensible, and mechanically compressible. Silks also display interesting thermal and electromagnetic responses, particularly in the UV range for insect entrapment and form crystalline phases related to processing. Silk fibres combine reasonable modulus (about 16 GPa for cocoon silk) with a very high strain to failure (order 20%), which means they can absorb a considerable amount of energy upon failure. Although substitution of natural materials has been underway for several decades, natural silks remain commercially important because of their unique properties, as well as their relative environmental stability and biocompatibility, and consumer preferences. The industries should look at the utilization of silk fibre in total for innovative marketable products of modern society's application and appreciation, which facilitate its all-round development. Silks have historically been used in medicine as sutures over the past 100 years and are currently used today in this mode along with a variety of consumer product applications. The novel mechanical and visual features of silk fibres from silkworms and spiders have driven interest in this family of structural protein fibres for centuries. The ability to manipulate silkworms for domesticated production of silk fibre, the opportunity to exploit spider silks via genetic engineering, and future options to mimic the novel features of this family of protein fibres using synthetic approaches, continues to drive strong interest in these protein fibres. With growing applicability of these fibres in biomedical and consumer product applications, this interest is likely to continue to expand.

Though, the traditionally, silk is used for making textiles, the new approaches must be attempted to extend the application of silk in nutritional, cosmetic, pharmaceutical, biomedical and bioengineering, automobile, house building and art craft products. During the past decade a great deal of progress has been made in understanding silk genetic and protein structures. On the other hand, materials scientists have long been fascinated by structure-function relationships in silk proteins. Since the exploration of biomaterial applications for silks, aside from sutures, is only a relatively recent advance, the future for this family of structural proteins to impact clinical needs appears promising.

11.7 Sources of further information and advice

Recommended text books are mentioned in the references. The periodicals such as *Indian Silk, Indian Journal of Sericulture, Sericolgia* can be referred for information on Indian sericulture, silk technology and other related topics. Occasional articles are found in *Journal of Applied Polymer Science, Journal of Textile Institute, Textile Research Journal* and *Indian Journal of Fibre and Textile Research.* Additional information on sericultural aspects of silk, silk fabric manufacture, dyeing, printing and processing aspects are available online on www.indiansilk.kar.nic.in, www.fao.org.

References

Akira, M., Kim, H.J., Tsai, I.Y., Wang, X., Cebe, P., Kaplan, D.L., 2006. Silk, Hand Book of Fibre Chemistry. Taylor & Francis Group, LLC.

Altman, G., Diaz, F., Jakuba, C., Calabro, T., Horan, R.L., Chen, J., Lu, H., Richmond, J., Kaplan, D.L., 2003. Silk-based biomaterials. Biomaterials 24, 401–416.

Altman, G.H., et al., 2003. Silk-based biomaterials. Biomaterials 24, 401–416.

Baruah, G.C., Talukdar, C., Bora, M.N., 1991. Indian J. Phys. 65B (6), 651–654.

Becker, N., Oroudjev, E., Mutz, S., Cleveland, J.P., Hansma, P.K., Hayashi, C.Y., et al., 2003. Molecular nanosprings in spider capture-silk threads. Nat. Mater. 2, 278–283.

Bell, F.I., McEwen, I.J., Viney, C., 2002. Fibre science: supercontraction stress in wet spider Dragline. Nature 416, 37.

Bhat, N.V., Ahirrao, S.M.,J., 1983. Appl. Polym. Sci 28, 1273–1280.

Bhat, N.V., Ahirrao, S.M., 1985. Text. Res. J. 55 (1), 65–71.

Bhat, N.V., Nadiger, G.S.,J., 1980. Appl. Polym. Sci. 25, 921–932.

Cao, Y., Wang, B., 2009. Biodegradation of silk biomaterials. Int. J. Mol. Sci. 10, 1514–1524.

Carboni, P., 1952. Silk, Biology, Chemistry and Technology. Chapman & Hall Ltd, London.

Chen, Z., Kimura, M., Suzuki, M., Kondo, Y., Hanabusa, K., Shirai, H., 2003. Synthesis and characterization of new acrylic polymer containing silk protein. Fibre 59 (5), 168–172.

Colonna–Cesari, F., Premilat, S., Lotz, B., 1975. J. Mol. Biol. 95, 71.

Craig, C.L., 1997. Evolution of arthropod silks. Annu. Rev. Entomol. 42, 231–267.

Craig, C.L., Hsu, M., Kaplan, D., Pierce, N.E., 1999. A comparison of the composition of silk proteins produced by spiders and insects. Int. J. Biol. Macromol. 24, 109–181.

Dalpra, I., Freddi, G., Minic, J., Chiarini, A., Armato, U., 2005. De novo engineering of reticular connective tissue *in vivo* by silk fibroin nonwoven materials. Biomaterials 26, 1987–1999.

Das, S., 1996. Studies on Tasar Silk. PhD thesis. IIT, Delhi.

Drukker, B., Hainsworth, R., Smith, S.G., 1953. J. Text. Inst. 44, T420.

Fraser, R.D.B., MacRae, T.P.S., 1973. In: Conformation in Fibrous Proteins. Academic Press, New York.

Hakimi, O., Knight, D.P., Vollrath, F., Vadgama, P., 2007. Spider and mulberry silkworm silks as compatible biomaterials. Composites Part B 38, 324–337.

Hardy, J., Scheibel, T.R., 2010. Composite materials based on silk proteins. Prog. Polym. Sci. 1093–1115.

Harris, B., 1999. Engineering Composite Materials. The Institute of Materials, London.

Iijuka, E., Itoh, H., 1997. Int. J. Wild Silkmoth Silk 3, 37.

Iizuka, E., Kawano, R., Kitani, Y., Okachi, Y., Shimizu, M., Fukuda, A., 1993a. Indian J. Seric. 32, 27.

Iizuka, E., Okachi, Y., Shimizer, M., Fukuda, A., Hashizume, M., 1993b. J Seric 1, 1.

Iizuka, E., Okachi, Y., Shimizer, M., Fukuda, A., Hashizume, M., 1993c. Indian J. Seric. 32, 175.

Iizuka, E., Vegaki, K., Takamatsu, H., Okachi, Y., Kawai, E., 1994. J. Seric. Sci. Jpn. 63, 64.

Iizuka, E., 1996. Size dependency of the physical properties of *Bombyx* silk. Journal of Sericultural Science of Japan 65 (2), 102–108.

Jin, H.J., Park, J., Valluzzi, R., Cebe, P., Kaplan, D.L., 2004. Bio-material films of *Bombyx mori* silk fibroin with poly ethylene oxide. Biomacromolecules 5, 711–717.

Jolly, M.S., Sen, S.K., Sonwalker, T.N., Prasad, G.K., 1979. Non-mulberry silks. In: Rangaswami, G., Narasimhanna, M.N., Kashivishwanathan, K., Sastri, C.R., Jolly, M.S. (Eds.), Manual on Sericulture. Food and Agriculture Organization of the United Nations, Rome, pp. 1–178.

Kaplan, D.L., 2004. "Silk" in EPST, Vol. 11, pp. 841–850.

Li, M., Ogiso, M., Minoura, N., 2003. Enzymatic degradation behavior of porous silk fibroin sheets. Biomaterials 24, 357–365.

Lucas, F., Shaw, J.T.B., Smith, S.G., 1960. Comparative studies of fibroins. I. The amino acid composition of various fibroins and its significance in relation to their crystal structure and taxonomy. J. Mol. Biol. 2, 339–349.

Magoshi, J., mizuide, M., Magoshi, Y., 1979. J. Polym. Sci. 17, 515–520.

Mahadevappa, D., Halliyal, V.G., Shankar, D.G., Bhandiwad, R., 2001. Mulberry Silk Reeling Technology. Oxford and IBH publication Co., India.

Manohar Reddy, R, 2009. Innovative and Multidirectional Applications of Natural Fibre, Silk – A Review. Academic Journal of Entomology 2 (2), 71–75.

Marsh, R.E., Corey, R.B., Pauling, L., 1955. An investigation of the structure of silk fibroin. Biochim. Biophys. Acta 16, 1–34.

Mauney, J.R., Nguyen, T., Gillen, K., Kirker, C., Gimble, J.M., Kaplan, D.L., 2007. Engineering adipose-like tissue *in vitro* & *in vivo* utilizing human bone marrow and adipose-derived mesenchymal stem cells with silk fibroin 3D scaffolds. Biomaterials 28, 5280–5290.

Minagawa, M., 2000. In: Nobumasa, H. (Ed.), Structure of Silk Yarn, Vol-I. Oxford & IBH Publishing Co. Pvt. Ltd, New Delhi, pp. 185–208.

Sehnal, F., 2008. Prospects of the practical use of silk sericins. Entomol. Res. 38, 1–8.

Sen, K., Murugesh Babu, K., 2004. Studies on Indian silk. I. Macrocharacterization and analysis of amino acid composition. J. Appl. Polym. Sci. 92, 1080–1097.

Sheu, H.S., Phyu, K.W., Jean, Y.C., Chiang, Y.P., Tso, I.M., Wu, H.C., et al., 2004. Lattice deformation and thermal stability of crystals in spider silk. Int. J. Biol. Macromol. 34, 325–331.

Spring, C., Hudson, J., 2002. Silk in Africa. University of Washington Press, Seattle.

Tsukada, M., Freddi, G., Minoura, N., Allara, G., 1994. Preparation and application of porous silk fibroin materials, J. Appl. Polym. Sci. 54, 507–514.

Vepari, C., Kaplan, D.L., 2007. Silk as a biomaterial. Prog. Polym. Sci. 32, 991–1007.

Further reading

Allmeling, C., et al., 2006. Use of spider silk fibres as an innovative material in a biocompatible artificial nerve conduit. J. Cell Mol. Med. 10, 770–777.

Altman, G.H., et al., 2002. Silk matrix for tissue engineered anterior cruciate ligaments. Biomaterials 23, 4131–4141.

Arcidiacono, S. et al,. Aqueous processing and fibre spinning of recombinant spider silks. Macromolecules 35, 1262–1266 49

Ayoub, N.A., Garb, J.E., Tinghitella, R.M., Collin, M.A., Hayashi, C.Y., 2007. Blueprint for a high-performance biomaterial: full-length spider dragline silk genes. PLoS One 2, e514.

Bini, E., et al., 2004. Mapping domain structures in silks from insects and spiders related to protein assembly. J. Mol. Biol. 335, 27–40.

Casem, M.L., Turner, D., Houchin, K., 1999. Protein and amino acid composition of silks from the cob weaver, *Latrodectus hesperus* (black widow). Int. J. Biol. Macromol. 24 (2–3), 103–108.

Champion de Crespigny, F.E., Herberstein, M.E., Elgar, M.A., January, 2001. Food caching in orb-web spiders (Araneae: Araneoidea). Naturwissenschaften 88 (1), 42–45.

Cunniff, P.M., Fossey, S.A., Auerbach, M.A., 1994a. Polym. Adv. Technol. 5, 401.

Cunniff, P.M., Fossey, S.A., Auerbach, M.A., Song, J.W., 1994b. Silk polymers: materials science and biotechnology. In: American Chemical Society Symposium Series, vol. 544, p. 34. http://pubs.acs.org/books/publish.shtml.

Exler, J.H., et al., 2007. The amphiphilic properties of spider silks are important for spinning. Angew. Chem., Int. Ed. Engl. 46, 3559–3562.

Garb, J.E., et al., 2007. Expansion and intragenic homogenization of spider silk genes since the triassic: evidence from mygalomorphae (tarantulas and their kin) spidroins. Mol. Biol. Evol. 24, 2454–2464.

Gosline, J.M., Denny, M.W., DeMont, M.E., 1984. Nature 309.

Gosline, J.M., Guerette, P.A., Ortlepp, C.S., Savage, K.N., 1999. The mechanical design of spider silks: from fibroin sequence to mechanical function. J. Exp. Biol. 202, 3295–3303.

Gould, P., December 2002. Exploiting spider's Silk. Mater. Today 42–47.

Hayashi, C.Y., et al., 1999a. Hypotheses that correlate the sequence, structure, and mechanical properties of spider silk proteins. Int. J. Biol. Macromol. 24, 271–275.

Hayashi, C.Y., Shipley, N.H., Lewis, R.V., 1999b. Int. J. Biol. Macromol. 24, 271.

Horan, R.L., et al., 2005. In vitro degradation of silk fibroin. Biomaterials 26, 3385–3393.

Huemmerich, D., et al., 2004. Novel assembly properties of recombinant spider dragline silk proteins. Curr. Biol. 14, 2070–2074.

Huemmerich, D., et al., 2006. Processing and modification of films made from recombinant spider silk proteins. Appl. Phys. A 82 (2002), 219–222.

Jin, H.J., Kaplan, D.L., 2003. Mechanism of silk processing in insects and spiders. Nature 424, 1057–1061.

Kaplan, D., et al. (Eds.), 1994. Silk Polymers: Materials Science and Biotechnology. American Chemical Society.

Kaplan, D.L., Mellow, C.M., Arcidiacono, S., Fossey, S., Senecal, K., Muller, W., 1997. Protein Based Materials. Birkhauser, Boston, pp. 104–107.

Ko, F.K., Kawabata, S., Inoue, M., Niwa, M., Fossey, S., Song, J.W., 2005. www.web.mit.edu/course/3/3.064/www/slides/Ko_spider_silk.pdf.

Lombardi, S., Kaplan, D.L., 1990. The amino acid composition of major Ampullate gland silk (dragline) of *Nephila clavipes* (Araneae, Tetragnathidae). J. Arachnol. 18, 297–306.
Madsen, B., Vollrath, F., 2000. Naturwissenschaften 87, 148.
Madsen, B., Shao, Z., Vollrath, F., 1999. Int. J. Biol. Macromol. 24, 301.
Mukhopadhyay, S., Sakthivel, J.C., 2005. J. Ind. Text. 35, 91.
Nadiger, G.S., Bhat, N.V.,J., 1985. Appl. Polym. Sci. 30, 4127–4136.
Randolph, V., 1992. Lewis, Spider silk: the unraveling of a mystery. Acc. Chem. Res. 25 (9).
Scheibel, T., 2004. Spider silks: recombinant synthesis, assembly, spinning, and engineering of synthetic proteins. Microb. Cell Factories 3, 14.
Sponner, A., Unger, E., Grosse, F., Weisshart, K., 2005. Differential polymerization of the two main protein components of dragline silk during fibre spinning. Nat. Mater. 4 (10), 772–775.
Van Beek, J.D., et al., 2002. The molecular structure of spider dragline silk: folding and orientation of the protein backbone. Proc. Natl. Acad. Sci. U.S.A. 99, 10266–10271.
Warwicker, J.O., 1956. The crystal structure of silk fibroins. Trans. Faraday Soc. 52, 554–557. https://doi.org/10.1039/TF9565200554.
Warwicker, J.O., 1960 Dec. Comparative studies of fibroins. II. The crystal structures of various fibroins. J Mol Biol. 350–362. https://doi.org/10.1016/s0022-2836(60)80046-0.

Fibre plants of arid regions of North Africa

12

Meriem Kaid-Harche, Abderrezak Djabeur
Laboratoire des Productions Valorisations Végétales et Microbiennes, Faculté des Sciences de la Nature et de la Vie, Université des Sciences et de la Technologie d'Oran Mohamed Boudiaf B.P., Oran, Algeria

12.1 Introduction

The use of plant fibres is booming along with scientific and technical progress. Among the fibre plants, flax, hemp and ramie find applications in different fields (textile, paper, biocomposite etc.) Vaney and Chambard (1917); Gay and Monrocq (1972) Roudier (2012) Mosiniak and Prat (2005). Their expansion is growing every year. In North Africa, other fibre plants grow but they are poorly known and poorly valued. In Algeria, fibre plants occupy a large area. They grow in a hinge region at the edge of the Tell and the pre-Saharan, constituting a barrier against desertification. These species belong to the family Poaceae.

Stipa tenacissima L. (Alfa) is widespread on the highlands; the range is estimated to be 3 million ha but covers only 450,000 ha. For example, the evolution of plant communities in the Alfa-dominated steppe (*Stipa tenacissima*) located in the southern Algiers region was assessed via a diachronic study that addressed floristic surveys carried out in two timescales. The results show a regression of the vegetation, which results in a change in their floristic composition, accompanied by a reduction of vegetal cover and contribution of *Stipa tenacissima* in the vegetal carpet with an impoverishment in floristic richness (Chaouch Khouane et al., 2018); it is also present on the coastal mountains. It covers areas where rainfall occurs in 400 mm isohyets in the North and 100 mm in the South but does not tolerate saline soils. The production of Alfa can reach 10 tons of dry matter per hectare (DM/ha), but the exploitable part is of the order of 1000−1500 kg DM/ha (Nedjraoui, 1981, 2003). On the pastoral level, several authors (Nedjraoui, 1981; Aidoud and Touffet, 1996; Kadi Hanifi et al., 2005) confirm its low energy value (0.3−0.5 UF 5/kg DM).

Lygeum spartum L. is a circum-Mediterranean plant; it grows in Southern Spain, Italy and Portugal in Sicily and Sardinia. In Africa, it grows in Morocco, Egypt, Tunisia and Algeria. The area of extension is in relation with its great ecological plasticity; in fact, this species colonizes the sands, and the clay soils often loamy and slightly salty; it covers the areas where the precipitation is between the isohyets 200 and 300 mm.

Handbook of Natural Fibres. https://doi.org/10.1016/B978-0-12-818398-4.00014-1

Aristida pungens is a perennial grass and cosmopolite; it fixes the sand in the sub-arid high-plateau region, the arid region and Sahara. It was reported in Hoggar, Egypt, Mauritania and South Africa (Battandier and Trabut, 1902).

From a dynamic point of view, these groups constitute a transitional stage between the steppe with the sagebrush and sagebrush steppe (Aidoud-Lounis, 1997).

They play an important role in the fight against erosion and soil erosion (Le Houérou, 1985). As a result, this species is considered an important element in combating the phenomenon of desertification. The morphology of Alfa clumps contributes to retaining the fine particles of soil transported by the wind which are deposited there by raising the substratum (Djebaili, 1984; Le Houérou, 1996).

12.2 Biology of the three species

12.2.1 Stipa tenacissima *L. (Alfa)*

The stem is a branched subupright rhizome formed by an internode and between very short nodes. At the base of the nodes, axillary buds that initiate vegetative units (leaves and apical bud, which gives the plant a tufted habit) exist. The leaves can reach between 1 and 1.50 m. The leaf is wound on itself on its upper side; the inner epidermis is hairy and it limits the evaporation; the biological cycle of the leaf is 2 years. Alfa has a fasciculate root system; it is formed by a very dense network of roots and rootlets, the roots sinking into the ground up to 50 cm deep (Fig. 12.1).

Figure 12.1 *Stipa tenacissima* L. (a) field of *Stipa tenacissima* L. in natural conditions, (b) *Stipa tenacissima* L. - detail of a plant (scale represents 50 cm).

According to the work of Boussaid (2013), the clump of Alfa *Stipa tenacissima* plant contributes through its form, in the retention at its base, of the fine particles of soils, transported by the wind and which are deposited there by raising the substratum. The fine particles are thus entrapped, added to the leaves of *Stipa tenacissima*, constitute an ecological biotope that attracts a very varied fauna of polyphagous predators and insects, cellulolytic microorganisms, and allows for the development of annual species, thus improving the pastoral quality of the courses.

The *Stipa tenacissima* inflorescences are sought by livestock especially during times of famine. The *Stipa tenacissima* is an important genetic resource for its hardness is well adapted to the drought and presents enormous possibilities of exploitation. Due to its ecological conformability, Alfa contains a gene pool 'markers very interesting for research in plant breeding and genetic engineering.

Stipa tenacissima flowers between April and May, the flower stalks reach more than 1 m, the inflorescences are loose and they carry caryopses protected by the glumes and lemmas, the caryopses are measured as 5–8 mm long and 1 mm on average wide. Maturation of caryopses takes place between May and June. The fruits are usually carried away by the wind, which explains why regeneration in situ by seeds is rare (Fig. 12.2). The multiplication is essentially vegetative. In the greenhouse, seedlings yielded clumps with leaves up to 2 m in length after 2 years of cultivation and the flower stalks produce viable caryopses (Harche, 1978) (Fig. 12.3).

12.2.2 Lygeum spartum *L. (Sparte)*

Lygeum spartum is the second perennial Poaceae after *Stipa tenacissima* which grows in Algerian arid and semiarid region. The wearing of the plant is similar to that of Alfa; it differs in flowering by its inflorescence called spathe (Fig. 12.4).

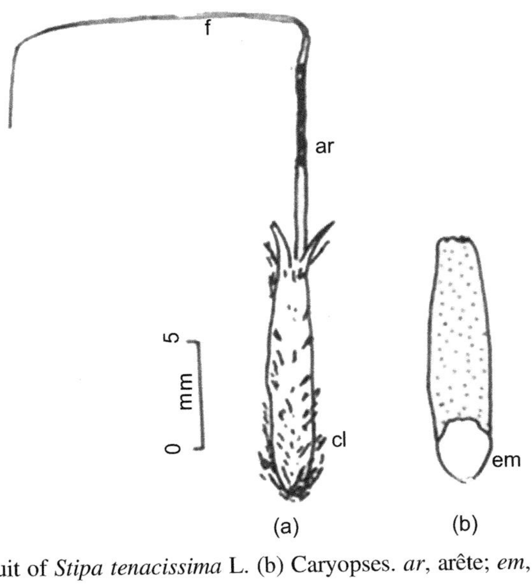

Figure 12.2 (a) Fruit of *Stipa tenacissima* L. (b) Caryopses. *ar*, arête; *em*, embryo.

Figure 12.3 The first development of *Stipa tenacissima* L.: 6 months, 1 year, 15 months, 2 years (from left to right).

Figure 12.4 View of a clump of *Lygeum spartum* (photograph taken in 2015 in Wilaya d'El Bayadh, Algeria).

The underground stem is a straight rhizome formed by nodes and between short nodes; at the nodes, there exist adventitious buds which initiate the adventitious roots of the axillary buds which initiate the leaves; the whole constitutes a vegetative unit (Fig. 12.5).

The Sparte have fasciculate root system, thanks to its anchoring as it stabilizes the sand. The anatomical structure of the root recalls that of *Stipa tenacissima* with the exception of the cortical parenchyma which is of lacunous type (Louerrad, 2017) (Fig. 12.6).

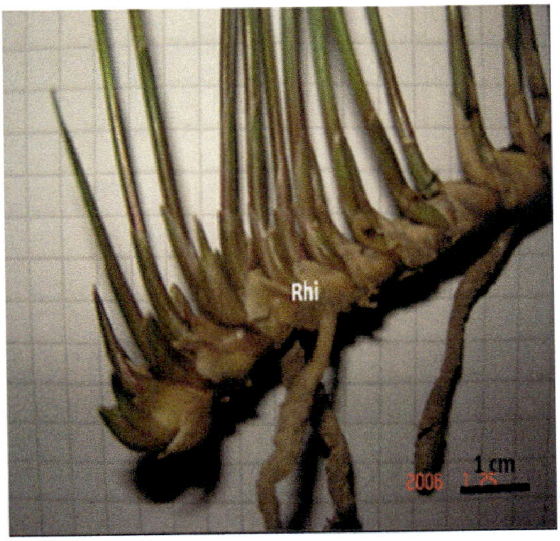

Figure 12.5 Rhizome (Rhi) of *Lygeum spartum* (Djabeur, 2007).

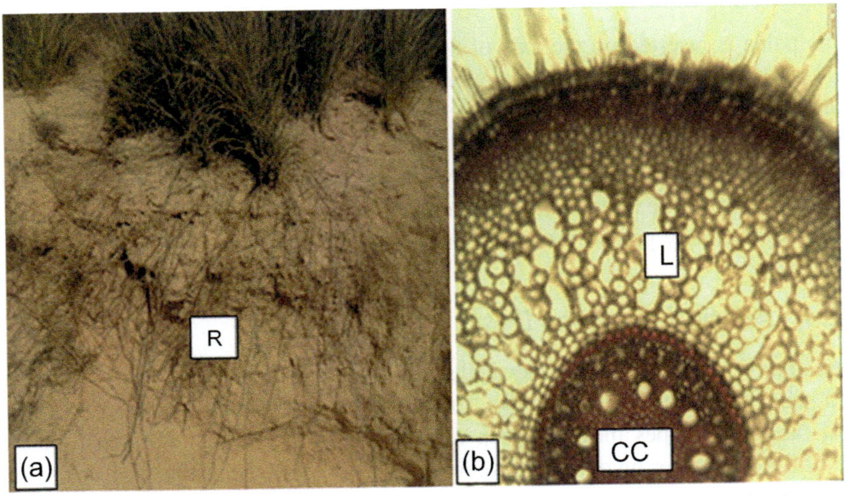

Figure 12.6 (a) Profile in the soil where *Lygeum spartum* grows. (b) Cross-section in a root. *R*, root; *CC*, central cylinder; *L*, gap (Louerrad, 2017).

During the month of May, the flower stems are formed. The inflorescence is a non-sessile uniflorous spikelet, composed of two lemmas that protect the floral pieces, three stamens and an ovary. The external lemmas are called lemmas. The internal lamella are called paleole. The lemma is inserted on the axis of the inflorescence and comprises two parts:

- Short lower part not exceeding 3 mm at the young stage; it is thick and welded with the lemmas of the other flowers. It is covered with hair. At maturation, the hairs become longer. This part as well as the hairs gradually lengthen during the development of the inflorescence.

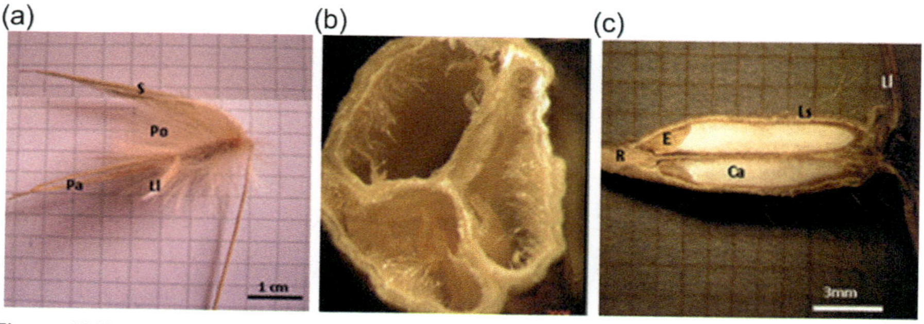

Figure 12.7 (a) Infructivity of *Lygeum spartum.* (b) Cross-section in a three empty cells infructescence. (c) Longitudinal section in two full cells infructescence. *E*, embryo; *Ca*, caryopses; *R*, rachis; *Li*, free Lemmas; *Ls*, solder Lemmas; *Po*, hair; *S*, spathe.

After fertilization, the coalescence of basal parts of lemmas forms an oblong, hairy tube 5–7 mm long and comprising as many compartments as there are flowers in the inflorescence (Djabeur, 2007) (Fig. 12.7)

- Membranous free upper part 2–4 mm long at the young stage. In the adult stage, it elongates or remains short; it is a distinctive genetic criterion in *Lygeum spartum* populations (Djabeur, 2007; Boughanmi-Abdedaim, 2010).

The palea is inserted on the axis of the spikelet, is totally membranous and transparent, is a kind of rolled sheet in which edge remains free. It is always oriented towards the external side of the inflorescence, that is, towards the lemmas. It comprises a short part of 2–3 mm at the young stage. It covers the ovary. Its median region is welded with the welded base of lemmas while its edges remain free. This part extends at the same time as the lemma, the free part, erected from 25 to 30 mm surrounding the stamens so that only part of the style emerges at the top.

12.2.3 Aristida pungens *Desf.*

Aristida pungens has a branched elevated subhorizontal rhizome, which gives the plant a tufted habit; the leaves are needle-like, the extremity is pointed and pungent, the roots are of fasciculate types they are deep and on the surface which stabilizes the sand (Fig. 12.8). The flowering occurs between March and April, the inflorescences are loose panicles, the spikelet has a flower, the fruit is a subfusiform caryopsis 5 mm long by 1.2 mm, it is protected by a lemma convoluted around the paleola bearing a ridge with three feathery branches—one median and two lateral (Quézel and Santa, 1962).

12.3 Anatomy of *Stipa tenacissima, Lygeum spartum* and *Aristida pungens* leaves

The *Stipa tenacissima* fibres leaves are differentiated from the early stages of development, they have a meristematic origin, the first fibres are located under the lower epidermis, the fibres continue their extension in the direction of the conductive bundle

Figure 12.8 View of *Aristida pungens* in natural condition fructification stage.

surrounding it and regain the upper epidermis. An adult leaf is multilobed and each rib contains fibres (Harche, 1984) (Fig. 12.9).

The fibrous tissue of *Lygeum spartum* is similar to that of *Stipa tenacissima*. Leaves from 48-day-old plants show in cross-section the location of subepidermal meristematic cells, which are differentiated into fibres. The fibrous tissue continues its development towards the conductive bundle surrounding it and prolongs its extension towards the upper epidermis. The same anatomy is found in *Aristida pungens* (Fig. 12.10). We can conclude that the limb of the three species possesses common characteristics: it is formed by several ribs, the upper epidermis is hairy, the fibrous tissue is important, these species have specific adaptations to water and thermal stress.

12.3.1 Fibre morphometry

The morphometric study of the fibres requires a treatment of the fragments of sheets beforehand. The samples of leaves of 1–2 cm are incubated in 110 V oxygenated

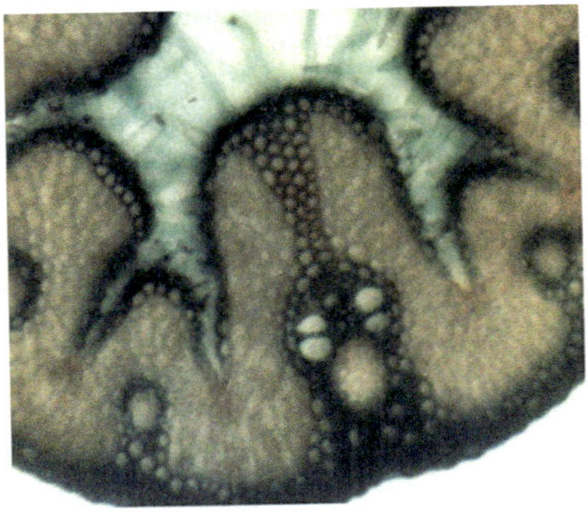

Figure 12.9 A transverse section in the young *Stipa tenacissima* leaf (3 months).

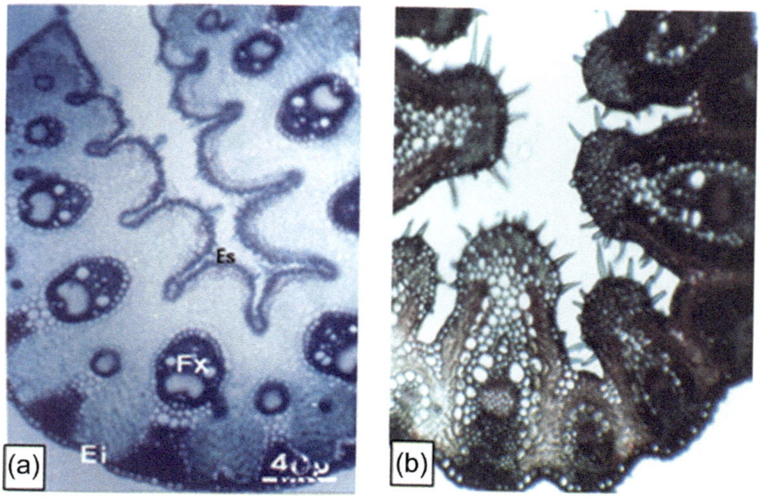

Figure 12.10 Transverse sections. (a) *Lygeum spartum* leaf, (b) *Aristida pungens* leaf.

water and acetic acid at equal volume for 24 h in the oven at 70°C. Then the mixture is drained, fibres are observed in photomicroscope, the fibres dissociation protocol is as follows:

Leaf pieces – Maceration in H_2O_2/acetic acid (v:v) 24 h at 70°C – Washing with water – Colouration with safranine (24 h) – Washing and agitation – Observation and measure of fibres.

The fibres are elongated and have morphological diversity. They are tapered at the ends, rounded, truncated or bevelled. The average fibre length is 0.38 mm to *Stipa tenacissima*, 1.70 mm to *Aristida pungens* and 1.72 mm to *Lygeum spartum* (Fig. 12.11).

12.3.2 Histochemistry and biochemistry of the fibre cell walls

The fibre walls are thick; the lignin-specific stains (the Maule reaction and the phloroglucinol hydrochloric acid reaction) reveal the presence of syringyl and coniferyl radicals.

Biochemical analysis of the cell wall shows the presence of cellulose, hemicellulose and pectin in all three species with varying contents (Table 12.1. Cell wall composition of three species).

Stipa tenacissima, Lygeum spartum and *Aristida pungens* are rich in polysaccharides. Cellulose represents about 50%. Hemicellulose represents about more than 20% in *Stipa tenacissima* and *Lygeum spartum*. Pectin is weak but higher in *Aristida pungens*.

Cellulose constitutes the microfibrillar framework of the wall. It is a macromolecule formed of β 1–4 glucose embedded in a matrix consisting of hemicelluloses of the xylan type, pectins in a small proportion and lignins. The global lignin determination according to the Klason technique shows that *Stipa tenacissima* contains 16.5%,

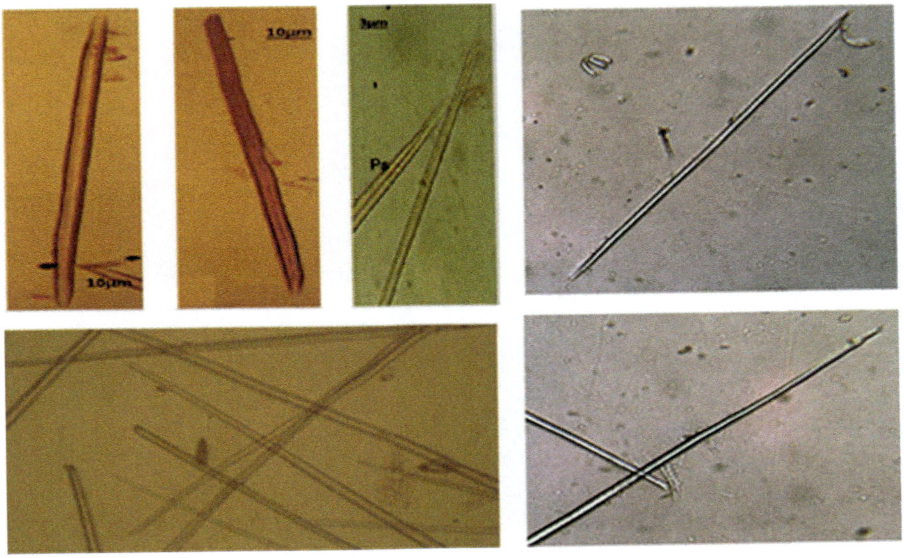

Figure 12.11 Morphology of fibres of the three species.

Table 12.1 Cell wall composition of three species.

Cell wall composition	*Stipa tenacissima*	*Lygeum spartum*	*Aristida pungens*
Cellulose	58%	47%	48%
Hemicellulose	28%	22.6%	19.4%
Pectin	0.95%	1.12%	2.24%

Lygeum spartum contains 17% and *Aristida pungens* contains 13%. Lignins result from the condensation of three monomers coumaryl alcohol (H), coniferyl (G) and sinapyl alcohol (S) constituting a three-dimensional macromolecule (Monties and Lapièrre, 1981), lignin permeate cell wall polysaccharides making it difficult to observe the structure of the wall.

Methylamine, considered as a mild extractant of lignins, was used to solubilize the lignins and reveal the cellulosic fibrillary framework (Harche, 1985). In the case of Alfa fibres, methylamine solubilizes 35% of the lignins after the treatment for 105 h. The application of ultrastructural cytochemistry techniques (PATAg) on ultrafine sections (Thiery, 1967) and MET observation equipped with a goniometric plate have made it possible to visualize the microfibrillar cellulosic strata and the cell wall architecture of the fibres. Foliar fibrous tissue of *Stipa tenacissima* offers the architectural diversity. (Fig. 12.12) and (Fig. 12.13). Several construction models exist, in regular or irregular hoop, in fishbone (Harche, 1985). This architectural diversity is probably one of the quality factors of the paper fibre of *Stipa tenacissima*. Similar models of parietal construction exist in *Lygeum spartum* and *Aristida pungens* fibres (Harche et al., 1990, 1992).

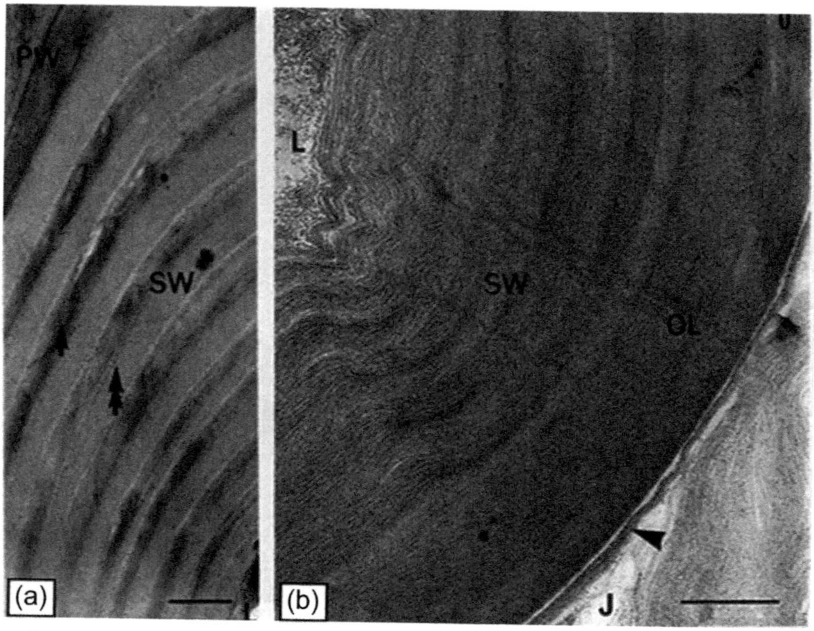

Figure 12.12 Bundle sheet fibres of *Stipa tenacissima*. Cell wall observed to MET. (a) Control. (b) After treatment by methylamine (105 h). *L*, lumen; *OL*, outer layer; *SW*, secondary wall; *J*, cell junction. The bar on micrographs represent 1 μm.

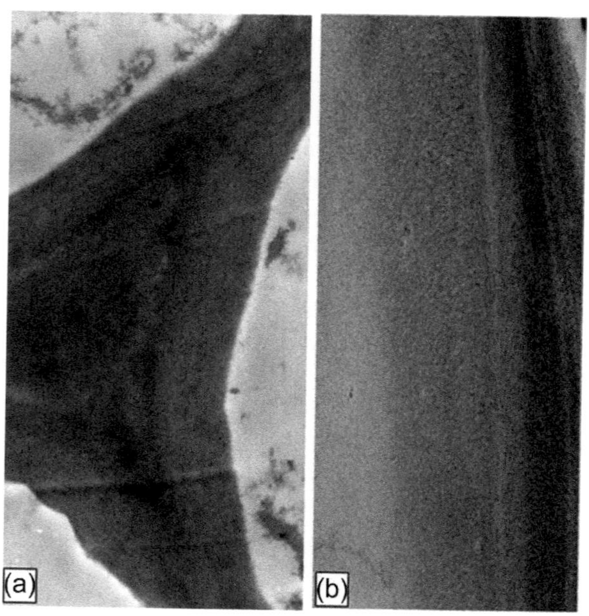

Figure 12.13 Subepidermal fibres of *Stipa tenacissima*. Cell wall observed to MET. (a) Control. (b) After treatment by methylamine for 105 h.

12.3.3 Valuations

Of the three species, it is *Stipa tenacissima* which is the best valued indeed. *Stipa tenacissima* (Alfa) is known as a source of noble raw material for the manufacture of paper. The exploitation and use of the Alfa goes back to the highest antiquity. Pline evoked the manual methods of harvesting *Stipa tenacissima* in Spain. The true growth of Alfa began in the 19th century when, thanks to the Routledge process (1836), the Scots used it as a raw material in the manufacture of pulp.

Stipa tenacissima is a paper plant in Algeria, known for its thick-walled fibres and cellulosic nature; in his work on the Alfa (Trabut, 1889), he wrote that this species provides a paper of excellent quality 'extra strong'. The leaves of Alfa were harvested in different regions of the Algerian highlands, baled and then transported by boat to France or England, and they were transformed into pulp.

In Algeria, the processing of Alfa (*Stipa tenacissima*) goes back to the beginning of the 19th century, the Cellunaf factory of Baba Ali was the first Alfa-based pulp mill established in Algeria. Another manufactory was built in the Mostaganem (North Western of Algeria). Fibres were processed into pulp mixed with hardwood pulp or softwood imported for pulp production (Fig. 12.14).

Alfa is also an artisanal plant; it is used for making carpets, trash, etc (Fig. 12.15).

The fibres of the Alfa have been used in the manufacture of orthopaedic prosthesis; recent work has revealed their textile potential (Dellal, 2012). In case of scarcity, this species is grazed by the sheep flock. In the context of the exploitation of biosourced materials in applications with high added value, *Stipa tenacissima* (Alfa) fibres are used as reinforcement elements in biocomposites based on polymer matrices (Sood and Dwived, 2018).

12.4 Seeds germination

Today, the area of Alfa (*Stipa tenacissima*) has decreased; the main causes are as follows: the lack of culture, overgrazing, increased drought. With this in mind and to regenerate these plants, work was conducted on the physiology of the germination of the three species.

Figure 12.14 Different stages of harvesting the *Stipa tenacissima* leaves mowing collecting and baling (from left to right).

Figure 12.15 Examples of articles made from *Stipa tenacissima* leaves.

The influence of several factors has been investigated. The study of germination is conducted with caryopses stored in the freezer (−15 to −20°C) or at room temperature (20–25°C). Before being germinated, caryopses are chosen according to their health status. The germination tests are carried out with caryopses wrapped in their lemmas (dressed caryopses) and with bare caryopses.

For caryopses of *Aristida pungens*, only dressed caryopses were tested. Caryopses of *Lygeum spartum* dressed for the study of sprouting are stripped of their hair to get better contact with soaked cotton.

Each germination test consists of caryopses dressed or naked, germinated in Petri dishes of 10 cm diameter, containing cotton soaked in distilled water or various solutions (gibberellic acid or polyethylene glycol). Hundred caryopses (50/box) were used in each germination test. The tests are carried out in the dark at temperatures ranging from 5 to 30°C for *Lygeum spartum*, and 5 to 40°C for *Stipa tenacissima*. In white light, only optimal germination temperatures of each species were used. Germination tests were performed in the presence of gibberellic acid (GA_3 10^{-3} M); that is, watering in this case is by the GA_3 solution instead of distilled water. For other experiences, we used polyethylene glycol-6000 (PEG-6000) solutions at different water potentials (−0.2, −0.5, −1.0, −1.5 and −2.0 Mega Pascal MPa-). Germination is considered to be carried out as soon as the radicle lengthens and pierces the teguments (Evenari, 1957; Côme, 1970).

The results showed that the caryopses of the three species germinate in a wide thermal range from 10 to 40°C. The thermal optimum germination varies from one species to another: it is 15–20°C in *Stipa tenacissima* L, 20–25°C in *Lygeum spartum* and 20–35°C in *Aristida pungens*. The caryopses of *Stipa tenacissima* prefer low temperatures, whereas those of *Aristida pungens* can prefer high temperatures.

The teguments of the nonmatured caryopse of the three species influence germination, with no doubt, by their inhibitory effect. This effect is accentuated in the caryopses of *Lygeum spartum* and *Aristida pungens*. The freshly harvested caryopses of *Stipa tenacissima* germinate better than postmatured caryopses, suggesting that the physiology of integumentary dormancy in caryopses of this species may not be the same as in the caryopses of the other two species *Lygeum spartum* and *Aristida pungens*. Indeed, the bare caryopses germinate better whatever their mode of conservation, whereas the clothes do not germinate or germinate with difficulty.

The dry preservation reduces the inhibitory activity of the seed coat and thus eliminates the dormancy of the covered caryopses by widening of the thermal range which allows a good germination. Even the low dormancy of bare caryopses is eliminated during dry storage, which makes germination more favourable.

The light has an inhibitory effect on the sprouting process of *Lygeum spartum* caryopses and *Aristida pungens* as long as the radicle has not yet pierced the tegument of the naked caryopse or lemmas; they are responsible for negative photosensitivity. The structures surrounding the embryo also have a role in this phenomenon. The caryopses of the three species germinate in a wide thermal range from 10 to 40°C. The thermal optimum germination varies from one species to another: it is 15–20°C in Alfa, 20–25°C in Sparte (*Lygeum spartum*) and 20–35°C in *Aristida pungens*. *Stipa tenacissima* caryopses prefer low temperatures for germination, whereas *Aristida pungens* caryopses prefer high temperatures.

The teguments of the nonmatured caryopses of the three species influence germination, with no doubt, by their inhibitory effect. This effect is accentuated in the caryopses of *Lygeum spartum* and *Aristida pungens*. The freshly harvested caryopses of *Stipa tenacissima* germinate better than postmatured caryopses, suggesting that the physiology of integumentary dormancy in caryopses of this species may not be the same as in the caryopses of the other two species.

Indeed, the bare caryopses germinate better whatever their mode of conservation, whereas the clothes do not germinate or germinate with difficulty.

The dry preservation reduces the inhibitory activity of the seed coat and thus eliminates the dormancy of the covered caryopses by widening of the thermal range which allows a good germination. Even the low dormancy of bare caryopses is eliminated during dry storage, which makes germination more favourable.

The light has an inhibitory effect on the sprouting process of *Lygeum spartum* caryopses and *Aristida pungens*, as long as the radicle has not yet pierced the tegument of the naked caryopsis or lemmas; they are responsible for negative photosensitivity. The structures surrounding the embryo also have a role in this phenomenon. However, a *Stipa tenacissima* seed does not exhibit this sensitivity to light.

These results confirm that the climatic and hydrous conditions of the soil control the process of germination of these species in natural conditions.

12.5 Conclusion

These plants grow in a region that is the steppe, which includes 8 million ha. Although traditionally pastoral (sheep rearing) with a nomadic population, this region has benefited from large means for the introduction of new development techniques to meet the nutritional needs of increasing sheep numbers and to settle the population (introduction of new species such as imported cereals or fodder plants, increased water drilling to meet the water requirements of introduced species, inadequate mechanization of industrial species to meet the needs of the cellulose industry plant. professionalism and consultation in decision-making). The state of research (Bedrani and Nedjaroui, 2008) highlights a change in the chemical nature of the soils of the steppe (salinization) which makes it uneducated, a disappearance of 25% of the Alfa area and a change in the steppic floristic composition and an advance of the sands (desertification). These upheavals are not without impact on the exodus to the interior or coastal cities, despite the efforts made in the construction of universities in the cities.

To preserve this biodiversity and to curb the weakening of the ecosystems of these three species, it is important to proceed with their regeneration by vegetative propagation and by sowing. These techniques are mastered (Djabeur, 2007; Djabeur et al., 2010). Conservation of caryopses, viability of caryopses, dormancy, dormancy emergence, optimal germination conditions and pretreatments that improve germination capacity are known (Djabeur et al., 2008 and Djabeur et al., 2010). Rhizome propagation is a faster method, it was already practised by foresters in natural conditions in the 19th century. These farming techniques will contribute to the development of these plants, which have several interests:

(1) Economic, as a major source of fibres (cellulose industry, value-added biocomposites);
(2) Ecological, soil fixation and stabilization of sands, and therefore the fight against desertification, thanks to their well-developed root systems,
(3) Artisanal, return to the use of *Stipa tenacissima* fibres, in the manufacture of handicrafts as in the past. This will open up jobs for the local population and therefore contribute to the rural exodus. If these actions are carried out, they will be part of the success of the sustainable development of the steppe regions.

References

Aidoud, A., Touffet, J., 1996. La régression de l'Alfa (*Stipa tenacissima* L.), graminée pérenne, un indicateur de désertification des steppes algériennes. Secheresse (Montrouge) 7, 187–193.

Aidoud-Lounis, F., 1997. Le complexe alfa-armoise-sparte (*Stipa tenacissima* L., *Artemisia herba-alba Asso*, *Lygeum spartum* L.) des steppes arides d'Algérie : structure et dynamique des communautés végétales. Thèse de Doct. Univ. d'Aix-Marseille, Marseille, p. 198.

Battandier, A., Trabut, L., 1902. In: Tunisie (Ed.), Flore analytique et synoptique de l'Algérie et de la. Alger, pp. 347–353.

Bedrani, M., Nedjraoui, D., 2008. La désertification dans les steppes algériennes : causes, impacts et actions de lutte? https://www.researchgate.net/publication/30439266.

Boughanmi-Abdedaim, K., 2010. Etude de deux cytotypes de *Lygeum spartum* L. par approches pluridisciplinaires: Palynologie et Cytogénétique classique et moléculaire. Thèse de Doctorat en sciences, Université des Sciences et de la Technologie USTO-MB, Oran, p. 108.

Boussaid, M., 2013. Diversité des populations naturelles de *Stipa tenacissima* L. en Algérie : approches phénotypique, caryologique et moléculaire. Thèse de Doctorat en sciences Université des Sciences et de la Technologie USTO-MB, Oran.

Chaouch-Khouane, H., Belhamra, M., Aidoud, A., 2018. Evolution of plant communities of Alfa (*Stipa tenacissima* L.). Courrier du savoir 26, 107–118.

Côme, D., 1970. In: Les obstacles à la germination. Masson et Cie, Paris, p. 162p.

Dellal, M., 2012. Evaluation du potentiel textile des fibres d'Alfa (*Stipa tenacissima* L.): Caractérisation physico-chimique de la fibre au fil .Thèse. Université de Haute Alsace, p. 153p.
Djabeur, A., 2007. Physiologie de la germination des caryopses de trois Poacées vivaces : Alfa (*Stipa tenacissima*), Sparte (*Lygeum spartum*) et *Aristida pungens*) . polymorphisme morphologique, biochimique et moléculaire de quelques populations de *Lygeum spartum*. Thèse de Doctorat es- sciences. Université des Sciences et de la Technologie USTO-MB, Oran, p. 120.
Djabeur, A., Kaid-Harche, M., Catesson, A.M., 2008. Structure des infrutescences et capacité fructifère de deux cytotypes de *Lygeum spartum* de l'ouest algérien. Ecol. Mediterr. 34, 5–12.
Djabeur, A., Kaid-Harche, M., Côme, D., Corbineau, F., 2010. In: Environmental Control of Germination and Dormancy of Seeds of Two Cytotypes of Lygeum Spartum a Perennial Grass of Semi-arid and Arid Areas in Ecologia Mediterranea, vol. 36, pp. 89–98.
Djebaili, S., 1984. Steppe algérienne, phytosociologie et écologie. Office des publications universitaires (OPU), Alger, p. 182.
Evenari, M., 1957. Les problèmes physiologiques de la germination. Bull. Soc. Franç. Physiol. vég. 3, 105–124.
Gay, M.C., Monrocq, R., 1972. Identification des fibres textiles naturelles par examen microscopique. BCIN, pp. 16–22, 56925.
Harche, M., 1978. Contribution à l'étude de l'Alfa (*Stipa tenacissima* L.) d'Algérie: Germination, croissance des feuilles et différenciation des fibres. Thèse Doct. 3éme Cycle. Univ. Sci. Technol. LILLE, p. 78p.
Harche, M., 1984. Origine et différenciation des fibres sous-épidermiques de la feuille d'Alfa (*Stipa tenacissima* L.). Ann. Sci. Nat. Bot., Paris, Tome 6, 207–226.
Harche, M., 1985. Différenciation et structure pariétale des fibres foliaires de l'Alfa (*Stipatenacissima* L.). Thèse Doct. Etat Univ, Paris VI, p. 88.
Harche, M., Chadli, R., Catesson, A.M., 1990. Diversity of cellulose microfibril arrangements in the walls of *Lygeum spartum* leaves. Ann. Bot. 65, 79–86.
Harche, M., Mekhaldi, A.E.K., Catesson, A.M., 1992. Structure et architecture pariétale des tissus foliaires de *l'Aristida pungens* Desf. Bull. Soc. Bot. Fr. 139, 127–134.
Kadi-Hanifi, H., Sadji, A., Amghar, F., 2005. The impact of anthropic action and aridity on the pastoral production in the *Stipa tenacissima* L. steppes of Algeria. Options Méditerranéennes, Sér. A 67, 61–66.
Le Houérou, H.N., 1985. La régénération des steppes algériennes. Rapport de mission, de consultation et d'évaluation. Ministère de l'Agriculture, Alger, p. 19.
Le Houérou, H.N., 1996. La régression de Stipa *tenacissima* L. graminée pérenne, un indicateur de désertification des steppes algériennes. Secheresse (Montrouge) 7, 87–93.
Louerrad, Y., 2017. Histologie racinaire et analyses des polyphénols et polysaccharides pariétaux de deux cytotypes de *Lygeum spartum* Thèse de doctorat. Université des Sciences et de la Technologie USTO-MB, Oran, p. 100.
Monties, B., Lapierre, C., 1981. Données récentes sur l'hétérogénéité des lignines. Pysiol. Vég. 19, 327–348.
Mosiniak, M., Prat, R., 2005. Les fibres textiles d'origine végétale. http://www.snv.jussieu.fr/bmedia/textiles/index.html.
Nedjraoui, D., 1981. Teneurs en éléments biogènes et valeurs énergétiques dans trois principaux faciès de végétation dans les *Hautes Plaines* steppique de la wilaya de Saida. Thèse de Doctorat de 3°cycle, Univ. USTHB, Alger.
Nedjraoui, D., 2003. Le profil fourrager de l'Algérie. FAO, p. 30p.

Quézel, P., Santa, S., 1962. Nouvelle Flore de l'Algérie et des régions désertiques méridionales. Centre National de la Recherche Scientifique CNRS, Paris, p. 1170p.

Roudier, A., 2012. Analyse multi-échelle du comportement hygro-mécanique des fibres de lin. Thèse de Doctorat en sciences de l'ingénieur , Univ. Blaise Pascal-Clermont II France.

Sood, M., Dwived, G., 2018. Effect of fiber treatment on flexural properties of natural fiber reinforced composites: a review. Egypt. J. Pet. 27, 775–783. https://doi.org/10.1016/j.ejpe.2017.11.005http://www.univ-usto.dz/theses_en_ligne/index.php?lvl=notice_display&id=2320.

Thiery, J.P., 1967. Mise en évidence des polysaccharides sur coupes fines en microscopie électronique. J. Microscopie. 6, 987–1018.

Trabut, L., 1889. In: Etude sur l'Halfa. Jourdan. Alger, p. 90.

Vaney, C.L., Chambard, P., 1917. Méthodes d'examen des fibres, vol. 63. publications de la société linnéenne de Lyon, pp. 27–36.

Further reading

Aidoud, A., Le Floc'h, E., Le Houérou, H.N., 2006. Les steppes arides du nord de l'Afrique. Secheresse (Montrouge) 17, 19–30.

Chadli, R., 1990. Contribution à l'étude du Sparte, *Lygeum spartum* L. : Germination; croissance des feuilles; différenciation et structure pariètale des tissus foliaires; essai d'obtention de pâte papetière à partir des fibres foliaires. Thèse de Magister, Université d'Oran, p. 96p.

Djebaili, S., 1988. Connaissances sur l'alfa (*Stipa tenacissima*). Biocénoses 3, 43–52.

Harche, M., Catesson, A.M., 1988. Cell wall architecture in Alfa (*Stipa tenacissima* L.) fibres. IAWA. Bull. 6 (1), 61–69.

Harche, M., Tollier, M.Y., Monties, B., Catesson, A.M., 1991. Caractérisation comparée des constituants (polyosides, lignines et acides phénoliques) des parois *Stipa tenacissima, Lygeum spartum* et *Aristida pungens*. Cellul. Chem. Technol. 25, 11–17.

Marion, J., 1958. L'Alfa, matière première pour l'industrie papetière. Rev. For. Fr. (Nancy) 345–347.

Pouget, M., 1980. Les relations sol-végétation dans les steppes Sud-algéroises. Travaux et documents d'Orstom, Paris, p. 569.

Robert, D., Roland, J.C., 1989. In: Doin (Ed.), Caractéristiques et stratégie évolutive des plantes. Tome 1 organisation cellulaire, p. 265. Paris.

Mineral fibres: basalt

13

Zhishen Wu [1], Xin Wang [1], Jianxun Liu [2], Xingfen Chen [2]
[1]School of Civil engineering/National and Local Unified Engineering Research Center for Basalt Fiber Production and Application Technology, Southeast University, Nanjing, Jiangsu Province, China; [2]National and Local Unified Engineering Research Center for Basalt Fiber Production and Application Technology, Southeast University, Nanjing, Jiangsu Province, China

13.1 Introduction to basalt fibres

Continuous basalt fiber (CBF) is a continuous basalt fiber made from natural volcanic ore (such as andesite basalt, andesite basalt and diabase), melted at high temperature (1500 °C) and pulled out through leak plate. CBF has a golden-brown appearance and belongs in the group of inorganic nonmetallic fibres.

CBF has excellent properties such as lightweight, high strength, high temperature resistance, corrosion resistance, oxidation resistance, radiation resistance, heat insulation and sound insulation (Dalinkevich et al., 2009). The mechanical properties of stabilized CBF exceed those of glass fibre. The strength, stiffness and durability are equal to or exceed those of aramid fibre (AF). CBF can replace the common carbon fibre (CF) in many aspects, and its high temperature resistance is even comparable with that of CF. It is a cost-effective fibre material. Together with CF, AF, ultra−high-molecular-weight polyethylene fibre, polyphenylene benzodioxazole fibre and polyphenylene sulfide fibre, CBF is called a 'high-tech' fibre.

CBF textile yarn can be made into roving, chopped yarn, fibre cloth, fibre felt and other products. As reinforced materials, these products can be made into fiber reinforced composites (FRP) with various properties, such as composite reinforcement materials, composite plates and composite grids. They have a wide range of uses in the aerospace, automotive and shipping industries, as well as in civil transport, energy and environment, chemical fire protection, national defense and other fields.

CBF has a plentiful and inexpensive source of raw materials, simple production process, excellent comprehensive performance, wide range of applications, long industrial chain and good price performance. The basalt fibre industry is a new material industry and a new resource industry. The development of this industry is conducive to the comprehensive development of basalt and other volcanic mineral resources, the use of high value-added products and the protection of the environmental through green materials, which in turn are conducive to sustainable development in a circular economy. However, each CBF production process has strict requirements. Accelerating industrial technological innovation and breaking through major technical bottlenecks are the driving forces for the healthy and rapid development of the CBF industry.

Handbook of Natural Fibres. https://doi.org/10.1016/B978-0-12-818398-4.00015-3

In 1922, the Frenchman Paul proposed basalt fibre manufacturing technology in American patent OS 1,438,428, but there was no substantial production. To meet the needs of military industrial development, the United States, the Soviet Union and other countries began to research CBF, especially in the research and development (R&D) of industrialization. In the 1950s, the Soviet Union's glass composites and glass fibre institutes developed basalt fibres. Later, under the orders of the former Soviet Union, the Ministry of the Building Materials Industry of Ukraine established the 'scientific research and production consortium of insulation materials', referred to as 'the consortium', where the main task was to develop CBF and production line of CBF and its products. The research laboratory of the consortium began to develop the CBF production process in 1972. In 1985, Ukraine took the lead in achieving the industrialized production of CBF, and all the products were used in the former Soviet Union's defense and military industry, aerospace and aviation fields. After the collapse of the Soviet Union in 1991, the Soviet Union's research results were declassified and turned from military to civilian use. However, the collapse of the Soviet Union slowed the development of the CBF industry in the following decades. In the past decade, Russia, Ukraine and Georgia have increased their investment in R&D, and their CBF industry has developed rapidly (Dalinkevich et al., 2009). Among the representative manufacturers is Kamenny Vek, which was established in 2002 in Dubna (a government laboratory) in Moscow. Kamenny Vek is the largest CBF producer in Russia, combining the technology development and market coverage capabilities of Basaltex in Europe and Showa Highpolymer in Japan.

The United States and Japan have studied CBF and issued CBF batch production technology patents, but these countries have not strategically promoted the development of CBF industry. Several major glass fibre companies in the United States, such as Owens Corning, have independently researched and developed CBF. Owing to the high technical difficulty of producing basalt fibre and in consideration of the company's own development, they essentially abandoned the CBF development project. As early as 2000, Japan established a joint venture with Toyota in Kiev, Ukraine, to produce CBF products used to make materials for automobile silencers, which were all sold back to Japan. However, owing to the national strategy of developing CF, Japan did not produce basalt fibre at home but focused on CF and high-performance glass fibre. In spite of this, Japan has been conducting research on CBF ore raw materials, production and preparation technology and penetrating agents and has mastered the key production technology of CBF. At present, although there is no large-scale CBF production in Japan, CBF is always produced in the domestic laboratory furnace. In addition, Japan has paid great attention to the application market of CBF and has been studying and exploring the application technology of CBF in automobile, civil engineering, structural reinforcement and repair, high temperature resistance and other fields.

Since the 1970s, the China Building Materials Academy and Nanjing Fiberglass Research and Design Institute have studied the production technology of CBF but have not achieved industrial production. Since 2002, China has learnt from the CBF production technology of Russia and Ukraine, has developed this technology with China's independent intellectual property rights, and has achieved industrialized production in 2003. The industrialization of CBF in China was initially achieved by private enterprises, which started from the production of CBF and progressed to the application research of CBF with domestic universities and research institutes such as Southeast University, Donghua University and Harbin Institute of Technology.

With new applications and increased demand for CBF, a fresh look at the production technology is required to remove bottlenecks and increase the scale of production. China's schools and institutes are studying the production process and equipment design. Under such circumstances, China's first national basalt fibre engineering research centre, the National and Local Joint Engineering Research Center for Basalt Fibre Production and Application Technology, was established in 2011 and has become an important source of technological innovation in China's CBF industry.

As an emerging industry, the CBF industry is still in its primary stage and the global intellectual property and rights competition characterized by the patent application protection of CBF and composite materials-related technologies is increasingly fierce. At present, Russia, Ukraine, China, Georgia, Germany, Belgium, Austria and other countries have manufacturers, but the capability is concentrated in Russia, Ukraine and China. There are 15 CBF manufacturers in China and more than 10 CBF manufacturers in other countries. In 2014, the total sales volume of CBF yarns abroad was approximately 4000 tons, and the output of CBF in China was nearly 4500 tons. The total production of CBFs has exceeded that of other countries.

13.2 Production of basalt fibres

13.2.1 Screening of basalt ore raw materials

The raw material of CBF is basalt. Basalt here refers not only to basalt ore but also to basalt, basalt andesite, andesite, diabase and other igneous rocks that can produce CBF. The 'basalt' mentioned in this book refers to basalt in a broad sense.

Basalt is the single-component ore raw material provided by nature. The mineral facies and composition of basalt in different areas vary, as do the properties of CBFs produced in different areas. Although basalt is widely distributed on the earth, not all basalt is used to produce CBFs. Therefore, before the production of CBF, the rock must be screened to select the igneous rock suitable for production.

Igneous rocks are rocks formed by magma intruding into the crust or cooling crystallization after eruption from the surface. According to the location of solidification, scientists categorize igneous rocks into two types: volcanic rock and plutonic rock. Volcanic rock refers to all the products ejected from magma (flow and fragmentation of rock debris called igneous debris) onto the earth's surface. Plutonic rock is rock that has solidified underground.

Geologists systematically classify and name igneous rocks according to their structure, colour, minerals present in the rocks and chemical composition. During field sampling, rocks are usually classified and named according to their structure. Once the rocks are brought back to the laboratory, they can be chemically analysed, classified and named.

The geological characteristics of igneous rocks (such as structure, chemical composition and mineral composition) are the basis for selecting rock materials suitable for the production of CBFs.

13.2.1.1 Classification of igneous rocks

Igneous rocks have different characteristics, including melting temperature, viscosity (flow resistance), chemical composition and mineral type, quantity and size. As most igneous rocks are of microcrystalline, cryptocrystalline and vitreous structure, mineral identification from thin sections is difficult. The simplest and most discriminating rock classification method is according to the chemical composition, which has the following advantages:

1. The chemical composition of the rock reflects the composition of the parent magma and provides information about the source of the rock.
2. The chemical composition of the magma determines the minerals and proportion of minerals to be crystallized.
3. A group of hypothetical minerals that can crystallize from magma of the same chemical composition (called the standard norm) is useful for comparing rocks.
4. The chemical composition can change continuously, which is beneficial to the identification of different rocks.

13.2.1.2 Classification method based on silica content

Silicon dioxide (SiO_2) is one of the most important oxides in igneous rocks, and the regular variation of SiO_2 content is therefore the main basis for the classification of igneous rocks. Igneous rocks are divided into ultrabasic rocks ($SiO_2 < 45$ wt%), basic rocks (SiO_2: 45–53 wt%), intermediate rocks (SiO_2: 53–63 wt%) and acidic rocks ($SiO_2 > 63$ wt%) according to the SiO_2 content (weight%).

Ultrabasic rock

This is a large class of igneous rocks with SiO_2 content less than 45% such as peridotite and kimberlite. The rock is dark in colour (dark green or black) and is relatively dense (3.1–3.6 g/cm^3). It is characterized by unsaturated SiO_2, FeO and more components of MgO (up to 15–24 wt%) and is almost entirely composed of ferro-magnesite minerals. The mineral composition is mainly olivine and pyroxene, followed by amphibole and biotite, without quartz. Common metallic minerals in ultrabasic rocks include magnetite and chromite. Such minerals are rarely distributed in the crust, accounting for only 0.4% of the exposed area of igneous rocks. Ultrabasic rocks are mainly of plutonic origin and are rarely found in epigenetic rocks and extrusive rocks.

Basic rock

This is a large class of igneous rocks with SiO_2 content of 45−53 wt% such as gabbro, diabase, and basalt. It features a SiO_2 content higher than that of ultrabasic rocks but still unsaturated and a CaO, FeO and MgO content of approximately 20%; the rock therefore contains a large quantity of basic plagioclase and dark minerals, without quartz. The mineral composition is mainly pyroxene and basic plagioclase, followed by olivine, amphibole and biotite. Metallic minerals in basic rocks often include titanomagnetite or ilmenite. The colour is darker than ultrabasic rock and intermediate rock. Such minerals are more widely distributed in extrusive rocks (such as basalt) than plutonic rocks. Basic rocks have a transitional relationship with ultrabasic rocks and intermediate rocks.

Intermediate rock

This is a large class of igneous rocks with SiO_2 content of 53−63 wt% such as diorite and andesite. The SiO_2 content is saturated or nearly saturated, the Fe_2O_3, FeO, MgO and CaO contents are lower than those of the basic rocks, and the K_2O and Na_2O contents are increased to 5%−6%. Therefore, the dark minerals decrease while the light minerals increase, and the ratio of the two is 1:2. Among the mineral components, the dark minerals are dominated by amphibole, followed by pyroxene and biotite, and the light minerals are dominated by intermediate plagioclase, sometimes including orthoclase and a very small quantity of quartz (below 5%). Such minerals are widely distributed, and extrusive rocks (such as andesites) are much more common than plutonic rocks. Intermediate rocks have a transitional relationship with basic rocks and acidic rocks.

Acidic rock

This is a large class of igneous rocks with SiO_2 content greater than 63 wt% such as granite and diorite. It features a saturated SiO_2 content, but the Fe_2O_3, FeO, MgO and CaO contents are not high. It has a large number of light-coloured minerals, mainly quartz, potassium feldspar and acid plagioclase. The dark-mineral content is very low, only approximately 10%, and is usually composed of biotite.

13.2.1.3 Continuous basalt fibre industry with common volcanic rock

Gabbro

The SiO_2 content in gabbro is 45%−52%, and the ferro-magnesite mineral content is 40%−90%. The main mineral components are pyroxene (such as common pyroxene, diopside and purpurine) and calcium-rich plagioclase. The secondary minerals are olivine, amphibole, biotite, quartz, orthoclase and iron oxides. The colour is generally dark grey, black grey or grey green, with other colours. Gabbro has full crystalline, coarse, medium-grained structure and massive structure. It can be divided into gabbro, syenite or syenite and olivine gabbro, according to its mineral composition. It occurs as stocks, sills, strains, basins, lids and dikes. It is believed that basalt magma crystallizes deep in the earth. Gabbro is characterized by a specific gravity of 2.8–3.1 g/cm^3, low porosity, weathering resistance and compressive strength of 2000–2800 kg/cm^2, and it is a good construction material.

Anorthosite

This basic plutonic rock, also known as the labradorite, consists of 90% basic plagioclase (usually perthite or calcite feldspar). Dark minerals (such as pyroxene, amphibole, and biotite) do not exceed 10%. Anorthosite sometimes contains small quantities of sulfides of potassium feldspar, titanomagnetite, rutile, iron and copper. Plagioclase is grey white and smoky grey, with holocrystalline semi-automorphic coarse-giant grain structure and massive structure. Anorthosite is often associated with gabbro occurring as veins or as independent intrusions. It is generally believed to be the product of the differentiation of deep underground basic magma.

Diabase

The SiO_2 content in diabase is 45%–52%. The composition is similar to that of gabbro. The main mineral composition is basic plagioclase and pyroxene, sometimes also containing olivine and magnetite. Plagioclase is usually automorphic, early crystallization and disorderly arrangement, with dark mineral pyroxene distributed in it. This structure is the diabase structure. If the plagioclase occurs in porphyritic structures, the rock is diabase porphyrite.

Freshly exposed diabase is dark grey or grey green and becomes lighter after weathering. It has a holocrystalline granular structure, often with typical diabase structure and dense massive structure, sometimes with vesicular or amygdaloidal structure. Diabase usually occurs in intrusions such as dikes or sills in bedrock. Diabase is generally believed to be formed by the crystallization of basaltic magma at shallow depths, and some of the diabase belongs to the eruptive facies. Diabase is characterized by a specific gravity of 2.8–3.0 g/cm^3 and compressive strength up to 2000 kg/cm^2 and can be polished well. It is commonly used as a building material and paving material.

Basalt

Basalt is the most widely distributed basic extrusive rock, with a SiO_2 content ranging from 45% to 52%, and is mainly composed of basic plagioclase and monoclase pyroxene, followed by orthopyroxene, peridot and amphibole. Pyroxene in basalt often has a complete crystalline, short columnar structure; plagioclase in basalt has a fine needle-like nondirectional arrangement. The structure is usually fine grain to cryptocrystalline or vitreous and is sometimes medium grained. In general, basalts have porphyritic texture, vesicular or amygdaloidal structure. Basalt is commonly black or brown-black; the weathered surface appears reddish brown. Basalts usually has holocrystalline-semicrystalline cryptocrystalline to fine-grained textures with dense massive, stomata or almond structures. Basalt often occurs in a large area of rock flow or crust, and sometimes in dikes, beds, stocks, basins, caps, dikes, etc. Basalt is characterized by a specific gravity of 2.7–3.3 g/cm^3, porosity of 1.28%, weathering resistance, compressive strength up to 3000 kg/cm^2 and hexagonal columnar joint development; it is easy to exploit. Basalt is divided into tholeiitic and alkaline basalt, the characteristics of which are shown in Table 13.1.

Basaltic andesite

Basaltic andesite is an intermediate extrusive rock with a SiO_2 content of 52%–57%. It is a volcanic rock with feldspar components from andesite and mafic minerals in basalt and is cryptocrystalline or semicryptocrystalline with a structure similar to that of andesite.

Andesite

Andesite is an intermediate extrusive rock. The SiO_2 content varies in the range of 57%–63%. Oversaturated SiO_2 occurs in 98.5% of andesites, and the common mineral quartz appears (mostly less than 15% content). The porphyritic crystals are mainly plagioclase, pyroxene (common pyroxene, purpurine), amphibole and biotite. Pyroxene is black or grayish green and short columnar, whereas amphibole is acicular. The matrix is mainly composed of interwoven structure and andesite structure (vitreous interwoven structure) and composed of plagioclase (feldspar, medium feldspar-dominated), pyroxene, chlorite and andesite glass; alkaline feldspar and quartz are rare, with only individuals filling in the intergranular space. Andesites are mostly dark grey or black.

Table 13.1 Characteristics of tholeiite and alkaline basalt.

	Tholeiite	**Alkaline Basalt**
Substrate	Usually fine-grained, intergranular structure	Usually fairly coarse, intergranular or diabase structure
	No olivine	Olivine is common
	Clinopyroxene = ordinary pyroxene	Titanium-iron Ordinary Pyroxene (Red)
	Orthogonal (perilla) pyroxene is common	Orthopyroxene is uncommon
	No alkaline feldspar	Potential occurrence of interstitial alkaline feldspar or feldspar-like rocks
	Gap-filling glass and/or quartz are common	Gap-filling glass is rare, quartz is not.
Phenocryst	Olivine is rare	Olivine is common
	Orthopyroxene is uncommon	Orthopyroxene does not
	Plagioclase is common	Plagioclase is relatively rare
	Clinopyroxene is grey-brown ordinary pyroxene	Clinopyroxene is titaniferous ordinary pyroxene (red)

Dacite

Dacite is intermediate-acid extrusive rock. The chemical composition and mineral composition of andesite are between those of andesite and rhyolite, usually with vitreous, vitreous interwoven or felsitic structure, with porphyritic and rhyolitic structure. Porphyritic crystals are mostly composed of intermediate plagioclase, some alkaline plagioclase, and sometimes contain a small quantity of quartz. The matrix is fine feldspar and quartz. The rocks are light in colour, grey or off-white. Dacite is often associated with rhyolite, trachyte and andesite.

Rhyolite

The SiO_2 content in rhyolite is greater than 69%. The degree of crystallization is not high, except in a few porphyritic rocks, and the matrix is mostly vitreous. The porphyry is mainly composed of potassium feldspar and quartz. The crystal shape is square and plate-like, with glassy luster and joints. The rocks are grey, pink or brick red.

13.2.1.4 Selection of raw materials for production of continuous basalt fibre ores

The basalt raw materials for the production of CBFs have strict chemical and mineral composition requirements, and not all basalt materials can be used to produce CBFs.

The types and contents of the major mineral components vary as follows (percentage by weight):

plagioclase 45%–60%, pyroxene 15%–30% (for basalt and some basalt andesite) or quartz 15%–27%, orthoclase 10%–18%, plagioclase 30%–50% (for some basalt andesite, andesite and rhyolite).

The chemical and mineral components of basalt are the basis of CBF production, determining the melting performance, fibre-forming performance, physical and chemical properties, production process, production equipment and product quality of the CBF. With a natural single-component raw material, the CBF production process has the following characteristics:

1. The chemical composition and mineral composition of the basalt ore from different geographic regions vary, and not all basalt can produce CBF.
2. The chemical composition and mineral composition of basalt ores fluctuate, and the key to guaranteeing the quality and performance of CBF products is to resolve the fluctuation of raw material composition of the ores.
3. As basalt is a single component of natural silicate raw material, CBF production generally does not need other ingredients, but the homogeneity of raw materials is demanded.
4. The mineral components in basalt determine the melt properties, fibre-forming properties, melt homogeneity and crystallization properties of CBFs. In the case of sufficient homogenization of the basalt glass melt, the physical and chemical properties of the fibre are determined by the chemical composition.
5. During the cooling process of volcanic magma eruption, the chemical reactions between components of basalt have been completed, so the silicate formation stage is not required. This is different from the melting process of glass, that is, the process in which various raw

materials form glass through silicate chemical reaction. This determines the difference between the CBF production process and the glass fibre process.

6. The melting of basalt is also a process in which silicate crystals (mineral compositions) open their lattices to form a glass network. Generally speaking, the lattice opening process requires much more energy than the process of glass melt forming silicate, which is one of the reasons for the high melting temperature of basalt.
7. As basalt glass melt has a high wettability, it is easy to product 'diffuse flow' at the leakage outlet wall of the leakage plate. If the 'diffuse flow' is severe, it may affect the wire-drawing operation.
8. Basalt glass melt has a strong tendency to crystallize, with an upper limit of crystallization temperature generally greater than 1250°C.
9. Basalt glass melt has a high hardening speed and cooling speed, a narrow fibre-forming temperature range, and varies sharply on viscosity.
10. The ferric oxide content in basalt is high, and the blackness coefficient of the basalt glass melt is close to 0.9; the dielectric property of the melt is therefore poor, which has a significant impact on the melting and drawing processes. However, because of the high iron oxide content in basalt, CBF products have excellent thermal insulation performance and chemical corrosion resistance.

13.2.2 Melting process of basalt ores

The process of basalt ore raw material melting at high temperature to form homogeneous, stable and suitable basalt glass liquid for fibre forming is called basalt melting.

Basalt glass melting is an important step in CBF production. The crystallization, homogenization and defects of basalt glass (such as stones or streaks), as well as the quality, performance stability, output, qualification rate, production cost and kiln life of CBF are closely related to the melting of basalt glass. A full understanding of the melting process and melting mechanism of basalt is necessary to formulate a reasonable melting system and production process and to obtain high-quality CBF.

The homogeneity and stability of basalt melt in the process of basalt melting have an important influence on the drawing process and fibre quality stability. Basalt ore mining, crushing and particle size control, as well as furnace structure, refractory materials, thermal process, melting system and other factors have an important impact on the homogeneity of basalt melt. In addition, the study of the basalt melting mechanism can effectively guide the adjustment and progress of the production process.

13.2.2.1 Melting process of basalt glass

The processing of raw materials used for the production of CBF can be roughly divided into four stages, in the entire process from basalt ore to melting: crushing and homogenization of basalt ore, formation of basalt melt, clarification and homogenization of basalt glass and cooling of basalt glass.

Comminution and homogenization of basalt ores

The particle size of basalt ore is within a certain range, which can promote the uniform mixing of ore raw materials, accelerate the melting of basalt melt and improve the homogeneity and quality of the basalt melt. Therefore, the bulk basalt ore must be crushed and homogenized.

The crushed basalt ore should be generally massive with a particle size of no more than 3 mm. Massive basalt can reduce the loss of molten materials. After grinding, the surface area of the raw material increases and the dispersion increases, which correspondingly increases the contact area between the ore particles, accelerates their melting, improves the melting speed of the ore and is beneficial to the homogenization of the glass liquid.

The crushed basalt ore must be mixed evenly to keep the chemical and mineral composition of the ore uniform. This is beneficial in improving the homogeneity of basalt melt and ensuring the stability of CBF quality, physical properties and chemical properties.

Basalt melt formation stage

Basalt ore begins to melt in the continuous heating process. The fusible eutectic mixture (such as diopside and calcite) first melts. With the generation of liquid phase, the melting of other minerals is promoted. At this stage, basalt melt has high viscosity, poor fluidity and poor homogeneity of glass liquid; the basalt melt is therefore uneven in chemical composition, mineral composition and properties. The formation of basalt melt is commonly in the range of 1250 ~ 1450°C.

Basalt melt stage of clarification and homogenization

As the temperature of the basalt melt continues to rise, the viscosity of the basalt melt decreases, and the dissolved bubbles (carbonates, sulfates and gases carried in the raw material) in the basalt melt rise to the surface of the melt and break. As there are few bubbles in basalt melt, the phenomenon at this stage is not as obvious as with glass fibre.

When the basalt glass melt solution is kept at high temperature for a long time, the viscosity of the solution is further reduced, which is conducive to the diffusion and convection of the solution at this stage and causes the chemical composition and mineral composition of solution to gradually become uniform. Basalt glass melt clarification and homogenization occur in the range of 1500–1700°C.

Basalt glass cooling stage

The homogeneous basalt glass solution is cooled until it reaches the forming temperature of CBF (when the viscosity of the basalt glass solution is approximately 10^3 Pa s). Crystallization of the basalt glass melt should be avoided during the cooling process.

13.2.2.2 Changes and reactions of components of basalt glass

In the process of basalt melting, the components of raw materials are interlaced with solid-phase reaction, liquid-phase reaction and gas-phase reaction, which are very complex reactions and changes.

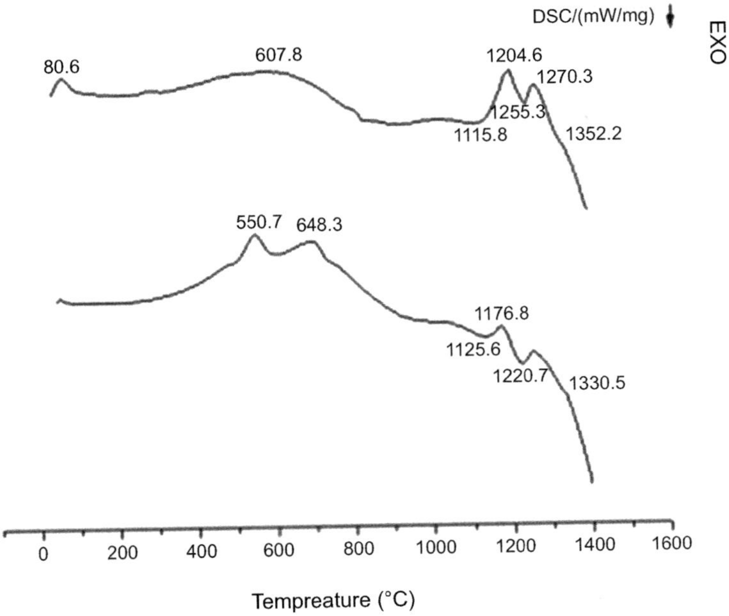

Figure 13.1 Differential scanning calorimetry (DSC) curve of basalt glass.

The changes in the basalt melting process can be observed by differential scanning calorimetry (DSC) (Fig. 13.1). There are four endothermic peaks and two exothermic peaks in the DSC curve of the basalt sample. At the first endothermic peak at approximately 100°C, adsorbed water is driven off in the process. The second endothermic peak at 550–680°C is mainly a semiconductor-type endothermic transformation (alpha quartz to beta semiconductor-type transition temperature of 575°C) and carbonate decomposition endothermic. The third endothermic peak at 1145–1220°C has several causes: one is the endothermic decomposition of orthoclase and sulphate crystal transformation or decomposition heat; and the second is heat absorption from plagioclase and diopside beginning to melt and producing liquid phase. The fourth endothermic peak in the 1265–1330°C range is mainly the endothermic basalt ore melting. Regarding the two exothermic peaks, the first exothermic peak in the 980–1200°C range is mainly due to the exothermic oxidation of magnetite and ilmenite. The second exothermic peak in the 1220–1320°C range is the highest crystallization temperature of basalt glass (the upper limit of the crystallization temperature).

13.2.2.3 Heat transfer of basalt glass melts

The diathermal property of basalt glass melt has an important influence not only on the basalt melting but also on the drawing process. Basalt glass melts contain a large number of pigmentation oxides (FeO, Fe_2O_3) and have a dark colour. The absolute blackness of basalt glass melt is close to 0.2, whereas the blackness of typical transparent glass is 0.9. The low thermal absorption coefficient of basalt glass melt has an important impact on the melting process and homogenization.

The cooling rate of basalt glass melt is different from that of ordinary glass melt. At the beginning of cooling, the temperature of the basalt melt surface layer drops much faster than that of the glass, resulting in rapid cooling of the surface layer. However, at a depth of 17−19 mm below the melt, the hardening rate of basalt melt is 1.5-2 times slower than that of alkali-free glass melt. The large temperature gradient in a basalt melt is due to its poor transmittance.

13.2.2.4 Characteristics of basalt glass melting process

Basaltic rock is the product of the hardening of natural silicate magma. According to the characteristics of rock materials, the basalt melting process has the following characteristics:

1. As basalt is a natural multicomponent raw material, the ore raw material does not need to be batching.
2. The basalt has completed the silicate reaction in the magmatic formation process, and therefore no silicate formation stage is required in the basalt melting process.
3. The high iron ion content in basalt and the poor thermal conductivity of iron ions result in high temperature differences and poor thermal permeability of basalt glass. Iron ions greatly influence the hardening speed of glass. FeO can greatly increase the hardening speed of the outer layer of the glass melt while decreasing the hardening speed of the inner layer, and as a result, basalt glass varies sharply on viscosity and has a narrow fibre-forming temperature range. FeO may also be reduced to elemental iron, which reacts with platinum−rhodium to form hypoeutectic substances, causing 'poisoning' of the leakage plate.
4. The basalt iron oxide content is high, the glass melt colour is dark and the diathermy is poor. If heating is only on the surface of a glass melt, the temperature difference between the upper and lower parts of the melt results in poor homogeneity of the melt.
5. Basalt ores have crystals in them that cool at different rates. When basalt glass is cooled, the faster-cooling crystals are smaller and fewer, whereas the slower-cooling ones are larger and more abundant. Basalt fibres crystallize readily.
6. If there are refractory minerals in the basalt ore, these refractory minerals are not conducive to the homogenization of basalt melt, and the basalt crystallizes readily in the cooling process.
7. The melting of mineral crystals in basalt ores requires more energy than for other ores, which means that basalt ores require higher melting temperatures and longer melting time.
8. The temperature range of basalt ore from liquid phase to vigorous melting is narrow and the change happens quickly.

13.2.2.5 Melting mechanism of basalt glass

The basalt melting process takes the basalt ore through a series of reactions and involves changes in the composition of the basalt glass.

According to the melting process of basalt and the structure of basalt glass, the melting mechanism of basalt glass is as follows: at high temperature, the lattice of framework-like or chain-like silicate mineral components (crystals) is opened to form a framework-like or chain-like structure composed of silicon-oxygen tetrahedron, aluminium-oxygen tetrahedron, silicon-oxygen hexahedron and other polyhedrons, which pass through the apex (Bridge oxygen) connects with each other to form an irregular three-dimensional network structure.

In the melting process of basalt, the higher the temperature and the longer the melting time is, the more intense the lattice destruction will be in the material. Therefore, there will be fewer regions for the ordered arrangement of atoms in the basalt glass and the number of disordered structures will increase. At a sufficiently high melt temperature, the number and size of locally ordered structural regions in the obtained basalt glass are greatly reduced, that is, the amorphous degree of the glass increases. In other words, the homogeneity of the basalt glass is greatly increased.

13.2.2.6 Basalt ore melting equipment

The melting of basalt ore is conducted in a kiln, in a process similar to that of glass. However, owing to the poor thermal conductivity and strong corrosion resistance of basalt melt, its kiln is quite different from the glass kiln. The structure of the melting furnace, the selection and quality of refractory materials, the thermal method and other factors directly affect the basalt melting process and the quality of the molten basalt glass. According to structure and capacity, kilns can be divided into crucible kilns and pool kilns. According to the heating mode, kilns can be divided into three categories: full flame kiln, full electric kiln and flame-electric combined kiln.

Crucible kiln

As the scale of CBF production is relatively small in the initial stage, the furnace is relatively small, similar in size to a crucible; this method is therefore also called the 'crucible method'. The advantages are a small furnace, small investment, convenient local process adjustment, and suitability for small batch products and special products. Disadvantages are a single furnace, low thermal efficiency, high energy consumption, low production efficiency, fibre product quality fluctuation and high comprehensive cost.

The production capacity of the 'crucible method' is very low. Generally, the output of the crucible kiln with a 400-hole leakage plate is 60–100 tons per year, and the life of the kiln is relatively short, generally 3–6 months. A low melting rate and high energy consumption result in a small production scale and high production cost.

Pool kiln

Corresponding to a pot furnace, pool kiln refers to a kiln with multiple bushings for fibre forming. In terms of structure, the kiln is a large-scale pool kiln that is generally several times the size of a crucible kiln and has large output, strict temperature control and stable product quality. It consists of the following main parts: a feeding pot, melting part, neck, cooling part (working part), regenerator, flue and chimney. Each part has its own functions and features, and at the same time the parts are closely connected and becomes a dynamic melting system for high-quality basalt melts.

All-electric melting furnace

Electric energy is used to heat the basalt glass melt in the all-electric melting furnace.

The theory of melting the basalt in the all-electric furnace is based on the principle that the basalt melt has electrical conductivity at high temperature (caused by the movement of alkali metal and alkaline earth metal ions). The electric energy is input into the melt and converted into heat energy to heat the basalt glass internally. Usually, an electrode is inserted into the molten basalt glass (generally half-way down the molten glass). Under the action of the electric field, the glass liquid generates heat energy through the movement of alkali metal and alkaline earth metal ions.

The characteristics of the all-electric melting furnace are as follows:

(1) It provides a high melting temperature in the interior and middle of the basalt melt, and heat energy is transferred to the upper and lower parts of the kiln. The heat dissipation is low, heat loss is low, heat utilization rate of the kiln is high (up to 80%) and energy consumption is reduced.
(2) The furnace body is small, occupying a small area and the fabrication and baking times for the furnace are short.
(3) Electrical equipment is used in each system in the production. Automatic control, high production stability, good control and stable quality are therefore easy to achieve.
(4) Electric energy is a clean energy source. No waste gas is produced during the production process, which supports clean production.
(5) As basalt has a high iron oxide content, the thermal conductivity of basalt glass melt is poor, which readily causes great temperature stratification in the kiln. The melting of basalt in the all-electric furnace can greatly reduce such a temperature difference and promote homogenization of the basalt glass melt. For dark glass liquid, such as basalt glass, electric furnaces are the preferred solution.

In 2014, Prof. Wu Zhishen from Southeast University in China led a scientific research team to independently design and study the world's first 'deep liquid level all-electric melting pool kiln' (Wu & Liu, 2014, in press). A series of in-depth studies was conducted on key technologies such as the structure of the all-electric furnace, refractory material structure and molybdenum electrode structure. The kiln had a successful one-time operation, in which the depth of liquid level was adjustable in the range of 600–1000 mm, the temperature fluctuation in the range of 140–460 mm was only 16°C, the internal temperature of the furnace was very uniform and the unit power consumption of the precursor fibre was reduced to approximately 2.5 kW h/kg, which showed strong energy-saving characteristics. The molybdenum electrode can be pushed or pulled out freely, and the life of the kiln can reach more than 2 years. This technology has laid a solid foundation for the large-scale industrial production of 10,000 tons of CBFs.

Flame furnace

The flame furnace melts basalt by heating the basalt raw materials with natural gas and combustion air. Russia and Ukraine use mainly flame furnaces to produce CBFs.

The characteristics of the flame furnaces are as follows:

(1) The heating by the flame furnace is on the surface of the basalt melt. The heat dissipation area around the kiln is large, heat loss is high, heat energy utilization rate is low and energy consumption is relatively high.
(2) Owing to the surface heating and the poor thermal permeability of basalt melt, a great temperature stratification occurs in the melt, which easily causes the melt to be inhomogeneous.
(3) The fuel used in the flame furnace is natural gas. If the atmosphere is improperly controlled, the Fe^{2+}/Fe^{3+} ions in the basalt melt are easily reduced to elemental Fe. The reaction between Fe and Pt readily causes platinum leakage plate poisoning.
(4) The fuel in the flame furnace will generate large number of waste gas, causing environmental pollution.

Combined gas and electricity furnace

Combined gas and electricity furnace melting is a type of heating method that combines electric energy and natural gas combustion energy. It combines the advantages of the all-electric furnace and the flame furnace.

At present, the kiln in China producing CBFs is mainly a single crucible. With the improvement of CBF production technology and the increase in production, the gas–electric combination furnace will be widely used in pool kiln production.

13.2.3 Drawing process for basalt fibre

CBF is manufactured by forming a basalt glass melt into fibrous shape through a platinum–rhodium drawing leakage plate. The fibres with desired diameter are then prepared by a wire-drawing machine at high speed. In the fibre-manufacturing process, the molten basalt glass is extruded through nozzles of the drain plate and pulled by the wire-drawing machine at a constant speed to solidify into a CBF precursor with a certain diameter. The CBF fibre formation is also an important part of the CBF production process, including the design of leakage plates, the position lines layout of the fibre-forming process and the selection and arrangement of the coolers. The drawing process is shown in Fig. 13.2.

The key equipment for the CBF forming process is the platinum–rhodium drawing leakage plate. The factors affecting the fibre formation are the coordinated control of the leakage plate temperature, the melt forming temperature and crystallization temperature, as well as the melt hardening speed and tension control. As the basalt melt is black glass, its blackness coefficient is high and its thermal permeability is poor, resulting in poor uniformity of the internal temperature of the melt. At the same time, because of the high crystallization temperature of the basalt melt, the crystallization is rapid, which makes fibre formation at the leakage plate difficult. This is also a key technology in the CBF industry. Scientific researchers and production engineers in this field in countries worldwide are concentrating their efforts on such problems.

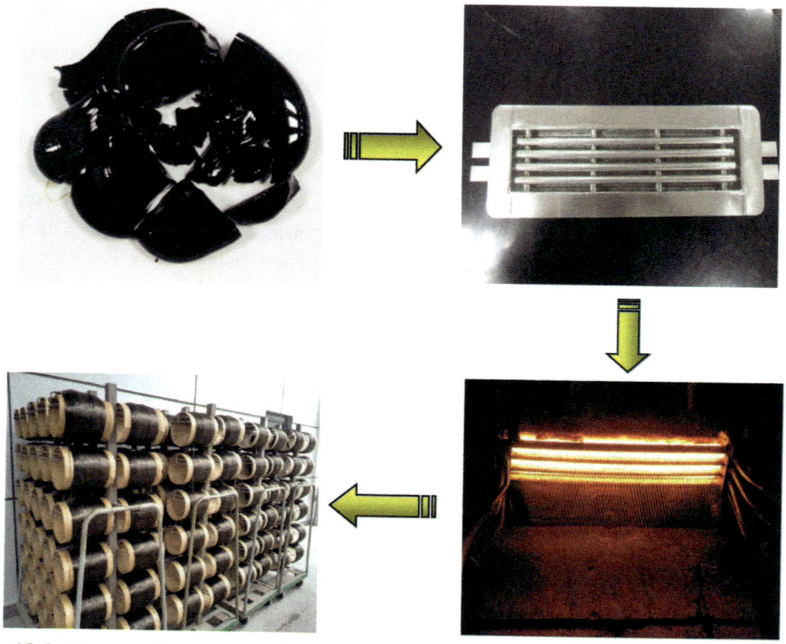

Figure 13.2 Process of continuous basalt fibre manufacturing.

13.2.3.1 Principle of continuous basalt fibre formation

The forming principle of basalt fibre is the same as that of glass fibre. The high-temperature viscous glassy melt is drawn from the leakage nozzle in a drop shape, and then the basalt fibre is drawn and solidified into a continuous fibre of a certain diameter by the wire-drawing machine at a set speed. At the lower part of the nozzle outlet, owing to the balance of the surface tension of the glassy melt and the drawing force, a crescent-shaped part with a gradually thinning diameter is formed, which is called the filament root. The distance from the nozzle outlet to the final diameter of the fibre is called the fibre-forming line. The interval containing the fibre line is called the fibre-forming zone. The distance from the nozzle outlet to the winding point of the fibres on the wire-drawing machine is called the wire-drawing line, which can be arbitrarily specified longer or shorter depending on the technological requirements. During the fibre-forming process, the fibre-forming line is always much shorter than the drawing line.

Whether the filament root and filament-forming line are stable or not is the key to obtaining direct and uniform high-quality fibres and reducing breakage during drawing. Experienced, industry practitioners believe that, on the basis of high-quality melt, as long as several major process parameters (liquid level, leakage plate temperature and drawing speed) are relatively stable, the drawing operation is stable. However, many years of practice have proved that the situation is not so simple. To stabilize the drawing process, in addition to the high-quality melt and these parameters being very important, cooling conditions, drawing ratio, air flow control and other factors have a great impact on the fibre-forming process.

From the rheological point of view, when the basalt melt flows in the leak nozzle, the viscous fluid is deformed by simple shearing under forced conditions, and the filament root is drawn into a glass fibre, in this type of deformation, the viscous free-flowing strand is subjected to uniaxial tension stress. The deformation is unequal over the cross-section and is not isothermal.

The fluctuation of the monofilament diameter of the CBF, fluctuation of the precursor diameter, breakage rate, and spinnability are all related to the fibre-forming process.

13.2.3.2 Drawing bushing of continuous basalt fibre

The forming process of basalt fibre is similar to that of glass fibre, both of which rely on a platinum–rhodium alloy leakage plate. The leakage plate used in the manufacturing of basalt fibre has been improved based on the platinum–rhodium leakage plate used for ordinary glass fibre.

In the glass fibre industry, the platinum–rhodium alloy leakage plate is an important piece of drawing equipment, and is usually used in conjunction with a glass fibre drawing furnace. At the time of fibre drawing, the glass liquid exiting the leakage plate flows out through the leakage nozzle at the bottom of the leakage plate and is drawn into a glass fibre having a certain fibre diameter by the wire-drawing machine. The common drawing leakage plate is generally longer than 350 mm. The leakage nozzles have a straight-hole shape and are arranged in a staggered manner on the bottom plate; the leakage plate designed in this manner is suitable for drawing ordinary glass frit with low melting point. For basalt melts with high melting point, poor thermal permeability, high crystallization temperature and fast crystallization, such a design will result in a large number of broken fibres and silk waste, which not only is highly labour-intensive for workers but also gives a low yield of finished product.

Current problems in the research on leakage plates for basalt fibre drawing are as follows:

(1) Materials: Measurements are lacking on the mechanical and electrical properties of materials at high temperatures, especially for oxide dispersion-strengthened materials. At the same time, devices are lacking that are suitable for measuring high-temperature mechanical properties of platinum-based materials, especially for the creep rate. Few in-depth studies exist on high-temperature strengthening mechanisms of materials, and research progress on new low noble metals and on antioxidation and creep-resistant materials has been slow.

(2) Material processing technology: The research on material manufacturing technology is not comprehensive enough. For example, properties of materials prepared by the diffusion-enhanced platinum internal oxidation process are found to be superior to those of materials with the same composition prepared by the powder metallurgy process. In addition, research is required on the influence of processing or annealing texture on the high-temperature mechanical properties of materials, the time and conditions of internal oxidation and the welding of oxide dispersion-strengthened materials.

(3) Leakage plate design aspect: At present, the design of the leakage plate is mainly based on an empirical method and simple thermoelectric calculation. There is no integral combination of experimental data, practical experience, finite element analysis, computer simulation technology and design optimization method. Little research has been reported on the structural improvement and optimization of existing leakage plates.

The leakage plate is key in CBF formation. The uniformity of the fibre diameter and the stability of the drawing operation are determined by the leakage plate. The working temperature of the leakage plate is generally approximately 1280–1320°C and the plate is easily deformed, owing to long-time erosion by high-temperature basalt melt. The leakage plate must be kept structurally stable and not easily deformed over a long time period at high temperature; these are considerations when selecting the material and design of the leakage plate. The platinum–rhodium alloy leakage plate used at present in the glass fibre and CBF industries has been studied and explored for more than 10 years.

At the beginning of the CBF industry, 200-hole and 400-hole drawing leakage-plates were commonly used. In China, 1200 holes were used for industrial production. With the increase in the quantity of basalt glass liquid in the kiln, especially the pool kiln, the drawing leakage plate evolved to large leakage plate with 1600, 2000 or even more holes.

The leakage plate is one of the most important pieces of equipment in basalt fibre production and is directly related to the formation quality and production efficiency of the fibre. In particular, the bottom plate of the leakage plate is prone to high-temperature creep deformation during basalt fibre drawing, thereby severely affecting drawing operation. To improve the service life of the leakage plate, general measures are to increase the strength of the bottom plate material, for example, to increase the rhodium content in the platinum–rhodium alloy or to increase the dispersion of nano-zirconia. At the same time, fabricating the leakage plate is difficult owing to the enhancement of the mechanical properties of the material. One measure to increase the life span is to change the structure of reinforcing bars to delay the deformation of the bottom plate. Generally, L-shaped ribs or ribs with tubular reinforcement (corundum tubes inserted into the tubes) are used, but in actual use, the effect on delaying the deformation of the bottom plate is minimal.

The National and Local Joint Engineering Research Center of Basalt Fiber Production and Application Technology of the Southeast University in China provides a high-service-life reinforcing rib and leakage plate for producing CBFs. The reinforcing rib is arch shaped and includes a steel column and arc reinforcement with upward protrusion at the top. The steel column is welded on the lower surface of the reinforcement; an arch hole is formed between two adjacent steel columns, which can effectively reduce the high-temperature deformation of the bottom plate.

13.2.3.3 Electrode

A molybdenum electrode is generally used in the basalt electrofusion technology. The molybdenum electrode has good high-temperature structural strength and mechanical strength at 1700°C, high-temperature oxidation resistance, good electrical conductivity, low thermal expansion coefficient, corrosion resistance, long service life, good processing performance and other advantages, and it is widely used in the glass and basalt fibre industries. The service life of the molybdenum electrode is directly related to the service life of the kiln.

The molybdenum electrode mainly has a strip, plate or rod shape. Strip and plate molybdenum electrodes have good high-temperature structural strength and the surface current density distribution tends to be more uniform, but they are difficult to replace during use, and the service life of molybdenum electrode is therefore usually the service life of the kiln. The rod-shaped molybdenum electrode has good high-temperature strength and can be continuously pushed into a kiln cycle; it is mostly used in large- and medium-sized electric melting furnaces.

The molybdenum electrode is not easily oxidized at low temperature and is easily oxidized at high temperature. Molybdenum electrode begins to oxidize at 400°C, producing nonvolatile MoO and MoO_2, which are adhering oxides layer, and therefore prevent further oxidation of the molybdenum. However, the oxide layer is oxidized to MoO_3 quickly at 500–700°C. Therefore, the molybdenum electrode must have protective measures to prevent the volatilization of molybdenum electrode during operation.

The protection methods for the molybdenum electrode are mainly water cooling (either direct water cooling or indirect water cooling), inert gas protection and antioxidation coating on the surface of the electrode.

13.2.3.4 Drawing machine

The drawing machine is one of the main pieces of equipment for CBF production and is also one of the factors influencing basalt fibre diameter and quality. At present, most CBF production uses an automatic barrel-changing drawing machine. The diameter of the head of the machine is generally 200–310 mm. The machine can draw a raw wire cylinder on one head, or it can use a separate drawing technology, to draw two original wire cylinders at the same time. The automation is helpful to stabilize the drawing operation. Moreover, as the barrel-changing time is generally less than 10 s, the silk waste produced by barrel-changing can be reduced, such that the wire-drawing yield reaches more than 95%.

The current national Chinese standard JB/T 7910–1999 stipulates the following for the wire-drawing machine: machine type, basic parameters, technical requirements, test method, acceptance criteria, marking, packaging, transportation, storage and manufacturing guarantee.

13.2.4 Surface treatment of continuous basalt fibres

In the CBF drawing process, it is necessary to coat the surface of the basalt fibre with a special surface treatment agent called a sizing agent. The sizing agent is a type of organic mixture, mainly composed of a film-forming agent, lubricant, coupling agent, and other organic emulsion in an aqueous solution. If the fibre surface is not coated with sizing agent, the floss and broken filament may be caused by mutual friction of single fibre filaments, such that the drawing operation cannot be performed. Moreover, hundreds or thousands of single filaments cannot then constitute a complete bundle and cannot be further processed into a variety of products, such as crepe or basalt cloth. Finally, CBFs cannot then be combined very well with the substrate (such as resin, or asphalt) to make fibre-reinforced composite materials. Therefore, the sizing agent is very important for the production and application of CBF (Fig. 13.3).

13.2.4.1 Function and mechanism of sizing agent

The sizing agent can effectively change the surface defects and properties of the fibre, which determines the performance of fibre and the final properties of the fibre-reinforced composite. The main function of the sizing agent is to connect the fibre with the matrix and transfer stress in the interface layer. The functions are summarized as follows:

(1) Clustering and bonding

The sizing agent can make the monofilaments in the CBF precursor bond into a bundle, giving them good cohesion, reducing loose or broken filaments, and facilitating plying, unwinding and textile processing. In the process of short cutting, the integrity of fibre bundle is maintained.

(2) Lubrication and protection

The sizing agent gives the CBF precursor good lubricity, smoothness and protection and reduces the filaments generated by mechanical abrasion in the process of drawing, drying, winding, unwinding and weaving.

Figure 13.3 Continuous basalt fibre sizing agent.

(3) Processing properties

The sizing agent can provide certain processing properties for CBFs, such as short cut, stiffness, bundling, banding, dispersibility and textility.

(4) Fibres and matrix compatibility

The sizing agent gives good compatibility between the CBF and the matrix, which results in good performance by the composite material.

13.2.4.2 Surface properties of continuous basalt fibre

The CBF is cylindrical and has a smooth surface. CBF is made of natural basalt, crushed and melted and then drawn through a leakage plate. After the basalt melt flows out of the leakage plate, the effect of surface tension shrinks the surface area of the melt to the smallest cylinder before cooling and solidification, and this shape is maintained during solidification. Therefore, the CBF has a smooth cylindrical appearance.

The metal ions on the surface of the CBF are associated with protons or hydroxyl groups from air or water owing to the coordination unsaturated. This results in hydroxylation of the surface or negative ions on the surface of the fibre that are anionic and have affinity for cations and groups.

Table 13.2 shows that the surface tension of CBF is approximately 60 mN/m, whereas those of glass fibre and of CF are approximately 40 and 20 mN/m, respectively. The surface tension and surface energy of CBF are higher than those of glass fibre and CF, and the chemical inertness of CBF is high. Therefore, the surface tension and surface energy of the CBF coated with sizing agent should be larger. In the sizing agent formulation of CBF, the bonding component with high polarity and surface energy should be selected, and the surface energy of the sizing agent should be reduced as much as possible by lubricants and other oils. At the same time, owing to the strong polarity of the basalt fibre surface, a component with larger polarity must be used in the sizing agent to give the fibre product surface a higher surface energy, such that the matrix resin can fully infiltrate the fibre surface.

As the surface energy of the CBF is larger than that of the sizing agent emulsion, this emulsion can be well dispersed on the surface of the basalt fibre. When the fibre coated with sizing agent is dried, only the organic coating remains after water evaporates. At this time, the surface tension of the fibre decreases to approximately 45 mN/m. The surface tension of the common matrix material is 35–45 mN/m (for example, the surface tension of polyester resin is 35 mN/m, and the surface tension of epoxy resin and vinyl resin is approximately 42 mN/m). In the production of resin-based basalt composites, as the surface energy of the resin is lower than that

Table 13.2 Surface tension of three kinds of fibres.

Uncoated fibre	Basalt fibre	Glass fibre	Carbon fibre
Surface tension (mN/m)	60	40	20

of the organic coating of sizing agent, the resin can effectively immerse the fibre to ensure the mechanical properties of the composite.

13.2.4.3 Components of sizing agent

The main components of the sizing agent are coupling agents, film-forming agents, lubricants, antistatic agents and other auxiliary components, such as pH regulator, plasticizer and cross-linking agent. The coupling agents in the sizing agent react with the fibre surface to form a strong chemical bond. At the same time, they have good adhesion with the resin matrix and act as a bridge. The film-forming agents have good adhesion to the surface of the fibre, and good solubility with the resin matrix, and produce a curing reaction. The lubricant acts reduce friction. Antistatic agents can effectively reduce the electrostatic effect of processing. Lubricants and antistatic agents must be capable of dissolving or diffusing into the resin matrix to minimize interference with the interfacial adhesion of the sizing agent or other components.

13.2.4.4 Classification of sizing agents

CBF sizing agents are mainly divided into two categories: reinforced sizing agents and reinforced textile sizing agents.

Reinforced sizing agents

The reinforced sizing agents can be used to reinforce thermosetting resins (such as epoxy, polyester, vinyl, phenolic resins), thermoplastic resins (such as polyurethane, polyethylene, polypropylene, nylon, PBT, PET and ABS), asphalt, cement and rubber. The reinforced sizing agent has the most variety and classification. In terms of the secondary processing form of fibre, the agents are divided into chopped and continuous fibre sizing. In terms of reinforcing materials, they are classified into thermosetting plastics and thermoplastic plastics. In terms of the hardness requirements of the fibre, they can be divided into soft-yarn and hard-yarn sizing agents. In terms of compatibility between the sizing agent and the reinforced matrix, the agents can be divided into fast-dissolving, medium-dissolving and slow-dissolving sizing agents.

A sizing agent is used in the composite of a reinforcing sizing agent and a matrix fibre. The composition of the sizing agent includes a coupling agent, which acts as a bridge between the fibre and the matrix resin, such that the resin and the fibre surface can produce a chemical bond or physical adsorption or other bonding behaviour, thereby improving the mechanical, physical, chemical, electrical, ageing resistance and other properties of the composite.

Different types of sizing agents give special properties for the processing and production of fibrous products and FRP products. The reinforced sizing agent must impart the following characteristics to the original filament:

(1) There is good adhesion between fibre and polymer, which is mainly achieved by the coupling agent.
(2) The fibre has good bundling and permeability, which is mainly achieved by the film-forming agent.

(3) The fibre is protected in various processing procedures, and friction is reduced, which are mainly achieved by lubricants.
(4) The fibre has an antistatic property, achieved by adding antistatic agent.
(5) The fibre has good processing properties, such as zonation, stiffness, softness, and cutting, through careful design of the composition and content of the sizing agent.

Reinforced textile sizing agents

Textile wetting is mainly used in fibre textile products. The main function is to protect the fibres in the twisting, plying and weaving process from wear, keeping the fabric smooth, flat and wool-free. As the textile sizing agent must ensure the good textile processing performance of the fibre, the sizing agent must have good bundling, lubricity, film-forming and antistatic properties. The textile sizing agent can be classified into amylose and paraffin sizing agents. In particular, the starch type sizing agent is generally used in the United States, Japan and Western Europe, whereas the paraffin sizing agent is mainly used in China. The raw material of the amylose sizing agent is nontoxic, has a good bundling effect on glass fibres and creates very few loose filaments during weaving. However, the starch and oil in the starch sizing agent will hinder the combining of glass fibre and matrix resin. After the woven fabric is formed, the starch sizing agent can be removed by hot cleaning. The process of hot cleaning causes not only a large loss in fibre strength but also an increase in cost. The affinity and permeability of the paraffin sizing agent are poor and severely affect the mechanical properties, ageing and water resistance of the composite.

For the above reasons, the textile sizing agent has been enhanced to improve not only the performance of fibre textile processing (drawing, twisting, plying and weaving) but also the penetration and bonding force between the fibre and matrix. The key technology of the sizing agent is the film-forming agent. The film-forming agent can be polyurethane resin, epoxy resin or acrylic resin. Generally, polyurethane resin is selected because the film formed by polyurethane resin is elastic and has strong adhesion, which can give good bundling protection to the fibre. At the same time, because polyurethane resin is a strong polar molecule, it has good compatibility and permeability with all resin matrix.

13.2.4.5 Preparation process of sizing agent

The preparation of the sizing agent has an important influence on its application. The concentrations of the single-component raw materials of the sizing agent are relatively high. If these raw materials with high concentration are directly mixed together and then diluted with water, a gel or precipitate is readily produced. In addition, the silane coupling agent must be hydrolyzed sufficiently; otherwise, its application and even the drawing process are affected.

The preparation process of the sizing agent includes the following:

(1) Hydrolysis of coupling agent: The amount of water added into the hydrolysis container is 20–30 times that of the coupling agent, and acid (acetic acid, citric acid, or other organic acid) is added to adjust the pH value to 3–4. With slow stirring (ensuring that no bubbles

are stirred into the water), the silane coupling agent is added slowly (usually in the form of a fine lines thread). The mixture is stirred for more than 30 min until the solution is clear, and the surface does not appear oily, which means the hydrolysis is completed. After hydrolysis, the pH value of the solution must not be greater than 6. If it is too high, the solution should be supplemented with acid until the pH value is adjusted to 5–6.

(2) Film-forming agent dilution: Several types of film-forming agents are fully diluted with 2–5 times the volume of deionized water.
(3) Lubricant dilution: Several types of lubricants are diluted sufficiently with 2–5 times the volume of deionized water (hot water if necessary to dissolve the lubricant).
(4) The antistatic agent is dissolved by adding deionized water.
(5) Final mixture: A small quantity of deionized water is added to the dispensing container of the sizing agent and stirred. Then, the diluted film-forming agent, lubricant, antistatic agent and coupling agent are successively poured into the sizing agent container while continually stirring. Finally, the remaining water is added, and the solution is stirred until well mixed.

The prepared sizing agent should be used as soon as possible, and the general service period should not exceed 72 h. The main purpose is to prevent the coupling agent from losing reactivity after self-polymerization.

13.2.4.6 Drying process of sizing agent

The CBF precursor coated with the sizing agent must be thoroughly dried before use. The purpose of drying is to discharge water and to make the sizing agent part of the fibre surface film, namely, to promote a chemical reaction in the single group of raw material in the sizing agent, to form a film. The drying process of the sizing agent has an important influence on the properties and applications of the fibre.

The drying process of the sizing agent is generally divided into two stages:

(1) The first stage is at 90–110°C for 2–6 h. This stage is the dehumidification stage, namely, expulsion of the water.
(2) The second stage is 110–130°C for 12 h. This stage is the film-forming stage, in which a single component of raw material undergoes a chemical reaction to form a protective film.

The sizing agent drying mechanism on the surface of the CBF is a dynamic process, as shown in Fig. 13.4. When the sizing agent is coated on the surface of the fibre, the sizing agent adheres to the surface of the fibre in the form of particles. When the agent-coated fibre is dried in a heating furnace, the particles diffuse slowly until a protective film is formed on the fibre surface.

The diffusion rate of the sizing agent is related to the formula of the sizing agent and the drying regime. In the sizing agent system, the components of coupling agent can react in the process of wetting agent emulsion and microwave drying process, the other components should be stable and do not react with other substances, to ensure the bundling and flexibility of basalt fibre. The higher the lubricant content is in the sizing agent, the faster the diffusion of the sizing agent will be at the same temperature. However, it is not the case that the higher the lubricant content is, the lower the interfacial adhesion of the composites will be, when the other components are unchanged. In the case of the same agent formulation, the higher the drying temperature is, the

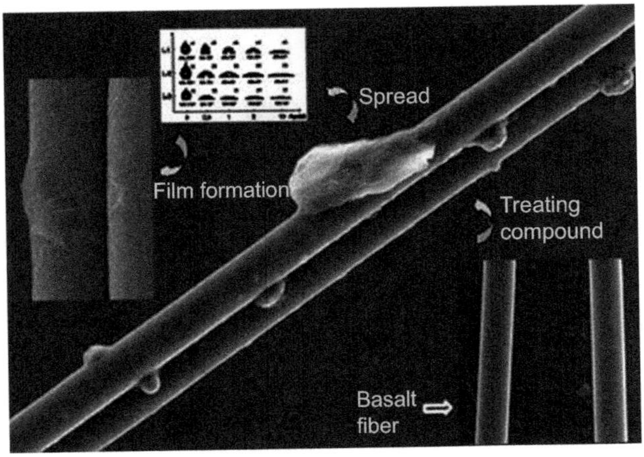

Figure 13.4 Drying mechanism of the sizing agent.

faster the diffusion of the sizing agent will be. If the drying temperature of sizing agent is too high, the moisture migration will be accelerated and the temperatures inside and outside the precursor cylinder will be inconsistent. If the temperature is too low, the film surface of the fibre will not be dense, will not play a protective role and will readily create wool or loose silk. If the drying time is too short, the water cannot be fully discharged, the film surface is insufficient and the quality of precursor is affected; if the drying time is too long, the fibre surface yellows.

For different types of sizing agent, the drying process is slightly different. The drying process of the same agent at different times (such as summer and winter) also varies. The drying process of the sizing agent should be adjusted periodically according to the drying effect, production condition and climate of the fibre.

13.2.5 Processing technology of basalt fibre products

Common basalt fibre products include basalt fibre roving, textile yarn, basalt fibre chopped yarn, bulk yarn, cloth, and felt, and basalt composite materials (Fig. 13.5).

Generally, the CBF precursor is not directly used but is processed into various products such as yarn, cloth, felt, fabric and tape. FRP composite products made of these products and a resin matrix are used in various fields (Fig. 13.6).

13.2.5.1 Chopped yarn

CBF chopped yarn is made with the CBF precursor, untwisted roving or pretreated precursor, or short untwisted roving. According to different application requirements, the chopped length of yarn is generally from 3 to 100 mm.

CBF chopped yarn is used to reinforce cement concrete, asphalt concrete and thermosetting plastics. It can also be made into needle felt, surface felt, chopped precursor felt, nonwoven fabric and roofing felt (or thin felt). The manufacture of dough molding compounds is also possible.

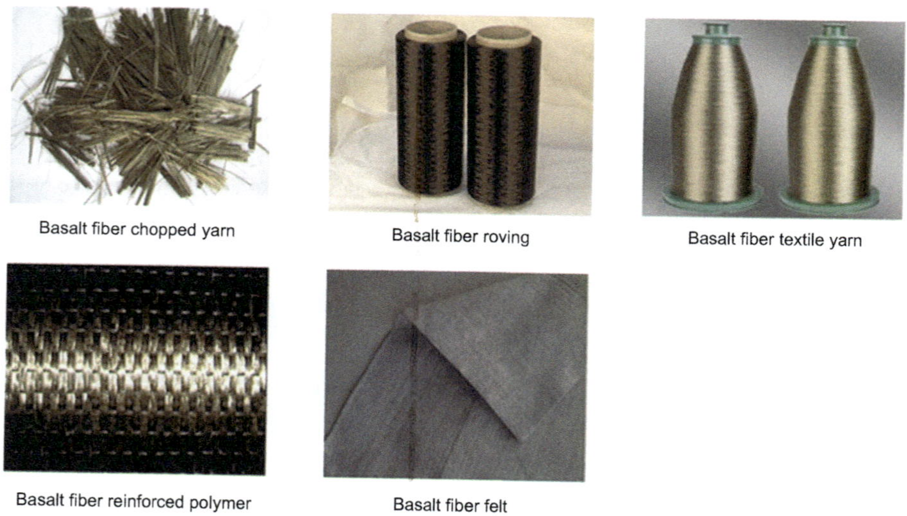

Figure 13.5 Continuous basalt fibre products.

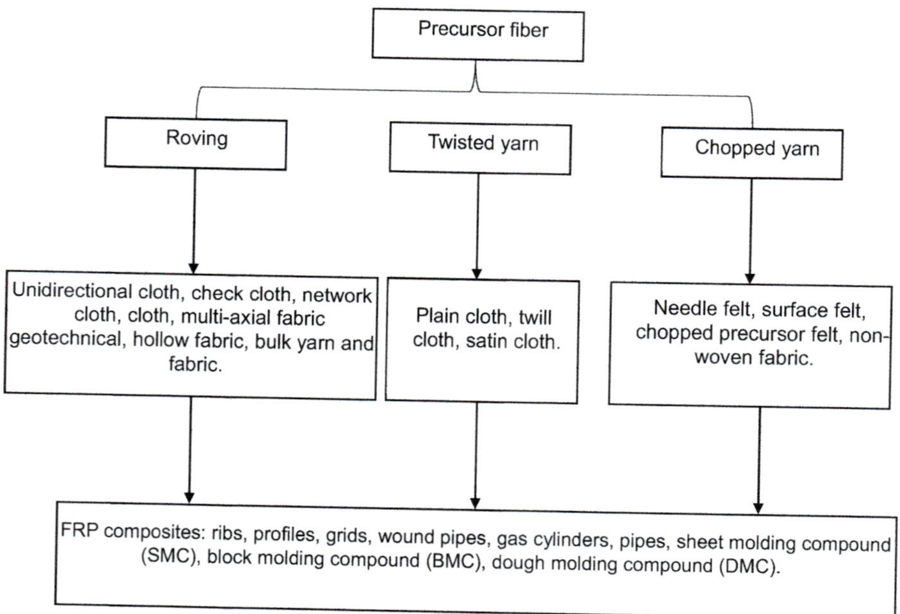

Figure 13.6 Continuous basalt fibre products.

13.2.5.2 Roving

CBF roving is a type of basalt fibre product that is made up of untwisted single parallel or multiple parallel precursors. The diameter of the monofilament is generally $\geq$13 μm. CBF roving is divided into direct roving and ply roving, and the linear density is generally 300 to 4800 tex (grams per kilometre).

CBF roving is suitable for weaving, spraying, pultrusion, molding, winding and chopping. It can be made into unidirectional cloth, woven cloth, grid cloth, geotextile and also composite with resin to make composite profiles, plates, ribs, grids, pipes and winding pipes.

13.2.5.3 Twisted yarn (textile yarn)

CBF twisted yarn (textile yarn) is a yarn obtained by twisting and doubling multiple basalt fibre precursors. The diameter of the monofilament is generally ≤ 11 μm. Textile yarns can be generally classified into weaving yarns and other industrial yarns. Weaving yarn mainly consists of bobbin yarn and bottle shaped cheese yarn.

CBF twisted yarn (textile yarn) is used to make various basalt fibre cloths, such as plain, twill and satin cloths. It can also be used to manufacture cord, sewing thread and electrical insulation materials.

13.2.5.4 Roving fabric

Unidirectional cloth: Unidirectional cloth is made of CBF roving in one direction (0 or 90°) and is stitched with organic fibre in the other direction. The characteristics of the unidirectional cloth are as follows: it has high strength in a single direction; as the roving strands are straight and parallel to each other, and the use of the fibre strength is maximized, the cloth surface is not easy to deform and resin permeability is good, and the type of yarn, the angle of arrangement, the number of layers, and the density of warp and weft can be determined according to requirements, which leads to good designability and miscibility, the cloth can provide greater material selection and design freedom. As a reinforcing material, unidirectional cloth is suitable for winding, pultrusion, hand lay-up and other forming processes. Unidirectional cloth is widely used in building reinforcement, sport equipment, hulls, and wind power blades, and in the aerospace industry.

Multiaxial fabric: Multiaxial fabric is a type of multilayer fabric that is a three-dimensional network structure composed of fibre bundles. CBF roving is sewn with organic fibre in different directions (0°, 90°, ±30°–90°). For example, if the roving is single warp or single weft, it is uniaxial fabric. When the roving is 0° and 90°, +45°, and −45°, or +80°, and −80°, the fabric is biaxial; at 0°(90°) and ±45°, or 0°(90°) and ±80°, the roving is triaxial. Multiaxial fabric has the advantages of structural integrity, design flexibility, good tear resistance and strong interlaminar shear force. Multiaxial fabrics are used in composite reinforced materials, which can be used in aircraft, spacecraft, automobiles, ships and armoured vehicles.

13.2.5.5 Twisted yarn fabric

Continuous basalt twisted yarn fabric is made of basalt fibre twisted as warp and weft yarn—a fabric woven from warp and weft yarn. According to the interweaving rules of warp and weft yarns, twisted yarn fabric is divided into plain, twill, satin and gauze cloths.

Plain weave: This is fabric in which the warp and weft yarns are interwoven at regular intervals (yarns up and down). Plain weave is characterized by compact structure, firm texture and smooth surface.

Twist fabric: Yarn for 2 up and 1 down twill, and 45° left oblique woven cloth. This fabric features less warp and weft interweaving, a smaller gap between warp and weft yarns, and closely arranged yarns. The fabric has higher density, higher strength, soft and loose structure and better elasticity than plain fabric.

Satin fabric: This fabric consists of two adjacent warp or weft yarns with individual points of organization uniformly distributed but discontinuous. As the single weaving points in satin fabric are covered by two adjacent warp or weft floats, the satin fabric surface is smooth, the interlacing points are fewer and the yield of warp and weft yarns is less.

13.2.5.6 Basalt fibre felt

Needle punched felt: Continuous basalt chopped yarn is carded and needled, which causes the fibres to entangle and reinforce each other to form needle punching felt with a certain strength. The needle punching felt is a three-dimensional microporous structure with disordered interlacing of monofilaments, such that it has the characteristics of small space and high porosity. Needle punching felt has good filtration, dust removal, sound absorption, shock absorption and other properties.

Chopped stand mat: This is a nonwoven fabric formed by bonding undirected, evenly distributed chopped fibres together. Its characteristics are moderate hardness, good flexibility, easy to completely penetrate, few voids, easy to process and excellent mold filling and mold covering.

Continuous strand mat: The continuous strand mat is made of CBF strands with nondirectional circular and random, mutual overlapping. It relies on the mechanical interlocking effect of mutual overlapping between the fibrils or a small quantity of binder bonding and is characterized by isotropy, good fluff, good adhesion and high strength.

13.3 Properties of continuous basalt fibre

13.3.1 Basic mechanical properties

13.3.1.1 Tensile strength of basalt fibre

Tensile strength refers to the maximum bearing stress of a material before it is broken. It indicates the ability of a material to resist a large extent of plastic deformation. It is an indicator of the mechanical properties of CBF. Fibre material is an ideal reinforcing material because of its high tensile strength. Table 13.3 shows the tensile strength of different fibres.

Table 13.3 Tensile strength of different fibres.

Fibre	Continuous basalt fibre*	Carbon fibre	Aramid fibre	S-glass fibre*	E-glass fibre*
Tensile strength/ MPa	3100–4840	3500–6000	3200–3700	4020–5000	2800–3800

Note: * refers to the tensile strength of the new ecological monofilaments.

Effect of structural homogeneity on tensile strength of fibres

The micro-inhomogeneity of the fibre structure reduces the tensile strength of fibres. When subjected to external forces, stress concentration will occur at defects in the basalt fibres, resulting in internal cracks that greatly reduces the tensile strength of the fibres. If the homogeneity of basalt fibres is better, the micro-inhomogeneous structure, chemical inhomogeneity and microdefects in basalt fibres will be reduced, and the tensile strength will be increased.

Militky's research has shown that CBFs break because of the heterogeneity of fibres (possibly near small crystals). The narrow melting temperature range of basalt glass leads to poor homogeneity of the glass melt, the high viscosity reduces melt fluidity and the narrow temperature range of fibre forming leads to a rapid increase in viscosity, which will affect the homogeneity of basalt glass, thereby affecting the tensile strength of the basalt fibre.

Chen Xingfen et al. analyzed the relationship between the mineral composition, chemical composition, melting temperature, viscosity, crystallization temperature and homogeneity of the melt and the tensile strength of a basalt fibre monofilament. The chemical composition, mineral composition, melting property and processing parameters of high-strength CBFs were determined. A study on improving the tensile strength of CBFs by mixing basalts was conducted. By introducing high-SiO_2-content basalt into low-SiO_2-content basalt, the tensile strength of basalt fibres can be improved and the melting properties of basalts can be optimized. The tensile strength and melting properties (viscosity, crystallization upper limit temperature and fibre-forming temperature range) of blended basalt fibres were tested and compared with the properties of natural single-component basalt fibres. The effects of chemical composition, mineral composition, and homogeneity of melt on the tensile strength of basalt fibres were discussed, which provides basic research for improving the stability and tensile strength of CBFs.

The initial liquid temperature and hemispheric point temperature of the basalt mixture decrease slightly, which is conducive to the melting, clarification and homogenization of basalt glass and to the improvement in tensile strength of basalt fibres.

Effect of surface microcracks on tensile strength of fibres

Griffith believed that the failure of glass began with surface microcracks, and with the crack propagation, the whole sample was broken. Owing to the existence of microcracks, the tensile strength of fibres decreases.

There are two reasons for the surface microcracks of basalt fibres. According to Zack, each fibre is subjected to a certain stress under tension during wire drawing. When the stress acts on the hardened fibre shell, surface microcracks will occur. For glass whose inner layer is still in a plastic state, the stress is dispersed and the glass does not crack easily. Increasing the hardening speed of glass and reducing the tension of fibres during forming are beneficial for reducing the microcracks on the fibre surface and improving the strength of fibres.

The mechanical damage and chemical corrosion on the fibre surface form surface microcracks. Coating the fibre surface with a protective film can reduce mechanical damage on the surface.

13.3.1.2 Elastic modulus of basalt fibres

Elastic modulus is a physical quantity that characterizes the relationship between the stress and strain of materials and represents the resistance of materials to deformation. The fibres generally obey Hooke's law at low and normal temperatures; the law can be expressed by the following formula:

$$E = \frac{\sigma}{\varepsilon}, \tag{13.1}$$

where σ is the stress, ε is the relative longitudinal deformation, and E is the elastic modulus.

In addition, the dynamic properties of materials (such as elastic modulus, loss modulus and tangent angle of mechanical loss) can be calculated using acoustic parameters and density of materials. The elastic modulus of CBFs is measured by the acoustic method in Russia. That elastic modulus is calculated according to the following formula:

$$E = c^2 \rho, \tag{13.2}$$

where c is the speed of sound, ρ is the density of fibres and E is the elastic modulus.

Table 13.4 shows the speed of sound transmission in fibres obtained at different drawing speeds. The data in this table show that the fibres obtained at higher drawing

Table 13.4 Elastic modulus of continuous basalt fibers determined by acoustic method.

Drawing Speed/ $m \cdot min^{-1}$	Average Diameter of Fibers/μm	Speed of Sound Propagation/ms^{-1}	Elastic modulus/Gpa
1580	15.0	5632	90.8
1760	14.0	6010	104.0
2070	11.0	6196	110.6
3060	9.0	6285	113.8

speed have higher sound propagation speed, and the dynamic modulus of elasticity of these fibres as calculated by Eq. (13.2) is therefore also higher.

When the dynamic elastic modulus of CBFs with different diameters was measured, the elastic modulus was found to decrease with increasing diameter.

The elastic modulus of CBFs is determined by the chemical composition and glass network structure. SiO_2 and Al_2O_3 make the glass network structure tight and increase the elastic modulus. Na^+, K^+ and Ba^{2+} oxides with large ionic radius and low charge are not conducive to the increase in elastic modulus. Ions with small ionic radius and strong polarization ability (such as Li^+, Be^{2+}, Mg^{2+}, Al^{3+}, Ti^{4+} and Zr^{4+}) can improve the elastic modulus. When K_2O replaces Na_2O, the elastic modulus of glass increases slightly. The effects of various oxides on elastic modulus are as follows: $CaO > MgO > B_2O_3 > Fe_2O_3 > Al_2O_3$.

The estimation of elastic modulus of glass based on glass composition has been studied. Some empirical formulae have been proposed since the 1920s and 1930s. As the elastic modulus of glass was mainly determined by a static method, the error in the empirical calculation formula derived from this method was more than 10%, and the elastic modulus of only some silicate glasses with narrow glass composition could be calculated. Since the appearance of the dynamic method for measuring glass elasticity in the 1950s, ample elastic data of silicate glasses have been obtained, and some calculation methods have been proposed. The representative Makishima method and Mackenzie method were developed in the 1970s. The Makishima method can calculate the elasticity of glass, but the calculation error is very large (5%–10%); the Mackenzie method has a calculation error of 2%–3% but gives only some properties of 11 oxides. Both methods are cumbersome.

Temperature resistance of basalt fibres

The so-called high-temperature resistant fibre often refers to fibres that can be used for a long time at temperatures above 180°C, that is, the physical and mechanical properties can be maintained at this high temperature or be obtained after a long service time at high temperature. The performance is as follows:

(1) No change in size at high temperature.
(2) High softening point and melting point.
(3) High ignition point.
(4) High decomposition temperature.
(5) Retention of general characteristics after long-term exposure to high temperature.
(6) Generally necessary properties for fibre products, such as softness, elasticity and processing properties.
(7) Flame retardant or nonflammable properties.

Basalt fibres have excellent flame-retardant properties and fully meet the above requirements. Basalt fibres have a wide range of service temperatures. The minimum and maximum service temperatures are −260 and 900°C, which is a much wider range than that of E-fibres (−60 to ~450°C) and para-aramid fibres (−196 to ~204°C). As for CFs, the oxidation resistance is poor, and CO_2 and CO are produced at 300°C. Some studies have shown that the strength of basalt fibres can be maintained at

85% after breaking when they work at 400°C, and at 80% of the original strength after breaking when they work at 600°C. In addition, if basalt fibres are pretreated at 780 ~ 820°C, they can work at 860°C without shrinkage, whereas the strength of mineral wool with good temperature resistance can be maintained at only 50% or 60%. Glass wool is completely destroyed after breaking.

To study the high-temperature resistance, CBFs were heated at 100, 200 and 400°C for 2 h in air and then cooled naturally at room temperature. The results for the measured tensile strength are shown in Fig. 13.7. The tensile strength of CBFs decreases with increasing temperature. When the temperature is in the range of 25–200°C, the tensile strength decreases slightly. However, when heated at 400°C for 2 h, the tensile strength decreases significantly but can still be maintained at more than 60%.

Temperature-resistant continuous basalt fibres

(1) Effect of composition on temperature resistance of CBFs

1. Effect of SiO_2 and Al_2O_3 on temperature resistance of CBFs

To fundamentally improve the high-temperature resistance of CBFs, the composition of the basalt raw materials must be changed. That is to say, basalt raw materials suitable for preparing temperature-resistant basalt fibres should be found or be injected with oxides such as SiO_2 or Al_2O_3. The high-temperature resistance of CBFs can be improved by improving the connection degree of the glass network structure.

By adding different quantities of SiO_2 and Al_2O_3 to basalt raw materials, the temperature resistance of CBFs increases, as shown in Table 13.5 and Fig. 13.8. The strength retention rate of CBFs is 63.73% at 400°C and 40.55% at 500°C. With the addition of SiO_2 and Al_2O_3, the strength retention of CBFs increases by approximately 4%. The fibres do not adhere when heated at 900°C for 2 h, as shown in Fig. 13.8.

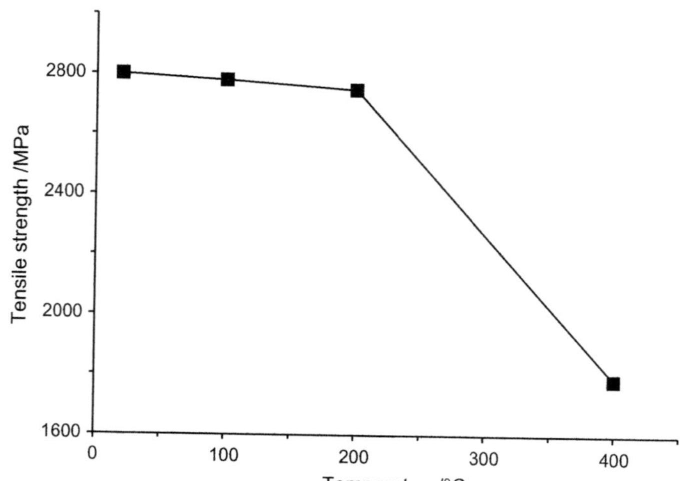

Figure 13.7 Temperature resistance of universal continuous Basalt fibers.

Table 13.5 Temperature resistance of CBFs treated at 400 and 500°C.

Sample	Strength retention rate/%	
	400°C	500°C
B0	63.73	40.55
B1	64.19	41.96
B2	65.78	42.22
B3	66.71	43.07
B4	67.76	44.33

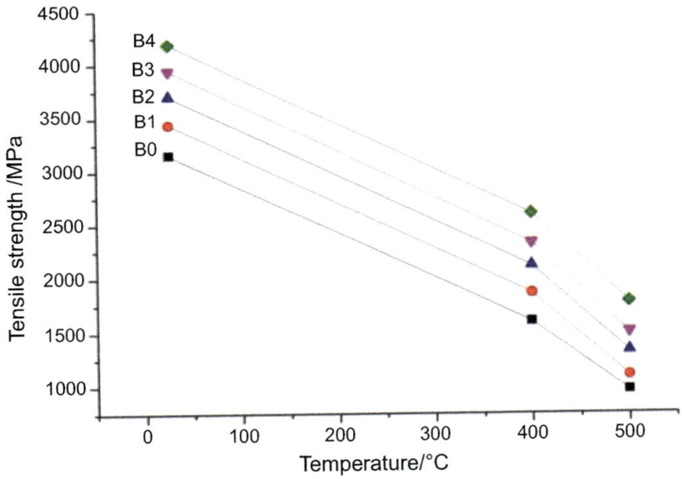

Figure 13.8 Tensile strength of samples at different treatment temperatures.

2. Effect of oxidant on temperature resistance of CBFs

Owing to the aggregation of Fe^{3+}, the temperature resistance of CBFs is improved. The internal polarization of Fe_t^{3+} on the skeleton of the glass network can make the network connect tightly. Adding an oxidant such as CeO_2 to basalt can increase Fe^{3+}, thereby improving the temperature resistance of CBFs, as shown in Table 13.6 and Fig. 13.9.

CeO_2 can react with Fe^{2+} at high temperature to produce Fe^{3+}. CeO_2 changes from a high-valent oxide to a low-valent state. Specifically, the following chemical reactions occur during the melting process of glass:

$$2CeO_2 + 2FeO = Fe_2O_3 + Ce_2O_3$$

Proper heat treatment of basalt fibres in the range below the softening temperature can improve the high-temperature resistance of CBFs and change basalt fibres from a glass state to a glass-ceramic state, that is, basalt glass-ceramic fibres. The heat

Table 13.6 Temperature resistance of samples treated at 400 and 500°C.

Sample	Room Temperature	400 °C		500 °C	
	Strength/ MPa	Strength/ MPa	Strength Retention rate/%	Strength/ MPa	Strength Retention rate/%
BF0	3200	1600	50.01	965	30.16
BF1	3253	1703	52.36	1058	32.54
BF2	3420	1835	53.66	1153	33.72
BF3	3835	2125	55.43	1327	34.62
BF4	3151	1658	52.63	1036	32.89

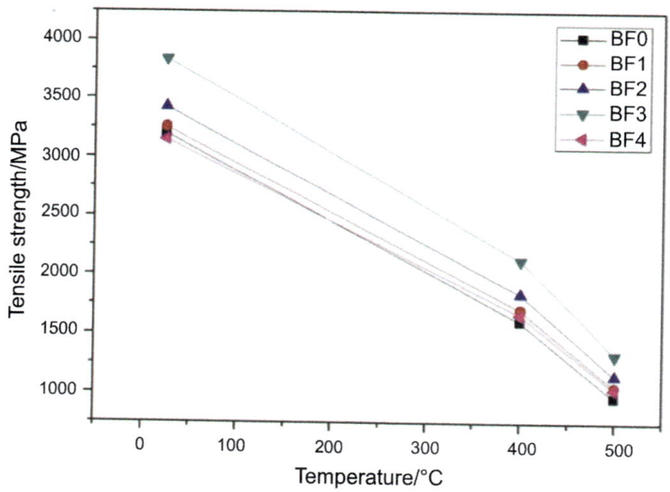

Figure 13.9 Tensile strength of different CeO_2 samples treated at different temperatures.

treatment process produces a polycrystalline phase in the basalt fibres, which can improve the high-temperature resistance of CBFs.

3. Effect of coating on temperature resistance of CBFs

Coating the CBF surface with thermostable organic compounds such as polysiloxane, phenylmethyl polysiloxane and other organic polymers can improve the thermostability. A shell-like protective layer is created on the surface of fibres by the coating, which can improve the temperature resistance of CBFs.

13.3.2 Corrosion resistance (chemical stability of basalt fibres)

To date, basalt fibres have been widely used in many materials such as fibre-reinforced composites, friction materials, shipbuilding materials, thermal insulation materials, automotive industry materials and high-temperature filter fabrics. Basalt fibres perform well in building materials because basalt fibres have high tensile strength and good alkali resistance. Furthermore, basalt fibres belong to the silicate group and can have good compatibility with cement mortar and other concrete matrix materials. In addition, basalt FRPs can resist acid and alkali corrosion, moisture, radiation and ageing. It can therefore be used to transport alkaline and other special liquids.

(1) Effect of OH- on alkali resistance of basalt fibres

The corrosion reaction of basalt fibres is mainly caused by the fracture of the Si–O tetrahedron in a strong alkaline environment. This is mainly due to the chemical reaction between the Si–O tetrahedron and OH-. With a further increase in OH- concentration, the surface of the fibres is exfoliated in layers, and the fibres change from smooth cylinders to massive scales. The corrosion mechanism is also different under different alkali conditions. With the weak alkali Ca(OH)2, the basalt fibres undergo local corrosion and shedding on the surface. CBF is a type of silicate fibre in which the Si–O–Si or Si–O–Al bonds, which constitute the framework of the glass network, readily react with the OH- in an NaOH solution. Fracture of the glass network skeleton promotes the dissolution of Si, K, Al and other elements and ultimately leads to fibre fracture or a great reduction in mechanical properties. Therefore, to prepare alkali-resistant basalt fibres, several aspects should be taken into consideration. First, an ore that can produce alkali-resistant basalt fibres is sought from volcanic rocks. Second, the composition can be adjusted by mixing two or more basalts to meet the requirements of producing alkali-resistant basalt fibres. Third, a special coating treatment is applied to the surface of basalt fibres to meet the requirements.

The alkali resistance of ordinary CBFs was studied by using 2 mol/L NaOH solution and other test media. CBFs were corroded in 2 mol/L NaOH solution at 80°C for 2, 4, 6, 8, 10 and 12 h. The quality loss and strength loss of basalt fibres were measured.

Figs. 13.10 and 13.11 show that the mass loss rate and strength loss rate of ordinary basalt fibres increase with increasing treatment time in an NaOH medium. With the increase in treatment time, the alkaline medium reacts with the basalt fibres to form a corrosion product layer, and then some cracks perpendicular to the fibres appear on the surface of that layer. The corrosion medium enters between the corrosion product layer and the new surface through cracks. The stress between the two layers lead to the shedding of the corrosion product layer, and OH- acts not only with the Si–O tetrahedron in the fibre but also with the Al–O tetrahedron or Al–O hexahedron, which causes further damage to the fibre skeleton. At the same time, the dissolution reaction of the free Al–O group can reduce the content of Al element, resulting in damage to the basalt fibre space structure and the loss of the corrosion layer, as shown in Fig. 13.12.

(2) Effect of chemical composition on alkali resistance of CBFs

1. Effect of Fe_2O_3 on alkali resistance of CBFs

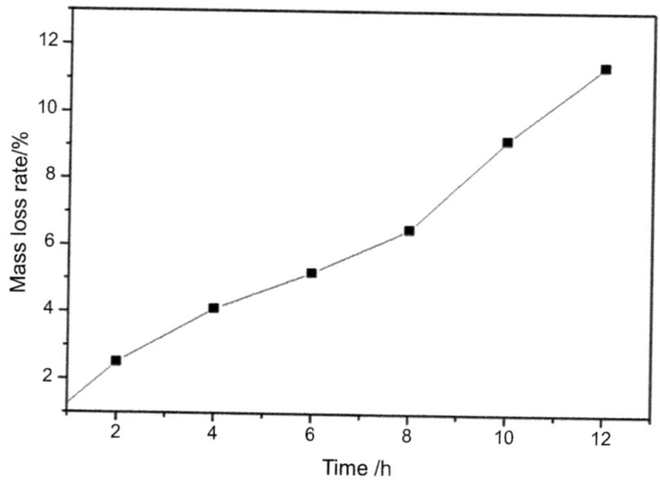

Figure 13.10 Mass loss rate of CBFs in NaOH solution at different time.

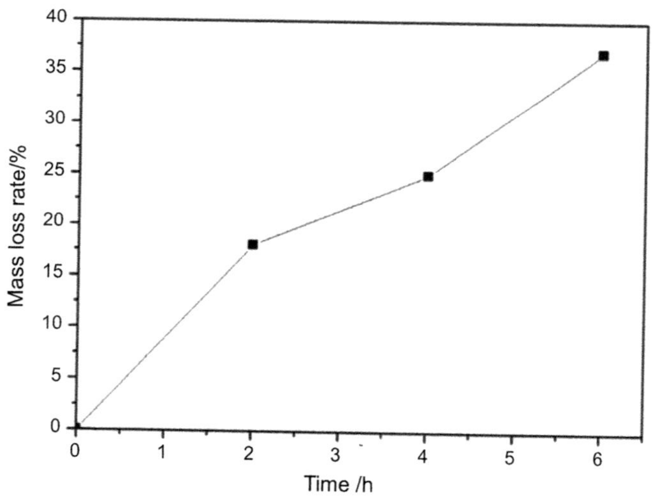

Figure 13.11 Fracture strength loss rate of CBFs in NaOH solution.

From the perspective of composition, the biggest difference between basalt fibre and glass fibre is that basalt fibre contains FeO and Fe_2O_3, whereas glass fibre does not. According to the table of solubility products, for $Fe(OH)_3$ $Ksp = 4.0 \times 10^{-38}$. In contrast, alkali-resistant glass fibre contains ZrO_2, and $ZrO(OH)_2$ has a lower solubility product ($Ksp = 6.3 \times 10^{-49}$) than $Fe(OH)_3$. Under the action of alkali solution, Fe_2O_3 on the surface of the basalt fibre will be transformed into a colloidal substance of $Fe(OH)_3$, which is dehydrated and polymerized to form a dense film on the glass surface to prevent the corrosion of the basalt fibre by alkali solution and hydrate. Therefore, increasing the Fe_2O_3 content is expected to improve the alkali resistance of basalt fibre. However, as FeO and Fe_2O_3 have different conversion forms at different

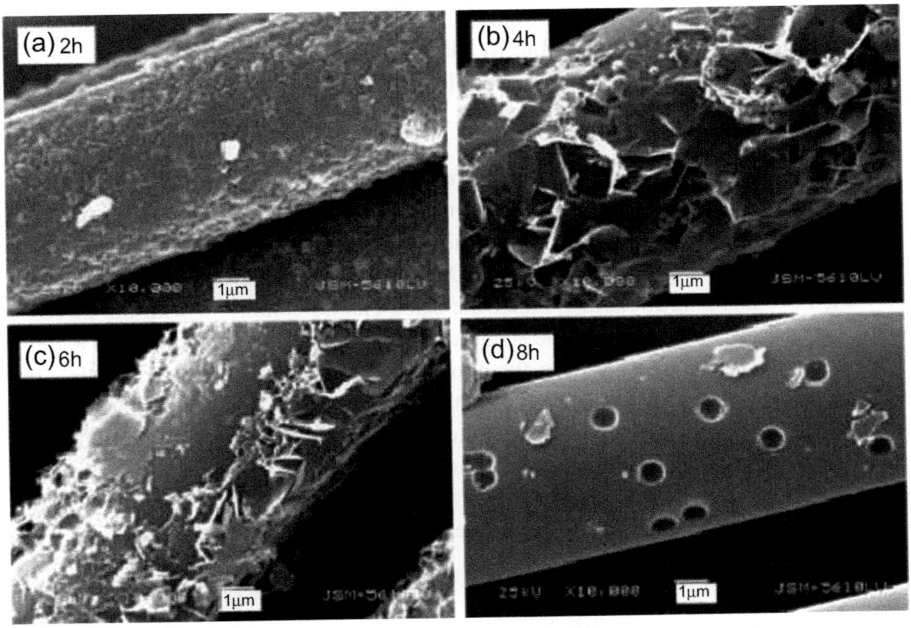

Figure 13.12 Morphology of CBFs in NaOH solution at different treatment time.

temperatures, it is difficult to know whether the full quantity of Fe_2O_3 exists or is reduced at high temperatures. Some scientists, represented by Professor Weyl from the University of Pennsylvania in the United States, performed a systematic study and believed that the oxidation state of iron has the following dynamic equilibrium relationship:

Low temperature, strong alkalinity and oxidation →

$$Fe^{2+} \longleftrightarrow Fe^{3+}$$

← High temperature, strong acidity and reducibility.

Because pure FeO does not exist in nature, when the temperature reaches 1475°C, Fe_2O_3 will partially decompose and give off oxygen to produce Fe^{2+}. So, Fe^{2+} and Fe^{3+} always coexist in silicate glass, just as the melting conditions are different, more or less. Therefore, the concept that Fe_2O_3 can improve the alkali resistance of basalt fibre requires further experiments and verification.

2. Effect of ZrO_2 on alkali resistance of CBFs

Regarding the alkali resistance of glass fibre subjected to an alkali solution, ZrO_2 on the fibre surface is transformed into a $ZrO(OH)_2$ gel and forms a dense film on the glass surface through dehydration polymerization, thereby preventing the corrosion of the glass fibre by the alkali solution and hydrate. Therefore, to improve the alkali resistance of CBF, zircon sand was added into its basalt raw material, and a specific

ZrO_2 content was introduced. Through the protection of ZrO_2 on the surface of the fibre, the corrosion of the basalt fibre by OH^- was reduced.

Strength retention and surface morphology of CBFs with ZrO_2 after erosion in NaOH solution of 2 mol/L for 1 h, 3 and 5 h at 98.5°C are shown in Figs. 13.13–13.15.

The above figures show that ZrO_2 can improve the alkali resistance of CBFs. Zr^{4+} exists in the form of a $[ZrO_6]$ octahedron in the zirconium silicate melt. Zr^{4+} is separated by at least two silica tetrahedrons. The Zr–O–Si bond strength is high and not easily destroyed at high temperature. ZrO_2 belongs to an octahedral coordination ($N = 8$); Zr^{4+} has a large radius. Under the same charge condition, with the increase of cationic coordination number, the accumulation of silica groups is enhanced and the viscosity is increased. The ZrO_2 on the glass surface is converted into a $Zr(OH)_2$ gel and forms a dense film on the glass surface through dehydration and polymerization to make its structure more compact, thereby preventing the corrosion of the glass fibre by hydrates such as those found in alkali solution. ZrO_2 can inhibit the precipitation of Si and Al ions in the basalt fibre and reduce the output of corrosion substances on the fibre surface, and in this manner hinder further erosion by the alkali solution.

(3) Effect of coating on temperature resistance of CBF

By coating the surface of CBF with a nano-SiO_2/epoxy composite coating or zirconia and titanium dioxide coating, the basalt fibre can be protected from erosion to a certain extent or undergo a small amount of erosion. Nano-SiO_2 particles prepared by the sol-gel method are doped with basalt fibre in the coating, which is referred to as 'coated modified basalt fibre' for short. Fig. 13.16 shows the topography of the insulation by scanning electron microscope (SEM) with no coating and with coating on basalt fibre, after exposure to 5% NaOH medium at 40°C for 24 h.

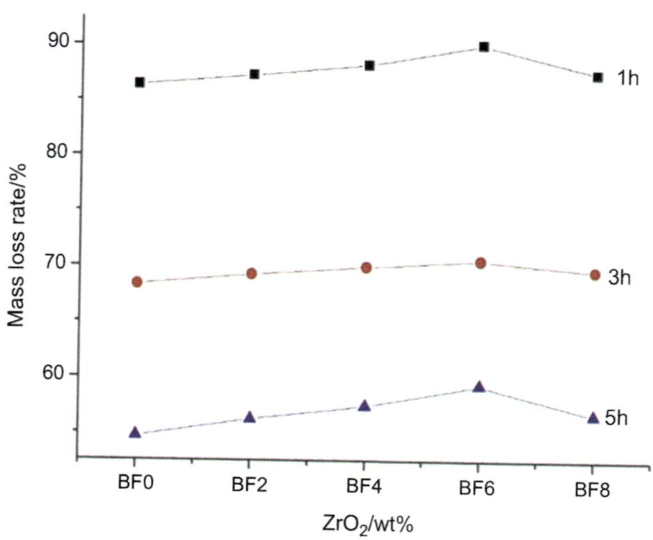

Figure 13.13 Strength retention rate of CBFs with different ZrO_2 contents after alkali corrosion.

Figure 13.14 Surface morphology of CBFs with different ZrO_2 contents after 1 h alkali corrosion ((a):Uncorroded, (b):0wt%, (c):2wt%, (d):4wt%, (e): 6wt%, (f):8wt%.)

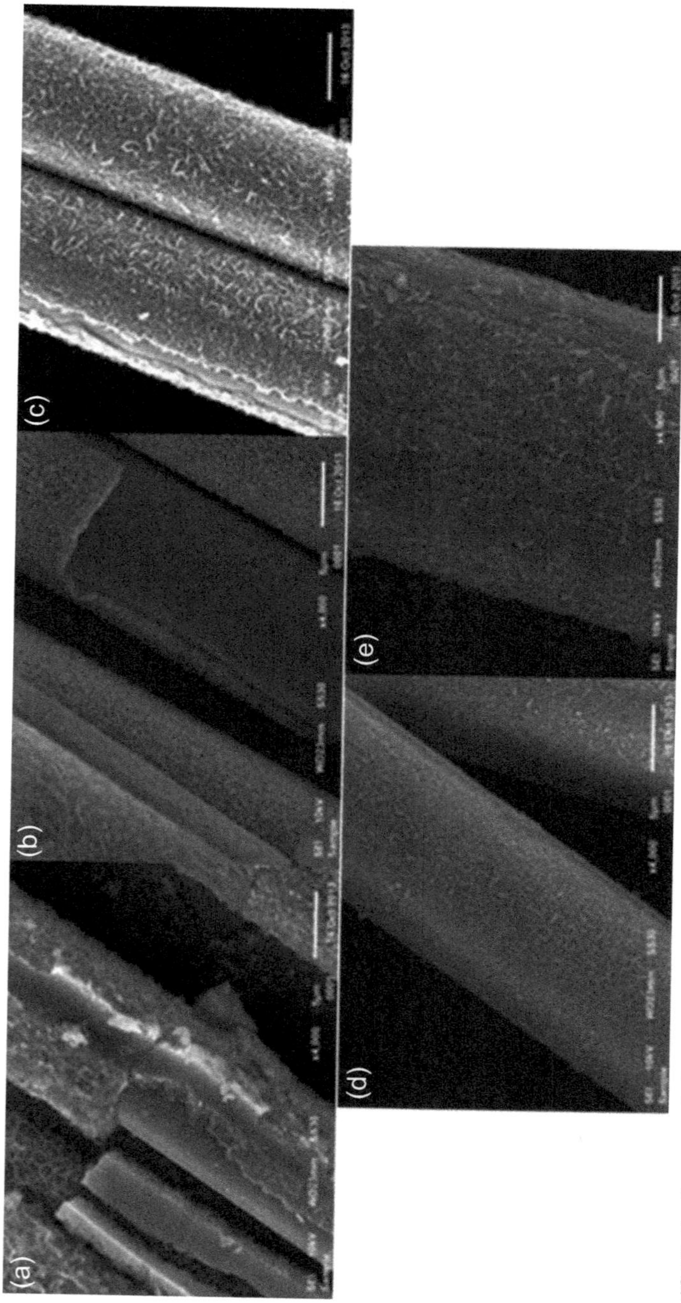

Figure 13.15 Surface morphology of CBFs with different ZrO_2 contents after 5 h alkali corrosion ((a):0wt%, (c):2wt%, (d):4wt%, (e):6wt%, (f): 8wt%.)

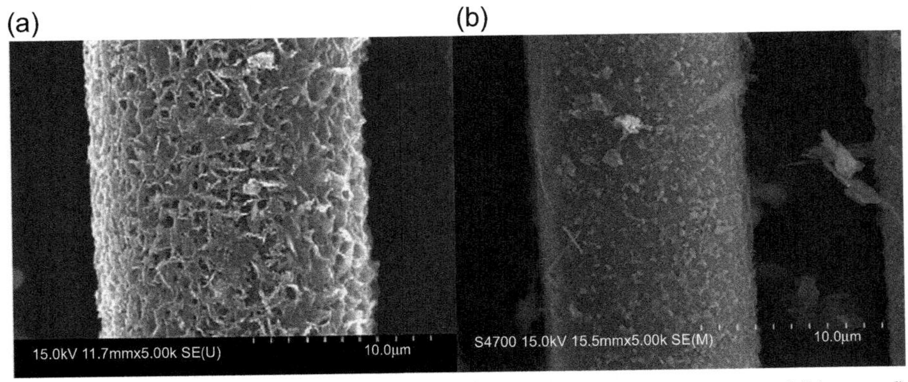

Figure 13.16 SEM photographs of basalt fibers after alkali corrosion ((a): uncoated (b): coated).

As shown in Fig. 13.16(a), after the alkali corrosion of the unmodified basalt fibre, its matrix was severely corroded, and a layer of white flake reaction product was generated on the surface. The base of basalt fibre modified by surface coating was not damaged after alkali corrosion. A layer of reactant protective film was formed on the surface of the basalt fibre modified by the coating to protect the base from further alkali corrosion. The reason is that the coating first hinders the diffusion of water molecules to the surface of the fibre to some extent and the existence of SiO_2 nanoparticles then makes the diffusion of water molecules to the surface of the fibre more difficult and the diffusion channel more tortuous. Second, SiO_2 particles in the coating hinder the absorption of OH^- in the process of OH^- reaching the surface of the fibre with water molecules. That is, SiO_2 particles react with OH^- entering the coating. The resulting products accumulate in the coating and form a new protective layer, which makes the fibre matrix avoid further erosion of OH^- and the strength decline rate of the fibre matrix is significantly controlled. The higher the content of SiO_2 particles in the coating, the stronger the alkali corrosion resistance.

13.3.3 Long-term mechanic properties

13.3.3.1 Creep performance

Owing to the potential advantages in prestressing application, the creep behaviour of basalt fibre–reinforced polymer (BFRP) is specially investigated.

Strain development

The typical strain curves with respect to time (ε-T) of BFRP are plotted in Fig. 13.17. Under the stress level of 70%, 68% and 65% all of the specimens fractured with the increase of strain. An initial rapid increment of creep strain can be observed during the first stage. Then, the rate of strain increase gradually decreases with respect to time and finally goes into the second stage characterized by a stable and slow rate

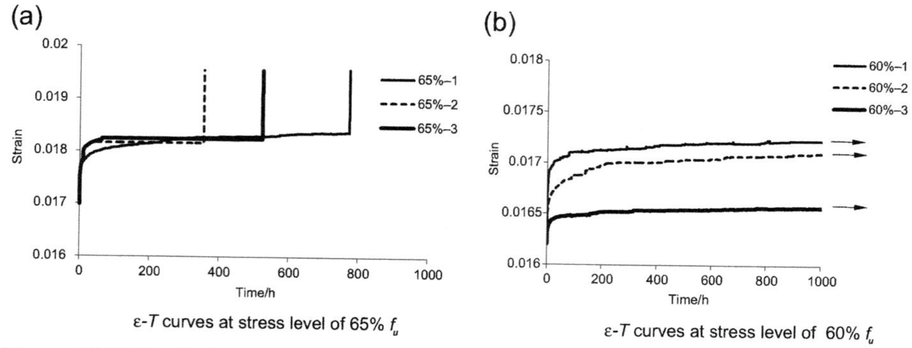

Figure 13.17 Typical **ε**-T curves of BFRP (Wang et al., 2014). (a) *ε*-*T* curves at stress level of 65%f_u, (b) *ε*-*T* curves at stress level of 60%f_u.

of creep increment. The second stage tends to be a long-term stage and specimens at some stress levels (e.g., 50% f_u) are capable of maintaining the second stage without a further rapid strain increase stage, the so-called creep rupture stage. For the specimens under higher stress level (e.g., 68% f_u), the third stage takes place with the sudden increase of strain up to failure. It should be mentioned that the *ε*-*T* curves of the specimens at the stress level of 70% f_u shows obviously different three stages featured by a shorter second stage where creep rupture occurred after a short time of sustained constant load. The typical three stages of creep development are evident for the specimens at the stress levels of 68% f_u and 65% f_u. For the specimens under even lower stresses such as 60% f_u and 50% f_u, their *ε*-*T* curves show only the first and second stages of strain increase, in which the symbol '→' denotes the specimens without creep rupture after 1000 h. Overall, it can be seen that with the increase of stress level, the scatter of the *ε*-*T* relationship will become more significant.

Creep rate

The creep rate defined by the ratio of the increased strain up to the end of the second stage to the initial strain of specimens at different stress levels are shown in Table 13.7 and Fig. 13.18. It is shown that under low stress levels, the creep rates maintain a relatively low level, but the variation is large, whereas the creep rates increase greatly but the relevant variation decreases with respect to increasing stress level. This observation

Table 13.7 Creep rate of BFRP specimens (Wang et al., 2014).

Stress level	50%	60%	65%	68%	70%
Mean value	3.58%	3.70%	6.51%	14.19%	15.17%
Standard deviation (SD)	1.33%	1.45%	0.96%	0.90%	0.54%
Coefficient of variation (CV)	37%	39%	15%	6%	4%

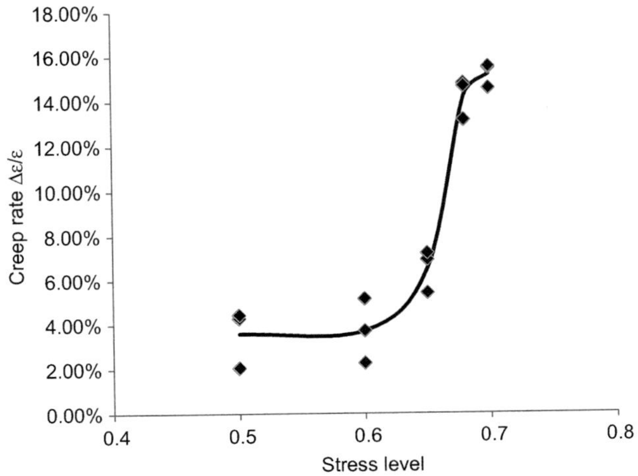

Figure 13.18 Relationship of creep rate and stress level (Wang et al., 2014).

demonstrates the mechanism of defect expansion inside the FRP as described above. Under the low stress level, the expansion of defect in the FRP is still in progress so that the large differences are observed. However, under high level of stress, all of the specimens ruptured within 1000 h indicating defect expansion is sufficient for each. Thus, relatively low variations are found.

Residual tensile strength

For the specimens that did not fail after 1000 h of sustained load, further loading up to failure was performed. The results of all six specimens under the stress levels of 50% f_u and 60% f_u are shown in Table 13.8 , which shows that the residual strength of these specimens maintains a high level close to the static tensile strength. Furthermore, the variation of residual strength is limited to a much lower level compared with the static tensile strength results. This further proves the creep deterioration mechanism described previously. It shows that insufficient expansion of defects in the FRP will not substantially affect the tensile behaviour of FRP, and on the other hand, the stress redistribution of fibres in the FRP in the first stage of creep development benefits the simultaneous load carrying capacity of fibres in the FRP resulting in lower variation induced by the nonuniformity of fibres.

Table 13.8 Residual tensile strength of specimens without creep rupture (Wang et al., 2014).

Number	60% f_u	50% f_u
Mean value of strength retention	93.8%	95.0%
CV	0.61%	0.32%

Prediction of creep rupture stress

To determine the creep rupture stress, the experimental data of sustained stress and rupture time are analyzed using a semilogarithmic graph. A linear fitting is calculated from the data using the least-square method. The results are shown in Fig. 13.19.

By substituting $T = 10^6$ h into the equation in Fig. 13.19, $Y_c = 0.59$ is obtained indicating that BFRP tendons can sustain a stress of 59% of its tensile strength without creep rupture during the service life time of approximate 114 years. The accuracy of the prediction represented by $R^2 = 0.9$ is also acceptable.

Considering the reliability analysis, the lower limit value for the confidence band according to the prescribe life time (10^6 h) is calculated as 0.52 (the value can be enhanced to 0.54 through pretension treatment as mentioned below), which means with a reliability of 95%, the BFRP tendons can carry a sustained stress of 52% of its tensile strength without creep failure within 114 years. This result has only a reduction of 12% compared to the prediction by mean values indicating a relatively small scatter of rupture time of BFRP tendons under different levels of stresses. It should be mentioned that in ACI-440.1R the creep rupture stresses of glass fibre–reinforced polymer (GFRP), AF-reinforced polymer (AFRP) and CF-reinforced polymer (CFRP) are 0.2 and 0.3 and 0.5, where a safety factor of 1/0.6 is considered. With the same safety factor, the creep rupture stress of BFRP tendons can be only 0.35 based on the mean value prediction of 0.59, which is much less than the predicted value of 0.52 based on statistical analysis. Thus, it apparently indicates that the limitation of creep rupture stress is relatively conservative in ACI-440.1R. In the future, with the accumulation of creep data of BFRP tendons, the creep rupture stress of BFRP tendons will become more accurate and evident. Based on the analysis in this study, the creep rupture stress of BFRP tendons is recommended to be 0.52 for practical application. It should be noticed that a large number of data are still necessary for a more reliable and representative advice for the creep rupture behaviour of BFRP because of the limited existing data.

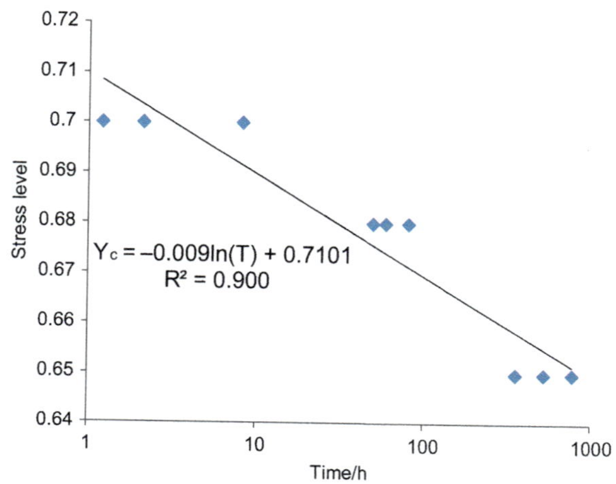

Figure 13.19 The approximation line for Y_c-logT (Wang et al., 2014).

Table 13.9 Creep rupture stresses (Wang et al., 2014).

Types	CFRP	AFRP	GFRP	BFRP
Creep rupture stress	0.70 f_u	0.55 f_u	0.29 f_u	0.54 f_u

The creep rupture stresses of several types of FRP are shown listed in Table 13.9, in which, the values of CFRP, AFRP and GFRP refer to the clauses in ACI-440.4R and the research on GFRP by Yamaguchi, and the value of BFRP is based on the study of Wang et al. (2014). In terms of creep rupture stress, GFRP has the lowest value, whereas CFRP possesses the highest value among the several types of FRP. Because creep rupture stress determines the upper limit of design stress, a high creep rupture stress is necessary for the prestressing component.

Creep strain rate control

Several methods were proposed to control creep behaviour of FRP composites, which mainly focused on the hybridization of different fibres. For instance, Maksimov and Plume (2001) explored the creep behaviour enhancement by FRP hybridization, which show creep strain of AFRP can be well restricted by hybrid with GFRP. Sasaki (2008) investigated the hybridization of GFRP and CFRP, which revealed that the creep strain of GFRP can be lowered from original 5%–3.5%. The similar mechanism of hybridization with higher modulus fibres plays an important role. Moreover, Acha et al. (2007) considered that creep deformation was directly affected by interfacial properties, which can be improved through the addition of coupling agents or the chemical modification of the fibres. In addition, the modification of the matrix in FRP was also investigated to control creep strain. Li et al. (2012) revealed that adding carbon nanotube in the resin can reduce creep strain rate of CFRP by an appropriate amount. Although above studies can control creep behaviour of FRP composite, they are still restricted by some practical reasons, such as low performance to cost ratio, stability and quality control of overall mechanical properties.

Aside from control methods from materials themselves, it may be a practical and effective method through external pretension referred to the treatment of low relaxation of high strength steel wires. The pretention process with tempering was widely adopted to lower the relaxation of high strength steel wires, in which the mechanism is to optimize the microstructure of steel (Libby et al., 1990). For FRP composites, as no creep was observed for the tensioned fibres themselves while the significant creep occurred in the resin (Ascione et al., 2012), it indicates if the fibres in FRP are sufficiently straight, the creep strain of the entire FRP material will be very limited (Zhou and Yang, 1985). However, due to the limit of production technology, the unevenness of fibre roving such as local bending and skewness, cannot be avoided during the process of pultrusion, which is the main cause of the rapid stress increment in the first stage of FRP tendon creep (Everett, 1996; Paluch, 1996). As revealed in Table 13.9, the residual strength of BFRP specimens experienced a sustained load of 1000 h can achieve more stable results compared to the original strength before creep test. This phenomenon indicates that through sustained loading the fibres in FRP can be straightened, which results in a more uniform load carrying of each fibre and enhancement of overall stability of strength. Furthermore, the research by Spathis and Kontou (2012) also showed that rate of creep strain is related to the internal stress in polymer material. With creep deformation of resin, the internal stress can be relieved by resin relaxation and transferred to fibres (Soudki, 1998). Thus, it can be estimated that the unevenness of fibres can be adjusted by means of applying pretension along the direction of longitudinal fibres. The resin in FRP continues to deform by creep when the FRP is subjected to a sustained load, which provides a possibility for the fibres to interact in the resin. During this process, fibres tend to be straightened due to the creep of resin and the previous unevenness of fibres can be adjusted. The mechanism on the improvement of internal fibre distribution in FRP through pretension process is shown in Fig. 13.20. Owing to the improved collaboration of fibres after pretension, the creep rate of BFRP can be significantly controlled (Fig. 13.21(a)), and the creep rupture stress of BFRP increases from 0.52 f_u to 0.54 f_u (Fig. 13.21(b)).

13.3.3.2 Fatigue performance

BFRPs and the related hybrid FRPs have good macroscale fatigue properties. Wu et al. (Zhou and Yang, 1985) found that the fatigue behaviour of BFRP was comparable to that of the conventional GFRP, whereas the fatigue behaviour of BFRP can be significantly enhance by hybridization with CFs, as shown in Fig. 13.18. Wu et al. (2010a) studied fatigue behaviour of BFRP and the related hybrid FRP tendons and found that basalt or carbon hybrid composites have higher fatigue life than the steel wire and can achieve integrated high performance in long-span cable-stayed bridges. Colombo et al. (Wang et al., 2013) carried out an experimental study on the static and fatigue properties of basalt fibre-reinforced vinylester and epoxy matrices. They demonstrated that the basalt fibre-reinforced epoxy has better fatigue behaviour than that reinforced vinylester. Dorigato and Petretti (Colombo et al., 2012) studied the fatigue properties of CF-reinforced, glass fibre– and basalt fibre–reinforced epoxy braided composites. It shows that the fatigue strength of BFRP composites is similar to that of GFRP

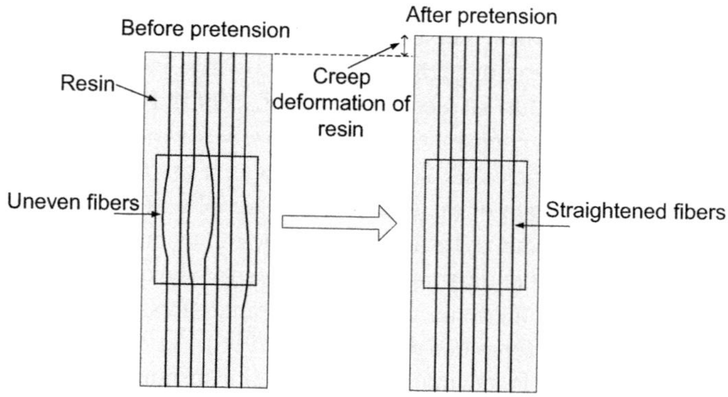

Figure 13.20 Mechanism of creep strain control by pretension (Wang et al., 2016a).

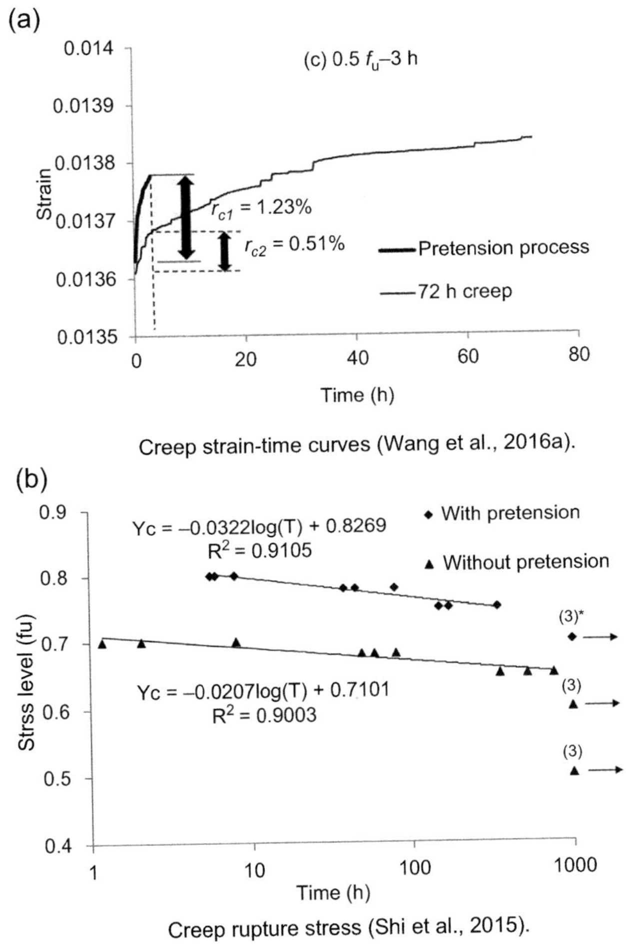

Figure 13.21 72-h Creep strain rates and creep rupture stress before and after pretension. (a) Creep strain-time curves (Wang et al., 2016a). (b) Creep rupture stress (Shi et al., 2015).

composites, but the loss of elastic modulus is less and the energy dissipation performance is stronger. In view of the small price difference between the two, both are far less than CFs, basalt fibres have good prospects for development (Fig. 13.22).

For BFRP tendons, the fatigue limit of BFRP tendons can be predicted to be 0.53 f_u (stress amplitude 0.04 f_u) (Dorigato and Pegoretti, 2012) or 0.7 f_u (stress ration R = 0.8) (Shi et al., 2017b) based on reliability analysis and stress amplitude commonly used in civil engineering structures. The fatigue limit is close to CFRP and much higher than AFRP and GFRP. The application prospect of BFRP as a high-efficient prestressing material is proved.

Fig. 13.23 shows the S–N fatigue curve of BFRP under different stress ratio (R = minimum fatigue stress/maximum fatigue stress). The degradation rate of the expected fatigue life of BFRP increases as the stress ratio decrease. It reveals that

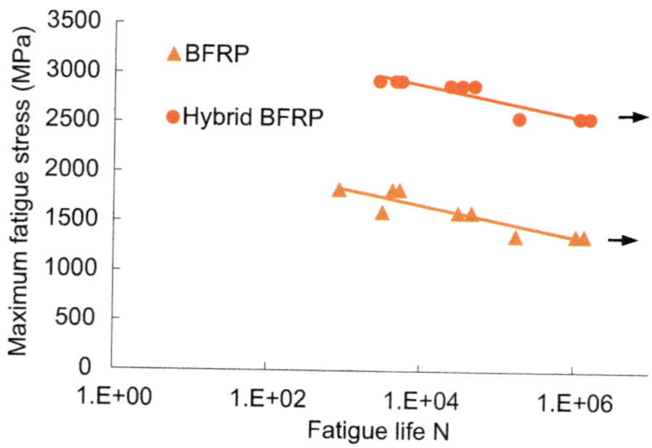

Figure 13.22 S–N fatigue curve of BFRPs and the carbon fibre hybrid BFRPs.

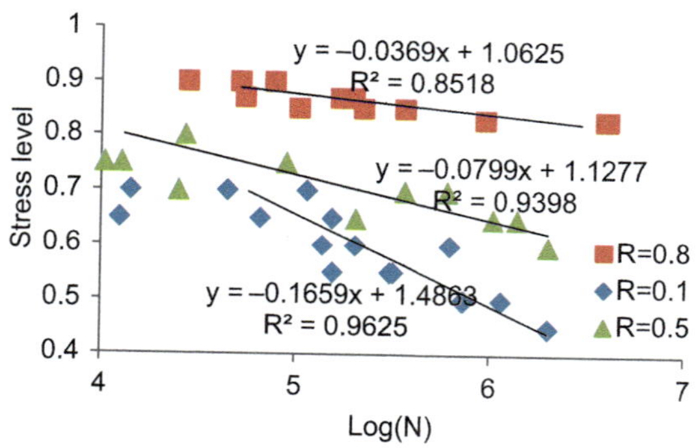

Figure 13.23 Linear stress level-log N_f curves of BFRP.

the fatigue failure has a different damage pattern from that of static failure. The degrees of damage pattern difference increase with the decrease of stress ratio from stress ratio 0.8 to 0.1.

The fatigue failure varies among the fatigue specimens of different stress ratio. Fig. 13.24 shows three typical failure surface at the stress level of 75% for specimens of R = 0.8, 0.5 and 0.1, respectively. It is clearly seen that the control mode of damage is discontinuous interface debonding at the stress level of 75% for R = 0.8. However, the failure modes shift into fibres wearing out and even breaking when the stress ratios R decrease.

13.4 Application of basalt fibres

Basalt fibres (in roving, textile yarn and chopped yarn) can be made into fibre sheet, mat, rope and other fibre products. Basalt fibres and their products can be used as reinforcements to make various excellent composite materials (bars, plates, profiles,

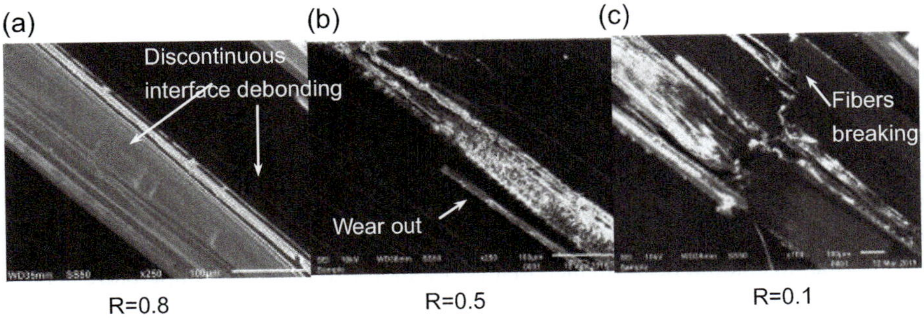

Figure 13.24 Failure pattern at the stress level of 75% for specimens of R = 0.8, 0.5 and 0.1. (a) R=0.8. (b) R=0.5. (c) R=0.1.

grids, cables and prepregs), as shown in Fig. 13.25. Basalt fibre products and their composites can be widely used in the fields of transportation, structural engineering, marine engineering and automobiles.

13.4.1 Applications in transportation

13.4.1.1 Short basalt fibre–reinforced asphalt pavement

In resolving the problems of conventional asphalt pavement, the addition of chopped fibres to asphalt is an effective solution. Fibre materials used in asphalt pavement must have the following properties: (1) a wide working temperature range; when mixing with asphalt at high temperature, the performance of fibre materials should not undergo thermal degradation, react with asphalt chemically or dissolve, which can effectively change the low-temperature hardening and high-temperature

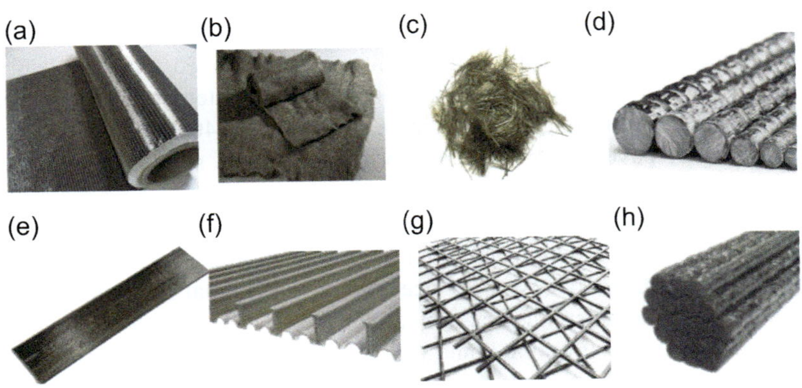

Figure 13.25 Basalt fibres and their products. (a) Fiber sheet. (b) Fiber mat. (c) Chopped yarn. (d) Composite bar. (e) Composite plate. (f) Composite profile. (f) Composite grid. (h) Composite cable.

softening of asphalt, and the materials should adapt to various high- and low-temperature pavement environments; (2) excellent mechanical properties that can improve the strength and toughness of the asphalt mixture; (3) good chemical stability, no chemical reaction with asphalt when producing the asphalt mixture and good adaptation to the working environment of various pavements; (4) good antiageing performance, with no ageing occurring in the asphalt mixing process and the pavement working process; (5) good water stability, inhibiting the oxidation ageing of asphalt and reducing the water damage of asphalt pavement; (6) good electrical and thermal insulation performance that is conducive to the high-temperature stability of the asphalt oil film and the prevention of electrochemical corrosion of the asphalt film; and (7) high performance–price ratio because of the huge quantity of short-term fibres used in asphalt. On the premise of guaranteeing the above properties, cost-effective fibre materials are advantageous. Considering the above factors, chopped basalt fibres have become the best choice for asphalt pavement reinforcement material. Compared with other fibres, there are some shortcomings; for example, CFs have excellent mechanical properties (but poor oil absorption and high price). Although lignin fibres have good oil absorption, their antiageing ability is poor. The mechanical properties of glass fibres in the alkaline environment of asphalt become severely degraded, except in the case of basalt. The brittleness of mineral fibres other than rock fibres is obvious.

The main technical indicators of short-cut basalt fibre-reinforced asphalt pavement include basalt fibre fineness, surface characteristics, length, dosage and mix ratio. A series of studies shows that the surface of short-cut basalt fibre used in asphalt pavement should be smooth, straight and pollution-free and an oil-friendly modification should be conducted to make it suitable for the asphalt system; the diameter of a single filament is 13–17 μm and the nominal lengths of the fibre are 6, 9 and 12 mm. Usually, the basalt fibre content should be 0.3%–0.4% of the total mass. The Marshall test method is used in asphalt mix design. The determination of the key technical indicators further optimizes and improves the durability, stability and crack resistance of basalt fibre–reinforced asphalt pavement.

The antifatigue-cracking ability of asphalt pavement mixed with basalt fibre is increased by a factor of 2–6, the anticracking ability at low temperature is increased by 15%–30%, the antirutting ability is increased by 25%–40% and the anti-water-damage ability of pavement is greatly improved. To date, short-cut basalt fibre–reinforced asphalt pavement has been applied to the Qiqihar–Daqing Expressway; Jilin Huinan–Baishan Expressway; Jiangsu Province S333 Provincial Highway, Yizheng Section Reconstruction and Extension Project; Mintai Avenue construction project in Yizheng City, Jiangsu Province; Beijing–Shanghai Expressway Special Treatment Project and other practical projects, as shown in Fig. 13.26. The pavement operation in these projects is good, and there are generally no pavement problems.

13.4.1.2 Reinforced concrete structures with BFRP bars

BFRP bars have the characteristics of high tensile strength, lightweight, nonmagnetism, good corrosion resistance, good fatigue resistance and a thermal expansion

Figure 13.26 Concrete structure reinforced with BFRP bars.

coefficient close to that of concrete, which can fundamentally solve the problem of insufficient durability of concrete structures. The tensile strength of CF-reinforced polymer (CFRP) tendons is as high as 1500–2500 MPa, but the cost is high, and the thermal expansion coefficient is quite different from that of concrete; it is therefore difficult to use CFRP tendons as reinforcing materials for concrete structures on a large scale. The tensile strength of BFRP tendons can reach 1000–1800 MPa, which is approximately 1.5 times the strength of ordinary glass fibre-reinforced polymer (GFRP) tendons. The elastic modulus of BFRP tendons is slightly higher than that of GFRP tendons, and the thermal expansion coefficient of BFRP tendons is similar to that of concrete. BFRP is the best choice of concrete reinforcement materials. In view of the application of BFRP bars in concrete, the author's team has developed various types of BFRP bars (such as longitudinal bars and stirrups) to meet the performance requirements of different structures, and has developed the related technology of reinforcing bars, such as coiling, bending and lapping, to meet the site construction needs (Fig. 13.2). In 2015, BFRP bars were used in the concrete structure of an island reef in Sansha City, South China Sea. The comprehensive cost of the structure increased by no more than 20%. The structure required no maintenance or management, and the service life of the structure was longer than 50 years.

Considering the advantages of BFRP bars, and aiming at the high transportation cost of structural materials (such as river sand and fresh water) and the vulnerability of internal bars to chloride ion corrosion, the author's team developed concrete reinforced by BFRP bars (Fig. 13.26) for sea sand or coral reef sand, which fundamentally solved the durability problem of coastal highway engineering. However, some problems occur such as low ductility and insufficient stiffness in the reinforced concrete structure with all BFRP bars. Therefore, the author developed the technology for a hybrid configuration of BFRP bars with reinforcement. BFRP bars with special ribs (including different shapes of winding ribs and integrated deep ribs) are arranged on the outside of the steel bars, and the stable sliding section in the bond–slip curve of the BFRP bars is used to delay the opening of cracks. The corrosion medium has difficulty penetrating into the surface of the internal steel bar, which ensures the durability of the structure, and the stiffness and ductility of the structure are significantly improved because of the reinforcement.

The BFRP-reinforcement hybrid configuration is an effective means to achieve good repairability of the structure. The linear elasticity of BFRP tendons enables them to provide secondary stiffness for the structure after yielding. The residual deformation of the structure after unloading is significantly less than that of an ordinary reinforced concrete structure under the same displacement. The stable slip section in the bond–slip curve of BFRP tendons can ensure that the structure does not collapse when the deformation is too large (the limit stage in Fig. 13.5) (Paluch, 1996). To achieve this effect, two preconditions must be satisfied. First, through reasonable design, the actual bond strength of fibre-reinforced polymer (FRP tendons should not exceed the corresponding bond stress when they break because once the FRP tendons break, the restriction on slip will disappear completely. Second, the FRP tendons must have a stable slip section to ensure both ductility and performance of the structure.

BFRP bars can also be used for pavement reinforcement to achieve long-distance continuous reinforcement (Fig. 13.26). They do not need overlap and welding as traditional bars do. The main construction technology involves only the binding of longitudinal and transverse bars, which shortens the construction period by 60% and improves the construction efficiency by a factor of 5. Compared with the same period of construction of ordinary concrete, bridge deck cracks are reduced by 30%, reducing maintenance costs; in addition, BFRP bars are lightweight, convenient to transport, easy to use and simple to construct, thereby saving labour costs. At present, continuous BFRP-reinforced pavement has been successfully applied in some domestic pavement projects such as the Yangjiang Highway and the Zhangshi Highway.

13.4.1.3 Basalt fibre cloth/BFRP plate reinforcing bridges and roadbeds

In addition to applications in the reinforcement of beams, slabs and columns, basalt fibre cloth can be used for roadbed reinforcement. The fibre cloth used for structural reinforcement is expensive and unsuitable for large-scale reinforcement of a roadbed. Therefore, the basalt fibre cloth material used in roadbed reinforcement is mainly low-cost polyester-basalt fibre nonwoven fabric. This type of material mixes discontinuous basalt fiber and polyester fibre and is produced by a spraying and setting process that can make full use of the excellent mechanics of basalt fibre. The material has the good oil absorption and toughness of polyester fibres. The fibre cloth coated with asphalt thermal binder can disperse the stress between subgrade and asphalt pavement, restrain and delay reflection cracks of the asphalt pavement caused by subgrade cracks, increase the flexural tensile strength of pavement by approximately 50% and prolong fatigue life by a factor of 3 compared to that without reinforcement. In addition, the performance of polyester-basalt fibre nonwoven fabric in preventing road cracks is better than that of polyester-glass fibre nonwoven fabric. Especially in a low-temperature road environment, polyester-glass fibre nonwoven fabric breaks because of reciprocating vehicle loading, whereas polyester-basalt fibre nonwoven fabric remains intact. To date, polyester-basalt fibre nonwoven fabrics have been applied in the engineering of the Huancheng Highway in Zhangjiakou, Yuxian County, Hebei Province, greatly improving the crack resistance of the asphalt pavement and prolonging the service life of the road.

13.4.1.4 Roadbed reinforced by basalt fiber geotechnical cloth

In addition to applications in the reinforcement of beams, slabs and columns, basalt fibre cloth can be used for roadbed reinforcement. The fibre cloth used for structural reinforcement is expensive and unsuitable for large-scale reinforcement of a roadbed. Therefore, the basalt fibre cloth material used in roadbed reinforcement is mainly low-cost polyester-basalt fibre nonwoven fabric. This type of material mixes discontinuous basalt fiber and polyester fibre and is produced by a spraying and setting process that can make full use of the excellent mechanics of basalt fibre. The material has the good oil absorption and toughness of polyester fibres. The fibre cloth coated with asphalt thermal binder can disperse the stress between subgrade and asphalt pavement, restrain and delay reflection cracks of the asphalt pavement caused by subgrade cracks, increase the flexural tensile strength of pavement by approximately 50% and prolong fatigue life by a factor of 3 compared to that without reinforcement. In addition, the performance of polyester-basalt fibre nonwoven fabric in preventing road cracks is better than that of polyester-glass fibre nonwoven fabric. Especially in a low-temperature road environment, polyester–glass fibre nonwoven fabric breaks because of reciprocating vehicle loading, whereas polyester–basalt fibre nonwoven fabric remains intact. To date, polyester–basalt fibre nonwoven fabrics have been applied in the engineering of the Huancheng Highway in Zhangjiakou, Yuxian County, Hebei Province, greatly improving the crack resistance of the asphalt pavement and prolonging the service life of the road.

13.4.2 Bridge/tunnel reinforced by BFRP grid

The CBF grid has the advantages of light and thin material, strong integrity, good durability, nonmagnetism and remarkable reinforcement effect. It can be used for the reinforcement of new structures and is widely used for the repair and reinforcement of tunnels, bridge structures and concrete bidirectional slabs; the reinforcement of underwater structures such as for ocean ports and wharfs; and the reinforcement of hospital floors, scientific research laboratories and observation stations (Figs. 13.27–13.31).

13.4.3 Applications in marine engineering

In addition to having excellent mechanical properties, CBF material can have good seawater resistance, heat resistance and sound insulation properties. CBF-reinforced composite bar (BFRP bar) has been widely used especially for buildings, bridges, highways and parking lots directly exposed to seawater and sea breeze in coastal areas. Continuous BFRP bar–reinforced concrete structures in salt lakes and coastal water

Figure 13.27 BFRP grid continuous production technology.

have longer life than those with steel bars. CBF nonwoven fabric can reduce the weight of a ship, and the cost is the same as that of E-glass fibre nonwoven fabric. It can be widely used in hulls, engine room insulation and superstructures in marine engineering.

13.4.4 Applications in automobiles

The lightweight automobile is the development direction of the automobile industry. The performance of CBF composites is better than that of steel and their weight is one-quarter that of steel. The application of composites in automobiles can greatly reduce the load of automobiles, reduce energy consumption and greatly improve the performance of automobiles. As shown in Fig. 13.30, CBF composites can be used to make an automobile wheel, intake pipe, door, bracket, chassis, cockpit, engine cover, reserve box cover, transmission shaft, rearview mirror and muffler. They can also be used to manufacture automobile muffler insulation and sound absorption materials, heat insulation gaskets, heat insulation covers, heat insulation thin plates, clutch discs and brake pads. A CBF honeycomb board can be made into a train carriage board, which not only reduces the weight of the carriage, but also is a good flame retardant material.

13.4.5 Application in energy and environmental engineering

Basalt fibre is a high-performance inorganic fibre with a diameter of several microns. It has attracted increasing attention in environmental engineering as an environmentally friendly (eco-friendly) carrier material. For this section, the progress of basalt fibre biocarrier research was reviewed. The characteristics of basalt fibre and its advantages in water treatment were elucidated. The fundamentals of a new basalt fibre bionest wastewater treatment technique were overviewed. The characteristics of ideal biocarriers for biological contact oxidation were discussed, which indicated that modification of

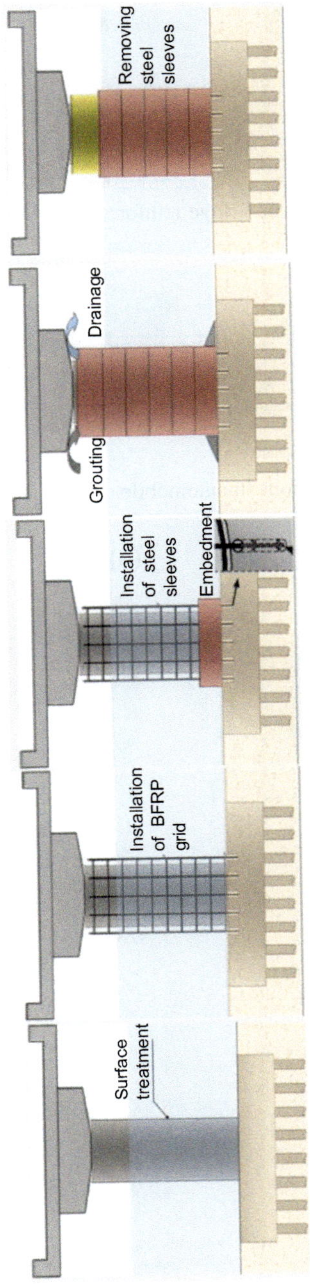

Figure 13.28 FRP grids reinforcement technology for underwater structures.

Figure 13.29 Nanjing Yangtze River Bridge reinforcement with BFRP grid.

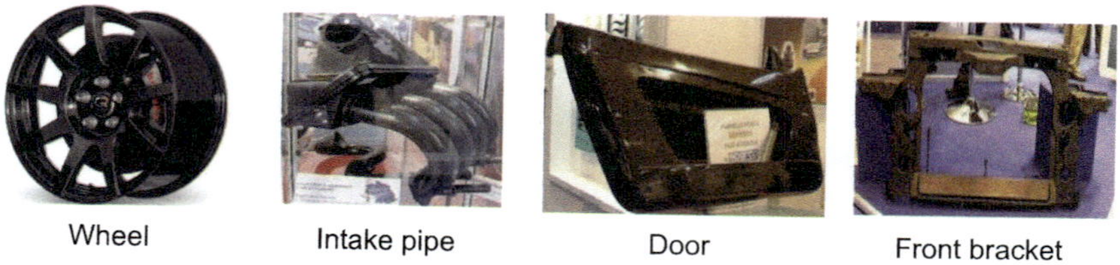

Figure 13.30 Basalt fibre applications in automobile. Wheel. Intake pipe. Door. Front bracket.

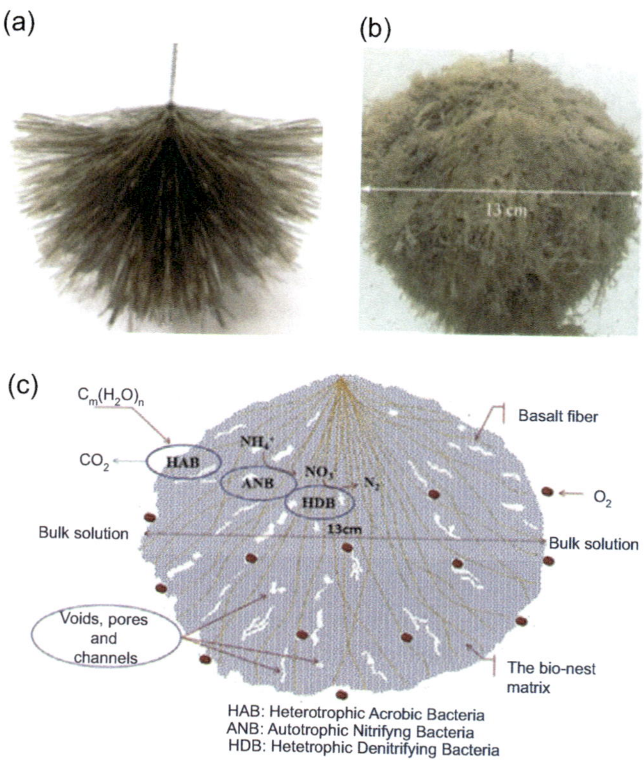

Figure 13.31 Basalt fibre packing structure: (a) basalt fibre bundle packing; (b) basalt fibre 'biological nest'; (c) 'biological nest' internal structure (Zhou and Zhou, 2017; Wu et al., 2016).

current basalt fibre material is necessary to improve the hydrophilicity, positive surface charge and bioaffinity. Methods reported in the literature to prepare modified basalt fibre, such as wetting agent coating, coupling agent grafting, surface etching and plasma processing, were reviewed.

13.4.5.1 Water purification mechanism of basalt fibre

Basalt fibre materials have gradually received attention in the field of environmental protection, especially as a microbial carrier in water treatment processes. At present, most of the sewage treatment at home and abroad is based on biological treatment technology. The traditional activated sludge method has a large area, poor impact load resistance and large sludge production. The biological contact oxidation method uses carriers dispersed in water to fix microorganisms, which can improve the contact area with the sewage greatly. In addition, it has the advantages of high energy efficiency, impact load resistance, small floor space and operation management. Therefore, it has been widely used in sewage treatment. The filler or carrier is the technical core of the biological contact oxidation, and it directly affects the treatment effect of the reactor. Biofillers can be divided into fixed, suspended and suspended according to the state in wastewater. The materials used commonly include polyolefin (Zhao et al., 2016), polyamide (Zhang, 2017), activated carbon (Cheng et al., 2017), zeolite (Nie et al., 2018) and CF (Miazga-Rodriguez et al., 2012a). Among them, the fibre material has the advantages of large specific surface area, ease of film hanging, no clogging and good oxygenation performance. However, the production process of organic synthetic fibre and CF has problems such as the discharge of polluting gas and wastewater. Disposal of organic waste materials would cause environmental stress.

The micron-sized basalt fibre of a monofilament has a diameter of approximately 7 ~ 21 μm (Zhou and Zhou, 2017) and is light weight; it has a large specific surface area (Fig. 13.31(a)), high tensile strength, high elastic modulus and elongation at break and good flexibility. As a filler, it is dispersed and entangled in a space cage structure under the action of hydraulic force and can aggregate and enclose a large number of microorganisms to form spherical microbial aggregates. Wu et al. (Zhou and Zhou, 2017) define this structure as a 'biological nest' (Fig. 13.31(b)).

Biofilms on the surface of conventional biofillers are heterogeneous. As the thickness increases, the composition of the flora and the type of respiration are layered. The scattering and entanglement of basalt fibre under the action of hydraulic force gives the biological nest an irregular pore structure inside, which affects the transmission and consumption of dissolved oxygen and nutrients. Therefore, an aerobic–anoxic–anaerobic layered structure also appears. The biological nest diameter is several orders of magnitude larger than the biofilm thickness, and the mass transfer and layered structures are more complex. Wu et al. (Zhou and Zhou, 2017) analyzed the heterogeneity, structural stratification and respiration type stratification of the physical structure and pointed out that the biological nest was rich in microbial flora.

They found aerobic bacteria such as nitrifying bacteria and nitrosomonas, facultative anaerobic flora such as flora and denitrifying bacteria and absolute anaerobic flora such as sulphate-reducing bacteria. They determined experimentally the distribution of dissolved oxygen, chemical oxygen demand (COD), ammonia nitrogen and total nitrogen in a biological nest of basalt fibre and examined the activity of the internal flora. According to the pore structure of the biological nest and the stratification of mass transfer, the distribution of the biological nest was as shown in Fig. 13.31(c) (Wu et al., 2016).

Generally, the thickness of the active biofilm formed by the extracellular polymeric substances after the microbes are attached to the surface of the filler is approximately 2 mm. The biological nest formed by the fibrous filler can form activated sludge with a diameter of more than 10 cm according to the length of the filler. Its biomass is greatly increased. The multilayer structure of aerobic–oxygen–anaerobic layers is formed inside the biological nest. The synergistic effect of different populations of microorganisms inside can achieve nitrogen and phosphorus removal while degrading organic matter and effectively shorten the sewage treatment process. The rich microbial phase and complete food chain help reduce sludge production (Zhou and Zhou, 2017). Ni et al. (Wu et al., 2016) conducted an experiment on the water treatment effectiveness of the basalt fibre biological nest reactor. The results show that the COD removal rate can reach 90%, and the reactor has a high denitrification capacity with an efficiency of 60.10% ~ 82.07%. Therefore, the application of basalt fibre filler in water treatment is a new type of water treatment technology with high efficiency, low cost, low land occupation and low sludge production.

13.4.5.2 Water treatment application of basalt fibre filler

The suspended biofiller prepared by basalt fibre has the advantages of environmental protection, easy membrane hanging, abundant biomass and high diversity of functional flora. It has broad prospects in water treatment applications and has achieved good results in the following applications:

Urban wastewater treatment

Chu et al. (Ni et al., 2018a) used basalt fibre as a filler to treat simulated urban wastewater. The experimental results showed that simulated urban wastewater with COD concentration of 500 mg/L could be treated under the condition of combined microbial preparation and bicyclic basalt fibre filler. The removal efficiency of COD could reach 97.22%. Yang et al. (Chu et al., 2013) treated municipal wastewater with composite microbial preparation and basalt fibre filler. The COD concentration decreased from 485.99 to 59.4 mg/L and the removal rate reached 87.78%. The ammonia nitrogen concentration decreased from the original 37.89 to 13.78 mg/L, and the removal rate was 63.63%. The total phosphorus concentration decreased from 6.15 to 1.09 mg/L, and the removal rate was as high as 82.28%. The effluent could reach the national first-class B emission standard.

Printing and dyeing wastewater treatment

Jiang et al. (Yang, 2013) treated printing and dyeing wastewater with basalt fibre as biofilm carrier. During the treatment process, the membrane speed and microbial activity were high. The COD removal rate reached 70% after only 5 days. Luo et al. (Jiang et al., 2016) used TiO_2 on basalt fibre felt to treat methyl red–simulated printing and dyeing wastewater. The basalt fibre felt was immersed three times in Ti solution, and the loading ratio of TiO_2 was 4.83%. At this time, the degradation rate of methyl red was 98.58%. Qi et al. (Luo et al., 2014) treated printing and dyeing wastewater with basalt fibre filler. Under Hydraulic Retention Time (HRT) = 15 h, the COD removal rate reached 67% and the Total Phosphorus (TP) removal rate reached 72%.

Water body repair

Jiang et al. (Qi et al., 2016) analyzed and compared the basalt fibre filler and biological rope filler treatments while simulating the purification of contaminated lake water. Under optimal conditions, the removal efficiencies of basalt fibre and biological rope on COD were 93% and 88%, respectively, and the NH^{4+}–N removals were 90% and 88.7%. This showed that basalt fibre was a better filler to treat COD, ammonia nitrogen and other pollutants and that the sewage purification performance was better than with biological rope filler. Huang et al. (Jiang, 2015; Huang et al., a, b) used basalt fibre in constructed wetlands to exert the synergistic effect of plant matrix microbes to purify sewage, whereas basalt fibre as a biofilm carrier could strengthen microbial action. The method has many advantages such as low construction cost, simple operation and good water discharge effects.

13.5 Environmental friendliness of basalt fibres during their life cycle

13.5.1 Environmentally friendly measures for basalt mining

The composition of basalt is similar to that of gabbro and the main mineral components are basic plagioclase and clinopyroxene. The related elements and minerals are copper, iron, zinc, cobalt, Iceland stone and agate. Basalt can be used as raw material for high-quality casting stone and is excellent stone material. Basalt belongs to the group of basic volcanic rocks. Its main components are silica, aluminium trioxide, iron oxide, calcium oxide and magnesium oxide (with a small quantity of potassium oxide and sodium oxide). Among them, silica is the most abundant, accounting for approximately 45%–55%. Basalt has high durability and many joints, most of which are hexagonal. It has a volume density of 2900–3300 kg/m^3, compressive strength of 100–300 MPa, high brittleness and strong weathering resistance. The colour of basalt is usually black, dark brown or dark green. Because of its dense texture, its specific gravity is greater than that of ordinary granite, limestone, sandstone and shale.

Basalt mining methods can be generally classified into four categories based on the main mining process (that is, the method of separating stones from ore bodies): sawing, splitting, controlled blasting and special cutting (mainly flame cutting and water jet cutting). At present, the latter three are commonly used in the exploitation of basalt ore in China. The following is a brief introduction to the commonly used mining methods.

(1) Controlled blasting method: The mechanism involves high-pressure gas produced by simultaneous explosions in row holes causing rock mass to split along the line of the row holes to achieve separation. This method has been widely used to date.
(2) Machine-type breaking method: This is divided into three methods: cutting, rolling and grinding. During rock breaking, the rock-breaking sand machine enters the basalt material and then moves to the front rock mass, showing a compact nucleus. A larger mass of debris occurs around the compact nucleus. The main reason for machine fracture not being widely used in hard rock is the severe wear and tear of instruments. The wear level depends on the content and particle size of hard mineral deposits (The main mineral is quartz) in the rock.
(3) Water jet cutting: This is divided into two types: low-pressure high flow and high-pressure low flow. In the former, the pressure does not exceed 2.1 MPa; the method is mostly used in hydraulic mining or coal mining. In the latter, the pressure can reach several hundred MPa; the method is used to cut basalt materials. In addition, a pulsed jet technology has been developed that can effectively crack strong rock without high power. At present, the instantaneous pressure reaches 5.6 GPa. Because of high energy consumption and a huge machine structure, the high-pressure water jet technique is now mostly used to assist in rock breaking by roadheaders and open-pit cone drills.
(4) Thermodynamic fracturing method: In this method, a high temperature gradient is formed in the basalt material and thermal stress is formed because of the different thermal expansion coefficients of the rock components, which causes the rock mass to spall or crack. Rock with more quartz has a better result.

13.5.2 Environmentally friendly measures

(1) Air quality control
  ① Dust in mining: Wet operation is adopted, personnel management is strengthened and mining activities are avoided in windy weather.
  ② Dust in ore crushing and processing: A water mist spraying system is set up, a spraying wet operation is conducted during ore crushing and the crushing area is screened to minimize dust production;
  ③ Dust in automobile transportation: Shielding is used for transportation vehicles, while driver management is strengthened, loading volume is controlled, vehicle speed is controlled and vehicle washing is enhanced.
(2) Noise abatement

The impact of mine production on the environment is further reduced by controlling explosive charges, adopting low-noise blasting methods, arranging reasonable blasting operation schedules, strengthening the maintenance of mining machinery and equipment, avoiding the increase in mechanical noise caused by poor performance of equipment, and slowing down or banning the sound of transport vehicles when passing through residential areas

(3) Solid waste treatment

Stripped surface soil and waste rock generated in the development process are piled up in the abandoned slag yard and the mining control area is regularly backfilled and greened every year according to the mining plan. Overhaul of mine production equipment is entrusted to external units.

(4) Wastewater treatment

① Rainfall in rainy season in mining area: A sedimentation tank is built and drainage ditch water is collected from the mine and waste yard for sedimentation and reuse.

② Wastewater from ore flushing: A three-stage sedimentation tank is set up in the processing area to treat and reuse ore flushing wastewater after sedimentation.

13.5.3 Environmentally friendly measures for crushing basalt ores

The crushing process of basalt ore consists of the following stages:

(1) Preliminary stage: primary fragmentation

The raw material passes through a silo, is evenly fed by a vibrating feeder and is transported to a jaw crusher for primary crushing. After primary crushing, the raw material and the fine material screened by the vibrating feeder are transported using a belt conveyor to the transfer stack.

(2) Phase II: medium and fine fragmentation

The conveyor belt evenly conveys the materials of the intermediate heap to the cone crusher for fine crushing and for screening by the circular vibration screen after crushing. Larger particles are sent back to the cone and broken again, medium-sized materials are sent to the sand machine for fine crushing and shaping and qualified granular materials are cleaned by a sand washer as final products.

The process has large production capacity, stable operation and uniform discharge levels, and dust is easily generated. Spraying with clean water is generally used to reduce the dust content in the production line. In addition, factory buildings around the production line can be constructed to collect in an organized manner the dust produced.

13.5.4 Environmental friendliness of basalt fibre production process

The production process of basalt fibre produces less waste and less pollution to the environment. When discarded, the product can be directly transferred into the ecological environment without any harm and is therefore a type of truly green and environmental protection material. The production process route is short and concise: Feeding → Melting → Drawing. The waste silk from the production process can also be collected, cut short, melted back to the kiln and then redrawn.

13.5.5 Environmentally friendly measures for preparation of basalt fibre composites

In the contemporary field of fibre-reinforced composite materials, glass fibre, CF, AF and basalt fibre are commonly used. For synthetic material such as CF and AF, inevitable harmful chemical scraps and waste are generated during their fabrication, and the enormous energy consumption of their production not only burdens the environment but also increases the cost and limits their popularization. Likewise, the extra fibrous fragments and dusts from production of glass FRP are found to be tremendously pernicious to human body in the long term and can hardly be decomposed and dissipated. In contrast, the BFRP is treated as a relatively healthy and natural material. Its eco-friendly characteristics are exhibited both in the production and the application stage.

Nowadays, automatic enclosed production lines are formed for BFRP materials. Owing to the inorganic, nonmetallic features of basalt rock, the fabrication of basalt fibre itself would not generate any deleterious material. As the fibres are impregnated and cured with resin, irritating volatile organic compounds (VOCs) could be produced because of the features of resin. With a custom-made gathering system composed of a blowing machine and ventilating flue, those VOCs are gathered and treated through a specific purification machine that can achieve zero discharge of harmful gases. Common purification methods include absorption, combustion and nonthermal plasma (NTP) as shown in Fig. 13.32.

The adsorption method is a physical method that takes advantage of adsorption when the VOCs come into contact with an adsorbent surface. Adsorption technology mainly includes adsorption by fixed bed, fluidized bed, mobile bed, pressure swing and power conversion. Its high purification efficiency is exhibited in the purification of low-concentration organic gas. This method can be used individually or in

Figure 13.32 BFRP production line and its waste gas collection and purification system.

combination with other methods. When the adsorption reaches saturation, the adsorbed material can be removed from the surface of the adsorbent by heating or other means to restore original state. Through desorption, organic wastes can be recovered and the adsorbent regenerated. Commonly used adsorbents are activated carbon, silica gel, zeolite, active alumina and molecular sieves. As silica gel, active alumina and molecular sieves first absorb the water vapour in the exhaust gas when adsorbing and desorption is difficult to desorb, activated carbon is often used as the adsorbent in the purification of organic exhaust gas because of its larger porosity, specific surface area and better adsorption effect. At present, granular activated carbon is widely used in the purification of organic waste gas in the production of BFRP.

Combustion is defined as the process of oxidizing high-concentration VOCs into a harmless gas that is much more convenient to treat. The energy generated by combustion can be further utilized. There are three combustion methods: direct, thermal and catalytic. Direct combustion is the direct burning of organic waste gas as fuel, suitable for organic waste gas with a higher concentration of combustible components. In this manner, the heat released during combustion is enough to compensate for the heat lost to the external environment, and the combustion zone is maintained at a sufficient temperature of approximately 1100°C for continuous combustion. Thermal combustion is suitable for purification of combustible organic gas whose concentration is relatively low. Auxiliary fuel is always required for this method, which leads to high cost. Catalytic combustion uses catalysts to purify organic waste gas at lower temperatures. The exhaust gas is preheated to 300–450°C and mixed before being transferred to the catalyst layer. This method requires less auxiliary fuel and equipment.

The NTP method has attracted growing interest in the last decades because it offers the opportunity to destroy or convert several air pollutants with high efficiency and a medium–low amount of energy. Plasma is defined as an ionized gas, the so-called fourth state of matter, in which the charges of ions and electrons are balanced. NTP has the potential to treat high flows for both low (<100 ppm) and high (>1000 ppm) concentrations of pollutants. In NTP, electrons and their surroundings are not in thermal equilibrium. Electric discharges in the gas heat the electrons instead of the gas itself, and the resulting high-energy electrons, active radicals and ions promote numerous chemical reactions in the ionized zones that are created. Although NTP is a very promising development for a variety of applications, it still has drawbacks including low energy efficiency, incomplete oxidation of the pollutants undergoing treatment and the formation of undesired by-products.

During the cutting of BFRP, some dust can be emitted. Through a dust gathering machine, the dusts can be recycled and reused. In general, assisted by multiple methods, eco-friendly production of BFRP can be achieved.

13.5.6 Recycling of waste BFRP

Landfilling and incineration are the two main conventional methods for disposing of waste BFRP. Such methods not only waste land resources; BFRP also does not degrade in 50–100 years after deep burial, which causes great potential harm in the ecological environment. Incineration treatments will also cause secondary pollution because the burning smoke contains toxic and harmful substances such as sulphur dioxide, nitrogen oxides, chlorinated hydrocarbons and dioxins. Meanwhile, owing to the excellent corrosion resistance and thermal stability of basalt fibre, it is almost impossible to decompose the intact basalt fibre wrapped in resin matrix through landfilling and incineration, and therefore its potential values in secondary utilization are wasted. At present, new environmentally friendly recycling methods have emerged that mainly include mechanical, thermal and chemical methods.

Mechanical recycling is a physical technique in which waste materials are recycled into new raw materials without changing the basic structure of the materials. Recycling of FRP composite materials mechanically started in the 1970s and is widely used for thermoset composite materials. This technique involves a sequence of processes including sorting, cutting, shredding, crushing, rotary cutting, milling and cycloning. The mechanical recycling technique is more suitable for GFRP and partially for CFRP; it presents a few advantages as it recover both fibres (short and long) and resins without any input or output of hazardous materials. This method can obtain fibre and resin material with a simple approach and no toxic or harmful substances are generated. However, the structure of the fibres is damaged in the process of recycling. As a result, long fibres cannot be recovered and the mechanical properties of the fibre are also significantly reduced. The potential value of recycled products is highly dependent on the size of the particles.

The thermal methods mainly consist of high-temperature pyrolysis, fluidized-bed pyrolysis and microwave pyrolysis. For the high-temperature pyrolysis method, using heat from air or an inert gas, the resin matrix can be decomposed into hydrogen, methane, hydrocarbon gas and low-molecular-weight organic compounds. No chemical reagents are used in this method, which prevents secondary pollution to the environment. The products of composite decomposition can further provide fuel for burning gas, but poisonous gas will be released during the process of recycling. The method of fluidized-bed pyrolysis emphasizes the use of air as fluidized gas in a fluidized-bed reactor to decompose the matrix components of composite materials. Through flow of air heated to high temperature, the heat generated in the recovery process can be fully extracted, and the fibre material is obtained by cyclone separation. Fluidized-bed pyrolysis enables the recycling of polluted composite material, especially chopped fibre. The microwave pyrolysis method takes advantage of the fact that CF is an electric conductor that can absorb microwaves. The main advantage of the microwave technique is that the material is heated in its core such that the transfer of thermal energy is fast, which helps to minimize heat losses to the surroundings and reduces energy consumption. This method is not suitable for basalt fibre, which is an insulator.

Chemical methods aim at converting the resin matrix in the waste FRP to small molecules by chemical reagents, to dissociate the FRP and recover fibre. The atmospheric-pressure dissolution method and supercritical-fluid dissociation method are two common chemical methods. With the atmospheric-pressure dissolving method, the resin material can be abstracted from the FRP and high-performance long fibres can be retained, but a large number of chemical reagents are required, which results in secondary pollution to the environment. The excellent solubility, diffusion ability, solvation ability and other special properties of supercritical fluids are conducive to the degradation reaction. In environmental ecology fields, supercritical fluid is often used in waste plastics recycling, biomass transformation and polymer degradation. Current research work shows that supercritical fluid can effectively recover the high-performance long fibre from FRP; at present, the main reaction media to recover CF are supercritical water, supercritical n-propyl alcohol and supercritical methanol.

13.6 Conclusions and future trends

Because of the characteristics and advantages of basalt fibre, it is considered as a green and sustainable material for our future development. Basalt fibre is known as a green industrial material of this century. In addition, basalt fibre has no toxic reaction in air or water and is nonflammable. Because of its high rigidity, low elongation or high elongation at break, it has become a choice of useful reinforcement materials at present. In the future, more basalt fibres with lighter, stronger and better corrosion resistant properties, like making hollow fibre, can be developed. Hollow fibre membranes refer to membranes which are like fibres and have self-supporting function. It is a kind of asymmetric membrane whose dense layer can be located on the outer surface of the fibres, such as reverse osmosis membrane, or on the inner surface of the fibres (such as microfiltration membrane, nanofiltration membrane and ultrafiltration membrane).

For the application technology of the series BFRP, we should focus on the following three aspects in the future: vigorously promote the large-scale application of existing mature technologies, including chopped fibre-reinforced asphalt and BFRP bars/BFRP grids reinforced bridge/tunnel; continuously improve new technologies and accumulate experience in demonstration projects, including short fibre-reinforced cement-based composite materials, BFRP anchor, intelligent BFRP reinforcement and reinforced structure, BFRP profile-concrete composite structure; constantly explore new application technologies and make technical reserves for more extensively structural requirements, including BFRP cable long-span bridge structure, BFRP profile structure, thermoplastic resin matrix based BFRP, prefabricated structure and BFRP piles, etc.

Basalt fibre is a green microscale inorganic flexible fibre material with good chemical stability, thermal stability, mechanical properties and hydraulic properties and can be used as a high-performance carrier in water treatment. Basalt fibre disperses in water to form a cage structure, which can gather and enclose a large number of microorganisms to form a spherical microbial aggregate with a diameter of more than 10 cm, that is, a 'biological nest'. 'Biological nest' has the advantages of large biomass and high activity and has a multimembrane structure (aerobic layer-oxygen layer-anaerobic layer). At present, basalt fibre filler has been gradually researched in urban wastewater treatment, printing and dyeing wastewater treatment and water body repair, and the results show that nitrogen removal and organic removal are highly efficient. The new biofiller should have good microbial adhesion, hydraulic properties, mechanical strength and environmental protection. Basalt fibre needs to be improved in hydrophilicity, surface weak positive polarity and roughness to improve its biological progeny. More research on the modification of water-treated biofillers for basalt fibre should be conducted. At present, the research and development, production and application of basalt fibre are still in the preliminary stage. In the future, systematic research on the preparation of modified basalt fibre biofiller and the mechanism and application of 'biological nest' technology are needed.

Index of acronyms

ABS Acrylonitrile butadiene styrene copolymers
AF Aramid fibre
AFRP Aramid fibre–reinforced polymer
BFRP Basalt fibre–reinforced polymer
CBF Continuous basalt fibre
CF Carbon fibre
CFRP Carbon fibre–reinforced polymer
COD Chemical oxygen demand
CV Coefficient of variance
DSC Differential scanning calorimetry
FRP Fibre-reinforced polymer/plastic
GFRP Glass fibre–reinforced polymer
HRT Hydraulic retention time
NTP Nonthermal plasma
PBT Polybutylene terephthalate
PET Polyethylene terephthalate
R&D Research and development
SD Standard deviation
SEM Scanning electron microscope
TP Total phosphorus
VOC Volatile organic compound

References

Acha, B.A., Reboredo, M.M., Marcovich, N.E., 2007. Creep and dynamic mechanical behavior of PP–jute composites: effect of the interfacial adhesion. Compos. A Appl. Sci. Manuf. 38 (6), 1507–1516.

Ascione, L., Berardi, V.P., D'Aponte, A., 2012. Creep phenomena in FRP materials. Mech. Res. Commun. 43, 15–21.

Cheng, Y., Tian, L., LV, W., et al., 2017. Research progress of flame retardant PA 66[J]. China Synth. Resin Plast. 34 (1), 81–84, 90(in Chinese).

Chu, J., Yang, X., Jiang, S., 2013. Treatment of city domestic sewage by combining microbial strains and basalt fiber[J]. Chin. J. Environ. Eng. 7 (11), 4372–4376 (in Chinese).

Colombo, C., Vergani, L., Burman, M., 2012. Static and fatigue characterisation of new basalt fibre reinforced composites[J]. Compos. Struct. 94 (3), 1165–1174.
Dalinkevich, A., Gumargalieva, K., Marakhovsky, S., Soukhanov, A., 2009. Modern basalt fibrous materials and basalt fiber-based polymeric composites[J]. J. Nat. Fibers 6 (3), 248–271.
Dorigato, A., Pegoretti, A., 2012. Fatigue resistance of basalt fibres-reinforced laminates[J]. J. Compos. Mater. 46 (15), 1773–1785.
Everett, R.K., 1996. Quantification of random fiber arrangements using a radial distribution function approach. J. Compos. Mater. 30 (6), 748–758.
Huang, J., Xiao, J., Ma, Y., et al. Constructed Wetland with Long Basalt Fiber Composite fille [P]r:China, CN201810255479.1.2018-07-06 (in Chinese).
Huang, J., Xiao, J., Ma, Y., et al. Constructed Wetlands with Multi-Stage Basalt Fiber Bundle Packing[P]: China, CN201810256298.0.2018-07-06 (in Chinese).
Jiang, D., 2015. Research on Water by In-Site Purification Technology Based on Basalt Fiber Carrier[D]. Jiangsu University, Zhenjiang (in Chinese).
Jiang, X., Wu, C., Wu, Z., et al., 2016. Treatment of printing and dyeing wastewater through biological contact oxidation process using basalt fiber as carrier[J]. Ind. water Treat. 36 (1), 78–82 (in Chinese).
Li, Y., Shen, M., Chen, W., et al., 2012. Tensile creep study and mechanical properties of carbon fiber nano-composites. J. Polym. Res. 19 (7), 1–8.
Libby, J.R., 1990. Steel for Prestressing, Modern Prestressed Concrete. Springer, US, pp. 11–44.
Luo, D., Xu, C., Gao, H., et al., 2014. Preparation and photocatalytic performance of TiO_2/basalt fiber mat composite[J]. Bull. Chin. Ceram. Soc. 33 (10), 2493–2497 (in Chinese).
Maksimov, R.D., Plume, E., 2001. Long-term creep of hybrid aramid/glass-fiber-reinforced plastics. Mech. Compos. Mater 37 (4), 271–280.
Miazga-Rodriguez, M., Han, S., Yakiwchuk, B., et al., 2012. Enhancing nitrification at low temperature with zeolite in a mining operations retention pond[J]. Front. Microbiol. (3), 271–280.
Ni, H.C., Zhang, X.Y., Zhou, X.T., et al., 2018. Feasibility of using basalt fiber as biofilm carrier to construct bio-nest for wastewater treatment[J]. Chemosphere 212, 768–776 (in Chinese).
Nie, X., Liu, C., Zhong, J., et al., 2018. Research progress on regeneration of coal—based granular activated carbon used in water treatment[J]. Rmal Power Gener. 47 (3), 1–11 (in Chinese).
Paluch, B., 1996. Analysis of geometric imperfections affecting the fibers in unidirectional composites. J. Compos. Mater. 30 (4), 454–485.
Qi, Y., Wu, Z., Xu, X., et al., 2016. Application of different fillers to biological contact oxidation process ofdyeing and finishing effluents[J]. Dye. Finish. (8), 16–19 (in Chinese).
Sasaki, T., 2008. Experimental Study on Long-Term Performance of FRP and FRP Bonding. Technique Master thesis Ibaraki University, Japan.
Shi, J., Wang, X., Wu, Z., et al., 2015. Creep behavior enhancement of a basalt fiber-reinforced polymer tendon[J]. Constr. Build. Mater. 94 (6), 750–757.
Shi, J., Wang, X., Wu, Z., Zhu, Z., 2017b. Fatigue behavior of basalt fiber-reinforced polymer tendons under a marine environment[J]. Constr. Build. Mater. 137, 46–54.
Soudki, K.A., 1998. FRP reinforcement for prestressed concrete structures. Prog. Struct. Eng. Mater. 1 (2), 135–142.
Spathis, G., Kontou, E., 2012. Creep failure time prediction of polymers and polymer composites. Compos. Sci. Technol. 72 (9), 959–964.

Wang, X., Shi, J., Liu, J., et al., 2014. Creep behavior of basalt fiber reinforced polymer tendons for prestressing application. Mater. Des. 59, 558–564.

Wang, X., Shi, J., Wu, Z., et al., 2016b. Fatigue behavior of basalt fiber-reinforced polymer tendons for prestressing applications[J]. J. Compos. Constr. 20 (3).

Wang, X., Shi, J., Wu, Z., Zhu, Z., 2016a. Creep strain control by pretension for basalt fiber-reinforced polymer tendon in civil applications. Mater. Des. 89, 1270–1277.

Wang, X., Wu, Z., Wu, G., Zhu, H., Zen, F., 2013. Enhancement of basalt FRP by hybridization for long-span cable-stayed bridge[J]. Compos. B Eng. 44 (1), 184–192.

Wu, Z., Jiang, S., Zhou, X., et al., 2016. The theoretical analysis for the formation and waste-water treatment mechanism of "Bio-nest"[J]. Adv. Environ. Prot. 6 (4), 61–68 (in Chinese).

Wu, Z., Wang, X., Iwashita, K., Sasaki, T., Hamaguchi, Y., 2010a. Tensile fatigue behaviour of FRP and hybrid FRP sheets[J]. Compos. B Eng. 41 (5), 396–402.

Zhishen Wu, Jianxun Liu. Movable Electric Machine Components of Continuous Basalt Fiber Porcelain Kiln and Their Replacement Method, National intellectural property administration, PRC, ZL20160504082.2 (In Chinese).

Yang, X., 2013. Study on Treating Municipal Wastewater by Combination of Compound Microorganism and Packing[D]. Jiangsu University, Zhenjiang (in Chinese).

Zhang, Y., 2017. Application of PO polyolefin composite material in anticorrosion of containers [J]. China High-Tech Enterp. (4), 44–45 (in Chinese).

Zhao, X., Wang, X., Wu, Z., Zhu, Z., 2016. Fatigue behavior and failure mechanism of basalt FRP composites under long-term cyclic loads[J]. Int. J. Fatigue 88, 58–67.

Zhishen Wu, Jianxun Liu, 2014. A Kiln and Heating Method for Mass Production of Continuous Basalt Fibres. National intellectural property administration. PRC. ZL2014101308819 (In Chinese).

Zhou, Z., Yang, Y., 1985. Preliminary discussion on creep mechanism of fiber reinforced plastics. Glass Fiber Reinf. Plast. Compos. 4, 012 (in Chinese).

Zhou, Y., Zhou, C., 2017. Study on application of new material carbon fiber in water treatment [J]. Chem. Manag. (12), 213 (in Chinese).

Further reading

Chen, X., Zhang, Y., Hui, D., Chen, M., Wu, Z., July 17–23, 2016. Development of melted process of continuous basalt fiber. In: The 24th Annual International Conference on Composites or Nano Engineering (ICCE-24) (Haikou, China).

Chen, X., Zhang, Y., Wu, Z., et al., 2017. Study of melting properties of basalt based on their mineral components. Compos. B Eng.[J] 11 (6), 53–60.

Chen, X., Zhang, Y., Wu, Z., et al., 2017. Improving the tensile strength of continuous basalt fiber by mixing basalts. Fibers Polym[J] 18 (9), 1796–1803.

Cheng, Y., Tian, L., Lv, W., Cui, Y., 2017. Research progress of flame retardant PA 66[J]. China Synth. Resin Plast. 34 (1), 81–84, 90.

Chu, J., Yang, X., Jiang, S., 2013. Treatment of city domestic sewage by combining microbial strains and basalt fiber[J]. Chin. J. Environ. Eng. 7 (11), 4372–4376.

Da-Jun, L., Cai-Yun, X.U., Hong-Jiao, G., et al., 2014. Preparation and photocatalytic performance of TiO_2/basalt fiber mat composite[J]. Bull. Chin. Ceram. Soc. 33 (10), 2493–2497.

Fiore, V., Scalici, T., Di Bella, G., Valenza, A., 2015. A review on basalt fiber and its composites. Compos. B Eng. 74, 74–94.

Huang, J., Xiao, J., Ma, Y., et al. Basalt Fiber Covered Vertical Flow Constructed Wetlands[P]: China, N201810255499.9.2018-07-06 (in Chinese).
Jia, X.U., 2009. Recycling technology of composite material[J]. Fiber Reinf. Plast. Compos. 26 (4), 100–103.
Jian-Xun, L., Mei-Rong, C., Zheng-Guo, J., et al., 2015. Determination of total iron and Fe^{2+} in basalt[J]. Spectrosc. Spectr. Anal. 35 (8), 2312.
Jianbin, S., Changcheng, M.A., Lihong, C., et al., 2013. Preparation and performance of basalt fiber reinforced PVDF/PP/PETG blends[J]. China Plast. 27 (11), 34–38.
Jiang, D., 2015. Research on Water by In-Site Purification Technology Based on Basalt Fiber Carrier [D]. Jiangsu University.
Kouparitsas, C.E., Kartlis, C.N., Varelidis, P.C., et al., 2002. Recycling of the fibrous fraction of reinforced thermoset composites[J]. Polym. Compos. 23 (4), 682–689.
Lee, S.H., Choi, Kim, J.S., et al., 2011. Circulating flow reactor for recycling of carbon fibre from carbon fibre reinforced epoxy composite[J]. Korean J. Chem. Eng. 28 (2), 449–454.
Liu, J., Jiang, M., Wang, Y., et al., 2013. Tensile behaviors of ECR-glass and high strength glass fibers after NaOH treatment[J]. Ceram. Int. 39 (8), 9173–9178.
Liu, J., Yang, J., et al., 2018. Effect of SiO_2, Al_2O_3 on heat resistance of basalt fiber. Thermochim. Acta 660, 56–60.
Liu, J., Yang, J., Huo, H., et al., 2017. Study on the effect of different Fe_2O_3/ZrO_2 ratio on the properties of silicate glass fibers[J]. Adv. Mater. Sci. Eng. 2017 (10), 1–7.
Miazga-Rodriguez, M., Han, S., Yakiwchuk, B., et al., 2012. Enhancing nitrification at low temperature with zeolite in a mining operations retention pond[J]. Front. Microbiol. (3), 271–280.
Moeser, K., 2014. Enzymatic. Degradation of epoxy resins: an approach for the recycling of carbon fiber reinforced polymers[J]. Adv. Mater. Res. 1018, 131–136.
Nageh, A.M., Wang, X., Wu, Z., et al., 2015. Basalt fiber reinforced polymer grids as an external reinforcement for reinforced concrete structures[J]. J. Reinf. Plast. Compos. 34 (19), 1–13.
Ni, H.C., Zhang, X.Y., Zhou, X.T., et al., 2018. Feasibility of using basalt fiber as biofilm carrier to construct bio-nest for wastewater treatment[J]. Chemosphere 212, 768–776.
Pickering, S.J., 2006. Recycling technologies for thermoset composite materials-current status [J]. Compos. Appl. Sci. Manuf. 37, 1206–1215.
Qi, Y., Wu, Z., Xu, X., et al., 2016. Application of different fillers to biological contact oxidation process of dyeing and finishing effluents [J]. Dye. Finish. (8), 16–19.
Qi-Lin, Z., Yi-Feng, G., Fei, L.I., 2014. Current research and development of the application of the pre-tightened tooth connection[J]. Fiber Reinf. Plast. Compos. 12, 52–55.
Rodolfo, M.I., Mitsuru, S., Motonobu Armando, T., et al., 2015. Carbon fiber recovery using water and benzyl alcohol in subcritical and supercritical conditions for chemical recycling of thennoset composite materials[J]. J. Mater. Cycles Waste Manag. 17, 369–379.
S, hi, J., Wang, X., Huang, H., Wu, Z., 2017. Relaxation behavior of prestressing basalt fiber-reinforced polymer tendons considering anchorage slippage. J. Compos. Mater. 51 (9), 1275–1284.
Sayed, A.M., Wang, X., Wu, Z., 2013. Modeling of shear capacity of RC beams strengthened with FRP sheets based on FE simulation. J. Compos. Constr.[J] 17 (5), 687–701.
Shi, Q., 2003. Development and application of continuous basalt fibers abroad. Glass fiber (4), 27–31.
Shi, J., Wang, X., Wu, Z., et al., 2017. Relaxation behavior of prestressing basalt fiber-reinforced polymer tendons considering anchorage slippage. J. Compos. Mater.[J] 51 (9), 1275–1284.
Shi, J., Wang, X., Wu, Z., Zhu, Z., 2015. Creep behavior enhancement of a basalt fiber-reinforced polymer tendon. Constr. Build. Mater. 94, 750–757.

Singha, K., 2012. A short review on basalt fiber. Int. J. Text. Sci. 1 (4), 19–28.

Song, J., Chen, L., Ma, C., Liu etc, J., 2013. Influence of basalt fibers on the strengthening and toughening of PE-HD/bamboo composites[J]. China Plast. 27 (7), 48–51.

Tang, Y., Wu, Z., Yang, C., et al., 2010. A new type of smart basalt fiber-reinforced polymer bars as both reinforcements and sensors for civil engineering application[J]. Smart Mater. Struct. 19 (11), 115001.

VivekDhand, Mittal, G., You Rhee, K., Park, S.-J., Hui, D., 2015. A short review on basalt fiber reinforced polymer composites. Compos. B Eng. 73, 166–180.

Wang, X., Shi, J., Wu, Z., et al., 2015. Effectiveness of basalt FRP tendons for strengthening of RC beams through the external prestressing technique[J]. Eng. Struct. 101, 34–44.

Wang, X., Xu, P., Wu, Z., et al., 2015. A novel anchor method for multi-tendon FRP cable: concept and FE study[J]. Compos. Struct. 120, 552–564.

Wang, X., Xu, P., Wu, Z., et al., 2015. A novel anchor method for multitendon FRP cable: manufacturing and experimental study[J]. J. Compos. Constr. 19 (6).

Weirong, D., Masatoshi, K., Shurou, Y., et al., 2002. An approach to chemical recycling of epoxy resin cured with amine using nitric acid[J]. Polymer 43 (10), 2953–2958.

Wu, Z., Fahmy, M.F.M., Wu, G., 2011. Damage controllable structure systems using FRP composites[J]. J. Earthq. Tsunami 5 (3), 241–258.

Wu, Z., Jiang, S., Zhou, X., et al., 2016. The theoretical analysis for the formation and wastewater treatment mechanism of "Bio-Nest"[J]. Adv. Environ. Prot. 6 (4), 61–68.

Wu, J., Tang, X., Liu, J., 2014. Determination of horizontal insertion depth of molybdenum electrodes [J]. Glass Fiber (5), 13–17.

Wu, Z., Wang, X., Wu, G., June 2012. Advancement of structural safety and sustainability with basalt fiber reinforced polymers. In: Proc. CICE 2012 6th International Conference on FRP Composites in Civil Engineering – Rome, Italy, pp. 13–15.

Wu, G., Wu, Z., Hu, X., Jiang, J., 2007. Current Research and Development of the Application of Basalt Fiber in Civil Engineering[C]. National FRP Academic Exchange, pp. 410–415.

Wu, Z., Wu, G., Wang, X., Hu, X., Jiang, J., 2009. New progress in R&D of basalt fibers and BFRP in infrastructure engineering. Ind. Archit. 1–14.

Wu, Z., Yuan, H., Asakura, T., et al., 2005. Peeling behavior and spalling resistance of bonded bidirectional fiber reinforced polymer sheets[J]. J. Compos. Constr. 9 (3), 214–226.

Wu, Z., Zhang, D., Karbhari, V., 2010. Numerical simulation on seismic retrofitting performance of reinforced concrete columns strengthened with fibre reinforced polymer sheets[J]. Struct. Infrastruct. Eng. 6 (4), 481–496.

Xia, J., Chundu, W., Zhiren, W., et al., 2016. Treatment of printing and dyeing wastewater through biological contact oxidation process using basalt fiber as carrier[J]. Ind. Water Treat. 36 (1), 78–82.

Xian-qi, H.U., LUO, Y.-F., Nian, S., 2004. Basalt continuous fiber and its reinforcing composite material[J]. Hi-Tech Fiber Appl. 4 (2), 1–5.

Xian-Qi, H.U., Nian, S., 2005. The applications of the CBF in war industry & civil fields[J]. Hi-Tech Fiber Appl. 6 (30), 7–13.

Xin, N., Chenglong, L., Junfeng, Z., et al., 2018. Research progress on regeneration of coal-based granular activated carbon used in water treatment[J]. Therm. Power Gener. 47 (3), 1–11.

Yang, X., 2013. Study on Treating Municipal Wastewater by Combination of Compound Microorganism and Packing[D]. Jiangsu University.

Yang, Y., Wang, X., Wu, Z., 2017. Damping behavior of hybrid fiber-reinforced polymer cable with self-damping for long-span bridges[J]. J. Bridge Eng. 22 (7).

Identification of natural textile fibres

14

R. Nayak[1], *S. Houshyar*[2], *A. Khandual*[3], *R. Padhye*[2], *S. Fergusson*[2]
[1]School of Communication and Design, RMIT University Vietnam, Saigon, Vietnam; [2]School of Fashion and Textiles, RMIT University, Brunswick, Australia; [3]Department of Textile Engineering, CET Bhubaneswar, Bhubaneswar, Orissa, India

14.1 Introduction

The identification of textile fibres is important for textile and apparel manufactures, fashion designers, automotive industries and forensic science. The identification methods, however common across disparate industries, are conducted very differently in each. Some methods are exclusively used by one group; others are shared, while some shared methods have greater or lesser utility for the analyst. Differences in physical and chemical properties are the basis for fibre identification. Generally, simple tests are enough to enable a fibre to categorize into the main groups of natural fibres where the differences in properties are relatively large. More complicated tests are essential to distinguish between fibres within a group. Many fibres can be identified by fairly simple tests needing only inexpensive equipment.

Although some new instrumental identification techniques originated with the technical developments, the old methods are often considered to be the best. Most laboratories will still rely on the simplest microscopical examination, which provides a range of analysis barely possible with any other method. Combined with spectroscopy, microscopy is the quintessential fibre identification tool. Instrumental methods of identification provide reliable means to distinguish between the main types of fibres according to their chemical structure and closely related fibres with only minor differences in properties.

The American Association of Textile Chemists and Colourist (AATCC) technical manual lists microscopy as useful method for fibre identification, whereas American Society for Testing and Materials (ASTM) reports infrared (IR) spectroscopy with additional physical properties as preferred method for fibre identification. AATCC also lists reaction of fibres to flame (such as melts near flame, shrinks from flame and burns in flame) as a test method for fibre identification. The amount of sample for analysis is nearly unlimited for producers, while it is minimal in the case of forensic analysis. Therefore, different methods are devised for fibre identification. For both the analysis the common parameters are determining the individual fibre's morphological, optical and chemical properties as distinctly as possible.

Handbook of Natural Fibres. https://doi.org/10.1016/B978-0-12-818398-4.00016-5

14.2 Natural textile fibres

A natural fibre is any fibre that exists as such in the natural state, such as cotton, wool or silk (Houck, 2009). Natural fibres differ from each other in cross-sectional (CS) shape, colour, surface contour, chemical structure as well as length and width (Ziabicki, 1976; Nayak et al., 2012). The natural textile fibres can be classified into three groups such as vegetable, animal and mineral according to their origin. But only the vegetable and animal fibres will be discussed in this chapter.

14.2.1 Vegetable fibres

Vegetable fibres are mainly collected from different parts such as seed (cotton, akund, kapok, coir); bast (flax, jute, hemp, ramie, kenaf) and leaf (sisal, abaca) of various plants. The vegetable fibres are also called cellulosic fibre as cellulose is the main component of the fibres. Additional information on vegetable fibres can be obtained from Chapters 1, 2, 3, 4, 5, 6, 7, 8, 9, 10, published in Vol. 1 of the Handbook of Natural Fibres. The identification of vegetable fibres includes the analysis of the size and relative thickness of cell wall; shape and thickness of lumen; the presence, type and distribution of dislocations and the direction of twist of the cellulose in the cell wall (Greaves, 1990). Additional information on vegetable fibres can be obtained from Chapters 2 and 3 of volume 1.

14.2.1.1 Seed fibres

Cotton is the most common seed fibre and accounts for approximately half of all textile fibres processed annually. The unique feature of cotton fibre is the presence of lumen and convolutions, which makes identification easier. Convolutions are twists in the fibre with many reversals along the fibre length, which are clearly visible in the polarized light of the microscope. Cotton can be distinguished from other fibres under the microscope between crossed polars (Perry et al., 1975). It remains substantially bright in all orientations, except that in the orthogonal positions dark bands cross the fibre at frequent intervals indicating the reversal points of the underlying spiral structure.

The appearance of cotton can be modified by specific chemical finishing to get desired result e.g., mercerization of cotton causes fibre swelling and untwisting. Therefore, the convolutions in mercerized cotton may not be easily visible (because of fibre swelling). Generally, the cross-section of cotton is kidney shaped and in case of mercerized cotton it is circular. Resin treatment and some other chemical finishes can alter the appearance of cotton fibre.

For preliminary identification of cotton, add a drop of cuprammonium hydroxide to the fibres mounted on a slide and examine under microscope (a) if the fibres dissolve readily and completely, it is scoured cotton or mercerized cotton; (b) if the fibres swell irregularly, forming beads and finally dissolve leaving a residue of particles that float away in the solvent, it is raw cotton and (c) if the fibres remain undissolved, it is chemically modified cotton. Iodine test can also be used to distinguish between mercerized and unmercerized cotton (Marsh, 1941). In this test cotton fibre is treated with a specific iodine solution, and if the fibre decolourizes immediately, it is unmercerized cotton. Mercerized cotton takes several hours for discolouration.

Other seed hairs such as akund, kapok and coir are readily distinguished from cotton by thin cell wall and absence of convolutions. Akund is fine, soft and lustrous but very weak like kapok. The appearance of akund under microscope is similar to kapok, but the base does not show the net-like thickenings seen in kapok. Kapok fibres are smooth, hollow, cylindrical, thin-walled, brittle and frequently bent over on it. Coir fibres are very stiff, dense and have distinct cross-section, which makes easier to identify by microscopy. Coir appears dark brown or opaque with very large, coarse ultimates under microscope.

14.2.1.2 Bast fibres

The colour of the bast fibres depend upon the retting conditions, e.g., the colour of water retted fibre is from white to yellow and the colour of dew retted fibre is from grey to dark silver. Flax has clockwise twist, polygonal cross-section of ultimates with thick walls and small lumina. The feature of flax fibre, which is common with other bast fibres is the presence of dark dislocations, often in the form of an 'X' under microscope that are roughly perpendicular to the longitudinal axis of the fibre. On the other hand jute has counterclockwise twist, microscopically it appears bundled and may have a yellowish cast. It has polygonal cross-section of the ultimates, which are angular with medium sized lumina.

The characteristic features of hemp differ from jute with more bundled ultimates, wider lumen and fewer nodes. Also the rounder and more flattened cross-section of hemp helps to distinguish it from jute. Hemp has counterclockwise twist like jute and may also have a brownish cast to it. Ramie has thick wall, flattened cross-section, very wide ultimates with the width ranging from 25 to 75 μm. Microscopically ramie has frequent, short dislocations, longer transverse striations and radical cracks may be present in cross-section. Kenaf, with an ultimate fibre length from 2 to 6 mm is more lustrous, harder and stronger than jute; and lighter in colour. The reader can find more pieces of information on bast fibres in Chapters: 3, 4, 5, 6, 7 and 18 of volume 1 and Chapters: 2, 4, 5, 6, 7, 11, 12, 17, and 19. of volume 2.

14.2.1.3 Leaf fibres

The irregular lumen size, acicular crystals, spiral elements and annular vessels are the features for identification of sisal. The cross-section of sisal looks like cut celery, and it has counterclockwise twist. The features for identification of abaca are uniform diameter of ultimates, waxy appearance, polygonal cross-section and counterclockwise twist. Abaca may contain spiral elements but often will have stegmata, which are visible as small crown-like structures. Sisal is distinguished from abaca by Billinghame's test (Perry et al., 1975).

14.2.2 Animal fibres

The animal fibres with protein as the main component include wool, hair and silk fibres. Although the chemical structure of hair fibres are related to wool, the physical characteristics are different, i.e., they differ in length, fineness, shapes and internal structure, which can be used for their identification. More pieces of information on wool are available in Chapters 11 and 12 (volume 1) and about silk in Chapters13 and 21 (volume 1) and in Chapter 3 (volume 2).

14.2.2.1 Wool

Wool is obtained from the hair of sheep and certain other animals, including cashmere from goats; mohair from goats, alpaca and animals in the camel family and angora from rabbits. Wool can be identified by careful microscopical examination of features such as CS outline, pattern formed by the scale margins, regularity of fibre diameter, type of medulla and thickness of cuticle. The morphology of these fibres is complex, consisting of a cortex covered by a cuticle of scales and in some instances a central medulla, often showing different patterns in different fibres. Finer wool fibres (e.g., merino wool) have no detectable medulla. The finest merino wool appears strikingly irregular in diameter when seen under microscope. The medulla in long wools and crossbred wools is simple unbroken, medium to narrow type.

At the beginning, the animal hair fibre to be identified should be confirmed that it is not human in origin. This is done from the banding characteristics (abrupt and profound colour changes along the hair shaft) of the animal hair. The complete identification of animal hair fibres mainly relies on the morphological features of the cortex, the size and shape of the scales and their pattern around the hair. The scale patterns may be visualized by mounting the hairs in a semipermanent mounting medium with a refractive index lower than that of keratin (1.55) or it can be done by a scanning electron microscope (SEM). The advantage of SEM is its very high resolution and relatively large depth of field, which enables a better discrimination of the surface characteristics. However, in SEM no internal features are visible, and due to the thin coating, the further examination of the fibre is not possible.

If the origin of the animal fibre is unknown, determining the diameter of the fibre assist in the identification (Houck, 2009). The medulla types and scale patterns vary in hairs of different species, which can be used for differentiation. Heat treatment and some chemical finish such as alkaline treatment, bleaching or chlorination can modify the structural characteristics of wool. Chlorination causes the scale margins gradually less prominent and may be complete removal with severe treatment.

Hamlyn et al. (1992) applied molecular biology techniques for the identification of animal hair fibres (wool, cashmere and mohair). In this technique, specific DNA probes were developed, which can distinguish between DNA isolated from closely related species. This technique is helpful for forensic analysis and biomolecular palaeontology. The chemical resistance of similar type of fibres such as wool, cashmere and mohair to solvents is very similar. Therefore, solubility test cannot be used to distinguish among these fibres. In these cases, scale height (Kusch and Arns, 1982; Wiedemann, 1987; Robson, 2000), scale pattern (She et al., 2001), scale frequency (Robson, 2000) and evenness of diameter (Langley and Kennedy, 1981) may be used in distinguishing between wool, cashmere and mohair fibres. She et al. (2001) used image processing and Artificial Neural Network (ANN) model to classify mohair and merino fibres on the basis of the scale pattern.

14.2.2.2 Silk

Silk is a natural protein fibre mainly produced by the larvae of insects that complete metamorphosis. The best type of silk is obtained from the cocoons of the mulberry silkworm (*Bombyx mori*). Unlike the wool fibres, silk fibres are generally easy to identify. The filaments of *B. mori* in gummed state are stuck together in pairs and covered with sericin. These silk filaments when degummed have smooth and uniform surface, semitransparent appearance and approximate to equilateral triangles with rounded apices cross-section.

Mulberry silk shows a triangular cross-section and a smooth surface, which markedly differ from the nonmulberry (wild) silk (Sen and Babu, 2004). The wild silk appears flat and ribbon-like with fine longitudinal lines. The cross-section of other silks can vary in shape, e.g., crescent-like for Anaphe and elongated wedge-shaped for tussah.

Spider silk is a protein fibre spun by spiders with mechanical properties superior to silkworm silk (Cunniff et al., 1994). It has a unique combination of strength, elasticity and resistance to compression, which is judged to be superior to that of synthetic fibres made of polyamide or polyester. Some specimens exhibit over 200% elongation and others possess tensile strengths equivalent to high performance fibres. Also compared with silkworm silk, it is more waterproof. In spite of many superior properties, spider silk has not been domesticated for textile applications because of difficulties associated in raising dense populations of spiders due to their solitary and predatory nature; reeling of a single fibre from the spider webs is difficult and the amount of silk generated by spiders is small in comparison to the silkworm cocoon silk (Mukhopadhyay and Sakthivel, 2005).

Artificial spider silk has already been spun successfully. The race is now on the perfection of spinning technology. The most promising methods till now include conventional solvent spinning of recombinant spider silk protein analogue (Du Pont Ltd) (O'brien et al., 1998) and extruding regenerated silk into a coagulation bath using a nanofabricated silicon spinneret (Seidel et al., 1998).

Care must be taken while identifying the silk fibre because of its similarities with nylon fibre in fibre diameter and infrared spectra. However, silk is less regular in appearance along its length than a nylon fibre. If difficulties are faced in identifying these fibres, place a short segment or cross-section of the fibre in question in a hot stage. Nylon will melt, and silk will not. Fluorescence microscopy may provide additional features based on the fluorescence of the dyes.

14.3 Identification methods

The identification of textile fibres is an important and challenging task. As there are many ways of identifying textile fibres, there is no strict routine test method followed by the textile analyst. Fibres are identified by one or more test methods in combination such as physical appearance, burning test, microscopic analysis, solubility test, staining test and physical property analysis. Although conventional approaches such as microscopy, burning test, staining test and solubility tests are easy to conduct, they have several limitations:

1. These methods are subjective and rely heavily on expertise.
2. These methods are destructive as there is fibre damage during the test.

14.3.1 Identification by physical appearance

It may be said with some justification that the simplest and most obvious way to identify a fibre is to look at it. As the physical appearances of fibres differ from each other, it can give a preliminary indication for fibre identification. Successful identification of fibres depends on experience and familiarity with the fibres. This method is subjective and often not sufficient for the complete identification.

14.3.2 Burning test

This is the first step in the fibre identification (BIS, 1981). This test relies on the behaviour of the fibre when exposed to a flame depending on the chemical composition of the fibre. In this method a small tuft of fibre is held by forceps for about 10 s in the flame of a burner and then removed (BIS, 1981). The behaviour of the fibre to the flame (whether it burns, shrinks, forms any bead, the type of smell emitted, nature of remnant ash) helps in identifying the generic type of the fibre. This method is still used in industry where the composition of an unlabelled bale of fibre needs to be identified. This method is very quick and can be performed almost anywhere. However, the test is subjective and destructive, consumes a considerable amount of material and can produce toxic fumes sometimes. Table 14.1 describes the burning behaviour of different natural textile fibres.

14.3.3 Microscopic analysis

Microscopical analysis is indispensable for positive identification of several types of cellulosic and animal fibres and also for distinguishing these fibres from man-made fibres (ASTM, 2000). The distinguishing features of each fibre though not enough for confirmatory identification, help to categorize into the correct group. The features include convolutions, lumen and reversal zones (cotton); fibre bundles with or without cross-markings (bast and leaf fibres); presence or absence of scales and scale margins (animal fibres) and smooth profile without scales or convolutions (silk). The other

Table 14.1 The burning behaviour of natural textile fibres.

Behaviour of fibre	Cellulosic fibres	Wool	Silk
When approaching flame	Does not fuse or shrink away from flame	Fuses and curls away from flame	Fuses and curls away from flame
When in flame	Burns quickly without melting	Burns slowly with some melting	Burns slowly with some melting
After removal of flame	Continues to burn without melting afterglow	Burns very slowly, usually self-extinguishing	Burns very slowly, sometimes self-extinguishing
Melts near flame	No	Yes	Yes
Shrinks from flame	No	Yes	Yes
Burns in flame	Yes	Yes	Yes
Continues to burn	Yes	Slowly	Slowly
Smell	Burnt paper	Burnt hair	Burnt hair
Appearance of ash	Small, light grey	Lumpy, blistered, irregular, black ash, brittle, breaks easily	Soft, round, black bead, brittle, pulverizes easily

More data about flammability and flame retardancy of natural fibres the reader can find in chapter 10, volume 2.
Adapted from Lyle, D.S., 1977. Performance of Textiles, John Wiley and Sons, New York, USA.

features include longitudinal shape, regularity of diameter and any distinguishing characteristics. But sometimes, the morphological differences among some species are not always sufficiently distinctive for precise identification because nature tends to use the same repertoire of cells in constructing the fibres of different plants (Goodway, 1987).

Microscopy has and will remain an important aspect of fibre examination (Palenik, 1999) primarily for both vegetable and animal fibres (Catling and Grayson, 1982; Brunner, 1974; Palenik, 1983; Cook, 1969). This method applies various microscopies such as light microscopy, stereomicroscopy, scanning electron microscopy and infrared microspectroscopy. The light microscopy usually provides information about the shape and interior structure of fibres, not about the external features. A stereomicroscope can record fibre characteristics such as size, crimp, colour and luster.

The external features of fibres can be observed by SEM with very high resolution and relatively large depth of field than the optical microscope. In SEM images the complete surface of a fibre can be seen in high detail, which gives a better discrimination of the surface characteristics. In this process, the fibres are made conductive by coating with a layer of conducting metal to allow the charge acquired from the incident electron beam to run to earth. The ability to portray surface topographies with such clarity has targeted the main application of SEM in fibre identification towards natural rather than synthetic. The IR microspectroscopy is widely used in forensic analysis.

A mounting medium (such as liquid paraffin) with a refractive index about a 10th higher than that of natural fibres can be used as a convenient method of enhancing contrast at the surface of the fibre so that surface details become distinctly visible in light microscopy (Perry et al., 1975). Microscopical analysis includes the longitudinal and CS examination. In longitudinal examination a small amount of fibre is placed on a glass slide with a suitable mounting medium and covered with a covering glass. The fibres are examined at a specific magnification, which provides a strong indication of the type of fibre present. If the longitudinal analysis fails for identification of a fibre, it is necessary to go for CS analysis. The CS of different types of natural fibres are different, and they vary in diameter, i.e., from 10 to 13 μm (cultivated silk) up to 40 μm (US wool).

In CS analysis, the CS of a fibre is prepared with a cross-sectioning device, placed on a glass slide with a suitable mounting medium, covered with a covering glass and examined at a specific magnification. The CS can be prepared by microtome, hand-sectioning (by the use of a cork, a C-clamp and a razor blade) and the plate technique (Wildman, 1947). For better quality, the hardy microtome is used for cutting the CS. In some cases despite of cutting the fibre CS, the CS can be inferred by slow focussing upon the fibre at fairly high powers. This method is called optical sectioning, which is quicker and much less tedious procedure than cutting CS. The longitudinal and CS images along with the features of different natural textile fibres are shown in Table 14.2.

Langley and Kennedy (1981), by using a combination of light microscopy and SEM examinations, suggested that the earlier criteria for identifying wool, mohair and cashmere were unreliable. The combination may be the correct and most reliable means of identification by microscopy, which reveals different characteristics of the fibres involved. However, Kadikis (1987) argued that the light microscopy reveals more information about the structure and internal characteristics of animal fibres than SEM, and hence the optimum instrument for such analysis. He also stated that the analyst working in this field should possess years of experience, a good visual memory, patience and the intellectual honesty to recognize when a fibre is difficult to be identified.

Both the light microscopy and SEM should be used in conjunction rather than being in competition, as each will have its advantages for circumstances. Extensions of the SEM technique include energy dispersive X-ray (EDX), scanning tunneling microscopy (Spurny and Stöber, 1981) and plasma etching (Alfy and Blakey, 1980). Microscopic analysis is the quickest, accurate and least destructive in determining the types of fibres. Forensic analyzers use first microscopy as preliminary identification and then other methods for confirmatory identification.

14.3.4 Solubility test

Solubility tests were widely used for fibre identification before the development of modern instrumental methods. The generic fibre types are identified by the solubility

Table 14.2 Microscopic views and features of natural fibres.

Longitudinal view	Cross-sectional view	Longitudinal features	Cross-sectional features	Fibre name
		Ribbon-like, flattened structure with frequent convolutions, sometimes changing directions; distinct but small lumen, containing protoplasm in raw fibre. Thin walls and few convolutions in immature fibres.	Flat, elongated or kidney bean–shaped, seldom round or oval; lumen as a line or oval. Some very thin-walled sections of immature or dead fibres.	*Cotton* (raw and bleached)
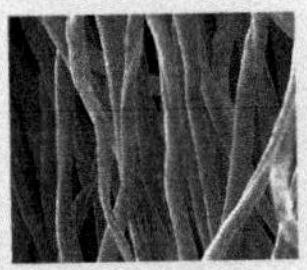		Ribbon-like fibres and fibre regions or less frequent depending on degree of mercerization; with greater part cylindrical and smooth; very small lumen or disappeared.	Most fibres round or oval (number depending on degree of mercerization); very small lumen as a small central spot or no lumen.	Cotton (mercerized)
	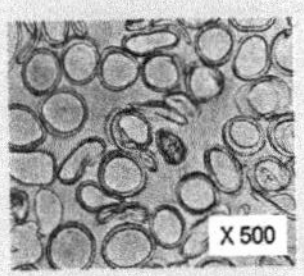	Smooth and cylindrical with rounded base; no convolutions or other structure except at the ends. When mounted in water, shows large, elongated air bubbles.	Circular or slightly elliptical; thin wall and large lumen.	*Kapok and akund*
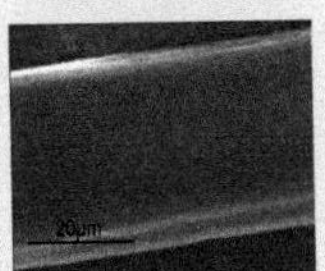	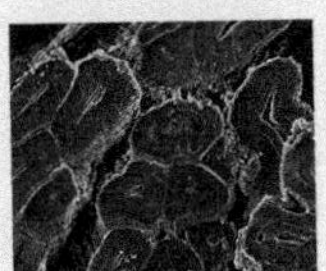	Fibre bundles with cross-markings, longitudinal and transverse fissures.	Bundles (and possibly some individual fibres).	*Ramie* (raw, before degumming)

Continued

Table 14.2 Continued

Longitudinal view	Cross-sectional view	Longitudinal features	Cross-sectional features	Fibre name
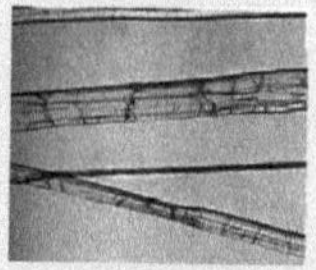	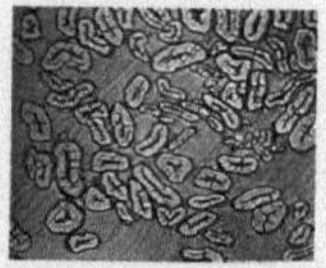	Isolated individual fibres, very broad and ribbon-like with infrequent twists; cross-markings, longitudinal and transverse fissures.	Long and narrow lumen or same shape as fibre section. Elongated polygons, often with curved side-lines, and sometimes rounded; thick wall, and radial fissures.	Ramie (degummed, and possibly bleached)
	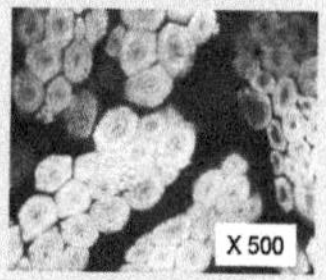	Fibre bundles, cross-markings, nodes, fissures, but otherwise smooth.	Shape and size of fibre bundles partly depends on preparation; ultimate fibres mainly sharply polygonal with narrow, round or oval lumen; also rounded oblong forms with larger lumen.	*Flax* (raw)
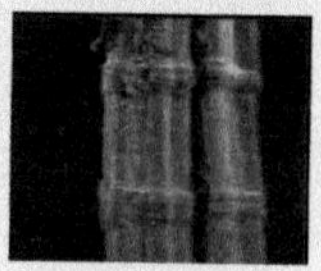		More or less isolated ultimate fibres depending on degree of bleaching; cross-markings, nodes, fissures but otherwise smooth.		Flax (bleached)
	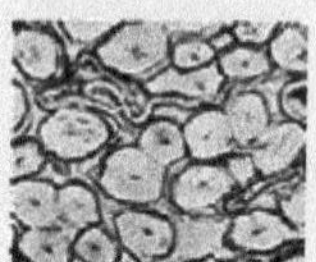	Fibres cylindrical, smooth, few cross-markings and nodes visible. Mixture of bundles and single fibres. Same as mercerized		Flax (mercerized) Flax (cottonized) (Crease-resistant)

 9	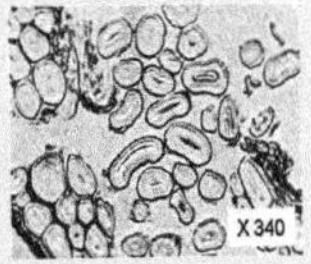	Similar to flax.	Similar to flax; often the lumen is as a mere line and indistinct.	*Hemp* (raw, bleached and cottonized)
	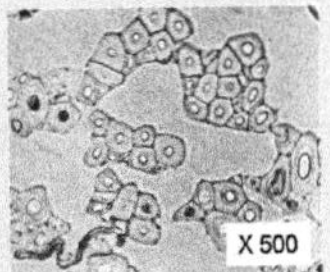 2	Fibre bundles, very rarely cross-markings; nodes or fissures; ultimate fibres (bleached or macerated) with lumen considerably varying in size along the same fibre.	Fibre bundles of varying size; ultimate fibres mainly sharply polygonal, some with rounded corners; lumen round to oval with very varying size.	*Jute*
 3	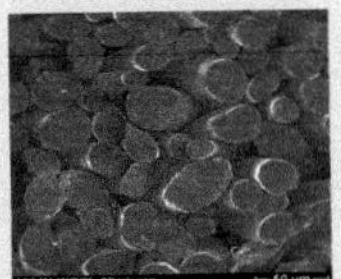 4	Irregular diameter and prominent scale margins. Medulla present in some medium and coarse fibres; may be fragmental, interrupted, or continuous.	Oval to circular, variable in diameter. Medulla (if present) is concentric and variable in size.	*Wool*
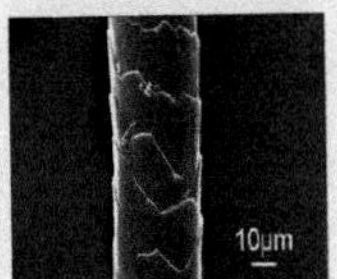 5	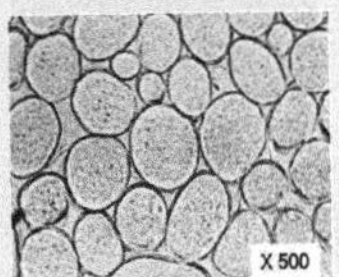 6	Regular diameter and smooth profile; scales very shallow; small vacuoles appearing black in some fibres.	Circular to oval, medulla occasionally fragmental or continuous in the coarser fibres.	*Mohair*
 7	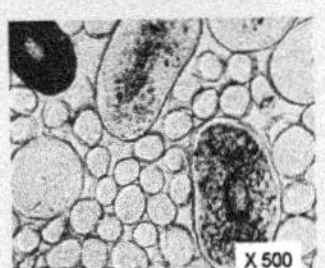 8	Fairly regular diameter with prominent scale margins. Medulla interrupted or continuous in coarse fibres.	Fine fibres: almost circular; coarse fibres: oval to circular, some flattened, often medullated.	*Cashmere*

Continued

Table 14.2 Continued

Longitudinal view	Cross-sectional view	Longitudinal features	Cross-sectional features	Fibre name
 9	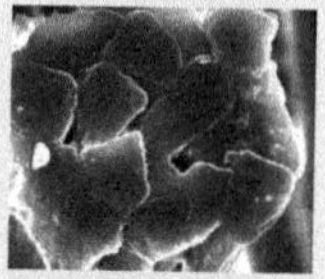	Fine fibres or filaments, cemented in pairs by silk gum. The gum layer is not always continuous.	Triangular, with rounded corners, in pairs.	*Bombyx silk* (raw)
	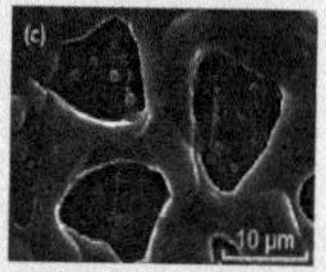	Fine fibres or filaments variable in diameter, single, smooth, nearly structure-less, sometimes flattened. Occasionally very fine fibres are seen, formed by the superficial splitting of the original fibre.	Separated fibres, triangular, rounded corners.	*Bombyx silk* (degummed)
	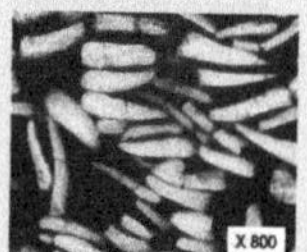 4	Flat irregular ribbons, usually separate, sometimes twisted and with longitudinal striations.	Very elongated triangles, usually separate.	*Tussah silk*

Adapted from Luniak, B., 1951. The Identification of Textile Fibres, Sir Isaac Pitman and Sons, London, UK; and Identification of Textile Materials, 1975, seventh Edition, The Textile Institute, Manchester, UK.

of the fibre in various reagents and comparing the data to the known solubility of several fibres. The reactivity of textile fibres to various solvents is different depending upon the chemical composition of the fibres. A suitable solvent can be selected for each fibre, which will dissolve only that fibre. Therefore, fibres which are soluble can be distinguished from those that are not. The list of commonly used solvents for different natural fibres is described in Table 14.3. Solubility tests are easy to perform and can be carried out in test tubes on macroscale or on the stage of a microscope (ideally used for just such test).

In test tube method, a small tuft of fibre is placed in a test tube containing the suitable solvent of the fibre. The fibres are identified by the solubility of the fibre. In microscopic method, a small amount of the fibre is placed under a cover slip with a drop of the appropriate solvent and the changes are observed under the microscope. For each solubility test, a fresh sample must be used to avoid any risk of cumulative damage effects. Great care must be taken to ensure that corrosive or harsh organic solvents do not contaminate any of the optical and mechanical parts of the microscope.

14.3.5 Staining test

In this method a small tuft of fibres are dyed with a mixture of dyestuffs for 1–5 min and washed thoroughly. The coloured fibres are compared with the known dyed samples or viewed carefully to identify the fibre. The chemical composition and reactivity of natural fibres differ from each other leading them to have affinities for specific stain or dye types. Therefore, mixtures of dyes have been developed for fibre identification, which produces known colours for individual fibre types. The best known stains for natural fibre identification are undoubtedly the Shirlastains (A and D) from Shirley Developments Ltd. (UK) (Ford and Warwicker, 1961) and Kayastain from Nippon Kayaku (Japan).

The satisfactory identification of natural fibres by a single identification stain 'Shirlastain A' is still widely used for identification of these fibres since it was developed in 1932. The identification between cotton and viscose obtained by Shirlastain A is not entirely satisfactory, and hence an improved distinction may be now obtained by new product, Shirlastain D.

In this method a small fibre tuft is stained in Shirlastain A at room temperature for 1 min, rinsed under cold tap water and compared with the colour chart for identification. The similar orange hues of chlorinated wool and degummed silk creates problem in distinguishing them. This can be avoided by staining a fresh sample in Shirlastain A at the boil for 1 min. Chlorinated wool appears black while degummed silk is dark, reddish brown. Although the same stain can be used two or three times, the recommendation is to use a fresh portion of the stain for each test.

The main limitation of staining test is that the fibre samples for staining should be undyed (pale-dyed samples are sometimes useable), wettable and free from any surface finish. Dyed fibres need to be stripped with strong reducing or oxidizing agents. In addition, any chemical damage to the fibre may also affect the colour of the fibre produced. Nevertheless, staining tests can be very useful particularly if combined with light microscopical examination.

Table 14.3 Solubility of natural fibres in different reagents.

	Sulphuric acid	Sulphuric acid	Sodium hypochlorite	Sodium hydroxide	Cuprammonium
Concentration (%)	60	75	0.25	10	-
Temperature (°C)	20	20	20	Boil	20
Time (minutes)	20	30	20	20	5
Cellulosic fibres	I	S	I	Sw	S
Wool	I	I	S	S	Sw
Silk	S	S	S	S	S

I, Insoluble; *S*, Soluble; *Sw*, Swell.
Adapted from Identification of Textile Materials, seventh ed. The Textile Institute, Manchester, UK, 1975 and Textile Laboratory Manual, vol. 5, third Edition, American Elsevier Publishing Company Inc.

14.3.6 *Identification from physical properties*

Various physical properties of unknown fibres such as density, moisture regain, refractive index and birefringence are compared with the reference standards from which the fibres are identified. The physical properties of natural fibres are described in Table 14.4.

14.3.6.1 *Gravimetric method (density)*

Different fibres have varying densities according to the polymer from which they are made, and this can be used as a diagnostic test. Generally, the densities of natural textile fibres range from 1.3 (wool) to 1.6 g/cm^3 (weighted silk), which may be used to confirm the identity of a fibre. Three methods such as density gradient column, pycnometer and a technique based on Archimedes principle are described in ASTM standard (ASTM, 2000).

The simplest approach for determining fibre density is by using a density gradient column. In a glass tube, a liquid of high density (pentachloroethane-1.71 g/cm^3) and a liquid of low density (xylene-0.87 g/cm^3) are used to form a linear density gradient region with the density increasing from the top to the bottom. The other commonly used liquids include ethanol/water mixture and varying strengths of salt solutions. Fibres of known density are used as calibration markers. Fibre of unknown density is allowed to sink in the tube until it reaches the region of equivalent density, i.e., where the density of the fibre matches that of the liquid. The density value of the fibre may be read off and compared with the standard values for identification (Table 14.4).

Thermogravimetric methods, titrimetric methods, kinetic measurements (rate of fall) and centrifuging techniques are variants of this method. The use of a single instrument to simultaneously identify cotton, wool, silk, viscose and synthetic fibres by thermogravimetry and differential scanning calorimetry has been reported (Crighton and Holmes, 1998; Perkins et al., 1966). In thermogravimetric method, the sample mass is monitored in a thermal balance as the temperature is raised linearly and the mass-temperature curves are plotted. The fibres are identified from the degree of mass change with temperature. In titrimetric method the volume of liquid adjusting the density is measured through a burette and the 'end point' of no movement of the sample is noted.

The gravimetric method requires very small sample, and the colour of the fibre do not interfere with the test. Also, this method is applicable to all fibre types, i.e., thermoplastic and nonthermoplastic. However, the possible interactions (mainly sorption or swelling) between the liquid and the fibre, as well as the trapped air bubbles must be avoided as this leads to a change in the density. This method cannot be used solely for confirmatory test as there is considerable overlapping in densities between different fibre types and difficulty in determining exact density values. Also, in all the density-based measurements, the temperature should be controlled accurately.

Table 14.4 Physical properties of natural fibres.

Fibre	Density (g/cm^3)	Moisture regain (%)	Refractive index $n_{\parallel}$	$n_{\perp}$	Birefringence Δn	Tenacity (g/den)	Extension at break (%)
Cotton	1.52–1.56	8.5	1.577	1.529	0.048	1.7–6.3	3-12
Flax	1.48–1.50	12.0	1.58–1.60	1.52–1.53	0.060	2.6-8.0	1.5-5
Hemp	1.48–1.49	12.0				3.0-7.0	1.5-5
Jute	1.44–1.50	17.0				2.0–6.3	1-2
Sisal	1.33	14.0					
Ramie	1.51–1.55	8.5	1.599	1.529	0.070	4.5–8.8	1.5-5
Wool	1.31	18.25	1.557	1.547	0.010		
Silk	1.32–1.60	11.0	1.591	1.538	0.053		

Remark: Physicomechanical properties of natural fibres, especially lignocellulosic fibres, depend on cultivation conditions and mainly on primary processing (especially on fibre extracting) methods.

Adapted from Houck, M.M., 2009. Identification of Textile Fibres, Woodhead Publishing Limited, CRC Press, USA.

14.3.6.2 Moisture regain

Different textile fibres can absorb and hold different amounts of moisture from the atmosphere depending on the chemical structure. The standard measure of the moisture termed as regain of natural fibres ranges from 8.5% (cotton) to 18.25% (wool). The standard regain (Table 14.4) of an unknown fibre can be determined which can help in identifying the fibre.

14.3.6.3 Refractive index

The refractive index of a fibre is directly related to its specific chemical composition. As the fibres are optically anisotropic, the refractive indices along and across the fibre axis will be different. Hence, the fibres are said to be birefringent. The refractive index is an extension of the microscopical method and can be used directly or in combination with melting and solubility behaviour for complete fibre identification. The birefringence (Δn) is the difference between the two refractive indices:

$$\Delta n = n_{\|} - n_{\perp},$$

where $n_{\|}$ and $n_{\perp}$ are the refractive indexes parallel and perpendicular to the fibre axis, respectively.

Refractive index and birefringence are the two most distinguishing features for the identification of a fibre's generic class. Although the refractive indices can be obtained by a light microscope with a polarizer, the most accurate method is to use monochromatic light and an interference microscope. The refractive indices ($n_{\|}$ and $n_{\perp}$) of undyed fibres can be determined by simply mounting the fibre in different liquids of know refractive index until the two are found, which causes the fibre to become invisible for polarized light along and across the fibre axis. The Refractive Index (RI) and birefringence values of different natural fibres are shown in Table 14.4.

The fibres dyed with dark shade require dye removal or stripping before their refractive index can be determined. The refractive index of some fibres (such as cotton and nylon) and the varieties of one group of fibre (wool and mohair) lie very close to one another, which necessitates the measurement of some other properties such as melting point and solubility for complete fibre identification. More training and experience are required in this method compared with other methods such as staining and solubility tests.

14.3.6.4 Other properties

Other properties not generally recognized as relevant in fibre identification (such as tensile strength, modulus and melting point) can also be used for identification of the fibres. Subramanian et al. (1983) proposed a method to distinguish between wool and mohair fibres by the examination of their tensile properties. The tensile strength and extension at break values of different natural fibres are shown in Table 14.4. The determination of melting point for fibre identification is mainly used for synthetic fibre identification as natural fibres do not melt. Some other properties such as the fibre length, the length of ultimates and fibre diameter (Table 14.5) can be used for fibre identification. In addition, the plant fibres can be distinguished by their chemical composition (mainly the cellulose and lignin content) as shown in Table 14.6. In addition, by testing flammability of fibres using cone-calorimeter and by estimating heat release rate, the fibres can also be identified.

There are significant differences in the burning characteristics of major natural fibres and hence can be used for their identification. One of the important properties used to compare the flammability of textile fibres is the limiting oxygen index (LOI). LOI is the quantity of minimum oxygen content (%) in nitrogen necessary to sustain candle-like burning. The LOI values for some important natural fibres are: Cotton-18, Silk-23 and Wool-25.

14.3.7 IR and Raman spectra analysis

The IR spectra of fibres are obtained by a double-beam spectrophotometer or by a Fourier transform infrared (FTIR) spectrophotometer. The identification of generic fibre types in question is made by comparing the absorption spectrum of these fibres with reference spectra of known samples done on the same instrument until a match is found. This technique is more valuable for distinguishing between different fibres of the same generic class (Tungol et al., 1990) and subclass (Tungol et al., 1991, 1993; Grieve et al., 1988; Bartick and Tungol, 1993). When the IR radiation interacts with a fibre, certain frequencies of energy are absorbed while others are transmitted or reflected depending on the functional groups (bonds) of the fibre which is used as a means for identification. Although the IR spectrometer is expensive, it offers an

Table 14.5 Length and diameter of natural fibres.

Fibre	Length of commercial fibre (mm)	Length of ultimate fibre (mm)	Diameter (μm)
Cotton	15–56		14–21
Jute	750–1500	0.8–6	5–25
Flax	700–900	13–60	12–30
Ramie	800	40–250	16–125
Kenaf	750–1500	2–6	12–36
Hemp	2500	5–55	16–50
Sisal	600–1000	0.8-8	100–400
Wool	25–300		18–55
Silk	Very long		8–15

Remark: Length and diameter of natural fibres, especially lignocellulosic fibres, depend on cultivation conditions and mainly on primary processing (especially on fibre extracting) methods.
Adapted from Identification of Textile Materials, seventh ed. The Textile Institute, Manchester, UK, 1975.

Table 14.6 Chemical composition (%) of plant fibres.

Fibre	Cellulose	Hemicellulose	Pectin	Lignin	Water soluble	Fat and wax	Moisture
Cotton	82.70	5.70			1.00	0.60	10.00
Jute	64.40	12.00	0.20	11.90	1.10	0.50	10.00
Flax	64.10	16.70	1.80	2.00	3.90	1.50	10.00
Ramie	68.60	13.10	1.90	0.60	5.50	0.30	10.00
Hemp	67.00	16.10	0.80	3.30	2.10	0.70	10.00
Sisal	65.80	12.00	0.80	9.90	1.20	0.30	10.00
Abaca	63.20	19.60	0.50	5.10	1.40	0.20	10.00

Turner, B.A.J. 1988. "The structure of textile fibres". In C. Jarman, Plant fibre and processing: A handbook, Intermediate Technology Publications, UK.

unambiguous means of chemically identifying unknown fibres. Furthermore, when only few fibres are available, this method is found to be the most valuable single test.

The IR spectrum of a sample is obtained by passing a beam of IR through the sample over a range of energies within the IR region; either a monochromatic beam or a Fourier transform instrument is used. In modern instruments, energy from an IR source is directed through the fibre by an interferometer. The track of the wavelength and the amount of light transmitted through the fibre at each wavelength is saved by a computer and the result is represented by a plot of wave number (cm^{-1}) against intensity (T%). The IR spectroscopy along with a microscope can be used for the complete analysis of a single textile fibre.

The natural fibres exhibit more intrasample variation in the absorption spectrum compared with synthetic fibres. The amount of twist in the fibres should be reduced by teasing apart and straightening the fibres before FTIR test. The IR absorption spectrum of various natural textile fibres is shown in Fig. 14.1. Table 14.7 shows the assignments of absorbance frequencies of natural fibres.

Sometimes the fibre may have to be flattened to reduce the effects of spectral distortion due to the diffraction caused by fibre's cross-section. A metal roller in combination with a clean glass microscope slide is used for flattening the fibre, which is then carefully transferred to the specimen holder. As the fibres are flattened during the test and the morphology is destroyed, the minimum fibre length should be used. In some cases, the available fibre is too short or too thin to obtain an FTIR spectrum by conventional means. In these cases, attenuated total reflectance (ATR) also known as internal reflection spectroscopy may be used. In ATR mode, the fibre is placed in tight contact with an IR transparent, high−refractive index crystal. IR spectrum is obtained by passing the IR radiation into the sample and crystal at angles greater than the critical angle. Recently, FTIR plays an important role in the forensic analysis of fibres (Grieve, 2000).

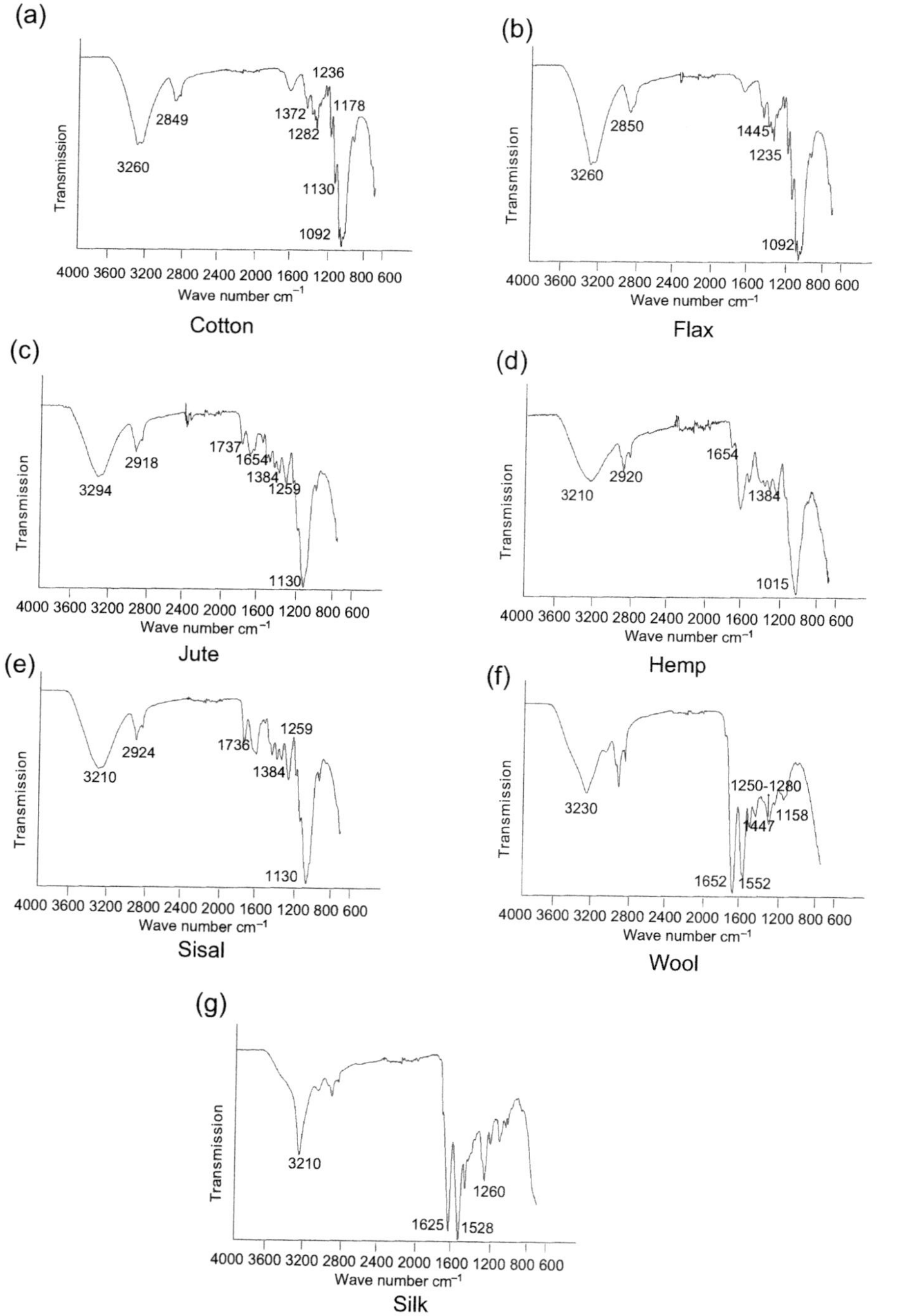

Figure 14.1 The infrared absorption spectrum along with the assignment of vibrations of various natural fibres (a) cotton, (b) flax, (c) jute, (d) hemp, (e) sisal, (f) wool and (g) silk.

Table 14.7 The FTIR assignments (absorbance frequencies) of natural textile fibres.

Cotton	Hemp	Flax	Jute	Sisal	Wool	Silk	Assignment
1042							C–O stretching
1092	1000–1162	1000–1162	1000–1162	1000–1162			C–C stretching
1130							Asym. Bridge C–O–C
					1158		C–N stretching
1178							Asym. Bridge C–O–C
1236		1235					O–H in-plane bending
					1250–1280		C–N stretching
			1259	1259			C–H bending
						1260	Absorption by amide III
1282							C–H stretching
1372	1384	1372	1384	1384	1447		C–H bending
		1445					$C-H_3$ asym. Stretching
					1552	1528	Absorption by amide II
					1652	1625	Absorption by amide I
	1654		1654	1654			C=C stretching
			1737	1736			C=O stretching
2849		2850					Sym. $C-H_2$ stretch
	2920		2918	2924			C–H vibration
3260	3200–3570	3200–3570	3200–3570	3200–3570	3200–3570	3200–3570	O–H stretching

Garside and Wyeth (2003) developed an ATR technique using FTIR for the identification of natural fibres and used the peak intensity ratio for their differentiation on the basis of relative lignin content. If the composition of cellulosic fibres is considered, the relative proportion of lignin with respect to other cellular components seems distinctive. According to harvesting time and degree of maturity of lignocellulosic fibrous plants, the lignin content in the fibre is changeable (liable). The authors applied ATR spectroscopy for the comparison of the lignin-to-cellulose content of the plant fibres for identification.

Raman spectroscopy is complementary to IR spectroscopy and is also a powerful tool that can be applied in routine fibre analysis following optical microscopy (OM) and microspectrophotometry (MSP) measurements. The fundamental principles of Raman spectroscopy and IR spectroscopy are different, the former being a scattering process while the latter is a pure absorption phenomenon. The spectroscopic selection rules for Raman and IR activities are different, and they produce complementary vibrational data. Many IR modes that are weak or not permitted in IR spectrum are very strong in Raman spectrum, e.g., molecular vibrations which involve nonpolar bonds are strong in the Raman spectrum, and those vibrations involving polar bonds are strong in the IR spectrum.

Raman spectroscopy can be conducted while the fibre remains mounted on a glass slide under cover slip as glass produces little response to Raman spectroscopy. But FTIR cannot be conducted with the fibres mounted to the glass slide due to the strong absorption of IR radiation by glass. Raman spectroscopy is very useful for the discrimination of dyed fibres and for complete fibre characterization sequence.

Raman spectroscopy was used to discriminate amongst untreated plant fibres on the basis of peak ratios derived from the associated C–H and C–O–C vibrations (Edwards et al., 1997). Raman spectroscopy has been used for the analysis (industrial and forensic) of fibres and has the potential as an analytical tool for fibre examination. The application of Raman spectroscopy for characterizing natural and synthetic fibres, and forensic analysis by European Fibres Group (Wiggins, 2003) has been reported. Raman spectroscopy can also be used to identify the main dye type present in a coloured fibre.

14.3.8 Other approaches

In several cases other techniques such as transmission electron microscopy (TEM), chromatography and pyrolysis can be used for the fibre analysis. These techniques along with some other methods (burning or solubility test) can be used for complete fibre identification. Unlike the light microscope and SEM, the TEM cannot be used to examine whole textile fibres directly. TEM is mainly confined in studying the fine internal structure of single fibres or parts of fibres and chemical reactivity. TEM is not directly used for fibre identification (Johnson, 1965) rather it is used indirectly by resolving the structure of the crystalline regions in the fibre (Sikorski, 1975).

Various chromatography such as gas, high-pressure liquid and thin-layer techniques can be used for the chemical analysis of textile materials (Weaver, 1984). Pyrolysis is mainly used for the identification of synthetic fibres. A method of qualitatively identifying a fibre without any elaborate equipment, with the aid of a simple solvent separation process combined with preliminary tests was suggested by Rist (Rist, 1987).

14.4 Practical approach

Microscopy and burning test are described (Garner, 1967) as the first step towards the fibre identification. An unknown fibre is investigated under the microscope for the longitudinal and CS appearance (Refer Table 14.2). Then the burning behaviour of the fibre is observed by holding a small tuft of the unknown fibre in forceps into the flames of a micro-burner for about 10 s (Refer Fig. 14.2). For the confirmation of the fibres, solubility test (Fig. 14.3) and staining test (Section 14.3.5) should be used. In staining test, a small fibre tuft is stained in Shirlastain A at room temperature for 1 min, rinsed under cold tap water and compared with the colour chart for identification. The ultimate fibre identification is done by repeating the above tests, carefully considering the inferences given on various methods of identification in Section 14.3 and considering the physical properties.

ASTM (ASTM, 2000) describes solubility test for the preliminary identification of the generic type of an unknown fibre. Alternatively, the IR spectra (obtained by an FTIR or a double-beam spectrophotometer) can be compared with the decision chart for the preliminary identification. The plant (native cellulose) and animal hair fibres are distinguished by the microscopical examination of longitudinal and cross-sections. Additional physical properties such as density, melting point, regain, refractive indices and birefringence are used for confirmatory identification.

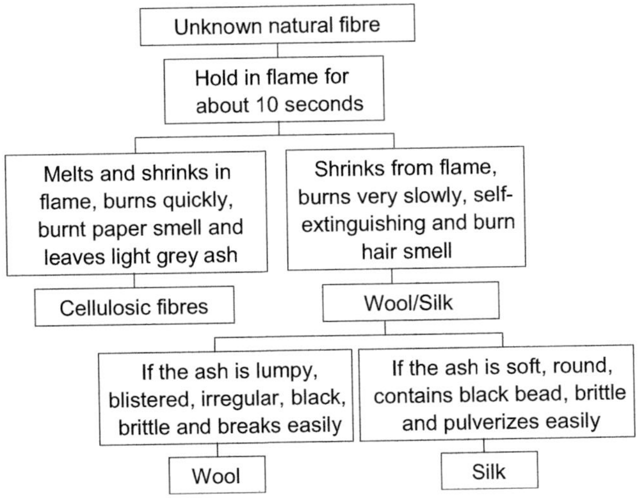

Figure 14.2 Scheme for identification of fibres according to burning test.

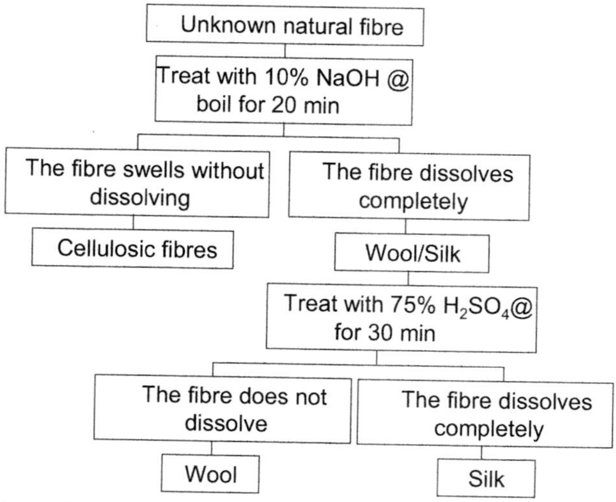

Figure 14.3 Scheme for identification of fibres according to solubility test.

14.5 Forensic analysis

Forensic analysis of fibres which rarely follows the industrial practice requires knowledge of fibre manufacturing processes and methods. Textile fibres are one of the most important forms of trace evidence in the context of criminal evidence. Forensic scientist can examine these trace evidence and compare them to a known fibre to discover possible common origins. The factors influencing the nature of fibre adherence to materials include type of fibre transferred, type of receiving material and the extent to which the receiving material is used after transfer.

In case of forensic analysis, apart from the murky origin and uncertain provenance, the sample size is limited (sometimes may be 1 or 2 fibres). Therefore, microscopy is the primary method of fibre identification. The preservation of the sample is also the concern. The focus of forensic scientists is fast nondestructive, selective and sensitive examination methods. In forensic analysis, FTIR and Raman spectroscopy are mainly used to determine the chemical composition of fibres, providing a generic fibre class and subclass.

14.5.1 FTIR and Raman spectroscopy

FTIR and ATR-FTIR are the widely used techniques used for the forensic identification of textile fibres by molecular structure characterization. They can also be used to identify the fibres present in the archeological textiles and differentiae among different textile fibres. These methods are fast, accurate and inexpensive, which is the main reason for their wide acceptance. Qualitative analysis and semiquantitative analysis, such as discriminant analysis and principal component analysis (PCA), are possible by using ATR-FTIR and chemometric tools. Furthermore, multicomponent fibre analysis can be performed by using these techniques.

Raman spectroscopy is an essential tool in many forensic laboratories in recent years because of good resolution, smaller sample size, ease of preparation and ability to use aqueous samples. It can be used for the identification of fibres, minerals, lipsticks and other evidences left by the criminals. Several studies have shown that Raman spectroscopy is one of the modern tolls to be used for the forensic analysis in recent times. Raman spectroscopy is appropriate for very small samples such as single fibres or very small amount of pigments or particles of 1 μm or less. Micro-Raman spectroscopy is also found to be an appropriate tool for forensic examination of dyes and pigments, which can also be used for textile fibres.

For example, Peets et al. (2017) identified and classified textile fibres using ATR-FTIR spectroscopy with chemometric methods. PCA and discriminant analysis was performed on the single-component and bicomponent textile materials. The PCA classification was suitable for single- to three-component fibres to identify each fibre. However, the limitations were the differentiation among the cellulose-based fibres (cotton, linen and in some cases viscose) were not possible, and distinguishing between silk and wool were partially possible (Peets et al., 2017).

Thomas et al. (2005) used Raman spectroscopy for the forensic analysis of black/grey and blue cotton fibres (reactively dyed) by varying the laser wavelength. The study has shown that this technique can be used for dyed fibre identification without interference from the cotton substrate, thereby, providing molecular information about the main dye used. However, caution must be taken in the choice of wavelengths, as the fluorescence and changes in sample spectra can occur with laser wavelength.

Wiggins et al. (2007) investigated the use of calculating the first derivative of the absorbance spectra of fibres as a tool for forensic analysis. The first derivative of the absorbance spectra provided more points of comparison that facilitated discrimination when the absorbance spectra produced for some samples were broad and featureless. Although it was found that the calculation of the first derivative can be a useful aid in the comparison of spectra, a high degree of caution is required while applying this method to fibres which exhibit a large intrasample variation in colour.

The application of the new combined micro-Raman and micro-X-ray fluorescence (XRF) spectrometer (PRAXIS apparatus) in the forensic analysis of fibres and other materials has been reported (Zieba-Palus et al., 2008). The studies confirmed the usefulness of the apparatus in criminalistic traces and the strong point of the instrument is that it provides a combination of two nondestructive methods in one instrument. The analysis is characterized by good repeatability, reproducibility and compatibility.

Lepot et al. (2008) analyzed five forensic fibre cases by Raman spectroscopy and found it is a good complementary method for MSP. The forensic analysis of dyed textile fibres by using chromatography, spectroscopy and mass spectrometry were reported in this work. For certain dye class, the absorption spectra in the visible region is a characteristic feature, which be subsequently identified by the Raman spectroscopy using a spectral library. This study showed that the Raman spectroscopy showed only little interference from the mounting resin and glass. Therefore, Raman spectroscopy is a powerful method that can be applied in routine fibre analysis following OM and MSP measurements. There are also other important works in forensic analysis by several researchers (Paulsson and Stocklassa, 1999; Causin et al., 2006; Miller and Bartick, 1978; Lang et al., 1986; Goodpaster and Liszewski, 2009).

Both the test methods FTIR and Raman spectroscopy require very small amount of sample and are nondestructive test. The samples used in these tests can be used for analysis of other properties. In several instances, FTIR and Raman work as complementary to each other. Some bonds that can't be detected effectively with FTIR can be detected with Raman and vice versa.

Although Raman spectroscopy has several advantages over other analytical tools with respect to simple fibre preparation and accurate identification, it suffers from some disadvantages such as fluorescence is the major concern assisted with the analysis of the fibre samples. Fluorescence occurs after a high-intensity visible or UV light excited the electronic molecular states. This influence can be reduced by changing the laser wavelength or application of metal colloids and surface enhanced Raman spectroscopy.

14.5.2 Analysis of dyed textile fibres

The identification of dyed textile fibre can be completed using spectroscopy, chromatography and mass spectrometry in addition to the microscopy. Microscopic analysis is an important aspect of fibre analysis as discussed in the previous section. For forensic analysis the stereomicroscope can be used to characterize the colour, dimension, crimp and luster of fibres. The dyed textile fibres can be easily identified by the polarized light microscopy as it can provide information on the polymer group by the rotation of incident polarized light put in the fibre. The description for identification of dyed textile fibres has been given in Fig. 14.4.

The measurement of dichroism of pigmented fibres can also help to identify the fibres for the forensic analysis (De Wael and Lepot, 2012). The use of OM and MSP with plane polarized light can measure the dichroic behaviour. For example, Wael and Lepot (De Wael and Lepot, 2012) observed strong dichroic behaviour in half of the samples with 20% showing weak dichroic behaviour. Regular dichroic effect was observed in 80% cases and inverse dichroic effect was observed in 20% of the cases. The dichroic behaviour strongly depends on the draw ratio and the crystal structure of the pigmented fibres.

14.5.3 Analysis of blended materials

In several instances, different types of textile fibres are mixed together to form blended yarn and fabric as the desirable properties that cannot be obtained by using a single fibre type is achieved in the blend. The use of blended fabrics is in rise, which has necessitated the development appropriate methods for the analysis of the blends. Of the identification methods described, burning test and quantitative method of blend analysis are used by many industries for fibre identification purposes in the blend. The burning test as described in Section 16.3.2 is used for the preliminary identification of the fibres. Once an idea has been made on the presence of the tentative fibres in

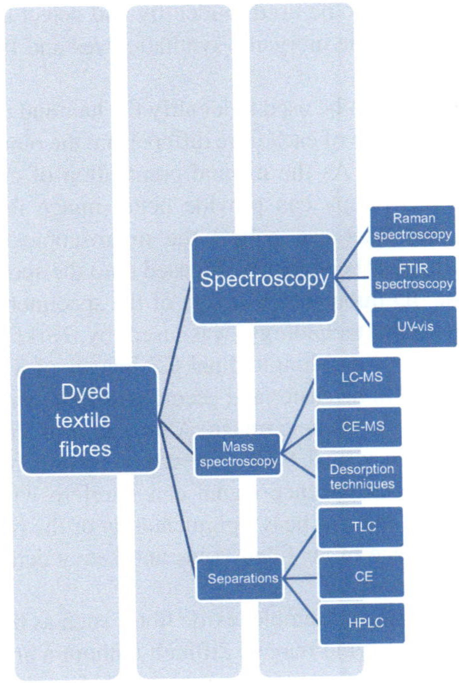

Figure 14.4 Methods used for the identification of dyed textile fibres.

the blend, the solubility test is performed to find the composition of the blend in the fibre mixture.

The solubility test has been described in Section 16.3.4. In blend analysis, 100 g of fabric is taken and conditioned in standard atmosphere. Appropriate solvent is used to dissolve one component of the blend. From the initial weight (100 g), the percentage of the fibre present in the blend is identified. In binary blends, the remaining percentage is the amount of the second fibre in the blend. In tertiary or higher blends, the other fibres are dissolved in suitable solvents and percentage of each component has been calculated from the initial weight till all the fibres are completed. Other approaches such as moisture regain, image analysis, physical properties, IR spectroscopy, fibre density and nuclear magnetic resonance. Several fashion brands provide the information on the blend of the fabric in the final garment (Nayak and Padhye, 2015a, b; Nayak and Padhye, 2015a; Nayak and Ratnapandian, 2018). However, it is essential to confirm these blends for exact reproducibility of the textile blends.

14.5.4 Other approaches

Other approaches such as the pyrolysis in combination with the gas chromatography and the mass spectroscopy can also be used for the forensic analysis of textile fibres form a small sample size. Scanning electron microscope (SEM) combined with the energy dispersive X-ray can also be used to identify and detect the elemental composition of the fibres containing the inorganic synthetic dyes and finishes applied to the fibres.

The confocal microscopy can be used to identify the hair and other textile fibres for forensic purposes. The CS shape of each fibre differs from the other, which can be used for the identification purposes. As the manual preparation of cross-section is rather difficult, the confocal microscopy can provide better image for identification. The confocal microscope with laser can collect the microscopical images with some additional capability. The laser beam can be focused onto the specimen, and by means of a photomultiplier, the 3D fluorescence image of the specimen can be recorded.

The confocal microscopy technology was used by Kirkbride and Tridico to generate the virtual transverse and longitudinal CS images of a wide range of fibres including hair. It was found to be easier and accurate as there was no need to prepare the fibre cross-sections for microscopic image. Although confocal microscopy can be successfully used to produce virtual transverse cross-sections from a wide range of fibres including hairs, autofluorescence signal can interfere with the image, which can be attenuated by the presence of heavy pigmentation or the presence of an opaque medulla in hairs. The presence of air-filled voids and heavy delustering agent on the fibres can produce wrong results.

In several occasions the drugs resemble textile fibres such as the cannabis represent tee flax fibre. Therefore, identifying may be difficult without a proper instrument. The use of gas chromatography and mass spectrometry can help to identify the fibres. For example, Broséus et al. (2010) used this technique to identify the textile fibres and differentiated from the illegal drugs. The compounds that distinguish between the fibres and the illegal drugs were identified by using a support vector machines. A classification rate above 99% was achieved with a false-positive rate less than 2%. This model successfully helped to differentiate between fibre and drug-type cannabis during the early growth stage.

The combination of FTIR and energy dispersive X-ray fluorescence (EDXRF) spectroscopy can be used to identify the textiles, buttons and metal garment components extracted during excavations. For example, Akyuz et al. (2014) used FTIR and EDXRF to investigate textiles and trims (buttons and metals) excavated from ancient Ainos (Enez in Turkey). The presence of partially degummed *B. mori* silk was confirmed using FTIR spectroscopy. Furthermore, the FTIR results also showed that the buttons were filled with some cellulosic fillings. The EDXRF spectroscopy results showed the silver-plated copper ornaments in the clothing. The SEM images showed that the textiles were almost degraded.

For the analysis of the historic materials various nondestructive techniques can be used. For example, Akyuz et al. (2014) used the combination of ATR-FTIR, EDXRF and scanning electron microscope (SEM) techniques, to identify the historical textiles from the imperial pavilion of the new mosque Eminonu-Istanbul (Turkey). The combined techniques identified linen fabric used in the preparation of the textiles. EDX findings showed that metallic gold was used for gilding in the original textiles. No gold gilding was observed in the repaired part of the textile. Both lead white [$2PbCO_3.Pb(OH)_2$] and gypsum [$CaSO_4.2H_2O$] were used in the paint layers of the original part of the textile. However, in the repaired part, the white pigment was mainly gypsum and contained only a small amount of the lead white. FTIR spectroscopy was used in quantifying the degradation of historic textiles in terms of the crystalline structure of cellulosic fibres.

14.6 Conclusions and future trends

Textile fibre identification is essential to know the composition of a textile material to reproduce the material, understand the composition or to compare among different materials. As there are a wide range of textile fibres, both natural and synthetic, it is a difficult task to uniquely identify these fibres from their visual appearances. Accurate identification of textile fibres is essential for fashion and textile industries, garment design houses and forensic science. This chapter has discussed various identification techniques of natural textile fibres from their appearance, physical and chemical properties and microscopic images. Although there have been several technological advancements in the fibre identification such as FTIR spectroscopy, Raman spectroscopy, EDX, confocal microcopy, gas chromatography and the mass spectroscopy, the traditional methods are prevalent in the industries as they are the cheapest alternative.

Recently, a wide range of analytical methods such as FTIR, Raman spectroscopy, electron microscopy, atomic force microscopy and vibrational spectroscopy are available apart from the OM for fibre identification. Yet as far as the analysis is concerned, the old methods are the best as they are cheaper and affordable by many large industries and Small to Medium Enterprises (SMEs). Microscopy still dominates in fibre identification by providing a range of analysis barely possible with any other methods. Combined with an analytical bench, such as IR or Raman, it is the quintessential fibre identification tool. Microscopy should be the first method of choice for any fibre scientist.

New developments in technical and medical textiles, microfibres and nanofibres and speciality fibres broaden and deepen the analytical world of fibre analyst. Some of the other new techniques include multiphoton fluorescence microscopy, optical coherence tomography and confocal microscopy. Although the first of the two instruments were developed especially for biosciences, it is likely that these will become more valuable in relation to fibre identification. Confocal microscopy is the advanced form of OM, which can measure lower than 200 nm (not feasible by OM).

Abbreviations

AATCC American Association of Textile Chemists and Colourist
ASTM American Society for Testing and Materials
ATR Attenuated total reflectance
CS Cross section
EFG European Fibres Group

EDX	Energy dispersive X-ray
EDXRF	Energy dispersive X-ray fluorescence
FTIR	Fourier transform infrared
HRR	Heat release rate
IR	Infrared
LOI	Limiting oxygen index
MSP	Microspectrophotometry
NMR	Nuclear magnetic resonance
OM	Optical microscopy
PCA	Principal component analysis
PLM	Polarized light microscopy
PPL	Plane polarized light
SEM	Scanning electron microscopy
SERS	Surface enhanced Raman spectroscopy
STM	Scanning tunnelling microscopy
SVM	Support vector machine
TEM	Transmission electron microscopy
XRF	X-ray fluorescence

References

Akyuz, S., Akyuz, T., Cakan, B., Basaran, S., 2014. Investigations of the historic textiles excavated from Ancient Ainos (Enez–Turkey) by multiple analytical techniques. J. Mol. Struct. 1073, 37–43.

Alfy, M.O.E., Blakey, P.R., 1980. J. Text. Inst. 71, 163.

ASTM, 2000. Standard Test Methods for Identification of Fibres in Textiles. ASTM, NC, USA.

Bartick, E., Tungol, M., 1993. Infrared microscopy and its forensic applications. Forensic Sci. Handb. 196–592.

BIS, 1981. Identification of Textile Fibres. Bureau of Indian Standards, New Delhi.

Broséus, J., Anglada, F., Esseiva, P., 2010. The differentiation of fibre-and drug type Cannabis seedlings by gas chromatography/mass spectrometry and chemometric tools. Forensic Sci. Int. 200, 87–92.

Brunner, H., 1974. The Identification of Mammalian Hair. Inkata Press, Melbourne, Aus.

Catling, D., Grayson, J., 1982. Identification of Vegetable Fibres. Chapman & Hall, Ltd, London UK.

Causin, V., Marega, C., Schiavone, S., Guardia, V., Marigo, A., 2006. Forensic analysis of acrylic fibers by pyrolysis-gas chromatography/mass spectrometry. J. Anal. Appl. Pyrolysis 75, 43–48.

Cook, J., 1969. Handbook of Textile Fibres I: Natural Fibres. Textile Book Service, Metuchen, NJ.

Crighton, J., Holmes, D., 1998. The application of thermal analysis to textiles. Part I: identification of fibre-blend components by using thermogravimetry. J. Text. Inst. 89, 198–207.

Cunniff, P., Fossey, S., Auerbach, M., Song, J., Kaplan, D., Adams, W., Eby, R., Mahoney, D., Vezie, D., 1994. Mechanical and thermal properties of dragline silk from the spider *Nephila clavipes*. Polym. Adv. Technol. 5, 401–410.

DE Wael, K., Lepot, L., 2012. Dichroism measurements in forensic fibre examination: Part 5—pigmented fibres. Sci. Justice 52, 161–167.

Edwards, H., Farwell, D., Webster, D., 1997. FT Raman microscopy of untreated natural plant fibres. Spectrochim. Acta A Mol. Biomol. Spectrosc. 53, 2383–2392.

Ford, J., Warwicker, J., 1961. Fibre identification by means of shirlastains A, D, and E. J. Text. Inst. Proc. 52, 617–619.

Garner, W., 1967. Fibres. American Elsevier Publishing Company Inc, NY, USA.

Garside, P., Wyeth, P., 2003. Identification of cellulosic fibres by FTIR spectroscopy: thread and single fibre analysis by attenuated total reflectance. Stud. Conserv. 48, 269–275.

Goodpaster, J., Liszewski, E., 2009. Forensic analysis of dyed textile fibers. Anal. Bioanal. Chem. 394, 2009–2018.

Goodway, M., 1987. Fiber identification in practice. J. Am. Inst. Conserv. 26, 27–44.

Greaves, P., 1990. Fibre identification and the quantitative analysis of fibre blends. Rev. Prog. Coloration Relat. Top. 20, 32–39.

Grieve, M., 2000. Back to the future-40 years of fibre examinations in forensic science. Sci. Justice 40, 93–99.

Grieve, M., Dunlop, J., Kotowski, T., 1988. Bicomponent acrylic fibers—their characterization in the forensic science laboratory. J. Forensic Sci. Soc. 28, 25–33.

Hamlyn, P.F., Nelson, G., Mccarthy, B.J., 1992. Wool-fibre identification by means of novel species-specific DNA probes. J. Text. Inst. 83, 97–103.

Houck, M.M., 2009. Identification of Textile Fibres. CRC Press, Fl, USA.

Johnson, D.J., 1965. Doctor of Philosophy. PhD Thesis. University of Leeds.

Kadikis, A., 1987. Text. Res. J. 57, 676.

Kusch, P., Arns, W., 1982. Schriftenriche des Deutsches Wollforschunginstitutes 87, 386.

Lang, P., Katon, J., O'keefe, J., Schiering, D., 1986. The identification of fibers by infrared and Raman microspectroscopy. Microchem. J. 34, 319–331.

Langley, K., Kennedy, T., 1981. The identification of specialty fibers. Text. Res. J. 51, 703.

Lepot, L., DE Wael, K., Gason, F., Gilbert, B., 2008. Application of Raman spectroscopy to forensic fibre cases. Sci. Justice 48, 109–117.

Marsh, J.T., 1941. Mercerising. Chapman & Hall.

Miller, J., Bartick, E., 1978. Forensic analysis of single fibers by Raman spectroscopy. Phys. Lett. 59, 68.

Mukhopadhyay, S., Sakthivel, J., 2005. Spider silk–providing new insights in the field of high performance materials. J. Ind. Text. 35, 91.

Nayak, R., Padhye, R., 2015a. Care Labelling of Clothing. Garment Manufacturing Technology. Elsevier.

Nayak, R., Padhye, R., 2015b. The Care of Apparel Products. Textiles and Fashion. Elsevier.

Nayak, R., Padhye, R., Fergusson, S., 2012. Identification of Natural Textile Fibres. Handbook of Natural Fibres: Types, Properties and Factors Affecting Breeding and Cultivation. Elsevier.

Nayak, R., Ratnapandian, S., 2018. Care and Maintenance of Textile Products Including Apparel and Protective Clothing: Apparel and Protective Clothing. CRC Press.

O'brien, J., Fahnestock, S., Termonia, Y., Gardner, K., 1998. Nylons from nature: synthetic analogs to spider silk. Adv. Mater. 10, 1185–1195.

Palenik, S., 1983. Light microscopy of medullary microstructure in hair identification. Microscope 31, 129–137.

Palenik, S., 1999. Microscopical examination of fibres. In: Robertson, J., Grieve, M. (Eds.), Forensic Examination of Fibres, 2 ed. CRC, New York.

Paulsson, N., Stocklassa, B., 1999. A real-time color image processing system for forensic fiber investigations. Forensic Sci. Int. 103, 37–59.

Peets, P., Leito, I., Pelt, J., Vahur, S., 2017. Identification and classification of textile fibres using ATR-FT-IR spectroscopy with chemometric methods. Spectrochim. Acta A Mol. Biomol. Spectrosc. 173, 175–181.

Perkins, R., Drake Jr., G., Reeves, W., 1966. DTA and TGA studies of flame-resistant fabrics. J. Appl. Polym. Sci. 10, 1041–1066.

Perry, D.R., Appleyard, H.M., Cartridge, G., Cobb, P.G.W., Coop, G.E., Lomas, B., Ritchie, G.G., Taylor, C., Welch, M.J., Farnfield, C.A., 1975. Identification of Textile Materials. The Textile Institute, Manchester, Uk.

Rist, D., 1987. Chemische faseranalyse. Textilveredlung 22, 368–373.

Robson, D., 2000. Animal fiber analysis using imaging techniques. Text. Res. J. 70, 116.

Seidel, A., Liivak, O., Jelinski, L., 1998. Artificial spinning of spider silk. Macromolecules 31, 6733–6736.

Sen, K., Babu, K.M., 2004. Studies on Indian silk. I. Macrocharacterization and analysis of amino acid composition. J. Appl. Polym. Sci. 92, 1080–1097.

She, F., Chow, S., Wang, B., Kong, L., 2001. Identification and classification of animal fibres using artificial neural networks. J. Text. Eng. 47, 35–38.

Sikorski, J., 1975. Structural studies of mammalian keratin. In: Atkins, E.D.T., Keller, A. (Eds.), Structures of Fibrous Biopolymers.

Spurny, K., Stöber, W., 1981. Some aspects of analysis of single fibers in environmental and biological samples. Int. J. Environ. Anal. Chem. 9, 265–281.

Subramanian, S., Phalgumani, G., Manjunatha, B., Sitaram, M., Shringapure, A., Bhatt, I., 1983. Assessment of crease recovery values of textile fabrics by different instruments. Indian J. Text. Res. 8, 16–22.

Thomas, J., Buzzini, P., Massonnet, G., Reedy, B., Roux, C., 2005. Raman spectroscopy and the forensic analysis of black/grey and blue cotton fibres:: Part 1. Investigation of the effects of varying laser wavelength. Forensic Sci. Int. 152, 189–197.

Tungol, M., Bartick, E., Montaser, A., 1990. The development of a spectral data base for the identification of fibers by infrared microscopy. Appl. Spectrosc. 44, 543–549.

Tungol, M., Bartick, E., Montaser, A., 1991. Analysis of single polymer fibers by Fourier transform infrared microscopy: the results of case studies. J. Forensic Sci. 36, 1027–1043.

Tungol, M., Bartick, E., Montaser, A., 1993. Forensic analysis of acrylic copolymer fibers by infrared microscopy. Appl. Spectrosc. 47, 1655–1658.

Weaver, J., 1984. Analytical Methods for a Textile Laboratory. AATCC.

Wiedemann, E., 1987. SAWTRI Bull. 21, 7.

Wiggins, K., 2003. The European Fibres Group (EFG) 1993–2002: "Understanding and improving the evidential value of fibres". Anal. Bioanal. Chem. 376, 1172–1177.

Wiggins, K., Palmer, R., Hutchinson, W., Drummond, P., 2007. An investigation into the use of calculating the first derivative of absorbance spectra as a tool for forensic fibre analysis. Sci. Justice J. Forensic Sci. Soc. 47, 9.

Wildman, A., 1947. The microscopy of textile fibres: aids to their identification. J. Text. Inst. Proc. 38, 468–473.

Ziabicki, A., 1976. Fundamentals of fibre Formation: The Science of fibre Spinning and Drawing. Wiley, New York.

Zieba-Palus, J., Borusiewicz, R., Kunicki, M., 2008. PRAXIS—combined [mu]-Raman and [mu]-XRF spectrometers in the examination of forensic samples. Forensic Sci. Int. 175, 1–10.

Testing of natural textile fibres

15

R. Harwood[1], *E. Smith*[2]
[1]Copernicus Textile Solutions Ltd., Desford, Leicestershire, United Kingdom;
[2]De Montfort University, Leicester, United Kingdom

15.1 Introduction

Testing of natural fibres is undertaken for a range of reasons and is a necessary aspect of the determination of fibre quality in respect to the suitability of fibres for particular end uses. The need for characterization of the physical properties of fibres, e.g., length, fineness, strength, elongation, colour, trash content, etc., are generally directly related to subsequent conversion into products, such as yarns, felts, woven and knitted fabrics, industrial and agro-textiles, and are related to the suitability of the fibres to provide the products which meet the end-user specifications. Fibre properties fall into two groups. First, those properties which are characteristics of the fibres themselves and which can be related directly to the quality of the fibre in specific areas of application and their performance in production processes. Fibre fineness and length are important examples of these intrinsic fibre properties: they are fundamental characteristics which affect, for example, fibre spinnability, the quality of yarns and the limit of yarn count that can be achieved. Second, natural fibres usually contain nonfibre contaminants generated in production or left over from the extraction processes, e.g., plant debris such as shives and cuticular matter, dust, grease, soil and vegetable matter. Such contaminants are important when determining fibre yield but can be of more fundamental importance when they limit the uses of fibres in target applications.

The use and usefulness of test data are, naturally, dependent on the fibre type and end use. Skill and experience play an essential part in the interpretation of the data within the diverse specialist areas of the natural fibre industries and on the particular testing required for production and quality control. Testing plays a very different role in research and development where the speed of generating test results, and the costs involved, can often be of less importance than the quality of the data generated. As a result, many of the test methods used in research have not been adopted for general use in the industrial environment.

Chemical testing of natural fibres offers a diverse set of tools which fall into two broad areas of application: testing for the extent and causes of 'chemical' damage and for characterization of fibres so that subsequent chemical and other wet processes can be optimized. However, this review focuses mainly on physical methods of testing and chemical methods are not considered here.

There are many sources which detail standard procedures for routine testing of fibres. The International Organization for Standardization (ISO) designates standards; the membership of the ISO comprises representatives from various national standards

Handbook of Natural Fibres. https://doi.org/10.1016/B978-0-12-818398-4.00017-7

organizations. In addition to publishing international standards, they also produce technical reports and technical specifications. The European Committee for Standardization (CEN) exists to develop European Standards (ENs) in various sectors to build a European internal market for goods and services and to position Europe in the global economy. National standards organizations include the British Standards Institution (BSI) and the German Institute for Standardization (DIN); both are members of ISO. ASTM, originally known as the American Society for Testing and Materials, predates both BSI and DIN and is an international standards organization in its own right, but it is not a national body in the USA; this is the American National Standards Institute (ANSI) in the USA. However, ASTM does develop standards in the USA in collaboration with volunteer committees across the world.

Note, it is not the intention of this chapter to be a catalogue of standards information but to focus on some of the more important testing methods, recent innovations and their uses. There are many excellent publications on textile testing, some of which are focused on natural fibres, which contain extensive information on test methods; some of these are listed at the end of the chapter.

15.2 Key issues in testing natural fibres

In comparison with man-made fibres, which are produced with known uniform properties, natural fibres lack consistency because of the fact they are produced under varying conditions. For example, variations in climate, soil type and local environment mean that fibre properties will vary between locations. For this reason, there is a need to test a range of fibre characteristics to assess their suitability for designated end uses.

15.2.1 Reasons for testing

By their very nature, natural fibres vary widely in their properties depending on a whole range of factors including agronomic, environmental, geographical and geographic variables, species varieties and breeds. Because of the international dimensions to trade and manufacturing, it is important to have recognized quality standards for raw materials. For cotton and wool, classification and grading standards have been established for many years, and there are comprehensive sets of international standards for testing and quality evaluation. However, manual grading is still commonplace in many of the smaller production countries of the world. Despite the existence of grading standards, processors will often undertake in-house testing of raw materials to optimize their processing and to confirm that input materials are appropriate for their requirements.

Quality classification systems exist for all fibres. Table 15.1 shows the factors on which the gradings of a range of fibres are based. Internationally recognized methods exist for testing the characteristics of many fibres, but for many of the low-volume fibres, quality is evaluated subjectively and testing is not automated and is hence labour-intensive; as a result, grading of fibres at source is generally carried out subjectively and manually by skilled graders; testing by end-users is often an essential part of their production activity.

Table 15.1 Grading factors for a range of fibres.

Fibres	Strength	Length/length distribution	Diameter	Fineness	Elasticity	Colour	Cleanliness
Cotton[a]		√	√	√		√	√
Linen flax	√	√		√		√	√
Short fibre		√	√	√		√	√
Flax							
Hemp	√	√		√			√
Ramie		√		√		√	√
Nettle	√	√	√	√		√	√
Coir							
Jute	√			√		√	√
Wool[a]	√	√		√	√	√	
Silk				√			

[a]Automated methods used routinely in some countries.

15.2.2 Sampling

As with almost all products obtained from animal or plant origins, natural fibres show a wide variation in their properties; therefore, proper selection of representative samples and statistical analysis are necessary to cope with the variation. Obviously, it is not possible, or reasonable, to sample the total amount of raw material; it would be too expensive and time-consuming, and furthermore, many tests are destructive. Therefore, samples of the material have to be taken for evaluation and the amount of material actually tested is often a very small proportion of the total. It is, however, imperative that the sample is an appropriate representation of the whole. For this reason, one must obtain an unbiased sample from the bulk material so that the properties to be evaluated are adequately representative of the properties of the whole. Test data can then be used with confidence. For example, if fibre diameter is the property being examined, the results should be an adequate representation of the bulk material and enable the prediction of performance in subsequent processes and products.

The total amount of a material under test can be seen as a collection of individual units (e.g., fibres) and in statistical terms this is termed the *population.* The characteristics of a population which differ between the individuals are termed the *variables;* for clean fibres, these will be properties such as fibre length, diameter and strength. A *sample* is a subset of a relatively small number of *individuals* selected to represent that population. Because of the variation found in natural materials, a sample must consist of a sufficient number of individuals for the data to be meaningful. For this reason, selection of the individuals comprising the sample should be done in such a way to ensure that a *random sample* is obtained and *sample bias* is avoided. A very variable sample will require a large number of individuals in the sample to obtain an adequate representation.

A *numerical* sample is defined as one where all the fibres in the population have an equal chance of selection. In a perfect numerical sample, the proportion by number of, say, long, medium and short fibres would be the same in the sample as in the whole population. However, such perfection is practically impossible and therefore a random sample is normally chosen. A random sample will have a *random error* associated with it, i.e., repeated measurements of the same property will produce different values scattered randomly around a central (mean) value. This error can be estimated by statistical analysis.

When the sample selection is influenced by factors other than chance, a sample ceases to be truly representative of the bulk and a *biased* sample results; bias may be attributed to the person doing the sampling. The effect of bias may be negligible or considerable and is characterized by test measurements being either consistently higher or lower than expected. Bias is also referred to as *systematic error* and cannot be detected by statistical analysis of test results.

Sampling methods for fibre will depend on the form in which the fibre is available, e.g., fibre in bales, sliver, carded web, yarn, etc., will demand different sampling techniques. The aim of sampling is to minimize random and bias/(systematic) errors. Fibre sampling will be affected by the degree to which fibres have been mixed or blended. If the bulk sample has been thoroughly mixed or blended, then the sample can be selected from one area with the assumption that fibres from all the original sources of supply are represented. However, if mixing was only partial, or if it is known that the material may vary in different regions of the bulk material, then samples should be taken from all parts of the bulk to ensure a representative test sample.

Fibre sampling methods are outlined below; a comprehensive explanation of different fibre sampling methods is given by Booth (1968) and Saville (1999).

15.2.2.1 Fibre sampling from bulk

When a contract is established between a buyer and a seller of fibre or yarn, it is usually a requirement that an independent third party determines the invoice mass of a consignment. The method of sampling to be used for commercial bales of fibre and containers of yarn is described in BS ISO 6741−2:1987. To assess fibre, yarn and fabric quality from bulk consignments, a range of different sampling methods will be required. BS EN12751:1999 describes sampling for acceptance testing to estimate without bias the desired property of the lot being evaluated.

15.2.2.2 Zoning

Zoning fibre in bulk form is likely to vary significantly from place to place throughout the bulk and the zoning method is used for selecting samples. A sample of fibre is taken from at least 40 different places, or zones, in the bulk material; these should be widely spaced apart. Each of these samples is divided into two subsamples and one subsample is discarded. The process is repeated with the remaining subsamples, i.e., each is divided into two and one half is discarded; the process is repeated again until about n/x fibres remain in the subsample ($n =$ total number of fibres required in the sample, $x =$ the number of original samples). Each of the 40 or so samples originally selected are treated; thus, the fibres remaining are collected together to give the required sample containing ‘n’ fibres. This method can be used for selecting samples from bulk raw cotton or other loose fibre.

15.2.2.3 Core sampling

The core sampling method was developed to obtain a laboratory sample representative of fibre in unopened bales of raw wool and may be used to assess the moisture content, grease and vegetable matter present and can be adapted for other fibre types. The density of the packaged fibre must be suitable for core sampling to take place: too loose or too dense and it is not applicable. A predetermined number of core samples are withdrawn from the bulk material by forcing a 60 cm tube with a sharpened tip into the bale. The tubes are sufficiently long to penetrate halfway into the bale, and coring is carried out from both ends and the sides of bales. All cores are combined when removed from the bale and stored in an air-tight vessel, and the weight of the bale and moisture content can be determined.

15.2.2.4 Fibre sampling from combed slivers, roving and yarn

Sampling sliver or yarn may lead to a length-biased sample as the fibres are laying approximately parallel to each other. This means that the longer fibres are more likely to be selected during sampling. Length-bias should be avoided where fibre length is the property to be measured, but length-bias will also impact on fineness and strength measurements as these often vary with fibre length. Length-bias can be overcome by preparing a numerical sample, or alternatively, a length-biased sample which allows for the bias to be incorporated into any calculation.

Textile testing is carried out for a wide range of particular reasons which depend on the specific circumstances and the information required. Commonly, testing is carried out to give an indication of performance of a material, to assess and compare qualities of materials, to estimate the content of useful fibre, to determine the effects of different processes or to see whether the material meets a particular specification. Whatever the purpose, it is imperative that test methods are reproducible. For example, when the same material is tested by a different operator or in a different laboratory, the results should be sufficiently similar, within experimental error, for the same conclusions to be obtained. Different operators may each have their own individual way of carrying out a test procedure, for example, there may be variations in how different people take measurements, or make instrumental adjustments; in all situations, the procedures used should be specified to minimize operator-based variations. It should also be borne in mind that when testing natural fibres, they are responsive to different atmospheric conditions and their properties may vary according to local temperature and humidity; standard conditions are commonly required for testing and should be specified (normally $20 \pm 2°C$ and $65 \pm 2\%$ RH). Similarly reproducibility in test procedures should be ensured between different laboratories and different dates. If test results vary significantly between laboratories, then the test method must be deemed unreliable. However, values obtained from repetitive testing of textile materials would not be expected to be identical and therefore statistical analysis needs to be carried out.

15.2.3 Standardization of test methods

To ensure reproducibility of test results, and facilitate trading in the global marketplace, test methods need to be standardized and, as mentioned earlier, most countries have their own standards organizations, e.g., the British Standards Organization, ANSI, ASTM International (USA) and Deutsches Institut für Normung (DIN, Germany). The European Committee for Standardization (CEN) comprises the national standardization bodies of 34 European countries with the remit to develop voluntary European Standards (ENs). These European standards are also national standards in each of the 34 member countries, and replace any conflicting national standards. The aim of CEN is to remove trade barriers for European industry and consumers and enable products to reach a wider market with lower development and testing costs. Similarly, on a global scale, the Vienna Agreement of 1991 is a joint agreement on technical cooperation between ISO and CEN for harmonization of international and European standards.

15.3 Test methods for natural fibres

Historically, the assessment of the 'quality' of natural fibres was carried out subjectively by skilled individuals, often aided by physical reference 'standards'. This approach is still commonly used in most of the less-developed fibre-producing countries. Today's international trade has led to the development, and wide acceptance, of internationally approved standard test methods including automated instrumental methods for the assessment of quality parameters. Standard methods are highly developed for cotton and wool, but for many other fibres there is still a long way to go before objective testing of natural fibre materials becomes the norm. The USA has led the development of test methods for cotton and, more recently, the principles for cotton trade 'harmonization' have been established (HS 52) as part of the harmonized commodity description and coding system. The methods developed for wool grading have been established primarily from research undertaken in Australia and New Zealand.

Some progress has been made for international grading for the quality of other natural fibres, but their importance in international trade is often localized and has not received the attention given to cotton and wool. That said, the current exploitation and studies of the potential uses of flax and other bast fibres require a standardization system, and this has been addressed by the FAO/ESCORENA European Cooperative Research Network on Flax and other Bast Plants. Procedures adopted for testing wool have been relatively easily adapted to other hair fibres, but most of the cotton tests that are based on objective methods have not been easy to apply to other cellulosic fibres. The need for the establishment of standard test methods for flax has been recognized in the main producing countries for many years, but the only real progress has come from ASTM International. ASTM standards for testing flax include ASTM D6961/D6961M-09 (2015)e1 Standard Test Method for Color Measurement of Flax Fiber; ASTM D7025-09(2015)e1 Standard Test Method for Assessing Clean Flax Fiber Fineness; and ASTM D7076-10(2015)e1 Standard Test Method for Measurement of Shives in Retted Flax.

The end use of fibres determines the relative importance of testing. For example, it is not necessary to test the quality of coir or jute using the rigour and range of methods applied to high-value fibres intended for use in fine woven or knitted apparel. Table 15.1 shows the more important characteristics which are important in the assessment of quality for a range of natural fibres.

15.4 Measuring the physical properties of natural fibres

Consideration of the physical properties of fibres such as fibre diameter, fineness and stiffness is imperative for quality control purposes. Fine fibres are used for the production of fine yarns and therefore are of high value. Similarly, the stiffness of a fibre impacts on its spinnability and therefore affects yarn and fabric properties. Fabric drape, handle, prickliness and softness are all dependent on the stiffness of a fibre used in its production.

15.4.1 Fibre diameter and fineness

Fibre fineness/diameter is one of the most important properties of fibres and is influential in determining end use. Fibre diameter is normally measured in micrometres (μm), commonly abbreviated to microns. The visual appearance and handle of fibre are influenced by the fineness and this will obviously have an impact on the value of the fibre. Generally, the value of a natural fibre for use in apparel and many domestic textiles increases with decreasing diameter, but for some industrial and technical textile applications, the relationship can be reversed. A simplified example is that, because there is a minimum number of fibres needed in the cross-section of yarns to provide strength and uniformity, fine yarns can only be satisfactorily made from fine fibres; fine yarns are more highly valued and are used in high-value products so, other factors being constant, fine fibres generally are of higher value than coarse, thicker fibres.

Fibre stiffness (resistance to bending) increases with increasing fibre diameter, i.e., the coarser the fibre, the stiffer it is. Bending rigidity of perfectly elastic fibres of circular cross-section is proportional to the fourth power of the fibre diameter ($B \propto d^4$); while most natural fibres are not perfectly circular, or perfectly elastic, the bending stiffness is still very sensitive to fibre fineness. The stiffness of a fibre affects the 'spinnability' and yarn properties; fibre stiffness also affects properties of fabrics such as drape, handle, prickliness and softness.

The diameters of natural fibres vary considerably; fibre diameter does not have a single absolute value and can range widely about the mean value even when fibre is from the same source; because of this, the use of single values to characterize fibre diameter (e.g., mean, median or airflow) is often inadequate for assessing diameter characteristics, for example, short flax fibres will always have a wide, skewed, distribution in diameter values even within batches from the same crop (see Fig. 15.1). Similarly, wool obtained from a single sheep will show variation but the distribution of diameters is less skewed (Fig. 15.2).

Another factor that needs to be considered when measuring fibre fineness is that the cross-section of different fibre types varies in shape; wool has an approximately circular cross-section, raw cotton has a collapsed tubular cross-section and bast fibres are often irregular polygonial in cross-section (Fig. 15.3(a)–(f)). Furthermore, natural fibres usually exhibit cross-section variability along their length.

Fibre diameter is not a particularly useful measurement in all circumstances, particularly when the deviations from circular cross-sections are significant and alternative ways have been developed to represent fibre fineness. Fibre fineness is often measured in terms of fibre linear density, e.g., decitex (mass in grams of 10,000 m of fibre) or denier (a historical term originally used to express the fineness of silk, defined as the mass in grams of 9000 m of fibre). Indirect assessments of fibre fineness were some of the earliest to be developed, especially for quality control purposes, and have been particularly valuable in the cotton and wool industries; the most common of these are the 'airflow' methods, based on the resistance of a standard mass of fibre to air flowing through the fibres.

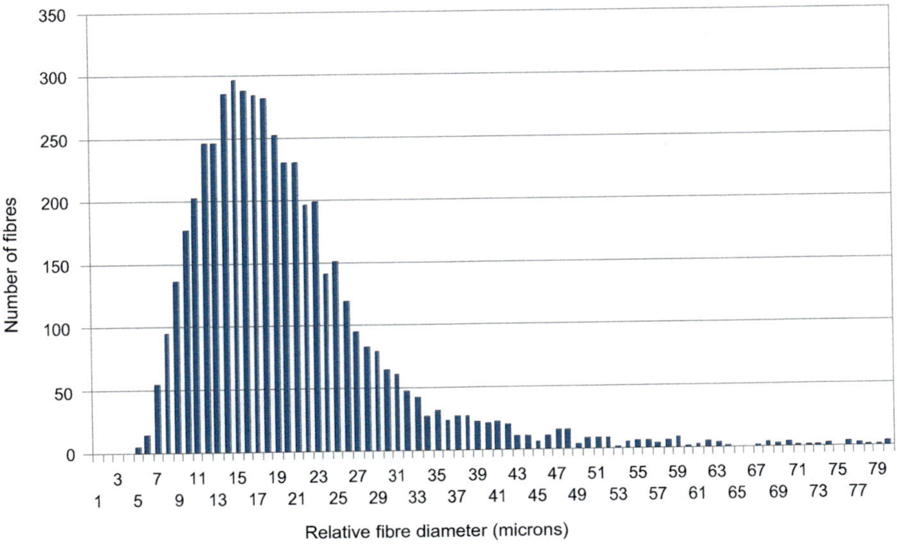

Figure 15.1 Fibre diameter distribution of short fibre flax (Laserscan).

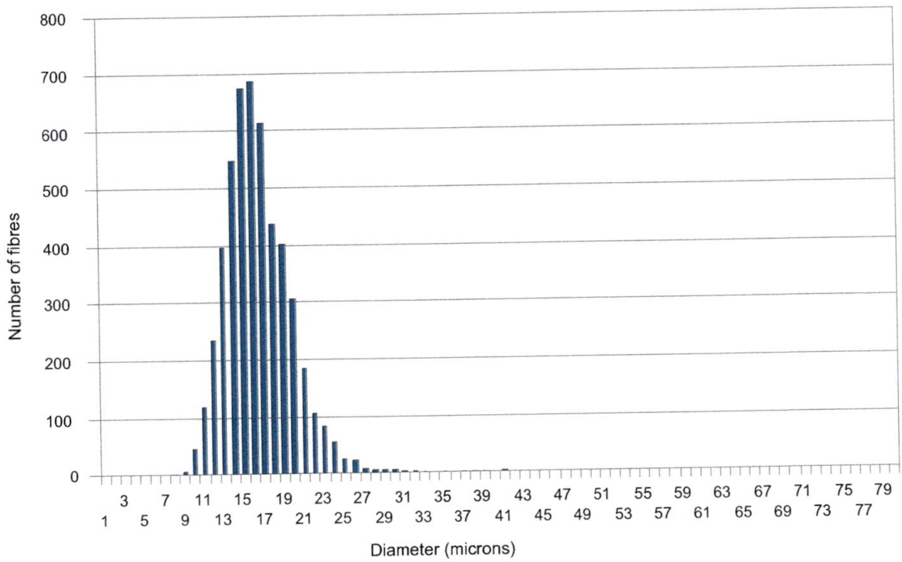

Figure 15.2 Relative fibre diameter of a wool sample (Laserscan).

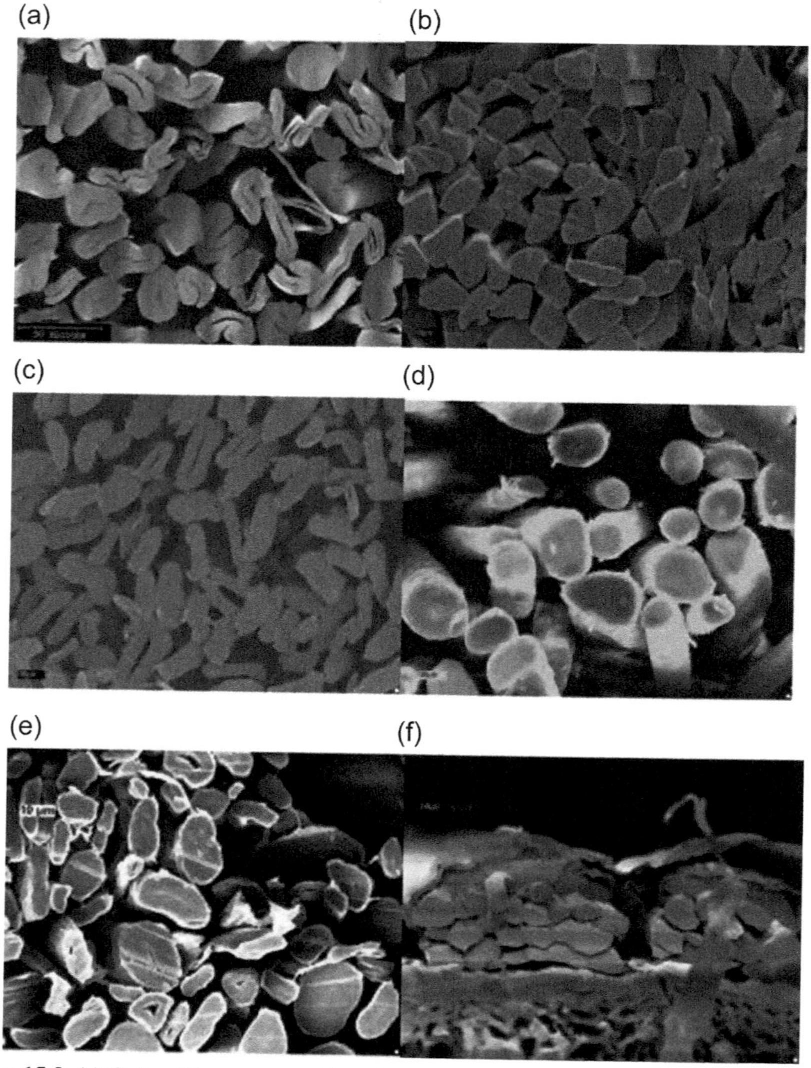

Figure 15.3 (a) Cotton fibre cross-section; (b) silk fibre cross-section; (c) ramie fibre cross-section; (d) wool fibre cross-section; (e) nettle fibre cross-section; and (f) flax fibre cross-section.

15.4.2 Methods for measuring fibre fineness

A series of 14 papers by Sommerville (2001–2007) reviewing methods of measuring fibre fineness are essential reading for those interested in this topic.

15.4.2.1 Cross-sections

Microscopy offers the only direct method for measuring the absolute characteristics of fibre cross-sections. Because of the irregularity of the cross-sections of most natural fibres, values of 'diameter' are usually an average based on a large number of 'diameter' measurements. Examination of fibre cross-sections is rarely used nowadays as it requires skill and experience, and is slow, expensive and highly labour-intensive.

15.4.2.2 Fibre diameter

The most common methods for measuring diameter are based on determining the width of short snippets of fibre using a projection microscope, but the processes are still laborious and are now only used for the characterization of 'standard' or 'reference' samples which are to be used as calibration standards for other methods.

Automation of diameter measurement for wool fibres has passed through many stages of development, reviewed by Sommerville (2001–2007), and today, there are a number of commercially available instruments which include the Optical Fibre Diameter Analyser (OFDA) (Australia), the Sirolan Laserscan (Australia), and the FibreShape (Switzerland) systems.

15.4.2.3 Optical fibre diameter analyser (OFDA)

The potential for using image analysis for automating the measurement of fibre diameter from microscope images was developed at the Australian Wool Testing Authority Ltd in the early 1980s. This led to the current range of OFDA systems produced by BSC Electronics Pty Ltd. OFDA 100 was originally developed to determine the average fibre diameter and standard deviation of wool fibres and is widely used by testing houses and mills across the world. The OFDA 100 system was developed for wool; its successor, OFDA 2000, determines a wide range of quality parameters including mean diameter (range 4–300 μm), standard deviation and curvature for fibre types which have a circular, or near-circular, cross-section including wool, cashmere, mohair, glass and synthetic fibres (Stobart and McColl, 2000). OFDA 4000 has been developed more recently as a single instrument to measure the fibre length distribution of wool fibre in top or sliver form in addition to the fibre diameter distribution. OFDA 5000 has been introduced for measurements on microfibres (diameter range 0.5–60 μm). OFDA 100 has been used for determining useful characteristics of other fibres including some bast fibres, but the data have to be interpreted to take account of the irregularity of the fibre cross-sections; nevertheless, the 'relative' fibre diameter data have proved very useful in a range of applications (Akin, 2010; Drieling et al., 1999; Grishanov et al., 2006; Wang and Wang, 2004).

The OFDA 2000 comprises an automated microscope which captures images of a moving sample of fibre snippets (normally 2 mm long) which are spread automatically onto a 70 mm square glass slide which is then scanned by the microscope. Software identifies and measures individual fibres to a resolution of 1 μm; the mean diameter and standard deviation are calculated to a resolution of 0.01 μm. The whole process takes a few minutes for, typically, 4000–20,000 measurements. The results are printed out as a histogram of the fibre diameter distribution and software generates values for coefficient of variation, prickle factor and spinning fineness (the last two relating only to wool and similar near-circular fibres).

15.4.2.4 Sirolan laserscan

Research on the use of photometry to determine fibre diameter was undertaken by the AWTA alongside the image analysis approach which led to the early OFDA machines. The principle is relatively simple: when a beam of parallel light (from a laser) shines on a photodetector, fibres at right angles to the beam of light will cast a shadow on the photodetector which, if the fibres are of constant length, will have an area directly proportional to the fibre diameter. The key to the success of this approach was the development of an optical discrimination system (Fig. 15.4), which ensures that only single-fibre snippets that fully intercept the laser beam are measured. Fig. 15.4 shows a schematic diagram of the principle of the 'optical discriminator'. Fibres to be measured must cross the central detector and two (but only two) of the outer detectors. In this way, only single whole fibres with an appropriate alignment are measured.

To measure fibre diameter using this principle samples should be appropriately prepared in terms of cleanliness and fibre separation (e.g., wool is scoured to remove grease and dirt, dried and conditioned). When the samples are in suitable condition, fibre snippets of 2 mm length are cut from an aligned random sample of fibre using a special guillotine. Fig. 15.5 illustrates the general features of the Sirolan Laserscan system: small quantities of the snippets are introduced into a premeasurement dispersion bowl containing an isopropanol/water mixture (which emulates standard moisture regain conditions) until the required number of measurements have been taken. The dispersion of fibres flows through a measurement cell where fibre snippets intersect a laser beam; those correctly aligned are detected and measured. Typically, 2000 to 10,000 measurements are taken and the software yields values of mean diameter, standard deviation, CV%, percentage of fibres less than 30 μm (comfort factor for wool), fibre curvature and a histogram (and data file) of the diameter distribution (Fig. 15.6) of fibre diameters up to 80 μm.

As with the OFDA systems, the Sirolan Laserscan was targeted at wool but has been successfully used for many other fibres including other protein fibres, flax and hemp (Grishanov et al., 2006; Lamb and Denning, 2004; Miao and Finn, 2008; Wang and Wang, 2004).

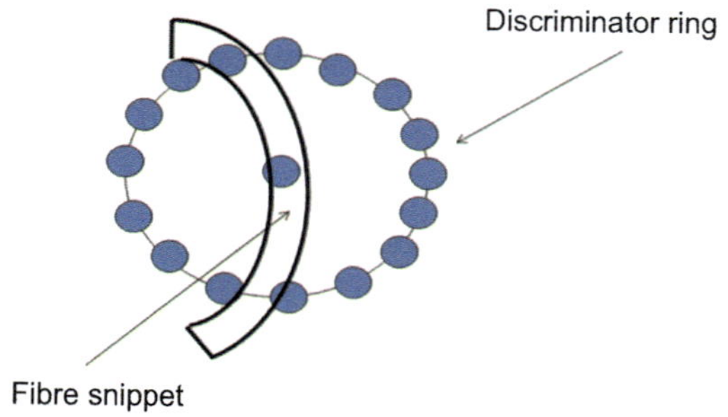

Figure 15.4 Principle of the Laserscan optical discriminator.
Courtesy CSIRO.

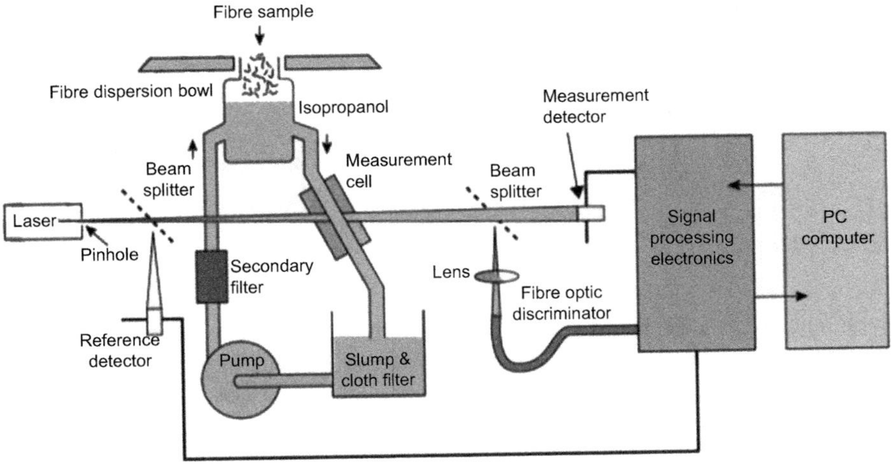

Figure 15.5 Principle of operation of the Sirolan Laserscan system.

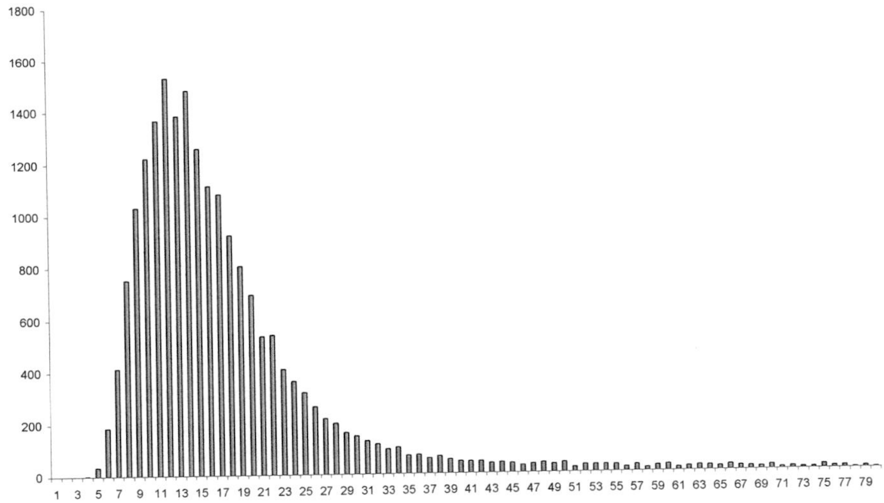

Figure 15.6 Distribution of relative fibre diameter of a cottonized flax sample (Laserscan).

15.4.2.5 *FibreShape*

FibreShape (Fig. 15.7) is a versatile quality control and characterization system which combines a high-resolution scanner with advanced image analysis software. The analysis of a scanned image provides mean fibre diameter, standard deviation and a histogram of diameter distribution; fibre length and shape information can also be determined. The FibreShape system is available with either manual or automatic sample preparation; it is a relatively low-cost system useful for quality control measurement of a wide range of characteristics for many natural fibres (Schmid et al., 2002; Schmidt et al., 2010; Mussig and Amaducci, 2018). FibreShape offers

Figure 15.7 FibreShape instrument.
Courtesy FibreShape.

a rapid characterization process, but the resolution is lower than with other methods, although there is good general correlation. Fibres of a fineness range from 5 μm up to 5000 μm (5 mm) can be measured (using appropriate software, fibre lengths from 2 μm to 30 cm can also be measured).

The performance of automated diameter measurement systems need to be checked and recalibrated using standard reference materials. For the OFDA and Laserscan machines, this is achieved using standard wool tops, where the mean fibre diameter and diameter distribution have been previously determined by direct measurement using a projection microscope (available from INTERWOOLLABS), but other reference materials may also be used. In some instances, recalibration can be more troublesome, e.g., returning the machine to the manufacturer for recalibration.

The Laserscan, OFDA and FibreShape systems have many uses in the natural fibre industry beyond wool characterization as they enable the measurement of mean fibre diameters and diameter distributions for quality specification, quality analysis, research and development for many fibres.

Flax and hemp have received a lot of attention in recent years particularly towards modification of flax and hemp tow and short fibre crops by enzyme and chemical 'retting' to produce cottonized fibre for use on short-staple spinning systems. The main issue is that flax and hemp fibre are not, as harvested and decorticated, single fibres but a mixture of single, elementary, fibres and bundles of elementary fibres held together by interfibre cement. During the carding and spinning processes, these bundles partially split into single fibres, but a proportion of bundles will remain intact. Enzyme and chemical retting treatments have been shown to improve bast fibre separation; high-energy ultrasound and shock waves have also proved successful for cottonizing bast fibres (Akin et al., 2001; Brown, 1984; Fila et al., 2001; Henriksson et al., 1997; Harwood et al., 2008b; Sirghie et al., 2005). Instruments that produce histograms of fibre distribution are particularly useful for observing the effects of retting and cottonization treatments as the changes in the diameter distribution give a quantified picture of the process.

Although flax fibre cross-sections are irregular polygons, they can be considered to be approximately circular. Changes in diameter distributions assist the quantification of cottonization treatments. Attempts have been made to produce estimates of single flax fibre diameter and the proportion of multiple fibre bundles in a sample; Grishanov (2006) produced software to assist this process. Diameter distributions are 'deconvoluted' and the output is an estimate of the numbers of single (ultimate) fibres and their mean diameter along with estimates of the number of fibres made up of two, three, four and five or more ultimate fibres. Although diameter values are output as microns, values are on a 'relative' diameter scale as fibre, and fibre bundle, cross-sections deviate significantly from circularity.

Workers at De Montfort University used this approach for determining the effectiveness of the retting process (Harwood et al., 2008a,b; Horne et al., 2008, 2010). Application of the deconvolution software to diameter distributions enables comparison of different fibre samples, irrespective of the degree of retting. Deconvolution software can provide the following additional data:

- Estimates of percentages of single fibres present in a sample, and fibre bundles containing two to five fibres.
- The average deviation of deconvoluted data from original data.
- The mean and standard deviation of ultimate fibre diameter.

Fig. 15.8 shows the theoretical distributions of diameter for average fibre diameter of 20μ and standard deviation $\sigma = 2\mu$ for single fibres and fibre bundles consisting of two to five fibres. The distributions are labelled (i)–(v), respectively.

15.4.2.6 Airflow method for measuring fibre fineness

The ease with which a fluid will flow through a porous mass is dependent on the surface area of the pores; the higher the surface area per unit volume, the greater is the resistance to fluid flow. The research which led to the adoption of airflow methods for the textile industries was laid down in the late 1930s and 1940s with the first Micronaire instrument produced in 1947. Considerable evolution followed in the next decade.

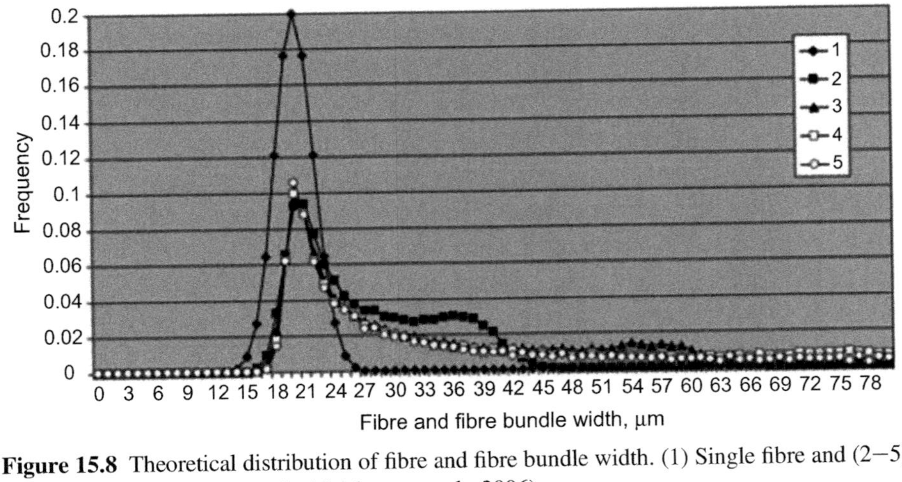

Figure 15.8 Theoretical distribution of fibre and fibre bundle width. (1) Single fibre and (2–5) two to five fibres in the bundle (Grishanov et al., 2006).

It was shown that the flow of air through a standard plug of fibres is proportional to the difference in pressure across the plug (ΔP) and the diameter of the spaces between the fibres squared (d^2):

$$\text{flow} = K \cdot \Delta P \cdot d^2$$

Using two principles – that at a given pressure difference, the airflow through a uniformly distributed mass of fibres is determined by the total surface area of the fibres, and fine fibres have higher specific surface areas (surface area per unit mass) than coarse fibres – two instrument types evolved (Anderson, 1954). The first type of instrument relates the pressure drop at a standard rate of air flow to d^2 ($d^2 \infty \Delta P^{-1}$). The second instrument using a constant air pressure applied to the fibre mass determines the flow of air through the sample: K is a constant so we can either relate d^2 to the flow (by keeping ΔP constant) or to ΔP (by keeping the flow constant) or to the drop in pressure (keeping the applied pressure constant). $d^2 \infty$ flow. All modern instruments are based on one of these scenarios.

The airflow concept was adapted to develop instruments for classifying wool and cotton fibre fineness and they are the standard quality control methods for assessing the fineness of these fibres (Fig. 15.9). Results for wool are expressed as the mean diameter in microns and for cotton as the mean fibre linear density in micrograms per inch (Micronaire value). The advantages of airflow systems are their wide acceptance by industry, the low cost of equipment and ease of operation. However, the data are limited in their quality; the equipment available is generally specific to either wool or cotton and results are sensitive to sample preparation. Use of airflow methods with

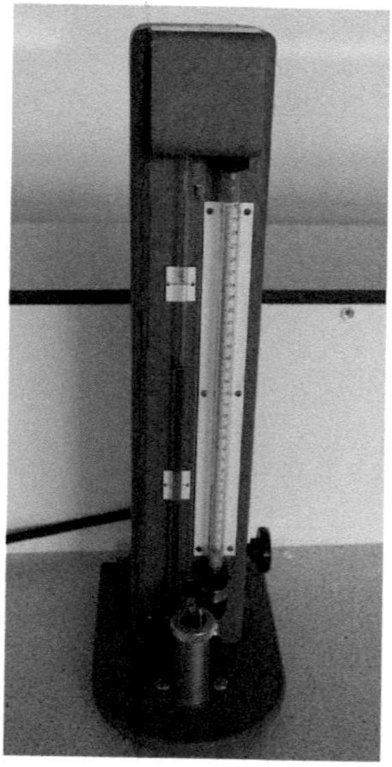

Figure 15.9 WIRA fibre fineness meter. (WIRA Instrumentation Ltd, Bradford, UK.).

other fibres is commonplace but only as a guide to relative fineness (Faughey et al., 2000; Sinha, 1970; Sinha and Bandyopadhyay, 1968) with no information about fineness property distribution (Mussig et al., 2010). One of the ASTM standards, for measurement of flax fibre fineness, is based on the airflow principle (ASTM D7025-09(2015)e1).

15.4.2.7 Operation

A fibre sample of known, standard, weight is compressed into a cylinder of a known, standard, volume and exposed to an air current at a known pressure. The rate of airflow through the fibre mass is measured using a rotameter which is calibrated to correlate with fibre fineness (using standard reference fibres whose fineness has been measured by a direct method), e.g., microns for wool, micrograms per inch for cotton.

The surface area of a fibre (length × circumference) is proportional to its diameter but for a given weight of sample, the number of fibres will increase with fibre fineness so that the specific surface area is inversely proportional to fibre diameter. The fibre diameter will therefore determine the ratio of airflow to differential pressure. For fibres of approximately circular cross-section and constant overall density such as unmedullated wool (BS 3183:1968), the estimate of fineness obtained by the airflow method shows a good correspondence to the average fibre diameter as determined by projection microscopy.

The same method can be used for measuring cotton fineness, and is usually carried out on raw cotton opened and blended using a laboratory blender or Shirley analyzer. However, results will be affected by the maturity of the cotton fibres as well as their fineness. For this reason, the results are expressed in arbitrary Micronaire units (Micronaire is the trade mark of the Sheffield Corporation) (BS 3181−1:1987a).

15.4.3 Fibre length and length distribution

The length of natural fibres varies considerably, even in samples of the same provenance, whether animal or plant. The fibre length is the second most important property of a fibre, fineness being the first. Wool is a relatively long fibre and fibre length will vary between different breeds of sheep; fine wool sheep breeds (e.g., Merino and Charollais) typically have wool of 4−10 cm, whereas long-wool breeds (e.g., Lincoln and Cotswold) have wool up to 30 cm long. Cotton is a much shorter fibre with Sea Island cotton fibres around 50 mm in length, whereas most are much shorter and some Indian cotton may be less than 12 mm in length.

The importance of fibre length lies in its processing, especially spinning. Generally, the longer a fibre, the easier it is to process. However, fibre-processing machines are designed to operate on a narrow range of staple lengths, and, even within a particular range of lengths, some machine adjustments must be made to optimize the process for the specific material being processed. Therefore, it is important that the raw material should not show much variance from the established standard length.

This variation in fibre length of the raw material also needs to be controlled during the combing processes where short fibre is combed out (combing waste). The quantity of short fibre present will impact on the waste produced and hence on the economics of the manufacturing. Longer fibre also assists in the cohesion of rovings and slivers as longer fibres will overlap more for any given level of twist and, consequently, less twist is needed to produce satisfactory slivers from long fibre.

With long fibre, there will be fewer fibre ends in a given length of yarn and yarn evenness improves with increasing fibre length. A longer fibre will produce a higher strength yarn for the same level of twist. Fibre length and fineness are quite strongly correlated in the cases of cotton and wool: for wool, the longer the fibre the coarser the wool; for cotton, the opposite is true as longer cotton fibres are generally finer than short cottons. Fibre length for wool and cotton is usually expressed as staple length: in simple terms, the staple length of wool is the average overall length of the fibres in their naturally crimped condition. For cotton, the staple length is between the mean length and the maximum length.

15.4.3.1 Fibre diagram machine

The fundamental principle of this machine is that it produces a fibre length diagram which is similar to that obtained from a traditional ‘hand draw’, when a fibre sample is hand laid on to a velvet board so that the length distribution can be observed. A fibre beard is produced in which all the ends of the fibres are aligned and the beard is then passed between two parallel plates that act as a capacitor. The change in capacitance is used to calculate the number of fibres present as a function of length, the distribution of fibre lengths is reported with the standard deviation and CV%. The maximum fibre length that can be detected using this type of instrument is approximately 300 mm.

Results are presented as Hauteur and Barbe fibre lengths. Hauteur (H) is the mean fibre length biased by the cross-section (linear density) of the fibres; Barbe (B) is the mean length biased by fibre weight. Hauteur and Barbe are both measured in mm and have the relationship:

$$B = H\left(1 + \left(\frac{CVH}{100}\right)^2\right)$$

where CVH is the coefficient of variation of Hauteur in %.

BS 5182:1975 describes the standard test for the measurement of wool fibre length.

15.4.3.2 Fibre length distribution tester

Testing is carried out on a beard of fibre produced by drawing a comb over a sample of the fibre. A typical beard of flax fibres removed from the comb after testing is shown in Fig. 15.10.

The beard is drawn over an illuminated sensor where the changes in optical density enable the fibre length to be calculated. The length distribution data are generated at intervals defined by the operator. Length measurements determined using a fibre length distribution tester are limited to fibres up to 80 mm long.

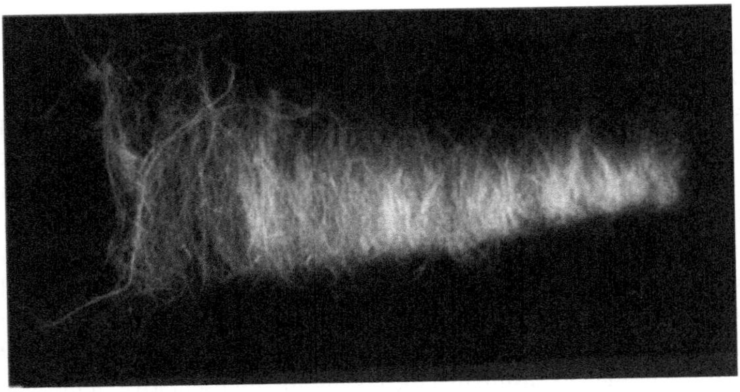

Figure 15.10 A beard of flax fibres.

The Almeter system for automated fibre length measurement consists of two separate instruments; a fibre preparation machine (Fibroliner) to prepare the fibre beard and a measuring instrument (Almeter) to scan the fibre beard calculating the mean fibre length and distribution as both Barbe and Hauteur. The OFDA 4000 incorporates fibre length measurement with fibre diameter measurement in a single machine with an integrated fibre aligner.

15.4.4 Fibre strength

Because of the variability found in natural fibres, testing the strength of these fibres is time-consuming and problematic. The strength of individual fibres shows a high degree of variability, and for this reason, a large number of fibres have to be measured to ensure statistical reliability. It is difficult to handle and mount single fibres in the jaws of a strength testing machine, especially if fibres are particularly fine. For this reason, tests are often carried out on fibre bundles.

To measure single fibre strength, a universal tensile tester is often used. Universal tensile testing equipment is widely available from many testing machinery suppliers, for example, Uster, SDL Atlas, Instron. The principle is very simple: individual fibres are clamped vertically between two jaws and extended at constant rate of elongation. The applied force (stress) is measured continuously as the elongation (strain) increases; a complete stress strain diagram may be produced, and alternatively, the maximum force required to break the fibre may be displayed along with the elongation at maximum stress and/or at the breaking point. The strength of fibres is normally expressed as 'tenacity', i.e., the maximum force (Newton) required to break the fibre divided by the linear density (tex) at the point of rupture.

There are instruments available that are specifically designed for single fibre strength measurement which are particularly useful for some fibre types (WIRA single fibre strength tester, the Lenzing Vibrodyn, Textechno Fafegraph HR, SDLATLAS Tinius Olsen H1KS). British standards BS EN ISO 5079:1996 and BS 4029:1978 describe the determination of breaking force and elongation at break of individual fibres, and tensile elastic recovery, respectively. The ASTM has published a standard test method for evaluation of the tensile properties of single textile fibres (ASTM D3822/D3822M-14).

The testing of bundles of fibres takes less time and involves fewer test samples than testing individual fibres. The use of manual instruments requires training and experience to produce valid results, but bundle strength has assumed greater importance than single fibre strength tests. The operating principle is to determine the breaking force required to rupture a carefully prepared thin ribbon of fibres which are clamped at very short gauge length; the weight of the fibres in the ribbon is determined and, so long as the fibre length is known, the tenacity of the fibres can be calculated (detailed instructions are always provided in the operating manuals). Careful sampling and multiple tests are needed to ensure that the results are representative of the whole batch of fibre. The 'Stelometer' (derived from STrength and ELOngation METER) is the instrument most commonly used for bundle strength testing of natural fibres and is available from a wide range of manufacturers. Semiautomated methods are commonly used and readily available, particularly for cotton testing; the operative is required to select and prepare the samples, usually with the aid of a 'fibrosampler' or other combing device to prepare a beard of fibres which is then inserted for automated testing (the USTER HVI System).

15.4.5 Fibre colour

Fibre colour will vary between different fibres and will be dependent on the processing to which it has been subjected. Colour can be measured instrumentally using a reflectance spectrophotometer. The technique requires controlled conditions of illumination, sample size and background. The principles of reflectance spectrometry are well established (McDonald and Rigg, 1980). As fibre samples exhibit variations in texture and colour, measurement using reflectance spectrophotometry requires careful sampling:

- Ensure samples are taken from different parts of the material and that duplicate measurements are taken to obtain a mean reflection spectrum.
- Ensure the sample is sufficiently thick to be opaque, thus preventing light transmitting through the sample and reflecting back from the sample holder.
- Rotate the sample 90 degrees between measurements and take a mean value of four measurements to minimize any direction effects caused by the texture of the fibre sample.

By far the most common system used for colour measurement is that developed by the CIE (Commission Internationale de l'Eclairage). The colour of a fibre sample is expressed in terms of three-dimensional colour coordinates, represented by the CIE L*a*b* colour space system, by which all surface colours can be represented.

Assessments of 'Brightness', 'Whiteness' and 'Yellowness' of textile materials are often used to monitor changes during processing. Caution is needed when using measurements calculated from reflectance data with natural fibres, particularly bast fibres, as calculated rankings are often significantly different from visual rankings.

15.4.6 Optical microscopy

Optical microscopy can provide a direct image of plant structure and in conjunction with staining methods can be used to study the arrangement of fibres in the plant stem and the presence of noncellulosic constituents. There are two types of optical microscopy relying on either reflected or transmitted light. Reflected light microscopy is used to study the surface of a sample. A sample can be simply placed under the microscope lens, or may be cut and polished into a flat surface to allow compositional and structural data to be collected.

Transmitted light microscopy requires samples that are optically transparent or translucent so that light can pass through them. Fibre samples are sufficiently thin for study without any preparation; however, other plant material will often require thin sections to be studied. Many techniques for preparing thin cross-sections are discussed in the literature (e.g., Greaves et al., 1995; Steedman, 1960; Stoves, 1957). To avoid distortion when cutting a cross-section, a microtome can be used with the sample embedded in a supporting medium which may be a wax or polymer resin. A wide range of resins are available: the use of Spurr's resin has proved particularly successful for bast fibres, as it is an embedding medium that is low in viscosity and useful for plant materials which are difficult to infiltrate (Akin et al., 1996; Dérue et al., 2002; Donaldson et al., 2001). Another resin that may be used for the same reasons is LR white (Fraser et al., 1982).

As with reflected light microscopy, no direct chemical information is obtained from transmitted light microscopy. However, the use of specific staining techniques enables information on chemical composition to be gained. For example, the distribution of pectic substances in flax has been studied using a 0.02% aqueous solution of ruthenium red. Similar studies for lignin (Fig. 15.11) use acid phloroglucinol (Akin et al., 1996) and toluidine blue (Fraser et al., 1982). Such studies have shown that the concentration of lignin in the fibre bundles was significantly lower than in the xylem of the stem.

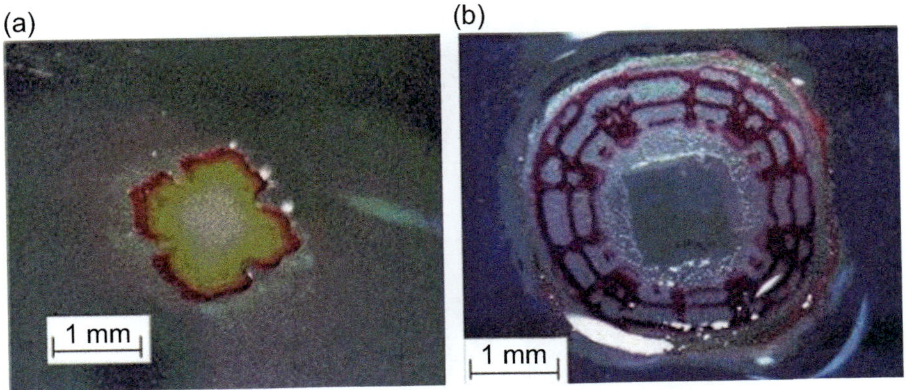

Figure 15.11 Cross-section of *Urtica dioica* stems (stained with phloroglucinol to highlight lignin: (a) upper stem; (b) lower stem.

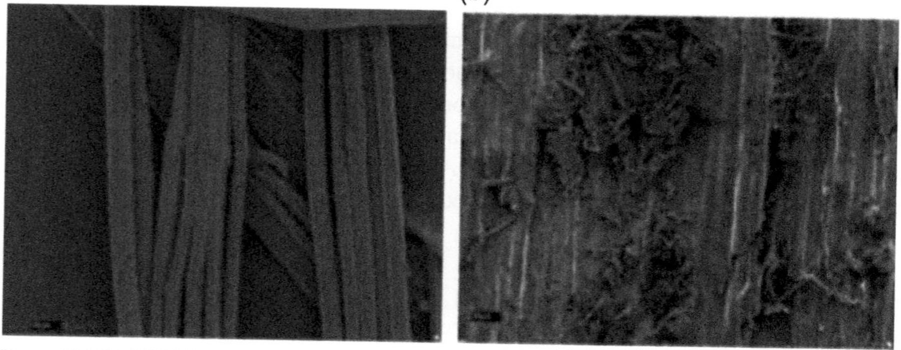

Figure 15.12 SEM images: (a) chemically retted flax fibres; (b) outer surface of field retted flax straw.

15.4.7 Electron microscopy

There are two forms of electron microscopy. Scanning electron spectroscopy (SEM) is a technique that has been available since the mid-1960s and is used for the study of surfaces. It can be used to study internal structures if a section through a sample is examined. Transmission electron microscopy (TEM) is an older technique dating from the 1930s, and is able to investigate internal structure. Although some elemental contrast can be obtained from SEM arising from variations in atomic number throughout a material, it is not suitable for studying the distribution of lignin or pectic substances in plant structures. However, it can be used to study the surface of fibres and fibre bundles. Fig. 15.12(a) shows an SEM image of chemically separated flax fibres and can be compared with Fig. 15.12(b) where there is clear evidence of fungal activity on the outer surface of flax straw which been retted in the field. Fig. 15.13 shows the effect of retting on the inner surface of field retted flax straw. The separation

Figure 15.13 SEM image showing inner surface of field retted flax straw.

Figure 15.14 SEM image showing fibre separation in field retted flax straw.

of individual fibres as a result of field retting is well illustrated by the SEM image of a cross-section of the flax stem in Fig. 15.14. The technique can be used for a visual evaluation of differences between fibres and related materials.

Modern SEM instruments often include energy dispersive spectroscopy (EDS) analysis. SEM-EDS allows elemental analysis of the surface and immediate subsurface of materials. This technique is useful for examining the uniformity and durability of surface treatments on textile fibres (Shen et al., 2017). SEM-EDS is particularly useful for metal analyses but can, theoretically, be used for all elements below sodium in the periodic table.

TEM is used for the study of internal structures and sample preparation is important as ultrathin sections must be used. To permit good imaging, samples are stained with heavy metal salts, usually lead or uranium, in the same way as for other microscopic techniques. The very high resolution of the method enables detailed structural information to be obtained.

15.4.8 Cleanliness

Cleanliness affects the cost of fibre materials; low-trash-content materials normally commanding higher prices than materials with higher trash contents. Although there are manual test methods for trash and other nonfibre for cellulosic fibres, automated methods are usually used. Instruments are available from a number of manufacturers (e.g., Uster and SDLAtlas) which allow the easy determination of trash, dust, microfibre and some other fibre contaminants. Dust is an important contaminant particularly with bast fibres as it can cause environmental health and processing problems in spinning and knitting mills. Test systems can also be used to determine the 'cleanability' of the materials, a major consideration for the processing of fibres. Micro-Dust and Trash analyzers such as the Suessen MDTA 3 and Uster instrument range were developed for cotton testing but are used for a range of other 'fine' fibre types.

Output data include the quantity of useful fibre and trash, dust, microfibre and nep content in fibre samples. The data are obtained by separation of the components by mechanical and airflow systems. Useful fibre is collected as a partially aligned sliver which can then be used for small-scale spinning trials. Textechno have recently introduced the MDTA 4, based on the MDTA 3 but also offers fibre length measurement of the collected clean fibre using digital image processing.

As an example, MDTA data were used in the evaluation of different flax accessions grown for short fibre to be used in high-value textile end uses (Harwood et al., 2008a). The aim of the work was to progressively select the 'best' accessions through evaluation of fibre obtained from straw from replicated trial crops harvested over three growing seasons. The selection criteria were initially based on the optimal properties for spinning on the cotton system (Table 15.2).

Five flax accessions were finally selected as having the optimum properties for spinning on the cotton system and were successfully blended with cotton to produce commercial 26 Nm (i.e., 38 tex) yarns with flax content up to 50% (Harwood et al., 2008a,b; Horne et al., 2010).

For protein fibres, the contaminants in raw materials are primarily grease and adhering soil which are effectively removed in raw fibre scouring before any further processing. Relatively simple laboratory test methods simulate commercial scouring processes and allow the clean fibre yield to be determined with supplementary standard tests for monitoring the quality of scouring processes (Wool Testing Authority, 2019).

15.4.9 Fibre length, strength, elongation and Micronaire (HVI)

The Uster HVI (high volume instrument) provides data relating to average fibre length, percentage of short fibres, average fibre tenacity, average elongation at break and fibre fineness (Micronaire value). The instrument was developed for cotton and is the standard measurement system used by the international cotton trade. The system can be used very effectively for cottonized flax and other fine bast fibres, but results obtained will be 'relative' rather than absolute values (Harwood et al., 2008a; Horne et al., 2010).

Table 15.2 Target specification for flax fibre on the cotton system.

Property	Target specification
Fibre length	Mean length 25–30 mm
Short fibre	Not more than 15%
Strength	Not less than 28 g tex^{-1}
Trash	Total trash content less than 10%
Fibre fineness	Micronaire value less than 7
Colour grade	The whiter the better

15.4.10 Sensory factors

After evaluation of the quality and performance characteristics of natural fibres, it follows that their effectiveness in end-use products must be assessed. While the uses of natural fibres encompass a hugely diverse range such as automotive, medical to fashion and clothing, it is worth mentioning here that the promotion of natural fibres in clothing has been addressed in studies of their comfort and physiological influence. Kawabata was a pioneer in the objective evaluation of fabric composed of both man-made and natural fibres. Fabric handle is clearly a subjective assessment, based on an individual perception of comfort values such as smoothness, stiffness and softness. Kawabata recognized that the stimuli causing a psychological response of handle are determined by the physical and mechanical properties of the fabric and developed the KES-F system whereby both subjective and objective measurements are correlated (Bishop, 1996; Kawabata and Niwa, 1989; Postle and Dhingra, 1989). Later research carried out by Zimniewska et al. (2004) investigated the physiological aspects of natural vs. man-made fibres.

15.5 Chemical properties

Chemical analysis of bast fibres using traditional 'wet' methods are mostly based on work undertaken on flax by the Linen Industries Research Association (LIRA) and, later, the Lambeg Industrial Research Association, from 1919 until it closed in 1993. A special collection of LIRA publications can be viewed at the Irish Linen Centre at Lisburn museum. Chemical testing of flax and hemp is undertaken by CELC (Confédération Européenne du Lin et du Chanvre), and by members of the European Cooperative Research Network on Flax and other Bast Plants. A comprehensive review of all chemical methods of testing of natural fibres is clearly beyond the scope of this chapter. However, key sources on chemical testing, analysis and properties are listed at the end of this chapter.

15.5.1 Chemical extraction of bast fibre

To evaluate various quality parameters of bast fibres, it is often necessary to extract these fibres from the plant as cleanly as possible. Many methods have been employed by previous workers to release the fibres from bast plants. Pott (2003) suggested that treatment with water at 160–180°C would separate flax fibre from straw by removal of pectin, hemicellulose and lignin similar to the steam explosion process (Nebel, 1995). Van den Oever et al. (2003) developed an autoclave process for small-scale fibre extraction. Other researchers have investigated the use of chemicals such as sodium chlorite and sodium hydroxide (Cook, 1984) as well as pectolytic enzymes (Akin et al., 2007; Militky et al., 2000). Enzyme systems used in conjunction with emulsifiers and ultrasound have also been studied (Sirghie and Van Langenhove, 2004). However, on a laboratory scale, the use of sodium hydroxide for fibre extraction has been found to produce the cleanest fibres and can be used for flax, hemp, ramie and nettle fibre

(McCormick, 2006). The laboratory method described here can be applied to either decorticated fibre or plant stems where decortication has not been carried out or is impractical. The fundamental principle is that a known mass of dry sample is heated in sealed stainless steel beakers containing 0.5 M NaOH and 0.025 M EDTA solution at 130°C for a time dependent on the plant material. After neutralizing with dilute acetic acid and rinsed with dilute ammonium hydroxide (1 mL.880 ammonia/l) and finally rinsing with water, the sample is dried, reweighed and the percentage fibre content calculated. Figs. 15.15(a) and 15.16(a) illustrate SEM images of flax tow and hemp tow fibre respectively showing the fibre bundles before separation. Figs 15.15(b) and 15.16(b) illustrate SEM images of flax and hemp fibre respectively separated to their ultimates by chemical extraction.When using a pressurized fibre-processing kier, the rate of sodium hydroxide extraction of bast fibres can be monitored by measuring alkali use by titration and dry solid content using a microwave moisture analyzer on liquor samples taken at intervals during the processing.

15.5.2 Alkali solubility

The degree of retting is of general importance for bast fibres and is generally associated with removal of pectins and other noncellulose materials. Determination of alkali-soluble materials is a useful test when following removal from the parent plant bast fibres will contain varying amounts of noncellulosic materials, typically pectins and hemicelluloses. Some materials are soluble in boiling water and further amounts are removed in normal wet processes. Alkali extraction can be used to gauge the effectiveness of a particular fibre processing method and when it is important to know how much soluble material is present. The method described would typically be used for bast fibres such as flax, hemp, ramie and nettle. Samples of fibre are boiled in 0.5 M Na_2CO_3 solution for 2.5 h and the loss of mass, compared with the loss when boiled in deionized water, is the 'alkali solubility'. The alkali solubility is calculated as follows:

$$\text{Alkali solubility}(\%) = \frac{100 \times (a - b)}{a} - \frac{100 \times (c - d)}{c}$$

where a = initial mass of fibre, b = mass of fibre after extraction with 0.5 M Na_2CO_3, c = initial mass fibre to be extracted with boiling deionized water and d = final mass of deionized water specimen.

15.5.3 Cellulose content (cuprammonium fluidity)

Natural retting and the use of chemicals and enzymes for fibre extraction and processing are likely to cause some degradation to the cellulose molecules by hydrolysis or oxidation. This may be manifest in an undesirable decrease in fibre strength. It is a well-known fact that pure cellulose dissolves in cuprammonium solution to give a viscous fluid and that any reduction in strength is accompanied by a nonlinear fall in the viscosity of the solution. If, however, the reciprocal of viscosity, which is termed

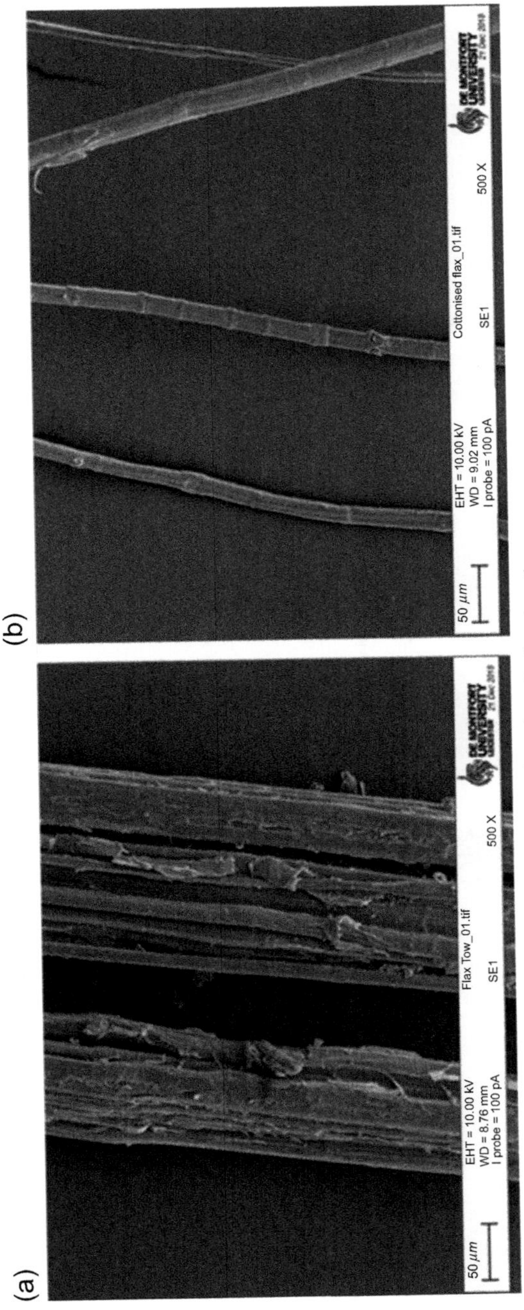

Figure 15.15 SEM images (a) flax tow fibre; (b) flax fibre after chemical extraction.

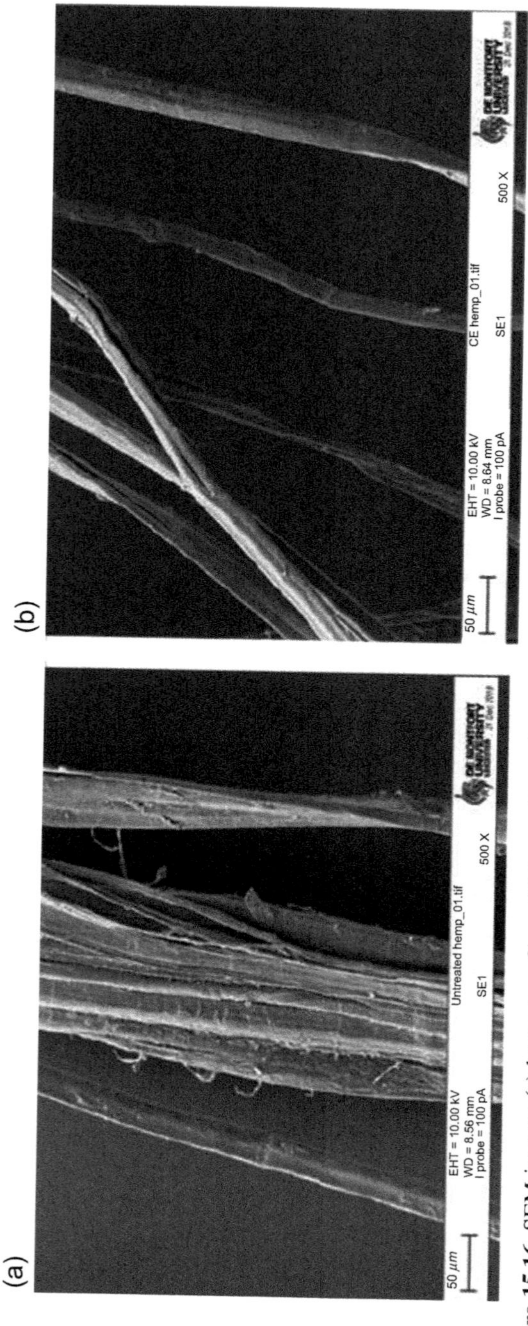

Figure 15.16 SEM images (a) hemp tow fibre; (b) hemp fibre after chemical extraction.

'cuprammonium fluidity', is used, this increases linearly with the percentage loss in strength over a wide range.

This method (BS 2610:1978a) has been widely used for cotton and regenerated cellulosic fibres but is more difficult to use for other natural cellulosic fibres because of the high, and variable, concentrations of noncellulosic components. However, the method has been adapted for flax by pretreating to remove the noncellulose materials (BS 3090:1978) and has been used on some other fibres.

15.5.4 Lignin content

Lignin is a complex 3D branched copolymer comprising, primarily, a mixture of three aromatic alcohols in a network of phenylpropene basic units. Lignin is a very diverse polymer and is hard to study which accounts for the large number of different analytical methods used. Lignin is responsible for strength and rigidity in plants; it also controls diffusion processes and provides protection from microbial attack. The quantitative estimation of lignin is very dependent on the method used; furthermore, the distribution of lignin within plant tissues is localized and is mainly associated with cell walls. The nature of the lignin present in plants also varies with growth stage; the degree of polymerization and cross-linking tends to increase with maturity (Blackburn, 2006). Lignin is important as it affects many aspects of fibre production, processing and properties. The situation is very complex. Lignin is important during plant growth, e.g., it influences the recovery of flax from lodging (Sharma, 1986); mature plants are more difficult to ret than crops harvested earlier and so lignin has a negative effect on fibre extraction. The presence of lignin has a negative effect on the quality and the commercial value of flax fibre (Sharma, 1986). As a result, investigation of the concentration, distribution and properties of lignin can be important at all stages of production and processing.

There are various methods for the determination of lignin. The choice of method will depend on whether quantitative or qualitative information is being sought and the type of plant material being investigated. Chemical methods include the Klason method (Theander and Westerlund, 1986) and the acetyl bromide assay (Hatfield et al., 1999). Instrumental methods include infrared spectroscopy (Himmelsbach et al., 1999), nuclear magnetic spectroscopy (Ralph et al., 2001), gas chromatography (Sonada, 2001) and Raman spectroscopy. With over 40 different methods to be found in recent literature, it is easy to see why the International Lignin Institute (ILI) has included lignin characterization and analysis their activities (ILI, 2011).

The distribution of lignin in plant tissues is most easily appreciated by simple staining of cross-sections. Fig. 15.11 shows an example of the sort of variation that is found when sections are stained with phloroglucinol and clearly demonstrates that the concentration of lignin in plant tissues varies dramatically across and along plant stems. Microspectroscopic methods, particularly infrared and Raman spectroscopy, have been successfully used to explore the distribution and nature of lignin and other components in plant cross-sections (e.g., Himmelsbach et al., 1999; Stewart et al., 1995).

15.5.4.1 Klason and acetyl bromide determination of lignin

The Klason method is described here because it has been extensively used for measurement of lignin concentrations in bast fibres. It is based on the observation that carbohydrates can be hydrolyzed by sulphuric acid to leave an acid-insoluble lignin residue (acid-insoluble lignin) which can then be determined gravimetrically. The procedure is carried out using a two-step hydrolysis of the carbohydrate components in which a sample of homogenized plant material is treated with a 72% solution of sulphuric acid at room temperature, and then a secondary hydrolysis is performed by diluting the mixture and refluxing (Browning, 1967). Owing to this harsh chemical treatment, the solid residue remaining is unlikely to be representative of the lignin as it occurs naturally in the plant but attempts have been made to characterize Klason lignin (KL) (Hatfield and Fukushima, 2005; Reeves, 1993; Reeves and Galletti, 1993).

The main criticism of the Klason method is that the lignin residue can be contaminated by nonlignin components. Inclusion of a preextraction step using an isotropic mixture of ethanol and benzene will remove waxes and cutaneous matter (Monties, 1989). Other sources of contamination are degradation products from proteins and carbohydrates. A preextraction step with 'acid detergent' to remove proteins results in a lower estimate for lignin concentration – acid detergent lignin (ADL). Hatfield et al. (1994) compared the insoluble residues produced by the KL and the ADL methods and concluded that KL is a better estimate of the total lignin concentration in forages than ADL. Other workers have made similar observations (Fukushima and Hatfield, 2004; Jung et al., 1999).

The quantitative determination of acid-insoluble lignin by either the KL or ADL method is commonly used in lignin chemistry, but its use for bast fibres is questionable. Many researchers have found the results to be unreliable and not reproducible for samples with low lignin content (Day, 2005); flax fibre falls into this category. Love et al. (1994) found the KL content of flax fibre to be $3.2 \pm 0.6\%$ which they believed overestimated the true lignin content of 0.9% as determined by nuclear magnetic resonance (NMR) spectroscopy.

The estimated concentrations of lignin in natural cellulosic fibres reported in the literature vary widely within each fibre and between fibres.

The acetyl bromide assay of lignin is widely used as a rapid and sensitive method for determining lignin concentration in small sample sizes (Johnson et al., 1961) and the sensitivity of the method makes it suitable for analyzing materials in which the lignin concentration is low, e.g., flax fibres. Acetyl bromide acetylates the free hydroxyl groups on the alkyl side chains and substitutes a bromine atom on the alpha carbon (Hatfield and Fukushima, 2005). The absorbance of the resulting solution is measured at 280 nm and the concentration of lignin assessed from a calibration curve using a suitably prepared lignin standard (Fukushima and Hatfield, 2001). Results from acetyl bromide assays are higher than obtained from the Klason or the ADL methods.

The wood pulp industry has a well-established method for measuring the bleaching required for wood pulp. For pulp bleaching, the severity needed is directly related to the lignin content of the pulp. The method provides a 'Kappa' number which is approximately proportional to the lignin content. Kappa number is determined by

ISO 302:2004 and is based on the back-titration of an excess of potassium permanganate used to bleach pulped material; the results are expressed as the Kappa number in the range of 1–100. Kappa number has been used for investigating lignin concentrations in bast plant materials used in papermaking (e.g., Danielewicz and Surma-Slusarska, 2010) and in their work on bast fibres Fischer and Müssig (2010) have used the Kappa number method extensively.

A useful summary table which illustrates the lignin contents measured for a wide range of natural cellulosic fibres has been produced by Müssig et al. (2010). No one method is ideal for all cases, and the limitations of each method have to be borne in mind when results are interpreted. A succinct review of methods for chemical analysis is included in the work of Blackburn (2006).

15.6 Instrumental methods

Standard laboratory instrumental methods are used for the determination of a host of fibre characteristics, including study of the chemical composition of fibres, their surface morphology, thermal and flammability behaviour, among many others. Such methods enable fibre quality and end-use behaviour to be assessed. This section provides examples of the methods used by various workers and serves as a source of information for further reading.

15.6.1 *Differential scanning calorimetry*

Differential scanning calorimetry (DSC) is a method used to study the thermal transitions that occur in materials as they are heated. The principle is based on measuring the energy required to steadily heat a sample at a steady rate (or emitted during cooling at a steady rate) relative to a reference material. A sample of the material under study is placed in a cell and an empty cell serves as a reference. As each cell is heated, the difference in heat input for the 2 cells is measured as the temperature increases (or the difference in heat emitted during cooling is measured). In this way, the heat capacity of the material can be calculated:

$$\text{Heat flow} = \frac{q}{t}$$
$$\text{Heating rate} = \frac{\Delta T}{t}$$
$$\text{Heat capacity} = \frac{(q/t)}{(\Delta T/t)} = \frac{q}{\Delta T} = Cp$$

Fibres are physically complex and contain both crystalline and amorphous regions. Heating such a sample will cause a glass transition to occur in the amorphous region which is the reversible transition from the brittle state into a molten state. The temperature at which this occurs (Tg, the glass transition temperature) can be observed from the DSC curve (Fig. 15.17).

When the temperature rises to above the glass transition temperature, a polymer will become more mobile until it reaches the optimal temperature for crystallization to take place. The molecules will arrange themselves in an extremely ordered fashion, and the

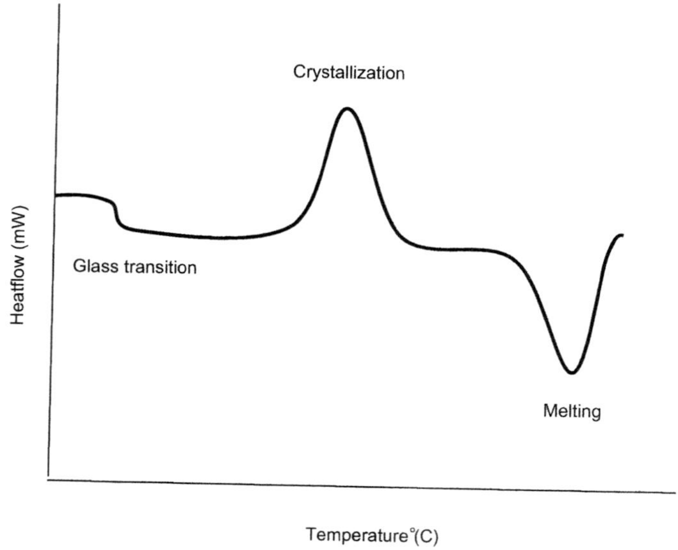

Figure 15.17 Features of a DSC curve. *DSC*, Differential scanning calorimetry.

sample will give off heat. This is observed as a reduction in heat flow. The area under the peak represents the latent heat of crystallization. Application of more heat will cause the crystals to fall apart and the material to melt.

DSC is therefore a method whereby significant information relating to the crystalline and amorphous components of a polymer can be gained.

There are many examples of the use of DSC in the study of natural fibres. DSC normally forms part of a suite of analytical methods and has proved especially useful for studying natural fibre composites. Again there are many examples of such work but the work of Sgriccia and Hawley (2007) where DSC is in the characterization of such composites is typical. Other illustrations are those of Islam et al. (2010) and Baiardoet al. (2004). Moran et al. (2008) utilized the technique in the extraction of cellulose from sisal fibre for use as a nanofiller for biodegradable matrices.

15.6.2 Thermogravimetric analysis (TGA)

For the evaluation of the thermal and flammability behaviour of natural fibres and products made therefrom, a cone calorimeter may be used. This enables testing of small samples of various materials in the condensed phase to be studied. Ignition time, mass loss rate, combustion products, heat release rate and other parameters associated with burning properties can be determined. This method has been widely used for example in the study of such properties of flax/polypropylene composites, e.g., Schartel et al. (2003); Manfredi et al. (2006). Many workers have found thermogravimetric analysis (TGA) as a useful tool for the study of thermal degradation behaviour of natural fibres. The principle of operation of the technique is the measurement of mass change with temperature. For example, compositional and structural information, as well as thermal degradation behaviour, of chemically treated hemp cellulose was studied by Ouajai and Shanks (2005), Titok et al. (2006) describe a thermogravimetric analysis of flax fibre bundles. Another example of how TGA may be used is a study of the pyrolysis products of flame retardant cotton fabrics carried out by Zhu et al. (2004). Majdanac and Teodorovic (1987) reported on the kinetics of the thermal decomposition of cotton cellulose. TGA enabled the thermal degradation and fire resistance of natural fibre composites to be investigated (Manfredi et al., 2006). The cationic modification and dyeing of ramie fibre is another area where TGA has been employed (Liu et al., 2007).

15.6.3 Spectroscopic methods

Spectroscopy in various forms has become a routine tool for investigating the structures and properties of natural fibres. The general principles and techniques are described in a wide range of sources. All forms of spectroscopy are based on the principle of measuring the quantity of energy absorbed or emitted when a sample is exposed to electromagnetic radiation or high-energy particles. Common forms of spectroscopy include the usual ultraviolet (UV), visible, infrared (IR), Raman and fluorescence methods and their Fourier transform (FT) variants; flame ionization methods (for metal analyses) including atomic adsorption, atomic emission, atomic fluorescence and later variants such as plasma emission; and onwards into NMR, mass spectrometry (MS), matrix-assisted laser-de-sorption ionization–time of flight (MALDI-TOF) MS and X-ray spectroscopy.

The literature available on the use of spectroscopic methods is very extensive and cannot possibly be adequately reviewed here. For this reason, the following is a very brief selection which illustrates some of the applications of these methods to natural fibres during recent years, and will provide a starting point for further literature research.

Work by Lang et al. (1986), Howell and Davies (1991), Edwards et al. (1997) and Garside and Wyeth (2003) are examples of investigations into the identification of natural fibres using IR transmission and reflectance techniques, FT-IR and Raman spectroscopy for the identification of a wide range of natural fibres. Meijer et al. (1995) used visible spectroscopy for the determination of pectin contents of flax.

A paper by Akin et al. (1996) describes the analysis of fibre and core fractions of flax by gas–liquid chromatographic methods, ^{13}C CPMAS NMR spectrometry, electron microscopy and UV absorption microspectrophotometry. Results give an insight into the structure and composition of the cell walls in relation to quality and end uses. Mwaikambo and Ansell (1999, 2002) used wide-angle X-ray diffraction, DSC, FT-IR and SEM methods to investigate the chemical and surface morphology of chemically treated hemp, sisal, jute and kapok fibres for improvement of fibre to resin bonding in natural composite materials.

It is known that the degree and method of retting flax stems will have a direct effect on the fibre quality. Morrison et al. (2000) evaluated the composition of dew-retted and water-retted flax fibres by chemical and mass spectral analyses to determine their chemical differences. Phenolics, waxes, cutin and carbohydrates were determined by gas liquid chromatography. Pyrolysis mass spectrometric analysis differentiated water- and dew-retted fibres. A later study by Himmelsbach et al. (2002) investigated changes in a range of chemical components of flax during retting using FT-IR microspectroscopic analysis of cross-sections. Other workers (Mooney et al., 2001) used high-performance anion-exchange chromatography and MALDI-TOF MS to study the sugars and oligomers extracted from flax during chemical and enzymatic retting processes.

Fibre surface properties and their interface with resin matrices are clearly important for the natural fibres composites industry. Modifying the fibre surface can improve hydrophilicity, wetting and adhesion properties. Singh et al. (2000) used diffuse reflectance FT-IR and other techniques to study the surfaces of modified sisal.

Sharma and Reinard (2004) explored the potential of visible and near-IR spectroscopy as a tool for the rapid quality assessment of flax fibres during the spinning process. They successfully focused on assessment of the quality of input fibre and changes in fineness at all stages from the doubling (blending) through to the roving stage. It was concluded that assessment using visible and near IR methods could allow fineness and spinning quality of flax to be predicted.

15.6.4 DNA analysis

DNA analysis has been considered to identify fibre type. ISO 18074:2015e specifies a testing method for DNA analysis to identify animal fibres such as wool, cashmere and yak. The standard states that identification of fibres by his method may be difficult if the fibres were processed under severe conditions or high temperatures especially for textile products, due to damage to the mitochondrial DNA. Dunbar and Murphy (2009) considered if rope fibres contained cells from the plants of origin, DNA analysis could be used to identify the fibre used in a rope and was able to differentiate by this method five different fibre types used in ropes, namely hemp, sisal, abaca, flax and jute. Applied DNA sciences (ADNAS) has developed a DNA testing technique to differentiate Pima cotton (*Gossypium barbadense*) from upland cotton (*Gossypium hirsutum*).

15.7 Future trends

The nature of international trade and the trend towards high levels of process automation are causing fibre quality to be described by results of objective testing methods. The grading of quality against specifications has been well established for many years for cotton and wool fibres but for most other fibres the production volume has been too low and qualities too varied for grading systems to be established.

The move away from qualities being defined relative to reference samples is relatively slow and very dependent on the fibre type and the end uses. The increased use of natural fibres in composites is an example of an end use which is influencing moves towards improved specifications; as the applications become more refined, raw material input to meet performance or processing criteria is becoming more commonplace.

The USA has led the trend for standardization of test methods with the work on flax and there is a growing trend for suppliers to provide test data, albeit restricted in diversity at present, alongside physical samples. It can be expected that measurements of fibre parameters such as fineness, length, strength, trash content, colour and useful fibre will be expected by consumers of a wider range of fibres in the near future.

In the research areas, many of the established methods for evaluating fibre characteristics and properties were very labour-intensive. For the major fibres, cotton and wool-type fibres, many test methods have become highly automated. For most natural cellulosic fibres, there has been little advance in automating methods for the evaluation of physical properties except for fibre diameter and length.

15.8 Sources of further information and advice

For further information and advice the reader is directed to the following publications:

Advances in Wool Technology (2009). Ed. N.A.G. Johnson and I.M. Russell. Cambridge: Woodhead Publishing.

ASTM D-6798-02 Standard Terminology Relating to Flax and Linen (2018). West Conshohocken, PA: ASTM. Basic Principles of Textile Coloration (2001). A. Broadbent. Bradford: Society of Dyers and Colourists. Bast and Other Plant Fibres (2005). R.R. Franck. Cambridge: Woodhead Publishing. Handbook of Natural Fibres. Vol. 1: Types, Properties and Factors Affecting Breeding and Cultivation (2012). Ed. R. Kozłowski. Cambridge: Woodhead Publishing.

Colour for Textiles: a User's Handbook (1993). W. Ingamells. Bradford: Society of Dyers and Colourists. Handbook of Natural Fibres. Vol. 2: Processing and Applications (2012). Ed. R. Kozłowski. Cambridge: Woodhead Publishing.

Industrial Applications of Natural Fibres, Structure, Properties and Technical Applications (2010). Ed. Jorg Mussig. Chichester: John Wiley.

Industrial Crops and Uses (2010). Ed. B.P. Singh. Oxford: CABI

Natural Fibres, Biopolymers, and Biocomposites (2005). Ed. A. K. Mohanty, M. Misra and L. T. Drzal. London: Taylor & Francis.

Physical Properties of Textile Fibres, fourth ed. (2008).W. E. Morton and J. W. S. Hearle. Cambridge: Woodhead Publishing.

Physical Testing of Textiles (1999). B. P. Saville. Cambridge, Woodhead Publishing. Principles of Textile Testing, third ed. (1968). J. E. Booth. London: Butterworths.

Textiles for Sustainable Development (2007). Ed. R. Ananadjiwala, L. Hunter, R. Kozłowski and G. Zaikov. New York: Nova Science Publishers.

Textile Laboratory Manual: Additional Methods, third ed. (1967). W. Garner. London: Heywood Books.

Textile Science: An Introductory Manual, fourth (revised) impression (1958). J.T. Marsh. London: Chapman & Hall.

Appendix: Abbreviations

AA	Atomic absorption spectroscopy
AE	Atomic emission spectroscopy
AF	Atomic fluorescence spectroscopy
ASTM	American Society for Testing and Materials
BSI	British Standards Institution
CIE	Commission Internationale de l'Eclairage
CPMAS	Cross-polarization/magic angle spinning
DIN	German Institute for Standards
DSC	Differential scanning calorimetry
EDS	Energy dispersive spectroscopy
EN	European standards
FT	Fourier transform
FT-IR	Fourier transform infrared spectroscopy
HPAEC	High-performance anion exchange chromatography
HVI	High volume instrument
IR	Infrared spectroscopy
ISO	International Organization for Standardization
LIRA	Linen Industries Research Association
MALDI-TOF	Matrix-assisted laser-de-sorption ionization—time of flight
MDTA	Microfibre dust and trash analysis
MS	Mass spectroscopy
NMR	Nuclear magnetic resonance spectroscopy
OFDA	Optical fibre diameter analyzer
SEM	Scanning electron microscopy
TEM	Transmission electron microscopy
TGA	Thermogravimetric analysis
UV	Ultraviolet spectroscopy
VIS	Visible spectroscopy

References

Akin, D.E., 2010. 'Flax − ASTM standardisation and harmonisation'. In: Mussig, J. (Ed.), Industrial Applications of Natural Fibres: Structure, Properties and Technical Applications. John Wiley, Chichester, pp. 371−380.

Akin, D.E., Gamble, G.R., Morrison, W.H., Rigsby, L.L., Dodd, R.B., 1996. Chemical and structural analysis of fibre and core tissue from flax. J. Sci. Food Agric. 72, 155−165.

Akin, D.E., Foulk, J.A., Dodd, R.B., McAlister III, D.D., 2001. Enzyme-retting of flax and characterization of processed fibers. J. Biotechnol. 89, 193−203.

Akin, D.E., Condon, B., Sohn, M., Foulk, J.A., Dodd, R.B., Rigsby, L.L., 2007. Optimisation for enzyme retting of flax with pectate lyase. Ind. Crops Prod. 25, 136−146.

Anderson, S.L., 1954. The airflow method for measuring wool fibre fineness. J. Text. Inst. 45, 312−316.

ASTM, 2014. ASTM D-3822/D-3822M-14 Standard Test Method for Tensile Properties of Single Textile Fibers. ASTM, West Conshohocken, PA.

ASTM, 2015a. ASTM D-6961/D6961 M-09 (2015) Standard Test Method for Color Measurement of Flax Fibre. ASTM, West Conshohocken, PA.
ASTM, 2015b. ASTM D-7025-09(2015)e1 Standard Test Method for Assessing Clean Flax Fibre Fineness. ASTM, West Conshohocken, PA.
ASTM, 2015c. ASTM D-7076-10(2015)e1 Standard Test Method for the Measurement of Shives in Retted Flax. ASTM, West Conshohocken, PA.
Baiardo, M., Zini, E., Scandola, M., 2004. 'Flax fibre—polyester composites'. Compos. Appl. Sci. Manuf. 35 (6), 703—710.
Bishop, D.P., 1996. Fabric sensory and mechanical properties. Text. Prog. 26 (3), 1—62.
Blackburn, B.M., 2006. 'Investigating Changes in the Lignin of Flax (*Linum usitatissimum*) during Maturity', PhD Thesis. De Montfort University.
Booth, J.E., 1968. Principles of Textile Testing. Butterworths, London.
Brown, A.E., 1984. *Epicoccum nigrum*, a primary saprophyte involved in the retting of flax. Trans. Br. Mycol. Soc. 83 (1), 29—35.
Browning, B.L., 1967. Methods of Wood Chemistry. John Wiley, New York.
BSI, 1978a. In: Method of Test for the Determination of the Cuprammonium Fluidity of Cotton and Certain Cellulosic Man-Made Fibres, vol. 2610. BSI, BS, London, p. 1978.
BSI, 1987a. In: Methods for Determination of Cotton Fibre Properties by the Airflow Method: Determination of Micronaire Value by the Single Compression Airflow Method. BSI, London, 3181-1:1987.
Cook, J.G., 1984. Handbook of Textile Fibres I: Natural Fibres, fifth ed. Merrow Publishing, Darlington.
Danielewicz, D., Surma-Slusarska, B., 2010. Processing of industrial hemp into papermaking pulps intended for bleaching. Fibres Text. East. Eur. 18 (6), 110—115. Datacolor Color Tools QC User's Guide and Reference version 1.2.1.
Day, A., Ruel, K., Neutelings, G., Crônier, D., David, H., Hawkins, S., Chabbert, B., 2005. Lignification in the flax stem: evidence for an unusual lignin in bast fibres. Planta 222 (2), 234—245.
Dérue, C., Gibouin, D., Verdus, M.C., Lefebvre, F., Darty, M., Ripoll, C., Thellier, M., 2002. Appraisal of SIMS applicability to boron studies in plants. Microsc. Res. Tech. 58, 104—110.
Donaldson, L., Hague, J., Snell, R., 2001. Lignin distribution in coppice poplar, linseed and wheat straw. Holzforschung 55, 379—385.
Drieling, A.R., Baumer, J., Mussig, J., Harig, H., 1999. Testing Strength, Fineness and Length of Bast Fibres, vol. 42. Technische Textilien, pp. 261—262.
Dunbar, M., Murphy, T.M., 2009. DNA analysis of natural fiber rope. J. Forensic Sci. 54 (1), 108—113.
Edwards, H.G.M., Farwell, D.W., Webster, D., 1997. FT Raman microscopy of untreated natural plant fibres. Spectrochim. Acta A Mol. Biomol. Spectrosc. 53 (13), 2383—2390.
Faughey, G.J., Sharma, S.S., McCall, R.D., 2000. Determining fibre fineness in flax using derivative thermogravimetric analysis, scanning electron microscopy, and airflow methods. J. Appl. Polym. Sci. 75, 508—514.
Fila, G., Manici, L.M., Caputo, F., 2001. *In vitro* evaluation of dew-retted flax by fungi. Ann. Appl. Biol. 138 (3), 343—351.
Fischer, S., Schenzel, K., Fischer, K., Diepenbrock, W., 2005. Applications of FT Raman spectroscopy and micro spectroscopy characterising cellulose and cellulosic biomaterials. Macromol. Symp. Spec. Issue Cellul. Cellul. Deriv. 223 (1), 41—56.

Fraser, T.W., Courtney, A.D., Harvey, B.M.R., 1982. Preharvest retting of flax: a light microscope study of the effects of glyphosate treatment on the maturation of stem tissue. Ann. Appl. Biol. 101, 533–537.

Fukushima, R.S., Hatfield, R.D., 2001. Extraction and isolation of lignin for utilisation as a standard to determine lignin concentration using the acetyl bromide spectrophotometric method. J. Agric. Food Chem. 49 (7), 3133–3139.

Fukushima, R.S., Hatfield, R.D., 2004. Comparison of the acetyl bromide spectrophotometric method with other analytical lignin methods for determining lignin concentration in forage samples. J. Agric. Food Chem. 52, 3713–3720.

Garside, P., Wyeth, P., 2003. Identification of cellulosic fibres by FTIR spectroscopy: thread and single fibre analysis by attenuated total reflectance. Stud. Conserv. 48 (4), 269–275.

Greaves, P.H., Saville, B.P., 1995. Microscopy in Textile Fibres. Oxford University Press in association with The Royal Microscopical Society, Oxford.

Grishanov, S.A., Harwood, R.J., Booth, I., 2006. A method of estimating the single flax fibre fineness using data from the Laserscan system. Ind. Crops Prod. 23 (3), 273–287.

Harwood, J., McCormick, P., Waldron, D., Bonadei, R., 2008a. Evaluation of flax accessions for high value textile end uses. Ind. Crops Prod. 27 (1), 22–29.

Harwood, R., Nusenbaum, V., Harwood, J., 2008b. 'Cottonisation of Flax', Proceedings of International Conference on Flax & Other Bast Plants, Saskatoon. Canada, 21–23 July.

Hatfield, R., Fukushima, R.S., 2005. Can lignin be accurately measured? Crop Sci. 45, 832–839.

Hatfield, R.D., Jung, H.-J.G., Ralph, J., Buxton, D.R., Weimer, P.J., 1994. A comparison of the insoluble residues produced by the Klason lignin abd acid detergentprocedures. J. Sci. Food Agric. 65 (1), 51–58.

Hatfield, R.D., Grabber, J., Ralph, J., Brei, K., 1999. Using the acetyl bromide assay to determine lignin concentrations in herbaceous plants: some cautionary notes. J. Agric. Food Chem. 47 (2), 628–632.

Henriksson, G., Akin, D.E., Rigsby, L.L., Patel, N., Eriksson, K.E.L., 1997. Influence of chelating agents and mechanical pretreatment on enzymatic retting of flax. Text. Res. J. 67 (11), 829–836.

Himmelsbach, D.S., Khahili, S., Akin, D., 1999. Near-infrared-Fourier-transform-Raman microspectroscopic imaging of flax stems. Vib. Spectrosc. 19, 361–367.

Himmelsbach, D.S., Khahili, S., Akin, D., 2002. The use of FT-IR microscopic mapping to study the effects of enzymatic retting of flax (*Linum usitatissimum* L.) stems. J. Sci. Food Agric. 82, 685–696.

Horne, M., Harwood, R., McCormick, P., Harwood, J., 2008. The commercial production of short-fibre flax for cottonisation. In: Proceedings of International Conference on Flax & Other Bast Plants, Saskatoon, Canada, pp. 129–136, 21–23 July.

Horne, M.R.L., Waldron, D., Harwood, J., Harwood, R.J., 2010. The production and extraction of flax fibre for textile fibres. J. Biobased Mater. Bioenerg. 4 (2), 98–105.

Howell, H.E., Davis, J.R., 1991. Qualitative identification of fibers using NIR spectroscopy. Text. Chem. Colorist 23 (9), 69–73.

HS 52 Cotton, Including Yarns and Woven Fabrics Thereof. US Harmonized System Codes.

International Lignin Institute, 2011. www.ili-lignin.com/.

Islam, M.S., Pickering, K.L., Foreman, N.J., 2010. Influence of accelerated ageing on the physico-mechanical properties of alkali-treated industrial hemp fibre reinforced poly(lactic acid) (PLA) composites. Polym. Degrad. Stab. 95 (1), 59–65.

Johnson, D.B., Moore, W.E., Zank, L.C., 1961. The spectrophotometric determination of lignin in small wood samples. TAPPI 44 (11), 793–798.

Jung, H.-J.G., Varel, V.H., Weimer, P.J., Ralph, J., 1999. Accuracy of Klason lignin and acid detergent lignin methods as assessed by bomb calorimetry. J. Agric. Food Chem. 47, 2005–2008.

Kawabata, S., Niwa, M., 1989. Fabric performance in clothing and clothing manufacture. J. Text. Inst. 80 (1), 19–50.

Lamb, P.R., Denning, R.J., 2004. Flax: Cottonised Fibre from Linseed Stalks. Report for the Australian Government, Rural Industries Research and Development Corporation. Publication No. 03/123.

Lang, P.L., Katon, J.E., O'Keefe, J.F., Schiering, D.W., 1986. The identification of fibers by infrared and Raman microspectroscopy. Microchem. J. 34, 319–331.

Love, G.D., Snape, C.E., Jarvis, M.C., Morrison, I.M., 1994. Determination of phenolic structures in flax fibre by solid-state 13C NMR. Phytochemistry 35 (2), 489–491.

Majdanac, L.D., Teodorovic, M.J., 1987. The influence of supramolecular structure on the kinetics of thermal decomposition of cellulose. Acta Polym. 38 (12), 661–666.

Manfredi, L.B., Rodriguez, E.S., Wladyka-Przybylak, M., Vazquez, A., 2006. Thermal degradation of fire resistance of unsaturated polyester, modified acrylic resins and their composites with natural fibres. Polym. Degrad. Stab. 91 (2), 255–261.

McCormick, P., 2006. 'Development of a Method of Extraction of Flax Fibre Suitable for Laboratory Testing', MPhil Dissertation. De Montfort University.

McDonald, R., Rigg, B., 1980. 'Publication sponsored by the society's colour measurement committee-XVI'. J. Soc. Dye. Colour. 96, 587–589.

Meijer, W.J.M., Vertregt, N., Rutgers, B., van de Waart, M., 1995. The pectin content: a measure of the retting and rettability of flax. Ind. Crops Prod. 4, 273–284.

Miao, M., Finn, N., 2008. Conversion of natural fibres into structural composites. J. Text. Eng. 54 (6), 165–177.

Militky, J., Bazjik, V., Kremenakova, D., 2000. Selected Properties of Cottonised Flax, 461 17. Textile Faculty of Liberec, Liberec, Czech Republic.

Monties, B., 1989. Lignins. In: Harborne, J.B. (Ed.), Methods in Plant Biochemistry. Academic Press, New York, pp. 113–157.

Mooney, C., Stolle-Smits, T., Schols, H., de Jong, E., 2001. Analysis of retted and non retted flax fibres by chemical and enzymatic means. J. Biotechnol. 89, 205–216.

Moran, J.I., Alvarez, V.A., Cyras, V.P., Vazquez, A., 2008. Extraction of cellulose and preparation of nanocellulose from sisal fibers. Cellulose 15, 149–159.

Morrison III, W.H., Archibald, D.D., Sharma, H.S.S., Akin, D.E., 2000. Chemical and physical characterisation of water- and dew-retted flax fibers. Ind. Crops Prod. 12, 39–46.

Müssig, J., Amaducci, S., 2018. Scanner based image analysis to characterize the influence of agronomic factors on hemp (*Cannabis sativa* L.) fibre width. Ind. Crops Prod. 113, 28–37.

Fischer, H., Müssig, J., 2010. Bast fibre processing and uses. In: Singh, B.P. (Ed.), Industrial Crops and Uses. CABI, Oxford, pp. 326–348.

Müssig, J., Fischer, H., Graupner, N., Dreiling, A., 2010. Testing methods for measuring physical and mechanical fibre properties (plant and animal fibres). In: Müssig, J. (Ed.), Industrial Applications of Natural Fibres. John Wiley, Chichester, pp. 267–309.

Mwaikambo, L.Y., Ansell, M.P., 1999. The effect of chemical treatment on the properties of hemp, sisal, jute and kapok for composite reinforcement. Die Angew. Makromotekulare Chem. 272, 108–116.

Mwaikambo, L.Y., Ansell, M.P., 2002. Chemical modification of hemp, sisal, jute, and kapok fibers by alkalization. J. Appl. Polym. Sci. 84, 2222–2234.

Nebel, K.M., 1995. New processing strategies for hemp. J. Int. Hemp Assoc. 2 (1), 6–9.

Ouajai, S., Shanks, R.A., 2005. Composition, structure and thermal degradation of hemp cellulose after chemical treatments. Polym. Degrad. Stab. 89 (2), 327–335.

Postle, R., Dhingra, R.C., 1989. 'Measuring and interpreting low-stress fabric mechanical and surface properties. Part III: optimization of fabric properties for men's suiting materials'. Text. Res. J. 59 (8), 448–459.

Pott, G.T., 2003. 'Reduction of Moisture Sensitivity in Natural Fibres', 2nd International Conference on Eco-Composites, 1–2 September. Queen Mary College, University of London, UK.

Ralph, J., Lapierre, C., Marita, J.M., Kim, H., Lu, F., Hatfield, R.D., Ralph, S., Chapple, C., Franke, R., Hemm, M.R., Van Doorsselaere, J., Sederoff, R.R., O'Malley, D.M., Scott, J.T., MacKay, J.J., Yahiaoui, N., Boudet, A.-M., Pean, M., Pilate, G., Jouanin, L., Boerjan, W., 2001. Elucidation of new structures in lignins of CAD- and COMT-deficient plants by NMR. Phytochemistry 57, 993–1003.

Reeves, J.B., 1993. Infrared spectroscopic studies on forage and by-product fibre fractions and lignin determination residues. Vib. Spectrosc. 5 (3), 303–310.

Reeves, J.B., Galletti, G.C., 1993. Use of pyrolysis-gas chromatography/mass spectrometry in the study of lignin assays. J. Anal. Appl. Pyrolysis 24, 243–255.

Saville, B.P., 1999. Physical Testing of Textiles. Woodhead Publishing, Cambridge.

Schartel, B., Braun, U., Schwarz, U., Reinemann, S., 2003. Fire retardancy of polyproylene/flax blends. Polymer 44 (20), 6241–6250.

Schmid, H.G., Mussig, J., Gerardi, H., 2002. Image scanning for measurement of cotton fibre width. In: ITMF International Committee on Cotton Testing Methods. Proceedings of the General Assembly, Bremen, 12–13 March 2002.

Schmidt, G.P., Palazuelos, M., Schmid, H.G., Powers, K.W., Mussig, J., 2010. Characterisation of Particle-Fibre Mixtures Using Quantitative Image Analysis (Unpublished manuscript).

Sgriccia, N., Hawley, M.C., 2007. Thermal, morphological, and electrical characterisation of microwave processed natural fibre composites. Compos. Sci. Technol. 67 (9), 1986–1991.

Sharma, H.S.S., 1986. Effect of glyphosate treatment on lignification of fibres of some flax cultivars. Ann. Appl. Biol. 108, 114–115.

Sharma, H.S.S., Reinard, N., 2004. Evaluation of visible and near-infrared spectroscopy as a tool for assessing fiber fineness during mechanical preparation of dew-retted flax. Appl. Spectrosc. 58, 1431–1438.

Shen, J., Smith, E., Chizyuka, M., Prajapati, C., 2017. Development of durable shrink-resist coating of wool with sol-gel polymer processing. Fibers Polym. 18, 1769–1779.

Singh, B., Gupta, M., Verma, A., Tyagi, O.S., 2000. FT-IR microscopic studies on coupling agents: treated natural fibres. Polym. Int. 49, 1444–1451.

Sinha, N.G., 1970. The effect of moisture on jute-fibre fineness determined by an air-flow method. J. Text. Inst. 61 (2), 93–95.

Sinha, N.G., Bandyopadhyay, S.B., 1968. An air-flow method for the determination of the fibre fineness of jute and mesta. J. Text. Inst. 59 (3), 148–156.

Sirghie, C., Van Langenhove, L., 2004. 'Ultrasonic – Chemical – Enzymatic Treatment: A New Method for Flax "cottonisation"', COST Action 847, Textile Quality and Biotechnology, WG 2 Bioprocessing of Bast Fibres, Abstracts, Maribor, Slovenia, 26–27 February.

Sirghie, C., Turcu, F.D., Popa, N., 2005. Process for Obtaining Fine Fibres from Flax and Hemp Rags. Patent No: RO119961.

Sommerville, P., 2001–2007. Fundamental Principles of Fibre Fineness Measurement. Australian Wool Testing Authority Ltd Newsletters. Available at: www.awta.com.au/Documents/ResearchPapers/Reviews/Fibre_Fineness_Measurement_Fundamentals.pdf.

Sonada, T., Ona, T., Yokoi, H., Ishida, Y., Ohtani, H., Tsuge, S., 2001. Quantitative analysis of detailed lignin monomer composition by pyrolysis-gas chromatography combined with preliminary acetylation of the samples. Anal. Chem. 73, 5429–5435.

Steedman, H.F., 1960. Section Cutting in Microscopy. Blackwell Scientific, Oxford.

Stewart, D., McDougall, G.J., Baty, A., 1995. Fourier-transform infrared microspectroscopy of anatomically different cells of flax (*Linum usitatissimum*) stems during development. J. Agric. Food Chem. 43, 1853–1858.

Stobart, R.H., McColl, A., 2000. A Comparison of Fiber Measurement Methods to Obtain Average Fiber Diameter of Fibers Other than Wool. IWTO, Commercial Technology Forum Paper CTF04, Christchurch, New Zealand.

Stoves, J.L., 1957. Fibre Microscopy: Its Technique and Application. National Trade Press, London.

Theander, O., Westerlund, E.A., 1986. Studies on dietary fibre. 3: improved procedures for analysis of dietary fibre. J. Agric. Food Chem. 34, 330–336.

Titok, V., Leontiev, V., Shostak, L., Khotyleva, L., 2006. Thermogravimetric analysis of the flax bast fibre bundle. J. Nat. Fibers 3 (1), 35–41.

Van den Oever, M.J.A., Bas, N., Van Soest, L.J.M., Melis, C., Van Dam, J.E.G., 2003. Improved method of fibre content and quantity analysis and their application to flax genetic diversity investigations. Ind. Crops Prod. 18, 231–243.

Wang, H.M., Wang, X., 2004. Evaluation of the fineness of degummed bast fibers. Fibres and Polymers 5 (3), 171–176.

Wool Testing Authority, http://www.wtaeurope.com/services/greasy-and-scoured-wool.php/. accessed September 2019.

Liu, Z.T., Yang, Y., Zhang, L., Liu, Z.W., Xiong, H., 2007. Study on the cationic modification and dyeing of ramie fibre. Cellulose 14 (4), 337–345.

Zhu, P., Sui, S., Wang, B., Sun, K., Sun, G., 2004. A study of pyrolysis and pyrolysis products of flame retardant cotton fabrics by DSC, TGA and Py-GC-MS. J. Anal. Appl. Pyrolysis 71 (2), 645–655.

Zimniewska, M., Kozłowski, R., Rawluk, M., 2004. Natural vs. man-made fibres: physiological viewpoint. J. Nat. Fibers 1 (2), 69–81.

Argyropoulos, D.S., 1998. Advances in Lignocellulosics Characterisation. Tappiu Press, Atlanta: GA.

ASTM, 2018. ASTM D-6798-02(2018) Standard Terminology Relating to Flax and Linen. ASTM, West Conshohocken, PA.

Bonatti, P.M., Ferrari, C., Ficher, B., Grippo, C., Torri, G., Costentino, C., 2004. Histochemical and supramolecular studies in determining quality of hemp fibres for textile applications. Euphytica 140 (3), 243–244.

BSI, 1968. In: Method for the Determination of Wool Fibre Diameter by the Airflow Method, vol. 3183. BSI, London, p. 1968.

BSI, 1975. In: Measurement of the Length of Wool Fibres Processed on the Worsted System, Using a Fibre Diagram Machine, vol. 5182. BSI, BS, London, p. 1975. ISO 2646:1974.

BSI, 1981. Textiles – Cotton Fibres: Evaluation of Maturity – Microscopic Method. BSI, London. BS ISO 4912:1981.

BSI, 1996. Textiles: Fibres – Determination of Breaking Force and Elongation at Break of Individual Fibres. BSI, London. BS EN ISO 5079:1996.

BSI, 1999. Textiles: Sampling of Fibres, Yarns and Fabrics for Testing. BSI, London. BS EN12751:1999.

Faughey, G.J., Sharma, H.S.S., 2000. A preliminary evaluation of near infrared spectroscopy for assessing physical and chemical characteristics of flax fibre. J. Near Infrared Spectrosc. 8 (1), 61–69.

Gamble, G.R., Snook, M.E., Henriksson, G., Akin, D.E., 2000. Phenolic constituents in flax bast tissue and inhibition of cellulase and pectinase. Biotechnol. Lett. 22 (9), 741–746.

Himmelsbach, D.S., Akin, D.E., 1998. Near-infra-red Fourier-transform Raman spectroscopy of flax *(Linum usitatissimum* L.) stems. J. Agric. Food Chem. 46 (3), 991–998.

ISO, 2015. Textiles–Identification of Some Animal Fibres by DNA Analysis Method–Cashmere, Wool, Yak and Their Blends. Geneva. ISO 18074:2015(E).

Jahn, A., Schroder, M.W., Futing, M., Schenzel, K., Diepenbrock, W., 2002. Characterisation of alkali treated flax fibres by means of FT Raman spectroscopy and environmental scanning electron microscopy. Spectrochim. Acta A 58 (10), 2271–2279.

Jonoobi, M., Harun, J., Shakeri, A., Misra, M., 2009. Chemical composition, crystallinity, and thermal degradation of bleached and unbleached kenaf bast (*Hibiscus cannabinus*) pulp and nanofibres. BioResources 4 (2), 626–639.

Kessler, R.W., Quint, B., Kessler, W., Ullmeyer, D., Urgerer, P., 1993. Quality estimation of flax by modern instrumental methods. In: Kozłowski, R. (Ed.), Flax in the World. Institute of Natural Fibres, Rome, pp. 79–92.

Morrison III, W.H., Akin, D.E., Archibald, D.D., Dodd, R.B., Raymer, P.L., 1999. Chemical and instrumental characterisation of maturing kenaf core and bast. Ind. Crops Prod. 10 (1), 21–34.

Peetla, P., Schenzel, K.C., Diepenbrock, W., 2006. Determination of mechanical strength properties of hemp fibres using near-infrared Fourier transform Raman microspectroscopy. Appl. Spectrosc. 60 (6), 682–691.

Roberts, B., Kessler, R.W., 2011. FAO-quality round robin test on flax. Personal Commun. R. Kozłowski.

Sgriccia, N., Hawley, M.C., Misra, M., 2008. Characterisation of natural fiber surfaces and natural fiber composites. Compos. Appl. Sci. Manuf. 39 (10), 1632–1637.

Sharma, H.S.S., Faughey, G.J., 1999. Comparison of subjective and objective methods to assess flax straw cultivars and fibre quality after dew-retting. Ann. Appl. Biol. 135, 495–501.

Sharma, H.S.S., Kernaghan, K., 1988. Thermogravimetric analysis of flax fibres. Thermochim. Acta 132, 101–109.

Steadman, R.G., 1997. Cotton testing. Text. Prog. 27 (1), 1–63.

Van der Velde, K., Kiekens, P., 2002. Thermal degradation of flax: the determination of kinetic parameters with thermogravimetric analysis. J. Appl. Polym. Sci. 83, 2634–2643.

Wang, H.M., Postle, R., Kessler, R.W., Kessler, W., 2003. Removing pectin and lignin during chemical processing of hemp for textile applications. Text. Res. J. 73 (8), 664–669.

Yelle, D.J., Ralph, J., Frihart, C.R., 2008. Characterisation of nonderivatized plant cell walls using high-resolution solution-state NMR spectroscopy. Magn. Reson. Chem. 46 (6), 508–517.

Part Two

Improving natural fibre production through breeding and cultivation

Cotton breeding

16

Mohamed Negm
Cotton Research Institute, Giza, Egypt

16.1 Taxonomy

The word cotton is used to describe cultivated species of the genus '*Gossypium*' (family *Malvaceace*) that has about 50 species of which only four are cultivated on commercial scale: two diploids (2n = 26), *G. arboreum* L. (A2A2) and *G. herbaceum* L. (A1A1), and two allotetraploids (2n = 52), *G. hirsutum* L. (AADD) and *G. barbadense* L. (AADD).

The diploid species are called as old world, or desi cotton (in India and Pakistan), while the allotetraploids species are commonly known as new world cottons, (Wendal et al., 1995). More than 97% of the cotton produced in the world comes from *G. hirsutum* L. and its intraspecific commercial cotton hybrids. Production of diploid species is limited to a few countries in Asia while *G. barbadense* is grown in Africa, Asia and Americans.

Numerous methods have been employed to classify cotton species (*Gossypium* spp.) based on morphology (Fryxell et al., 1992) including meiotic behaviour (Menzel, 1954) and karyotype (Beasley, 1940); genetic and molecular methods have also been used (Wendel, 1995; Iqbal et al., 1997). The diploid species in *Gossypium* are morphologically diverse and are represented genetically by eight different genome types including the A, B, C, D, E, F, G, and K genomes (Endrizzi et al., 1985; Stewart, 1995). Except for the A genome, all the diploid genomes are composed of wild species. These wild species represent a rich genetic resource for cotton improvement (Stewart, 1995).

Kohel and Lewis (1984), reported that the finding of cotton remains from Nubia (Egypt), and Mohenjo-Daro (Pakistan), dating from about 2700 B.C. gave credence to the idea that cottons (*G. arboreum*) were domesticated in these areas which already knew the productive agriculture and the arts of spinning and weaving. The origin of *G. herbaceum* is possibly from Southern Africa and the Sahel of North Africa and was introduced to the Near East in ancient times where it was subsequently domesticated. Later, *G. herbaceum* spread from the Northern Africa to Arabia, Baluchistan and perhaps Sind (Pakistan). For New World species, based on archeological contexts dating from about 2500–1750 B.C. in Ancon-Chillon district of Peru, the centre of diversity for *G. barbadense* is northwestern South America. This district appears to be the site where cottons belonging to *G. barbadense* were domesticated. About

Handbook of Natural Fibres. https://doi.org/10.1016/B978-0-12-818398-4.00018-9

G. hirsutum, archeological remains point to Mexico, however, the oldest evidence, from about 3500–2300 B.C. shows that *G. hirsutum* was domesticated in at least two, or perhaps three, separate areas in the region. Historical evidence indicates that in the early days of human civilization, three principal countries of the east have been distinguished by their distinctive type of fabrics; China as the land of silk, Egypt as the country of flax (and linen), and India of cotton.

Thus, it is India which has always been regarded as the country of cotton growing and manufacturing, while, on the other hand, Egypt has always been regarded as the country of flax - growing and of the linen industry.

16.1.1 Species of Gossypium

There are 43 diploid species with 2n = 26 chromosomes which have been classified into seven genomes from A to G, and seven tetraploid species with 2n = 52 chromosomes with genome designation AD (Table 16.1). Apart from four cultivated species, the remaining 46 are wild species. Currently, India is only country where, all the four cultivated species are grown commercially.

16.2 Morphology

16.2.1 Germination of cotton seed

The cotton growing season varies from 100 to over 190 days depending on climatic conditions and plant species. Seed cotton germination is epigeal and is sown at between 2 and 4 cm depth and rather deeper in dry conditions and shallower when soil is wet. Cotton has a taproot system that grows quickly and may reach a depth of 20–25 cm before the emergence of a seedling above the ground, as shown in Figs. 16.1 and 16.2, (Govila and Rao, 1969; Balock, 2004). The depth to which it penetrates is roughly equal to the height of the stem. For the adult plants, the lateral roots arise freely from the upper thicker part of the taproot and run more or less horizontal and constitute the bulk of the feeding roots when the soil is moist but penetrate much more deeply when the soil is dry (Bird, 1981). Excessive moisture, soil pH and soil hardness (Farbrother, 1972) affect the development of the root system.

16.2.2 Stem and branches

The cotton plant consists of vertical stem carrying two kinds of branches; sympodial or fruiting branches and monopodial or vegetative branches. Vegetative branches and the main stem do not bear flowers directly but produce fruiting branches (Mauney, 1966). The number of monopodial branches may vary from zero to reaching as many as 10. A higher number of monopodial branches shape the plant to a bushy structure which is not desirable for machine picking. No more monopodial branches can be formed once

Table 16.1 Cotton species genomes and distribution.

Species	*Genome*	**Distribution**
DIPLOID (2n = 26)		
1. G. africanum	A	Africa
2. G. herbaceum (cultivated)	A_1	Afghanistan
3. G. arboreum (cultivated)	A_2	Indo-Burma, China and Arab
4. G. anomalum	B_1	Africa
5. G. triphyllum	B_2	Africa
6. G. barbosanum	B_3	Cape Verde
7. G. capitis-viridis	B_4	Cape Verde
8. G. sturtianum	C_1	Australia
9. G. nandewarense	$C_{1\text{-}n}$	Australia
10. G. robinsoni	C_2	Australia
11. G. australe	C_3	Australia
12. G. pilosum	"C"	Australia
13. G. costulatum	C_5	Australia
14. G. populifolium	C_6	Australia
15. C. cunninghamii	C_7	Australia
16. G. pulchellum	C_8	Australia
17. G. nelsonii	C_9	Australia
18. G. enthyle	"C"	Australia
19. G. londonderriense	"C"	Australia
20. G. marchantii	"C"	Australia
21. G. exiguum	"C"	Australia
22. G. rotundifolium	"C"	Australia
23. G. fryxellii	"C"	Australia
24. G. binatum	"C"	Australia
25. G. nobile	"C"	Australia
26. G. thurberi	D_1	America
27. G. armouianum	D2-1	America
28. G. harknessii	D2-2	America

Continued

Table 16.1 Cotton species genomes and distribution.—cont'd

Species	*Genome*	**Distribution**
29. G. klotzschianum	D3-K	America
30. G. davidsonii	D3-d	America
31. G. aridum	D4	America
32. G. raimondii	D5	America
33. G. gossypioides	D6	America
34. G. lobatum	D7	America
35. G. trilobum	D8	America
36. G. laxum	D9	America
37. G. turneri	"D"	America
38. G. stocksii	E1	Arabia
39. G. somalense	E2	Arabia
40. G. areysianum	E3	Arabia
41. G. incanum	E4	Arabia
42. G. longicalyx	F1	Africa
43. G. bickii	G1	Australia
TETRAPLOID (2n = 52)		
44. G. hirsutum (cultivated) *(AD)*	(AD)1	America
45. G. barbadense (cultivated)	(AD)2	America
46. G. tomentosum	(AD)3	Hawai
47. G. lanceolatum	(AD)	America
48. G. mustelinum	(AD)	America
49. G. darwinii	(AD)	America
50. G. caicoense	(AD)	America

Compiled from Gotmare, V., Singh, P., Tule, B.N. http://www.cicr.org.in/pdf/wild_species%20.pdf.

a sympodial branch appears on the main stem. The cotton plant usually produces 16–18 fruiting branches with two to five fruiting nodes on each branch (Jenkins et al., 1990), as shown in Fig. 16.3. Of the total possible fruiting sites, only, small fractions mature as productive bolls. In general, cotton matures bolls on position one more often than on other positions. The bolls closer to the main stem are desirable for better quality. The fruiting branches at nodes nine through 14 produces the bulk of lint and only a small percentage of bolls on nodes higher than fruiting branch position 16 matures (Buxton and Kerby, 1981).

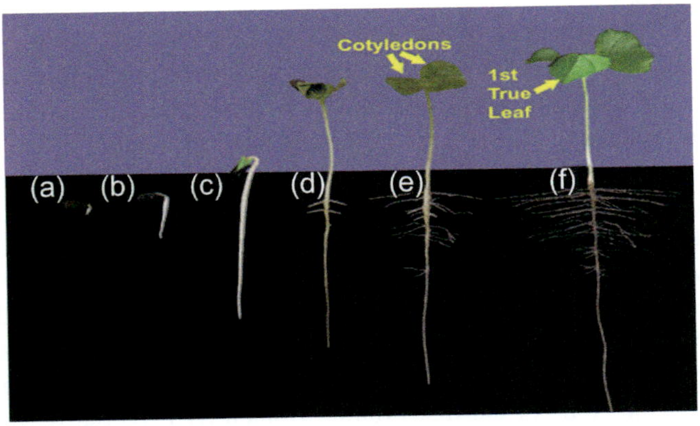

Figure 16.1 Seed cotton germination.

Figure 16.2 Cotton root system.

Figure 16.3 Cotton stem and branches.
Source: http://cotton.tamu.edu/General%20Production/cotplantgrows.pdf.

16.2.3 Leaves

The cotton plant has two types of leaves, those which arise from the main stem are referred to as main stem leaves, while those that arise from the fruiting branch are referred to as subtending leaves. Leaves are arranged in a spiral of 3/8 way round the circumference of the stem. They are broad and lobed with three to five lobes. The vegetative branches show the same direction of its spiral as the main stem (George, 2007). The development of the buds at each node largely depends on environmental conditions, close spacing, low fertility and drought. The first of the buds to develop at each main stem node is the axillary bud producing either fruiting branches or vegetative branches which depends on conditions sometimes before the branches appears (Mauney, 1966). The second and third bud at each node rarely develops below the Node of the first Fruiting Branch (N F B) but if conditions are favourable, a second bud will develop. This often produces a short fruiting branch or vegetative branch, independent of species and variety. High temperatures increase the NFB (George, 2007). The main leaf is palmate, but the leaf shape and size depend on species and varieties. The cotyledons are between an opposite pair of simple leaves with short stalks. They are the first to be formed and are rather fleshy, glabrous and usually expand to a width of 4 cm within 3 days. The terminal bud of loosely furled leaves protects the growing point as it develops from between the cotyledons (George, 2007).

16.2.4 Reproduction biology of Gossypium species

Gossypium species have a complete hermaphroditic, solitary, axial flower that begins to from four to 5 weeks after planting (Macfarlane et al., 2002). The floral buds form epically and flowers approximately 25 days after they first appear. Flowering follows a distinct pattern with the first flower opening low on the plant at first position of the fruiting branch and approximately 3 days later a flower appears and opens on the next higher fruiting branch at the same position. About 6 days after the first flower on that branch, the second flower will open. The same pattern continues provided the plant continues to grow actively. Flowers open at dawn and remain open for a single day (Liogier, 1994).

16.2.4.1 The flower

Cotton has a complete flower with multicarpellary syncarpous superior ovary. The flower bud, called square, begin to form at the initiation of the fruiting branch. The first square produced on a fruiting branch is referred to as a first-position square. As this square develops the portion of the fruiting branch between the main stem and the square elongates. An auxiliary meristem also develops adjacent to the square. The auxiliary meristem produces a second position square and subtending leaf (Qayyum et al., 1992). The flower first becomes visible as a small green cone closely hard pressed to the fruiting branch. Three bracts or bracteoles, joined at their bases, surround the buds. The bracteoles are heart shaped forming a pointed canopy over

Figure 16.4 Cotton flower.
Source: http://cotton.tamu.edu/General%20Production/Georgia%20Cotton%20Growth%20and%20Development%20B1252-1.pdf.

the bud (Austin, 1998). Within the bracts is an inconspicuous green calyx as described in Fig. 16.4. The day before the flowering, the conical bud starts to enlarge rapidly (Mauney, 1968; Merideth and Wells, 1984). The petals double their length in 24 h and early on, the following day the corolla opens widely to expose the stamina column and the stigma. The floral parts are epicalyxes (E). 3–4, calyx (K) five undulate, petals (P) 5, androecia's (A) 100–150, gynoecia's (G) 3–5 (Macfarlane et al., 2002). The corolla is white or slightly cream colour found in upland cotton. The day after anthesis the petals turn pink and the next day or two they drop off to reveal the young green boll.

16.2.4.2 *Flowering pattern*

The flowering data provides a convenient method of studying the sequence of development of the cotton plant. The cotton flower opens early in the day and petals start to turn pink by evening (Mitchison, 1971); Macfarlane et al. (2002). The next day the fertilized 1-day old flowers can easily be distinguished by their dark pink colour from nearby open flowers. Both the vertical flowering interval (VFI), between successive fruiting branches on the main stem, and horizontal flowering interval (HFI), between successive flowers on the fruiting branch follow an orderly pattern in the same cotton plant (Tharp, 1960). Seasonal average of HFI range from 5.26 to 8.5 days in the United States and VFI is about 3 days (Mauney, 1968). In the West African cotton producing countries, the average interval ranges from 2 to 3 days for VFI and 5.5–6.0 days for HFI. The differences are caused by differences in temperatures. The ratio increases during the season from between 2.1 for the early flowering to 2.6 for the late flowering. This ratio appeared to be specific for genetically different cottons. The day of the first flower indicates the earliness of the crop and marks the start of the flowering period (Iziko, 2002). It is a resultant of number of factors besides

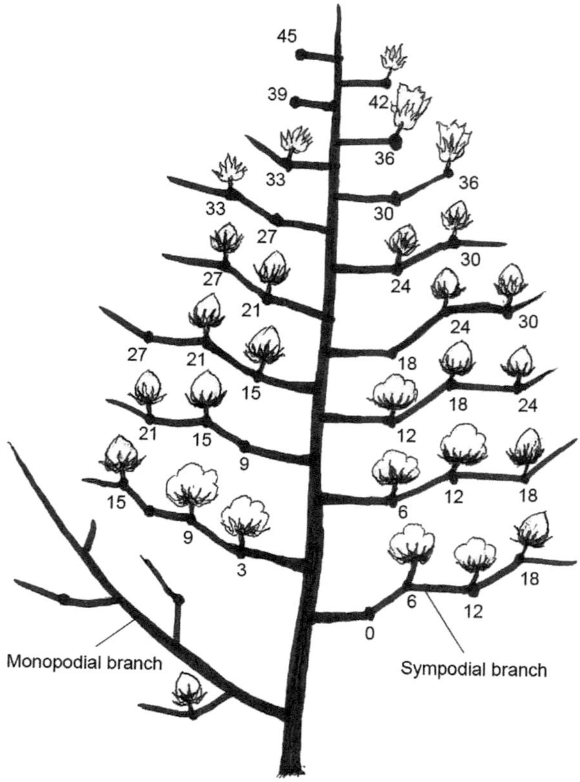

Figure 16.5 Flower pattern.
Source: Image drawn by Keshav Kranthi, based on Oosterhuis, D.M., 1990. Growth and development of a cotton plant. In: Miley, W.N., Oosterhuis, D.M., eds, Nitrogen Nutrition in Cotton: Practical Issues, Proc. Southern Branch Workshop for Practicing Agronomists. Publ. Amer. Soc. Agron., Madison, WI.

sowing data, temperature, water supply and NFB, species, varieties and day and night temperatures.

The above diagram 'Fig. 16.5' showed the development of fruiting points on a cotton plant. Each number indicates the day on which a fruiting point was formed.

16.2.4.3 Pollen grain and pollination

Soon after anthesis, (dehiscence) the spiny round pollen grains are shed, the pollen grains are relatively large, heavy and not easily dispersed by wind (Jenkins, 1992). Cotton is a facultative self-pollinator and an opportunistic out crosser, when insect pollinators are present, (Oosterhuis, 1990; Macfarlane et al., 2002). Fertilization of ovules occurs about 12–30 h after pollination as shown in Fig. 16.6. Out-crossing occurs only rarely and almost all seeds in the boll are formed as a result of self-pollination (John and David, 1995).

Figure 16.6 Cotton pollination.
Source: ICAC Research Associated Program, Kranthi, K.R., 2012. Bt Cotton, Question and Answer. The Secretary Indian Society for Cotton Improvement, (ISCI) c/o Central Institute for Research on Cotton Technology. Matunga, Mumbai.

16.2.4.4 Fruit development

The growth and development of cotton fruit, bolls begin immediately following fertilization although the most rapid period of growth occurs between 7 and 18 days (Jenkins et al., 1990). During development 'Fig. 16.7', the bolls are spherical to ovoid and green in colour. Maximum boll size is achieved about 25 days after fertilization with full maturity later. Mature bolls are thick, leathery and dry rapidly to become brittle and brown (Igbal et al., 2006; Bariola et al., 1981). Such bolls split open revealing seed cotton composed of seeds and fibres attached to seed coat. The cotton boll is a dehiscent schizocarp with three to five locules each containing about eight seeds (Percy, 2006). About 100 bolls or more may be needed to gather a reliable boll weight

Figure 16.7 Fruit development.

average. The fruit usually ripens from 50 to 70 days after fertilization, with the time varying with species, variety, temperature and humidity. In a field there are greater variations from plant to plant and from boll to boll which appears to be independent of date of flowering and position of the boll on the plant. Temperature accounts for 96% of the variation in the boll period. Night temperatures have greater effect on flowering and boll development than the day temperature (Gipson and Johan, 1970).

The boll reaches full size half way through the boll period. At the same time, the seeds inside the boll it reach full size and the lint grown to their full length with thickening of the fibres well started. The boll wall is pitted at outside with gossypol glands sucked as the boll dries out at maturity, with the outer wall of each loculus opening outwards exposing the lint and the seeds (Ahmad et al., 2008).

16.2.4.5 The seed morphology

Cotton is grown primarily for its fibre, which are produced by epidermal cells of the seed coat. Prior to ginning and delinting the seed coat bears two types of fibres, long lint fibres valued by the textile industry and short fuzz fibres known as linters used in various products. The cotton seed measures about 10 −4 mm and weighs about 80 mg (George, 2007). Beneath the epidermal cells, which produce the seed hairs, there is a thin outer seed coat with an inner epidermis. This has a different origin from that of the inner seed coat but the two are fused together (Christiansen et al., 1960; Merideth and Wells, 1984).

16.2.5 Intraspecific and interspecific hybridization

Natural gene transfer within adjacent G. *hirsutum* plants varieties can occur but at relatively low frequencies (Llewellyn and Fitt, 1996). It is estimated that cross-pollination between cotton plants in adjacent rows does not exceed 1%−2% of seeds formed. Fertile progeny is also produced when *G. hirsutum* is cross-pollinated with G. barbadense (Brubaker et al., 1999). This potentially provides another ready means by which G. hirsutum genes may be spread in the environment. Although cross-pollination may occur, cotton is considered as self-pollinated crop (Niles, 1970). However, there are no morphological barriers to cross-pollination based on flower structure. Hybrid vigour in cotton has been observed in interspecific crosses as well as in crosses between varieties within a species (Fryxell, 1969; Jenkin et al., 1990; Stroman, 1961). Research has shown that crosses between *G. barbadense* and *G. hirsutum* could be more productive than either parent. Because of the differences in the characteristics of the lint of the two species, it frequently has objectionable qualities in the hybrid. This problem is less likely to arise in intraspecific hybrids where considerable hybrid vigour has also been seen.

16.3 Cotton breeding

The Objectives of Cotton Breeding vary among cotton growing countries and have also been different at different times though the main focus of achieving higher yield has never been compromised. Once a commercial variety is released for commercial cultivation that becomes a standard to develop newer varieties. Not lint, which is the net target product, but seed cotton yield has often been the criteria to measure yield except in countries where farmers sell lint after having it custom ginned. Other than yield, the priority for fibre quality characteristics has also been never ignored. The demand from the textile sector has guided breeders to priorities fibre quality parameters i.e. fibre length, micronaire, strength and others. Once a breeder has developed a high yield breeding line, testing a promising line for resistance to insect pests and diseases and assessing its agronomic requirements is a normal of every breeding program. Additional desirable features like plant type and early maturity are always preferred without scarifying the priority for higher yield and improved fibre quality over a standard variety or varieties.

16.3.1 Cotton breeding of yesterday

Only three methods of breeding have been employed throughout the world, i.e.

Introduction,
Selection and
Hybridization.

I. Introduction

Acclimatization played a much greater role in the development of introduced cotton germplasm. Plant introduction usually mean the transfer of a new germplasm from one country. The imported germplasm has to be tested for production and desirable features, as mentioned above, at several locations and several seasons. The germplasm line found most suitable is adopted as a commercial variety without any selection. The reliance on introductions has diminished since it has been understood that there is a science behind the carryover of characters and they cannot merely be manipulated based on production conditions. Introductions do not have an impressive history of success, although varieties developed in one part of the world do have a chance to excel in performance when grown under more suitable conditions within a country or across countries. Introduction is a quick and least expensive method to introduce new varieties. Introduction also provides for adoption of new non-existing features. A major risk in introducing new varieties from other countries is the importation of new pests and diseases along with new genotypes.

II. Selection

Natural mutations occur giving rise to variability in the existing populations, would that be a commercial variety grown for years. Selection is choosing a desirable plant or plants (and bulking them or not) that look different from the bulk of the population and harvesting the seed of the plant separately to produce progeny in the next generation and screen them for commercial adoption.

16.3.1.1 Selection methods

A. Mass selection

The mass selection is selecting better-looking plants every year from the existing population and bulk the similar looking plants and grow then next year. The same process of selecting and bulking seed continues until a favoured population of similar looking plants is achieved for commercialization. Mass selection is a rapid, simple and less expensive of plant-breeding procedures that provides for superiority over the previously grown variety from which selections have been made. However, neither drastic improvement is expected, nor introduction of newer features can be hoped. There is no doubt that in crops like cotton variability within an existing population does not occur so often.

B. Pedigree selection

In pedigree method, detailed records of the origin of the selected individual plant or lines are maintained until a desirable line is finalized for commercial adoption.

C. Hybridization

This is probably the most obvious and a scientific way to improve production based on improving the genetic background of cultivation material. Cotton production itself was initiated by introduction in the Indian subcontinent; it initially failed and then succeeded.

Selection from within a population, having not been able to produce desirable results, brought hybridization to the forefront as a breeding tool. The material from where selections were made often comprised adopted/commercially grown varieties. The major limitation to the selection method of developing varieties has been a lack of enough variability, as selection had to rely either an existing variability in the population resulting from natural out-crossing or natural mutations. Drastic deviations from existing populations were not expected though usually there was no fear of adaptation issues.

Because of the limitations described above, efforts were made to induce variability in the existing homozygous populations via mutagenesis. Chemical mutagens were employed without much success. However, radiation was extensively used in a hit and miss fashion in many countries in the 1960s and 70s. Gamma rays were more effective than other sources of radiation on cotton for desirable mutations. A number of different doses were tried, and it was found that 30 and 35 krad induced chromosomal changes with a minimum number of deleterious mutations. Pakistan probably benefitted the most from gamma radiation in the form of developing heat-tolerant variety NIAB 78. There are many different ways of using radiation to induce beneficial mutations, including radiating wet/soaked seed, pollen grains, using varying doses, and using different sources of radiation; all have been tried. Scientists have concluded that the high number of deleterious effects and undesirable linkages induced by radiation rendered this technique too costly and inefficient to continue, (Chaudhry, 2016).

The hybridization procedure co-existed parallel to introduction and selection for a short while after the discovery of the Mendelian laws of segregation and independent assortment, but soon took over as the only method employed on large scale for many reasons.

It is true that hybridization is a long process involving crossing of two parents, selection from segregating populations starting from F2 generation onward, and attaining genetically pure breeding lines before they are carried forward for commercialization.

Nevertheless, hybridization continues as a modern method of developing varieties, but the practice relies mostly on the experience and judgement of breeders, rather than science. In other words, breeding is practiced more as an art of selecting better plants that will continue performing better in subsequent generations and will become commercial varieties.

Unfortunately, breeders lacked a precise test that would ensure success to a higher degree. A big team of breeders attempting a large number of crosses has a higher chance for success because they were able to screen more combinations/segregation generations. Nevertheless, there is no guarantee that a more qualified or bigger team will develop a superior variety in a shorter time.

A complicated genetic control of various characters, negative correlations among desired characters, and other factors complicated and restricted the flow of varieties. There were gene combinations that every breeder desired to introduce into varieties, but they were not successful because they could not pick and choose genes carrying the desired characters. Genetic purity of merging line could only be trusted but not verified.

Many efforts were made to overcome some of the constraints to successful breeding of superior lines. The efforts included using back-crossing to retain or transfer a specific character, it is the same technique now used to transfer biotech genes. Single crosses, double crosses, varieties crossing with segregation populations and many more options have been tried, but only back-crossing proved its worth. The development of varieties with single, double and even triple biotech gene transfers is possible because of the experience learnt from conventional breeding.

The segregating populations could be carried forward via two ways I.e. Mass selection or Pedigree selection, as discussed above. Here Pedigree selection is discussed in detail which it the norm in most breeding programs around the world.

16.3.1.2 Crossing and handing segregation population

Breeders usually make crosses between two high yielding parents that have all the desired characters existing between the two parents. It is hoped that the segregating population with emerge with a segregate that not only have the good qualities of both combined in them but rather excel in some respects over both the parents in which one parent has a unique character and the objective is to introduce that character into the commercial stream.

Which variety is better for use as a male parent and which selected variety should be used as a female parent has no value as the reciprocal crosses have not shown significant differences. Nevertheless, the varieties which are known to be severely attacked by bollworms or exhibit high rate of shedding are not to be used as female parents, and varieties with strong growth habit are preferably chosen as female parents in order to secure sufficient successful hybrid vigour. Once a cross is made (and F0 seed is grown next year), the first filial generation (F1) plants flowers are selfed, and carefully picked to avoid any mixing. The F_1 plants are supposed to be similar behaving in the field. The F_1 seed, which is heterogenous, is grown to produce segregating F_2 population. The F_2 population has the maximum variability and provides an opportunity for a breeder to search for combinations on his hit list.

The segregating F_2 population is very critical for a breeder wherein he has to use is visual analytical skills as all the seed from the F_2 population cannot be planted next year. The breeder has two options, either to following the bulk method wherein he will take a portion of the F_2 seed and continue following the same procedure until (F_6 or F_7) when he is sure that the population has reached a level of purity. The breeder can select single plants for further testing or initiate single plant selection at one or two generations earlier. It all depends on the facilities available to handle segregation populations. However, the breeder is risking to lose good combination plants in the delay to make single plant selection. A modified form of this bulk selection procedure could also be employed where in single plant could be initiated at early stages, but the selected plants are bulked each year to produce the next generation. It is not desirable to test segregating bulk populations for fibre quality as the character expression is stabilized yet. The options discussed in this paragraph differentiate the bulk method followed in the Selection as a method to develop new varieties and hybridization/ dealing with segregating populations.

The second generation (F2) population, where segregation is at maximum, must be grown on a reasonable area unlike the F_1 population. The Pedigree method of selection in a hybridization programme provides for the opportunity to trace back the performance of each generation plants and go for early discard of plants and lines. However, it is cumbersome to select, maintain and test all single plants for yield and various other priorities characteristics. The single plant selections from the F_2 population could be subjected to fibre testing. The same could be repeated in the subsequent generations i.e. select single plants, grow them in progeny rows and continue growing progenies until a purity is reached. The Pedigree selection provides for to trace back the performance of a particular generation to its origin in F_2 generation, which is great asset for a breeder to trust his selection. How many plants should be selected in each generation and what should be the size of the progeny rows all depends on the manpower and facilities available. A typical breeding programme follows the procedure mentioned below.

Emasculation, pollination and collecting hybrid seed (F0)	Parents	Year 1
Planting single seed, normal spacing selfing (if required) and picking bulk seeds.	F_1	Year 2
- Plant a population of 500–2000 plants initiate monitoring the populations from early growth stage, select single plants, harvest single plants separately. - Measuring single plant seed cotton testing for GOT, boll weight, seed index, and fibre quality parameters. - Scrutinize the data for all aspects and the best 30–60 single plants for planting next year.	F_2	Year 3
- Grow 20–40 plants from each selected plant in a row or small plot. - Monitor progenies of among themselves and single plants within progenies. - Select progenies with a high number of better performing plants. - Select single plants from within better performing progenies and from even progenies not performing as good. - Discard the rest of the population. - Single plants selections are tested for yield and fibre quality. - Select the best 25–50 families and or single plants. Discard the rest of population	F_3	Year 4
Repeat the F_3 generation process and 20–25 best performing families.	F_4	Year 5
Selection and data analysis continue as in the previous generation.	F_5	Year 6
- Similar behaving plants and/or families, depending upon the field performance and lab data, can be bulked to start early yield testing trials. The yield trial could have no replication for 1 year. Bulking of progenies and single plants could be delayed for another year or two if the field data or the lab data show unacceptable variation for any of the tested parameters.		
Yield trials must be replication and single plants/families must be put to more extensive testing.	F6	Year 7

Regional or zonal trials at multiple locations (throughout the potential growing areas) are started.

The selected families can be named as strains for further **testing and evaluation.** From now on the yield trials must be conducted against a standard variety. The strains only perform better that a standard variety or varieties be carried forward. Some strains might show variation within the population even at this stage, those must be discarded or rogued of from off type plants. Confirming the performance year after years becomes very critical from this stag on. Testing of strains must be continued for multiple years until it is established that the population is not morphologically segregating and quality parameters have been established. Permanent name is assigned at this stage that will be used in the commercial market. Variety, a group of morphologically similar behaving plants, will be known by this name. The breeding programme is a continuous process and each year a number of crosses are made, thus a flow of promising varieties is always obtained.

16.3.1.3 Seed production

New varieties are formally registered or should be registered for record keeping with the registration authority. Many countries still have a legal variety approval process without which no variety can be grown on commercial scale. The variety registration and approval process is consistent and in line with practices in most countries except a few countries. As breeding and seed products shift to the private sector strict self-certification, high quality maintenance standards and intense competition among seed companies all work to assure the supply of quality seed to growers. In most countries, registration and approval are in the purview of government agencies and departments. Approved varieties are notified and promoted, usually by respective breeders, whether in the private or public sector. Once approved, varieties are generally not denotified in many countries and even if denotified, they are not immediately taken out of production. Varieties taken out of production, replaced by other varieties or no longer desired for planting for any reason should be denotified.

Commercial seed production and distribution is an integral component of a successful breeding program. The success of any variety, developed locally or introduced from another country, will depend on having the right planting seed system. The planting seed that is made available to growers must be pure and have good germination properties. Quality seed maintenance adds to the active life of a variety. It can be achieved only if seed production, certification and distribution of planting seed for cotton are managed. In most countries, seed production and distribution is spread across organizations/departments/ministries within the government and the private sector. Breeders develop varieties and are responsible for maintaining and producing seed that will be used for multiplication purposes. Consequently, only the subsequent seed generations (generation 1, 2 or 3) reach the farmers' fields. The planting seed that is used by farmers for sowing a crop has different nomenclature among countries although the seed may have the same purity standard and is of the same generation level. The international Cotton Advisory Committee undertook a study in 2013 on the nomenclature used with the objective to popularize a common nomenclature across countries and regions. The International Cotton Advisory recommends that following names be used for various stages of planting seed (ICAC, 2013).

16.3.1.4 Breeder seed

This will be the seed produced or supplied by breeders for multiplication purposes only. Seed may be multiplied by the public or private sectors, but under the supervision of a breeder. Breeder supervision will be the certification, if required. Purity of the seed will be 100%.

16.3.1.5 Foundation seed

This is the seed produced from the crop of Breeder Seed. Foundation seed will be used to multiply on a large scale to supply growers with seed for the next generation. External certification by a neutral agency or self-certification will be mandatory. Purity must be at least 99%.

16.3.1.6 Certified seed

This is the seed produced from Foundation Seed and it is the seed that will be supplied to growers. It is recommended to maintain purity close to 99%.

16.3.1.7 Registered seed

Seed produced from certified seed by a farmer under his own supervision for self-planting or to supply to other farmers is known as registered seed. It should carry the label 'not certified' and its purity should be no less than 98%.

Commercial seed is not recommended for planting. It will go to oil production or livestock feed.

16.3.1.8 Commercial cotton hybrids

Heterosis is the superiority of F1 hybrids in one or more characters over the parental values. In current stage, heterosis and hybrid vigour are used as synonyms and interchangeable.

Generally, positive heterosis is considered desirable, but for some characters negative heterosis is also desirable. In cotton, negative heterosis for plant height, micronaire value, maturity duration and gossypol content is desirable because it shows superiority over the parents. The true heterosis differs from pseudo-heterosis (luxuriance). When talking of commercial cotton hybrids productivity increase over commercial varieties is the main consideration, no doubt other parameters particularly fibre quality must be at least equivalent to commercial varieties if not better. Producing commercial cotton hybrids presents two steps and both are important for utilization of heterosis in yield or superiority in quality. The steps are explained in the following sections.

Identification of combiners

Commercially grown cotton varieties have a high yielding ability. To utilize hybrid vigour, such F_1 crosses should be found which are capable of producing higher yield over all commercial varieties of the area. For identification of parents, the specific combinability is more important than the general combinability. Parents must be identified which, when crossed, express an acceptable degree of heterosis in yield without loss in quality. The performance of the hybrid has to be consistent over many years. Many countries including Australia, Israel, Pakistan and United States tried to find specific combiners that were more productive than commercial varieties, but they could not. Only India found such combiners and has grown commercial cotton hybrids on millions of hectares for decades. Indian production conditions are not the subject here but there could be other reasons for the success of hybrid cotton in India. India in now considering moving to straight varieties.

Production of hybrid seed

Cotton, having been having like a self-pollinated crop, makes it quite expensive to produce hybrid seed every year for large scale commercial cultivation. Hand emasculation and pollination are not feasible for many apparent reasons. The cytoplasmic male sterility and genetic male sterility have been tried and they have their own limitations in terms of lack of good restorers and 1:1 segregation respectively. Gametocides were also trying in the United States in the 1980s and early 1990s but they were also not successful for producing hybrid seed. The work on gametocides was stopped in 1995.

The International Cotton Advisory Committee published a good review of the commercial cotton hybrid work in 1997 where in situation in many countries including Australia, Egypt, India, Israel, Pakistan, Peru, Sudan, Vietnam, United States and Uzbekistan has been discussed (ICAC, 1997). Hybrid cotton did not succeed (outside India) because of failure to find combiners that consistently gave significantly higher yield over commercial varieties and high cost of F_1 seed production. Interspecific diploid hybrids were also tried without any success for the same reasons. It is not possible to extend hybrid cotton beyond F_1 generation because of the variation in fibre quality in a segregation population.

16.3.2 *What breeding is currently going through?*

Breeding is increasingly conducted in the private sector, although it will take many more years until this shift is complete. The public sectors in many countries see private sector breeding as a challenge to their authority and have sometimes been reluctant to cooperate.

Often times, public sector breeding programs have been judged in terms of the number of varieties released for commercial production and the area planted to such varieties. Some breeders fear that financial support from governments will dissipate if they surrender the right to commercialize varieties. However, the shift to private sector breeding is a reality, and this shift allows public sector resources to be better utilized in other areas of research that require higher attention than received in the past.

Breeding programs, whether in the private sector or public, are faced with issues that will have long-term consequences if proper measures are not taken. Constraints are often openly discussed, but solutions are not given, or if solutions are mentioned, pathways are not shown to resolve them. Hence the elimination of constrains is not expected soon, and the consequences are potentially severe and long lasting.

The major issues confronting cotton breeding programs are:

- The genetic base of current varieties is narrow and getting narrower. Only a limited circle of varieties is hybridized to produce newer varieties, hence the breeders end up with only a slight change of achieving improvements over existing varieties. Individual companies or public sector breeding teams are using their own varieties for the sake of maintaining their known and accepted stylized series. They are doing so to have a higher success rate in commercializing new varieties than they would have if they attempted crossing two diverse genotypes. So, breeders themselves have fused this problem into their breeding programs, and they are the ones who must change their approaches.

- Germplasm exchange has almost disappeared. Germplasm availability is the extent of freedom that a breeder has to obtain and use any genetic material that exists in or outside a country. The result of stringent plant variety protection in each country is restricting breeders' rights to freely use any genetic material. These restrictions inhibit further development of innovations.
- The advent of biotechnology and intellectual property laws has increased the market for improved seed. These changes made it easier for commercial breeders/companies to be rewarded and to recover the cost of their investments. But, at the same time, intellectual property protections limit the ability of farmers or rival breeders to reproduce seed. Private seed companies have benefitted the most from the expansion of the seed industry to more than $2 billion US dollars. Public sector breeders and programs have not been compensated for their work to the same degree as private programs, resulting in shrinkage of public sector breeding programs.
- The focus in cotton breeding has shifted to short-term objectives to achieve quick returns. Budgets for long-term fundamental scientific research are shrinking, which is not in the ultimate best interest of breakthrough achievements.

The rise of molecular genetics, particularly with the commercialization of biotech crops, has accelerated the shift toward private sector plant breeding. Many scientists believe that as conventional breeding is replaced by biotechnological approaches, public sector programs and institutions will retreat from classical plant breeding. This, in turn, has a negative effect on breeding education at universities and the research work at public institutions (Roupakias, 2014). Biotechnology is relatively new and can be done in large centralized laboratories; there has been a rapid expansion of biotechnology research. That research is essential, but there has also been a reduction in public sector plant breeding efforts, which could result in a lack of progress in the development of elite germplasm and effective commercial cotton cultivars (Constable, 2015). Conventional breeding and molecular genetics are complementary, and both are needed to fundamentally improve cotton varieties.

16.3.3 Future trends: cotton breeding of tomorrow

Cotton breeding is in a high-transition stage in which the way that genetic principles are applied is changing. It is quite possible that 'conventional breeding' together with transgenic breeding, complimented with even newer developments in biotech approaches, will find a common name. The name for the new breeding approach is not known, but it might be something like 'directed breeding,' wherein the breeder will have a specific, predetermined target, and the breeder will hybridize with certainty. Future breeders will not be working with hit and miss trial methods and with an unlimited wish list of targets. Rather, breeders will have a source for the anticipated outcome.

The long process of selection from F2 to many later generations, currently followed for the sake of producing a homozygous population, has to be shortened. The production of haploid plants, and the doubling of chromosomes in the cotton genome, have long been targeted, but additional options may become available with the progress we are making. The other time-consuming aspect of breeding is multi-location performance testing of varieties under varied sets of agronomic practices. This also needs to be changed. It is not economical to test candidate varieties at many locations, under various sets of agronomic situations, and then make selections based just on normal conditions.

Cotton, especially *G. barbadense*, is highly sensitive to photoperiodic complexities and thermal conditions. Researchers have talked about 'global varieties' that will perform equally well anywhere in cotton producing areas. Imagine if a good team of breeders is allowed to focus on innovative developments rather than breeding varieties based on luck and not having to wait 12–14 years to know the fate of their breeding lines.

The biggest change is, of course, going to come from biotechnology applications. Over the next 20–30 years, breeding will be employed to transfer specific new traits into current varieties. Quoting many researchers, Abdurakhmonov (2013), stated that the 21st century's 'omics' science and innovative genomics tools are the most promising approaches, in combination with contemporary cotton breeding knowledge and strategies. The strategies include:

Decoding of cotton genomes and the mapping and characterization of the genetic basis of complex traits (referred to as quantitative trait loci-QTLs) that provide better exploitation of existing genetic diversity of cotton germplasm and gene pools, and a widening of the genetic diversity of commercialized cotton cultivars using modern marker-assisted selection (MAS), marker-assisted backcross selection (MABS) and genomic selection (GS) programs.

The use of marker-assisted technology can be easily employed if a particular feature is controlled by a small number of genes, and their impacts are least influenced by ambient conditions. Unfortunately, many genes control lint yield and fibre quality properties, each inducing only a small effect. For example, Shen et al. (2011) stated that advanced-backcross quantitative trait locus (QTL) analysis of an interspecific *G. hirsutum* × *G. barbadense* population, showed that 28 fibre length QTLs were identified including qFL-chr1 on chromosome one of the A-sub genome. The *G. barbadense* allele at this QTL contributed to longer fibres and explained up to 24% of the phenotypic variance. Managing these quantitative traits is even more difficult because of the number of genes involved. DNA markers associated with QTLs for improved fibre quality such as length, strength and uniformity will be explored for Pima cotton. It might take many years to overcome challenges, such as the simultaneous improvement of yield and fibre quality, but molecular technologies will certainly accelerate the process of improving the cotton genome. Malik et al. (2014) have presented a good review of the role of molecular markers in cotton genetic improvement, including future prospects for the practical utilization of new molecular technologies. An overview (of literature) of genetic diversity studies in cotton using different kinds of markers, i.e. RAPDs, SSRs, AFLPs, ISSRs (Inter Simple Sequence Repeats) and SNP, showed that thousands of Upland and barbadense populations, and a number of diploid cultivated and uncultivated species genotypes, have been studied in the United States, China, India, Pakistan and many other smaller cotton producing countries. Enormous work is going on to develop more efficient DNA markers for plant breeders and geneticists for developing cultivars of cotton in more efficient ways. It is hoped that SNPs markers will have big impact on molecular assisted selection and mapping studies in the future due to an abundance of sophisticated detection systems that will be developed.

16.4 International exchange of germplasm

Lack of fluent exchange of germplasm is hurting cotton breeding programs everywhere in the world including developed countries. Unlike other field crops that are taken care of by the Consultative Group on International Agriculture Research (CGIAR), cotton does not have an international institute/centre on cotton research. Such a centre could have many benefits for the entire cotton community including all producing and consuming countries. Efforts have been made but the current scenario is not conducive to initiate a new research institute on a crop that is most grown in the developing countries. Cotton is grown as a fibre crop, but cotton seed oil is a big source of edible oil in many countries, which further increases the value of the crop. Time may not be right for a 'consuming institute/centre' that undertakes basic research, but an option is available to establish an international centre that undertakes applied research in fields that could make is self-supporting. Seed money will be anyway needed as will be public and private sector backing. An easiest thing that may not even require any funding is to upgrade an existing research centre to be recognized as an international germplasm centre responsible for maintaining and facilitating exchange of cotton germplasm in the world.

16.5 Conclusion, future trend

Cotton is still largely under genetic control, and breeding programs strive to raise and increase cotton yields and keep up with the industry demands for quality. Fortunately this has been achieved. However, fibre quality also depends on the environment and on crop management. Crop management for high yields is fully compatible with high fibre quality, but the environment is not always predictable or manageable. Hence there will always be some fibre quality variability coming from the field, which must be understood, evaluated and valued by the cotton market.The extensive research done on cotton became more formal and was easier to understand after it was discovered that there are genes that carry a blueprint of the characters to be expressed under a given set of growing conditions. Such discoveries, unimaginable in the early years of cotton research, were severely questioned and remained shelved for about half a century. The theory of evolution did not satisfactorily address many concerns, and it was practically impossible to give up the long-held belief in the inheritance of acquired characters. Fortunately, however, the law of inheritance of characters and the independent assortment of genes were rediscovered and applied.

16.6 Fundamental books and websites about cotton breeding

- Australian Government, 2008. The Biology of *Gossypium hirsutum* L. and *Gossypium barbadense* L (www.ogtr.gov.au). [An on line article that describes taxonomy, morphology, growth and growing conditions of *Gossypium hirsutum* L. and *Gossypium barbadense* L.].
- ISAAA, 2017. Global Status of Commercialized Biotech/GM Crops in 2017: Biotech Crop Adoption Surges as Economic Benefits Accumulate in 22 Years. ISAAA Brief No. 53. ISAAA: Ithaca, New York.
- Smith, C.W., Cothren, J.T., 1999. Cotton: Origin, History, Technology and Production. John Wiley and Sons., New York. [A book dedicated to detailed presentation of cotton history, technology, and production].
- Supak, J.R., Snipes, C.E., 1991. Cotton Harvest Management: Use and Influence of Harvest Aids. The Cotton Foundation, Memphis, TN. [A book that presents techniques, methods and aides for a successful cotton harvesting].
- Wakelyn, P.I., Chaudhry, M.R., 2010. Cotton: Technology for the 21st Century. International Cotton Advisory Committee. Washington DC20006 US.
  www.icac.org
  www.cottoninc.org
  www.cotton.com
  www.icracotton.org
  www.isaaa.org

Acknowledgement

The author would like to express his deepest thanks and gratitude to Dr. MD. Rafiq Chaudhry, Former Head of Technical Information Section, International Cotton Advisory Committee-ICAC, for his invaluable advice and great help and revision.

References

Abdurakhmonov, I., 2013. Role of genomic studies in boosting yield. ICAC Rec. XXXI (4), 2013.

Ahmad, W.M.U., Khan, M.R., Parveen, U., Shah, S.A., 2008. Genetic variability and correlation analysis in upland cotton. Sarhad J. Agric. 24, 195–200.

Austin, N., Khan, M.D., Khan, M.A., Ahmad, M., 1998. Genetic studiesof cotton (*Gossypium hirsutum* L.). I. Combining ability and heterosis studies in yield and yield components. Pak. J. Sci. Ind. Res. 41, 54–56.

Balock, M.J., 2004. Genetic variability and heritability estimates of some polygenic traits in Upland cotton. Pakistan. J. Sci. Ind. Res. 42 (6), 451–454.

Bariola, L.A., Henneberry, T.J., Hittock, D.L., 1981. Fruiting of Upland and Pima cotton with different planting dates. Agron. J. 73, 711–715.

Beasley, J.O., 1940. The origin of American tetraploid *Gossypium* species. Am. Nat. 74, 285–286.

Bird, L.S., 1981. Cotton seed and germination stand establishment. Beltwide Cotton Prod. Res. Conf. 3, 18–21.

Brubaker, C.,L., Bourland, F.M., Wendel, J.F., 1999. In: cothren, S.W.C., Cothren, J.,T. (Eds.), Cotton: Origin, History, Technology and Production. John Wiley and Sons, Inc, New York, USA, pp. 3–32.

Buxton, D.R., Kerby, T.A., 1981. Competition adjacent fruiting forms in cotton. Agron. J. 73, 867–871.

Chaudhry, M.R., 2016. Cotton breeding: developments and opportunities. In: 33rd International Cotton Conference Bremen. March 16 - 18, 2016.

Christiansen, M., Moore, R., Rhyne, C., 1960. Cotton seed quality preservation by hard seed coat characteristics which restrict internal water uptake. Agron. J. 52, 2–3.

Constable, G.A., 2015. Cotton Breeding and Physiology Research in Australia. Presented at 75th Plenary Meeting of the ICAC, Mumbai, India. December 6-11, 2015. Available at: www.icac.org.

Endrizzi, J.E., Turcotte, E.L., Kohel, R.J., 1985. Genetics, cytology, and evolution of *Gossypium.* Adv. Genet. 23, 271–375.

Farbrother, H.G., 1972. The field behavior of Gezira clay soil under irrigation. Cotton Grow. Rev. 49, 1–27.

Fryxell, P.A., 1969. A classification of *Gossypium* L. (Malvaceae). Taxon 18, 585–694.

Fryxell, P.A., Craven, L.A., Stewart, J.M., 1992. A revision of *Gossypium* sect. Grandicalyx (Malvaceae), including the description of six new species. Syst. Bot. 17, 91–114.

George, A., 2007. The Principles of Plant Genetics and Breeding, pp. 87–95.

Gipson, J.J., Johan, H.E., 1970. Temperature variety interrelationship in cotton, seed development and composition. Agron. J. 47, 26.

Gotmare, V., Singh, P., Tule, B.N. http://www.cicr.org.in/pdf/wild_species%20.pdf.

Govila, O.P., Rao, C.H., 1969. Studies on the *in vitro* germination and storage of pollen. J. Palynolognol. 5, 37–41.

Iqbal, J., Aziz, N., Saeed, N.A., Zafar, Y., Malik, K.A., 1997. Genetic diversity of some elite cotton varieties by RAPD analysis. Theor. Appl. Genet. 94, 139–144.

Iqbal, J., Reddy, G., El-zik, K.M., Pepper, A.E., 2006. A genetic bottleneck in the evolution under domestication of upland cotton *Gossypium hirsutum* L. examined using DNA. Genetics 103, 547–554.

Iziko museums of Cape Town, 2002. Gossypium, vol. 5. Cotton. Biodiversity explorer. http://www.museums.org.za/bio/plants/malvceae/gossypium.htm.

Jenkins, J.N., Mccarty, J.C., Parrot, W.L., 1990. Effectiveness of fruiting sites in cotton yield. Crop Sci. 30, 365–369.

Jenkins, J.N., 1992. Cotton in: Traditional Crop Breeding Practice: An Historical Review to Serve as a Baseline for Assessing the Role of Modern Microbiology Organization for Economic Cooperation and Development (OECD), Paris, France, pp. 61–70. http://www.oecd.org/dataoecd/57/48/1946204pdf.

John, M.P., David, A.J., 1995. Breeding Field Crops, fourth ed., pp. 156–158

Kranthi, K.R., 2012. Bt Cotton, Question and Answer. Publisher: The Secretary Indian Society for Cotton Improvement, (ISCI) c/o Central Institute for Research on Cotton Technology, Adenwala Road, Matunga, Mumbai 400 019.

Kohel, R.J., Lewis, C.F. (Eds.), 1984. Cotton. Am. Soc. of Agronomy, Madison, U.S.A.

Liogier, H.A., 1994. Descriptive Flora of Poerto Rico and Adjacent Islands, vol. 3. Editorial de la University de Puerto Rico, Rio Piedras p461p.

Llewellyn, D., Fitt, G., 1996. Pollen dispersal from two field trial of transgenic cotton in the Namoi valley. Aust. Mol. Breed. 2, 157–166.

Macfarlane, T.D., Watson, L., Marchart, N.G., 2002. Flora Base: The Western Australia Flora. Western Australia Herbarium Department of Conservation and land management. Government of Western Australia. http://www.florabasecalm.wa.gov.au.

Malik, W., Ashraf, J., Zaffar Iqbal, M., Khan, A.A., Abdul Qayyum, Ali Abid, M., Noor, E., Ahmad, M.Q., Abbasi, G.H., 2014. Molecular markers and cotton genetic improvement: current status and future prospects. Sci. World J. 2014. http://www.hindawi.com/journals/tswj/2014/607091/.

Mauney, J.R., 1966. Production of fruiting point. In: Hoover, E., Porter (Eds.), Advances in Production and Utilization of Quality Cotton: Principles and Practices. Iowa State University Press, pp. 256–261.

Mauney, J.R., 1968. Morphology of the cotton plant. In: Hoover, E., Porter (Eds.), Advances in Production and Utilization of Quality Cotton: Principles and Practices. Iowa State University Press, pp. 25–40.

Menzel, M.Y., 1954. A cytological method for genome analysis in *Gossypium*. For. Genet. 40, 214–223.

Merideth, R., Wells, R., 1984. Comparative growth of obsolete and modern cotton cultivars: II. Reproductive Dry Matter Partitioning. Crop Sci. 24, 863–868.

Mitchison, J.M., 1971. The biology of cell cycle. In: Upland Cotton, the Day after Anthesis the Petal Turns Pink and the Next Day or Two it Drops off to Reveal the Young Green Boll. Cambridge Univ. Press, London, p. 153.

Niles, G.A., 1970. Development of plant type with special adaptation to narrow-row culture, 6-7 January. In: Brown, J.M. (Ed.), Proc. Beltwide Proc., Houston, TX. National cotton council, Mephis TN, pp. 63–64, 1970.

Oosterhuis, D.M., 1990. Growth and development of a cotton plant. In: Miley, W.N., Oosterhuis, D.M. (Eds.), Nitrogen Nutrition in Cotton: Practical Issues, Proc. Southern Branch Workshop for Practicing Agronomists. Publ. Amer. Soc. Agron., Madison, WI.

Percy, R.G., Cantell, R.G., Zhang, J., 2006. Genetic variation for agronomic and fibre properties in an introgreted recombinant inbred population of cotton. Crop Sci. 46, 1311–1317.

Qayyum, S.M., Chondhry, N.A., Ansari, M.M., Baig, M.A., Memmon, M.I., 1992. Correlation and regression analysis among yield and its economic character, in upland cotton *G. hirsutum L.* Agric. Vetinary Sci. 8 (12), 28–34.

Roupakias, D.G., 2014. Germplasm availability and development of superior cotton cultivars. In: Presented at 74th Plenary Meeting of the ICAC, Thessaloniki, Greece. November 2-7, 2014. Available at: www.icac.org/getattachment/mtgs/Plenary/73rd-Plenary/Agenda/BS2-Roupakias-(1).pdf.

Shen, X., Cao, Z., Singh, R., Lubbers, E.L., Xu, P., Wayne Smith, C., Paterson, A.H., Chee, P.W., 2011. Efficacy of qFL-Chr1, a Quantitative Trait Locus for Fiber Length in Cotton (*Gossypium* spp.) (2011) Crop Science: 51:2005–2010. Available at: http://nespal.org/peng_lab/papers/Shen%20et%20al.%202011.pdf.

Stewart, J.M., 1995. Potential for crop improvement with exotic germplasm and genetic engineering. In: Constable, G.A., Forrester, N.W. (Eds.), Challenging the Future: Proceedings of the First World Cotton Research Conference. CSIRO, Canberra, Australia, pp. 317–327. Brisbane, Australia, 13–17 February 1995.

Stroman, G.N., 1961. An approach to hybrid cotton as shown by intra and interspecific crosses. Crop Sci. 1, 363–366.

Technical Information Section, ICAC, 1997. Commercial cotton hybrids. ICAC Rec. XV (2), 1997.

Technical Information Section, ICAC, 2013. The planting seed industry. ICAC Rec. XXXI (3), 2013.

Tharp, W.H., 1960. Cotton seed composition-relation variety. Maturity and environment of the plant. In: Bailey, A.E. (Ed.), Cotton Seed and Cotton Seed Products. Inter-science Publishers, Inc., New York, pp. 117–156.

Wendel, J.F., 1995. Cotton. In: Simmonds, N.W. (Ed.), Evolution of Crop Plants. Longman Scientific & Technical, Essex, UK, pp. 358–366.

Wendal, J.F., Brubaker, C.L., Percival, A.E., 1995. Genetic diversity in *Gossypium hirsutum* and the origin of Upland cotton. Am. J. Bot. 79, 1291–1310. https://doi.org/10.230/2445058.

Further reading

Technical Information Section, ICAC, 1998. Breeding for yield improvement needs changes. ICAC Rec. XVI. No. 1, 1998.

Developments in fibrous flax and linseed breeding and cultivation

17

M. Pavelek, M. Vrbová - Prokopová, E. Ondráčková, M. Ludvíková, M. Griga
AGRITEC Plant Research Ltd., Šumperk, Czech Republic

17.1 Introduction

Decrease of flax areas at the beginning of the 21st century has been stopped, and now the world flax and linseed areas are relatively stable. Now there is the tendency to replace flax areas by linseed areas. For more information about flax see chapters: 6 in volume 1, and chapters 4, 5 and 6 of volume 2.

17.1.1 World flax and linseed areas, biggest growers and producers and European flax and linseed areas

World flax and linseed areas are relatively stable and fluctuate between 1,812,423 ha in 2010 and 3,110,775 ha in 2015 (FAOSTAT ....). The world's biggest growers and producers are Russia (710,277 ha in 2016), China (294,700 ha in 2016), India (293,000 ha in 2016), Canada (338,000 ha in 2016), Kazakhstan (633,615 ha in 2016) and the USA (148,152 ha in 2016; Table 17.1). Total flax and linseed areas of the biggest growers and producers vary between 1,743,564 ha (in 2010) and 2,755,095 ha (in 2015). In the beginning of 2015, there is a little bit decrease in flax and linseed areas (Table 17.1, FAOSTAT ...).

Traditional west European fibre producers France, Belguim, the Netherlands representing approximately 60% of the world market stopped the decrease of flax areas in 2011, 2012, 2013 and increased their flax areas from 18. 291 ha in 2013 to 39,707 ha in 2016. The other biggest growers and producers in the framework of the rest of EU countries are Poland, Czech Republic and Slovakia. However, their flax and linseed areas are more variable and several last years are exclusively represented by linseed only. Flax breeding and growing in the Czech Republic was definitely finished in 2010, and from 2010 all areas are represented by linseed. Flax and linseed areas in Poland fluctuated between 1234 ha in 2012 and 7457 ha in 2016, in Czech Republic between 1481 ha in 2016 and 4291 ha in 2009 and in Slovakia between 2116 ha in 2011 and 978 ha in 2013 (Table 17.1, FAOSTAT ...). Since 2013, there has been a certain increase of linseed areas.

Handbook of Natural Fibres. https://doi.org/10.1016/B978-0-12-818398-4.00019-0

Table 17.1 Harvesting areas and production quantity – world's, biggest growers, EU countries 2009–2016.

	2009 HA	2009 PQ	2010 HA	2010 PQ	2011 HA	2011 PQ	2012 HA	2012 PQ	2013 HA	2013 PQ	2014 HA	2014 PQ	2015 HA	2015 PQ	2016 HA	2016 PQ
World	2,104,734	2,157,621	2,034,388	1,812,423	2,212,293	2,134,648	2,456,642	2,024,669	2,248,043	2,273,440	2,631,816	2,654,232	3,009,392	3,110,775	2,764,340	2,925,282
Russia	144,342	10,262	226,442	178,213	191,533	437,236	558,294	341,565	438,349	299,768	441,475	365,088	636,271	523,534	710,277	672,691
China	336,930	318,135	324,400	352,812	322,100	358,641	317,860	390,505	312,890	398,809	310,000	387,000	294,517	361,189	294,700	361,569
India	407,900	169,200	342,000	153,700	359,000	147,000	323,000	152,000	296,000	147,000	293,070	141,730	284,000	155,000	293,000	125,000
Canada	623,300	930,100	353,300	423,000	273,200	368,300	384,400	488,900	422,100	730,700	620,800	872,500	646,000	942,000	338,000	579,000
Kazakhstan	58,400	47,650	225,200	94,610	309,670	273,077	369,629	157,878	384,338	295,021	556,152	419,957	622,545	491,458	633,615	561,771
USA	127,070	188,550	169,160	230,030	70,010	141,783	135,980	147,280	69,610	85,250	122,220	161,750	184,540	256,420	148,520	220,480
Ukraine	46,800	37,300	56,300	46,800	58,700	51,100	52,900	41,400	41,980	31,920	37,600	46,630	66,700	76,380	68,700	92,160
Belorussia	48,086	10,033	46,762	10,399	49,981	13,143	35,675	8664	29,024	7005	22,395	7229	20,422	6876	22,025	6499
TOTAL	1,792,828		1,743,554		191,533		2,177,738		1,994,291		2,403,712		2,755,095		2,508,837	
France	9633	20,614	18,086	35,756	16,351	30,547	12,077	23,742	8510	16,147	12,083	16,561	22,308	42,594	23,292	45,197
Belgium	11,227	7393	14,000	8600	10,160	5689	10,581	8082	7900	7400	10,000	7000	12,000	8000	14,000	9000
The Netherlands	2163	2499	1422	1422	2156	1195	2077	1765	1881	1981	1983	1388	2405	2405	2415	1787
TOTAL	23,023		33,508		28,667		24,735		18,291		24,066		36,713		39,707	
Great Britain	28,782	54,175	43,838	71,894	36,453	71,083	27,788	41,730	34,498	62,131	14,538	39,253	15,000	29,000	274,000	48,000
Germany	4100	4000	6859	7000	4600	6000	4600	6000	3700	6000	4200	6000	5000	7000	4200	6000
Spain	2498	2262	33,000	58,000	13,000	10,000	22,000	10,000	8000	6000	153,000	114,000	251,000	179,000	249,000	179,000
Austria	535	642	669	847	669	906	691	871	601	626	824	1035	1047	1200	1289	1734
Sweden	990	1880	1896	2360	1462	2300	8800	1240	4840	9600	6560	1100	700	1400	837	1680
Denmark	620	196	574	196	568	191	562	186	545	181	526	176	507	171	489	166
Italy	3	2	3	2	3	2	3	2	3	2	3	2	3	2	3	2

Ireland	200	100	800	800	1300	1500	300	300	100	200	400	600	300	300	100	200
Portugal	200	200	400	200	500	400	300	200	200	200	300	300	365	316	233	219
Latvia	1624	2082	3413	3379	2162	2611	1234	1602	1470	2049	2386	4124	4499	6097	7457	8287
Lithuania	2631	4291	4094	3928	2475	3433	1683	2402	1513	2066	1813	2390	1599	2127	1481	2237
Poland	1444	3876	2114	1672	2116	2331	1053	1228	978	1155	1302	1334	1391	1651	1869	2338
Czech R.	838	1099	1608	1817	1447	2626	3953	3553	3042	1046	1621	2567	2177	3638	1904	3159
Romania	103	150	180	213	290	59	81	58	82	58	81	58	80	58	78	58
Bulgaria	910	948	354	339	527	547	330	420	700	700	640	540	602	692	596	655

Explanations: *HA*, harvesting area; *PQ*, production.

17.2 Key issues of fibre flax breeding and cultivating

In this section, the basic sources and steps of breeding work are discussed, starting with European flax genetic resources followed by the breeding methods that are generally used.

17.2.1 Flax genetic resources – national inventories

Successful breeding work depends on the range of genetic diversity kept in the respective flax national collections (national inventories) of individual European countries followed by the breeding methods used. The first step to make an inventory of the European flax gene pool was made at the ad hoc meeting of flax collection holders within the newly established Industrial Crops and Potato Network at the last European System of Cooperative Research Networks in Agriculture (ESCORENA) network by European Cooperative Programme for Crop Genetic Resources (ECP/GR) held in Prague, on 7–8 December 2001 (Maggioni et al., 2001). The structure of the respective national inventories – Bulgaria, Czech Republic, France, Germany, Hungary, the Netherlands, Poland, Romania, Russia, Ukraine – and the methods of the work with flax genetic resources were presented (Maggioni et al., 2001).

The stocktaking of European flax genetic resources in 2001 showed the following status. Bulgarian flax collection is kept in two institutions. In Sadovo, 945 accessions were maintained in 2001. The evaluated 450 accessions showed that 23% are linseed, 29% flax and 46% of intermediate type. Landraces and old cultivars create 58%, advanced cultivars 15% and breeder's lines 14%. Except for *L. usitatissimum,* there are 12 wild species in the collection: *Linum altaicum, Linum austriacum, Linum bienne, Linum flavum, Linum grandiflorium, Linum humile, Linum perenne, Linum punctatum, Linum setaceum, Linum strictum, Linum trigynum* and *Linum viscosum* (Shamov, 2001). In addition, 283 accessions are maintained at the AgroBioInstitut (ABI) in Kostinbrod. Of these, 178 were received from IGR–Sadovo, 29 from other Bulgarian institutions, 71 from foreign institutions and five are own breeding materials. These accessions were originated from Europe, Asia, Africa and America. Only 6.4% are of Bulgarian origin. Cultivars represent 54% of the collection, landraces and primitive cultivars 27.5%, breeding lines 15%, wild forms 1% and genetic stocks 2.5% According to the plant type, fibre flax represented 31% of the collection, linseed 35%, combined type 32% and other types 2%. 259 accessions were described for 15 morphological traits, five biological and four yielding characters, according to International Union for the Protection of New Varieties of Plants UPOV (1991, 1995) and to the International Flax Database (IFDB) descriptors (Pavelek, 1994, 1995) and other descriptors for the species *L. usitatissimum* L. (Rykova et al., 1987, 1989). Linseed and combined type accessions were also analyzed with biochemical methods for oil and proteins content, Cd accumulation and fatty acid composition of oil according to standards of the International Standard Organization (ISO). Fibre quality evaluation of flax varieties was also made according to the Bulgarian standards (Balabanova and Atanasov, 2001).

The Czech flax national collection has been managed since the 1960s by Agritec Ltd., Sumperk. It is one of the largest European collections, with 2011 accessions of flax, linseed, intermediate and wild types in 2001. Based on geographical representation, the main part of the collection is of European origin, followed by a number of accessions from Africa, the USA and Australia. The flax national collection is described according to the Descriptor List (Pavelek, 2001a) by 22 passport and 55 special descriptors covering morphological traits and biological and yielding characteristics (Pavelek, 2001b).

The French flax and linseed germplasm collection contains 1650 accessions with about 50 Institut National De La Recherche Agronomique (INRA) lines. It is maintained by the INRA flax breeding laboratory. The genetic resources were collected from 42 different countries, but mainly from the USA (360 accessions), Argentina (327), Russia (192) and France (108). The collection is composed of 426 accessions of linseed, 273 of flax and 207 of intermediate *Linum*. Other genotypes seem not to have an agricultural interest but could be used as parents for breeding (Nôžková et al., 2014). Characterization is completed, and impurities or heterogeneous materials are detected. A catalogue of all the accessions, including their passport data, is available on Internet at http://www.inra.fr/Internet/Produits/Lin/index.htm. About 800 genotypes have almost fully been described for 17 qualitative and 14 quantitative traits (Fouilloux et al., 2001).

The German flax collection was previously kept in the two German genebanks (Federal Centre for Breeding Research on Cultivated Plants, Braunschweig, Germany [BAZ] Gene Bank/Braunschweig and Institute of Plant Genetics and Crop Research, Gatersleben, Germany [IPK] Genebank/Gatersleben) in 2001. A total of 2304 accessions of *Linum* plant genetic resources (PGRs) were maintained as follows – BAZ: 621 and IPK: 1683. Almost 95% of these accessions (2181) belong to *L. usitatissimum* as the only crop species of the genus. This species actually constitutes (with the exception of four as yet undetermined accessions) the entire BAZ collection, whereas in the IPK collection, 85 entries from at least 25 other species of the genus *Linum* are also maintained. With respect to the origin of the material, Russia/the former Soviet Union and Germany with 184 and 177 accessions, respectively, are the major donor countries, followed by Italy (87), Hungary (75), France (72), Iran (67) and Portugal (58). In total, 67 countries are listed as donors (BAZ: 32, IPK: 67). Regarding the continents, Europe with 1147 accessions is the origin of almost 50% of the entries, whereas Asia (179), America (154), Africa (82) and Australia (2) donated the second half, together with 745 accessions of unknown origin. In the active collection of the BAZ, modern varieties (72%) and breeding lines (21%) prevailed, whereas only few landraces (6%) were represented. In the IPK collection, modern varieties and breeding lines form a major part as well (total 44%). Less than 1% landraces and 5% wild material are present, whereas the sample status of 50% of the accessions is unknown. About equal amounts of oil (37%) and fibre (45%) types were maintained at Braunschweig, only 3% were classified as combination types. In Gatersleben, however, combination types dominate with 49%, followed by 27% fibre and 7% oil types; the usage of 17% of the entries is unknown. Concerning characterization and evaluation, the respective BAZ activities were focused on inflorescence and agronomic and seed trait. The evaluation activities for the genus *Linum* include 46 descriptors. In the IPK genebank, *Linum* accessions were also characterized for a majority of the IFDB descriptors, the respective data not yet being available in an electronic format (Dehmer et al., 2001). Nowadays all accessions from BAZ have been moved to the IPK, which is now the only institution responsible for the whole German germplasm stock (Dehmer et al., 2001).

The Hungarian national *Linum* collection, which is over 45 years old, is part of the national collection and is composed of 409 accessions. 94% of the *Linum* collection belongs to the species *L. usitatissimum.* The analysis of *L. usitatissimum* by subtaxa showed a wide range of variation. 80% of the collection derives from Hungary (40.3%), former Czech and Slovakia (15.9%), Germany (13.2%), Great Britain (3.2%), Romania (2.7%), France (2.4%) and Poland (2.2%). 71% of the collection is composed of advanced cultivars. Characterization and evaluation have already been carried out on 307 accessions, equivalent to 75% of the *Linum* collection (Simon, 2001).

The Dutch *Linum* collection was adopted by the Centre for Genetic Resources, the Netherlands (CGN) in 1995, after a group of four private Dutch breeding companies encouraged the institute to take action for the proper conservation of this valuable collection. In 1996, an arrangement for cooperation in the field of maintenance and evaluation was established between the breeding companies and CGN. The collection included 974 accessions in 2001 and is divided into fibre flax, linseed, intermediate flax and wild species.

A minimal descriptor list was developed based on the descriptors for *Linum* of the IFDB of the FAO European Cooperative Network on Flax. The present CGN descriptor list includes 16 descriptors that can be divided in two groups:

- 10 mandatory descriptors: These traits should be always recorded during the regeneration.
- Six optional descriptors: These traits can be recorded when sufficient time is available or at the occurrence of diseases for which the character can be optimally scored in the field.

In the framework of a multidisciplinary fibre research programme, a core collection of fibre flax was developed. The core was generated over the period from 1998 to 2001 and the development was conducted after different stages, resulting in the final core collection of 84 accessions (Soest, 2001).

The Polish Flax Collection managed in former Institute of Natural Fibres (INF), nowadays Institute of Natural Fibres and Medicinal Plants (IWNiRZ) Poznań, Poland, comprises 864 accessions: 48 accessions of wild species, 29 landraces or primitive cultivars, 588 advanced cultivars, 102 breeder's resources and 97 accessions of unknown status. The IWNiRZ collection holds also unique Poland-originating accessions. All genotypes present at the IWNiRZ genebank have been characterized as far as genetic and economic qualities are concerned. The flax collection is also described according to resistance to diseases, especially the most dangerous one – Fusarium wilt. The most important task is the improvement of the protection and preservation of flax and hemp genetic resources. The main aim of the IWNiRZ genebank is to help breeders in their work by increasing availability of genetic resources for breeding and research and by screening the core collection for properties important for breeders (Rutkowska-Krause, 2001).

The Romanian *Linum* national collection is represented by 3845 accessions, mainly of *L. usitatissimum* (98%), with another 20 species. Only 220 accessions are considered as duplicates. 2161 accessions are oil varieties, 1068 are fibre type, 71 are intermediate and 545 unknown. The institutions holding flax collections in Romania are the Research Institute for Cereals and Industrial Crops of Fundulea (2880 accessions), the Suceava genebank (520), the Agricultural Research Station of Livada (420) and the Agricultural Research Station of Simnic (25). The National Genebank of Suceava has the responsibility to preserve 'ex situ' the Romanian agrobiodiversity. About a third of the collection has Romanian origin, from which a small part is represented by old cultivars (Strajeru, 2001; Vasile, 2001).

The Russian flax national collection located in the N. I. Vavilov Research Institute of Plant Industry (VIR) in St Petersburg, Russia consists of 5521 accessions and represents the biggest European collection. It covers a wide range of diversity. The collection includes all three flax types: fibre, oil, usually called linseed, and intermediate type, which can be used for both purposes. Some wild species are also included in the collection. The collection covers the whole area where flax is cultivated. More than half of the collection is represented by local folk bred varieties. Also several commercial varieties, lines and other breeding material are conserved. The main part of the collection is composed of accessions originating from the former Soviet Union Republics. The main donors of genetic material are Russia, Uzbekistan, Ukraine and Tajikistan. The majority of these accessions are landraces, which now can be found in this collection only. Practically all flax commercial varieties ever bred on the territory of the USSR are maintained. Breeding lines and donors of different agronomic characters are widely represented too. Apart from the former Soviet Union, flax genetic material from 58 foreign countries is included in the collection. Some countries, such as India and Ethiopia, are represented mainly by local accessions. For some other countries such as, for example, Germany and the USA, the collection is represented mainly by commercial varieties and breeding material. A genetic collection consisting of about 250 inbred lines and lines of lower generations are conserved at the VIR. These lines carry different morphological and agronomic characters. This collection includes different phenotypes of traits such as colour and shape of flowers and seeds, resistance to rust, duration of vegetative period, plant height and others. For many of them, genetic control of the character has been identified (Brutch, 2001).

The Ukrainian flax collection contained 1042 samples in 2001. The collection was only established after 1992 because all material necessary for carrying out scientific research and breeding work were previously obtained from the VIR. The collection, based in Glukhiv, is currently composed of samples from 45 countries, the largest amount from Russia (256), 117 from Sweden, 54 from the Netherlands, 75 from Czech Republic and Slovakia, 52 from Ukraine, 52 from the USA, 34 from Argentina, 30 from France, 23 from Poland, 23 from Hungary and 61 from Germany. So far, 870 accessions have been characterized and a number of samples have been singled out for their useful properties, such as 132 for fast ripening, 154 for seed yield, 81 for straw yield, 76 for fibre content in the stalks, 42 for plant height, 114 for good spinning capacity of fibre and 41 for resistance to lodging. All the materials have also been evaluated for their resistance to diseases, and five carriers of resistance to Fusarium wilt have been identified. Characterization of the collection is carried out and includes information about country of origin, originating institute, authors, botanical species, intraspecific taxon, life cycle, type of development, ploidy, value of the sample, availability, etc. Unfortunately, this work is still far from completion (Virovets et al., 2001).

The Italian flax collection is maintained by CRA-Instituto Sperimentale per le Colture Industriali (CRA-ISCI). It comprises 380 accessions, mainly originating from European countries. A database (MS Access) for both passport and descriptive data on the collection, according to the IFDB structure, is available for internal use (Mandolino, 2006).

Part of the present Latvian flax collection has been built by repatriation of Latvian material from Russian and German institutes. The Latvian flax genetic resources are maintained by two institutes: the Latvian Genebank of Cultivated Plants and the Latgale Agricultural Science Centre. The former maintains 26 accessions from Latvian origin in both base and active collections (Grauda, 2006).

The Portugal flax collection maintained by Banco Português de Germoplasma Vegetal/Direcção Regional de Agricultura de Entre Douro e Minho (BPGV/DREM) consists of 147 accessions. Estação Agronómica Nacional (EAN Genebank) maintains eight accessions. All accessions originate from Portugal (De Sousa, 2006).

The Slovakian flax genetic resources are stored at the Genebank of Slovak Republic, which is the part of the Research Institute of Plant Production in Piestany. It consists of 170 accessions, of which 145 accessions are both in base and active collections and are safety duplicated at RICP Praha-Ruzyne. The passport data of 136 accessions are computerized and ready to be submitted into the IFDB (Nožková, 2006).

The Lithuanian collection and storage of flax genetic resources was started in 1994. The collection of *L. usitatissimum* L. in Lithuania consists of 922 accessions in the active collection of which 51 accessions are stored under long-term storage conditions. At the Field Crops Coordinating Centre in Lithuanian Institute of Agriculture in Akademija, Kėdainiai district, 51 accessions of flax are stored in long-term storage facilities. The active flax and linseed collection is stored and investigated at the Upytė Research Station of the Lithuanian Institute of Agriculture (Jankauskienne, 2006).

Nowadays the work with flax collections is in progress, the traditional methods of evaluation based on passport and special descriptors analysis are still replenished by molecular methods of characterization to rationalize the work with collections, to create core collections as well as to identify and to distinguish the respective accessions. Various DNA markers have been widely used for diversity analysis in plants, including randomly amplified polymorphic DNA (RAPD) molecular markers, inter-simple sequence repeat (ISSR), amplified fragment length polymorphism (AFLP) and simple sequence repeat (SSR). All these methods have also been applied to study flax germplasm diversity (Cloutier et al., 2009; Diederichsen and Fu, 2006; Everaert et al., 2001; Fu, 2002, 2005; Fu et al., 2002a; Fu et al., 2002b; Vromans, 2006; Wiesnerová and Wiesner, 2004). One of the last developed molecular methods is inter-retrotransposon amplified polymorphism being also often used for germplasm collections analysis (Antonius-Klemola et al., 2006; Kalendar and Schulman, 2006; Smýkal et al., 2008; Vukich et al., 2009).

17.2.2 International flax database

"Convention on biological diversity" (UNCED, 1992) and "Global Plan of Action" (FAO, 1990) are the main documents for conservation, maintenance and utilization of PGRs. This work is coordinated by FAO and International Plant Genetic Resources Institute (IPGRI), (currently Bioversity International) in Rome, Italy. The Global Plan of Action is focused on worldwide networks creation resulting in PGR protection, conservation and utilization. Via IPGRI, the ECP/GR has been operating since 1980 and nearly all European countries are involved (Maggioni et al., 2001). The activities on PGRs are carried out through 10 networks in the respective working groups in the framework of action VII (2004–08) where also the creation of the international crop databases is considered to be one of the FAO and IPGRI networks' activities. These central crop databases are accessible via the European Internet Search Catalogue (European Plant Genetic Resources Search Catalogue with Passport Data on Ex Situ Collections Maintained in Europe [EURISCO]) (Maggioni et al., 2001). The EURISCO web catalogue automatically receives data from the national inventories. It effectively provides access to all *ex situ* PGR information in Europe and thus facilitates locating and accessing PGRs. EURISCO is hosted at and maintained by the IPGRI on behalf of the Secretariat of the ECP/GR. EURISCO and the list of FAO/IPGRI Multicrop Passport Descriptors (MCPDs) were accepted by the European database managers in 1996 for data exchange (Lipman et al., 1997) and have also been used as a starting point for the IFDB development.

The system of flax germplasm evaluation, description and characterization is not uniform either in Europe or in the USA and Canada and differs according to the breeders' requests in individual countries. Regarding the different evaluation and characterization systems used in the individual European genebanks as well as overseas and in order to exploit the richness of the flax genetic diversity for breeding activities, it is necessary to have a good, unique system of evaluation, description and characterization, which should be uniform in European countries at least. Therefore the IFDB can be considered as an initial input for this aim (Pavelek, 1995).

The IFDB has been managed and coordinated by the Agritec Ltd. since 1993 in the framework of FAO ESCORENA Flax and other Bast Plants Network (FAO FOBPN) (Pavelek, 1995, 1997; 1998a, b; Pavelek et al., 2001) and then since 1999 in the framework of IPGRI Coordination Group Network for Sugar, Starch and Fibre Crops (CGN-SSFC), and nowadays in Sugar, Starch, Fibre Crops & Aromatic Plants Network (CGN SSFC&APN) at Bioversity International. It presently includes passport data of 8385 accessions of 11 collections from 10 countries in an ACCESS IFDB structure. From an analysis of the delivered data, it can be concluded that approximately 37% of the accessions are unique. It also revealed large differences in the percentage of descriptors filled in among the collections. The total European gene pool is estimated at 27,437 accessions, maintained at 16 genebanks. The majority of these accessions are maintained in Russia, Romania, Germany, Czech Republic and France. In EURISCO, approximately 17,175 *Linum* ssp. accessions are recorded of 31 institutes from 21 countries (Pavelek, 2009).

A new IFDB passport and E/CH structure after the meeting of Flax and Hemp Group by Sugar, Starch and Fibre Crops Network (Wageningen et al., 2005; Bas et al., 2006) has been adopted and demonstrated in the database managers meeting in Quedlinburg in 2009 (Pavelek, 2009). It is based on MCPDs and consists of 17 passport descriptors, six additional passport descriptors and seven recommended E/CH descriptors (Pavelek, 2009). Based on 7934 entries (except 451 entries from Bulgaria), contemporary analysis of IFDB MCPDs was carried out and the results are shown in Table 17.2.

Contemporary status of IFDB concerning the entries of the individual European genebanks and the structure according to the type of use is presented in Tables 17.3 and 17.4.

Table 17.2 Survey of International Flax Database Multicrop Passport Descriptors filled in.

Multicrop passport descriptors	Filled in (%)
ACCENUMB	40.34
ACCENAME	72.37
INSTCODE	100.00
ACQDATE	36.36
DONORCODE	68.39
DONORCODECTY	30.19
DONORNUMB	47.32
ORGCTY	83.80
GENUS	100.00
SPECIES	100.00
USETYPE	64.19
PLOIDY	53.38
ORIGIN	57.60
CHARMATER	51.94
PEDIGREE	33.97
GROWTHHAB	59.97
LIFETYPE	60.81
COLLYEAR	56.40
COLLSITE	38.76
LONGSITE	44.23
LATITSITE	44.59
ELEVSITE	31.62
COLLNUMB	33.10
COLLDATE	27.83
COLLSRS	42.05
AVAILAB	56.78
ALTERDATE	53.49

Table 17.3 Contemporary status of International Flax Database (IFDB) at the beginning of 2004.

European genebanks – databases included	Number of accessions
Previous IFDB	1385
IPGR Saadovo – Bulgaria	(453[b])
CGN Wageningen – the Netherlands (from Internet)	747 (974[a])
CAS Latgale region – Latvia	15
IPK Gatersleben – Germany (from Internet)	1680 (2304[a])
ENMP Elvas – Portugal	87
Suceava, Livada – Romania	499 (247[a])
VIR St. Petersburg–Russia	1486
VNIIL Torzhok – Russia	243
INF Poznan – Poland	200
ABI Tapioszele – Hungary	410
IBC Glukhiv – Ukraine	1182
Total	7934 (8986[a])

[a]All entries are described by passport data and partly by describing data.
[b]All entries are described by passport and describing data.

Table 17.4 Structure of contemporary status of International Flax Database according to the type of use.

Use type	Number and % of accessions
Fibre	3700 (46.63%)
Linseed	825 (10.40%)
Combined type	509 (6.42%)
Others	77 (0.97%)
Not specified	2823 (35.58%)
TOTAL	7934

From Table 17.4 especially it is obvious that nearly one-half of the IFDB is created with fibre flax, followed by linseed. Unfortunately, more than one-third is not specified because this descriptor was not filled in.

Based on geographical analysis, it is understandable that most maintained accessions are of European origin Fig. (17.1). The respective part of Asia's accession origin is kept in the VIR St Petersburg, Russia collection and also an insignificant part of

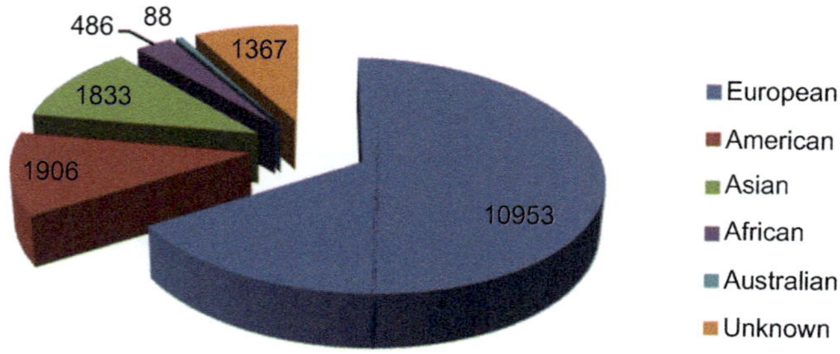

Figure 17.1 Structure of IFDB according to the geographical origin.

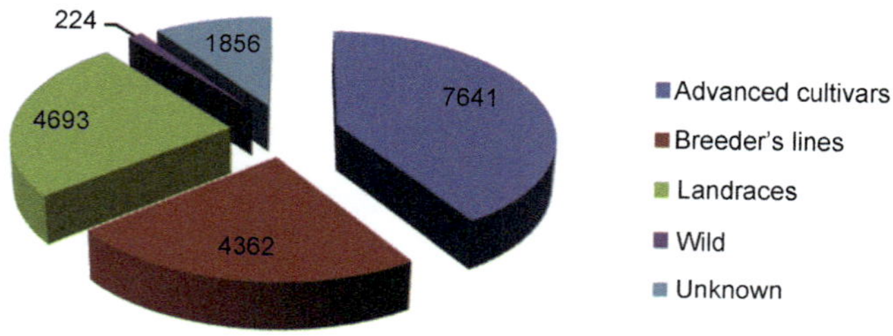

Figure 17.2 Structure of IFDB according to the origin.

accessions belongs to the category 'unknown' where the geographical representation is not identified – for example, in genebanks IPK Gatersleben, Germany, and CGN Wageningen, the Netherlands.

The more significant variability was identified in the category 'origin of the maintained material' (Fig. 17.2). The biggest part in most European genebanks is created by the advanced cultivars, followed by the breeding lines. This trend is marked in the genebanks, which are often located in the research or breeding institutes especially interested in breeding activities and the development of their own origin varieties. The category 'landraces' is maintained in the genebanks with a long history like in St Petersburg – Russia, Sadovo – Bulgaria and Sumperk – Czech Republic, where the base of collection was just established on old flax and linseed materials at the beginning. Categories 'breeder's lines' and 'landraces' are in the IFDB and practically comparable. The category 'wild material' is registered in individual genebanks only in Sadovo – Bulgaria, IPK Gatersleben – Germany and INF Poznan – Poland, and it depends on the genebank manager's decision whether this category becomes the part of *L. usitatissimum* collection or not. The respective part belongs also to the category 'unknown', which is nearly 57% in ISCI Bologna – Italy, 37% in IPK Gatersleben – Germany and 30% in IBC Glukhiv – Ukraine.

Because in the IFDB there are now 8000 incomplete records and the total European gene pool consists of approximately more than 27,000 accessions, it is necessary to adopt another 19,000 records into the IFDB structure in the future (Pavelek, 2004).

17.3 Methods of flax and linseed breeding

L. usitatissimum L. species is a self-pollinated crop, and its genetic improvement can be carried out through conventional breeding methods of hybridization and selection on the one hand or utilizing new techniques such as haploidy, interspecific hybridization, mutation, tissue culture and transformation on the other. Commercial breeding of flax started at the end of the nineteenth century (Vromans, 2006). Nowadays, several breeding methods are available, but the pedigree method is the most common one used in flax breeding (Salas and Friedt, 1995).

Breeding work is carried out in the agreement with breeding aims and it is focused on two types of use, fibre flax and linseed. Content of fibre in the stem and yield of fibre per hectare are the main characteristics of flax and the subject of improvement by breeding methods. Long and short fibres are used in textile industry; waste products such as flax shive and tow are used in building, car and air industries for furniture, paper and fuel production; chaff and seeds in the food industry and flax oil for technical use – varnish, colours and paint production. Genetic resources for fibre content and fibre yield improvement are known, and the breeding work is based mainly on traditional methods. Breeding of linseed varieties is especially directed to the qualitative parameters of the seed. It means fatty acid composition, higher content of lignans, predominantly secoisolariciresinol, and lower content of cyanogenic glycosides predetermine the different directions of utilization of linseed and thus increasing linseed yield. More emphasis is placed on the complex use of the whole plant, both in terms of seed processing and the use of fibre for technical purposes. Thus, the linseed varieties compensate a certain extent of the flax area loss in last years. Primarily, mutation breeding for different fatty acid composition has been used.

An indispensable precondition for genetic and breeding research is good knowledge and targeted work with genetic resource collections as mentioned above. This includes both intraspecies variability of certain crops (old and modern cultivars, landraces and experimental exactly defined lines – mutants, transloci lines) and variability of wild forms or forms of a certain genus often crossed with cultural forms. Genetic resource collections demonstrate gene pools of all spectrum traits and characters, which can be utilized based on hybridization. A very close genetic link between requested and nonrequested characters can be considered a large problem by incorporating some abilities from the wild species into cultural forms resulting often in the loss of cultural phenotype and its cultural characteristics. It is often very difficult based on backcrossing to renew the former phenotype. Very important for breeding work (choice of parental genotypes) is the fact that these parental genotypes are equipped with the requested genes on the genetic background of the yielding line.

Hybridization followed by progeny selection is still the main breeding method in flax and linseed breeding programmes. Intraspecies breeding is predominantly used.

17.3.1 Conventional methods of breeding

Conventional methods include intraspecies intervarietal crossing, pedigree method of breeding and single seed descent (SSD) method:

basic cross: A×B
topcross: A×B, A×C, A×D, A×E
saturating cross: (A×B)×C
backcross: (A×B)×B, (A×B)×A
convergent cross: (A×B)×(C×D)×(E×F)
different types of diallel crosses: A×B, A×C, A×D, B×C, B×D, C×D

17.3.1.1 Breeding in European countries

Flax and linseed genetic resources collections for breeding aims are studied, and the choice of initial useful parental patterns is frequently discussed. Tadesse et al. (1997) studied crosses between high × high– and low × high–yielding flax components, resulting in the recommendation that hybridization of low- and high-yielding flax lines may be useful for increasing genetic variability and obtaining high-yielding flax lines. Based on the results (Pavelek, 2001c), also crosses of high × high–yielding flax components can be successful when the genetic diversity of parental lines is large. In case of Venica variety (SL-988 × Viking) registered for growing in the Czech Republic in 2001 (Pavelek, 2001c), parental lines were based on completely different varieties as follows: SL-988 (Regina × Toržokskij), Regina (Reina × /Reina × Fibra), Toržokskij (M-34 × Tomskij-10), Viking (Natasja × Silva) and Natasja (Wiera × Mapun). Fibre content in the stem and its quality represented by thinness and divisibility can be considered main criteria for all European countries focused on flax breeding. Selection for fibre content should be effective (Fouilloux, 1988), and fibre content is positively influenced by the recessive gene cumulation (Fouilloux et al., 1991). Nevertheless, the evaluation of the correlation relationship between fibre content and qualitative characters of the highest breeding stages is very problematic (Kate, 1991).

Based on these traditional methods all nine – two flax and seven linseed – cultivars registered for growing in the Czech Republic were bred. From this, all flax and five linseed varieties are of the Czech origin. The situation is similar in other European countries such as France, the Netherlands, Poland, Romania, former countries of Soviet Union, Ukraine, Lithuania, Latvia and Russia. There are often simple or sophisticated cross combinations derived from the West European and East European cultivars (A × B; /A×B/×C; /A×B/×/B×C/). Also interspecies hybridization was used with the species *Linum crepitans* L. to include earliness into the *L. usitatissimum* L. Success in this direction was achieved, although these early maturing types were not used in further breeding. The effectiveness of interspecies crossing is usually low, but it could be increased using the method of embryo tissue culture – cultivation of unmatured isolated embryos in in vitro conditions. Breeding of resistant lines is generally carried out on provocation fields infected by pathogens sometimes followed by tests at the stage of seedlings. Also Trouvé (1996) indicated that using a few high-performing genotypes for flax breeding over the past century resulted in a significant loss of genetic variability. Recurrent selection breeding is suggested for flax to avoid genetic stagnation and to reach genetic population variability (Trouvé, 1996). Good yield of the stem and fibre is positively correlated with the vegetation period. Vegetation period length, course of vegetative and generative phases as well as fibre content are the good inherited characters (Pavelek, 1985, 1991a; 1991b). Beckmann and Kromer (1995) consider fibre content, tensile strength, percentage elongation and Young's modulus as selection criteria for breeding of flax varieties. Khotyleva et al. (1997) in agreement with Fouilloux et al. (1991) and results (Pavelek, 1980, 1982) reported the content of fibre is influenced by additive effects and also incomplete dominance and superdominance.

In the second half of the last century, the breeding work was focused on increase of yielding potential. A lot of new fibre flax and also linseed varieties were bred in traditional European countries such as France, the Netherlands, Czech Republic, Poland and Russia (Ivanova and Shamov, 1996; Pavelek, 2001c; Pavelek and Tejklová, 2005; Scheer-Triebel et al., 1997; Todorov and Lukipudis, 1997) and also in other countries such as India and the USA (Foster et al., 1997, 1998; Geleta, 1999; Pradhan et al., 1999; Stephens, 1996; Yadav et al., 2000).

Nowadays the breeding work is predominantly directed on the characters stabilizing the yield, for example, resistance to fungi diseases, unfriendly outside factors, and qualitatively new content of matters. New methods of breeding work are looked for based on mutagenesis (Rakouský et al., 1998; Tejklová, 1995, 2002, 2008), molecular approaches (method of isoenzyme spectra – Krulíčková et al., 2002; Pošvec and Krulíčková, 1999; Yurenkova et al., 1992), anther culture, mutation breeding, RAPD analysis (Fu et al., 2003; Mansby et al., 2000), and methods of transformation, making it possible to solve the problem when the possibilities of genetic variability are limited and exhausted. Fibre content and resistance to *Fusarium* and scorch are the most important breeding goals for fibre flax (Fouilloux, 1988), whereas seed yield, fatty acid composition and resistance to rust (*Melampsora lini)* and *Fusarium* are important for linseed breeding (Rowland, 1998). *Alternaria linicola, Fusarium oxysporum* f. sp. *lini, Septoria linicola, M. lini, Phoma exigua* var. *linicola* and *Botrytis cinerea* can be considered most harmful for both flax and linseed throughout the world (Ondřej et al., 2008), and the resistance breeding work is carried out in most European countries. Intensity of occurrence and harmfulness of the respective diseases are strongly influenced and modified by geographical conditions. Recently, powdery mildew and pasmo have been widespread causing local disease epidemics. Maintaining resistance to rust and wilt and breeding for resistance to pasmo and powdery mildew are the objectives in developing new varieties.

Complex *Alternaria* + *Oidium* + *Melampsora* occur in subtropical conditions especially, whereas *Fusarium* + *Septoria* + *Melampsora* + *Oidium* are located in dry arid areas. In the northwest European countries, there are *Alternaria* + *Botrytis*; in southwest areas and south countries, complex *Fusarium* + *Oidium* predominates.

Pasmo is a widespread harmful flax disease. The causal organism of pasmo is *S. linicola*. This fungus attacks above-ground parts of flax and overwinters in the soil on infected flax stubble. Flax is susceptible to pasmo from the seedling stage to maturity. High moisture and warm temperature favour the dinase. Epidemics can occur early in the season, but it is most readily recognized on the flax stems in the early ripening stage as irregular bands of brown mottling alternating with bands of green. Pasmo can cause defoliation and premature ripening. Depending on the earliness and severity of the infection, pasmo reduces seed yield as well as seed and fibre quality.

A. linicola is a serious disease of linseed in England, Ireland, Denmark, the Netherlands, Belgium and Normandy. The loss of the yield reaches 10%–60% (Chauhan and Srivastava, 1975). This disease is spread by the seeds or the rest of damaged plants. The main symptoms on its host include poor germination, damping-off of seedlings, leaf spotting and a form of head blight in the capsules, which can result in a loss in yield and a reduction in oil quality. Although the attack occurs initially at the seedling stage, disease symptoms are rarely discernible on the growing crop until capsule formation is almost complete later in the season.

Information concerning the genetics of resistance is disunited. The resistance influenced by one major gene was demonstrated by Kalia et al. (1965) and Evans et al. (1995, 1997), whereas polygenic resistance was found by Klose et al. (1993).

Powdery mildew (*Oidium lini*) is a common and widespread foliar disease of flax in most flax-growing areas of the world. Occurrence of *O. lini* was detected in 50%–60% of cases in the last century in England and Germany especially connected with the decrease of yield in the range 5%–20%. In India (Pavgi and Singh, 1965) and China, the loss reached 30%–60%. Generally it occurs in Europe, Australia, Asia and America. The symptoms are characterized by a white powdery mass of mycelia that start as small spots and rapidly spread to cover the entire leaf surface. Heavily infected leaves dry up, wither and die. Early infections may cause severe defoliation of the flax plant and reduce seed yield and quality. Some flax varieties are resistant or moderately resistant to this disease. Taxonomy and origin are still not completely investigated. It is supposed to result from the existence of two or three species: *Erysiphe polygoni, Erysiphe cichoracearum* and *Sphaerotheca lini.* Heritability of resistance is being studied in India only. Formerly, one gene of resistance (R1) was detected (Badwal, 1975), in 2000, two genes, and in 2002, three genes of resistance (R1, R2, R3) were demonstrated (Rashid and Duguid, 2002). Based on the results reached in the Czech Republic, it was stated that the genetic background of the resistance to *O. lini* is unclear, is valid for the set of tested varieties and could be different depending on the choice of varieties (Ondřej et al. 2008).

F. oxysporum f. sp. *lini, Fusarium avenaceum* and *Fusarium equiseti* are considered to be the most serious diseases of flax and linseed, together called Fusarium wilt. It is spread over the world in semiarid or arid areas especially. This disease is one of the limiting factors of flax and linseed growing in the USA and Canada (Kommendahll et al., 1970). Thanks to the programme of resistant breeding in North America; all commercial flax and linseed cultivars are characterized as resistant or moderate resistant (Kenaschuk and Rashid, 1993; Kenaschuk et al., 1996). Resistant flax and linseed cultivars were also developed in Europe (Kroes et al., 1998, 1999). Wide pathogenic variability in *F. oxysporum* f. sp. *lini* has been reported. Many different biotypes and pathogenic types of *F. oxysporum* f. sp. *lini* have been detected and isolated (Armstrong and Armstrong, 1968; Kommedahl et al., 1970) from different regions in the world: Argentina, the USA, Canada, Australia, India, Japan (Borlaug, 1945; Kroes et al., 1999, Muir and Westcont, 2003; Houston and Knowles, 1949; Milikan, 1945, 1948; Sharma and Mathur, 1971; Tochinai and Takee, 1950). Some varieties are more susceptible than others. Early infections may kill flax seedlings shortly after emergence, whereas delayed infections cause yellowing and wilting of leaves, followed by browning and death of plants. Resistance is probably influenced by the minor genes, and it has a polygenic character. Investigations carried out in eight countries on nine localities during 1995–96 showed that population virulence in the USA and Canada was very strong, medium in France, Germany, Czech Republic and Russia, and very weak in Belgium and the Netherlands. The most resistant varieties were AC Linora, AC Emerson and Atalante. The most susceptible were Ocean, Barbara, Liflora and McGregor varieties (Kroes, 1997). In the Czech Republic, 80 linseed breeding lines were tested in field conditions in 2005. 47.9% of the lines showed medium resistance, whereas 52.1% were medium susceptible. In 2007, 55 breeding lines were evaluated. 41.8% of the lines were resistant including standard varieties Atalante and Jantar. Medium resistance was detected in 25.6% lines; also in linseed varieties including Bajkal and Lola. 20% lines showed medium susceptibility with linseed varieties Flanders, Lindor, Recital and Kaolin. Susceptibility was detected in 7.2% of the lines of varieties Niagara, Baladin and Alaska, and high susceptibility was found in 5.4% of the lines including Astral, Jupiter and Ocean varieties (Ondřej et al., 2008). The varying levels of resistance or susceptibility are evidence of the polygenic character of the resistance, and it can be used by the breeding work on the provocation fields (Ondřej et al., 2008).

Breeding work is carried out according to breeding aims. For flax and linseed, these are presented in Table 17.5.

Through the breeding aims, the following target parameters should be reached:

Flax (*Linum usitatissimum* L.)

Table 17.5 Breeding aims for flax and linseed.

Flax	Linseed
• High resistance to lodging	• High resistance to lodging
• High resistance to pathogen complex	• High resistance to pathogen complex
• Middle vegetation period	
• Average unretted stem yield	• Middle vegetation period
• High content of fibre in the stem	• Low stem yield
• High yield of fibre per hectare	• High seed yield
• New quality: low linolenic acid content	• High content of fat in the seeds
	• High fat yield per hectare
	• New quality: low linolenic acid content

Flax ideotype

- resistance to lodging: 9 p;
- resistance to pathogen complex: 8 p;
- middle vegetation period;
- yielding potential of unretted stem reached in trials: 7–8 t ha^{-1};
- yielding potential of the seeds reached in trials: 1.10–1.30 t ha^{-1}
- long fibre content potential reached in trials: 22%–25%;
- total fibre content potential reached in trials: 39%–41%;
- long fibre yielding potential reached in trials: 1.25–1.40 t ha^{-1} and
- total fibre yielding potential reached in trials: 2.50–3.0 t ha^{-1}.

Cultivars will be enhanced by

- low linolenic acid content: less than 5%;
- increasing absorption and accumulation of Cd and Pb and increased toleration to Cd and Pb and
- resistance to the herbicide Basta.

Linseed ideotype

- resistance to lodging: 9 p;
- resistance to pathogen complex: 8 p;
- middle vegetation period;
- seed yielding potential reached in trials: 2.30–2.40 t $ha.^{-1}$;
- fat content potential reached in trials: 42%–46% and
- fat yielding potential reached in trials: 980–1000 kg ha^{-1}.

Cultivars will be enhanced by

- low linolenic acid content: less than 5%;
- increasing absorption and accumulation of Cd and Pb and increased toleration to Cd and Pb and
- resistance to the herbicide Basta.

17.3.1.2 Breeding in India, the USA and Canada

India, the USA and Canada are the biggest growers and producers of linseed in the world (*Situační a Výhledová Zpráva*, 2009). A list of linseed areas and production figures in 2003–07 is presented in Table 17.6.

It is obvious from the table that the production in India has an extensive character in comparison with the USA and Canada, and according to Gill (1987), the seed yield is very low. Therefore the main breeding goals especially in India and also in other overseas countries (Gill, 1987) are

- improvement of the seed yield;
- improvement of content and quality of oil;
- resistance to diseases and pests and
- early-maturing cultivars development.

Table 17.6 Survey of linseed areas (ha) as well as the production (mill. t) in India, the USA and Canada, 2003–07.

Country	2003		2004		2005		2006		2007	
	Area	Production	Area	Production	Area	Production	Area	Production	Area	Production
India	450,100	176,700	476,500	196,500	448,700	169,700	436,800	172,500	426,000	167,000
USA	235,900	264,830	206,800	263,360	386,480	500,280	310,400	279 00	141,235	149,963
Canada	728,400	754,400	518,000	516,900	732,600	990,600	785,200	988,800	524,000	633,500

The breeding approaches are similar to those in Europe including conventional breeding methods of selection and hybridization and also new techniques such as male sterility, haploidy, interspecific hybridization, mutation and tissue culture (Gill, 1980).

17.3.1.3 Breeding in India

Pure line selection

This method was used at earlier stages of linseed improvement (Gill, 1987), and many of the old linseed varieties originating from India were pure lines, e.g., NP 12, NP 121 and NP 124.

Hybridization

Hybridization has been used to combine different characters in one material since the beginning of the last century (Gill, 1987). This method was often used for incorporating the resistance from different sources into the hybrids. The pedigree method was the most used method at the time (Gill, 1987).

Pedigree method of breeding

The pedigree method of breeding, commonly used from the 1930s, (Culbertson, 1954; Grahan and Roy, 1924; Hayes et al., 1955; Kenaschuk, 1975), includes three main steps:

- crossing between parental varieties possessing the characters to be combined;
- growing the progenies of the selected types and
- keeping accurate records in order that the individual plants can be traced from one generation to the next.

A detailed description is presented by Gill (1987). The pedigree method of breeding has been used for developing the improved linseed varieties in India.

Modified pedigree method

This method is based on the SSD after hybridization followed by the homozygosity process for several generations. Usually in F_5 generation, each progeny is maintained in bulk (Brim, 1966). Choice of the individual plants can be applied for the traits with high level of heritability (rust resistance for example), while the characters of complex disposition (like yield or other less heritable characters) are easier to detect after the homozygous lines establishment. This method can be carried out when the parental varieties are very well adapted to the outside conditions.

Backcross method

This method is recommended when desirable genes need to be incorporated into the genotype, for example, resistance to diseases into the susceptible varieties. The method was described by Kenaschuk (1975), and it was used not only for resistance incorporation but also for yellow seed incorporation into the brown seeded variety (Gill, 1964).

Recurrent selection

Recurrent selection by linseed can be applied when undesirable linkages need to be overcome. This method was successfully used in cotton (El-Adl and Miller, 1971; Miller and Rowlings, 1967), soybean (Hanson et al., 1967), wheat and barley (Redden and Jensen, 1974) and tobacco (Matzinger and Wernsmann, 1968).

Mutation breeding

Mutation breeding has a quite long history in flax and linseed, and chemicals or rays were often used to improve requested characters. Favourable mutations are reported for many different traits by various authors. Improvement of oil content was demonstrated by Larter et al. (1965), Rath and Scharf (1968), Srinivasachar and Malík (1971), Seetharam (1972) and Srinivasachar et al. (1972). Other improvements include seed yield (Bari, 1971); iodine value (Harpstead, 1961; Rath and Scharf, 1968; Srinivasachar and Malík, 1971; Srinivasachar et al., 1972); seed weight (Seetharam, 1972) and height and maturity (George and Nayer, 1973).

Interspecific hybridization

All the above-mentioned linseed breeding centres in India predominantly used the pedigree method of breeding. Nevertheless, interspecific hybridization has also been used to improve and extend the genetic variability of fatty acid composition (Gill, 1966; Yermanos, 1966). Interspecific hybridization, however, is difficult due to the different numbers of chromosomes reaching 8, 9, 10, 12, 14, 15, 16, 18, 30 and also more than 30 haploid chromosomes. The method was only successful when crossing the species with the same number of chromosomes as in the case of *L. usitatissimum* L., for example, which could incorporate resistance to the rust *M. lini.* Some species such as *Linum grandiflorum, Linum perenne* and *Linum austriacum* are self-incompatible, and crossing between them and *L. usitatissimum* species was not successful (Ghai, 1966; Gill, 1966).

Hybrid linseed

Production of hybrid varieties accompanied by vigour or heterosis effect is connected with knowledge about floral morphology and the mode of pollination. Although heterosis by flax ranged from 40% (Carnahan, 1947) to 231% (Dubey and Singh, 1969) flax and linseed hybrid production is not so usual as in maize, sugar beet, sorghum, onion or pear millet. The main reason is that the flower of the male sterile flax has a small corolla which fails to open to allow cross-pollination. The problem of the male sterile flax was investigated by Dubey and Singh (1965, 1966), Kumar and Singh (1970, 1972) and Kumar (1971). In India, especially, a lot of studies were devoted to the investigation of combining ability of flax or linseed varieties in diallel crosses permitting also the evaluation of heterosis effect. Positive and significant heterosis for seed yield per plant at 40 crosses and fibre yield at 13 crosses was confirmed by Pant and Mishra (2008). This effect is often derived from the cumulative effect of the additive genes, dominance, or nonallelic interaction. Hence, the success of any selection or breeding programme depends on precise estimation of various components

(Mohammadi et al., 2010). The mutual relations of the genes expressed by the general combining ability (GCA) or specific combining ability (SCA) influence manifestation of the respective characters. GCA by flax and linseed is connected with traits such as plant height, number of capsules per plant, weight of capsules per plant and seed yield (Popescu et al., 1999), whereas nonadditive effects expressed by SCA were confirmed by Sood et al. (2007) for number of the seeds per capsule, 1000 – seed weight. Sometimes both additive and nonadditive gene actions were found (Bhateria et al., 2006).

In India there are different linseed breeding centres located in different states focused on various characters. Eight states in India – Bihar, Delhi, Gujarat, Madhya Pradesh, Punjab, Uttar Pradesh, Rajasthan and West Bengal – are involved in linseed breeding (Gill, 1987). Linseed breeding in India is very well organized, and it has an old history. The research programmes on linseed were strengthened in 1947 by the Indian Central Oilseeds Committee (ICOC), later under the project 'Intensification of Regional Research on Cotton, Oilseeds and Millets' (PIRCOM) in 1958. Because the results of breeding work were not as good as expected, the Indian Council of Agricultural Research started the All India Coordinated Research Project on Oilseeds (AICRPO) in 1968 (Gill, 1987) using international approaches. As a result of selection and hybridization, several different linseed varieties were developed from time to time as reviewed below. A detailed breeding work in the respective states above mentioned was described in the first edition of the Handbook of Natural Fibres (Pavelek et al., 2012)

17.3.1.4 Breeding in Canada

Canada is the world's largest producer of oilseed flax, with Saskatchewan producing about 80% of the flax grown in Western Canada with a current production of 1.041 mt (Flax Council of Canadaweb sites, production in 2005/2006). There are three breeding programmes in Canada focused on the development of brown-seeded flax types and yellow-seeded low-linolenic linseed types under the common name 'Solin'. The first two programmes called 'Agriculture' and 'Agro-Food' Canada are carried out in Morden and Manitoba. The third one focused on the development, and breeding of agricultural crops is undertaken at the University of Saskatchewan in Saskatoon (Flax Council of Canada, Saskatchewan Flax Development Commission). The first two programmes started in 1900, and based on them, the first flax fibre varieties were developed: Diadem, Ottawa 770 B, Ottawa 829 C and Novelty. In the 1950s varieties Linott, Raja and Rocket were bred and in the 1960s varieties Dufferin, McGregor, Norlin, NorMan, AC Linora, AC McDuff, AC Emerson, AC Carnduff and AC Lighting were developed. Breeding work at the university in Saskatchewan started in the 1920s, and from this period the varieties Royal and Redwood 65 originated. In 1974 the breeding programme was prolonged and updated, and based on this, the varieties Vimy, Somme, Flanders, CDC Normandy and CDC Valour have been bred.

Mutation breeding was the main breeding method used by both the Australian and Canadian breeding programmes which started development of linseed varieties with completely different fatty acid composition in comparison with traditional composition standard in the species *L. usitatissimum* L. Using X-ray irradiation, the linseed variety Redwood 65 was bred, and from it, the linseed variety Dufferin registered in 1979 was derived (Micke et al., 1985). Via chemomutagenesis, new fibre genotypes resistant to wilt (*F. oxysporum)* were developed and in Australia the genotypes with various fatty acid composition completely unique to *L. usitatissimum* L. species with lower linolenic acid content (from 60% to ca. 30%) were created. Based on mutual crossing of these types with lower linolenic acid content quite new types have been bred with unique composition of fatty acids not known in *L. usitatissimum* L. species till this time (Green, 1986a, 1986b). The linolenic acid content is less than 3% while linoleic content was raised from approximately 15%–70% (Dribnenki and Green, 1995). From these types, many other low linolenic linseed varieties were derived under the name 'Linola' and commercialized at the beginning of the 1990s (Dribnenki and Green, 1995; Green, 1992). The first linseed variety originated in the framework of this programme was Linola TM947, a late-maturing yellow-seeded variety with very high oil content in the seeds and good resistance to lodging (Dribnenki et al., 1999). The other one is Linola TM2047 (Solin 2047) (Dribnenki et al., 2003). This variety is immune against strains of rust (*M. lini)* of North American origin and oil content is higher than in Linola TM947. The third one is Linola TM2090 (Solin 2090) (Dribnenki et al., 2004), which is good in oil content and seed yield and is also immune to North American strains of *M. lini* and relatively resistant to Fusarium wilt *F. oxysporum* f. sp. *lini.* The last two varieties were produced by the Canadian company Agricole United.

Nowadays the breeding work in Canada is focused on dual purpose flax *L. usitatissimum* L., to determine whether the plant of linseed can be used for both oil and fibre (Nandy and Rowland, 2008). For dual purposes (seed/oil, straw/fibre), it is recommended to use crosses linseed × linseed or linseed × flax lines rather than flax × flax ones (Foster et al., 1997, 1998). Recombinant inbred lines (RILs) based on crosses of linseed × linseed or linseed × flax lines showed positive genotypic correlations resulting in increasing the chances of extracting RILs producing both high seed/oil and straw/fibre yields (Foster et al., 2000).

Current studies have shown no tight linkage between stem fibre content and oil traits indicating the possibility of developing dual purpose varieties (Foster, 1997). It is known that product quality and productivity can be improved by a better understanding of the material structure. To increase the fibre yield and identify superior varieties for fibre production, it is very important to know in detail the ultrastructure of the fibre-bearing stem tissues (Roland and Vian, 1991). To combine both fibre and oil in one plant, RILs were studied using light microscopy and molecular markers to screen the parents and the RIL population to identify polymorphic markers for the identification of possible dual purpose varieties (Nandy and Rowland, 2008).

17.3.1.5 Breeding in the USA

Almost all the activities connected with flax and linseed are concentrated in North Dakota. North Dakota produces over 90% of the flax seed in the USA. The value of the flax seed crop in North Dakota is estimated at $45 million per year. During recent years, the USA has been a net importer of flax seed. At present, the only flax breeding and genetics programme in the USA is situated at the North Dakota Agricultural Experiment Station. In fact, the programme at NDSU is one of only three in North America. The value and market for flax seed as a healthy food continues to develop. A recent news release indicated that a major baby food manufacturer will be adding an enriched omega-3 product. This could require a significant quantity of flax seed. Research continues to support flax seed as a healthy choice for heart health and reduced cancer risk.

The primary objective is to develop and evaluate genetic material to improve yield potential while maintaining resistance to pests, maintaining oil content and oil quality and maintaining other agronomic characteristics for potential cultivars. Because producers have historically planted later than would be expected to produce greatest yields, a part of the breeding effort will be devoted to evaluation of a delayed seeding date. With the interest in flax as a human food, a minor effort will continue to evaluate material with a yellow seed coat colour, which is preferred for 'eye appeal'.

Main breeding goals are as follows:

- to develop flax cultivars with desirable agronomic characteristics; seed yielding ability; quantity and quality of oil;
- to reach tolerance to wilt and pasmo and
- resistance to known North American races of rust.

The breeding program is divided into two major parts: (1) rust resistance for two genes conditioning resistance (M^3 P^3) and (2) use of other genes for resistance (either as one gene or multiple). The two parts would require different groups of crosses. In addition, a minor effort will be to continue to evaluate golden (yellow-seeded) flax as preferred in the human food market. Most of the breeding programmes, as in India or Canada (Kenaschuk, 1975), will follow the pedigree breeding scheme. The details of the planned breeding methods to be used are as follows:

- Cross and F_1 generation: crosses between selected parents are made in the greenhouse, and the F_1 plants are grown in a greenhouse.
- F_2 generation: from 500 to 3000, F_2 plants are grown the following year in the greenhouse and evaluated for rust resistance. Resistant plants are grown to maturity.
- F_3 generation: progenies of rust-resistant F_2 plants are grown in the field, and selection is made on agronomic appearance for the best plants within the best family. Normally, four plants are selected from each selected family.
- F_4 generation: plants are evaluated for rust resistance in the greenhouse and only lines homozygous for resistance will be planted in the field the following year. Four to six plants will be selected from selected families.
- F_5 generation: seed from selected F_4 plants are planted as hills. Selected hills will be advanced.
- F_6 generation: lines are grown in field trials for yield evaluation. In addition, data on other characters such as flowering date, plant height and wilt tolerance are collected. In later generations, yield testing will be conducted at multiple locations on selected lines. The most promising lines will be evaluated in 'tri-state' trials. All lines in the 'tri-state' nurseries will be evaluated for oil percentage, iodine value, height, maturity, lodging and pasmo as well as yield in both early and late seeded nurseries. Presently, the 'tri-state' nurseries are grown in ND, SD, and Manitoba, Canada. Each cooperating station grows two trials per

season. Selected lines from 'tri-state' nurseries are grown in regional trials in the flax growing area. As a final stage of evaluation, lines will be evaluated in Research Extension Centre trials and increased, named and released. Selected material of high priority will be rapidly advanced in a modified SSD method to 'pure' lines for yield evaluation.

Population improvement

The secondary objective to develop and maintain populations with useful genetic variability is necessary to support the development of improved cultivars. Much of this effort will centre on the health food aspect of flax. Yellow seed coat is preferred in human food products. In addition, concern has been expressed concerning the level of Cd in the seed of several plants including flax.

Several potential sources of yellow seed colour have been identified in the US collection of flax (Miller and Hammond, 1988). Several lines have been selected for crossing with low-accumulating Cd lines (additional breeding is planned after evaluation of Cd content in the crossed material). In addition, crosses have been initiated with selected advanced yellow-seeded experimental lines and adopted brown-seeded lines. The segregation of advancing generations will be followed to determine mode of inheritance of yellow seed coat colour and potential as varieties for the health food market. As the material is developed it may be important to evaluate other quality characteristics for health benefits (lignan content, etc.). Populations need to be available in the event of market change. Diverse populations will be maintained to attempt to meet changes in demand.

17.3.1.6 Breeding in China

Flax and linseed genetic resources

The flax breeding was begun from the collection of germplasm. The study of flax germplasm conservation began in the 1950s, and the main goal was germplasm collection and identification (Wang, 2008).

The first Catalogue of Flax Germplasm Resource of China has been compiled in 1978, and 570 accessions were included in it. It includes 408 local varieties and eminent lines from China, and 162 accessions from 15 oversea countries, including Russia, Switzerland, Sweden, Romania, and so on. After that, the *Catalogue of Germplasm of Main Bast Fiber Crops of China* was compiled in 1990. 138 flax accessions were embodied in it, and they had been embodied in *Catalogue of Flax Germplasm Resource of China*. The *Continuation* of *Catalogue of Germplasm of Main Bast Fiber Crops of China* was completed in 1995, 2113 flax accessions from Heilongjiang, Hebei and Inner Mongolia were embodied in it. The *Second Continuation* of *Catalogue of Germplasm of Main Bast Fiber Crops of China* was completed in 2000, 240 flax accessions were embodied in it. 170 of them are domestic flax varieties and new lines and 70 of them are from other countries. So far, there are about 4000 flax accessions in China, 2607 of them have been compiled in the *Catalogue of Flax Germplasm Resource of China* or *Catalogue of Germplasm of Main Bast Fiber Crops of China* (Wang, 2008).

The 2934 flax accessions which have been stored in the National Long-term Seed Genebank and National Medium-term Bast Fiber Crop Seed Genebank are offered by 13 academies or institutes from eight provinces. And the database of flax germplasm resources has been established.

Breeding work

In China, breeding work on flax began in 1950s with the collection of local cultivars and introducing of varieties from Europe and Japan. This not only solved the urgent demand but also increased the resources of germplasm. Huaguang No.1 and No.2 with the high yield were selected from the Japanese variety 'Beierna' in Heilongjiang (Wang, 2001). Those varieties have been spread and utilized, what broke the barrier of using only local cultivars.

In the 1960s, the introduction and identification of varieties was a primary objective, and varieties such as л-1120, which presented high yields, were spread.

In the 1970s, based on the increase in the resources of germplasm, the breeders began the work of cross breeding, mainly focused on single cross. At the same time, they developed the breeding through combination of cross breeding and mutation breeding, and some varieties such as Heiya No.2 and No.4 had been bred. This made yield, oil content and fibre content increase obviously.

Since 1980, breeders developed the breeding of high yield, high fibre, high oil and disease resistance and set up a lot of new methods and bred some new varieties such as Heiya No. 6, No.7 and No.8 and Shuanya No.5, which had the properties of high yield, good quality and disease resistance.

During the past nearly 60 years of breeding work, plenty of new varieties have been bred. This provided the following solutions to problems. First, yield had been greatly increased, as the stem yield of fibre flax increased from 1200 to 9000 kg ha^{-1}. The highest record of stem yield was 10,669.5 kg/ha, which was created by variety ZHONGZHUYAMA (Chinaflax) No.2 at Yunnan province (Wang et al., 2014). Second, diseases of flax have been controlled. For example, Heiya No.6 and Chinaflax No.4 have high wilt, rust and powdery mildew disease resistance. Third, quality was evidently enhanced. The long fibre content of new varieties such as Heiya No.8, 9, 10, 11, Shuanya No.5, ZHONGZHUYAMA (Chinaflax) No.3 and ZHONGZHUYAMA (Chinaflax) No.4 amounted to 18%–20%. It was a great achievement in the flax breeding.

Breeding centres and methods of breeding

About 10 institutes are engaged in flax (including linseed) breeding in China now. And many new methods were used to flax breeding. The following are the main methods.

Induced mutation Induced mutation was used from 1960s in China. And some fine flax lines with higher fibre content or early maturing were obtained. They were treated by $^{60}Co - \gamma$ ray of 20,000–50,000 Rad. And some flax varieties were bred by this method too. Besides γ-ray, some other inducing methods, such as chemical, antibiotics, and cosmic rays, were also used. The chemicals are useful for the mutation of flax. Three different EMS (Ethyl methanesulfonate) percentages (0.3%, 0.6% and 0.75%) were used to mutagenize 10,000 flax seeds.

Interspecific hybridization Interspecific hybridization is used for wild species gene incorporation into the genotype. If hybridization is not possible because of incompatibility of pollen and stigma, interspecific hybrids might be obtained by protoplast fusion techniques, followed by regeneration of the fusion product or by embryo-rescue techniques and other methods. Protoplast culture and isolated immature embryos of distant hybridization culture of flax lie in the study stage of medium and culture condition selection. Wang (2008) did the interspecific hybridization *L. usitatissimum* × *L. perenne*. The results show that repeat-pollination + plant growth regulator (GA3, NAA and 2,4-D) is an effective method to overcome interspecific cross-incompatibility of flax. The authenticity of interspecific hybrids was identified by SSR molecular marker (Zhang et al., 2012).

The utilization of male-sterility flax A sterile plant was discovered from oil flax (linseed) 'Yanza No.10' in the Inner Mongolian Academy of Agricultural Sciences in 1975. It was believed that it was a nuclear sterile type controlled by dominant single gene. It has been used in 18 Academies of Agricultural Sciences and has been transformed to fibre flax of a sterile type. The sterile flax was used in breeding means such as cross, backcross, and rotational crossing selection. A lot of good strains have been bred (Wang, 2001). Dang et al. (2000) used antibiotics to treat flax cultivars; by which, male-sterile mutant was obtained and the sterility can be inherited. The sterile type can be used to disposition of hybrid.

Ploidy breeding There are some researchers studying on the methods to double flax chromosome. The results are that the flax be soaked in water for 24–36h and then treated by colchicine. The colchicine concentration is 0.075%–0.2%, and treatment time is 12–24 h. The 12%–40% seeds can be doubled.

Normally, the time spent between a first cross and final new flax variety will be 12–15 years. For self-pollinating crops, such as flax, it has been argued that using a haploid generation might result in a considerably more efficient breeding program.

Another method for obtaining haploid flax plants is anther or microspore culture. Sun (1979) reported that the obtained three regeneration shoots from flax anther culture at 1978, and he obtained 21 shoots again at 1979. There were 11 shoots rooted among of them. They were confirmed to be haploids by test of chromosomes. The anther culture experiments showed that callus formation, expressed relative to the number of cultured anthers, varied from 19% to 31%; shoot regeneration on calli expressed relative to the number of regenerated shoots ranged from 1% to 13%, mainly depending on the genotype used. Song (1996) improved the anther culture, and plants regeneration rate was increased by lower temperature pretreatment of anther and double layers (one liquid and one solid) medium. Some excellent flax resources were created by anther culture. The first new fibre flax variety Shuangya No.13 was bred by the anther culture in China (Song, 2008).

Twin seeds often contain one diploid and one haploid plant. The haploids derived from twin seedlings are sterile, but chromosomes can be doubled to get plants and seeds for breeding. Kang et al. (2011) found that the rate of polyembryo seeds was more than 50% in F_1. And they have got the high-yield DH flax lines by this method. The highest straw yield was 8009 kg/ha (Kang et al., 2014).

Tissue culture Somatic cell regenerated plants had extensive hereditary variation, and this variation is stable hereditary in the early regeneration, and most parental characteristics can be retained. Sun (1983) reported that they obtained regenerative plants from the culture of flax stem tip, cotyledon and hypocotyl, and they have studied the influence of the light, temperature, medium and additament to the producing of callus and regenerative plant. The male sterility, susceptible to disease and short stem can be obtained by somatic cell culture. Chinese scientists have bred a flax variety by tissue culture (Ji et al., 2012). They obtained the callus from hypocotyls, then treated them by γ-ray, and then got the regeneration plants from the callus. After two generation selection, they got the new line 06-3. It is new flax variety Shuangya No.16, which was registered at 2012 (Ji et al., 2012).

Gene transformation Wang et al. (2004) reported on flax gene transformation by *Agrobacterium tumefaciens* in 1998, and the regenerative plants had been obtained by this method, the primary genetic transformation system mediated by *A. tumefaciens* was established. The Bar and *vhb* genes were transferred to flax mediated by *A. tumefaciens*. It was verified that target gene was transferred to genome of regenerative plants by molecular methods of polymerase chain reaction (PCR) in the same year.

17.3.1.7 Breeding in the Czech Republic

Flax breeding (*L. usitatissimum* L.) has a long-standing tradition in the Czech Republic since the beginning of the last century. At that time, mainly varieties of flax and flax industry had a relatively significant position during this period, supported by the processing industry. The situation began to change dramatically in the first years of the 21st century and especially after the entry of the Czech Republic to the EU on 1 May 2004. Due to unfavourable circumstances, particularly in the first decade of the 21st century (the elimination of subsidies for the production of flax, invasion of cheap linen and commodities from countries in Asia, especially China), there has been a decline in European flax areas, both in Western countries (France, the Netherlands, Belgium) and in Eastern European countries, including also the Czech Republic. However, for the first time in the history of the production of flax fibre in the Czech Republic, there was also inactivation of the processing factories, and in a few years, the complete disappearance of this sector of industry. The impact of these unfavourable circumstances has also negatively affected the breeding work of AGRITEC Ltd. Šumperk, which had to react quickly to this situation, and for this reason in 2010, the breeding of fibre flax varieties was definitively terminated. All attention was paid to breeding of linseed varieties with new qualities of the seed (differentiated FA content, higher content of lignans, lower content of cyanogenic glycosides) predetermining the different directions of utilization of linseed and thus increasing its yield. On newly registered varieties, much higher requirements are currently emphasized than it was in the past for the application of both the traditional types of flax and linseed. More emphasis is placed on the complex use of the whole plant, both in terms of seed processing and the use of fibre for technical purposes. Thus, the linseed varieties compensate a certain extent of the flax areas loss and in this respect the Central Institute for Testing in Agriculture in Brno also responded very flexibly and dynamically and undertook a system for testing the utility value of varieties in the registration procedure, especially in terms of quality parameters of the seed and in the SDO testing system, which also assesses the stem's technological properties. This trend was also captured by AGRITEC Ltd. Šumperk, so that three new varieties of linseed RACIOL, AGRIOL and AGRAM were bred and successfully registered in the Czech Republic from 2011 to 2017.

RACIOL linseed variety is middle early-maturing, blue-flowering variety with the low-to-middle high plants, resistant to lodging, fat content is middle high. Fatty acid content (FA) has been changed by mutation breeding. Alpha linolenic acid content had been decreased to 30%, linoleic acid content had been increased to 40% (Fig. 17.3). Iodine number is moderately high. Advantages of this linseed variety are a favourable change in the ratio of essential fatty acids in oil, a unique composition of FA not yet known at *L. usitatissimum* L., high seed and fat yield and yellow seed colour. This variety is recommended to be used in bakery industry for pastry, for dough mixing as a rich source of essential omega-3 fatty acid, for edible oil in cold kitchen. Mouldings can be used as components in food mixtures.

AGRIOL linseed variety is late-maturing, blue-flowering variety with yellow colour seeds. Thousand seed weight is low to medium, the seed as well as fat yield is above average, 15%–20% higher compare with standard varieties, and fat content is high. Fatty acid composition had been changed in direction to very low alpha linolenic acid content and very high linoleic acid content (Fig. 17.4). Combination of very high seed and fat yield and yellow seed colour are considered main advantages of this linseed variety. AGRIOL variety is thought to replace former variety AMON. It can be used in bakery industry for pastry, for edible oil in cold kitchen. Mouldings can be used as components in food mixtures.

AGRAM linseed variety is middle early-maturing, blue-flowering variety with brown colour seeds, and stem length is low to medium with resistance to lodging. This variety excels higher resistance of roots and stem basis to diseases. Iodine number is middle high (165). Fatty acid composition is similar to RACIOL variety. Alpha linolenic content had been decreased to 30%, and linoleic acid content had been increased to 40%. AGRAM variety can be considered brown seed variation of RACIOL variety. Main advantages of this variety are: unique fatty acid composition and also very low cyanogenic glycosides content. AGRAM variety is recommended for edible oil production, for rational human nutrition, for use in bakery industry for pastry, for dough mixing.

17.3.1.8 Conclusion

Requested characters can be found in another species; however, it is difficult to transmit them due to hybridization barriers. Also induced mutagenesis is nearly at the top of

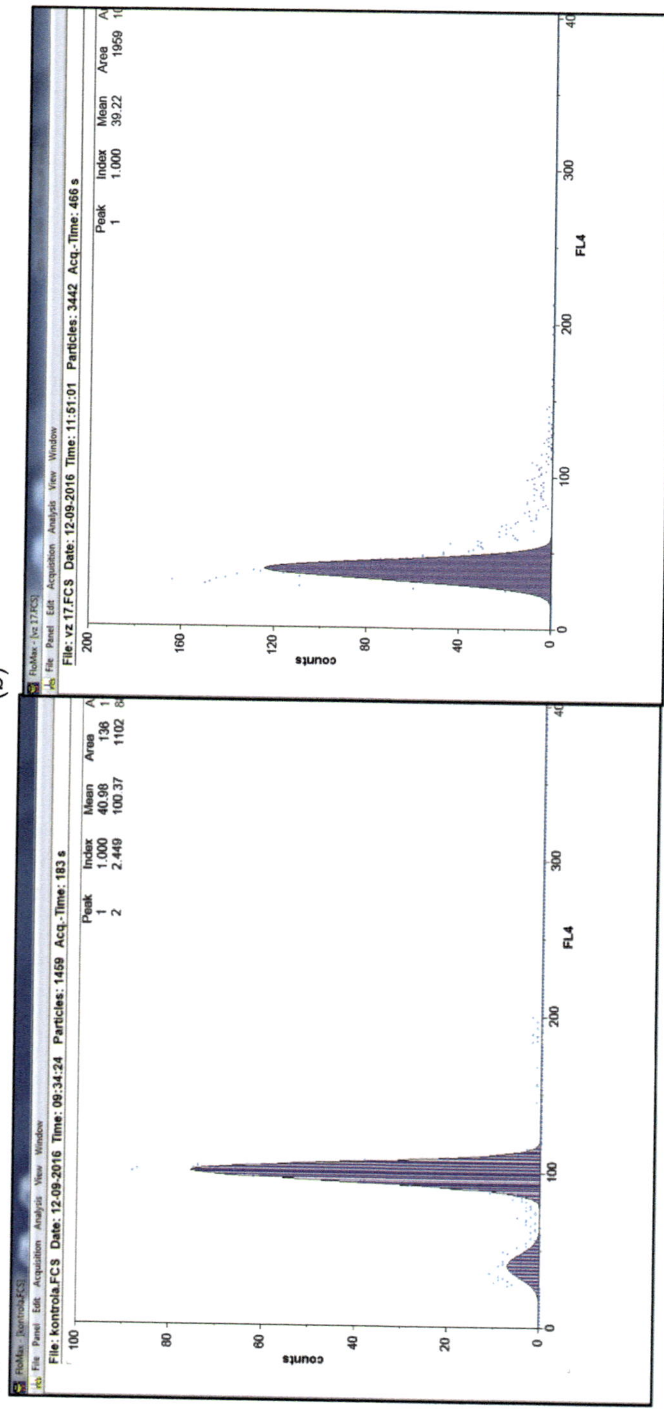

Figure 17.3 Fatty acid composition in RACIOL linseed variety compare to standard linseed varieties.

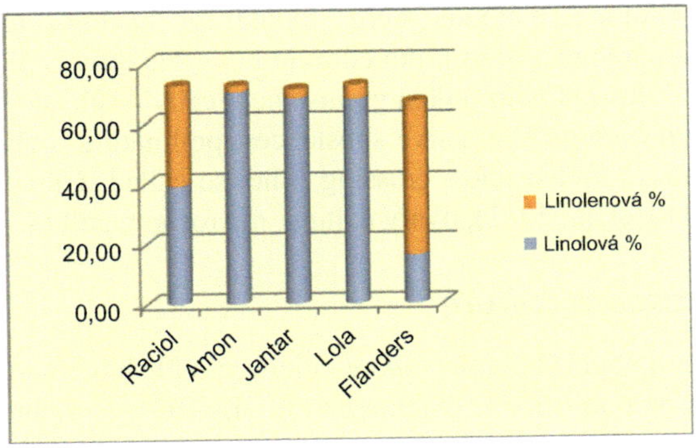

Figure 17.4 Alpha linolenic acid content in linseed variety AGRIOL compare with standard variety AMON.

its utilization (mutagenesis is a causal phenomenon of statistically less effective requested large populations with high working demands). However, the technologies used in modern industry require new varieties equipped with new characteristics. Sustainable agriculture requires more ecologically friendly ways of growing and for systems of phytoremediation of contaminated soils. A revolutionary development might view the plants as 'factories' for the synthesis of specific compounds. Thus the traditional methods are likely to be augmented or overtaken by methods of cellular and genetic engineering and molecular biology.

17.3.2 Unconventional methods of breeding

17.3.2.1 Haploid production in flax

Crop improvement is based on the extension of genetic variability followed by selection of suitable phenotypes with desired combination of features, and then reduction to eliminate genetic variability to meet the conditions of distinctness, uniformity and stability in newly developed varieties. Extension of the variability for desirable properties is induced by intrageneric or intergeneric hybridization, mutation induction, somaclonal variation utilization or somatic hybridization. Genotype stabilization is achieved either by the classical breeding methods (pedigree selection, bulk breeding) or by using haploids. The second method is very effective: completely homozygous genotypes, nonsegregating in any trait, are obtained in one generation. For comparison, it took even than 2 decades from the mutation induction to the official registration of the first low-linolenic variety Linola 947 using the classical breeding method (Dribnenki and Green, 1995).

The oldest method of haploid production is the polyembryonic method (Rosenberg, 1974). Some genotypes are able to produce twins in one seed, from which mostly one is diploid and one is haploid, but in a very low frequency and, moreover, induction of twinning into breeding materials is a very lengthy procedure.

Some biotechnological methods were studied for haploids and subsequently double-haploids (dihaploids [DH]) production in flax (Poliakov et al., 1994): pollination of emasculated flowers with pollen grains chemically or physically treated, treatment of plants with biologically active substances, pollinating diploid flax cultivars with tetraploid ones, interspecies crossing (the so-called Grandiflorum method described in Poliakov et al., 2001), ovary culture, microspore culture or anther culture.

17.3.2.2 Microspore culture

The first information about flax microspore culture establishment and androgenic regenerated plants was published in Nichterlein et al. (1991a). Nichterlein and Friedt (1993) reported detailed methodology of shoot regeneration from microspore-derived embryoids and microcalli in microspore culture. The highest frequency of regenerated plants was obtained in microspore cultures of the hybrid Atalante × Szegedi 62 (F1) at 30°C. From 61 microspore calli 65.6% produced shoots. Shoots were rooted on MS medium containing indolylacetic acid (IAA) (0.1 $mg.L^{-1}$), silver nitrate (1 $mg.L^{-1}$) and Gelrite (6 $mg.L^{-1}$). From 113 inoculated shoots, 44 surviving plants were obtained from the more responsive hybrid. The majority of the plants were haploid. After colchicine treatment, 24% of the initially haploid plants produced seeds.

However, the overall efficiency of plant regeneration is still not suitable for practical breeding (Obert et al., 2009).

17.3.2.3 Ovary culture

The first results of experiments with ovary culture and callogenesis in ovary culture were reported by Bartošová and Pret'ová (2003). Bartošová et al. (2003a,b) presented callogenesis and shoot regeneration in culture of isolated ovaries. After that several other works followed. Obert et al. (2005) reported the highest frequency of callusing ovaries: 64% in genotype AC Emerson on N6 medium with growth regulators as for the anther culture. Ploidy of callus cells was $2n$, $4n$ up to $16n$. Regeneration from ovary calli was realized via bud and shoot formation.

Bartošová et al. (2005) reported shoots obtained in ovary culture, which were diploid. Callus induction and regeneration rates in ovary culture were genotype-dependent (Bartošová et al., 2006). Flow cytometry analysis showed $2n$ in yellow–green calli of genotypes Viking and AC Emerson and $4n$, $6n$, $8n$ and $16n$ in dark green calli.

Burbulis et al. (2007b) reported effect of genotype, growth regulators and level of sucrose on callus induction in ovary culture.

Ovary culture is at the beginning of its development, and it is not yet sufficiently effective to be used in flax and linseed breeding.

17.3.2.4 Anther culture

Anther culture is the most promising, most studied and most successful method of haploid and DH production in flax. Microspores developing in anthers have a reduced number of chromosomes (n). They develop into pollen grain coming through the several developmental stages. In the uninuclear stage, they are able to change their direction of development in specific in vitro conditions and haploid plants can grow up from them. Such plants are not fertile, they must be treated with some mitotic active substance to double their chromosome set. In many cases, however, spontaneous diploids are caused by endomitosis. Progenies of DH plants are completely uniform in genotype and they cannot segregate in any character. The absence of a dominant effect in complete homozygotes makes selection of desired genotypes easier (Poliakov et al., 1994). That is why using of anther culture can reduce the time needed for obtaining pure lines.

Anther culture seems to be an excellent method for breeding interval abbreviation commonly. It has been developed progressively since the 1960s in different plant species. Guha and Maheswari (1964) reported production of Datura haploid plants by culturing anthers. Anther culture in flax has been studied since the 1970s.

The first experiments on flax anther culture were described in Sun (1979). Soon afterwards, Sun Hong Tao and Fu Wei Dong obtained complete plants from F1 hybrids through the callogenesis in isolated anthers, bud regeneration in calli, shoot development from buds, and shoot rooting (Sun and Fu, 1981). This method is called indirect regeneration in contrast to embryogenesis, which is named direct regeneration. Cytological tests confirmed microspore origin of regenerated plants because haploid number of chromosomes was detected.

17.3.2.5 Anther culture as an object of research

A system of indirect plant regeneration from the microspores in flax anther culture was assumed and further improved by other researchers in different laboratories around the world. Experiments were conducted on flax and linseed as well. There are no differences among those two cultural types, for fibre (flax) and for seeds (linseed) production, in *in vitro* cultures. The first results of experiments with flax anther culture from individual laboratories were reported: Nichterlein et al. (1989a) (Institute of Agronomy and Plant Breeding, Justuts-Liebig-University, Giesen, Germany), Poliakov (1991) (All Russian Flax Research Institute, Torzhok, Russia), Tejklová (1992) (Agritec, Research, Breeding and Services Ltd Šumperk, Czech Republic), Rogalska and Rutkowska-Krause (1994) (Institute of Natural Fibres, Poznan, Poland), Pret'ová et al. (1995) (Institute of Plant Genetics and Biotechnology SAS, Nitra, Slovak Republic), Lassaga and Bretón (1997) (Estacion Experimental INTA, Paraná, Argentina, Fac. de Ciencias Agropecuarias, Universidad Nacional de Entre Ríos, Paraná, Argentina), Chen et al. (1998a, 1998b, 1998c, 1998d) (Agriculture and Agri-Food Canada, Morden Research Centre, Morden, Canada), Kurt and Evans (1998) (Department of Agronomy, Faculty of Agriculture, University of Ondokuz Mayis, Samsun, Turkey), Soroka (2004) (Institute of Oilseed Crops, Ukrainian Academy of Agricultural Sciences, Zaporozhye, Ukraine), Song et al. (2004) (Raw Material and Industry Institute of Flax, Heilongjiang Province, Shuangcheng 150,111, China), Burbulis et al. (2005a; 2005b) (Laboratory of Genetics and Biotechnology, Faculty of Agronomy, Lithuanian University of Agriculture, Kaunas, Lithuania) and Sun (2009) (Agronomy College of Inner Mongolia Agricultural University, Huhhot 010,019, China).

17.3.2.6 Method optimization – study of factors influencing efficiency of method

Researchers have solved the problems associated with each step of anther culture: callus induction, determination of ploidy in callus, bud regeneration, shoot elongation, shoot rooting, determination of ploidy in plants, colchicine treatment and transfer of plants into soil. The whole protocol is demonstrated in Table 17.7., Figs. 17.5–17.17. The aim of their effort was to develop a universal highly effective method of DH line production from microspores via anther culture applicable to large flax and linseed breeding material.

Table 17.7 Protocol of anther culture in flax and linseed according to Tejklova (2008).

Donor plant growing	Donor plants for anther collection are grown, if possible, at lower temperatures (daily minimum temperature of 5–10°C, maximum daily temperature of 15–20°C). If the plants are grown in vegetation containers, it is recommended to use basic fertilizer before planting: 0.857 g NH_4NO_3 + 1.717 g $Ca(H_2PO_4)_2.H_2O$ + 1.665 g K_2SO_4 per container with a capacity of 5 L. The sowing density is 3 × 3 cm. During the phase of rapid growth – additional fertilizing should occur once every 14 days with full fertilizer, such as Super Wuxal (AGROBIO, Opava) according to instructions. Microspore regeneration ability is increased by growing donor plants on short days, so it is recommended for the cultivation of flax in field conditions, when plants begin to bloom in mid-June, to shorten the light period about a week before bud collection for about 8–9 h by covering plants with black canvas cover Fig. 19.5.
Flower bud collection	Cut the buds with a length of 5–6 mm and a width of about 1.8 mm in linseed and a length of about 4–5 mm and a width of about 1.5 mm in flax (that is: cut the buds of maximal size, when it does not have coloured petals), remove any adjacent leaves (Fig. 19.6). The most successful regeneration is in the anthers from the first few buds on the plant at the bud development stage.
Bud sterilization	Buds are briefly rinsed with 70% ethanol and then with distilled water, pour 1% chloramine B (Bochemie as Bohumin), shake for 3 min., pour chloramine B off, pour fresh solution of chloramine B, again for 3 min on a shaker and finally (from now on in sterile conditions – in a sterile box, with sterile instruments). Pour off chloramine B and rinse buds thoroughly three times with sterile distilled water (distilled water, sterilized in an autoclave for 15 min at 121°C).

Table 17.7 Continued

Extirpation of anthers	Grasp the flower bud with tweezers, cut off about 1 mm segment of the calyx from the stem side of the bud with a scalpel (so as not to damage the anthers), push aside sepals and petals and take out the anthers without filaments (Fig. 19.7). The colour blooming genotypes should have whitish petals. Anthers are put on the surface of agar medium AC (for preparation see Appendix 1) in Petri dishes for callogenesis induction (maximum of 50 anthers in a Petri dish with diameter of 10 cm, with 20 mL of culture medium), close the dish with parafilm (parafilm M, Pechiney, Chicago). Sterilization of buds and isolation of the anthers must be carried out quickly, as anther tissues are negatively influenced by lack of nutrition and the presence of sterilizing agents, which is difficult to completely rinse out of the flower buds.
Callus induction	Extirpated anthers are cultured on AC medium (Table 17.8), 1 week in the light (16 h photoperiod 22/20°C day/night) and then transferred for 1 week in darkness at 21°C and then back to light. In pollen sacs, calli from microspores start to develop, anthers are cracked and calli continue to increase in volume (Fig. 19.8). Callus induction is the highest in the third to fifth week after the anthers extirpation.
Bud regeneration	Transfer green callus 2–5 mm in diameter (Fig. 19.9) to regeneration medium AB (Table 17.8; 5 pcs per 100 mL; Erlenmayer flasks with 25 mL medium) and cultivate them on 16 h photoperiod (22/20°C day/night). Every 3 weeks transfer callus to fresh medium of the same composition, remove necrotic (browning) parts and divide a large callus into pieces with a diameter of 5 mm. A callus may also arise from somatic tissue – from pollen sacs or from filament (Fig. 19.9) Regenerants from callus of such origin are of donor plant genotype and the production of doubled haploids is not therefore possible. Regeneration of buds in the callus (Fig. 19.10) occurs mainly during the first three subcultures on regeneration medium.
Long-term shoot-tip culture	Developing clusters of buds the size of about 3 mm are maintained on medium L (Table 17.8). Periodically after 2–4 weeks, transfer explants to fresh medium, while removing necrotic parts and callus. Use developing shoots for rooting. Cultivation on 16 h day, 22/20°C day/night. Higher temperatures caused stronger callogenesis. In these conditions, new buds are constantly set up and shoots develop (Fig. 19.11)

Continued

Table 17.7 Continued

Culture activation	If culture is maintained so long that the production of new buds and shoots is reduced, it is appropriate to include either 1-week subculture on medium L with BAP increased to 1 μM, or 2 times 3-week subcultures on medium L with BAP increased to 0.1 μM, and then again to keep the culture on L medium.
Shoot rooting	2- to 3-cm-long shoots of normal appearance (not vitrified) are successively picked from the long-term multiple shoot culture (19.12 a), cut off with a sharp scalpel and injected vertically (not diagonally) into rooting R medium (Table 17.8). Often shoots are rooting on MS medium without growth regulators. Shoots usually develop enough roots within 2 weeks (19.12 b)
Transfer into nonsterile conditions	Soil for regenerated plants should be fine, sandy, with a small proportion of humus: mix of sandy soil, sand and peat in a ratio of 6:2:1, pH around 5.5. Rooted shoots without callus on the bases must be thoroughly rinsed to remove rest of agar from roots, then immersed for 1 min into 0.15% Previcur and planted into pots with sterilized soil (1 h in autoclave at 121°C). Transferred plantlets have to be covered with a transparent cover (glass, plastic). Plantlets are cultivated in a conditioned chamber (fluorescent tubes, 16 h day, 8 h night, 12,000 lux, 22°C). After enrooting plants (about 7th day after planting), the cover is partly lifted, and the next day, when the plants do not wither, it is completely removed (Fig. 19.13)
Growing of plants to maturity	To flower, flax and linseed plants need to be grown on long days (15–16 h) and with sufficient light intensity (above 10,000 lux) and adequately watered and fertilized.
Colchicine treatment	Plants regenerated from anther culture, in which haploid chromosome number was determined using flow cytometry or which show signs of haploid (a subtle plant, stomata reaching only two-thirds of the size of control diploid genotype, plant is not setting bolls) are decapitated, the stem is bent, and the cutting end of the stems is immersed in a 1% colchicine solution in a 1.5-mL test tube with about 1 mL of solution. The test tube is attached to the stem with the foil (Fig. 19.14); After the upper leaves get faded (it takes about 16 h), the test tube is removed. Branches growing from auxiliary buds near the decapitated stem top are usually dihaploid and set seeds (Fig. 19.15); Treatment with colchicine is not carried out in spontaneous diploids (Fig. 19.16)

Figure 17.5 Day-shortening for donor plants with black nonwoven textile. (a) donor plant without covering, (b) covered plant.

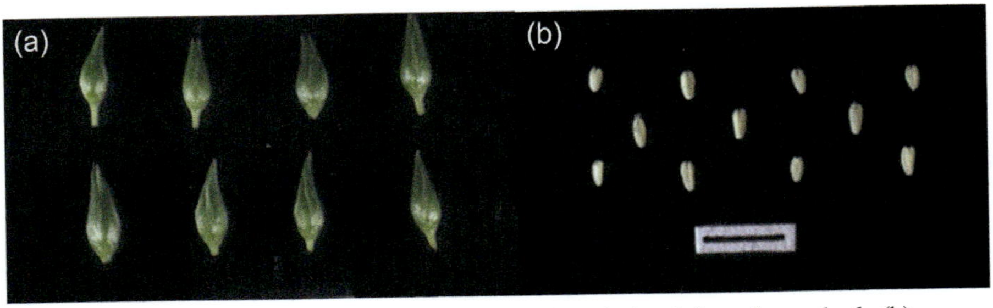

Figure 17.6 Buds ready for sterilization (a) and anthers isolated from flower buds (b). Bar = 5 mm. Callus development from microspore on induction medium.

Figure 17.7 Calli suitable for transfer to regeneration medium.

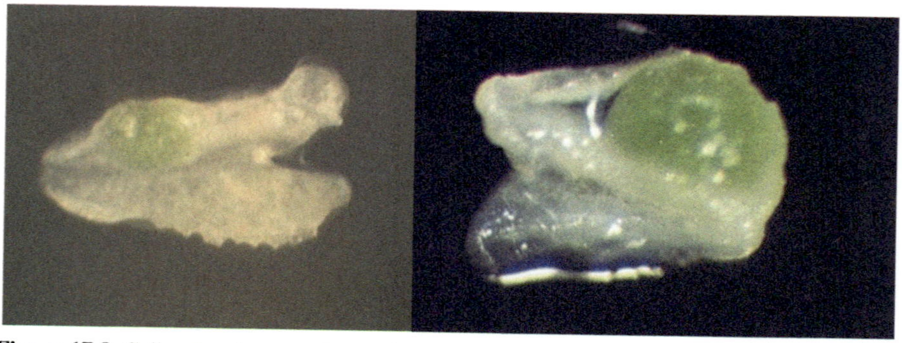

Figure 17.8 Callus development from microspore on induction medium.

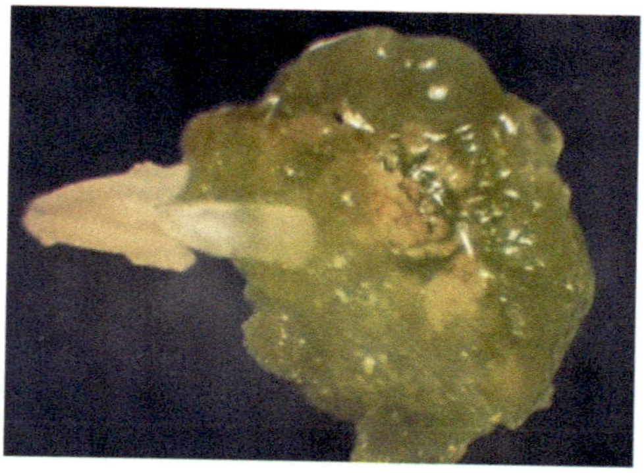

Figure 17.9 Callogenesis on filament. Callus originated in somatic cell.

Figure 17.10 Bud regeneration in anther callus.

Figure 17.11 Shoot development from regenerated buds. Bar = 1 cm.

Figure 17.12 Cut shoots ready for rooting (a) and rooted shoot after 2 weeks on medium R (b).

Figure 17.13 Rooted shoots are transplanted into soil to grow into flowering and set the seeds.

Figure 17.14 Colchicine treatment.

Figure 17.15 Dihaploid branches developed after colchicine treatment.

Figure 17.16 Haploid (left) and spontaneously diploid (right) regenerant.

Figure 17.17 GUS histochemistry in tissue culture of flax – blue sections demonstrate transformed tissues; of course, stable transformation must be confirmed by repeated histochemical and namely molecular (DNA) subsequent tests (which have been done).

Over the years, many experiments have been undertaken for the identification of factors affecting individual steps of anther culture.

17.3.2.7 Genotype

Anther culture is developed for linseed and flax varieties (Nichterlein et al., 1989b). Production of anther calli was generally higher in flax genotypes (Tejklová, 1996), and conversely, bud regeneration was higher in linseed genotypes (Tejklová, 1998). Bud regeneration was lower in hybrids having flax genotypes as one parent, compared with linseed ones (Rakouský et al., 2001). Genotype of donor plants is shown to be an important factor influencing regeneration capacity of anthers (Nichterlein et al., 1991a). Nichterlein et al. (1991b) noted the various responses of linseed varieties to cultivation conditions at all phases of plant regeneration (callogenesis, bud and shoot regeneration, shoot rooting).

Callogenesis and organogenesis were higher in F1 than in parental varieties (Poliakov et al., 1994; Rogalska and Rutkowska-Krause, 1994); the highest frequencies of both characteristics were reached in variety Orshanski-2. Chen et al. (1998b) described the highest callus induction (63.5%) and overall efficiency of regeneration (12%) in one F1 linseed hybrid. Higher frequency of bud regeneration per isolated anthers in F3 hybrids compared with pure lines was demonstrated in Rakouský et al. (2001), as well, but in Tejklová (1996), differences among genotypes in callus induction were more significant than differences between linseed and flax varieties or between varieties and hybrid populations.

Differences among genotypes in anther callus production were described by Vigier and Bretón (2005), Gonzales and Bretón (2005) and Burbulis et al. (2005a) (response namely on growth regulator combinations). Variation for anther culture response was significant both within and between the hybrids and their parents (Burbulis and

Blinstrubiene, 2006). On the other hand, Tejklová (1992) reported that from 15 flax and linseed genotypes, only linseed Duferin was without any callus induction. Frequencies of callogenesis were similar in all genotypes.

Wielgus and Mankowska (2007) and Wielgus et al. (2008, 2009) reported differences in callus production in anthers of F1 from different combination crossing of linseed. Anthers from some crossings were not able to form callus, whereas anthers from combinations Linola with Szafir, Opal and Bionda were characterized by higher than average capacity to form callus.

Genotype had a significant effect on callus induction and shoot regeneration from anther-derived calli of linseed and flax (Chen et al., 1998b; Poliakov et al., 1994). Cultivars significantly affected callus induction rate both on solid and liquid medium (Kurt and Evans, 1998). Chen et al. (1999) demonstrated a strong genotype effect on callus induction and shoot regeneration in anther culture in their study on 44 flax genotypes with diverse genotypic background. Some genotypes responded well to media with different levels of growth regulators (Chen and Dribnenki, 2002; Chen et al., 2003).

Significant differences in callus induction in anthers and plant regeneration among genotypes were observed by Rutkowska-Krause et al. (2003, 2004) and Obert et al. (2003, 2004). Only one of four varieties produced shoots in the anther culture reported by Burbulis et al. (2005a).

Areco was a more responsive genotype compared with Marina (Tejklová, 1998). Callus induction and bud regeneration, as well, were higher in Areco. Sun (2009) and Lassaga et al. (2010) compared regeneration capacity in different linseed genotypes, and Burbulis et al. (2009) selected four responsible genotypes from nine studied linseed varieties in their experiment. Results illustrated that genetic background is important in callus induction and shoot production in the anther culture.

A strong genotype effect on callus induction and shoot regeneration was described in Burbulis et al. (2009). They suggested determining specific combinations of growth regulators for each genotype. On the other hand, the callus induction and plant regeneration in anther culture from both types of flax (for fibre and for seed production), highly homozygous varieties, and hybrid populations, as well, were obtained on the same culture media.

17.3.2.8 Donor plants

The genotype and the physiological condition of donor plants as well as the developmental stage are crucial for haploid induction (Obert et al., 2004).

17.3.2.9 Physiological condition of donor plants

Influence of donor plant ontogenetic stage on anther culture was demonstrated in Bretón and Lassaga (1999). Better responsiveness of anthers collected from plants in the ontogenic phase bud development compared with anthers from plants in the late-flowering phase was observed by Tejklová (1996), when callogenesis in anthers from the first harvest reached 9% and from the second one, 2 weeks later, only 0.7%. Callus induction was higher in anthers harvested from donor plants at the beginning of the flowering period than from anthers harvested later.

Lassaga et al. (2010) used anthers from plants in ontogenetic stage fully flowering for high anther callus production, but very low bud regeneration from calli was obtained. Anthers harvested from plants at early-flowering stage were more responsive than anthers from plants at a late-flowering stage (Chen et al., 1998b).

Lower sowing density and additional fertilizing of donor plants caused higher callogenesis in anthers cultured in vitro (Tejklová, 1998).

17.3.2.10 Temperature of donor plants cultivation

Lower temperature in cultivation room for donor plants (14°C in daytime, 8°C at night) supported callogenesis and organogenesis in the anthers (Nichterlein et al., 1991b).

Significant differences in callus induction were noticed when anthers were collected from plants cultivated in the field after cold days in comparison with anthers collected after hot days (Tejklová, 1996). Low temperature probably delays pollen-grain development, so that more microspores are in the right development stage for callusing or endogenous conditions in anthers are more convenient for microspore callusing in low temperature.

Lowered cultivation temperature of donor plants for about 4 days before anther collection improved anther responsiveness in experiments in which donor plants were grown on the medium-long days (Tejklová, 1996).

Donor plants were grown in the same light conditions but either in lower (7–13/13–25°C day/night) or in higher (13–20/25–31°C day/night) temperature. Callogenesis in anthers harvested from donor plants from lower temperature was 7 times higher (Tejklová, 1998).

Precondition of donor plants influenced callogenesis in subsequently isolated anthers (Burbulis et al., 2005a). Anthers from donor plants grown at 18/14°C were more responsive than those grown at 22/18 degrees (day/night).

17.3.2.11 Day length of donor plants cultivation

Nichterlein et al. (1991b) cultivated donor plants in a 16-h photoperiod. Tejklová (1992) presented the significant influence of donor-plant cultivation condition on callus induction in the anther culture. The highest callogenesis was obtained in anthers from plants cultivated during the early spring in a greenhouse, the lower frequencies were in anthers from plants cultivated from the end of March to the first half of June in vegetation halls and the lowest frequencies were in anthers from plants cultivated from the second half of April to the beginning of July in the field. In the next experiment, callogenesis in anthers harvested in the field was higher than in anthers from the greenhouse (Tejklová, 1996). Both experiments have in common that the conditions for growing donor plants differed in temperature and day-length. High temperature and long photoperiod of donor plants growing both decreased anther responsiveness in the anther culture. Including a low-temperature period in culture system for donor plants before anther collection did not increase the callogenesis in anthers probably because the plants were grown on a long day (Tejklová, 1996). Day-length of donor plants notably influenced callogenesis and organogenesis in cultured anthers. Anthers harvested from plants grown in short day (8 h) were more responsive compared with anthers from plants grown on a 16-h day (Tejklová, 1998).

Study on the influence of photoperiod shortening for donor plants on flax anther culture confirmed the importance of this factor. Anthers from plants grown on the short day (8-h photoperiod) were significantly more responsive than anthers from the long day (16-h photoperiod) (Rakouský et al., 2001).

Rutkowska-Krause et al. (2004) cultivated donor plants either in the greenhouse (winter vegetation period; 16-h day) or in the vegetation hall (summer period; about 16-h day). Temperature was not defined. Statistical analysis showed that differences in callogenesis and organogenesis between those two periods were not significant.

17.3.2.12 Colchicine treatment

In some experiments, colchicine treatment applied on donor plants before anther collection improved organogenesis in anther calli (Rakouský et al., 2001).

17.3.2.13 Bud size

Mononuclear microspores were observed in the flax flower buds with a diameter of 1.32–1.78 mm and with a length of 4.08–4.78 mm, depending on genotype, testing place and year (Poliakov et al., 1995), and with diameter of 1.3–1.8 mm and length of 4.0–4.5 mm (Rutkowska-Krause et al., 2003, 2004). Lassaga et al. (1998) found the higher response in linseed anthers having about 1.5 mm in length. Bud regeneration was achieved only from flower buds with mononuclear microspores but anthers with microspore tetrads or with dinuclei pollen grains produced only calluses in low frequency (Rutkowska-Krause et al., 2004). Uninuclear microspores were observed in floral buds of 1.3–2.3 mm diameter in European linseed genotypes with larger flowers (Nichterlein, 2003).

17.3.2.14 Anther pretreatment

Temperature

Low-temperature (4°C) pretreatment on inflorescences and isolated anthers of flax did not seem to lead to the increase in percentage of callusing and plant regenerating anthers (Poliakov et al., 1995; Rogalska and Rutkowska-Krause, 1994). Later Rutkowska-Krause et al. (1995) reported that refrigerator-temperature pretreatment on inflorescences before anther extraction reduced callogenesis induction in the anthers. Cold pretreatment on extracted anthers cultured on medium improved responses of anthers only when 12-h pretreatment was used on anthers of variety Belinka.

Cold pretreatment of isolated anthers (4°C, 1–5 days) inhibited anther callogenesis and storage of flower buds (4°C, 2–8 days) did not increase anther callogenesis (Tejklová, 1996). Three days' pretreatment of flower buds in the dark at 4°C significantly lowered the number of callus inducing anthers (Kurt and Evans, 1998). Also Tejklová (1998) detected that pretreatment on isolated anthers 1–5 days/4°C decreased callogenesis in anthers. Pretreatment 16–48 h at 38°C increased callus induction in anthers insignificantly. Temperature 41°C applied for any tested period decreased callus induction in anthers compared with control without any temperature pretreatment. Fourteen days' low-temperature (4°C) pretreatment decreased callogenesis and organogenesis in anther culture of cultivar Nike compared with nontreated anthers (Rutkowska-Krause et al., 2003).

Chen et al. (1998b) found out that 1-day culture in 35°C in the dark improved overall efficiency of regeneration in one F1 linseed hybrid but not in the second tested hybrid. Three days' pretreatment in different temperatures (6, 28 and 36°C) influenced irregularly anther callus induction in four genotypes and their hybrids (Bretón, 2001). Obert et al. (2003, 2004) presented callus induction in the anthers improved by cold pretreatment (7 days/8°C) but only in some genotypes. Dramatic increase of induction rate in flax anther culture when anthers were pretreated for 3 days at 4°C and afterwards kept for 1 day in 35°C was declared in Pret'ová et al. (2006).

Position of cultured anthers on culture medium

Anthers immersed into agar culture medium showed lower callus production than anthers placed on medium surface in all variants (Tejklová, 1996). Anther orientation on the culture medium surface showed no significant effect on callogenesis and regeneration (Chen et al., 1998b).

Light condition

Callogenesis in anthers cultivated 4 weeks after isolation in the dark in the same culture temperature was higher than in anthers cultivated in the 16-h photoperiod (Tejklová, 1996) but organogenesis in those calli was low. One-week culture of anthers in the dark was enough for better callus induction (Tejklová, 1998) and did not decrease the regeneration from anther calli in next subculture.

Two-week culture of anthers in darkness after isolation significantly increased callus induction and bud regeneration in the anther culture (Rutkowska-Krause et al., 2004). Cultivation of anthers in the dark during the first month of culture improved callus induction in anthers (Rutkowska-Krause et al., 1995).

Osmotic pressure

Preculture of anthers on medium containing 15% sucrose for 2–7 days before transfer to the same medium with 6% sucrose increased the frequency of microspore-derived shoot regeneration and reduced frequency of shoot regeneration from somatic tissues in the anther culture (Chen and Dribnenki, 2004). The same results were obtained if 6% sucrose and 9% nonmetabolize osmoticum polyethylene glycol were used instead of 15% sucrose.

17.3.2.15 Culture medium

Culture medium consistency

Overall lower frequency of callogenesis in anthers was recorded on a liquid culture medium compared to a solidified one (Kurt and Evans, 1998).

Basal medium – macro and micro nutrients

Nichterlein et al. (1989a) tested two basal media. MS with reduced ammonium nitrate to 10% was more convenient for the flax anther culture.

Nichterlein et al. (1991a, 1991b) tested different basal media with different levels of amino acids, growth regulators and carbohydrates, and with different pH as well. Effect of culture media combination on bud induction in anthers calli was obvious. Medium A22 with lowered level of ammonium nitrate, completed with naphthalene acetic acid (NAA), 6-benzylaminopurine (BAP) and sucrose for callus induction, and P20 with standard ammonium nitrate supplemented with zeatin (1 $mg.L^{-1}$) and maltose for bud regeneration were the best medium combination. Anthers from Atalante gives 31% bud regeneration in such culture conditions.

Sun et al. (1991) presented large experiments in which influence of macronutrient levels in different combinations on anther callus induction and shoot regeneration was studied. An optimal combination for callogenesis and following organogenesis was B5 medium with 2 $mg.L^{-1}$ (2 mg/L) of kinetin and 2 $mg.L^{-1}$ of IAA modified by double dose of potassium nitrate and calcium chloride, but although it was the best combination, frequency of organogenesis in isolated anthers was only 0.24%.

Significant influence of culture medium on callus induction in cultured anthers was also noticed by Tejklová (1992). The highest callogenesis (up to 18%) was reached in anthers on modified MS medium (10% NH_4NO_3) supplemented with 1.11 $mg.L^{-1}$ of 2,4-D and 0.21 $mg.L^{-1}$ of kinetin and nil callogenesis was obtained on the medium N6 with 1 $mg.L^{-1}$ BAP. Besides the influence of nitrogen level and form, influence of growth regulators, however, must be considered in this experiment. Effect of ammonium nitrate concentration on the flax anther culture was studied by Tejklová (1998). In the experiment with two levels of ammonium nitrate in callus-induction medium (10 or 100% of NH_4NO_3 in basal MS medium) and the same two levels of ammonium nitrate in bud-regeneration medium, combination of 10% in callus-induction medium with 100% in bud-regeneration medium was the most successful. Bud regeneration was 4% per isolated anthers and 66.67% per anther calli.

Pret'ová et al. (2001) presented sporadic callogenesis on media N6 and N&N. Obert et al. (2004) induced higher callus formation on media Mo, N6, MS and N&N supplemented with various combinations of growth regulators. Callus formation was the highest according to genotype.

Influence of $CuSO_4$ on callus induction in anthers was studied in Villamonte and Bretón (2004) and Bretón et al. (2004). Higher concentration (10 mM) of this salt in culture medium improved callogenesis in anthers compared with 1 mM $CuSO_4$.

Grauda et al. (2005) compared two basal media for flax anthers of four genotypes. Medium NLN 82 supplemented with 1 $mg.L^{-1}$ BAP and 0.05 $mg.L^{-1}$ NAA and 6% sucrose was better than MS medium with 1 $mg.L^{-1}$ BAP and 2 $mg.L^{-1}$ 2,4-D for embryo formation. Embryos formed green regeneration zones after transferring them on the regeneration medium.

Burbulis and Blinstrubiene (2006) obtained callogenesis higher than 30% on medium with reduced ammonium nitrate to 10% of MS basal medium. Anthers were cultured in the dark at 25°C. Bud (or shoot) regeneration was not presented.

Overall efficiency of regeneration was reduced on 2× salts medium while it was increased on 1/4 salts medium, compared to 1× salts medium (international Patent PTC/CA2002/001,722, Chen et al., 2003).

Growth regulators

Influence of four growth regulators on flax anther culture was described in Sun et al. (1980). From eight auxins which were added into callus-induction medium, 2,4-D, NAA and Picloram were more favourable for anther callus induction while callus was induced at lowest frequencies when IAA, IBA and TIBA were added into induction medium (Tejklová, 1996). Zeatin was more suitable for callogenesis in anthers and for bud regeneration in anther calli, as well, than BAP (Tejklová, 1998).

MT in callus-induction medium resulted in higher frequency of callogenesis in anthers compared with BAP in the same concentration. When calli from medium with MT were transferred on regeneration medium with either meta-Topolin (MT) or benzylaminopurine (BAP) in the same concentrations, bud regeneration was significantly higher on medium with BAP (Tejklová, 1998).

The type of auxin appeared to determine the organogenetic capacity of anther-derived calli. The combination of 2 $mg.L^{-1}$ of 2,4-D with 1 $mg.L^{-1}$ of BAP significantly increased the overall efficiency of regeneration by 42% (Chen et al., 1998a, 1998c). Regeneration medium with 4.5 μM zeatin had significantly higher percentage of calli forming shoots than the same basal medium with 0.01 thidiazuron (TDZ) (Chen et al., 1998b). Percentage of calli forming shoots increased as the TDZ concentration increased; however, shoots from higher TDZ concentrations were abnormal in phenotype and often died during the transfer into soil.

Soroka (2004) claimed that callus grew and developed better at BAP concentration of 2 $mg.L^{-1}$, compared with 4 and 6 $mg.L^{-1}$ in culture medium. Shoot and root regeneration was low and was not dependent on BAP level. Transfer of explants onto fresh medium stimulated de-differentiation of regenerated structures.

Gonzales and Bretón (2005) studied the influence of different levels of kinetin on callus induction in anthers. 2 $mg.L^{-1}$ of Kin was the best. Substitution of zeatin with kinetin in higher concentrations did not improve organogenesis in the anther calli (Gonzales et al., 2005).

For flax there are currently available of any progressive practices, which will allow the improvement and optimization of double-haploid production. According to low capability of anthers to regenerate haploid shoots, the influence of new derivatives onto experienced model of experiments, multiple shoot culture at first, was tested (Smýkalová et al., 2010). Recent studies on anther culture describes the BAP + NAA (2: 1 and 1: 2) and high concentration of the carbon source in the medium, with the level of genotype callogenesis dependent (Obert et al., 2004; Burbulis and Blinstrubiene, 2006). However, many promising breeding lines are recalcitrant (a situation where in in vitro conditions problematic regeneration occurs or regeneration does not occur at all). To get a larger number of regenerated plants in our experiments, there were applied the new derivative of BAP and regulator of cytokinin metabolism INCYDE (inhibitor of cytokinin degradation). For verification of these derivatives influence the initiatory medium with high concentration of NAA was used. Onto this medium, the hypocotyls from flax seedlings were placed and after 1 week they were onto medium with addend derivatives of hormones INCYDE and BAP9THP. After 8 weeks there were observed different growth parameters (number of buds, number of shoots, the shoot length). A trial was carried out on multivariate cultures, in which, in addition to the AGT breeding line, the hypocotyls of the Jitka, Tábor and Venica varieties were cultivated. The behaviour of these varieties in in vitro conditions is already described from previous experiments (Smýkalová, 2010), and it was therefore possible to compare the obtained data on growth parameters and to evaluate the potential effect of the genotype. In this case, it was again confirmed (in varieties sensitive and nonregenerative in vitro) that the addition of the two above-mentioned derivatives will significantly affect the ability of the shoots to regenerate the explant Table 17.8., Fig. 17.18. Then, the explants were placed on the media for 5 weeks, and the individual growth parameters were then evaluated (number of calluses, number of buds and number of shoots). Individual cultures differed significantly Table 17.9.

Based on these previously experiments, INCYDE and BAP9THP were applied to the media for the recovery of callus shoots, which were obtained by anthers cultivating. Anthers were placed according to the original methodology (Tejklová, 2008) on the AC medium, which proved to be calogenized. Subsequently, however, the obtained calluses were placed (after about 2 months on AC medium) on media with different composition: AB-I and AB-T (i.e., the original AB according to the method of anther flax culture – Tejklová, 2008, enriched by INCYDE or BAP9THP). Here again, the effect on subsequent shoot regeneration was confirmed, the growth regulator INCYDE had a statistically positive effect on this desired parameter Table 17.9.

The presence of haploid cells in the regenerated plants was verified using flow cytometry. The flow cytometric determination of the number of chromosomes was performed on individual selected shoots (always from one unique callus). Several regenerated shoots there were identify a haploid set of chromosomes Fig. 17.18. The new derivative of BAP (BAP9THP) and INCYDE demonstrates suitability for anther culture and induction of shoot regeneration from anthers. The effect of derivates on callus induction occurred with greater regenerative capacity, better quality yields. Simultaneously, there were observed different responses of flax genotypes depending on the type of in vitro culture and medium composition.

Vitamins

Medium containing 10 mg.L^{-1} of thiamine had a significantly higher overall efficiency of shoot regeneration than the medium with 0.1 mg.L^{-1} (Chen et al., 1998a,c). Similarly 1 mg.L^{-1} of thiamine in callus induction medium slightly decreased callus induction and significantly improved bud regeneration compared to standard (0.1 mg.L^{-1}) level of thiamine (Tejklová et al., 2008).

Carbohydrates

There were no significant differences in callus induction and bud regeneration between variants of 6% or 10% sucrose in medium for callus induction (Tejklová, 1996).

Table 17.8 Composition of culture media.

Medium	Macro-microelements	Vitamins ($mg.L^{-1}$)	Growth regulators ($mg.L^{-1}$)	Carbohydrates ($mg.L^{-1}$)	Others ($mg.L^{-1}$)	pH
AC	MS but only 165 $mg.L^{-1}$ NH_4NO_3	Nicotinic acid, 0.5	NAA, 1	Maltose, 30,000	Myo-Inositol, 100	5.5
		Pyridoxine HCl, 0.5	Meta-topolin, 2	Sucrose, 30,000	Glycin, 2	
		Thiamine HCl, 1			L-glutamine, 750	
					Agar Difco, 5500	
AB	MS	Nicotinic acid, 0.5	BAP, 1	Sucrose, 30,000	Myo-Inositol, 100	5.8
		Pyridoxine HCl, 0.5			Glycin, 2	
		Thiamine HCl, 1			L-glutamine, 375	
					Agar Difco, 5500	
L	MS	MS	NAA, 0.005 μM	Sucrose, 40,000	Myo-Inositol, 100	5.5
			BAP, 0.050 μM		Glycin, 2	
					Agar Difco, 5500	
R	MS	MS	NAA, 0.005 μM	Sucrose, 40,000	Myo-Inositol, 100	5.5
					Glycin, 2	
					Agar Difco, 5000	

AB, medium for bud regeneration in anther calli; *AC*, medium for callogenesis in anthers; *BAP*, 6-benzylaminopurine; *L*, long-term shoot tip culture; *MS*, medium according to Murashige and Skoog (1962); *NAA*, naphthyl acetic acid; *R*, rhizogenesis.

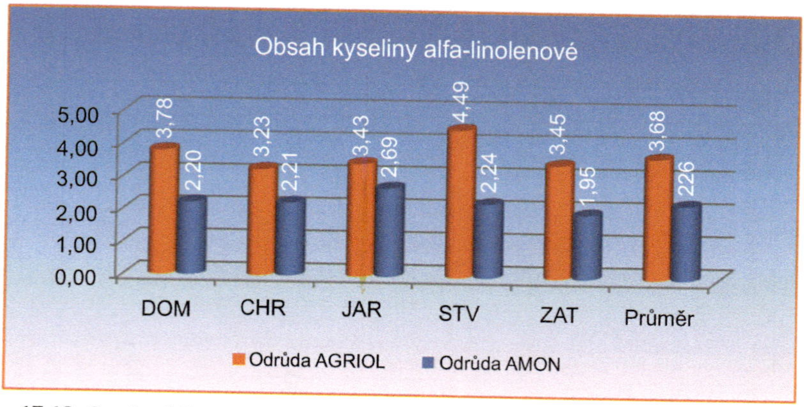

Figure 17.18 Graph of flow cytometry - (a) control and (b) haploid.

Replacement of half amount (30 g) of sucrose in callus-induction medium with the same amount of glucose decreased the callogenesis in anthers while replacement of half amount of sucrose with the same amount of maltose increased anther calli production (Tejklová, 1998).

Carbohydrates in the induction medium had a significant effect on both callus induction and shoot regeneration. Medium for callus induction containing 6% or 9% maltose showed the most overall efficiency of regeneration among five levels of maltose evaluated (Chen et al., 1998a). Level of sucrose in culture medium affected proportion of microspore-originated and somatic cell-originated regenerants. Frequency of spontaneous chromosome doubling appeared to depend on the sucrose concentration in the induction medium. In comparison with sucrose, lactose was found to increase callus induction from anthers; however, the effect of lactose on shoot regeneration appeared to be genotype-dependent (Chen and Dribnenki, 2002).

Rutkowska-Krause et al. (2004) described the highest callus induction on medium with lower level of sucrose (3%) or sucrose combined with glucose (5.5% of each); maltose was the worst carbohydrate. An increasing level of sucrose from 6% to 9% increased callus induction in linseed varieties Lirina and Barbara (Burbulis et al., 2005a) but significantly reduced frequency of responding anthers in their hybrids (Burbulis and Blinstrubiene, 2006). Bud (or shoot) regeneration was not presented.

17.3.2.16 Temperature and light conditions of anthers

Culture of anthers at 35°C for 1 day before transfer to 25°C in darkness significantly increased the overall efficiency of regeneration. Improved efficiency was attributed to the increased organogenetic capacity of anther-derived calli (Chen et al., 1998b).

Callogenesis was higher in anthers cultured in the dark, but organogenesis was higher in anther calli developed on a 16-h photoperiod (Tejklová, 1998). A dark period for 1 week was sufficient for callogenesis improvement (Tejklová, 1998).

Lassaga et al. (2010) incubated anthers for 10 days in darkness at 27°C and then in 16-h photoperiod and 25°C. After 25–30 days, frequency of anthers producing green

Table 17.9 Differences among individual cultures.

Medium	Jitka			Venica			Tábor			AGT 583/05		
	No of buds	No of shoots	Shoot length (cm)	No of buds	No of shoots	Shoot length (cm)	No of buds	No of shoots	Shoot length (cm)	No of buds	No of shoots	Shoot length (cm)
L	2.2 ± 0.2	0.7 ± 0.05	3.57 ± 0.2	3.2 ± 0.16	1.0 ± 0.05	2.50 ± 0.13	2.60.3±	0.4 ± 0.2	2.50 ± 0.12	3.3 ± 0.5	0.8 ± 0.04	3.88 ± 0.7
L-I	4.6 ± 1.3	3.9 ± 0.2	3.03 ± 0.15	5.4 ± 0.5	2.4 ± 0.12	2.08 ± 0.1	6.2 ± 0.4	3.9 ± 0.1	2.50 ± 0.5	10.4 ± 1.7	3.4 ± 0.17	3.71 ± 0.2
L-T	4.9 ± 1.5	3.7 ± 0.18	3.62 ± 0.5	8.8 ± 1.6	2.5 ± 0.13	2.40 ± 0.16	5.4 ± 0.7	2.8 ± 0.14	3.29 ± 0.15	9.5 ± 1.1	3.3 ± 0.18	3.82 ± 0.9
L-T9	8.8 ± 2.1	2 ± 0.1	2.45 ± 0.4	9.6 ± 0.9	1.5 ± 0.07	2.8 ± 0.14	6.7 ± 0.6	1.8 ± 0.09	2.6 ± 0.2	10.1 ± 2.5	2.9 ± 0.15	3.1 ± 0.2

calli was 17%–78%, according to genotype. Frequency of anthers with green callus was 41% in the cultivar Atalante. However, frequency of shoot regeneration was very low.

17.3.2.17 Conditions for higher callus induction decreased shoot regeneration

Conditions for higher callus induction were often connected with lower shoot regeneration as demonstrated in the previous item.

> The highest shoot regeneration rate was found in cross 2 (27.5%), which also showed the lowest callus-formation rate (Friedt et al., 1995).
> Higher bud regeneration related to calli was detected in variant with lower callus induction (Tejklová, 1996, 1998).
> High content of auxins (NAA and IAA) in initial medium induced higher callus production, but there was no bud regeneration in the calli (Rutkowska-Krause et al., 2004).

17.3.2.18 Shoot production and rooting

Lower growth regulator levels (0.02 mg.L^{-1} BAP plus 0.001 mg.L^{-1} BAP) were better for shoot production (Rutkowska-Krause et al., 2003). Culture medium for shoot-production in the callus explants producing buds with lower growth regulator levels (0.02 mg.L^{-1} BAP, 0.001 mg.L^{-1} NAA) was more productive than medium with higher levels both of them (1 mg.L^{-1} BAP, 0.05 mg.L^{-1} NAA) (Rutkowska-Krause et al., 2004).

The highest shoot rooting was detected on half-strength basal SH2 medium (Rutkowska-Krause et al., 2003). Lower concentrations of sucrose and glucose were more suitable for rooting of cut shoots. Optimal medium for shoot rooting was half-strength SH2 medium containing 10 g sucrose without or with 10 g glucose (Rutkowska-Krause et al., 2003, 2004). Chen et al. (2003) noted that medium for shoot regeneration and elongation containing 1% sucrose produced longer and more vigorous shoots than the same medium containing lower or higher concentrations of sucrose.

Tejklová and Rakouský (2005) improved rooting frequency and percentage of more viable rooted plantlets by Na_2MoO_4 enhancement in rooting medium.

17.3.2.19 Cytological analysis

Flow-cytometric analysis of calli developed in the anther culture demonstrated haploid, diploid, tetraploid, mixoploid (diploid and tetraploid or tetraploid and octoploid) cells after 3 months of culture (Rutkowska-Krause et al., 1996).

Rutkowska-Krause et al. (2001b) described ploidy in callus cells $2n$ and $4n$ after 3-month culture, and $2n$, $4n$ and $8n$ after 6-month culture in vitro.

Regenerated plants were diploid (mostly) or haploid (Nichterlein et al., 1991b). Microspore origin of plants regenerated from F1 was proved cytologically (haploid plants) or by monitoring of the homogeneity of progeny (in spontaneous DH lines)

(Friedt et al., 1995). Up to 50% of anther culture derived diploid plants originated from the somatic tissues of anthers which was demonstrated by segregation in their offspring.

Chen et al. (1998c,d, 2001) determined the microspore origin of anther culture derived plants using ISSR and RAPD markers. These molecular methods can differentiate plants originated from different microspores in one anther. About 80% of the regenerated plants originated from microspores and 60% of them were doubled haploids (Chen et al., 1998b).

Segregation in studied traits (colour of corolla, filaments and seeds) was found in five offspring from 13 diploid anther regenerants tested (Rakouský et al., 2001). That means 65.5% of regenerants were derived from microspores and 35.5% from the anther somatic cells.

Bartošová et al. (2005) demonstrated homozygosity, somaclonal variation and epigenetic modifications on the level of isozyme spectra of acid phosphatase and peroxidase in plants regenerated from microspores.

17.3.2.20 Anther culture applications in flax breeding

The importance of this method for flax breeding is highlighted in all contributions on anther culture of flax. The main benefit is to shorten the breeding period. Another application is outlined in the selection at the haploid level and mutation induction systems.

Sun et al. (1980) and Nichterlein et al. (1989b) outlined the first use of anther culture in flax breeding.

Friedt et al. (1995) regenerated doubled-haploid lines from hybrids from intraspecific crossing with the aim of raising alpha-linolenic content in linseed. They used 6679 anthers of the F1 donor plants of the three crosses. Callogenesis reached on average 20.6% and shoot regeneration 4.6% per isolated anthers. Finally 39 doubled-haploid lines were tested from which 4 lines produced more than 60% alpha-linolenic acid in fat, 50% lines showed a fat content superior to that of their cross-parents. Crosses were carried out in 1991, evaluation of doubled-haploid lines in field tests were carried out in 1994. Classical breeding (pedigree selections or bulk breeding) of autogamic crop takes about 15 years. These numbers clearly demonstrate a benefit of anther culture in flax.

Analysis of the progenies of plants regenerated from anthers showed that the doubled haploids (anther lines) are worse in agronomical traits than the initial variety but progenies in following generations in some cases exceeded the initial variety (Poliakov et al., 1994). Some anther lines were more resistant to Fusarium sp. or completely resistant to *M. lini* (Poliakov et al., 1994). The number of regenerants obtained in the anther culture significantly surpassed initial cultivars in the complex traits (Poliakov et al., 1998). Anther line BTL-10 was selected as the best genotype available in new flax variety establishment. On the other hand, Steiss et al. (1998) described pedigree lines showing higher performance than simultaneously developed DH linseed lines.

Straathof (1998) highlighted the importance of the anther culture for in vitro selection system establishment. The main advantage of using haploids in linseed breeding is the rapid development of completely homozygous lines within one generation and efficient means of genotypic selection (Chen et al., 1998c).

Rutkowska-Krause et al. (1998) evaluated anther lines derived from F2 generation. Some regenerated lines segregated plants with white and/or blue petals. This could be due to mutation induction during the period of plant regeneration (Rutkowska-Krause et al., 1998).

Anther lines showed significantly ($P = 0.01$) lower variability of total plant length, technical length, plant weight and seed weight of individual plants compared to initial material pedigree line Su 45/85, and flax variety Texa, as well (Tejklová and Pavelek, 2000). The anther culture is available for stabilization of breeding materials in flax.

Two rust resistance genes were inherited in expected Mendelian ratios in microspore derived populations developed via anther culture (Chen et al., 2001). Rutkowska-Krause et al. (2001a) observed gametoclonal variation in plants regenerated from anthers. Changes of morphological traits and resistance to Fusarium wilt were described. Some perspective lines with higher resistance to Fusarium wilt were obtained. Rutkowska-Krause et al. (2003) described method of flax Alba 2 development using the anther culture techniques. Tejklová et al. (2003, 2008) and Tejklová (2004) presented linseed anther lines overcoming the control varieties in seed yield and resistance to complex diseases in field trials. One of them, AGT 997/05, is going to be registered in CR under the name Raciol.

The first results with radiation of isolated flax anthers by gamma ray for mutations induction were published in Song et al. (2004). Suitable gamma ray dosage was determined.

17.3.2.21 Conclusion

Isolation of parts of the flax plant (hypocotyl, cotyledon, part of stem, leaf, anther) breaks the plant integrity causing plant tissue de-differentiation and growth of callus tissue in controlled conditions. Growth regulators are known to be the main factors controlling callogenesis and organogenesis in plants. Balance of growth regulators is apparently very fragile in flax. Markedly lower levels of growth regulators (auxins and cytokinins, as well) compared, for example, to levels used in multiple-shoot culture in pea, are able to induce tissue dedifferentiation and intensive callogenesis in explants. Differentiation of meristems in callus tissue and bud development from them is very hard to obtain. Probably the very high sensitivity of flax tissues to culture conditions is the reason why direct regeneration in flax has not been obtained yet. Burbulis et al. (2007a) assumed that differences in morphogenetic reaction of different linseed flax genotypes could be determined by the balance of endogenous hormones. Tejklová (1996) described flax variety Dufferin as a nonresponsive genotype in one experiment but it was responsive in the next experiment. Balance of endogenous hormones could be influenced by many factors; however, there has not yet been any study in this field of research. It could be interesting to detect, for example, growth regulator levels in shooting and nonshooting calli in flax, correlation between culture temperature of donor plants and endogenous hormones in buds or in anthers, and many other factors.

For obtaining plants originated from microspores, the interplay of multiple factors is needed: genotype, physiological condition of donor plant, developmental stage of microspores, desired sequence of changes in culture conditions, etc. However, it is possible that there are other factors influencing the outcome of anther culture, which have not been discovered yet. It is necessary to continue in experimental research for optimal factor combination allowing plant regeneration from microspores in the anther culture in the large breeding material at this time. Nevertheless, the method of anther culture for haploid production corresponding to the current level of knowledge has already been drafted and may be possible in breeding of flax.

17.3.2.22 Literature

In the following subsections, published articles of scientists and researchers working in the European as well as world scientific teams dealing with flax and linseed are cited.

Summarizing publications (reviews)

Summarizing publications with chapters on flax anther culture include: Bergmann and Friedt (1997); Nichterlein and Horn (2005); Millam et al. (2005); Song (2007); Burbulis et al. (2007a); Pret'ová et al. (2006, 2007); Obert et al. (2009) and Song (2009).

Patents

Doubled haploid production is considered a very remarkable method for flax breeding and it has been patented (US Patent 7297838, international Patent PTC/CA2002/001722).

Method descriptions

Although plant regeneration was obtained from different flax and linseed varieties, the frequency of callogenesis in anthers and bud regeneration in anther calli were irregular and usually too low to be used as a routine breeding method. However, the work of Chen et al. (1998c) presented 30% overall efficiency of plant regeneration from the anther culture under optimal conditions and this result is very promising for routine use of this method in flax and linseed breeding (Table 17.10).

The complete protocol for anther culture in linseed (and for flax, too) is described in Nichterlein (2003); Lassaga et al. (2004) and Rutkowska-Krause et al. (2004).

The optimized method of anther culture in flax is detailed in Tejklová (2008). The English version is summarized in Tables 17.7 and 17.11. Photographs of the individual steps in the method are shown in Figs. 17.1–17.12.

17.4 Modern methods in flax and linseed breeding

17.4.1 DNA molecular methods

Exploitation and characterization of flax genetic resources and evaluation of flax genetic variability are very important for flax germplasm management and breeding.

Table 17.10 Multiple shoot organogenesis observed in culture after application of cytokinin derivatives or modulators of their activity (after they were cultured on L induction medium − MS + 0.005 μM of 1 μM NAA + BAP).

L (control)	L-INCYDE	L-Roscovitine	L-BAP9THP	L-PEO-IAA
0.005 μM NAA 0.05 μM BAP	0.005 μM NAA 0.05 μM BAP 1 μM INCYDE	0.005 μM NAA 0.05 μM BAP 1 μM Roscovitine	0.005 μM NAA 0.05 μM BAP 1 μM BAP9THP	0.005 μM NAA 0.05 μM BAP 1 μM PEO-IAA

BAP, 6-benzylaminopurine; *INCYDE*, inhibitor of cytokinin degradation; *NAA*, naphthalene acetic acid.

Table 17.11 EU Flax and Linseed catalogue (182 items).

	Variety	Maintainer
1	Abacus	*UK 1361
2	Access	*FR S12606
3	Agatha	(*)FR S13491, NL 227
4	Agram	*CZ 1352
5	Agriol	*CZ 1352
6	Alizee	*FR 12171
7	Alpaga	*FR S12606
8	Altea	
9	Altess	*FR 12606, *UK 224
10	Ambre	*FR S12606
11	Amina	*BE 1479, *NL 712
12	Amon	
13	Andréa	
14	Angora	*FR S12606
15	Antello	*IT 475
16	Aquarius	*UK 1252
17	Aramis	*FR S12171
18	Aretha	*FR S13491, *NL 712
19	Aries	*FR S13491, *UK 1021
20	Arkéa	*FR S12606
21	Artemida	
22	Astral	
23	Atena	
24	Audriai	
25	Avian	*FR S13694, *NL 736
26	Axelle	
27	Batsman	*UK 620
28	Belinka	
29	Bilton	*UK 1063
30	Biltstar	*UK 1935
31	Bingo	*DE 147, *UK 23

Continued

Table 17.11 Continued

	Variety	Maintainer
32	Birdseye	*UK 1361
33	Blizzard	
34	Bolchoï	*FR S12171
35	Bonita	*NL 712
36	Boréal	
37	Boudica	*UK 1361
38	Bowler	*UK 23
39	Brighton	*UK 23
40	Brigitte	*DE 8864
41	Bukoz	*PL 893
42	Caesar Augustus	*NL 582
43	Calista	*FR S13491, *NL 712
44	Carina	*UK 1252
45	Chantal	*NL 736
46	Christine	*FR S13694
47	Cirrus	*FR S12171
48	Cledor	*FR S12606
49	Codruţa	*RO 1017
50	Comtess	*FR 13243
51	Cosmin	*RO 1017
52	Cristalin	*FR 13243
53	Cumulus	*UK 1361
54	Damara	*FR S13491, *NL 753
55	Dangiai	
56	Daniel	*UK 355
57	Dauréa	*FR S12606
58	Drakkar	*FR 12171
59	Duchess	
60	Eden	*FR 12171
61	Edita	*LT 32
62	Elan FD	*RO 1001

Table 17.11 Continued

	Variety	Maintainer
63	Elïxir	*FR S12171
64	Elodie	*NL 592
65	Empress	*UK 1882
66	Eurodor	*FR 12606
67	Evasïon	*FR S12171
68	Evea	*FR 13243
69	Faser	*UK 1361
70	Ferdinand	
71	Festival	*FR 12882
72	Filéa	*FR S12606
73	Flanders	*NL x, *UK 1361
74	Floral	*IT 417
75	Frima	
76	Galaad	*IT 417
77	Gemini	*UK 1021
78	GK Helga	*HU 149567
79	Glacial	*IT 417
80	Goldstern	*DE 4117
81	Heljä	*FI 6102
82	Hella	*DE 8864
83	Helmi	*FI 6102
84	Hivernal	*IT 417
85	Ilona	*NL 227
86	Ineke	*UK 1361
87	Ingot	*UK 1361
88	Inuit	*IT 1370
89	Jade	*FR S12606
90	Jan	*PL 893
91	Jantar	*CZ 231
92	Jantarol	*PL 611
93	Juliet	*DE 404, *UK 1303

Continued

Table 17.11 Continued

	Variety	Maintainer
94	Kaolin	*IT 417, *UK 6162
95	Kastyciai	
96	Lagoon	*UK 1782
97	Landfrau	*DE 9407
98	Laser	*UK 1361
99	Laura	*NL 227
100	Libra	*CZ 1207, *UK 1252
101	Lion	*UK 1252
102	Lirina	*EE 11, *LV 81, *RO 1001, *DE 39
103	Lisette	*FR S13694, *NL 736
104	Lola	*CZ 1207, *SK 448
105	Loréa	
106	LS Koral	*IT 417
107	Lucie	*NL 592
108	Luna	
109	Luncavăţ	
110	Lutea	*FR 12171
111	Magéa	*FR S12606
112	Malika	*FR S13491
113	Marmalade	*UK 1361
114	Marquise	*FR S12606
115	Marylin	*NL 736
116	McGregor	*IE 112
117	Melina	*BE 1479, *FR S13491, *NL 227
118	Merlin	*IT 417
119	Meteor	*UK 1361
120	Mistral	*IT 417
121	Modran	*PL 893
122	Nathalie	*FR S10772, *NL 592
123	Natural	*IT 417
124	Nike	*PL 893

Table 17.11 Continued

	Variety	Maintainer
125	Nikol	*HU 149567
126	Noemie	*NL 592
127	Novéa	*FR S12606
128	Octal	*UK 263
129	Oléane	*FR 12171
130	Olga	*FR S12171
131	Oliver	*UK 1236
132	Oliwin	*PL 611
133	Omégalin	*FR S12171
134	Orival	*IT 417
135	Ötztaler	*AT x
136	Paula	
137	Phoenix	*UK 1361
138	Pilgrin	*UK 1361
139	Podium	*FR S12171
140	Progress	*FR S12606
141	Proméo	*FR S12606
142	Raciol	*CZ 222
143	Rasa	*LT 32
144	Rina	*CZ 222
145	Rūta	*LT 32
146	Sara	*PL 893
147	Sartai	
148	Scorpion	*DE 1220, *UK 6064
149	Selena	*PL 893
150	Selin	*PT 2
151	Serenade	*DE 147
152	Serpent	*UK 1361
153	Sideral	(*)FR 12339, IT 417
154	Snaigiai	
155	Solal	*IT 417

Continued

Table 17.11 Continued

	Variety	Maintainer
156	Star FD	*RO 1001
157	Stélléa	*FR S12606
158	Sunrise	*UK 1361
159	Suzanne	*NL 736
160	Szafir	*PL 611
161	Taurus	*UK 1021
162	Télios	
163	Temida	
164	Toundra	*FR S12171
165	Valoal	*IT 417
166	Vasilelin	
167	Vega 2	
168	Végéa	*FR S12606
169	Venica	*CZ 222
170	Vesta	*FR S13491, *NL 712
171	Violin	
172	Vitalin	
173	Vivéa	*FR S12606
174	Volga	*IT 1370
175	VT50	*UK 6064
176	Windermere	*UK 1636
177	WPB Anette	*FR 14040
178	WPB Celeste	*FR 45162
179	WPB Felice	*FR 45162
180	WPB Pauline	*FR 45162
181	Zenith	*UK 1361
182	Zoltan	*HU 149567

There are different ways to evaluate flax genetic variation, such as morphological characteristics, isozymes and molecular markers. Morphological characters tend to be more quantitative and environmentally dependent while isozymes are limited in numbers. DNA markers' abundance, environmental intensivity and nontissue-specific characteristics are some of their advantages. They are useful for variety identification and evaluation of DNA variation. Different molecular markers including RAPD, restriction fragment length polymorphism (RFLP), AFLP and SSR have been developed to analyze flax genetic diversity. SSR markers are still very limited in flax: only 11 and 28 SSR markers have been reported in two independent variety identification studies. The first comprehensive study on the development and analysis of a large set of SSR markers in flax is from Cloutier (2009). The 635 alleles detected by the 275 polymorphic EST-SSRs were used to study the genetic relationship of 23 flax accessions. Four major clusters and two singletons were observed. Subclusters within the main clusters correlated with the pedigree relationship amongst accessions. The expressed sequence tags - signature sequence tags (EST-SSTs) developed herein represent the first large-scale development of SSR markers in flax. They have potential to be used for the development of genetic and physical maps, quantitative trait loci mapping, genetic diversity studies, association mapping and fingerprinting cultivars.

Fu (2005) assessed the geographic patterns of flax variability in the world collection of cultivated flax by RAPD markers. Sixteen RAPD primers were applied to screen 2727 flax accessions representing 63 countries and one group of unknown origin, and 149 RAPD bends were scored for each accession. Grouping the accessions into 12 major regions explained 8.2% of RAPD variation. Accessions from East Asia and European regions were most diverse, but accessions from the regions of the Indian subcontinent and Africa were most distinct. Accessions from the West Asia region were genetically more related to those from the Africa region and less to those from the Indian subcontinent region. These findings are significant for understanding flax domestication and also are useful in classifying intraspecific diversity of cultivated flax, establishing a core subset of the flax collection, and exploring new sources of genes for flax improvement.

To characterize germplasm at the DNA level, a variety of molecular techniques are currently available. Compared to alternative genetic marker systems, AFLPs are generally considered relatively powerful in germplasm analysis because of the high number of markers that can be generated per analysis. In the study from van Treuren et al. (2004), AFLPs have been used to identify redundant germplasm in the flax collection of the Centre for Genetic Resources, the Netherlands (CGN). Pairwise comparison of accessions was performed by ANOVA in order to identify redundant germplasm. Stepwise bulking of accessions until all remaining accessions were significantly different showed that the 29 accessions of breeder's lines could be reduced to 14. Only a small negative effect of this bulking approach on the among-population component of variance was observed, showing a reduction of 2.6%. This result is discussed in relation to improving the efficiency of collection management.

17.4.2 Flax transformation

Flax transformation has been described in detail in the first edition of this chapter. In this second edition the attention was especially focused on the last modern methods published after 2015 year. In the study from Beyaz et al. (2016), an efficient inoculation method was described which was based on increased osmotic pressure of explants in the stage of inoculation with bacteria. Vacuum infiltration is based on forming negative atmospheric pressure that allows the penetration of bacteria into the inner cell spaces and consequently increases the transformation efficiency. In these experiments, the highest transformation efficiency (88.37%) was observed when 7-day-old seedlings having cotyledon were dried in sterile cabin (under vacuum conditions) for 35 min.

To investigate the influence of the position the explants, three different hypocotyl explants were investigated according to the position where they were isolated from. After inoculation of *A. tumefaciens*, the seven-day-old hypocotyl section of the seedlings was divided into three parts labelled number 1 (part above the root), number 2 (part above number 1) and number 3 (part below the cotyledon leaves), and these parts were left to co-cultivate for 2 days. Then, hypocotyl segments were transferred to selected medium containing antibiotics. The presence of the npt-II (neomycin phosphotransferase II) gene in the transformants was confirmed by PCR analysis. It was observed that the highest transgenic plant number was obtained from the explants taken from the region labelled 1, which was followed by explants taken from the region labelled 2. No transgenic plants could be obtained from explants isolated from the region labelled 3. This study showed that not only the explant type but also the position the explant was excised from is very important in gene transfer via *A. tumefaciens* (Beyaz et al., 2017).

In 2018, Mandal et al. published experiments with the *Agrobacterium* – mediated genetic transformation through floral – dipping of buds and of friable embryogenic calli. For floral-dipping transformation, flower buds were infected with Agrobacterium strain LBA4404, involving a popular binary vector pCAMBIA1305.2. 76.6% of the buds showed distinct GUS expression after 48 h of Agrobacterium infection, especially in the reproductive. In experiments with calli, high proliferating calli were treated with acetosyringone and infected with *A. tumefaciens*. After shootlet regeneration, calli were used for GUS histochemical assay. In 77.5% calli, bluish spots were observed those confirmed GUS expression at the transient level.

In EU, the application of genetically modified (GM) flax encounters numerous legal restrictions. Thus, in the study from Dzialo et al. (2017) to optimize the new strategy of plant modification, plants developed through two methods were compared: the widely popular agrotransformation and the relatively recent strategy of treatment with oligodeoxynucleotides (ODNs). In summary, the acquired data revealed that the methylation status of plants upon ODN treatment as well as transgenesis was changed.

In novel study, Wróbel et al. (2019) described inserting of gene encoding PHA synthase from *Pseudomonas aeruginosa*. The gene was expressed in flax plants with the aim of modifying the mechanical properties of plants. Transgenic modified plants showed increases in the values of all measured parameters: the largest increase was noted for tensile stiffness, which was 2 to 3 times higher than in wild-type plants. The created plats also exhibited higher tensile strength and Young's modulus.

17.4.3 GMO flax (genetic engineered flax)

Since the commercialization of GM crops 13 years ago, the global area of GM crops has increased from 1.7 to 125 million hectares. Over 13 million farmers grew GM crops in 2008 compared to 12 million in 2007. Three new countries, Burkina Faso (cotton), Egypt (corn) and Bolivia (soy), planted GM crops for the first time, bringing the total number of countries growing GM crops to 25. The top 8 GM countries are the USA 62.5 mil. ha/2008 (alfalfa, canola, cotton, corn, soybean, squash, sugar beet, papaya), Argentina 21 mil.ha (cotton, corn, soybean), Brazil 15.8 mil.ha (cotton, corn, soybean), India 7.6 mil.ha (cotton), Canada 7.6 mil.ha (canola, corn, soybean, sugar beet), China 3.8 mil.ha (cotton, papaya, petunia, poplar, sweet pepper, tomato), Paraguay 2.7 mil.ha (soybean) and South Africa 1.8 mil.ha (cotton, corn, soybean). The International Service for the Acquisition of Agri-biotech Applications (ISAAA) expects that 15 or more countries will plant GM crops from now until 2015, bringing the total number of countries planting GM crops to 40. Asia, Eastern and Southern Africa, West Africa, North Africa and the Middle East are estimated to be the new adopters of GM crops. Countries in Eastern Europe and Latin/Central America may also adopt GM crops. Several new GM crops are expected between now and 2015, one being GM rice. The ISAAA states that pest and disease resistant rice is awaiting approval in China and Golden Rice, which could help curb deficiencies such as vitamin A, is expected to be available in 2012.

For 13 years, herbicide tolerance has consistently been the dominant trait. In 2008 herbicide tolerant GM crops (soybean, corn, canola, cotton and alfalfa) made up 63% of the global GM market. Stacked double and triple traits occupied 22% of the GM crop area and insect resistant varieties made up the remaining 15%.

GM flax is developed in agronomic traits in the first place. Herbicide tolerance is very important for weed control, as well as fungal resistance of flax for disease resistance, insect resistance for resistance against pests and stress tolerance to adapt to climate and local factors. Quality traits of flax are very interesting as well: modified composition of ingredients (oil composition), and enrichment with health-promoting ingredients, such as flavonoids (antioxidants) and omega-3 fatty acids (by transference of a marine algae to flax seed, the seeds then contain omega-3 fatty acids EPA and DHA, which are a feature of fat-laden saltwater fishes). These acids are assumed to have a preventive effect towards hypertension and specific degenerative diseases. Renewable resources are as also important in terms of modified composition (modified elasticity and thermoplastic characteristics of the flax seed fibre for the synthesis of biological degradable synthetic material), production of pharmaceutical agents (molecular farming usage of GM flax seed as a system to produce pharmaceuticals; to date, only experimental) and land reclamation = phytoremediation (heavy metal strained soil) – the plants are modified so that they are able to grow within this soil and can extract heavy metals and then accumulate them within the plant biomass.

Field trials with GM flax seed have been undertaken in the EU in three applications in three countries (Sweden, Poland and Czech Republic) in 2005–07. Traits studied in these experiments were oil composition, flavonoids content, elasticity (bioplastics), herbicide tolerance, insect resistance, insect and fungal resistance and heavy metal absorption.

No GM flax is currently grown. A herbicide-resistant GM flax was introduced in 2001, but was soon taken off the market because European importers refused to buy it. In September 2009 it was reported that Canadian flax exports had been contaminated by an unapproved, illegal, GM (GM) variety, known as Triffid. Since linseed derived from GM flax is not authorized in Europe, products containing even minimal amounts cannot be made commercially available. Transgene contamination of flax seed is used widely in the food industry, including bread, and as source of omega-3 fatty acids. On 10 September 2009, EU Rapid Alert System for Food and Feed (RASFF) reported finding an unapproved GM flax/linseed variety in cereal and bakery products in Germany. In Germany, the GM linseed has been found in at least seven states up to now. Baking ingredients contaminated with the GM linseed had been distributed to 15 German states and exported to other countries by a German company based in Hessen. Linseed is an ingredient in baked-goods and muesli. Consumption of products containing minute traces of GM linseed does not present a health risk. The GM flax variety FP967 (CDC Triffid) is not authorized for food or feed in the EU; it has tolerance to soil residues of sulfonylurea-based herbicides, and was developed by the Crop Development Centre at the University of Saskatchewan in Canada. Authorization of the variety was rescinded in Canada in 2001. Cultivation of CDC Triffid flax has since been banned. Canada supplies approximately 70% of the total flax/linseed in the EU annually. Because GM flax FP967 is not authorized in the European Union, there is zero tolerance for the variety. That means any raw material or flax/linseed derivative analyzed to be positive for FP967 is illegal and not marketable in the EU. The Canadian Grain Commission is investigating how the admixture of the GM flax in linseed products could have occurred.

17.5 New breeding techniques

New breeding techniques (NBTs) also named 'New Plant Engineering Techniques' are methods which can accelerate development of the new plant varieties with desired traits by modifying the DNA of the seeds and plant cells. These are fairly new methods which have been especially developed in the last decade and have rapidly evolved in recent years. These new techniques, often involve 'genome editing' whose intention is to modify DNA at specific locations within the plants' genes so that new traits and properties are produced in crop plants.

These methods are successfully used in Czech breeding programme via GM flax and linseed for phytoremediation.

As most cultivars of flax and linseed (*L. usitatissimum*) are capable of accumulating cadmium (Cd), they are suitable candidates for phytoextraction of the metals from contaminated soil (Vrbová et al., 2013). A certain limitation of this species in phytoextraction is due to retention of a substantial amount of Cd in roots of both flax and linseed cultivars, which restricts the technoeconomic suitability of flax to soil Cd concentrations ranging from 1 to 10 $mg.kg^{-1}$ (Angelova et al., 2004; Jiao et al., 2004; Bjelková et al., 2011). However, severely damaged agricultural soils often contain 10–30 mg Cd kg^{-1} and the surface soils in the vicinity of metal-processing plants may contain several hundred mg Cd kg^{-1} (Kabata-Pendias, 2011), which indicates the need of GM plants for phytoremediation. As was already stated, no "hyperaccumulators" were identified within the natural flax/linseed germplasm (Griga and Bjelková, 2013; Griga and Vrbová, 2011), and thus heavy metal (HM) tolerance/accumulation could be improved via genetic engineering.

17.5.1 History of flax transformation

The first transgenic flax cells were created in 1983 (Hepburn et al., 1983); the transformation was performed via *A. tumefaciens*. The first attempts aimed at transfer of whole unmodified T-DNA of *A. tumefaciens* and *Agrobacterium rhizogenes* were later changed for integration of model prokaryotic genes (Basiran et al., 1987; Mlynárová et al., 1994; Bretagne-Sagnard and Chupeau, 1996) and of specific genes coding economically important traits (Jordan and McHughen, 1988; McHughen, 1989; McHughen and Holm, 1995) into the flax genome. Flax/linseed transformation has been already reviewed in a number of journal papers or book chapters (Rakouský and Tejklová, 1999; McHughen, 2000; Pretová et al., 2007; Pavelek et al., 2012; Badere, 2014; Ludvı'ková and Griga, 2015). Although the first transgenic *L. usitatissimum* was obtained 30 years ago (Jordan and McHughen, 1988) flax/linseed transformation is not considered routine because it appears remarkably genotype-dependent (Zhan et al., 1988; Rakouský and Tejklová, 1999). Efficiency of flax transformation can be improved by modification of explant-Agrobacterium cocultivation period and injury of infected explant surface by removal of epidermis or sonication (Vrbová et al., 2013). Then the transformation protocol has been several times improved (Vrbová et al., 2013).

17.5.2 Gene delivery methods in flax/linseed transformation

Various methods to deliver genes into the flax genome have been developed over the years. The most frequently used procedure has been *Agrobacterium*-mediated hypocotyl transformation. Flax, like most dicotyledonous crop species, is suitable to gene transfer via *Agrobacterium* (Hepburn et al., 1983). Flax cells can be relatively easily transformed with *A. tumefaciens* and easily grown when the suitable inoculation/selection/regeneration procedure is applied (Jordan and McHughen, 1988). The first report of the regeneration of flax transformed by *A. rhizogenes* was described by Zhan et al. (1988). Another possibility is offered by the particle gun (biolistic) method, adapted to flax in the Crop Development Centre (University of Saskatchewan, Saskatoon, Canada). Wijayanto and McHughen (1999) documented a successful biolistic process to regenerate transgenic linseed. Bastaki and Cullis (2014) described the Agrobacterium transformation of flax via floral dip. Results showed that the floral-dip method could replace the previously used techniques for flax transformation because of its simplicity, low cost and high transformation rate.

17.5.3 Genetically modified flax for phytoremediation

As crop species including both Cd-accumulating and Cd-tolerant cultivars, flax is also gaining increasing attention for potential use in phytoextraction of Cd (Broadley et al.,

2001; Kos et al., 2003; Angelova et al., 2004; Shi and Cai, 2009; Smýkalová et al., 2010; Soudek et al., 2010; Hradilová et al., 2010; Bjelková et al., 2011; Najmanova et al., 2012). Regrettably, phenotypic variation within commercial flax and linseed cultivars (Smýkalová et al., 2010; Soudek et al., 2010; Bjelková et al., 2011) or germplasm resources (Bjelkova', unpublished data) does not appear high enough to justify the use of conventional breeding approaches for conditioning of phytoextraction plants. In general, the introduction of novel traits for the uptake and accumulation of pollutants into high biomass is proving to be a successful strategy for the development of improved phytoremediators. Recent research is directed to overexpression of genes whose protein products are involved in metal uptake, transport and sequestration or act as enzymes involved in degradation of hazardous organics or design systems containing special endophytes (Macek et al., 2012). The only fibre crop that has been GM to date is flax. The first successful result of genetic engineering of flax for HMs tolerance and accumulation (Vrbová et al., 2013) represents a promising approach to improve phytoremediation potential of fibre crops. In this study, in an attempt to enhance the phytoextraction capacity of *L. usitatissimum* through overproduction of an efficient heterologous Cd-binding peptide, a breeding line to constitutively express genetic fusion of α-domain of mammalian MT 1 (alpha MT1a) and uid A gene was engineered. The MTs are a family of cysteine-rich peptides capable of high-affinity coordination of HM ions (Coyle et al., 2002; Freisinger, 2008). In the last decade, many attempts were made to prepare transgenic plants bearing genes coding for metal-binding proteins of different origin—human MT, animal MT, and yeast MT genes (Bailey et al., 2003). Authors found increased Cd content in roots of transgenic tobacco expressing mammalian MT but decreased Cd content in leaves (Dorlhac De Borne et al., 1998). The performance of transgene products has been enhanced by implanting an additional HM-binding site (Macek et al., 2002). Vrbová et al. (2013) presumed that the production of αMT1a should lead to a greater capacity to detoxify and sequester Cd. In the study, an improved flax transformation protocol involving cocultivation of hypocotyl segments with partially removed epidermis with *Agrobacterium* suspension for 10 min in the presence of 200 $mg.L^{-1}$ cellulase was developed. As a result of successful transformation, hypocotyl explants of transformed line showed 1.7-fold higher biomass than those of parental flax line on media containing 15 mg Cd per litre, indicating that αMT1a transgene did confer higher Cd tolerance to engineered plants. Although even the best performing transgenic line did not exhibit increased stem Cd levels when grown in soils with natural Cd background (0.25 mg $Cd.kg^{-1}$), it accumulated 3.3- and 1.9-fold higher amount of Cd than the parental line from soils containing 20 and 360 mg $Cd.kg^{-1}$, respectively. The modified plants were genetically stable and had an improved ability to grow in contaminated soil (for field testing see below), extract HMs (particularly Cd), and accumulate them within the plant biomass, so they could be successfully used for reclamation and phytoremediation of HM strained soil. While the first laboratory and field tests of alpha-MT flax (officially started/permitted in 2009) were connected with Cd, recent research is extended to other metals/metalloids (Pb, As, Se; Cvečková et al., 2013, 2014, unpublished data). It has been well documented earlier that genetic engineering of tobacco, Arabidopsis, and Brassica species aiming at improved production of PCs or heterologous meta-binding proteins or peptides, including MTs, may further increase the ability of transgenic plants to tolerate and accumulate HM ions (Bhargava et al., 2012). Several lines of evidence suggest that overproduction of plasma membrane metal transporters responsible for loading of Cd into xylem may support the translocation of accumulated metal to shoots (Krämer, 2010). Data obtained by Vrbová et al. (2013) show that constitutive overproduction of αMT1a transgene increased the natural capacity of flax to accumulate Cd in stems. Earlier studies have revealed certain difficulty in formulation of general high-performance agrotransformation protocol for *L. usitatissimum*, which is being attributed to genotype-dependent predisposition of particular cultivar (Zhan et al., 1988; Rakouský and Tejklová, 1999). Apart from selection of appropriate genotype, the overall yield of T_0 transformants can be improved by enlarging the area of stably transformed tissue through promoting the Agrobacterium virulence and tissue invasion during the cocultivation period (Gelvin, 2003; Tzfira and Citovsky, 2006; Vrbová et al., 2013). The second transgene for production of heterologous metal-binding protein that was used for flax transformation via optimized Agrobacterium-mediated transformation protocol was CUP gene encoding MT with a high affinity for HMs (Griga et al., 2009; Griga and Vrbová, 2011). This transgene was also fused with uid reporter gene for easy and rapid detection of putative transformants (Fig. 17.17). Najmanova et al. (2007) also described the preparation of transgenic flax with enhanced metal tolerance. The aim of the work was to engineer flax to express the yeast glutamylcysteine synthetase (γ-ECS) and GS gene under the control of a strong, light-regulated rubisco promoter (RBcS). So far, they managed to clone these two genes under RbcS promoter into *Escherichia coli*. Overexpression of γ-ECS and GS genes may result in increased HM tolerance. Both are key enzymes in glutathion biosynthesis. Glutathion belongs to cysteine sulfhydryl-rich peptide thiols, which are believed to play important roles in the detoxification of many HMs such as arsenic, mercury and cadmium in plants.

17.5.4 General conclusion

As generally accepted, practical applications of GM technology in field crops were realized in the three subsequent/factual phases (which—of course—are carried out partly in parallel): (1) crop modification for better field cultivation (herbicide tolerance; resistance/tolerance against biotic and abiotic factors—virus, bacterial and insect pests, salt and drought tolerance), (2) product quality improvement (changed chemical composition of the seeds, fruits, tubers; also fibre characteristics in fibre crops) and (3) molecular farming (utilization of plants as 'cell factories' for the production of recombinant proteins of various determination—pharmaceuticals, bioplastics, edible vaccines). What about the risks of transgenic fibre crops with phytoremediation potential for the environment and biodiversity? Is the improved ability of HMs/metalloids accumulation an ecological/environmental/evolutionary advantage? Might GM fibre crops with such characteristics represent a danger for the environment? The answer is simple and evident – 'NO!' Of course, each example should be thoroughly evaluated case by case. The most progressed data on GM crops risk evaluation have been collected in flax (Jhala et al., 2008, 2009, 2011; Griga et al., 2008 and unpublished data) Despite the fact that culture flax has a number of related wild species—even with the same number of chromosomes (n = 15)—occurring in Eurasia and North America (Jhala et al., 2008, 2009; 2011; Ludvíkova' and Griga, 2015), the natural hybrids are not known. Thus, the unintended release of GM pollen represents a risk just for culture flax (Ludvíkova' and Griga, 2015) – the risk of contamination of breeders' or seed production fields.

17.6 Sources of further information and advice

The common catalogue of varieties of agricultural plant species (182 flax and linseed) is downloadable from the Health and Consumers: Plants Section of the European Commission website:

http://ec.europa.eu/food/plant/plant_propagation_material/plant_variety_catalogues_databases/search/public/index.cfm?event=SearchForm&ctl_type=A.

At present, there are 182 varieties of flax and linseed originating in the United Kingdom, France, Netherlands, The Czech Republic, Belgium, Italy, Germany, Poland, Romania, Lithuania, Hungary, Finland, Slovak Republic, and Austria (Table 17.11.).

Appendix: abbreviations

AFLP	Amplified fragment length polymorphism
ANOVA	Analysis of variance
BAP	6-benzylaminopurine
BAZ	Federal Centre for Breeding Research on Cultivated Plants, Braunschweig, Germany
CDC Triffid	a flax (*Linum usitatissimum* L.) variety tolerant to soil residues of triasulfuron and metsulfuron-methyl
CGN	Centre for Genetic Resources, the Netherlands
DH	dihaploids
EST-SST	expressed sequence tag – signature sequence tag
EURISCO	European Plant Genetic Resources Search Catalogue with Passport data on ex Situ collections maintained in Europe
IAA	Indolylacetic acid
INRA	Institut National de la Recherche Agronomique

IPK	Institute of Plant Genetics and Crop Research, Gatersleben, Germany
ISO	International Organization for Standardization
MS	Murashige and Skoog medium (1962)
NAA	Naphthalene acetic acid
RFLP	restriction fragment length polymorphism
SSR	Simple sequence repeat
TDZ	thidiazuron
UPOV	International Union for the Protection of New Varieties of Plants

References

Angelova, V., Ivanova, R., Delibaltova, V., Ivanov, K., 2004. Bio-accumulation and distribution of heavy metals in fibre crops (flax, cotton and hemp). Ind. Crops Prod. 19, 197–205.

Antonius-Klemola, K., Kalendar, R., Schulman, A.H., 2006. TRIM retrotransposons occur in apple and are polymorphic between varieties but not sports. Theor. Appl. Genet. 112, 999–1008.

Armstrong, G.M., Armstrong, J.K., 1968. Formae speciales and races of Fusarium oxysporum causing a tracheomycosis in the syndrome of disease. Phytopathology 58, 1242–1246.

Badere, R.S., 2014. Linseed improvement: a glance at the major developments in recent years. In: Makde, K.H., Pawar, S.E. (Eds.), Prof. A. D. Choudhary Commemoration—A Glimpse of Current Vistas in Plant Science Research. Edition I. Hislop College Publication Cell, Nagpur, pp. 75–92.

Badwal, S.S., 1975. Inheritance of resistance to powdery mildew in linseed. Indian J. Genet. Plant Breed. 35, 432–433.

Bailey, N.J., Oven, M., Holmes, E., Nicholson, J.K., Zenk, M.H., 2003. Metabolomic analysis of the consequences of cadmium exposure in Silene cucubalus cell cultures via 1H NMR spectroscopy and chemometrics. Phytochemistry 62 (6), 851–858.

Balabanova, A., Atanasov, A., 2001. 'Preservation, evaluation and utilization of *Linum* L. germplasm in the AgroBioinstitute Kostinbrod, Bulgaria – current status and strategy'. In: Maggioni, L., Pavelek, M., Van Soest, L.J.M., Lipman, E., compilers (Eds.), Flax Genetic Resources in Europe. Ad hoc Meeting, 7–8 December 2001, Prague. Czech Republic and International Plant Genetic Resources Institute, Rome, Italy.

Bari, G., 1971. Effects of chronic and acute irradiation on morphological characters and seed yield in flax. Radiat. Bot. 11, 293–302.

Bartošová, Z., Pret'ová, A., 2003. Induction of callogenesis in ovary and anther cultures of flax. In: Proceedings of the 10th Scientific Seminar, Piešt'any (Slovakia), November, pp. 25–28.

Bartošová, Z., Roux, N., Pret'ová, A., 2003a. Ovary culture in *Linum usitatissimum* L. In: Proceedings of the 11th International Conference on Plant Embryology, September 2003, Brno, Czech Republic, p. 116.

Bartošová, Z., Roux, N., Pret'ová, A., 2003b. 'Green plants regenerated from ovary culture in flax (*Linum usitatissimum* L.)'. Book of Abstracts. In: 5th International Symposium in the Series Recent Advances in Plant Biotechnology, Stará Lesná, 7–13 September 2003, p. 65.

Bartošová, Z., Obert, B., Takáč, T., Kormut'ák, A., Pret'ová, A., 2005. Using enzyme polymorphism to identify the gametic origin of flax regenerants. Acta Biol. Cracov. Bot. 47 (1), 73–178.

Bartošová, Z., Masar, S., Pret'ová, A., 2006. Flax plant regenerated from unpollinated ovules cultured in ovary segments. Acta Hortic. 725 (2), 869–871.

Bas, N., Pavelek, M., Mandolino, G., Carboni, A., Lipman, E., 2006. 'Report of a Working Group on Fibre Crops (Flax and Hemp)'. First Meeting, 14–16 June. Wageningen, The Netherlands.

Basiran, N., Armitage, P., Scott, R.J., Draper, J., 1987. Genetic transformation of flax (*Linum usitatissimum*) by *Agrobacterium tumefaciens*: regeneration of transformed shoots via a callus phase. Plant Cell Rep. 6, 396–399.

Bastaki, N.K., Cullis, C.A., 2014. Floral-dip transformation of flax (*Linum usitatissimum*) to generate transgenic progenies with a high transformation rate. J. Vis. Exp. 94, 52189.

Beckmann, A., Kromer, K.H., 1995. Evaluation of test standards for measuring the fibre content and strength values of flax. Zemed. Tech. 41 (3), 121–124.

Bergmann, R., Friedt, W., 1997. Haploidy and related biotechnological methods in linseed (*Linum usitatissimum* L.). In: Jain, S.M., Sopory, S.K., Veilleux, R.E. (Eds.), In Vitro Haploid Production in Higher Plants, vol. 5. Kluwer, Dordrecht, pp. 1–16.

Beyaz, R., Darcin, E.S., Aycan, M., Kayan, M., Yildiz, M., 2016. A novel method for high-frequency transgenic shoot regeneration via *Agrobacterium tumefaciens* in flax (*Linum usitatissimum* L.). J. Plant Biotechnol. 43 (2), 240–247. https://doi.org/10.5010/JPB.2016.43.2.240.

Beyaz, R., Aycan, M., Yildiz, M., 2017. Explant position effect on gene transformation to flax (*Linum usitatissimum* L.) via *Agrobacterium tumefaciens*. Period. Biol. 119 (3), 223–228.

Bhargava, A., Carmona, F.F., Bhargava, M., Srivastava, S., 2012. Approaches for enhanced phytoextraction of heavy metals. J. Environ. Manag. 105, 103–120.

Bhateria, S., Sood, S.P., Pathania, A., 2006. Genetic analysis of quantitative traits across environments in linseed (*Linum usitatissimum* L.). Euphytica 150 (1–2), 185–194.

Bjelková, M., Genčurová, V., Griga, M., 2011. Accumulation of cadmium by flax and linseed cultivars in field-simulated conditions: a potential for phytoremediation of Cd- contaminated soils. Ind. Crops Prod. 33, 761–774.

Borlaug, N.E., 1945. Variation and Variability of Fusarium Lini, University of Minnesota, vol. 168. Agricultural Experiment Station Technical Bulletin.

Bretagne-Sagnard, B., Chupeau, Y., 1996. Selection of transgenic flax plants is facilated by spectinomycin. Transgenic Res. 5, 131–137.

Bretón, A.M., 2001. Respuesta en la callogénesis in vitro de anteras de lino (*Linum usitatissimum* L.) para tres temperaturas de incubación. J. Basic Appl. Genet. Sociedad Argentina de Genética 14 (2), 100–101.

Bretón, A.M., Lassaga, S.L., 1999. Callogénesis in vitro de anteras de lino (*Linum usitatissimum* L.) según el memento de floración. In: 29° Congreso Argentino de Genética. 32° Congreso de Genética de Chile. III Jornadas Chileno Argentinas de Genética., Rosario. Libro de Actas del 29° Congreso Argentino de Genética. Sociedad Argentina de Genética, p. 402.

Bretón, A.M., Dittrich, A.A., Lassaga, S.L., 2004. Efecto del aumento de la dosis de sulfato de cobre en el medio de inducción sobre la callogénesis en cultivo de anteras de *Linum usitatissimum* L. J. Basic Appl. Genet. Sociedad Argentina de Genética 17, 101–102.

Brim, G.A., 1966. A modified pedigree method of selection in soybeans. Crop Sci. 6, 220.

Broadley, M.R., Willey, N.J., Wilkins, J.C., Baker, A.J.M., Mead, A., White, P.J., 2001. Phylogenetic variation in heavy metal accumulation in angiosperms. New Phytol. 152, 9–27.

Brutch, N., 2001. The flax genetic resources collection held at the Vavilov Institute, Russian Federation. In: Maggioni, L., Pavelek, M., van Soest, L.J.M., Lipman, E., compilers) (Eds.), Flax Genetic Resources in Europe. Ad hoc Meeting, 7–8 December 2001, Prague. Czech Republic and International Plant Genetic Resources Institute, Rome, Italy.

Burbulis, N., Blinstrubienė, A., 2006. Comparison of anther culture response among *Linum usitatissimum* L. cultivars and their hybrids. Acta Univ. Latv. Biol. 710, 131–138.

Burbulis, N., Blinstrubienė, A., Sliesaravičius, A., Venskutonienė, E., 2005a. Influence of genotype, growth regulators, sucrose level and preconditioning of donor plants on flax (*Linum usitatissimum* L.) anther culture. Acta Biol. Hung. 56 (3–4), 323–331.

Burbulis, N., Blinstrubiene, A., Venskutoniene, E., KatauskyteL, 2005. Organogenesis in callus cultures of Linum usitatissimum L. Acta Univ. Latv. Univ. Biol. 691, 129–135.

Burbulis, N., Blinstrubienú, A., Kuprienú, R., Sliesaravičius, A., Venskutonienė, E., 2007a. Optimization of linseed flax (*Linum usitatissimum* L.) in vitro cultures. Žemdirbyst/Zemdirbyste/Agriculture 94 (4), 120–128.

Burbulis, N., Blinstrubienė, A., Sliesaravičius, A., Kuprienė, R., 2007b. Some factors affecting callus induction in ovary culture of flax (*Linum usitatissimum* L.). Biologija 53 (2), 21–23.

Burbulis, N., Blinstrubienė, A., Kuprienė, R., Žilenaitė, L., 2009. Effect of genotype and medium composition on flax (*Linum usitatissimum* L.) anther culture. Agron. Res. 7 (Special issue I), 204–209.

Carnahan, H.L., 1947. Combining Ability in Flax (*Linum usitatissimum*). M.Sc. thesis. University of Minnesota, Minnesota, USA.

Chauhan, L.S., Srivastava, K.N., 1975. Estimation of loss of yield caused by blight disease of linseed. Indian J. Farm Sci. 3, 107–109.

Chen, Y., Dribnenki, P., 2002. Effect of genotype and medium composition on flax *Linum usitatissimum* L. anther culture. Plant Cell Rep. 21, 204–207.

Chen, Y., Dribnenki, P., 2004. Effect of medium osmotic potential on callus induction and shoot regeneration in flax anther culture. Plant Cell Rep. 23 (5), 272–276.

Chen, Y., Kenaschuk, E., Dribnenki, P., 1998a. High frequency of plant regeneration from anther culture in flax, *Linum usitatissimum* L. Plant Breed. 117, 463–467.

Chen, Y., Kenaschuk, E., Procunier, J.D., 1998b. Plant regeneration from anther culture in Canadian cultivars of flax (*Linum usitatissimum* L.). Euphytica 102, 183–189.

Chen, Y., Kenaschuk, E., Dribnenki, P., 1998c. Production and utilization of doubled haploids in linseed/linola breeding. In: Proceedings of the 9th International Congress on Plant Tissue and Cell Culture, Jerusalem, Israel, 14–19 June, p. 134.

Chen, Y., Hausner, G., Kenaschuk, E., Procunier, D., Dribnenki, P., Penner, G., 1998d. Identification of microspore-derived plants in anther culture of flax (*Linum usitatissimum* L.) using molecular markers. Plant Cell Rep. 18 (1–2), 44–48.

Chen, Y., Kenaschuk, E., Dribnenki, P., 1999. Response of flax genotypes to doubled haploid production. Plant Cell Tissue Organ Cult. 57 (3), 195–198.

Chen, Y., Kenaschuk, E., Dribnenki, P., 2001. Inheritance of rust resistance genes and molecular markers in microspore derived populations of flax. Plant Breed. 120, 82–84.

Chen, Y., Lin, L., Duguid, S., Dribnenki, P., Kenaschuck, E., 2003. Effect of sucrose concentration on elongation of shoots from flax anther culture. Plant Cell Tissue Organ Cult. 72, 181–183.

Cloutier, S., Niu, Z., Datla, R., Duguid, S., 2009. Development and analysis of EST-SSRs for flax (*Linum usitatissimum* L). Theor. Appl. Genet. 119, 53–63.

Convention on Biological Diversity (CBD, UNCED), 1992. http://www.cbd.int/doc/legal/cbd-en.pdf.
Coyle, P., Philcox, J.C., Carey, L.C., Rofe, A.M., 2002. Metallothionein: the multipurpose protein. Cell. Mol. Life Sci. 59, 627–647.
Culbertson, J.O., 1954. Breeding flax. Adv. Agron. 6, 174–178.
Cvečková, M., Větrovcová, M., Vrbová, M., Griga, M., 2013. In Vitro Testing of Potential for Selenium Accumulation by Linseed (*Linum usitatissimum* L.). Abstracts of Fourth Ann. Conf. COST Action FA 0905—Essential and Detrimental Trace Elements Entering the Food Chain via Plants. Norwegian University of Life Sciences, UMB, Norway, p. 61.
Cvečková, M., Vrbová, M., Pavelková, M., Větrovcová, M., Smýkalová, I., Griga, M., 2014. Transgenic linseed (*Linum usitatissimum* L.) with alphaHMT1a: the effect on the accumulation of several metal elements. In: Abstracts of 11th International Phytotechnologies Conference, Heraklion, Crete, p. 155.
Dang, Z.H., Zang, J.P., She, X.C., 2000. Induction male sterility in linseed by antibiotics treatment. Chin. J. Oil Crop Sci. 22 (1), 46–48.
De Sousa, T., 2006. 'National collection – status report, Portugal'. In: Bas, N., Pavelek, M., Maggioni, L., Lipman, E., compilers) (Eds.), ECP/GR, Report of a Working Group on Fibre Crops (Flax and Hemp). First Meeting, 14–16 June 2006, Wageningen, The Netherlands.
Dehmer, K., Frese, L., Freytag, U., Knupfer, H., Kurch, R., Schutze, G., 2001. Status report on the Linum collections in German genebanks. In: Maggioni, L., Pavelek, M., van Soest, L.J.M., Lipman, E., compilers) (Eds.), Flax Genetic Resources in Europe. Ad hoc Meeting, 7–8 December 2001, Prague. Czech Republic and International Plant Genetic Resources Institute, Rome, Italy.
Diederichsen, A., Fu, Y.B., 2006. Phenotypic and molecular (RAPD) differentiation of four infraspecific groups of cultivated flax (*Linum usitatissimum* L. subsp. *usitatissimum)*. Genet. Resour. Crop Evol. 53, 77–90.
Dorlhac De Borne, F., Elmayan, T., De Roton, C., De Hy, L., Tepfer, M., 1998. Cadmium partitioning in transgenic tobacco plants expressing a mammalian metallothionein gene. Mol. Breed. 4, 83–90.
Dribnenki, J.C.P., Green, A.G., 1995. Linola (TM) "497" low linolenic acid flax. Can. J. Plant Sci. 75, 201–202.
Dribnenki, J.C.P., McEachern, S.F., Green, A.G., Kenaschuk, E.O., Rashid, K.Y., 1999. Linola TM "1084" low-linolenic acid flax. Can. J. Plant Sci. 79 (4), 607–609.
Dribnenki, J.C.P., McEachern, S.F., Chen, Y., Green, A.G., Rashid, K.Y., 2003. Linola TM "2047" low-linolenic acid flax. Can. J. Plant Sci. 83 (1), 81–83.
Dribnenki, J.C.P., McEachern, S.F., Chen, Y., Green, A.G., Rashid, K.Y., 2004. Linola TM "2090" low-linolenic acid flax. Can. J. Plant Sci. 84 (3), 797–799.
Dubey, D.K., Singh, S.P., 1965. Mechanism of pollen abortion in three male sterile lines of flax (*Linum usitatissimum* L.). Crop Sci. 5, 121–124.
Dubey, D.K., Singh, S.P., 1966. Use of cytoplasmatic male sterility for the production of hybrid seeds in flax *Linum usitatissimum* L. Crop Sci. 6, 125–126.
Dubey, D.K., Singh, S.P., 1969. Extent of heterosis and problems of hybrid seed production in linseed (*Linum usitatissimum* L.). Balwant Vidyapeeth J. Agric. Sci. Res. 6, 83–87.
Dzialo, M., Szopa, J., Czuj, T., Zuk, M., 2017. Oligodeoxynucleotides can transiently up- and downregulate CHS gene expression in flax by changing DNA methylation in a sequence-specific manner. Front. Plant Sci. 8, 755. https://doi.org/10.3389/fpls.2017.00755.
El-Adl, A.M., Miller, P.A., 1971. Transgressive segregations and the nature of gene action for yield in an intervarietal cross of upland cotton. Crop Sci. 11, 381–384.

Evans, N., McRoberts, N., Hitchcock, D., Marshall, G., 1995. Assessing linseed (*Linum usitatisssimum* L.) resistance to *Alternaria linicola* using a detached cotyledon assay. Ann. Appl. Biol. 127 (23), 263–271.

Evans, N., McRoberts, N., Hitchcock, D., Marshall, G., 1997. Identification of the determinants of host resistance and pathogenicity in interactions between *Alternaria linicola* (Groves and Skolko) and *Linum usitatissimum* L. accessions using multivariate analyses. Ann. Appl. Biol. 130, 537–547.

Everaert, I., de Riek, J., de Loose, M., van Waes, J., van Bockstaele, E., 2001. Most similar variety grouping for distinctness evaluation of flax and linseed (*Linum usitatissimum* L.) varieties by means of AFLP and morphological data. Plant Var. Seeds 14, 69–78.

FAOSTAT 2016. In: http://www.fao.org/faostat/en/#data/QC.

Flax Council of Canada, 2006. http://www.flaxcouncil.ca/english/index.jsp?p=statistics2&mp=statistics.

Food and Agriculture Organization of the United Nations, 1996. Global Plan of Action for the Conservation and Sustainable Utilization of Plant Genetic Resources for Food and Agriculture.

Foster, R., Pooni, H.S., Mackay, I.J., 1997. Quantitative evaluation of *Linum usitatissimum* varieties for dual purpose traits. J. Agric. Sci. 129 (2), 121–124.

Foster, R., Pooni, H.S., Mackay, I.J., 1998. Quantitative analysis of *Linum usitatissimum* crosses for dual purpose traits. J. Agric. Sci. 131 (3), 285–292.

Foster, R., Pooni, H.S., Mackay, I.J., 2000. The potential of selected *Linum usitatissimum* L. crosses for producing recombinant inbred lines with dual purpose characteristics. J. Agric. Sci. 134 (4), 399–404.

Fouilloux, G., 1988. Breeding flax methods. In: Marshall, G. (Ed.), Flax: Breeding and Utilisation. Springer, Dordrecht, pp. 14–25.

Fouilloux, G., Trouve, J.P., Chaboche, I., Dhorne, D., 1991. Flax breeding for high productivity and high quality. In: Heredity of Fibre Content: Flax as a Fibre and Oil Bearing Crop. Proceedings of the FAO European Regional Workshop on Flax, Brno, Czechoslovakia, 18–20 June, pp. 46–63.

Fouilloux, G., Dorvillez, D., Blouet, F., 2001. 'The French flax and linseed germplasm collection – status 2001'. In: Maggioni, L., Pavelek, M., van Soest, L.J.M., Lipman, E., compilers) (Eds.), Flax Genetic Resources in Europe. Ad hoc Meeting, 7–8 December 2001, Prague. Czech Republic and International Plant Genetic Resources Institute, Rome, Italy.

Freisinger, E., 2008. Plant MTs-long neglected members of the metallothionein superfamily. Dalton Trans. 47, 6663–6675.

Friedt, W., Bickert, C., Schaub, H., 1995. In vitro breeding of high-linolenic, doubled haploid lines of linseed (*Linum usitatissimum* L.) via androgenesis. Plant Breed. 144, 322–326.

Fu, Y.B., 2002. Redundancy and distinctness in flax germplasm are revealed by RAPD. Plant Genet. Resour. 4, 117–124.

Fu, Y.B., 2005. Geographic patterns of RAPD variation in cultivated flax. Crop Sci. 45, 1084–1091.

Fu, Y.B., 2002a. Redundancy and distinctness in flax germplasm as revealed by RAPD. Plant Genetic Resources 4, 117–124.

Fu, Y.B., Peterson, G., Diederichsen, A., Richards, K.W., 2002b. RAPD analysis of genetic relationships of seven flax species in the genus *Linum* L. Genet. Resour. Crop Evol. 49, 253–259.

Fu, Y.B., Rowland, G.G., Duguid, S.D., Richards, K.W., 2003. RAPD analysis of 54 North American flax cultivars. Crop Sci. 43, 1510–1515.

Geleta, N., 1999. Performance of improved linseed varieties in western Ethiopia. Agri Topia 14 (2), 5.

Gelvin, S.B., 2003. Agrobacterium-mediated plant transformation: the biology behind the "Gene-Jockyeing" tool. Microbiol. Mol. Biol. Rev. 67, 16–37.

George, K.P., Nayer, G.G., 1973. Early dwarf mutant in linseed induced by gamma rays. Curr. Sci. 42, 137–138.

Ghai, B.S., 1966. Cytogenetical Study of Interspecific Hybrids in the Genus *Linum*. Ph.D. thesis. P.A.U. Hissar, p. 136.

Gill, K.S., 1964. The Species of *Linum*. Annual Crop Science Society Meeting, Tucson, USA.

Gill, K.S., 1966. Evolutionary Relationships Among *Linum* Species. Ph.D. thesis. University of California, Riverside, USA.

Gill, K.S., 1980. Objectives, breeding approaches and achievements in linseed (*Linum usitatissimum*). In: Gill, K.S. (Ed.), Breeding Oilseed Crops. Springer, Dordrecht, pp. 212–225.

Gill, K.S., 1987. Linseed. Indian Council of Agricultural Research, New Delhi.

Gonzales, C.I., Bretón, A.M., 2005. Evaluación de la callogénesis en el cultivo in vitro de anteras de lino utilizando tres concentraciones de furfuryl-6-amino purina en el medio de inducción. In: Segundas Jornadas de Difusión en Investigación y Extensión (INEX 2005), Concordia. Universidad Nacional de Entre Ríos.

Gonzales, C.I., Vigier, J.I., Bretón, A.M., 2005. Evaluación de callogénesis y regeneración en el cultivo in vitro de anteras de lino modificando el tipo y concentración de citocinina en los medios de cultivo. In: XIII Jornadas de Jovenes Investigadores de AUGM, Tucumán. Libro de Resúmenes XIII Jornadas de Jóvenes Investigadores de AUGM. Universidad Nacional de Tucumán, p. 72.

Grahan, R.J.D., Roy, S.C., 1924. Linseed (*Linum usitatissimum*) hybrids. Agric. J. India **1** 28–31.

Grauda, D., 2006. 'National collection – status report, Latvia'. In: Bas, N., Pavelek, M., Maggioni, L., Lipman, E., compilers) (Eds.), ECP/GR, Report of a Working Group on Fibre Crops (Flax and Hemp). First Meeting, 14–16 June 2006.

Green, A.G., 1986a. A mutant genotype of flax (*Linum usitatissimum* L.) containing very low levels of linolenic acid in its seed oil. Can. J. Plant Sci. 66, 499–503.

Green, A.G., 1986b. Genetic modification of seed fatty acid composition in *Linum usitatissimum* L. J. Aust. Inst. Agric. Sci. 52 (3), 175–176.

Green, A.G., 1992. The evaluation of Linola as a new oilseed crop for Australia. In: Proceedings of the 6th Australian Society of Agronomy Conference, Armidale, pp. 471–474.

Griga, M., Vrbová, M., 2011. Biotechnological approaches to improve heavy metal accumulation by flax/linseed plants for phytoremediation purposes. In: BIT's First Annual Congress of Environmental Biotechnology, Dalian, China.

Griga, M., Bjelková, M., 2013. Flax (*Linum usitatissimum* L.) and hemp (*Cannabis sativa* L.) as fibre crops for phytoextraction of heavy metals: biological, agro-technological and economical point of view. In: Gupta (Ed.), Plant Based Remediation Processes, Volume 35 of the Series Soil Biology, pp. 239–271. ISBN 978-3-642-35564-6.

Griga, M., Tejklová, E., Seidenglanz, M., 2008. Assessment of possible risks associated with release of genetically modified flax and linseed (*Linum usitatissimum* L) into environment of the Czech Republic. In: 10th International Symposium Biosafety of Genetically Modified Organisms: Past Achievements and Future Challenges. Welington, New Zealand, p. 105.

Griga, M., Vrbova, M., Horaček, J., Smýkal, P., Tejklová, E., 2009. Agrobacterium-mediated transformation of flax (*Linum usitatissimum* L.) with heavy metal binding protein genes. In: Abstracts Joint Annual Meeting AmericanSociety of Plant Biologists and Phycological Society of America, Honolulu, HI, P32012, pp. 242–243.

Guha, S., Maheswari, C., 1964. In vitro production of embryos from anthers of Datura. Nature 204, 497.

Hanson, W.D., Probst, A.H., Caldwell, B.E., 1967. Evaluation of population of soybean genotypes with implications for improving self- pollinated crops. Crop Sci. 7, 99–102.

Harpstead, D.D., 1961. The Effect of Gamma Ray Ant Thermal Neutron Irradiation on Dormant Flax Seeds as Measured by Oil Quality and Seed Yield. Ph.D. thesis. University of Nebraska, Lincoln, USA.

Hayes, H.K., Immer, F.R., Smith, D.C., 1955. Methods of Plant Breeding. McGraw-Hill, New York.

Hepburn, A.G., Clarke, L.E., Blundy, K.S., White, J., 1983. Nopaline Ti-plasmid, pTiT37, T-DNA insertions into flax genome. Mol. Appl. Genet. 2, 211–224.

Houston, B.R., Knowles, P.F., 1949. Fifty years survival of flax *Fusarium* wilt in the absence of *Fusarium* culture. Plant Dis. Rep. 33, 38–39.

Hradilová, J., Řehulka, P., Řehulková, H., Vrbová, M., Griga, M., Brzobohatý, B., 2010. Comparative analysis of proteomic changes in contrasting flax cultivars upon cadmium exposure. Electrophoresis 31, 421–431.

Ivanova, R., Shamov, D., 1996. Morphological and agronomic characteristic of introduced varieties of fibre flax. Rasteniev'Dni Nauki 33 (10), 35–37.

Jankauskienne, Z., 2006. 'National collection – status report, Latvia'. In: Bas, N., Pavelek, M., Maggioni, L., Lipman, E., compilers) (Eds.), ECP/GR, Report of a Working Group on Fibre Crops (Flax and Hemp). First Meeting, 14–16 June 2006, Wageningen, The Netherlands.

Jhala, A.J., Hall, L.M., Hall, J.C., 2008. Potential hybridization of flax with weedy and wild relatives: an avenue for movement of engineered genes? Crop Sci. 48, 825–840.

Jhala, A.J., Weselake, R.J., Hall, L.M., 2009. Genetically engineered flax: potential benefits, risks, regulations, and mitigation of transgene movement. Crop Sci. 49, 1943–1954.

Jhala, A.J., Bhatt, H., Topinka, K., Hall, L.M., 2011. Pollenmediated gene flow in flax (*Linum usitatissimum* L.): can genetically engineered and organic flax coexist? Heredity 106, 557–566.

Ji, Y.R., Liu, Y.F., Han, C.W., et al., 2012. Breeding report of new fiber flax variety Shuangya No.16. Plant Fiber Sci. Chian 34 (3), 109–111.

Jiao, Y., Grant, C.A., Bailey, L.D., 2004. Effects of phosphorus and zinc fertilizer on cadmium uptake and distribution in flax and durum wheat. J. Sci. Food Agric. 84, 777–785.

Jordan, M.C., McHughen, A., 1988. Glyphosate tolerant flax plants from mediated gene transfer. Plant Cell Rep. 7, 281–284.

Kabata-Pendias, A., 2011. Trace Elements in Soils and Plants. CRC Press, Boca Raton. FL9781420093681.

Kalendar, R., Schulman, A.H., 2006. IRAP and REMAP for retrotransposon-based genotyping and fingerprinting. Nat. Protoc. 1, 2478–2484.

Kalia, H.R., Chand, J.N., Ghai, B.S., 1965. Inheritance of resistance to Alternaria blight of linseed. J. Res. 2, 104–105. Punjab Agricultural University.

Kang, Q.H., Guan, F.Z., Wu, G.W., et al., 2011. Research and application of polyembryonic germplasm in flax(Linum). Plant Fiber Sci. Chian 33 (4), 179–182, 201.

Kang, Q.H., Ma, T.F., Guan, F.Z., et al., 2014. Breed. Eval. DIHAPLOID Strains Polyembryonic Flax 36 (1), 265–269.

Kate, R.M., 1991. Flax breeding in the past, prospects for the future. In: Flax as a Fibre and Oil Bearing Crop. Proceedings of the FAO European Regional Workshop on Flax, Brno, Czechoslovakia, 18–20 June, pp. 46–63.

Kenaschuk, E.O., 1975. Flax breeding and genetics. In: Harapiak, J.T. (Ed.), Oilseed and Pulse Crops in Western Canada. Western Co-operative Fertilizers, Calgary, pp. 203–221.

Kenaschuk, E.O., Rashid, K.Y., 1993. AC Linora flax. Can. J. Plant Sci. 73 (3), 839–841.

Kenaschuk, E.O., Rashid, K.Y., Gubbels, G.H., 1996. AC Emerson flax. Can. J. Plant Sci. 76, 483–485.

Khotyleva, L.V., Polonetskaya, L.M., Poskannaya, S.I., 1997. Genetic control of fibre and seed productivity determined from diallel crosses in linseed. Genetika Moskva 33 (6), 800–803.

Klose, A., Bauers, F., Paul, V.H., 1993. Pathogenicity of two isolates of *Alternaria linicola* on 16 cultivars of *Linum usitatissimum*. Bulletin OILB/SROP **16** 100–108.

Kommendahll, T., Christensen, J.J., Frederiksen, J.B., 1970. A half century of research in Minnesota on flax wilt caused by Fusarium oxysporum. Minn. Agric. Exp. Stn. Tech. Bull. 273.

Kos, B., Grčman, H., Leštan, D., 2003. Phytoextraction of lead, zinc and cadmium from soil by selected plants. Plant Soil Environ. 49, 548–553.

Krämer, U., 2010. Metal hyperaccumulation in plants. Annu. Rev. Plant Biol. 61, 517–534.

Kroes, G.M.L.W., 1997. Aspects of Resistance of Flax and Linseed (*Linum usitatissimum)* to *Fusarium Oxysporum* f. sp. *lini*. Ph.D. thesis. Wageningen Agricultural University, Wageningen, The Netherlands.

Kroes, G., Sommers, E., Lange, W., 1998. Two in vitro assays to evaluate resistance in *Linum usitatissimum* to Fusarium wilt disease. Eur. J. Plant Pathol. 104, 561–568.

Kroes, G.M.L.W., Loffler, H.J.M., Parlevliet, J.E., Keizer, L.C.P., Lange, W., 1999. Interactions of *Fusarium oxysporum* f. sp. *lini,* the flax wilt pathogen, with flax and linseed. Plant Pathol. 48, 491–498.

Krulíčková, K., Pošvec, Z., Griga, M., 2002. Identification of flax and linseed cultivars by isozyme markers. Biol. Plant. 45, 327–336.

Kumar, S., 1971. Haploid linseed. Agra Univ. J. Res. 40, 23–25.

Kumar, S., Singh, S.P., 1970. Inheritance of male sterility in some introduced varieties of linseed (*Linum usitatissimum* L.). Indian J. Agric. Sci. 40, 184–191.

Kumar, S., Singh, S.P., 1972. Inheritance of partial male fertility in linseed (*Linum usitatissimum* L.). Indian J. Agric. Sci. 42, 34–38.

Kurt, O., Evans, G.M., 1998. Anther culture potential of linseed (*Linum usitatissimum* L.): effects of genotypes and pretreatment on callus formation and differentiation. Turk. J. Agric. For. 22, 553–560.

Larter, E.N., Wenkeidt, A., Glore, R., 1965. "Redwood 65": an improved flax variety. Can. J. Plant Sci. 42, 515–516.

Lassaga, S.L., Bretón, A.M., 1997. 'Respuesta al cultivo in vitro de anteras de lino (*Linum usitatissimum* L.) con dos fuentes carbonadas a distintas concentraciones', XXVIII Congreso Argentino de Genética. In: En: Actas del Encuentro Latinoamericano de Bioética y Genoma Humano, San Miguel de Tucumán, Argentina,14–18 September 1997, p. 185.

Lassaga, S.L., Milisich, H.J., Sanchez, A.A., 1998. Linseed (*Linum usitatissimum* L.) anther culture: response at different pollen development stages. Correlations with morphological bud characteristics. Bast fibrous plants today and tomorrow. Nat. Fibres 2, 150.

Lassaga, S.L., Camadro, E.L., Bonell, M.L., Franzone, P., 2004. Diallel analysis of callus formation ability in linseed anther culture. Plant Breed. 123 (5), 502–504.

Lassaga, S., Bretón, A., Gieco, L., Milisich, H., Dittrich, A., 2010. Cultivo in vitro de lino (*Linum usitatissimum* L.). Ciencia, docencia y tecnología (Entre Ríos) 215–233. Universidad Nacional de Entre Ríos, 5/2010.

Lipman, E., Jongen, M.W.M., van Hintum, T.J.L., Gass, T., Maggioni, L., compilers), 1997. Central Crop Databases. Tools for Plant Genetic Resources Management. International Plant Genetic Resources Institute, Rome. Italy/CGN Wageningen, The Netherlands.

Ludvíková, M., Griga, M., 2015. Transgenic Flax/Linseed (*Linum usitatissimum* L.)—expectations and reality. Czech J. Genet. Plant Breed. 51, 123−141.

Macek, T., Macková, M., Pavlíková, D., Száková, J., Truksa, M., Singh Cundy, A., et al., 2002. Accumulation of cadmium by transgenic tobacco. Acta Biotechnol. 22, 101−106.

Macek, T., Sura-de Jong, M., Kotrba, M., Viktorova, J., Lovecka, P., Fiser, J., et al., 2012. Genetically modified plants designed for phytoremediation of toxic organic and inorganic contaminants. In: first ed.Anjum, N.A., Pereira, M.E., Ahmad, I., Duarte, A.C., Umar, S., Khan, N.A. (Eds.), Phytotechnologies: Remediation of Environmental Contaminants. CRC Press, Raton, FL, pp. 415−426. 617 pp. 9781439875186.

Maggioni, L., Pavelek, M., Van Soest, L.J.M., Lipman, E., compilers, 2001. Flax Genetic Resources in Europe. Ad Hoc Meeting, 7−8 December 2001, Prague. Czech Republic and International Plant Genetic Resources Institute, Rome, Italy.

Mandal, C., Dutta, S., Meena, K., Mandal, A.B., 2018. Efficient plantlet regeneration using hypocotyl explants and Agrobacterium-mediated genetic transformation in flax (*Linum usitatissimum* L.). Plant Cell Biotechnol. Mol. Biol. 19 (7&8), 358−371.

Mandolino, G., 2006. 'National collection − status report, Italy'. In: Bas, N., Pavelek, M., Maggioni, L., Lipman, E., compilers (Eds.), ECP/GR, Report of a Working Group on Fibre Crops (Flax and Hemp). First Meeting, 14−16 June 2006, Wageningen, The Netherlands.

Mansby, E., Diaz, O., Bothmer, R., 2000. Preliminary study of genetic diversity in Swedish flax (*Linum usitatissimum*). Genet. Resour. Crop Evol. 47, 417−424.

Matzinger, D.F., Wernsmann, A.E., 1968. Four cycles of mass selection in a synthetic variety of an autogamous species *Nicotiana tabacum* L. Crop Sci. 8, 239−243.

McHughen, A., 1989. Agrobacterium mediated transfer of chlorsulfuron resistance to commercial flax cultivars. Plant Cell Rep. 8, 445−449.

McHughen, A., 2000. Transgenic linseed flax (*Linum usitatissimum*). In: Bajaj, Y.P.S. (Ed.), Transgenic Crops I, Biotechnology in Agriculture and Forestry, vol. 46. Springer, Berlin, pp. 338−351.

McHughen, A., Holm, F., 1995. Development and preliminary field testing of a glufosinate-ammonium tolerant transgenic flax. Can. J. Plant Sci. 75, 117−120.

Micke, A., Maluszynski, M., Donini, B., 1985. Plant cultivars derived from mutation induction or the use of induced mutants in cross breeding. FAO Mutat. Breed. Rev. 3.

Millam, S., Obert, B., Pret'ová, A., 2005. 'Plant cell and biotechnology studies in *Linum usitatissimum* − a review'. Plant Cell Tissue Organ Cult. 82 (1), 93−103.

Miller, F.J., Hammond, J.J., 1988. Mutation rate of yellow to brown seeds in the line C.I. 3259 (U605), a yellow seeded flax cultivar with multiple gene resistance to rust. Proc. Flax Inst. U.S. 56−58.

Miller, P.A., Rowlings, J.O., 1967. Break up of initial linkage blocks through intermating in cotton populations. Crop Sci. 7, 192−204.

Millikan, C.R., 1945. Wilt disease of flax. J. Dept. Agric. Vic. 43, 305−313, 354−361.

Millikan, C.R., 1948. 'Studies of strains of *Fusarium lini*'. Proc. R. Soc. Vic. 61, 1−24.

Mlynárová, L., Bauer, M., Nap, J., Preťová, A., 1994. High efficiency Agrobacterium-mediated gene transfer to flax. Plant Cell Rep. 13, 282−285.

Mohammadi, A.A., Saedi, G., Arzani, A., 2010. Genetic analysis of some agronomic traits in flax (*Linum usitatissimum* L.). Aust. J. Crop. Sci. 4 (5), 343−352.

Muir, A.D., Westcott, N.D., 2003. Flax, the Genus Linum. Agriculture and Agri-Food Canada. Saskatchewan, Canada. Routledge, London and New York.

Najmanova, J., Mackova, M., Macek, T., Kotrba, P., 2007. Preparation of transgenic flax with enhanced metal tolerance. In: The 13th European Congress on 356 16. Transgenic Fiber Crops for Phytoremediation of Metals and Metalloids II. Subject Specific Studies, Biotechnology, Barcelona, Spain, Volume: Journal of Biotechnology, vol. 131 (Suppl. 2), ISSN 0168_1656.

Murashige, T., Skoog, F., 1962. A revised medium for rapid growth and bioassays with tobacco tissue cultures. Physiol. Plant 15, 473–497.

Najmanova, J., Neumannova, E., Leonhardt, T., Zitka, O., Kizek, R., Macek, T., et al., 2012. Cadmium-induced production of phytochelatins and speciation of intracellular cadmium in organs of *Linum usitatissimum* seedlings. Ind. Crops Prod. 36, 536–542.

Nandy, S., Rowland, G.G., 2008. Dual purpose flax (*Linum usitatissimum* L.) improvement using anatomical and molecular approaches. In: Proceedings of the International Conference on Flax and Other Bast Plants, Saskatoon, Saskatchewan, Canada, 21–23 July, pp. 31–39.

Nichterlein, K., 2003. Anther culture of linseed (*Linum usitatissimum* L.). In: Maluszynski, M., Kasha, K.J., Forster, B.P., Szarejko, I. (Eds.), Doubled Haploid Production in Crop Plants: A Manual. Springer, Dordrecht, pp. 249–254.

Nichterlein, K., Friedt, W., 1993. Plant regeneration from isolated microspores of linseed (*Linum usitatissimum* L.). Plant Cell Rep. 12, 426–430.

Nichterlein, K., Horn, R., 2005. Haploids in the improvement of linaceae and Asteraceae. 1. Linaceae. In: Palmer, C.E., Keller, W.A., Kasha, K.J. (Eds.), Haploids in Crop Improvement II. Springer, Dordrecht, pp. 277–283.

Nichterlein, K., Umbach, H., Friedt, W., 1989a. Investigation on androgenesis in breeding of linseed (*Linum usitatissimum* L.). In: Proceedings of the 12th Eucarpia Congress, Vorträge für Pflanzenzüchtg, vol. 15, pp. 25–13.

Nichterlein, K., Nickel, M., Umbach, H., Friedt, W., 1989b. Recent progress and prospects if biotechnology in breeding of linseed (*Linum usitatissimum* L.). Fat Sci. Technol. 91 (7), 272–275.

Nichterlein, K., Nickel, M., Umbach, H., Friedt, W., 1991a. New methods and recent progress in the breeding of flax. In: Proceedings of the European Regional Workshop on Flax, 2. Brno (CSFR), 18–20 June 1991, pp. 175–183.

Nichterlein, K., Umbach, H., Friedt, W., 1991b. Genotypic and exogenous factors affecting shoot regeneration from anther callus of linseed (*Linum usitatissimum* L.). Euphytica 58, 157–164.

Nožková, J., 2006. 'National collection – status report, Slovakia'. In: Bas, N., Pavelek, M., Maggioni, L., Lipman, E., compilers) (Eds.), ECP/GR, Report of a Working Group on Fibre Crops (Flax and Hemp). First Meeting, 14–16 June 2006. Wageningen, The Netherlands.

Obert, B., Dedičová, B., Hricová, A., Šamaj, J., Pret'ová, A., 2003. Flax anther culture: effect of genotype, cold treatment and media. In: Book of Abstracts, 5th International Symposium in the Series Recent Advances in Plant Biotechnology, Stará Lesná, 7–13 September 2003, p. 68.

Obert, B., Dedičová, B., Hricová, A., Šamaj, J., Pret'ová, A., 2004. Flax anther culture: effect of genotype, cold treatment and media. Plant Cell Tissue Organ Cult. 79, 233–238.

Obert, B., Bartošová, Z., Pret'ová, A., 2005. Dihaploid production in flax by anther and ovary cultures. J. Nat. Fibers 1 (3), 1–14.

Obert, B., Žáčková, Z., Šamaj, J., Preťová, A., 2009. Doubled haploid production in Flax (*Linum usitatissimum* L.). Biotechnol. Adv. 27 (4), 371–375.

Ondřej, M., Ondráčková, E., Tejklová, E., Pavelek, M., Matysová, B., 2008. Choroby olejného lnu a možnosti rezistentního šlechtění proti nim. In: Šlechtitelský Seminář 2008 – Významné Choroby hlavních hospodářských Plodin, Praha, 28 February, pp. 41–46.

Pant, S.C., Mishra, V.K., 2008. Heterosis over superior parents under diallel cross in linseed X flax (*Linum usitatissimum* L.). Indian J. Plant Genet. Resour. 21 (2).

Pavelek, M., 1980. Dědičnost vybraných znaků a vlastností lnu a vhodné hybridní kombinace pro tyto znaky. Len a Konopí 18, 43–47.

Pavelek, M., 1982. 'Rezistence lnu vůči chorobám, zvýšení odolnosti proti poléhání a zlepšení ranosti – úsek Dědičnost odolnosti proti předčasnému hnědnutí stonku (*F.avenaceum* Fr. Sacc.) ve čtyřřadovém křížení', *Závěrečná zpráva Šumperk*. VŠÚTPL 48 s.

Pavelek, M., 1985. 'Genetická analýza doby kvetení u přadného lnu'. Len a Konopí 20, 77–89.

Pavelek, M., 1991a. 'Utváření a průběh fenofází u odrůd a novošlechtění lnu s odlišnou délkou vegetační periody'. Len a Konopí 21, 69–79.

Pavelek, M., 1991b. Uniformita v nárocích na teplotu v některých fenofázích vegetační periody u lnu. Len a Konopí 21, 81–88.

Pavelek, M., 1994. Recent state of international flax database and future development', European cooperative network on flax, breeding research group. In: Report of Flax Genetic Resources Workshop, Second Meeting, Brno, 8–10 November 1994, pp. 57–63.

Pavelek, M., 1995. Further development of International Flax Data Base and special descriptors for more detail evaluation of agronomic and processing characters. In: Breeding for fibre and oil quality in flax', Proceedings of the Third Meeting of the International Flax Breeding Group, 7–8 November, 1995, St Valery en Caux, France, Centre technique pour l'étude et l'amélioration du lin (CETEAL), Paris, France, pp. 1–13.

Pavelek, M., 1997. Discussion for IFDB standard varieties. In: Information Bulletin of the FAO European Cooperative Research Network on Flax and Other Bast Plants, Institute of Natural Fibres – Coordination Centre of the FAO Network on Flax and Other Bast Plants, Poznan, Poland, vol. 1. Euroflax Newsletter, pp. 17–20, 7.

Pavelek, M., 1998a. International flax data base – poster'. ECP/GR steering committee meeting. In: Symposium on Implementation of the GPA in Europe, Braunschweig, Germany, 29 June–5 July, pp. 370–371.

Pavelek, M., 1998b. 'Analysis of current state of international flax data base – lecture'. In: Proceedings of the Bast Fibrous Plants Today and Tomorrow, Breeding, Molecular Biology and Biotechnology beyond the 21st Century, St Petersburg, Russia, 28–30 September, pp. 36–44.

Pavelek, M., 2001a. 'Status of national collections – Czech Republic'. In: Maggioni, L., Pavelek, M., van Soest, L.J.M., Lipman, E., compilers) (Eds.), Flax Genetic Resources in Europe. Ad hoc Meeting, 7–8 December 2001, Prague. Czech Republic and International Plant Genetic Resources Institute, Rome, Italy.

Pavelek, M., 2001b. International flax database. In: Maggioni, L., Pavelek, M., van Soest, L.J.M., Lipman, E., compilers) (Eds.), Flax Genetic Resources in Europe. Ad hoc Meeting, 7–8 December 2001, Prague. Czech Republic and International Plant Genetic Resources Institute, Rome, Italy.

Pavelek, M., 2001c. Flax *Linum usitatissimum* L.Venica. Czech J. Genet. Plant Breed. (CJGPB) 3.

Pavelek, M., 2004. Stocktaking of flax genetic resources in Europe. In: FAO Flax and Other Bast Plants Network Workshop, 'Bast Fibrous Plants for Healthy Life', Bosna I Hercegovina, Banja Luka, 24–28 October.

Pavelek, M., 2009. 'International flax database – status of recent development', lecture. In: CNG Meeting, Quedlinburg, 7–10 October 2009.

Pavelek, M., Tejklová, E., 2005. Survey of Czech breeding methods, development of the new Czech flax (*Linum usitatissimum* L.) variety Venica: the way of breeding and agronomical properties of flax and linseed cultivars registered in the Czech Republic. J. Nat. Fibers 1 (4), 17–36.

Pavelek, M., Tejklová, E., Horáček, J., 2001. Flax national collection, international flax database and breeding of flax, linseed and both types in the Czech Republic. In: Proceedings of the Second Global Workshop, 'Bast Plants in the New Millennium', 3–6 June, Borovets, Bulgaria, pp. 64–78.

Pavelek, M., Tejklová, E., Ondřej, M., Vrbová, M., 2012. Developments in fibrous flax breeding and cultivation. In: Kozlowski, R. (Ed.), Handbook of Natural Fibres, Properties and Factors Affecting Breeding and Cultivation, vol. 1. Woodhead Publishing Ltd, Oxford, pp. 393–468.

Pavgi, M.S., Singh, U.P., 1965. Parasitic fungi from North India. Mycopathol. Mycol. Appl. 27, 81–87.

Poliakov, A.V., 1991. Utilization of haploidy and tissue culture technique for obtaining new breeding material of fibre flax. In: Book of Abstracts. 2nd European Regional Workshop on Flax, 18–20 June 1991, Brno, Czechoslovakia, p. 54.

Poliakov, A.V., Loshakova, N.I., Krylova, T.V., Rutkowska-Krause, I., Trouve, J.P., 1994. Perspectives of haploids use for flax improvement (*Linum usitatissimum* L.). In: Kozłowski, R. (Ed.), Report of Flax Genetic Resources Workshop, 2nd Meeting of European Cooperative Network on Flax, Brno, Czech Republic, 8–10 November 1994, pp. 38–44.

Poliakov, A.V., Proliotova, N.V., Rutkowska-Krause, I., 1995. Some aspects of flax anther culture (*Linum usitatissimum* L.): cytology and pretreatment. Nat. Fibers 39, 21–28.

Poliakov, A.V., Pavlova, L.N., Alexandrova, T.A., Loshakova, N.I., Marchenkov, A.N., Rutkowska-Krause, I., 1998. Anther culture as a tool for production of new flax genotypes (*Linum usitatissimum* L.). Nat. Fibres 2, 210–213.

Poliakov, A.V., Egorova, E.G., Rutkowska-Krause, I., 2001. "Grandiflorum" method: a new effective way to produce doubled haploids of flax (*Linum usitatissimum* L.). Nat. Fibres 1, 121–126.

Popescu, F., Marinescu, I., Vasile, I., 1999. Combining ability and heredity of some important traits in linseed breeding. Rom. Agric. Res. 11–12, 33–13.

Pošvec, Z., Krulíčková, K., 1999. Identification of flax (*Linum usitatissimum* L.) cultivars by isozyme markers. In: Abstracts of the 29th Annual ESNA Meeting. Wye College, London, p. 155.

Pradhan, B., Mishra, A., Mishra, P.K., Mishra, A., 1999. Evaluation of linseed (*Linum usitatissimum* L.) varieties in the west central table land zone of Orissa. Environ. Ecol. 17 (1), 91–93.

Pret'ová, A., Bugárová, Z., Mlynárová, October 2–6, 1995. Biotechnology in flax'. Book of abstracts of symposium recent advances in plant biotechnology. Nitra, Slovak Republic 73–77.

Pret'ová, A., Obert, B., Hricová, A., 2001. Improvement of flax using innovative biotechnology approaches. Nat. Fibres 1, 146–150.

Pret'ová, A., Obert, B., Bartošová, Z., 2006. Haploid formation in maize, barley, flax, and potato. Protoplasma 228, 107–114.

Pret'ová, A., Obert, B., Bartošová, Z., 2007. Flax. In: Pua, E.C., Davey, M.R. (Eds.), *Transgenic Crops VI.* Biotechnology in Agriculture and Forestry Series. Springer, Heidelberg, pp. 129–148.

Rakouský, S., Tejklová, E., 1999. Methods and prospects of flax transgenesis. Czech J. Genet. Plant Breed. 35, 125–129.

Rakouský, S., Tejklová, E., Wiesner, I., Wiesnerová, D., Kocábek, T., Ondřej, M., 1998. T-DNA induced mutations and somaclonal variants of flax. Nat. Fibres 2, 244–246.

Rakouský, S., Tejklová, E., Wiesner, I., 2001. Recent advances in flax biotechnology in the Czech Republic. Nat. Fibres 1, 151–155.

Rashid, K.Y., Duguid, S.D., 2002. 'Genetic Resistance to Powdery Mildew in Flax', Canadian Phytological Society Annual Meeting. Waterhon Lakes National Park, Alberta. Available at: www.pubs.nrc.

Rath, L., Scharf, H., 1968. Oil content and degree of saturation of the oils and correlation between some characters in flax mutants. Theor. Appl. Genet. 38, 280–288.

Redden, R.J., Jensen, N.F., 1974. Mass selection and mating systems in cereals. Crop Sci. 14, 345–350.

Rogalska, S.M., Rutkowska-Krause, I., 1994. Anther culturability of some flax cultivars. In: Kozłowski, R. (Ed.), Report of Flax Genetic Resources Workshop, 2nd Meeting of European Cooperative Network on Flax, Brno, Czech Republic, 8–10 November 1994, pp. 45–51.

Roland, J.C., Vian, B., 1991. General staining and preparation of thin cells. In: Hall, J.H., Hawes, C. (Eds.), Electron Microscopy of Plant Cells. Academic Press, London, pp. 1–66.

Rosenberg, L., 1974. Haploidní rostliny Linum usitatissimum L. Len a konopí **2** 107–114.

Rowland, G.G., 1998. Growing Flax: Production, Management and Diagnostic Guide. Flax Council of Canada and Saskatchewan Flax Development Commission.

Rutkowska- Krause, I., 2001. The flax and hemp collection of the Institute of Natural Fibres, Poland. In: Maggioni, L., Pavelek, M., van Soest, L.J.M., Lipman, E., compilers) (Eds.), Flax Genetic Resources in Europe. Ad hoc Meeting, 7–8 December 2001, Prague, Czech Republic. International Plant Genetic Resources Institute, Rome, Italy.

Rutkowska-Krause, I., Mankowska, G., Poliakov, A.V., Proliotova, N.V., November 7–8, 1995. Plant regeneration through anther culture of flax (*Linum usitatissimum* L.). In: Breeding for Fiber and Oil Quality in Flax, Proceedings of the 3rd Meeting International Flax Breeding Research Group, St Valéry en Caux, France, pp. 83–90.

Rutkowska-Krause, I., Poliakov, A.V., Malepszy, S., September 25–28, 1996. Flow cytometric analysis of ploidy level in callus culture of flax (*Linum usitatissimum* L.). In: Proceedings of the 4th Workshop of FAO Network on Flax, Rouen, France, pp. 221–226.

Rutkowska-Krause, I., Mankowska, G., Poliakov, A.V., 1998. Somaclonal and gametoclonal variation of flax (*Linum usitatissimum* L.) in relation to breeding purposes. Nat. Fibres 2, 214–220.

Rutkowska-Krause, I., Poliakov, A.V., Mankowska, G., 2001a. Observation of heredity changes among plants regenerated in flax (*Linum usitatissimum* L.) anther culture. Nat. Fibres 1, 111–120.

Rutkowska-Krause, I., Malepszy, S., Mankowska, G., Poliakov, A.V., 2001b. Polyploidy as a stage characteristic for in vitro culture of flax (*Linum usitatissimum* L.). Nat. Fibres **2** 1–3.

Rutkowska-Krause, I., Mańkowska, G., Rutkowska, E., Rajewicz, M., 2003. Anther culture as a novel method of obtaining the new fibre flax variety ALBA 2. Euroflax 2/3, 11–14.

Rutkowska-Krause, I., Mankowska, G., Poliakov, A.V., 2004. Regeneration of androgenic flax (*Linum usitatissimum L.*) plants and their application in breeding programme. In: Mapping of European Germplasm for International Flax Database Creation, Use in Breeding for Different Flax and Linseed Varieties: Proceedings FAO ESCORENA Workshop, September 18–19, 2002. Natural Fibres, Special Edition 2004/01, Šumperk, Czech Republic, pp. 92–101. ISBN 80-902754-5-1.

Rykova, R., Kutuzova, C., Korneichuk, V., Rosenberg, L., Kovalinska, Z., Bondira, N., 1987. Širokij Unificirovannyj Klassifikator CEV Vida *Linum usitatissimum* L. Leningrad.

Rykova, R., Kutuzova, C., Korneichuk, V., Rosenberg, L., Kovalinska, Z., Bondira, N., 1989. Meždunarodnyj Klassifikator SEV Vida *Linum usitatissimum* L. Leningrad.

Salas, G., Friedt, W., 1995. Comparison of pedigree selection and single seed descent for oil yield in linseed (*Linum usitatissimum* L). Euphytica 83, 25–32.

Scheer-Triebel, M., Krukelmann, E., Heyland, K.U., 1997. Yield potential of dual purpose flax (*Linum usitatissimum* L.) for technical use depending on genotype and seed density. Angew. Bot. 71 (1–2), 24–30.

Seetharam, A., 1972. Interspecific hybridization in Linum. Euphytica 21, 489–495.

Shamov, D., 2001. Status of the Bulgarian national flax collection. In: Maggioni, L., Pavelek, M., van Soest, L.J.M., Lipman, E., compilers (Eds.), Flax Genetic Resources in Europe. Ad hoc Meeting, 7–8 December 2001, Prague. Czech Republic and International Plant Genetic Resources Institute, Rome, Italy, pp. 14–18.

Sharma, L.C., Mathur, R.I., 1971. Variability in first single spore isolates of *Fusarium oxysporum* f. *lini* in Rajasthan. Indian Phytopathol. 24, 698–704.

Shi, G., Cai, Q., 2009. Cadmium tolerance and accumulation in eight potential energy crops. Biotechnol. Adv. 27, 555–561.

Simon, A., 2001. Status of the Hungarian national *Linum* collection. In: Maggioni, L., Pavelek, M., van Soest, L.J.M., Lipman, E., compilers) (Eds.), Flax Genetic Resources in Europe. Ad hoc Meeting, 7–8 December 2001, Prague, Czech Republic and International Plant Genetic Resources Institute, Rome, Italy.*Situační a Výhledová Zpráva Len a Konopí,* Mze, Těšnov, Praha, Červen 2009, 44 [Situation and Outlook Report of Flax and Hemp, Ministry of Agriculture, Těšnov, Prague, June 2009, p. 44.

Smýkal, P., Hýbl, M., Corander, J., Jarkovský, J., Flavell, A.J., Griga, M., 2008. Genetic diversity and population structure of pea (*Pisum sativum* L.) varieties derived from combined retrotransposon, microsatellite and morphological marker analysis. Theor. Appl. Genet. 117, 413–424.

Smýkalová, I., Vrbová, M., Tejklová, E., Větrovcová, M., Griga, M., 2010. Large scale screening of heavy metal tolerance in flax/linseed (*Linum usitatissimum* L.) tested in vitro. Ind. Crops Prod. 32, 527–533.

Soest, L.J., 2001. 'Status of national collections – The Netherlands'. In: Maggioni, L., Pavelek, M., van Soest, L.J.M., Lipman, E., compilers (Eds.), Flax Genetic Resources in Europe. Ad hoc Meeting, 7–8 December 2001, Prague. Czech Republic and International Plant Genetic Resources Institute, Rome, Italy.

Song, S.M., 1996. The progress of the study on flax anther culture. China's Fiber Crops 18 (4), 4–6.

Song, S.M., 2007. Advances in flax haploid breeding. J. Heilongjiang August First Land Reclam. Univ. 3.

Song, S.M., 2008. The breeding of new fiber flax variety Shuangya No.13. Plant Fiber Sci. Chian 30 (4), 197–198, 216.

Song, S.M., 2009. Application of the biotechnology in breeding for flax herbicide resistance. J. Heilongjiang August First Land Reclam. Univ. 1.

Song, S.M., Tian, Y.J., Ji, Y.R., Xia, Z.M., Jiang, L.H., 2004. Preliminary study on the radiation of flax anther by G-ray. China's Fiber Prod. 4.

Sood, S., Kalia, N.R., Bhateria, S., Kumar, S., 2007. Detection of genetic components of variation for some biometrical traits in *Linum usitatissimum* L. in sub-mountain Himalayan region. Euphytica 155 (1–2), 107–115.

Soroka, A.I., 2004. Influence of nutrient medium composition on callusogenesis and regeneration in anther culture of oil flax. Cytol. Genet. 38 (2), 20–25.

Soudek, P., Katrušáková, A., Sedláček, L., Petrová, Š., Kočí, V., Maršík, P., et al., 2010. Effect of heavy metals on seed germination in twenty three cultivars of flax (*Linum usitatissimum* L.). Arch. Environ. Contam. Toxicol. 59, 194–203.

Srinivasachar, D., Malík, R.S., 1971. Gamma ray induced variability in the iodine value of linseed oil. Curr. Sci. 40, 298–299.

Srinivasachar, D., Seetharam, A., Malík, R.S., 1972. Combination of three characters (high oil content, high iodine value and high yield) in a single variety of linseed *Linum usitatissimum* L. obtained by mutation breeding. Curr. Sci. 41, 169–171.

Steiss, R., Schuster, A., Friedt, W., 1998. Development of linseed for industrial purposes via pedigree-selection and haploid technique. Ind. Crops Prod. 7, 303–309.

Stephens, G.R., 1996. 'Connecticut fibre flax trials 1992–1993'. Bull. Conn Agric. Exp. Stn. 932, 11.

Straathof, T.P., 1998. Regeneratie van somatische cellen en inductie van androgenese bij vlas ten behoeve van de ontwikkeling van een in vitro selectiesysteem. Prophyta 43 (2), 127–128.

Strajeru, S., 2001. The Romanian flax collection. In: Maggioni, L., Pavelek, M., van Soest, L.J.M., Lipman, E., compilers) (Eds.), Flax Genetic Resources in Europe. Ad hoc Meeting, 7–8 December 2001, Prague. Czech Republic and International Plant Genetic Resources Institute, Rome, Italy.

Sun, H., 1979. Preliminary report on anther culture of flax, 24. Ko' Hsueh Tung Pao Exue Tong Bao, pp. 948–950.

Sun, S.F., 2009. Key Technique to Increasing Callus Induction Frequency in Linseed Anther Culture', Inner Mongolia Agriculture Science and Technology 2009-03.

Sun, H., Fu, V., 1981. Induction of pollen plants in flax (*Linum usitatissimum* L.) and preliminary observations on performance of their progenies. Acta Genet. Sin. 8, 369–374.

Sun, H., Fu, W., Dong, L., Fu, W., Liu, S., 1980. Influence of IAA, NAA, 2,4-D and Kin on Flax Anther, Petal and Ovary Culture. China's Fibre Crop, pp. 241–243.

Sun, H.T., Fu, W.D., Dong, L.H., et al., 1983. Study on regeneration plants from tip of stem cotyledon and hypocotyl of flax. Chin. Sci. Bull. (20), 1332–1334.

Sun, H., Fu, W., Dong, L., Fu, W., Liu, X., 1991. The screening of best medium for flax anther culture by using variance analysis of multifactorial test. In: Proceedings of the European Regional Workshop on Flax 2. Brno, CSFR, 18–20 June 1991, pp. 184–195.

Tadesse, N., Lay, C., Dybing, C.D., 1997. Comparative seed yield performance of high-by-high and low-by-high crosses in flax. Plant Breed. 116 (6), 561–566.

Tejklová, E., 1992. The first results with anther culture of *Linum usitatissimum* L. In: Steiner, W., Haltrich, D. (Eds.), Biotechnology in Central European Initiative Countries. Federal Ministry for Science and Research, Graz, Austria, p. 51.

Tejklová, E., 1995. Induced mutagenesis in flax (*Linum usitatissimum* L.). In: Breeding for Fiber and Oil Quality in Flax, Proceedings of the 3rd Meeting of International Flax Breeding Research Group, St Valéry en Caux, France, pp. 42–50.

Tejklová, E., 1996. Some factors affecting anther cultures in *Linum usitatissimum* L. Rostl. Vyroba 42, 249–260.

Tejklová, E., 1998. Study of anther culture in flax (*Linum usitatissimum* L.). Nat. Fibres 2, 202–209.

Tejklová, E., 2002. 'Curly stem – an induced mutation in flax (*Linum usitatissimum* L.)'. Czech J. Genet. Plant Breed. (CJGPB) 38, 125–128.

Tejklová, E., 2003. 'Anther culture in flax breeding'. Book Abstract. In: The 11th International Conference on Plant Embryology, Plant Reproduction: From Mendel to Molecular Biology, Brno, 1–3 September 2003, p. 52.

Tejklová, E., 2004. Linseed breeding programme in Agritec Ltd. Nat. Fibres 1, 104–116.

Tejklová, E., 2008. Metodika indukce mutací u lnu setého (Linum usitatissimum L.) pomocí etylmetansulfonátu. Agritec s.r.o. Šumperk.

Tejklová, E., Pavelek, M., 2000. Comparison of variability in flax lines derived from traditional breeding methods and anther culture. Vortr. Pflanzenzüchtg 47, 57.

Tejklová, E., Rakouský, S., 2005. Biotechnology in flax and linseed breeding in the Czech Republic. In: Proceedings of the 11th International Conference for Renewable Resources and Plant Biotechnology, Narossa 2005 on CD, Poznaú.

Tochinai, Y., Takee, G., 1950. Studies on the physiologic specialization in *Fusarium lini* Bolley. J. Fac. Agric. Hokkaido Univ. 47, 193–266.

Todorov, T.S., Lukipudis, S., 1997. Effect of genotype in flax on stem quality in relation to regions of cultivation. Rasteniev'Dni Nauki 34 (2), 18–20.

Trouvé, J.P., 1996. Textile flax breeding facing market demand. C. R. Acad. Agric. Fr. 82 (8), 55–63.

Tzfira, T., Citovsky, V., 2006. Agrobacterium-mediated genetic transformation of plants: biology and biotechnology. Curr. Opin. Biotechnol. 17, 147_154.

UPOV, 1991. International Union for the Protection of New Varieties of Plants, Geneva, Technical Working Party for Agricultural Crops, Twentieth Session, Beltsville, USA, 13–17 May 1991. Working paper on the revision of test guidelines for flax, linseed (*Linum usitatissimum* L.).

UPOV, 1995. International Union for the Protection of New Varieties of Plants, Guidelines for the Conduct of Tests for Distinctness, Uniformity and Stability. Flax, Linseed (*Linum usitatissimum* L.).

van Treuren, R., van Soest, L.J.M., van Hintum, T.J.L., 2004. Marker-assisted rationalisation of genetic resource collections: a case study in flax using AFLPs. Theor. Appl. Genet. 103, 144–152.

Vasile, I., 2001. Progress in fibre flax breeding at the agricultural research station Livada, Romania. In: Maggioni, L., Pavelek, M., van Soest, L.J.M., Lipman, E., compilers) (Eds.), Flax Genetic Resources in Europe. Ad hoc Meeting, 7–8 December 2001, Prague. Czech Republic and International Plant Genetic Resources Institute, Rome, Italy.

Vigier, J.I., Bretón, A.M., 2005. Evaluación de la callogénesis en el cultivo in vitro de anteras de lino para genotipos derivados de dos líneas puras. In: Segundas Jornadas de Difusión en Investigación y Extensión (INEX 2005). Concordia. Universidad Nacional de Entre Ríos.

Villamonte, S., Bretón, A.M., 2004. Evaluación de la callogénesis en el cultivo de anteras de lino utilizando dos dosis de sulfato de cobre en el medio de inducción. In: Actas XII Jornadas de Jovens Pesquisadores da AUGM, Curitiba, Brazil, pp. 87–89.

Virovets, V., Loginov, M.I., Mukuvoz, V., Kozub, L.N., 2001. The Ukraininan fibre flax collection and related breeding activities. In: Maggioni, L., Pavelek, M., van Soest, L.J.M., Lipman, E., compilers) (Eds.), Flax Genetic Resources in Europe. Ad hoc Meeting, 7–8 December 2001, Prague. Czech Republic and International Plant Genetic Resources Institute, Rome, Italy.

Vrbová, M., Kotrba, P., Horáček, J., Smýkal, P., Švábová, L., Větrovcová, M., et al., 2013. Enhanced accumulation of cadmium in *Linum usitatissimum* L. plants due to over-production of metallothionein α-domain as a fusion to β-glucuronidase protein. Plant Cell Tissue Organ Cult. 112, 321–330.

Vromans, J., 2006. Molecular Genetic Studies in Flax (*Linum usitatissimum* L.). PhD thesis. Wageningen University, The Netherlands.

Vukich, M., Schulman, A.H., Giordani, T., Natali, L., Kalendar, R., Cavallini, A., 2009. Genetic variability in sunflower (*Helianthus annus* L.) and in the Helianthus genus as assessed by retrotransposon-based molecular markers. Theor. Appl. Genet. 119, 1027–1038.

Wageningen, T.N., Grauda, D., Rashal, I., Stramkale, V., 2005. The use of in vitro methods for obtaining of flax breeding source material. In: Proceedings of the 11th International Conference for Renewable Resources and Plant Biotechnology NAROSSA 2005 on CD. Poznań.

Wang, Y.F., 2001. The history and present situation and development direction of flax breeding in China. Nat. Fibers (Spec. Ed.) 2, 1–5.

Wang, Y.F., 2008. The present situation and suggestions about the study of flax germplasm resource in China. J. Life Sci. 2 (1), 67–71.

Wang, Y.F., Kang, Q.H., Liu, Y., et al., 2004. Study on flax genetic transformation mediated by *Agrobacterium tumefaciens*. J. Nat. Fibers 1 (1), 1–10.

Wang, Y.F., Qiu, C.S., Xue, S.D., et al., 2014. Breeding and demonstration of new flax variety ZHONGYAMA No. 2 breeding report. Plant Fiber Sci. Chian 36 (5), 229–233.

Wielgus, K., Mankowska, G., 2007. Anther culture of F1 generation obtained through crossing of Linla with different linseed cultivars. Biotechnologia 2, 84–89.

Wielgus, K., Mankowska, G., 2009. Estimation of callus formation capability of anthers obtained as a result of crossing different linseed cultivars. N. Biotech. 25 (1), S301.

Wielgus, K., Mankowska, G., Bialas, W., 2008. Estimation of callus formation capability of anthers obtained as a result of crossing Linola cultivar with other linseed cultivar plants. Biotechnologia 2, 84–89.

Wiesnerová, D., Wiesner, I., 2004. ISSR-based clustering of cultivated flax germplasm is statistically correlated to thousand seed mass. Mol. Biotechnol. 26, 207–214.

Wijayanto, T., McHughen, A., 1999. Genetic transformation of Linum by particle bombardment. In Vitro Cell. Dev. Biol. Plant 35, 456–465.

Wróbel-Kwiatkovska, m., Kropiwnicki, M., Zebrowski, J., Beopoulos, A., Dyminska, L., Hanuza, J., Rymowicz, W., 2019. Effect of mcl-PHA synthesis in flax on plant mechanical properties and cell wall composition. Transgenic Res. Feb 28 (1), 77–90. https://doi.org/10.1007/s11248-018-0105-y.

Yadav, R.K., Singh, P.K., Gupta, R.R., 2000. Phenotypic stability for yield and quality attributes in linseed (*Linum usitatissimum* L.). Crop Res. Hisar 19 (2), 301–304.

Yermanos, D.M., 1966. Variability in seed oil composition of 45 Linum species. J. Am. Oil Chem. Soc. 43, 546–549.

Yurenkova, S.I., Khotyleva, L.V., Zhuchenko, A.A., 1992. Comparative study on isoenzyme spectra of fibre flax cultivars. Dokl. Akad. Nauk. Belarusi 36, 473–475.

Zhan, X., Jones, D.A., Kerr, A., 1988. Regeneration of flax plants transformed by Agrobaterium rhizogenes. Plant Mol. Biology 11, 551–559.

Zhang, L.L., Bai, W., Mi, J., et al., 2012. Identification interspecific hybridization between cultivated flax and wild flax. Acta Agriculturae Boreali-Sinica 27 (Suppl. ment), 57–60.

Further reading

Dehmer, L., 2009. Status of National Collections – Germany, Oral Notification.

Desphande, R.B., 1950. A note on the investigations on breeding rust resistant linseed at the I.A.R.I. Indian J. Genet. Plant Breed. 10, 7–13.

Fu, Y.B., Diederichsen, A., Richards, K.W., Peterson, G., 2002c. Genetic diversity within a range of cultivars and landraces of flax (*Linum usitatissimum* L.) as revealed by RAPD. Genet. Resour. Crop Evol. 49, 167–174.

Gill, K.S., 1975. The future strategy of research in plant breeding. Sou Agalu 6, 22–27.

Gill, K.S., Singh, G., 1960. Study of exotic linseed varieties in search of new germplasm. Indian Oilseeds J. 5, 258–266.

Howard, G.L.C., Khan, A.R., 1924. Studies in Indian oilseed. No. 2, linseed. Mem. Dept. Agric. India Bot. Sci. 12, 135–181. L.

Kozłowski, R., Grabowska, L., Baraniecki, P., Mścisz, J., 1993. Recultivation by flax and hemp culture of soil polluted by heavy metals. Natural Fibres, Poznań, Poland. Go to next line.

Li, X.P., 1994. New flax variety Shuangya 1. Crop Genet. Resour. 4, 54.

Li, X.P., Jie, T.Y., Hua, Y.Y., Zhi, L.Q., Jun, G.X., Jiang, W.G., Yu, W.C., Tian, Y.J., Yin, Y.H., Li, Q.Z., Guan, X.,J., Wei, G.J., Wang, C.Y., 1997. Selection of a new flax cultivar Shuangya 5. China's Fibre Crops 19 (1), 7–8, 21.

Mańkowski, J., Kołodziej, J., Kubacki, A., Baraniecki, P., Pniewska, I., Pudełko, K., 2018. In: Malina, G. (Ed.), Wykorzystanie konopi włóknistych do rekultywacji i zwiekszenia żyzności gleb zdegradowanych – projekt LIFE11 ENV/PL/445; Remediacja, rekultywacja i rewitalizacja, (The use of hemp for reclamation and increasing the fertility of degraded soils - project LIFE11 ENV / PL / 445; Remediation, reclamation and revitalization), pp. 67–84. Poznań ISBN: 978-83-64959-05-9.

Meikle, R.D., 1977. Linum L. In: Meikle, R.D. (Ed.), Flora of Cyprus, vol. 1, pp. 317–324.

Murre, M., 1955. Vezelvlas. Meppel. Uitgeverij Ceres, The Netherlands. Mutant Variety Database FAO/IAEA: http://mvgs.iaea.org/Search.aspx?ID=308.

Nôžková, J., Remeselníková, K., Bjelková, M., 2014. Characterization and evaluation of flax seeds (Linum usitatissimum L.) on selected genotypes. J. Cent. Eur. Agric. 12 (1), 193–207.

Ockendon, D.J., Walters, S.M., 1968. Linum L. In: Tutin, T.G., Heywood, V.H., Burges, N.A., Moore, D.M., Valentine, D.H., Walters, S./M., Webb, D.A. (Eds.), Flora of Europea, 2. Cambridge University Press, Cambridge, pp. 206–211.

Qiang-He, S., Mi-Jun, Qiang, H.S., Mi, J., 1996. A new flax cultivar Ba Ya 5. Crop Genet. Resour. 1, 5.

Song, S.M., Tao, S.H., Dong, F.W., Bing, W.C., Hui, Y.Z., Song, X.M., Sun, H.T., Fu, W.D., Wu, C.B., Yuan, Z.H., 1996. Study on flax anther culture. China's Fibre Crops 18 (4).

Williams, I.H., 1988. The Pollination of Linseed and Flax, vol. 39. Bee World, pp. 145–152.

Zhang-Hui, D.W., Wang, Y.L., Zhang, H., Ding, W., Wang, Y.L., 1996. A preliminary study on utilization of dominant nuclear sterile flax in breeding. Acta Agric. Boreali Sin. 11 (2), 38–42.

Prevention of fungi and bacteria growth in natural fibres

18

Ryszard M. Kozłowski[1], *Judyta Walentowska*[2]
[1]Journal of Natural Fibers Editor-in-chief, coordinator of FAO/ESCORENA European Cooperative Research Network on Flax and other Bast Plants, former director of the Institute of Natural Fibres, Poznań, Poland; [2]Institute of Natural Fibres and Medicinal Plants, Poznań, Poland

18.1 Introduction

Natural fibres serve as an excellent raw material for 'green products'. They are biodegradable, fully recyclable and relatively cheap to produce, and are characterized by low specific weight, thermal and acoustic parameters with favourable values, high hygroscopicity and also by a tendency not to accumulate electrostatic charges on their surface, except wool.

In specific conditions, i.e., high humidity and temperature, natural fibres and textiles (nonwovens and fabrics) are susceptible to biodegradation. The biodeterioration caused by microorganisms affects finishing and insulation materials containing natural fibres, which are used in the automotive and construction industries. Protection against biodeterioration is of great importance not only in the apparel sector, but also in the paper industry and in libraries, archives and museums, where microbiological purity of air is essential. Hygienic finishing is also of utmost importance for textiles used by humans, especially those used for medical purposes.

In unprotected materials made from natural fibres, microorganisms cause the degradation mainly of cellulose, which is the basic structural component of natural fibres. This leads to a loss of strength in the material and causes the emission of odours, ultimately leading to a worsening of the microbiological purity of the air.

18.2 Key issues of fungal growth, especially mildew, in natural fibres

Microorganisms: mould, mildew, fungi, yeast and bacteria are found in environment, e.g., on our underwear, garments and on our bodies. Also, home furnishing such as carpets, shower curtains, mattresses and upholstery are susceptible to biodeterioration. High humidity and temperature cause fungi, especially mildew, decompose cellulose, that is, the main component of natural fibres.

Handbook of Natural Fibres. https://doi.org/10.1016/B978-0-12-818398-4.00020-7

18.2.1 Microorganisms causing biodeterioration of textile materials from natural fibres

Fungi, especially mildew, play the biggest role in the biodeterioration of textile materials made from natural fibres and in the aerobic degradation of cellulose. The optimal conditions for their growth are a relative humidity of 70%–90%, a temperature of 24–30°C and a pH of about 6.0. The most active fungi in this process are *Chaetomium, Stachybotrys, Alternaria, Verticillium, Trichoderma, Penicillium* and *Aspergillus.* Bacteria that prefer very high relative humidity (96%–99%) and pH 6.8–8.0 are less significant in this respect. The following bacteria strains are the main cellulose decomposing agents: *Cytophaga, Sporocytophaga, Bacillus* and *Clostridium* (Szostak-Kotowa, 2004).

Mildew is a group of fungi that play an essential role, both positive and negative, in many branches of economy. Mildew is characterized by a wide range of biochemical properties. The main applications of mildew are as follows:

- biotechnological production of antibiotics, enzymes, organic acids, lipids and hormones;
- food industry (meat processing, cheese production);
- proecological activity (biological processing of wastewater, composting of organic waste).

Mildew is able to produce a number of enzymes (amylase, pectinase, cellulase, proteinase, lipase, catalase), which can be used in the industrial areas mentioned above, demonstrating the positive impact of mildew on the economy. One example is cellulolytic enzymes – cellulases. Strains of *Trichoderma viride* and *Chaetomium globosum* fungi, which show cellulolytic activity, are used to test the efficiency of treatments applied on fabrics and coverings to protect them against fungi activity (Żakowska and Piotrowska, 2008).

However, mildew can also have a negative impact, decomposing food, paper, leather and wood. The materials most susceptible to biodeterioration are textiles: fibres, fabrics, nonwovens, floor coverings and carpets of both chemical and natural origin.

Microbiological decomposition of natural fibre textile materials affects both the raw material (i.e., the fibres) and fabrics and other items made of natural raw materials, including linen, hemp, cotton, jute, ramie, sisal and manila. Leather, paper and blended fabrics made of natural and chemical fibres such as polyamide, polyacrylonitrile, polypropylene, polyethylene and polyester are also affected by microbiological decomposition.

The mildew forms most often found in textiles are shown in Table 18.1 (Zyska and Żakowska, 2005).

18.2.2 Methods of determining the resistance of natural fibre textile materials to the action of mildew

In studies dealing with the resistance of textiles to mildew action, the following methods are the most commonly used:

- ***The agar method***, which allows the resistance to fungi of cellulose and cellulose derived products to be determined, and enables evaluation of antifungal finishing. This method is

Table 18.1 Mildew most often found in textiles.

Type of material	Fungus species
Fibres, fabrics and items made of natural raw materials	*Chaetomium globosum, Aspergillus niger*
Viscose fibres	*C. globosum*
Fibres, blended fabrics containing natural and chemical fibres	*A. niger, Penicillium funiculosum, C. globosum*
Leather	*A. niger, Penicillium wortmannii, Chaetomium cupreum*
Paper (including wallpapers, gypsum-cardboard panels, roofing board, library materials)	*Aspergillus flavus, Aspergillus fumigatus, Aspergillus versicolor, A. niger, Penicillium chrysogeum, P. funiculosum, C. globosum, Trichoderma viride, Cladosporium herbarum*

also used for testing the efficiency of textile finishing against selected test fungi and comparing various types of finishing in vitro. The following mildew strains are employed in this method, all of which are harmful to human health: *Aspergillus niger van Tieghem, Chaetomium globosum Kunze, Gliocladium virens Miller, Paecilomyces variotii Bainier, Penicillium ochrochloron Biourge*. The samples are subjected to incubation in a chamber, at a temperature of $29 \pm 1°C$ and relative humidity of $90 \pm 5\%$, for 4 weeks. After the test, mildew growth on the samples is evaluated both visually and with the use of a stereoscopic microscope according to a 6-degree scale (see Table 18.2). If no visible mildew is observed, the area of growth inhibition is tested on the agar near the sample. Then the breaking strength of the materials is tested by checking the breaking force (Standard EN 14119:2003).

Table 18.2 Evaluation of fungal growth on the samples.

Growth degree	Evaluation
0[a]	No visible growth evaluated microscopically
1[a]	No visible growth evaluated with naked eye, but clearly visible microscopically
2	Growth visible with naked eye, covering up to 25% of tested surface
3	Growth visible with naked eye, covering up to 50% of tested surface
4	Considerable growth, covering more than 50% of tested surface
5	Very intense growth, covering all tested surface

[a]If the fungal growth on agar around the tested sample is partially or completely inhibited, then the inhibition zone must be measured.

- ***The soil burial test,*** which is aimed at determining the resistance of textile materials containing cellulose to the action of microorganisms occurring in the soil. The test is used for comparison of protected and unprotected samples of the same material. The material tested is buried in soil with the required biological activity for a given period, after which the breaking strength is tested. The method allows a comparison of the breaking strength before and after the material is buried in the ground. The buried samples are subjected to incubation at a temperature of 29 ± 1°C and relative humidity from 90% to 100%, for a period of no longer than 9 days, up until the point at which the unfinished samples show approximately an 80% drop in breaking strength. The final assessment in the soil burial method also involves visual evaluation with a microscope that allows changes in colour, structural surface damage, loss of cellulose fibres and overall appearance of the sample to be observed in light (Standard EN ISO 11721−1:2001).

With regard to resistance to bacteria, there are two basic methods used:

- ***The quality screening method***, where the bactericidal properties of the material are tested using the following bacteria species that pose a health risk to humans: *Staphylococcus aureus* (gram-positive bacteria) and *Klebsiella pneumoniae* (gram-negative bacteria). Other suitable species can also be used, including the gram-positive bacteria *Corynebacterium xerosis, Bacillus licheniformis, Micrococcus flavus* and *Staphylococcus haemolyticus* and the gram-negative bacteria *Escherichia coli* and *Pseudomonas aeruginosa.* Before the test, 25 × 50 mm samples of the material are soaked in 0.2 mL of fresh synthetic sweat with an acidic pH. Then linear inoculation with the above-mentioned bacteria is carried out on the trypticase soy agar (TSA). The stripes of the tested materials soaked with synthetic sweat are placed on the TSA medium, perpendicular to the line of bacterial inoculation. Samples are placed in an incubator at a temperature of 37 ± 2°C for 24 h. After completion of the test, the growth inhibition zone is shown on the bottom side of the sample (Standard AATCC 147−1998).
- The ***quantity method,*** which tests the antibacterial activity of the material against the microorganism that shows the highest resistance. In this test four 20 × 20 mm samples are inoculated with 0.1 mL of bacterial suspension and then incubated at 37 ± 2°C. After 2 h, they are rinsed in a sterile physiological salt solution, and centrifuged for 15 min at a speed of 150 rpm. After this preparation, a series of dilutions are prepared and each is inoculated at 0.1 and 1 mL on a sterile Petri dish. The samples soaked in agar are incubated at 37 ± 2°C, for 24 h, and finally all growing bacteria colonies are calculated. After the test, both the bacteriostatic activity (growth inhibition) and the bactericidal activity of the textile materials protected with biocides are tested (Standard AATCC 100−1998).

18.3 Methods of preventing fungal growth, especially mildew, and bacteria in natural fibres

In high humidity conditions, natural fibres require protection against biodeterioration caused by fungi, especially mildew. The methods of preventing fungi growth involve different approach to the problem, for example, can be based on chemical and genetic modification of natural fibres, and use of chemical or natural biocides.

Antimicrobial agents can be imparted to textiles at the fibre stage or at the stage of finishing. Owing to ban on various chemicals used in the textile processing because of their toxic and carcinogenic properties, ecofriendliness is considered as very important nowadays.

The other important consideration is problem of the durability, wash and wear conditions, and competitive economy of these processes (Raichurkar et al., 2017).

18.3.1 Protection of natural fibre textile materials against growth of microorganisms

Of all microorganisms, mildew forms show the greatest ability to decompose cellulose in a natural environment. Mildew growth can occur during, for instance, a poorly controlled production cycle or as a result of incorrect storage of materials. The main factor causing mildew growth is high air humidity of over 70%. This study has shown that the main negative effect of microbiological decay of textiles is strength reduction. The bast fibres are characterized by low resistance: after 16 days of testing the tensile strength was reduced by 83%–96%. For cotton fabric with unit weight of 250 g/m^2, tensile strength was reduced by 82%–98% after 14 days of testing. For cotton fabric with unit weight of 910 g/m^2, the tensile strength reduced by 92% after 6 days of soil burial testing. Viscose fibres showed a complete loss of strength, whereas cotton fibres retained 50% of their initial strength. With cotton and viscose fabrics with a unit weight of 880 and 930 g/m^2, slower decomposition was observed using the soil test than that observed for cotton fibres. Manmade fibres and fabrics show higher resistance to mildew. After 240 days of testing the strength of polyacrylonitrile fibres was reduced by 5%–13%, and the elongation was also reduced by 15%–32%. After the same test period, the strength of polyester and polyamide fibres was reduced by 20% and 18%, respectively, and an increase in swelling and reduction in elongation was also observed. Tests conducted on floor coverings made of polyvinyl chloride on a natural fibre backing such as jute felt, flax-hemp backing, and also on backing made of polyester fibres and synthetic fibre waste showed no resistance to cellulolytic fungi after as little as 14 days of testing. Tests conducted on carpet backing made of natural and synthetic fibre mesh, and on a backing made of synthetic fibres and polyurethane foam, mostly showed no resistance to growth of mildew fungi after 4 weeks of testing. Only the floor covering on a polyurethane foam backing proved resistant to mildew growth (Zyska, 1999).

For this reason natural fibres and natural fibre products require protection against microbiological decomposition. Protection may be provided simply by coating the cellulose, by chemically or biochemically modifying the cellulose, or through the use of chemical inhibitors such as fungicides and biocides from plant origin.

18.3.2 Methods of protecting natural fibre textile materials against mildew growth and bacteria

Chemical and genetic modification of natural fibres

Cellulose, the main component of natural fibres, can be modified by chemical reaction so that it is no longer susceptible to microbiological attack.

Chemical modification of natural fibres includes the following methods, among others:

- *Acetylation* – the method is based on the reaction of lignocellulosic material with acetic anhydride at elevated temperatures, usually without a catalyst. The acetylation strongly reduces hydrophilicity and in fact can make the material hydrophobic. In addition, modification of this sort causes bulking of the cell wall and renders the material less susceptible to biological decay. Biological protection is afforded by lower equilibrium moisture content. Moreover, the cell wall polymer chemistry is modified, so that the fibres are not recognized by enzymes (Pott, 2004).
- *Mercerization* – this method leads to an increase in the amount of amorphous cellulose at the expense of crystalline cellulose. The most important modification that this method is designed to achieve is the removal of hydrogen bonding in the network structure. The following reaction proceeds as a result of treatment with alkalis: Fibre–OH + NaOH → Fibre–O-Na+ + H_2O. The effect of the treatment with NaOH depends on the concentration of the alkaline solution, its temperature and the duration of the treatment. Optimum conditions for mercerization increase the degree of bonding at the polymer–fibre interface and lead to a rise in the rupture stress of cellulose (Kozłowski and Władyka-Przybylak, 2004). These treatments are not fungicidal, but they are preservative and as such they compete with fungicidal treatments. More data about chemical treatment of natural fibres one can find in volume 2, Chapter 8. Improving the properties of natural fibres by chemical treatments.

Microorganisms: bacteria, mildew, fungi and yeasts impair the functional, aesthetic and hygienic properties of cotton textiles and their blends, causing, e.g., staining and degradation phenomena. Antimicrobial finishes prevent the growth of bacteria and fungi. There are two types of antimicrobial finishing: leaching and nonleaching type. Very popular antimicrobial finishing agent is Saraguard 5700. The hydrolysed silane (silanol) covalently bonds to fabric surface. This antimicrobial agent is introduced to cotton fabric by padding method. Cotton is padded with 20–40 g/L of Saraguard, with the pickup of about 70%, pH is adjusted to 5.0–6.0. After padding, fabric is dried at 160°C for 2 min. Saraguard 5700 shows good antimicrobial activity against *S. aureus*, *E. coli*, and fungal attack. Protected cotton fabric is resistant to 20 home launderings. The Saraguard 5700 plays a key role in hygiene textile market.

Studies on the protection of cotton and cotton–polyester fabrics (100% cotton, 52% cotton and 48% polyester, 33% cotton and 67% polyester) were conducted, with both raw and finished fabrics tested. The finishing covered: dressing removal, mercerization, bleaching, washing and applying end-use finish. The resistance of cotton and cotton–polyester fabrics to mildew growth was determined by a soil test. The degree of biodeterioration was assessed after 1, 2, 5, 8, and 11 weeks. The evaluation of the effect of microorganisms was based on the extent of reduction of tearing strength. Cotton fabrics, both raw and finished, had undergone complete degradation after only 2 weeks. The most resistant was the blend of 33% cotton and 67% polyester fibres (Szostak-Kotowa et al., 1988). In other studies, plain-wave 100% cotton fabric was finished with a nonformaldehyde containing product based on imidazolidinone: 1,3-dimethyl-4,5-dihydroxyethyleneurea. In pretreatment processes, the fabric was bleached in an H_2O_2 bath, mercerized in NaOH solution and neutralized with diluted

CH_3 COOH solution. The resistance of finished and unfinished cotton fabrics to the action of soil microorganisms was again determined by the soil burial test. After 12 days of testing, the unfinished sample was completely degraded. The finished sample was undamaged, although a colour change was observed on the sample surface. Smaller colour changes indicated that the finished samples of cotton fabric are less susceptible to biodeterioration than the unfinished samples (Tomšič et al., 2007).

Genetic modification of natural fibres is a new area of research which might allow cellulose fibres to be developed that are genetically resistant to the destructive action of microorganisms. The results of studies on the resistance of polyhydroxyalkanoates to mildew and bacteria are not currently available.

Use of chemical biocides

The contemporary antimicrobial agents in textiles are quaternary ammonium compounds, chitosan, triclosan, polyhexamethylene biguanide, TiO_2 nanoparticles, zinc nanoparticles, copper nanoparticles, noble metals (Ag, Au) nanoparticles and many naturally occurring agents. These agents according to their mechanism of antimicrobial activity, toxicity, durability and ecological accessibility can be divided into biocides and biostatic agents, leaching and bounding antimicrobials, controlled release and barrier forming agents, also good or poor washing resistance, respectively (Raichurkar et al., 2017).

Triclosan inhibits the growth of the microorganisms such as algae, bacteria and fungi. Processing temperature of triclosan is under 270°C; the application is for indoor end uses. The food content approval is in EU 5 mL- 5 mg/kg. Chemical composition of triclosan: 5 chloro2-(2,4 dichloro-phenoxy) phenol). Common names are Lexol 300, Irgasan DP 300. It is white powder or liquid, only slightly soluble in water, but soluble in methanol, ethanol, diethyl ether and strong basic solutions such as 1 M NaOH. It is used since 1972, is present in soaps (0.1%−1%.0%), deodorants, toothpastes, shaving creams, mouth washes, cleaning supplies, toys, beddings, socks and trash bags. Also, triclosan is used for plastic and composites production, for food packages and is approved by European Commission. Triclosan is stopping not only bacteria but also fungi and mildew, but triclosan is toxic to aquatic microorganisms, which is produced by Clarion company (documents of triclosan producer).

US Food and Drug Administration informed US companies that 6 September 2017 was the deadline by which they have to stop marketing of those products, citing concern about triclosan causing antibiotic resistance and lack of evidence that soaps containing triclosan were more active than soaps with water alone. Six best-selling antibacterial toothpastes containing 0.3% triclosan (3 mg/g toothpaste) US Environmental Protection Agency limit for chronic dietary exposure 0.3 mg triclosan per kg body weight/day (Hiolski, 2017).

Many different types of bound antimicrobials are currently used for textiles such as polyhexamethylene biguanide, Biosil, magnesium peroxides, chitosan, tobacco leaf extract, curcumin (1,7-bis (4-hydroxy-3-methoxyphenyl)-1,6-heptadiene-3,5-dione).

Unfortunately, most chemical biocides that have long been used in the textile industry pose environmental hazards and health risks. Cooper fungicides, e.g., copper naphthenate, copper hydroxynaphthenate and copper 8-quinolinolate, were widely used in the 1940 and 1950s. In the 1960s, copper borate solubilized for aqueous application in a complex with zirconium was developed. This was applied on a denim cotton fabric and subjected to a soil test. After 5 weeks of testing a 100% fabric strength preservation was observed. Similar fabric tests conducted in a 6-month trial showed a 60% reduction of strength, yet no visual change in appearance. For comparison, the cotton fabric samples retained only 25% of their initial strength and intensive growth of mildew was observed (Block, 1967). The current biocide market for the textile applications is presented in Table 18.3 (Libudzisz and Kowal, 2000; Szostak-Kotowa, 2005; Halford, 2006).

Tests were carried out on cotton fabrics protected by different biocides (Afrotin LC, Mystox ELN, Mystox WFA, Antiback MF, Antiback MFB, Fungitex OP, Fungitex ROP). The fabric samples were washed three times and exposed to *A. niger*. The best results were observed for fabric protected by Fungitex OP (benzimidazole derivatives), which showed no mildew growth after three washes (Seventekin and Ucarici, 1993). Tests were also carried out on flax fibre and nonwovens protected by Cupramina B, a market product, based on boron and copper compounds. High efficiency was obtained for these materials, which showed no mildew growth, no loss of breaking force and no change in odour. Cupramina B is characterized by a low toxicity for humans and warm-blooded animals and can be used in interiors designed for human habitation or long-term use. Consequently, it is widely used to protect natural nonwovens and wood for the automotive and building industries (Kozłowski and Walentowska, 2006; Walentowska and Kozłowski, 2006).

Table 18.3 Biocides used in the textile industry.

Trade name	Active ingredient
Acticide ® TC 08	Zinc pyrithione with carbendazim
Acticide ® TC 10	Encapsulated octylisothiazolinone
Tolcide C 30	2′(thiocyanomethylthio) benzothiazole
Preventol GD	2,2′-dihydroxy-5,5-dichlorodiphenylmethane
Preventol O extra	o-phenylphenol
Myacide	2-Bromo-2-nitropropane-1,3-diol
Cunilate 2419-75	8-Hydroxyquinoline copper
Giv gard DXN	6-Acetoxy-2,4-dimethyl-1,3-dioxane
Densil P	Dithio-2,2′-bis(N-methylbenzamide)
Mystox LPL	5-Chlorophenol laurynian
Fungitex ROP	bis(chlorophenylhydroxy)methane
Sanitized BSC	Thiobendazol
Quecophob LE 6 and LE 8	Fluorocarbon nanodispersion for external areas.

A study was carried out on the antifungal efficiency of chemical biocides for aerial disinfection by thermal fogging in libraries and archives. The fungi strains used were those most frequently found as contaminants in these environments, namely *A. niger, Aspergillus ustus, Aspergillus flavus, P. variotii, T. viride, C. globosum, Myrothecium verrucaria, Stachybotrys atra, Cladosporium herbarum* and *Penicillium chrysogenum,* as well as combinations of these. The following biocides were tested: econazole, orthophenylphenol, imazalil and thiabendazole. Orthophenylphenol and imazalil at concentrations of 3% and 5%, respectively, were the most efficient at protecting against both individual fungi strains and the mixture of 10 tested strains. The effect of thiabendazole on the properties of paper was also tested. No significant effect on the fibre strength or level of oxidation was observed. Thiabendazole at a concentration of 10% proved very efficient at both air purification and disinfection of surfaces. As well causing no damage to paper, thiabendazole is also known to cause no visible damage to varnished or painted surfaces and metal shelves. In addition, it does not leave greasy stains on surfaces or an unpleasant odour in rooms (Rakotonirainy et al., 1999).

Other related data can be found in Vol. 2, Chapter 21 'Antimicrobial natural fibres' of the Handbook of Natural Fibres.

Use of natural biocides

Currently are also explored many natural products – such as extracts from leaves, roots, stem, seeds, flowers and fruits, which exhibit antibacterial and antifungal properties.

Above-mentioned products can be used in native forms and as microcapsules to improve durability and controlled release of active substances (Thilagavathi and Bala, 2007). Fundamental data about microencapsulation are provided in the article of Sebastian Gouin entitled Microencapsulation industrial appraisal of existing technologies and trends (Gouin, 2004).

Of the natural biocides, which have become the centre of interest in recent years, the most important are plant extracts and oils and chitosan. The bactericidal and fungicidal properties of essential oils have been well-known for some time. They are particularly widely used for sanitary and cosmetic applications, in the pharmaceutical and food industries and for plant protection. Active substances (alkaloids, flavonoids, terpenes, tannins) that are present in essential oils such as those from thyme, oregano, clove, sage, tea tree, chamomile and mint have natural antimicrobial properties (Rios and Recio, 2005). Bakkali et al. (2008) claim there are about 3000 essential oils known to science, of which 300 are of economic importance; they are complex natural mixtures which can contain about 20–60 components at high concentrations (20%–70%) with other components present in trace amounts. For example, carvacrol (30%) and thymol (27%) are the major components of essential oil from *Origanum compactum,* while menthol (59%) and menthone (19%) are the main components of essential oil from *Mentha piperita.* These major components determine the biological properties of the essential oils. The influence of essential oils from sage, mint, hyssop, chamomile and oregano on the growth of gram-negative and gram-positive bacteria was studied.

In the case of sage, mint, chamomile and hyssop, bacteriostatic activity was observed. Oregano oil shows both bacteriostatic activity and bactericidal activity, most likely due to its high content of phenolic compounds (Marino et al., 2001).

Below are short characteristics of some natural herbal resources and derived essential oils.

Aloe vera – a skin care product that possesses antifungal and antibacterial properties that can be exploited for medical care application such as wound dressing and bioactive textiles, also to treat cotton fabrics and finishing of silk using BTCA as a cross-linking agent.

Azadirachta indica, commonly known as neem, nimtree or Indian lilac; neem extract was recognized as the most promising source of insect control, with antibacterial and medicinal properties. The active ingredients are from seeds, bark, leaves and roots. They are limonoids: azadirachtin, sallannin, and nimbin. Some work was reported on application as antimicrobial agent to cotton and cotton–polyester textiles.

Adzuki beans (*Vigna angularis*) exhibit antibacterial effect against *S. aureus, Aeromonas hydrophila* and *Vibrio parahaemolyticus,* and contains large amount of polyphenols including proanthocyanidins.

Clove oil (eugenol) – main product of *Syzygium aromaticum.* It is applied for finishing cotton textiles. Fastness of fabrics was improved by using dimethylol dihydroxyethyleneurea–based inbuilt catalyst (KVSI). It was proved that clove oil with concentration 0.5% shows 17 mm of zone inhibition of *S. aureus*, and *K. pneumoniae*, whereas cotton fabric treated with 1% clove oil (with KVSI) shows 47 mm zone inhibition against *S. aureus* – gram-positive bacteria (Raichurkar et al., 2017).

Eucalyptus oil (*Eucalyptus radiata*) is effective against infections caused by bacteria, fungi and viruses (Raichurkar et al., 2017).

Pomegranate (*Punica granatum*) contains polyphenols and flavonoids. Fabrics finished with microencapsulated polyphenols and flavonoids were recorded as effective against *S. aureus* and fungi *A. niger* (Raichurkar et al., 2017).

Prickly chaff flowers (*Achyranthes aspera)* manifests antimicrobial activity against both gram-positive and gram-negative bacteria. Treated cotton fabric was tested by parallel streak method for gram-negative bacteria *E. coli* (Raichurkar et al., 2017).

Tea tree (*Melaleuca alternifolia)* – oil from this tree consist over 100 different compounds and has antiseptic, dermatological and antifungal properties (Raichurkar et al., 2017).

Tulsi leaves (*Osmium basilicum*); main constituent is eugenol (70%), methyl eugenol (20%), carvacrol and caryophyllene. From ancient ages, tulsi leaves were used as antimicrobial, insecticidal agent and are also suitable for textile application. Methanolic extract of tulsi leaves were applied on cotton fabric by dipping method (Raichurkar et al., 2017).

Garlic (*Allium sativum*) – potential antimicrobial compound in garlic was synthesized as ajoene –0.1%–0.5%. Team of Thomas Wirth at Cardiff University developed a short and effective ajoene synthesis. It was obtained from 27% on the original milligram scale to 65% on a 100 g scale. The researchers made 169 g of ajoene which retains that distinctive pungent garlicky smell (Sanderson, 2018).

Biocides of plant origin such as extracts and essential oils have become the focus of research in various branches of economy. These biocides are used in plant protection, food processing and the cosmetics industry and in the production of packaging, medical, finishing and insulation materials containing natural fibres. They are also used to preserve museum exhibits and antique book collections. The use of essential oils from clove, cinnamon and oregano as antimicrobial solutions in paper packaging was studied. Cinnamon oil used in a paraffin coating totally inhibited growth of *A. flavus* (Rodriguez et al., 2007). The studies on the antimicrobial properties of mint oil in protecting food against infestation with *Aspergillus* spp. and *Penicillium* spp. fungi, and *E. coli, Bacillus* ssp. and *S. aureus* bacteria, have shown its antifungal and antibacterial effectiveness (Gulluce et al., 2007). Tests were carried out on the action of essential oil against the mildew most commonly present in libraries and archives, namely *A. niger, C. globosum, Penicillium frequentans* and *P. variotii.* The vapours of different essential oils (armoise, boldo, clove, eucalyptus, lavender, tea tree, thuja, wormseed) were investigated. The different compounds were preliminarily screened by a microatmosphere method. Of the essential oils, eucalyptus oil *(Eucalyptus globolus)* showed moderate antifungal activity and wormseed oil *(Chenopodium ambrosioides)* was the most effective at inhibiting mildew growth (Rakotonirainy and Lavédrine, 2005).

Studies were also conducted on the antifungal efficiency of mint oil and mint extract in the protection of flax fibre used as a raw material in the production of nonwovens and fabrics. Mint oil and extract were applied by padding from a methanol solution at 20% concentration. 5 mL of the solution was used on 1 g of dry flax fibre. The tested samples of flax fibre were exposed to the action of the following fungi mixture: *A. niger* van Tieghem, *Chaetomium globosum* Kunze, *Gliocladium virens* Miller, *P. variotii* Bainier, *Penicillium ochrochloron* Biourge. The flax fibre samples were allocated on an agar medium and inoculated with a suspension of the testing fungi. Samples were incubated at a temperature of $29 \pm 1°C$ and relative air humidity of 90%, for 4 weeks. After the test, an evaluation of the anti-fungal properties was carried out on the basis of visual assessment of the degree of mildew growth and by testing the specific strength of the flax fibre. Mint extract did not show anti-fungal properties. Mint oil showed slightly lower anti-fungal activity: fungi growth was not visible with the naked eye, but single traces of fungi were visible microscopically, and a slightly higher loss of specific strength was observed. The results are shown in Table 18.4 (Foksowicz-Flaczyk and Walentowska, 2008).

Table 18.4 Antifungal properties of flax fibre protected with biocides.

Biocide	Degree of fungal growth (scale from 0 to 5)	Loss of specific strength (%)
Mint oil	1	22
Mint extract	4	83
Control unprotected	5	100

The anti-fungal efficiency of grapefruit extract at protecting natural textiles was tested. The biocide was applied to flax, needled nonwoven by the padding method, by applying 5 mL of solution per gram of non-woven. This was repeated six times, after which the protected samples were dried for 24 h in ambient temperature. The protected samples were exposed to *Chaetomium globosum* Kunze, at a temperature of 29 ± 1°C and relative air humidity of 90%, for 14 days. The protective effect of the grapefruit extract was evaluated visually based on the degree of mildew growth on the samples and also by testing the relative loss of breaking force. Maximum efficiency was observed for these biocides used on flax nonwovens: no fungi growth and no loss of breaking force occurred. Unprotected nonwoven material underwent complete decomposition, with mildew growth observed on all sample surfaces, along with 100% loss of breaking force (Kozłowski and Walentowska, 2006).

Thyme oil was found to display high anti-fungal activity when used to protect linen–cotton blended fabric. Thyme oil was applied by padding from methanol solutions of 20%, 12% and 8%; 1.2 mL of solution was used per gram of the dry fabric. The samples were exposed to the action of the following fungi mixture: *A. niger* van Tieghem, *Chaetomium globosum* Kunze, *Gliocladium virens* Miller, *P. variotii* Bainier and *Penicillium ochrochloron* Biourge. The tested samples were allocated on an agar medium and inoculated with a suspension of testing fungi. The samples were incubated at a temperature of 29 ± 1°C and relative air humidity of 90% for 4 weeks. For all concentrations, no visible growth was observed under microscopic evaluation, and no loss of breaking force occurred. For comparison, control samples showed the highest degree of decomposition (fifth degree) and 100% loss of breaking force. The results are shown in Table 18.5 (Kozłowski et al., 2008).

An assessment of the antibacterial activity of some commercial textile products was conducted. Socks, briefs, shoe insoles, foot spray, shoe spray, foot powder and hygienic pads were tested for resistance to *S. aureus* and *E. coli.* For most products, some degree of bacterial inhibition was achieved, except for socks and briefs. Future work should be aimed at studying the use of herbal extracts and chitosan as ecological antimicrobial agents for finishing purposes (Thilagavathi et al., 2006).

Chitosan is a natural, nontoxic and biodegradable natural polysaccharide. The only difference between chitosan and cellulose is that the former has an amine group in the C-2 position, whereas in the latter, this position contains a hydroxyl group (Sivaramakrishnan, 2007). Chitosan is a deacetylated derivative of chitin, which is derived from marine shells and mollusks. The antimicrobial activity of chitosan is

Table 18.5 Antifungal properties of linen-cotton blended fabric protected with thyme oil.

Biocide	Degree of fungal growth (scale from 0 to 5)	Loss of breaking force (%)
Thyme oil 20%	0	0
Thyme oil 12%	0	0
Thyme oil 8%	0	0
Reference samples (with methanol)	5	100

also used for other purposes, including improving dyeing performance, as a deodorant, and as an antistatic agent. However, a major limitation in their application in the textiles field is that they are only effective against microorganisms at high concentrations, when they are able to form a film on the surface of the fabric, decreasing air permeability (Joshi et al., 2009).

18.4 Future trends

A new generation of biocides – with relatively low toxicity to human and environment – are subject of scientific research and set future trends. Application of plant derived biocides, modification by hydrochloride of PHMG - polyhexamethylene guanidine and silver nanoparticles open the possibility to obtain antimicrobial barriers for natural fibres.

18.4.1 Developments in use of biocides for the protection of natural fibre textile materials: environmental and health considerations

New regulations have certainly limited the application of many commonly used biocides. Therefore, there is growing demand for the replacement of old and sometimes environmentally unsafe compounds with newer more environmentally friendly alternatives. Furthermore, EU Directive 98/8/EC dealt with the introduction of biocidal products into the market and required the withdrawal of products based on substances that are exceptionally toxic to humans and the environment. The development of ecological biocides is especially important because of the REACH system (**R**egistration, **E**valuation and **A**uthorization of **Ch**emicals), introduced by the EU (EC Regulation No. 1907/2006 of the European Parliament and of the Council of 18 December 2006). The REACH specifies the principles of using chemical substances in order to increase environmental safety and eliminate health hazards.

Other compounds with low toxicity that are effective in protecting against biodeterioration are quaternary ammonium salts (QACs) which are widely known to be bioactive and have high antimicrobial activity. The first quaternary ammonium salt was obtained by Menschutkin in 1890 (Menschutkin, 1890). In 1916, Jacobs discovered that QACs had antibacterial properties (Jacobs, 1916). These compounds were first used in disinfection in 1935 (Domagk, 1935), and in 1977 for wood preservation (Butcher and Drysdale, 1977; Butcher et al., 1977). In 1997, Seddon discovered ionic liquids, which are a new group of QACs (Seddon, 1997). Ionic liquids usually consist of an organic cation and an inorganic or organic anion, which are liquids at temperature below 100°C. These salts have a number of interesting properties, namely a wide liquid range of about 300°C, nonvolatility, and ability to act as solvents for organic and nonorganic compounds, high thermal stability, high ionic conductivity, antielectrostatic properties and easy recyclability; as a result, they are of increasing interest. Studies have also proved that ionic liquids are characterized by antimicrobial activity (Pernak et al., 2003, 2004a). These compounds were successfully used in protection of cellulose materials like wood (Pernak et al., 2004b) and paper (Przybysz et al., 2005) against biodeterioration. The antimicrobial activity of ionic liquids strongly depends on their structure, and especially on the type of anion they contain. $(NO_3)^-$ and $(NO_2)^-$ salts are very effective agents against bacteria and fungi (Pernak et al., 2006). Among the quaternary ammonium compounds, quaternary silicones such as 3-trimethoxy-silylpropyl-dimethyloctadecyl ammonium chloride have been used as a durable odour preventative on socks. However, it is less than 90% effective on bacteria and has limited activity against fungi (Sivaramakrishnan, 2007).

Biosil is an antibacterial agent with a composition of quaternary ammonium salts with an organosilicone. This biocide is fixed to fibre or textile by graft-polymerization and it is not removed even after multiple launderings. It's worth to signalize that biosil is accepted to be used in inner wear, event in other products worn directly on the skin (Raichurkar et al., 2017).

The research on ecofriendly antimicrobial finishing of natural textiles has also covered silver nanoparticles. Metallic silver combined with zeolite and dispersed in the polymer before extrusion spinning provides a polyester fibre that can be intimately blended with cotton to produce a durable antimicrobial composite with excellent antibacterial and antifungal properties (Sivaramakrishnan, 2007).

Silver bromide – bacteria fighter

A new type of antibacterial coating could make it easier and cheaper to endow biomedical implants with bacteria fighting power of silver. Preparation of this product is done by adding Ag salts to the cationic copolymer: poly (4 vinylopiridine–CO–poly (4 vinyl N-hexapyridinum bromide). Using this AgBr containing product make the $Ag+$ ions more available and more active than elemental silver (because AgBr is sparingly soluble). This coating can kill gram-positive and gram-negative bacteria on any surface and also in solution form (Halford, 2006).

A research group, based mostly Genentech, report a class of compound called arylomycin, which can kill some of the most resistant gram-negative bacteria, as they tested derivatives of naturally occurring arylomycins and found one compound called GO 775, that is at least 500 times as potent as natural precursors against gram-negative bacteria such as *E. coli*, *K. pneumoniae* and considerably active against *Acinetobacter baumannii* and *P. aeruginosa* that cause hospital acquired infections. This compound also works against drug-resistant strains of this species and in animal model of infection: GO 775 gums up the peptidase. The enzyme has two key catalytic residues – a serine and lysine. When attacking similar enzymes drugs often target the serine. But the aminoacetonitrile warhead of GO 775 instead forms a covalent amidine with the lysine. This research was conducted by a team led by Christopher Heise of Genentech, published in *Nature* 2018, DOI: 10. 1038/s41586-018-0483-6. These reveals were described by Megha Satyanarayana in C&EN in 2018 (Satyanarayana, 2018).

The antimicrobial properties of chitosan and silver nanoparticles on cotton fabric were tested. The emulsion imparted a very good and durable antibacterial property (against *E. coli* and *S. aureus)* on cotton at low concentrations (Hu et al., 2005).

Scholz et al. (2005) tested the antibacterial and antifungal properties of fabrics consisting of SiO_2 fibres coated with precious metal physical vapour deposition layers using the magnetron sputtering technique. Layers of silver, copper, gold, platinum and platinum/rhodium were deposited on fabrics. Silver and copper were effective against bacteria *S. aureus* and *K. pneumoniae.* Other coating metals did not show any antibacterial effect. In the fungal test, the samples were exposed to *C. globosum, A. niger* and *Penicillium funiculosum.* Copper was effective against all mildew: no mildew growth was observed on the surface of fabric after copper treatment. Other metals did not have any antifungal effect: after treatment, mildew growth covered from 25% to 100% of fabric surface.

Bacteria and fungi colonize on surface of living/working spaces as well as on daily use items. There is now a plenty of evidence that in environment most microorganisms are to be found in organized 3-dimensional structures. A biofilm represent a sessile microbial community composed of bacterial and/or fungal cells growing on various surfaces, immersed in a cell produced extracellular matrix. Microbial cells grow in a form of microcolonies which are surrounded by water channels, providing nutrients and metabolic products. In biofilm formation, we can distinguish few successive stages. These highly specialized structures are characterized by a strong resistance to commonly used antibiotics and biocides. Use of higher concentration of biocides especially now is not recommended due to environmental and toxicity concerns.

Antimicrobial peptides as promising biocides

Other compounds that inhibit the growth of microorganisms are peptides. Because of their high efficacy in fighting numerous pathogens, antimicrobial peptides have promising potential for use in agriculture, medicine and the food industry. This is of particular importance as the pesticides traditionally used for plant protection and the antibiotics widely used in health care have caused an increase in the resistance shown by pathogens. However, before antimicrobial peptides can be used, a number of regulations must be passed and, in the case of medical applications, clinical trials must be run to prove their effectiveness.

Currently antimicrobial peptides represent one of intensively studied group of novel antimicrobials. Peptides: camel KWKLFKKIQAVLKVL-NH_2, citropin 1.1 GLFDVIKKASVIGGL-NH_2, lipopeptide 1 Palm-KK-NH_2, lipopeptide 2 Palm-RR-NH_2, temporin A FLPLIQRVLSQIL-NH_2, pexiganan GIGKFLKKAKKFG-KAFVKILKK-NH_2, omiganan ILRWPWWPWRRK-NH_2 in the research were synthesized manually by the solid-phase method using 9-fluorenylmethoxycarbonyl chemistry, next were purified by high-performance liquid chromatography and analysed by matrix-assisted laser desorption ionization−time of flight mass spectrometry. Tested microbial strains from Polish collection of microorganisms from Natural Science Institute (*Bacillus anthracis* 34F2, *Bacillus atrophaeus* ATCC 9372, *Enterococcus hirae* ATCC 10541, *E. coli* 11229, *E. coli* ATCC 25922, *K. pneumoniae* ATCC 4352, *Pantoea agglomerans* ATTC 21434, *Proteus mirabilis* PCM 543, *P. aeruginosa* ATCC 9027, *S. aureus* ATCC 25923, *S. aureus* ATCC 6538P, *Staphylococcus epidermidis* PCM 2118, *Streptococcus pneumoniae* ATCC 49619,

Streptococcus pyogenes PCM 465. Tested compounds have shown a diverse activity in MIC assay (lowest dose of biocide) performed for reference strains of bacteria. Investigated peptides have shown a significantly stronger activity against gram-positive bacteria. The most susceptible strain of bacteria was *S. epidermidis*. MIC value of lipopeptides, pexiganan and omiganan was 0.5–1 mg/L. Tested essential oils (thyme oil) with main component thymol and clove oil (main component eugenol) and triclosan also proved to be significant antibacterial agents (Kozlowski et al., 2013a,b; Sobczak et al., 2013; Sobczak et al., 2014; Nakonieczna et al., 2018).

Some more data about antimicrobial peptides and essential oils activity are presented in the article: New trends in area of application effective and environmentally friendly biocides in polymers and textile materials (Kozlowski et al., 2013a,b).

In this research were tested quaternary ammonium salts, more essential; oil (thyme, clove, tea tree, and geraniol, eucalyptus and hemp oil. Also was investigated biocide: 2-hydroxypropyl-3-piperazinyl-quinoline carboxylic acid methacrylate also called BCA, produced by South Korean company: Micro Science Technology Co Ltd It was a clear liquid, odour-less, pH 9–14, boiling point 100°C, and freezing point below 0°C, soluble in water.

In agriculture, positive results were obtained from tests on using one of peptides, namely polyoxin, in commercially available fungicides. Polyoxin inhibits the synthesis of chitin, the basic component of cell walls in fungi. Polyoxin B effectively inhibited the growth of mildew in fruit trees and polyoxin D zinc salt was used on peat soils as a fungicide against soil pathogens, for example *Rhizoctonia solani* (Keymanesh et al., 2009). In the near future, it would be interesting to test the viability of using antimicrobial peptides in the protection of natural fibre textiles against biodeterioration.

Another one of these peptides, cecropin B, was used to modify the *Bombyx mori* silk fibroin films (obtained from the giant silkmoth *Hyalophora cecropia)* by the carbodiimide chemistry method (Bai et al., 2008). As a result of this modification, a smaller contact angle was achieved, along with greater surface roughness in the silk fibroin films. Those properties could prove important in preventing the contamination of the silkworm fibre biomaterial.

We know that silk is a composition of two proteins: in crude silk, content of sericin is about 20%–30%. Sericin is a composition of 18 amino acids. It was discovered that sericin has antimicrobial, antioxidant and antitumour properties. Antimicrobial activity can be utilized to deliver numerous activities of textile, such as sportswear, outside clothes, underpart shoes, decoration textiles, upholstery, healing centre materials, wound care wraps, towel and wipes. Sericin can be also utilized for different creams, salve and shampoos (Saravanan, 2017).

A further natural peptide, lysozyme, is an enzyme obtained from chicken egg whites. Its action involves the decomposition of bacteria cell walls, especially of gram-positive bacteria. Lysozyme can be combined with other antibacterial compounds to use such combinations in medicine, e.g., for wound dressings or creams. In the Air Force Research Laboratory at Florida's Tyndall Air Force Base a new material was developed by combining lysozyme and silver acetate in methanol; the mixture was then exposed to light and tested microbiologically. The tests indicated that the following microorganisms were susceptible to the action of the mixture: *E. coli, S. aureus, B. anthracis* and *Candida albicans* (Halford, 2009).

Modified cellulose fibres containing covalent bounded biocidal preparations based on copolymer of polyhexametyleno guanidine and its reactive derivatives are protected against *S. aureus*. This implies the possibility of using them in protective wear in the health care field. PHMG has got a relatively wide spectrum of biocidal activity. It belongs to the so called AMMS group, or antimicrobial macromolecules, similar in structure and mode of action to natural proteins that fight infections, i.e., HDPS. Studies on the spectrum of activity of PHMG derivatives and their possible applications to natural and man made polymers are also being conducted in Poland.

18.4.2 Bioavailability of compounds used in the protection of natural fibre textile materials

The bioavailability of the compounds involved is key in ensuring the safe use of biocides (for protection against bacteria, fungi – including mildew – and insects) in fibre and textile applications. Bioavailability is related to the possible migration of these substances from textiles to the human body and their harmful effects on human physiology. Examples of substances with easy bioavailability are ZnO and Ag nanoparticles. Varsha et al. (2011) proposed an approach involving the attachment of chitosan molecules onto cotton cellulose fabric via coupling reaction. This induced covalent attachment following the incorporation of ZnO microparticles into the chitosan layer by an 'equilibration-cum-hydrothermal' approach. The fabric shows antibacterial efficiency against *E. coli*, indicating its potential for use in biomedical applications.

18.5 Conclusion

Natural fibres as green, renewable and environmentally friendly materials are successible to biodeterioration caused by bacteria, fungi and mildew, which in favourable conditions, e.g., at elevated temperature and high humidity decompose the main structural components of natural fibres, especially cellulose. There are plenty of fungi like *Aspergillus, Chaetomium, Alternaria, Trichoderma, Penicillium, Stachybotrys* and bacteria such as *Cytophaga, Sporocytophaga*, and *Clostridium* bacillus, which contribute to the deterioration of natural fibres. In this chapter are discussed the methods of determining resistance to action of mildew: the agar method, soil burial test and quality screening method. There are also discussed methods of preventing fungal growth, mildew and bacteria action by influence of antimicrobial agents and application of chemical and also many natural products such as extracts from herbal plants origin.

It is also considered protection of natural fibres facing the rules according to REACH system demands, future trends and problems of bioavailability of different biocides, used for protection of natural fibres and different goods made on their base. Natural biocides have now been rediscovered and have become a safe alternative to traditional compounds, leading to the whole area of biodeterioration prevention becoming more ecofriendly and increasingly harmless for human health.

The possibility of producing antimicrobial barriers (against bacteria and fungi, including mildew) for natural fibre textiles through the use of biocides of natural origin or chemical biocides of low toxicity allows for the elimination of toxic chemical biocides. Conducting research into improvements in protection against biodeterioration is inextricably bound up with the market demand and growing interest in functional natural fibres.

18.6 Sources of further information and advice

Bast and other Plant Fibres, ed. R. R. Franck. Cambridge: Woodhead Publishing, 2000.

Biodegradable and Sustainable Fibres, ed. R. S. Blackburn. Cambridge: Woodhead Publishing, 2005.

Enzymatic degradation of insoluble carbohydrates, ed. Saddler J.N. and Penner M.H., ACS Symposium Series 618, 1995.

Fungicides, ed. D. C. Torgeson, Volume 1: *Agricultural and Industrial Applications.* Volume 2: *Environmental Interactions.* New York: Academic Press, 1967.

International Biodeterioration and Biodegradation, Official Journal of the International Biodeterioration and Biodegradation Society.

Mikrobiologia Materiałów. Microbiology of Materials, ed. B. Zyska and Z. Żakowska. ISBN 83-7281-150-5. Technical University of Łódź, Poland, 2005.

Polymer Green Flame Retardants. Ed. Papaspyrides C.D. Kiliaris P. Flame Retardancy and Protection against Biodeterioration of Natural Fibers: State-of-Art and Future Prospects (2014), Chapter 23, In *Polymer Green Flame Retardants*, Kozłowski, R.M.; Muzyczek, M. and Walentowska, J.: Elsevier B.V., ISBN 978-0-44,453-808-6, Amsterdam 801-838.

Renewable Resources and Plant Biotechnology, ed. R. Kozłowski, G. E. Zaikov and F. Pudel. New York: Nova Science Publishers, 2006.

Textiles for Sustainable Development, ed. A. Rayesh, L. Hunter, R. Kozłowski and G. Zaikov. New York: Nova Science Publishers, 2007.

References

Bai, L., Zhu, L., Min, S., Liu, L., Cai, Y., Yao, J., 2008. Surface modification and properties of *Bombyx mori* silk fibroin films by antimicrobial peptide. Appl. Surf. Sci. 254, 2988–2995.

Bakkali, F., Averbeck, S., Averbeck, D., Idaomar, M., 2008. Biological effects of essential oils: a review. Food Chem. Toxicol. 46, 446–475.

Block Seymour, S., 1967. Application and Use of Fungicides as Industrial Preservatives. In: Torgeson, D.C. (Ed.), Fungicides, vol. 1. Academic Press, New York (10).

Butcher, J.A., Preston, A.F., Drysdale, J.A., 1977. Initial screening trials of some quaternary ammonium compounds and amine salts as wood preservatives. For. Prod. J. 27 (7), 19–22.

Domagk, G., 1935. A new class of disinfectant. Dtsch. Med. Wochenschr. 61, 829–832.

Foksowicz-Flaczyk, J., Walentowska, J., 2008. Eco-friendly antimicrobial finishing of natural fibres. Mol. Cryst. Liq. Cryst. 484, 207–212.

Gouin, S., 2004. Microencapsulation industrial appraisal of existing technologies and trends. Trends Food Sci. Technol. 15, 330–347 (Elsevier).

Gulluce, M., Sahin, F., Sokmen, M., Ozer, H., Daferera, D., Sokmen, A., Polissiou, M., Adiguzel, A., Ozkan, H., 2007. Antimicrobial and antioxidant properties of the essential oils and methanol extract from *Mentha longifolia* L. ssp. *Longifolia*. Food Chem. 103, 1449–1456.

Halford, B., 2006. Silver bromide bacteria fighter. Chemical & Engineering News (C&EN) 8. July 17, 2006.

Halford, B., 2009. Antimicrobials from egg whites and silver. Chem. Eng. News 38, 20 April.

Hiolski, E., 2017. Toothpaste compounds, including triclosan, build up in toothbrushes. Chem. Eng. News 95 (43). October 30, 2017, 5.

Hu, Z.G., Chan, W.L., Chan, K.W., Szeto, Y.S., 2005. New preparation of chitosan/silver nanocomposite and its antibacterial activity on cotton. Polym. Prepr. 46 (2), 767–768.

Jacobs, W.A., 1916. The bactericidal properties of the quaternary salts of hexamethylene-tetramine. J. Exp. Med. 23, 563–568.

Joshi, M., Wazed, A., Purwar, R., 2009. Ecofriendly antimicrobial finishing of textiles using bioactive agents based on natural products. Indian J. Fibre Text. Res. 34, 295–304.

Keymanesh, K., Soltani, S., Sardari, S., 2009. Application of antimicrobial peptides in agriculture and food industry. World J. Microbiol. Biotechnol. 25, 933–944.

Kozlowski, R., Dawgul, M., Jaskiewicz, M., Rajkiewicz, M., Tyszkiewicz, W., Sobczak, M., Kamysz, W., 2013a. Antimicrobial peptides as promising future biocides, 2013. In: Poster at 12th International Conference on Frontiers of Polymers and Advanced Materials (12th ICFPAM). The University of Auckland Business School, New Zealand, pp. 8–13. December.

Kozlowski, R., Rajkiewicz, M., Wertejuk, Z., Tyszkiewicz, W., Majewski, K., Kamysz, W., Syryjczyk, R., 2013b. New Trends in Area of Application Effective and Environmentally Friendly Biocides in Polymers and Textile Materials. In: Proceedings of the Conference ELASTOMERY 2013, XVth International Science and Technology Conference, Warsaw, October 23-25, 2013.

Kozłowski, R., Walentowska, J., 2006. Efficiency of biocides in protection of lignocellulosic non-woven against biodeterioration. Annals of Warsaw Agricultural University, Forestry and Wood Technology 59, 6–11.

Kozłowski, R., Władyka-Przybylak, M., 2004. Uses of natural fibre reinforced plastics. In: Wallenberger, F.T., Weston, N.E. (Eds.), Natural Fibres, Plastics and Composites. Kluwer Academic Publishers, Dordrecht, pp. 249–274.

Kozłowski, R., Muzyczek, M., Mieleniak, B., Walentowska, J., Flaczyk, J., Konczewicz, W., 2008. State of art in treatment of natural fibres against fire and biodeterioration. In: Textile Processing: State of the Art & Future Developments, the 5th International Conference of Textile Research Division, NRC, Egypt. Abstract of Book ISSN 1687–2126, 7–8.

Libudzisz, Z., Kowal, K., 2000. Mikrobiologia Techniczna (Technical Microbiology), Wydawnictwo Politechniki Łódzkiej, vol. 2. Technical University of Lodz, Poland, p. 212.

Marino, M., Bersani, C., Comi, G., 2001. Impedance measurements to study the antimicrobial activity of essential oils from *Lamiaceae* and *Compositae*. Int. J. Food Microbiol. 67, 187–195.

Menschutkin, N., 1890. Beiträgen zur Kenntnis der Affinitätskoeffizienten der Alkylhaloide und der organischen Amine. Z. Phys. Chem. 5, 589–601.

Nakonieczna, J., Wolnikowska, K., Ogonowska, P., Neubauer, D., Bernat, A., Kamysz, W., 2018. Rose Bengal-mediated photoinactivation of multidrug resistant pseudomonas aeruginosa is enhanced in the presence of antimicrobial peptides. Front. Microbiol. 9, 1949.

Pernak, J., Sobaszkiewicz, K., Mirska, I., 2003. Anti-microbial activities of ionic liquids. Green Chem. 5, 52–56.

Pernak, J., Sobaszkiewicz, K., Foksowicz-Flaczyk, J., 2004a. Ionic liquids with symmetrical dialkoxymethyl-substituted imidazolium cations. Chem. Eur J. 10, 3479–3485.

Pernak, J., Zabielska-Matejuk, J., Kropacz, A., Foksowicz-Flaczyk, J., 2004b. Ionic liquids in wood preservation. Holzforschung 58, 286–291.

Pernak, J., Śmiglak, M., Griffin, S.T., Hough, W.L., Wilson, T.B., Pernak, A., Zabielska-Matejuk, J., Fojutowski, A., Kita, K., Rogers, R.D., 2006. Long alkyl chain quaternary ammonium-based ionic liquids and potential applications. Green Chem. 8, 798–806.

Pott, G.T., 2004. Natural fibres with low moisture sensitivity. In: Wallenberger, F.T., Weston, N.E. (Eds.), Natural Fibres, Plastics and Composites. Kluwer Academic Publishers, Dordrecht, pp. 105–122.

Przybysz, K., Drzewińska, E., Stanisławska, A., Wysocka-Robak, A., Cieniecka-Rosłonkiewicz, A., Foksowicz-Flaczyk, J., Pernak, J., 2005. Ionic liquids and paper. Ind. Eng. Chem. Res. 44, 4599–4604.

Raichurkar, P.P., Nadiger, V.G., Ranjit, T., Nadiger, G.S., 2017. Antimicrobial agents for textile application. Colourage 43–50. November 2017.

Rakotonirainy, M.S., Lavédrine, B., 2005. Screening for antifungal activity of essential oils and related compounds to control the biocontamination in libraries and archives storage areas. Int. Biodeterior. Biodegrad. 55, 141–147.

Rakotonirainy, M.S., Fohrer, F., Flieder, F., 1999. Research on fungicides for aerial disinfection by thermal fogging in libraries and archives. Int. Biodeterior. Biodegrad. 44, 133–139.

Rios, J.L., Recio, M.C., 2005. Medicinal plants and antimicrobial activity. J. Ethnopharmacol. 100, 80–84.

Rodriguez, A., Batlle, R., Nerin, C., 2007. The use of natural essential oils as antimicrobial solutions in paper packaging. Part II', Progress in Organic Coatings 60, 33–38.

Sanderson, K., September 3, 2018. Synthesis success smells sweet. Chemical & Engineering News, (C&EN) 96 (35), 9.

Saravanan, N.M., 2017. Sericin used in medical textiles. Colourage 81–85. Oct 2017.

Satyanarayana, M., 2018. Attacking Gram-negatives bacteria in a new way. Chemical & Engineering News (C&EN), September 17, 2018, 96 (37), 9.

Scholz, J., Nocke, G., Hollstein, F., Weissbach, A., 2005. Investigations on fabrics coated with precious metals using the magnetron sputter technique with regard to their antimicrobial properties. Surf. Coat. Technol. 192, 252–256.

Seddon, K.R., 1997. Ionic liquids for clean technology. J. Chem. Technol. Biotechnol. 68, 351–356.

Seventekin, N., Ucarici, O., 1993. The damage caused by micro-organisms to cotton fabrics. J. Text. Inst. 84 (3), 304–313.

Sivaramakrishnan, C.N., September 2007. 'Antimicrobial finishes'. Colourage 50–54.

Sobczak, M., Debek, C., Oledzka, E., Kozłowski, R., 2013. Polymeric systems of antimicrobial peptides—strategies and potential applications. Molecules 18 (11), 14122–14137. https://doi.org/10.3390/molecules181114122. Open Access, available from: www.mdpi.com/1420-3049/18/11/14122. ISSN 1420-3049.

Sobczak, M., Kamysz, W., Tyszkiewicz, W., Debek, C., Kozłowski, R., Oledzka, E., Piotrowska, U., Nałecz-Jawecki, G., Plichta, A., Grzywacz, D., Furtak, E., 2014. Biodegradable macromolecular conjugates of citropin: synthesis, characterization and *in vitro* efficiency study. React. Funct. Polym. 83, 54–61.

Standard AATCC 100-1998. Antibacterial Finishes on Textiles: Assessment of Materials.

Standard AATCC 147-, 1998. Antibacterial Activity Assessment of Textile Materials: Parallel Streak Method.

Standard EN 14119, 2003. Testing Textiles – Evaluation of Micro-fungi Action.

Standard EN ISO 11721-1, 2001. Tesivaratiles – Determination of the Resistance of Cellulose-Containing Textiles to Micro-organisms: Soil Burial Test – Part 1: Assessment of Rot-Retardant Finishing.

Szostak-Kotowa, J., 2004. Biodeterioration of textiles. Int. Biodeterior. Biodegrad. 53, 165–170.

Szostak-Kotowa, J., 2005. 'Włókna i tkaniny' ('Fibres and fabrics'). In: Zyska, B., Żakowska, Z. (Eds.), Mikrobiologia Materiałów (Microbiology of Materials), Wydawnictwo Politechniki Łódzkiej. Technical University of Lodz, Poland), 83-7283-150-5, 89–136.

Szostak-Kotowa, J., Leśniak, A., Skrzypek, M., 1988. In: 'Biodegradacja Materiałów Bawełnianych I Bawełniano-Poliestrowych' ('Biodegradation of Cotton and Cotton-Polyester Fabrics') Zeszyty Naukowe Akademii Ekonomicznej (Scientific Papers of Cracow University of Economics), Kraków, vol. 275, pp. 67–77.

Thilagavathi, G., Bala, C., 2007. Micro-encapsulation of herbal extract for microbial resistance in healthcare textile. Indian J. Fibre Text. Res. 32, 351–354. September 2007.

Thilagavathi, G., Rajendrakumar, K., Rajendran, R., February, 2006. Antimicrobial effectiveness of some commercial textile products. Colourage 49–51.

Tomšič, B., Simončič, B., Orel, B., Vilčnik, A., Spreizer, H., 2007. Biodegradability of cellulose fabric modified by imidazolidinone. Carbohydr. Polym. 69, 478–488.

Varsha, T., Bajpai, S.K., Bajpai, M., 2011. Novel strategy for synthesis of ZnO microparticles: loaded cotton fabrics and investigation of their antibacterial properties. Journal of Engineered Fibers and Fabrics 6 (3), 73–81.

Walentowska, J., Kozłowski, R., 2006. Protection of Natural Fibres against Biodeterioration Caused by Mildew. Corrosion Protection, 9s/A/2006, PL ISSN 0473-7733, 328–331.

Żakowska, Z., Piotrowska, M., 2008. 'Mikroorganizmy w procesach biotechnologicznych – grzyby strzepkowe' ('Microorganisms in biotechnological processes – hyphae fungi'). In: Mikrobiologia Techniczna (Technical Microbiology), Wydawnictwo Naukowe PWN SA, Warszawa. Polish Scientific Publishers PWN, Warsaw, Poland), 978-83-01-15223-0, Vol. 2, Part VI, (Chapter 7), 150–177.

Zyska, B., 1999. Zagrożenia Biologiczne W Budynku (Biological Threat in a Building). Arkady, Warsaw, Poland, pp. 81–88, 83-213-41117-9.

Zyska, B., Żakowska, Z., 2005. 'Włókna i tkaniny' ('Fibres and Fabrics'), Mikrobiologia Materiałów (Microbiology of Materials), Wydawnictwo Politechniki Łódzkiej. Technical University of Lodz, Poland.

Further reading

Arnaud, C., 2018. Conjugate attacks Gram-negative bacteria. Chem. Eng. News 96 (15). April 09, 2018, 14.

Butcher, J.A., Drysdale, J.A., 1977. Relative tolerance of seven wood-destroying Basidiomycetes to quaternary ammonium compounds and copper-chrome-arsenate preservative. Mater. Org. (Berl.) 12 (4), 271–277.

Jayalakshmi, I., 2016. Impact of value addition anti-microbial functional finish on pure wool and wool blend materials. Colourage 41–44. November 2017.

Radulescu, H.C., Dinca, L., Ghituleasa, C., Joyce, E., Vinatoru, M., Beddow, J., Mason, T.J., Lazar, V., 2014. Conferring antimicrobial properties to fabrics by sonochemical embedding of nanoparticles. Industria Textila 65 (5), 217–252.

Saraf, N.M., 2015a. Guard Yourself with Silane Based Antimicrobial Agent against Microbes. Part I - Cotton. Colourage, pp. 49–51. February 2015.

Saraf, N.M., 2015b. Guard yourself with silane based antimicrobial agent against microbes. Part II- Polyester and polyester/cotton. Colourage 51–52. June 2015.

Sen, A., Sambhy, V., MacBride, M.M., Peterson, B.P., 2006. Chemical & Engineering News (C&EN), p. 8. July 2006.

Wild silks: their entomological aspects and their textile applications

19

Richard S. Peigler
Department of Biology, University of the Incarnate Word, San Antonio, TX, United States

19.1 Introduction

Wild silks are defined as any silks that are not mulberry silk, derived from *Bombyx mori*. Whilst mulberry silk accounts for over 95% of all silk in commerce today and historically, some of the wild silks play an important role in international commerce, and many of them have significance in the material culture of tribal peoples who collect cocoons and process them into textiles. Unlike mulberry silk, which is pure white, wild silks are beige, dark brown, golden brown, or pale green in their natural colours (Fig. 19.1). Although a few wild silks have been exploited locally in ancient Greece and historically in Mexico (possibly even pre-Columbian), Japan and China, today most are associated with Nigeria, Botswana, Namibia, South Africa, Madagascar, Indonesia, Bhutan, China, Japan and notably India. Eri silk, the great majority of which is produced in India, is also cultured in Bangladesh, Thailand, Indonesia, Korea, Brazil, Ethiopia, Kenya and Vietnam. Although the market for handwoven textiles is saturated globally, due mainly to internet sales and intense local competition, many wild silk products are brought home by travellers and sought by textile collectors. Some wild silks are now promoted for poverty alleviation and fair trade to support cooperatives of weavers by giving them access to international markets via the internet. Unfortunately, there are now counterfeit pieces in the trade, mostly composed of mulberry silk or spun cotton dyed beige or golden to look like the more desirable wild silks (Figs. 19.1–19.16).

In this chapter, I will limit coverage to those silks that are being processed into textiles today or were important well into the 20th century. As an entomologist, whose research interests focus on taxonomy of Saturniidae (Figure 19.1), I can provide here correct and reliable scientific names of the silkmoths, many of which still appear in the literature as misspelled or incorrect names. I will also cite statistics on current production and make some projections of what I believe the near future holds for these silks. Tables 19.1, 19.2 and 19.3 list the species treated in this chapter. *B. mori* (Linnaeus, 1758) belongs to the moth family Bombycidae, whereas the wild silk producers belong to the families Saturniidae, Lasiocampidae, and Notodontidae. Under a microscope, the fibres of silk of Saturniidae are seen to be flat like tape or ribbons

Handbook of Natural Fibres. https://doi.org/10.1016/B978-0-12-818398-4.00021-9

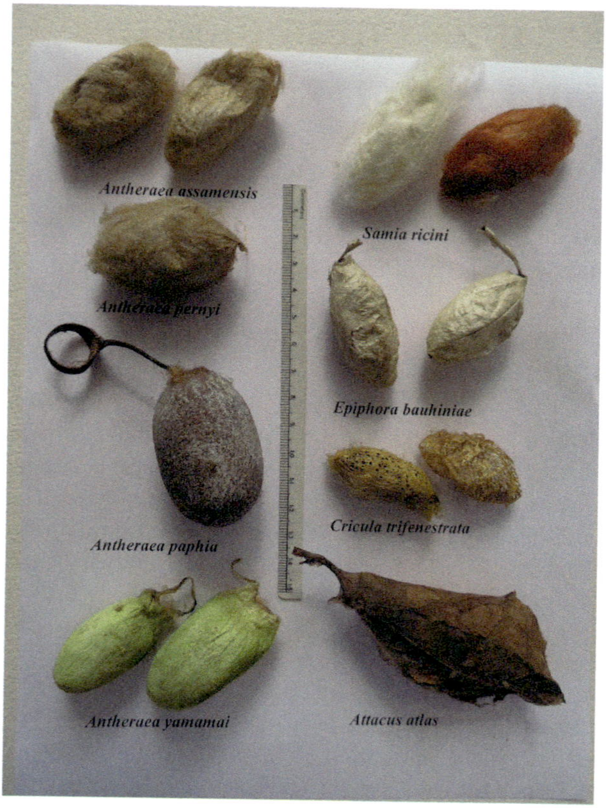

Figure 19.1 Cocoons of Saturniidae that are used in sericulture.

Figure 19.2 Muga silkmoth (*Antheraea assamensis*), pinned male.

Figure 19.3 Muga silkworm in Upper Assam, May 2017.

Figure 19.4 Muga silk on loom in Sualkuchi, Assam, May 2017.

(Travis and Peigler, 2016). Compared with mulberry silk, the silks of Saturniidae contain much less glycine, but more alanine and serine (Babu, 2013).

Wild silk textiles are produced in two main ways from cocoons (Fig. 19.1). The first is reeling, where the silk is unwound from the degummed cocoon as a single continuous thread. These fibres are then twisted together (throwing) to be woven or used in warps, sometimes as few as two or three single parallel threads. Textiles made from reeled silk are usually sheer and finely textured. Besides *B. mori*, the only silks that are commonly reeled are from the genus *Antheraea* (tensan, tussah, muga and tasar silks). The other method is called spinning, which entails pulling the silk off the degummed cocoon little by little and spinning it with a drop spindle, spinning wheel, or other machines. It can also be spun by hand by rolling the pulled silk over one's thigh or an inverted clay pot, and this technique of ghicha spinning is done with tasar

Figure 19.5 Students performing Bihu Dance in Titabar, Assam, wearing traditional costumes of muga silk, (girls' mekhela chaddars, boys' vests), May 2017.

Figure 19.6 Shirt of tasar silk woven in Jharkhand, tailored in Varkala, Kerala, India, in 2013.

silk. Spinning silk from cocoons is basically the same process that is used with wool and cotton, since all such fibres are not long enough to reel. Textiles woven from spun silk yarn are coarser and heavier than those made with reeled silk. A recent innovation called slivering renders spun silk into very fine thread, leading to textiles that are finely textured like reeled silk. For example, eri silk was never used for saris in India because

Figure 19.7 Shawls of tasar silk, all natural coloured. Clockwise from upper left: kantha embroidered in 2012, in Shantiniketan, West Bengal; Baluchari brocade, woven in 2009, in Bakreswar, West Bengal; from Kotpad, Odisha, Dongria Kondha tribe, tasar wefts, handspun cotton warps, maroon dye from *Morinda citrifolia*, 2016; ikat weave from Gopalpur, Odisha, lotus flower motifs, 2016.

Figure 19.8 Ghicha spinning tasar silk on clay pots, Central Tasar Research & Training Centre, Ranchi, Jharkhand, June 2017.

Figure 19.9 Tussah silkmoth (*Antheraea pernyi*), living female.

Figure 19.10 Tussah silkworms on oaks, Liaoning Province, China, June 2008.

it was too coarse and heavy, but slivered eri silk is now being made into saris and other garments that were traditionally made of reeled tasar and mulberry silks and cotton (Zethner et al., 2012). Except in China, virtually all wild silk textiles are handwoven. Indian textiles are almost always weft dominant, so that the warps are only visible as fringes at the ends of the uncut weaving. Those from Ethiopia, Indonesia, China, Namibia, Thailand, and Vietnam are usually a plain weave, but *Rothschildia* silk pieces from Argentina and Malagasy textiles are warp dominant (Figs. 19.17–19.33).

Finished threads, yarns, and cloth of wild silks are always shades of beige, brown, or golden when left in their natural colours. Many tribal peoples prefer these natural colours, but enhance their textiles with colourful brocade, embroidery, block printing, ikat weaving, beads, and even batik. In other cases, wild silks are dyed bright colours, before or after weaving, but never to the extent of mulberry silk. Cocoons are attached

Figure 19.11 Automatic loom making brocaded blanket of tussah silk, Liaoning Prov., June 2008.

Figure 19.12 Tussah silk skirt and sport coat.

Figure 19.13 Tablecloth, China, 1920s, natural coloured hand-reeled and handwoven tussah silk.

Figure 19.14 Scarves of tussah silk ("oak tasar") from India. With green dye from Uttarakhand, 2010; and natural coloured from Manipur, 2012.

Figure 19.15 Tensan silkmoth (*Antheraea yamamai*), living male.

Figure 19.16 Tensan silkworms, two colour forms.

Table 19.1 Wild silk producers in the family Saturniidae.

Species of silkmoth	Name of silk	Countries
Antheraea assamensis (Helfer 1837)	Muga	India
Antheraea paphia (Linnaeus 1758) (=*Antheraea mylitta*)	Tasar, tassar	India
Antheraea pernyi (Guérin-Méneville 1855)	Tussah	China
Antheraea yamamai (Guérin-Méneville1861)	Tensan	Japan
Samia ricini (Wm. Jones 1791)	Eri, endi	India, etc.
Attacus atlas (Linnaeus 1758)	Fagara, atlas	Indonesia
Cricula trifenestrata (Helfer 1837)	Cricula	Indonesia
Epiphora bauhiniae (Guérin-Méneville 1832)	Epiphora	Sudan
Rothschildia schreiteriana Breyer & Orfila 1945	Rothschildia	Argentina
Rothschildia maurus (Burmeister 1879)	Rothschildia	Argentina

to the host tree or shrub by a peduncle spun by the larva to firmly affix it to a twig (most Saturniidae), or sometimes by the side of the cocoon (*Cricula*, most Lasiocampidae). The caterpillars of others (Notodontidae treated here) construct a large silken bagnest attached in the host tree. This communal bagnest is occupied by sibling larvae that eventually form individual cocoons within it, the moths later emerging from the bagnest.

Some wild silks exist only in captivity (*Samia ricini*) like *B. mori* or are kept captive in most stages (*Antheraea pernyi, Antheraea assamensis, Antheraea yamamai*). A certain amount of husbandry is carried out with others (*Borocera* in recent times,

Table 19.2 Wild silk producers in the familyLasiocampidae.

Species of silkmoth	Name of silk	Countries
Borocera madagascariensis Boisduval 1833	Landibe, lande	Madagascar
Borocera cajani Vinson 1863	Landibe, lande	Madagascar
Borocera marginepunctata Guérin-Méneville 1844	Landibe	Madagascar
Europtera punctillata (Saalmüller 1844)	Landibe	Madagascar
Gonometa postica (Walker 1855)	Kalahari wild silk	Namibia, Botswana, South Africa
Gonometa rufobrunnea Aurivillius 1922	Kalahari wild silk	Namibia, Botswana, South Africa

Table 19.3 Wild silk producers in the family Notodontidae.

Species of silkmoth	Name of silk	Regions
Anaphe panda (Boisduval 1847)	Sanyan	Africa
Anaphe reticulata Walker 1855	Sanyan	Africa
Anaphe infracta Walsingham 1885	Sanyan	Africa
Epanaphe moloneyi (Druce 1887)	Sanyan	Africa
Epanaphe venata (Butler 1878)	Sanyan	Africa
Epanaphe carteri (Walsingham [1888])	Sanyan	Africa
Hypsoides radama (Coquerel 1855)	Landibe	Madagascar
Hypsoides antsianakana (Oberthür 1922)	Landibe	Madagascar
Hypsoides, several more species	Landibe	Madagascar

Antheraea paphia), where people put eggs on trees and protect caterpillars. In others, the cocoons or bagnests are collected in the field (*Gonometa, Attacus, Cricula, Epiphora, Rothschildia, Anaphe, Epanaphe,* traditionally *Borocera*) from wild populations of the silkmoths. It is always best if only empty cocoons are collected, so that wild silkmoth populations can be managed as a renewable resource.

Figure 19.17 Tensan silk products from Japan, scarf, coin purse, business card case.

Figure 19.18 Eri silkmoths (*Samia ricini*), living males, traditional eri chaddar in background.

Figure 19.19 Tray rearing of eri silkworms in Meghalaya, May 2017.

Figure 19.20 Washed eri silk floss, drying in sun near Guwahati, Assam, May 2017.

Figure 19.21 Eri silk sliver, Fabric Plus factory near Guwahati, May 2017.

19.2 Muga silk

Production of muga silk is limited the northeastern India, where this small industry is based on the Himalayan *A. assamensis*. (Fig. 19.2). It is believed that muga silk has been cultured for millennia based on archaeological samples found at Harappa in what is now Pakistan (Kenoyer, 2017). The muga farmers place newly hatched caterpillars on the host trees and try to protect them from ants, birds and other natural enemies as they mature. The full-grown caterpillars descend the trees to spin their cocoons on the ground but are intercepted by the muga farmers so they will spin their cocoons in captivity (Fig. 19.3). There are six generations per year, but only two of these produce cocoons with enough silk to be used commercially. Moths emerge

Figure 19.22 Miss Deep Tifull (Khasi tribe) hand spinning eri silk in Umden, Meghalaya, May 2017.

Figure 19.23 Shawls of eri silk, clockwise from upper left: Ethiopia, 2010; plaid weave from Umden, Meghalaya, Khasi tribe, 2017; block printed from Bangladesh, 2015; handspun yarns with natural dyes, Nong Ya Plong village, Khon Kaen Prov., Thailand, 2015.

and mate in captivity, and the resulting eggs are hatched to yield caterpillars to be placed on trees again. The host trees are primarily in the laurel family, two of the main ones being soalu (*Litsea monopetala*) and som (*Machilus gambeli*), the latter cited by its synonym *Machilus bombycina* in virtually all literature on muga silk (Seidel and Peigler, 2018). Production has been gradually increasing in recent years: 124 metric tonnes (MT) in 2010–11 to 166 MT in 2015–16. Assam produces over 85% of muga silk, followed by Meghalaya with 11% and smaller amounts from

Figure 19.24 Atlas moth (*Attacus atlas*), pinned female.

Figure 19.25 Jacket of *Attacus* silk from Yogyakarta, Java, made in May 2007.

Figure 19.26 *Epiphora bauhiniae*, living male from Senegal, 2006.

Figure 19.27 Landibe silkmoth (*Borocera cajani*), living female beside her cocoon, 2018.

Figure 19.28 Mrs. Josiane Rafanjanirina weaving landibe silk on her ground loom in Ambositra, Madagascar, May 2018.

tiny states bordering Assam (Nagaland, Arunachal Pradesh and Manipur). In any case, muga silk is highly esteemed in India and the demand always exceeds the supply, leaving little for export. The Central Silk Board actively promotes muga culture and provides healthy eggs to muga farmers at no cost.

Muga silk is famous all over India for its natural golden colour, so it is rarely dyed. The fabric (Fig. 19.4) is used in traditional clothing in northeastern India, especially the mekhela chaddar (skirt plus shawl) that Assamese tribal women wear during special occasions like the Bihu festival (Mazumdar, 2013). These garments are richly

Figure 19.29 Landibe silk scarves from Central Highlands, Madagascar.

Figure 19.30 Weavings from southwestern Madagascar, 1993, Tandroy tribe. Woman's beaded and brocaded loincloth of pure landibe silk (left) and man's loincloth, mostly handspun cotton, with weft twining (akosto) in landibe silk at each end.

brocaded in red, with smaller amounts of yellow, green, or blue. Men wear muga silk waistcoats (vests) and turbans (Fig. 19.5). Most such pieces are handwoven using reeled silk, but sleeveless Jawahar coats may be tailored from heavier cloth woven with spun yarns. As explained by Phukan (2010), muga has become so expensive that there are now a lot of counterfeit textiles sold as muga, which are usually mulberry or tasar silk dyed shades of gold. Many rural people and tribals now freely accept the substitute because it is more affordable and looks like genuine muga, a situation similar to the shift from sanyan silk to a cotton substitute in Nigeria discussed below.

Figure 19.31 Kalahari wild silk, 2018, two scarves from Leonardville, Namibia, and a cushion cover (cotton muslin with handspun appliqué of wild silk, natural colour) from Stellenbosch, South Africa.

Figure 19.32 Pinned silkmoths in Notodontidae. *Anaphe infracta*, *Anaphe reticulata*, *Epanaphe moloneyi*, *Epanaphe carteri*, all from western Africa; *Hypsoides antsianakana* female from Madagascar. All specimens courtesy of Dr. Stefan Naumann, Berlin.

Figure 19.33 Two ipele (skirt wrappers) composed of anaphe silk from Ondo, Nigeria, early 20th century. Yoruba tribe. Textiles supplied by Duncan Clarke, London.

In other cases, weavings have warps of acrylic, tasar silk, or mulberry silk, obscured by the muga wefts so that the finished piece is considered muga and is sold as such. In recent years, Japanese people have imported muga thread to be woven into obis, blended with mulberry silk. Muga silk has GI registration (Geographical Indication) status in Assam, and pieces labelled as such are guaranteed to be authentic. A sari composed of muga silk is a very expensive and prestigious possession. About 20 years ago, the Italian firm Anichini used muga silk in curtains, bedding and cushion covers, and such pieces are authentic. Historical pieces such as rumals from Bengal, Persia, and Kashmir occasionally have embroidered motifs composed of golden muga thread; I bought such a piece in London in 1995.

19.3 Tasar silk

Tasar silk is only produced in India where it has ancient origins dating back at least 4000 years (Kenoyer, 2017). Today India produces about 3200 MT of raw tasar silk but in recent years this amount fluctuated slightly above or below that amount. Although primarily produced by tribal peoples in eastern India (West Bengal, Bihar, Chhattisgarh, Jharkhand, Odisha and eastern Maharashtra) (Krishna Rao and Amarnath, 2006), this type of silk, also called kosa silk, is broadly known to the Indian public, so it is traditional for tasar silk to be worn in wedding ceremonies – saris of brides and kurtas of grooms. Tasar silk enjoyed a large export market in the 1960–1980s, bringing a lot of foreign currency to India; the fabrics were made into men's sport coats, neckties, and women's suiting by the apparel industry in Europe. In India, most tasar silk is made into kurtas, saris, men's shirts (Fig. 19.6), salwar kameez, scarves, and shawls, the latter which are the ubiquitous dupattas worn by women all over India, all these often in the natural beige and brown shades (Fig. 19.7). Beginning in the 1970s and increasingly in the 1980s, the demand in India exceeded the supply, so exports declined. This was due to overexploitation of the source by collecting too many cocoons everywhere they could be found. Now India imports as much Chinese tussah as possible to use as a substitute for tasar, and this is viewed as positive by keeping the handlooms running, but also negative since some in India view the trend as dumping by China of a cheaper commodity. New and vintage pieces of tasar are now listed by the thousands by internet merchants in India and even by the Indian diaspora in Europe and North America, where the term tussar is now almost universally used instead of tasar. This silk has been called tropical tasar to distinguish it from temperate tasar, but I consider the latter to be Himalayan cultures of *A. pernyi* (Peigler, 2012).

Tasar sericulture is based on the large moth *A. paphia*, which I believe was derived thousands of years ago from the Himalayan *Antheraea frithi* (Moore), although there is no consensus on this hypothesis (Peigler and Naumann, 2016), but I expect it will eventually be resolved by genomics. Although the synonymy of the names *A. paphia* and *Antheraea mylitta* (Drury) was established over two centuries ago (d'Abrera, 2012), many authors in India still use both names to distinguish varieties, or to separate the so-called wild populations from the cultured ones. The tasar silkworms feed on several trees including arjun (*Terminalia arjuna*) and sal (*Shorea robusta*) and eventually produce a huge cocoon. They are placed on the trees as newly hatched larvae, and as they mature, the farmers protect them from birds and other natural enemies with limited success.

There are some very distinctive and well-known categories of tasar textiles showing the natural brown and beige shades of the undyed silk. Women's tasar dupattas from West Bengal have ghicha-spun wefts and warps (Fig. 19.8), with colourful Baluchari repeating motifs at each end and matching butis brocaded in the main field. The large shawls from Kotpad, Koraput District, Odisha (formerly spelt Orissa), are much sought after, recognized by interlocking temple designs and various butis; the warps are handspun cotton, the wefts ghicha-spun tasar and the consistently maroon dyes come from the aal tree (*Morinda citrifolia*), also called Indian mulberry. From Odisha, we also see dupattas of reeled tasar silk with colourful ikat designs, mainly lotus flowers. Beautiful dupattas of reeled tasar from West Bengal may be richly decorated with kantha embroidery showing birds, flowers, or other designs, but most such Bengali dupattas are made of cotton or mulberry silk. Some tasar textiles have GI registration such as those from Bhagalpur and Kotpad (Fig. 19.7), and many of the famous pattachitra paintings from Odisha are on tasar fabric.

19.4 Tussah silk

Tussah silk, derived from *A. pernyi (Fig. 19.9)*, is a major industry in China and is well-known to westerners. Although tussah silkworms are cultured outdoors on oaks (*Quercus*) pruned to be large bushes (Fig. 19.10), they are kept in captivity as eggs, cocoons, and adult moths (Qin, 2011). Apparently, populations of *A. pernyi* do not exist in nature, and there is a lot of evidence that this species was derived by selective breeding thousands of years ago from the Himalayan *Antheraea roylei* (Moore), as detailed by Peigler (2012). Cultures of *A. pernyi* were also introduced decades ago to Ukraine, Japan, Korea and Romania, but these populations had to be maintained by humans. The populations of the Balearic Islands of Spain apparently have persisted in nature, but I suspect they have reverted to a wild '*roylei*' form. To test this hypothesis, Spanish authors need to collect wild cocoons, instead of only adults attracted to lights, since the papery cocoons of *A. roylei* are quite unlike the dense *A. pernyi* cocoons that are rich in silk. The so-called oak tasar or temperature tasar (dubbed by Indian workers as *Antheraea proylei*) that are now cultivated in the Himalayas of India (Manipur, Uttarakhand and Jammu and Kashmir) are also to be viewed as cultures of *A. pernyi*, as explained by Peigler (2012). There are more than 130 named varieties of *A. pernyi*, most of which are maintained at the Sericultural Research Institute of Liaoning [SRIL] (1994), where I visited for several days in 2008 (Fig. 19.11). Chinese researchers have developed varieties with pure white silk and have made great progress in controlling natural enemies (parasitoids, predators and pathogens) of tussah silkworms.

The natural colour of tussah silk is a pinkish beige or creamy beige, sometimes termed ecru. It can be darker brown in the Indian 'oak tasar' but I have a scarf from Manipur that is pale creamy beige (Fig. 19.14). Some tussah cocoons are reeled, but a majority are converted into spun yarn. Antique tablecloths (embroidered with white mulberry silk in dragon or flower motifs) (Fig. 19.13) and handkerchiefs (brocaded with colourful bands of mulberry silk at the edges and corners) were exported from China into the USA and UK by the thousands in the 1920 and 1930s. These pieces always consist of natural coloured beige silk that was hand-reeled and handwoven, and there is often ladder work or leno insertions in the tablecloths. Natural coloured beige robes and pyjamas from Japan also date back to the first half of the 20th century and are invariably printed with brightly coloured images of flowers, pagodas or dragons. All these examples can be found in antique markets and estate sales, and names such as pongee and Shantung were formerly applied to them. China produced a lot of heavy suiting fabric from tussah, especially in the 1970–1980s when tons of it were imported and used by famous designers like Eileen Fisher, Ralph Lauren, Geoffrey Beene and others (Fig. 19.12), and there are still merchants in Canada, Hong Kong, Italy and California who regularly offer tussah fabrics that they import from China in a variety of colours and weaves. However, in the last 25 or so years, most tussah silk has been exported from China to India where it is used as a substitute for the tasar silk in India that has been overexpl oited leading to declining production in the face of increased demand by Indian consumers (see under 'Tasar silk'). An agent in Guangzhou who connects sellers and buyers of tussah silk told me in 2008 that most of the tussah leaves the port of Dalian on ships headed directly to India. She explained that Chinese companies made far fewer textiles and finished garments because they could sell cocoons, yarn, and floss for greater profit directly to India. In any case, tussah silk remains an important commodity in Chinese export markets, but this silk has never been used in China for ethnographic or traditional clothing like most wild silks in other countries have. Today China produces about 23,000 MT of raw tussah silk, with a gradual increase each year over the previous year.

19.5 Tensan silk

Japanese people hold tensan silk in high esteem because it is indigenous to Japan and less than 0.5 MT of raw tensan silk is produced each year (Akai and Kuribayashi, 1990). This wild silk industry is based on *A. yamamai* whose larvae feed primarily on oaks (*Quercus*) (Figs. 19.15, 19.16). The moths fly in late summer or autumn, the eggs overwinter, larvae feed in the spring, and summer is passed in a green cocoon spun among the leaves in the host trees. The cocoon colour is retained in the finished threads and textiles as a beautiful light pale green.

The products are generally small, such as coin purses, business card cases, neckties, and altar cloths in which the limited supply of tensan silk is blended with greater amounts of mulberry silk, yet they are very expensive to purchase (Fig. 19.17). The silkworms are raised outdoors under large cages covering small pruned oak trees, as I observed at Ariake near Hotaka in 1994. The silk is reeled by hand in small filatures. In recent times, tensan production has become limited to very few places, primarily Shiga, Nagano, and Fukushima prefectures. Japanese researchers have produced a strain with light blue cocoons. To help meet the demand in Japan, tensan silk is also raised in North Korea and Hangzhou, Zhejiang Province, China.

19.6 Eri silk

The eri silk industry (ericulture) is based on *S. ricini*, which only exists in captivity (Fig. 19.18). The ancestral species is the Himalayan *Samia canningi* (Hutton), a moth with a compact, brown cocoon, attached to a twig by a strong peduncle. The cultivated species differs from the wild progenitor in several ways: in *S. ricini,* the larvae are disease resistant, large puffy cocoons are produced without a peduncle and the moths cannot fly. These exactly correspond to comparisons between *B. mori* which only exists in cultivation, and its wild ancestor *Bombyx mandarina* (Moore). In their detailed revision of the genus *Samia*, Peigler and Naumann (2003) suggested that the eri silkmoth has been cultivated for more than 2000 years. Historically, *Samia cynthia* (Drury 1773) in China has been used for its silk for centuries and into the 20th century, and *Samia pryeri* (Butler) was occasionally used in Japan. Many authors have treated the eri silkmoth as a subspecies of the ailanthus silkmoth, i.e., *Samia cynthia ricini*, an error still commonly seen in literature, and *Philosamia* is also not the correct name for this genus.

Eri silkworms are fed primarily on the leaves of castor bean (*Ricinus communis*) but other trees are also used including payam (*Tetradium fraxinifolium*, incorrectly cited as

Evodia fraxinifolia in virtually all literature on ericulture), cassava (*Manihot esculenta*), and species of *Ailanthus* and *Jatropha*. The larvae can be reared in pans on racks (Fig. 19.19) in a rearing house that is screened to keep out predators and insect parasitoids, or in cut bunches of castor leaves draped over a string or rod, often covered by netting. Mature larvae spin cocoons in baskets or bundles of twigs. The tribals in northeast India almost always eat the pupae, and such entomophagy is a major incentive for rural people to practise ericulture (Seidel and Peigler, 2018). The cocoons occur in two forms, brick red and the more common snow white; when processed these yield beige yarn and white yarn, respectively (Javali, 2011).

Many women in villages spin yarn from the cocoons (Fig. 19.22) and weave the textiles on their handlooms, usually as a plain weave, but sometimes weft dominant. The fabric can be tailored to make salwar kameez sets (tunic plus pants), shirts, skirts, etc., but most pieces remain uncut, ready to be used when taken off the looms. This can include the lungi or turban for men, but the majority of eri silk is made into narrow scarves (mufflers) and broad chaddars (shawls) (Fig. 19.23). Chaddars for men are larger and plain cream or beige, without colourful brocade or embroidery. Those for women are smaller and almost always have bands of red (plus black or green or blue) brocading at each end and small matching butis in the main field, displaying traditional Assamese or tribal motifs (Mazumdar, 2013). Eri silk is also used for blankets and quilting in northeast India. An interesting point about eri silk in Meghalaya is that some pieces are plaids, now a part of the textile arts of that small mountainous state where Scottish missionaries introduced plaid weaving in the 1770s. I was fortunate to buy several plaid scarves and chaddars of eri silk when I visited Sonidan, Meghalaya, in 2017 (Seidel and Peigler, 2018); they have natural dyes of red from stick lac, yellow from wild turmeric and black from iron.

Where ericulture has been adopted in other places (Ethiopia, Thailand, Vietnam, etc.), the finished products are usually scarves and shawls, and the projects, although limited, contribute to poverty alleviation via fair trade cooperatives. Some eri pieces from Ethiopia are noteworthy for their bright colours derived from natural dyes like indigo and madder (Fig. 19.23) and traditional weave patterns with leno insertions. Introduction of ericulture to Ethiopia from Japan in 2001 has been successful because the castor grows well there, and the silk is processed like Ethiopians have been spinning and weaving cotton for centuries (Badola and Peigler, 2013). In Bhutan, eri silk has been imported for many centuries as yarn or cocoons from India, and then made into traditional Bhutanese apparel by highly skilled weavers who used natural dyes and myriad of bright colours: the gho (robe) and kabné (shoulder cloth) for men, and the kera (belt), kira (skirt), and rachu (shoulder cloth) for women. Such pieces now have high market value outside of Bhutan among textile collectors, but many Bhutanese are now trying to stop the loss of their cultural heritage from sales of old textiles to outsiders. Most such pieces are composed of eri silk, although wool, cotton, etc., have also been used (Myers and Bean, 1994). Textile collectors and curators might also watch for intricately woven veils (marame, singular maramă) from southern Romania composed of white eri silk. Eggs of eri silkmoths were imported from North Korea to Romania in 1966 and ericulture persisted there until the 1980s, but marame are usually composed of mulberry silk.

Traditionally, ericulture has been tedious because the weavers had to hand-spin the yarns (Fig. 19.22)and the rearers sold their cocoons to middlemen for low prices. This is rapidly changing in Assam (where most of the world's eri silk is produced) due to an enlightened business model that empowers the rearers and weavers. Several companies, led by Fabric Plus headquartered in Guwahati, are now buying as many cocoons as the rearers can produce at higher prices and converting the silk into a high-quality yarn that is of uniform thickness (Figs. 19.20, 19.21). This machine-spun yarn is much easier and faster to weave, so the weavers are buying the yarn at fair prices and producing more textiles on their handlooms in their homes. Fabric Plus has won numerous awards from the Indian government and business groups for its practices that favour social justice over profit (Zethner et al., 2012). In May 2017, I was fortunate to spend several days in Guwahati visiting the factories and headquarters of Fabric Plus and learnt that the company is also promoting eri silk into new markets across India and abroad, even in high-end boutiques in Europe. India produces more than 5600 MT of raw eri silk annually and this is increasing every year.

Regarding so-called peace silk, also called ahimsa silk, in which it is possible to obtain silk from the cocoons without killing the pupae (simply because eri silk cocoons are spun into yarn after the moths emerge), eri silk is usually promoted as such, the sellers on websites employing terms cruelty-free and nonviolent to entice buyers, claiming that the moths are allowed to fly free into the forest. However, this notion of peace silk is in fact an invalid marketing ploy, since eri silkmoths cannot fly and the great majority of pupae are eaten by those who reared them. In cases where only empty cocoons are being harvested from the field, as in some projects for landibe silk (*Borocera*) and Kalahari wild silk (*Gonometa*) (see below), the term peace silk can rightfully be applied, but entomophagy of landibe pupae is still sometimes practised in Madagascar.

19.7 Attacus and Cricula silks

We know from reports by British authors working in India almost two centuries ago that silks from *Attacus atlas* and *Cricula trifenestrata* were exploited on small scales in the Bengal region, where tribals collected the cocoons in the forests, but the silk of neither is used in India today. Larvae of the atlas moths feed on many kinds of trees, and those of *Cricula* favour mango (*Mangifera indica*), cashew (*Anacardium occidentale*), avocado (*Persea americana*) and other trees. Species of *Attacus* and *Cricula* occasionally reach pest status on cultivated trees in India, the Philippines, Indonesia and other countries of Southeast Asia, but in most cases, the cocoons are destroyed by the farmers. It was against this backdrop that Japanese sericulturists helped launch a small but successful project in the 1990s to utilize these cocoons in Yogyakarta (Java, Indonesia), a city known for its textile arts.

The atlas moths (Fig. 19.24) are among the largest moths in the world, and their large cocoons are light to dark brown, wrapped in leaves, and suspended by a soft and often long peduncle. Long ago, the British called the silk by the term fagara silk. The Javanese call it atlas silk, not to be confused with some ikat fabrics of mulberry silk from Uzbekistan also called atlas silk. The yarns and fabrics are usually a medium brown colour, but can be golden brown or darker brown, depending on which kinds of leaves the caterpillars ate.

Moths of the genus *Cricula* are much smaller and resemble a dead leaf in shape and colour, and transparent spots in the wings enhance the camouflage by resembling holes in leaves (d'Abrera, 1998). The caterpillars have stinging spines that become incorporated into the cocoons, so these must be removed by careful washing before the silk can be processed further. The cocoons of *Cricula* are a metallic golden colour, but the processed silk yields dull or bright yellow textiles that are softer than those derived from *Attacus*.

The products from Java made from silk of atlas and *Cricula* moths include both woven textiles and pressed cocoons. The woven items are mainly scarves and shawls, but the fabric has also been used on clutch purses, cushion covers, tote bags, and women's shoes, whereas the pressed cocoons show up on lampshades, picture frames, and jewellery boxes. Occasionally the fabric is tailored into garments, as shown in the illustrated jacket (Fig. 19.25). The scarves and shawls are usually a plain weave, but sometimes have leno insertions or intricate patterns such as honeycomb or plaids. The project in Yogyakarta has been promoted for poverty alleviation, providing fair trade employment or supplementing incomes of local people. The finished products generally find their way into high-end boutiques in West Malaysia, Japan and cites in Indonesia, and can be bought by wholesalers on the internet.

19.8 Epiphora and Rothschildia silks

The African atlas moth, *Epiphora bauhiniae*, is a large and attractive moth that ranges widely in Africa from Kenya west to Senegal, southward to Namibia and South Africa (Fig. 19.26). The colourful caterpillars feed on buffalo thorn (*Ziziphus mucronata*) and Christ's thorn jujube (*Ziziphus spina-christi*) and then spin compact cocoons with smooth surfaces, each attached by a short peduncle to the hostplant. European colonialists working in Africa took note of this moth because of its conspicuous cocoons, and performed some small trials leading to combed floss and spun yarns. There are at least 12 additional species of *Epiphora* in sub-Saharan Africa (d'Abrera, 1998), all of which produce cocoons that have sericultural potential, but I found no reports that any of those have been exploited.

Hill-Thomas (2012) illustrated a handful of cocoons as 'small wild silk pods, unknown species' which I could readily identify as *E. bauhiniae* from her photograph. She wrote that although the Marka tribe weavers prefer the larger bagnests of the *Anaphe* (see below), some older women insisted that individual cocoons (*E. bauhiniae*) were also used when the larger ones were not available. Therefore, I believe that some vintage or antique textiles from West Africa will eventually be found to contain silk of *Epiphora*.

Recent efforts by Mohamed Tom Eltayb (2014) of Khartoum University advocated exploitation of silk from *E. bauhiniae* in Sudan, in hopes to provide employment opportunities to people in rural areas. Eltayb outlined the biology and life cycle of the moth in Sudan. His quantitative experiments consisted of trials on various host-plants, growth rates of the caterpillars, effects of rainfall on cocoon shell weight, degumming the cocoons, spinning the silk into yarn and various attempts to dye it. He also reported on the characteristics of the *Epiphora* silk, including tensile strength, filament length and thickness and fibre elongation. Eltayb's detailed treatise is available as a print-on-demand book.

The fact that elderly women in a few remote villages in northwestern Argentina could begin to collect cocoons and spin and weave the silk from a common moth demonstrates that small projects can spring up anywhere there is a reliable source of useable cocoons. Adriana Zapata and Graciela Jurado (2018) recently documented the exploitation of *Rothschildia schreiteriana* and *Rothschildia maurus* in the village of Ancasti, Department of Catamarca, Argentina, where women have been weaving scarves and knitting shawls from the silk of these two exquisite moths. There are at least 25 species in the genus *Rothschildia* ranging from southern Texas and southern Arizona down throughout much of South America, all of which produce large cocoons (d'Abrera, 1998).

19.9 Landibe silk

Landibe silk, called lande in southwestern Madagascar, is cultured primarily in the Central Highlands by weavers of the Merina and Betsileo tribes. For centuries, landibe has been the fibre of choice to make shrouds, which the Malagasy in the Central Highlands use to wrap the dead, and to rewrap them in elaborate ceremonies when they are exhumed and reinterned every few years. The people believe that they must keep their ancestors satisfied since they can bring good or bad luck to the living (Green, 2004). In southwestern Madagascar, landibe silk has been used to skillfully weave elaborate loincloths for women and men (Fee, 2004). The most comprehensive treatment of the sericulture and textile applications of landibe silk was provided by Gade (1985). Besides shrouds, landibe silk is primarily woven into lambas (scarves and shawls) that are very popular with the general populace, as mufflers in the cool winters of the Highlands and as fashionable scarves sold in high-end boutiques and handicraft shops all over the island. Moreover, cooperatives of women weavers have been formed to produce these scarves which are then sold on the internet with the help of European and American merchants, the profits mostly going back to the women (Figs. 19.28, 19.29). As such, these ventures qualify as poverty alleviation and fair-trade operations. Other landibe silk products include tote bags, clutch purses, beanie caps, and natural coloured silk embroidered onto shirts for men. In Madagascar, I found such items for sale in shops in hotels, airports, handicraft markets, and directly from the weavers themselves.

Landibe silk comes from cocoons of several silkmoths in the Lasiocampidae (Table 19.2) on the island (Peigler, 2004) that are collected in the forests and mixed and used interchangeably. The main ones appear to be *Borocera cajani* (Fig. 19.27) and *Borocera madagascariensis* but in her dissertation Tsiresy Razafimanantsoa (2016) identified a few more that are used. She detailed the biology of these moths, including hostplants, seasonal flight times, and population dynamics. The moths were formerly more common and widespread on the island, but extensive deforestation on the western half has led to large areas of savanna (Gade, 1985) so many silkmoth populations have been extirpated. However, on a night walk to see lemurs in Kirindy Forest Reserve, a tiny patch of original habitat near the west coast, I found an empty *Borocera* cocoon hanging on a small tree a metre above ground level. In the Central Highlands, there are now large patches of tapia forest (*Uapaca bojeri*), and today most landibe cocoons are collected on these small trees. The landibe caterpillars are covered with black, urticating spines that are incorporated into the cocoon by the spinning caterpillar, leaving them outwardly projecting, serving as a defence against predators. Bagnests of the *Hypsoides* (Table 19.3) have also been collected for silk in Madagascar, although I suspect this may no longer be the case.

The cocoons are collected by the thousands, degummed and handspun into yarn. Shrouds are usually left in their natural brown colour. Two or three woven panels are stitched together to make the piece wide enough to serve as a shroud. Some lambas and shrouds are very colourful and elaborate (Green, 2004). The fair-trade lambas usually consist of stripes formed by warp yarns. Although most Malagasy textiles are warp dominant, there are occasional plaids. The warp striping may be symmetrical but is usually asymmetrical. Colouration is usually imparted by vegetable dyes, leading to various earth tones, the most common being a reddish brown called mena, often accompanied by stripes in black or the natural (undyed) brown shades. Some scarves are now brightly coloured with synthetic dyes, but traditional colours are still preferred by most buyers in Madagascar and abroad. Older women in the coops who are skilled weavers can produce pieces with geometrical brocades. Glass or metal beads are sometimes incorporated near both ends of lambas (Fig. 19.30). All these Malagasy textiles have fringes at each end, usually knotted or braided. Most weavers also work with locally produced mulberry silk, and the narrow white stripes in landibe silk pieces are mulberry silk. Almost all weavers in Madagascar are women.

19.10 Kalahari wild silk

Exploitation of this wild silk stands unique because it does not have historical origins and is not a part of the cultural heritage of any groups of people. Although the French colonialists more than a century ago were aware that the moths in the genus *Gonometa* produce large cocoons, no commercial ventures were launched until the 1980s in Botswana, and soon after in neighbouring areas of South Africa and Namibia. The cocoons are collected in the field on small trees, and after degumming, the silk is handspun into yarn, sometimes dyed in bright colours. The main products are scarves and shawls, but I have seen a few other items like men's neckties and bowties and cushion covers (Fig. 19.31). These projects are almost entirely started and maintained to provide employment to people, so they qualify as fair trade and social justice enterprises. One additional benefit is that cattle inadvertently eat cocoons when browsing on low branches, and this can even lead to death if too many cocoons block the digestive systems in the cattle. Thus, there is an incentive to remove as many cocoons as possible in areas where cattle are raised.

Two species are used in this sericulture. Larvae of *Gonometa postica* feed on several woody legumes formerly placed in *Acacia*, but now classified in the genera *Vachellia* and *Senegalia*. Larvae of *Gonometa rufobrunnea* feed exclusively on mopane trees (*Colophospermum mopane*) and the range of the moth coincides with that of the tree. Caterpillars of both have stinging spines, and these are incorporated into the cocoon, as in the related *Borocera* from Madagascar. However, the oval cocoons of *Gonometa* are much harder and denser than the soft cocoons of *Borocera*. The South African ecologist Ruan Veldtman (Veldtman et al., 2002) has intensively studied both species of *Gonometa* and authored several publications about their ecology and biology. Populations fluctuate drastically over time, making it difficult for people to depend on these silkmoths to provide an ample supply of cocoons every year, so weavers often rely on locally grown mulberry silk to keep their looms operating. The wild silk projects were formerly in Francistown, Botswana, in the 1980s, but primarily in recent years in Leonardville, Namibia. Weavers in South Africa obtain cocoons from all three countries whenever possible.

19.11 Sanyan silk

West Africans have been exploiting local wild silks for centuries, and the resulting textiles and garments have great cultural significance. An important paper detailing the history and ethnography of this silk in Nigeria was published by McKinney and Eicher (2009), and this chapter extracts several points from that work. The tribes of Nigeria, notably the Hausa, Edo, and Yoruba, as well as the Marka in Burkina Faso (Hill-Thomas, 2012) and the Dogon in Mali (Douny, 2013), are some of the main ones using these silks. Traditionally, the women process the silk and spin it into yarns, and the men do all the weaving. In recent decades, women have also been weavers (Clarke, 2009a,b). The men weave narrow strips up to 10 cm wide and 12 m long which are then cut and tailored into garments that are held in high esteem and imply social status. The well-known regalia worn by men of high rank were often composed of this wild silk. They consist of a robe called an agbada in Yoruba (riga in Hausa), with matching cap and trousers. Elaborate embroidery on these pieces was also wild silk, usually golden or dull yellow in colour, in motifs like daggers and the endless knot. Sanyan is the name used by the Yoruba for wild silk. Other names in West Africa for local wild silks include tsamiya (Hausa), akpa-obuku (Igbo), and sapar ubele (Edo). Many wild silk pieces show the open work (holes) and floating wefts that are characteristic of West African textiles. Others are composed of subtle stripes in the warps, usually a grayish or brownish natural colour of the silk. Many of the narrow strips are tailored into the ipele, a wrap-around skirt worn by women on special occasions. Highly esteemed cotton textiles in Nigeria are alaari containing mulberry silk dyed a magenta and etu with blue derived from indigo dye, but sanyan is the most valued of the three.

There are several species of Notodontidae found across Africa that have been exploited for silk in West Africa, and they are related to the *Hypsoides* of Madagascar, mentioned above. Those on the mainland belong to the genera *Anaphe* and *Epanaphe* (Table 19.3) and the moths are much smaller than the Saturniidae and Lasiocampidae (Fig. 19.32). No comprehensive treatments of the African Notodontidae have been published since the work of Gaede (1928), but identification of the species used for silk is possible using his text and colour figures.

The moths lay masses of eggs on host trees, and the larvae live communally in a silken bagnest which they construct. They hide in the bagnest by day and venture out at night to feed on leaves. Eventually they make individual cocoons inside the bagnest. Most species have two generations per year. Urticating hairs of the caterpillars are incorporated into the bagnest so these must be thoroughly washed out before the silk can be spun into yarn. Female moths also have a mass of urticating scales on their abdomens that they wipe onto the mass of freshly laid eggs. The caterpillars in this group feed on many species of trees in the forest, some favourites being coffee berry (*Bridelia micrantha*), wild pear (*Dombeya rotundifolia*), obeche (*Triplochiton scleroxylon*) and the introduced tamarind (*Tamarindus indicus*). The obeche is an important timber tree that has been greatly overexploited, resulting in population declines of the silkmoths. However, the bagnests themselves have been overcollected so that few can now be found in many areas of West Africa. Bagnests that still contain caterpillars or pupae sell for higher prices in the local markets because people also eat the larvae and pupae, as well as use the silk.

As explained by McKinney and Eicher (2009), true sanyan silk has become very rare and expensive since about 1980, and only very wealthy people can commission a garment composed of it. Even then, the weavers now usually use handspun cotton for the warps, so that only the wefts are sanyan. African textiles are very collectible by westerners, so if collectors want to be certain to buy pieces composed of wild silk, they must pay high prices for ones made in the early 20th century or at least before the 1960s. As for people in Nigeria, they have accepted the fact that the 'sanyan' they now weave and wear is cotton that was dyed the same colour as the wild silk. Called kugu, this cotton substitute still holds the same significance and status as the silk. One final point to make is that sericulture in West Africa is not plagued by problems of child labour, and the weavers generally sell directly to buyers, so profits are not lost to middlemen, which has often been the case in India.

19.12 Conclusion and future trends

Wild silks are a diverse topic covering history, culture, ethnography, and entomology. Raising wild silkworms or collecting cocoons in the field is work-intensive but has persisted because it is often a part of the cultural heritage without regard to profitability. However, many girls are not following in the steps of their mothers and grandmothers by choosing careers that leave no time for weaving at home. Many boys in Northeast India do not wish to continue with muga farming as they watch their fathers work hard for little return. The populations of the silkmoths are often not treated as renewable resources when cocoons are overcollected. Deforestation and other habitat alteration also impact populations of the silkmoths negatively in many areas. Even climate change is a threat to some sericulture in the Himalayas because there is less water from glaciers and snow melt to keep rivers flowing that maintain silkmoth habitats in valleys below. Although it appears that tussah and eri silks have a bright future, I am not optimistic about most of the others persisting far into this century. In any case, it is probably advisable to collect, document, and preserve as many of these textile treasures as possible since they will become increasingly uncommon over the next decades. Loss of culture by indigenous peoples continues to be pervasive and relentless.

Acknowledgements

In my studies on wild silks, I have received significant and ongoing support from the following colleagues, each of them over many years: Kiran Badola, Dilip Barooah, Nahum Ben-Yehuda, Bhuban C. Chutia, Duncan Clarke, the late Bernard d'Abrera, Victor Das, Elodie Richard-Das, Kumud Ghosh, Harold Labrique, the late Claude Lemaire, Reeta Luikham, Toshimasa Mitamura, Stefan Naumann, Rolf G. Oberprieler, Jarek Ostrowski, Ulrich Paukstadt, Rodolphe Rougerie, Monimala Saikia, Ryan A. St Laurent and Ruan Veldtman. My colleague Tanja Stampfl and her husband Andre Martinez modeled tussah clothing, and my students Rowdy C. Zett and Matthew C. Seale (landibe scarves), Ricardo Mata (tasar shirt), and Cody L. Seidel (*Attacus* jacket) modeled wild silks. Funds from the Sister Joseph Marie Armer Endowed Chair (University of the Incarnate Word) enabled me to make study trips to India in 2017 and Madagascar in 2018. All photographs in this chapter were made by the author.

References

Akai, H., Kuribayashi, S., 1990. Tensan: Science & Technology. Science House, Tokyo, p. 274, 4 color plates. [in Japanese].

Babu, K.M., 2013. Silk: Processing, Properties and Applications. The Textile Institute & Woodhead Publishing, Cambridge, p. 182 xv.

Badola, K., Peigler, R.S., 2013. Eri Silk: Cocoon to Cloth. Bishen Singh Mahendra Pal Singh, Dehra Dun, p. 93.

Clarke, D., 2009a. Nigeria: Frauen weben – Männer weben. In: Gardi, B. (Ed.), Raffiniert und Schön: Textilien aus Westafrika. Museum der Kulturen Basel. Christoph Merian Verlag, pp. 100–129, 200.

Clarke, D., 2009b. Nigeria: women weave – men weave, pp. 100-129. In: Gardi, B. (Ed.), Woven Beauty: The art of West African Textiles. Museum der Kulturen Basel. Christoph Merian Verlag, p. 200.

d'Abrera, B., 1998. Saturniidae Mundi: Saturniid Moths of the World, Part 3. Goecke & Evers, Keltern, Germany, p. 171.

d'Abrera, B., 2012. Saturniidae Mundi: Saturniid Moths of the World, Part 2. Hill House Publishers, Melbourne & London, p. 174.

Douny, L., 2013. Wild silk textiles of the Dogon of Mali: the production, material efficacy, and cultural significance of sheen. Textiles 11 (1), 58–77.

Eltayb, M.T., 2014. The Potentiality of the Wild Silk Spinners in Sudan. Lambert Academic Publishing, Saarbrücken, p. 88.

Fee, S., 2004. Chapter 6: Textile traditions of southwestern Madagascar, pp. 92-111. In: Kusimba, C.M., Odland, J.C., Bronson, B. (Eds.), *Unwrapping the Textile Traditions of Madagascar*. Field Museum, Chicago. UCLA Fowler Museum of Cultural History, Los Angeles, p. 196.

Gade, D.W., 1985. Savanna woodland, fire, protein and silk in Highland Madagascar. J. Ethnobiol. 5 (2), 109–122.

Gaede, M., 1928. Thaumetopoeidae. plate 68. In: Seitz, A. (Ed.), The Macrolepidoptera of the World, The African Bombyces & Sphinges, vol. 14. Alfred Kernen Publisher, Stuttgart, pp. 395–400, 600 pp., 80 color plates.

Green, R.L., 2004. Chapter 5: Betsileo textiles: negotiating identity between the living and dead, pp. 73-91. In: Kusimba, C.M., Odland, J.C., Bronson, B. (Eds.), Unwrapping the Textile Traditions of Madagascar. Field Museum, Chicago. UCLA Fowler Museum of Cultural History, Los Angeles, p. 196, 2004.

Hill-Thomas, G., 2012. Silk in the Sahel: Tuntun and Marka Faso dan fani in northwestern Burkina Faso. Afr. Arts 45 (2), 58–69.

Javali, U., 2011. Eri Silk: Fibre, Yarn and Fabric Characterization. Lambert Academic Publishing, Saarbrücken, p. 229.

Kenoyer, J.M., 2017. Chapter 2: Textiles and trade in south Asia during the proto-historic and early historic period. In: Hildebtandt, B. (Ed.), Silk: Trade and Exchange along the Silk Roads between Rome and China in Antiquity. Oxbow Books, London & Philadelphia, pp. 7–26 xxi + 130.

Krishna Rao, J.V., Amarnath, S. (Eds.), 2006. An Introduction to Vanya Silks. Vanya: Wild Silks of India: A User's Compendium, vol. 1. Central Silk Board, Bangalore xii + 196.

Mazumdar, L., 2013. Textile Tradition of Assam: An Empirical Study. Bhabani Books and Fabric Plus Pvt. Ltd, Guwahati, Assam xvii + 194.

McKinney, E., Eicher, J.B., 2009. Unexpected luxury: wild silk textile production among the Yoruba of Nigeria. Textiles 7 (1), 40–55.

Myers, D.K., Bean, S.S. (Eds.), 1994. From the Land of the Thunder Dragon: Textile Arts of Bhutan. Serindia Publications, London. Peabody Essex Museum, Salem, 247.

Peigler, R.S., 2004. Chapter 10: The silk moths of Madagascar. In: Kusimba, C.M., Odland, J.C., Bronson, B. (Eds.), Unwrapping the Textile Traditions of Madagascar. Field Museum, Chicago. UCLA Fowler Museum of Cultural History, Los Angeles, 2004, 153–163, 196.

Peigler, R.S., 2012. Diverse evidence that *Antheraea pernyi* (Lepidoptera: Saturniidae) is entirely of sericultural origin. Tropical Lepidoptera Research 22 (2), 93–99.

Peigler, R.S., Naumann, S., 2003. A Revision of the Silkmoth Genus *Samia*. University of the Incarnate Word, San Antonio, p. 239, 148 color figures.

Peigler, R.S., Naumann, S., 2016. What exactly is *Antheraea paphia* (Linnaeus, 1758)? Atalanta 47 (3/4), 500–520.

Phukan, R., India, 2010. Muga Silk: Problems and Prospects of Muga Silk Industry of Assam. Verlag Dr Müller GmbH, Saarbrücken, p. 146.

Qin, L., 2011. Tussah Silkworm Science. Beijing Normal University, Beijing, p. 317 [in Chinese].

Razafimanantsoa, T.M., 2016. Vers à Soie: Bio-ecologie et Influence des Plantes Nourricières: Cas de *Borocera cajani* Vinson 1863 (Lasiocampidae Gonometinae) de Madagascar. Éditions Universitaires Européennes, Saarbrücken, p. 143 xiii +.

Seidel, C.L., Peigler, R.S., 2018. Entomological aspects of sericulture based on *Antheraea assamensis* and *Samia ricini* (Saturniidae) in Assam and Meghalaya. Tropical Lepidoptera Research 28 (1), 13–18.

SRIL [Sericultural Research Institute of Liaoning], The records of tussah varieties in China. Ser. Res. Inst. Liaoning, 1994. Shenyang, 274.

Travis, J.H., Peigler, R.S., 2016. A closer look at wild silks. Textiles 43 (2), 22–24.

Veldtman, R., McGeoch, M.A., Scholtz, C.H., 2002. Variability in cocoon size in southern African wild silk moths: implications for sustainable harvesting. Afr. Entomol. 10 (1), 127–136.

Zapata, A., Jurado Cazaux, G.B., 2018. Polillas y tejidos de seda en bosques nativos de Argentina. Revista Facultad de Ciencias Exactas, Físicas y Naturales 5 (1), 77–82.

Zethner, O., Koustrup, R., Barooah, D., 2012. Indian Ways of Silk: Precious Threads Bridging India's Past, Present and Future. Bhabani Books, Guwahati, Assam, p. 128.

New emerging natural fibres and relevant sources of information

20

Ryszard M. Kozłowski[1], *Maria Mackiewicz-Talarczyk*[2], *Jorge Barriga-Bedoya*[3]
[1]Journal of Natural Fibers Editor-in-chief, coordinator of FAO/ESCORENA European Cooperative Research Network on Flax and other Bast Plants, former director of the Institute of Natural Fibres, Poznań, Poland; [2]Institute of Natural Fibres and Medicinal Plants (IWNiRZ), Associate Editor Journal of Natural Fibers, Secretary of FAO/ESCORENA European Cooperative Research Network on Flax and other Bast Plants, Poznań, Poland; [3]The ŁUKASIEWICZ Research Network - Industrial Institute of Agricultural Engineering, Poznań, Poland

20.1 Introduction

Natural fibrous resources include lignocellulosic-like seed, bast, leaves, grass, bark and protein. Natural fibres are renewable, biodegradable, have excellent properties, such as good hygroscopicity and air permeability, and do not release substances harmful to health and to the environment. Mankind utilizes them since more than 7000 years B.C.

These lignocellulosic fibrous plants such as cotton, kenaf, jute, sisal, coir, ramie, flax, hemp, curaua, pineapple, bamboo and many other fibres are recently discovered and described in this chapter. The fibres can be extracted, processed, modified, functionalized, and used in production of textiles – woven, knitted, nonwoven, technical, and 3D textiles as well as a reinforcement and fillers for more environment-friendly biocomposites.

Those resources could be also explored for pulp and paper, and for obtaining agrochemicals.

Currently the production of most major natural fibres is expected to reach a level of about 40 mln tons/year in the middle of 21st century and is estimated to grow about 3% annually.

The research on the new fibres described in this chapter and analysed by novel physicochemical methods shows the new possibilities of fibres application (Kozlowski and Muzyczek, 2017).

In our opinion these up-to-now underestimated fibres will play an important role in the new world, which must fight with terrible pollution of our Earth and oceans caused by man-made polymers and fibres residues (Luong, 2017).

Handbook of Natural Fibres. https://doi.org/10.1016/B978-0-12-818398-4.00022-0

20.2 Characteristics and properties of recently discovered and described new natural lignocellulosic fibres

20.2.1 Seed fibres

The main representative of this group is cotton, largely cultivated and processed in the world, and described in two chapters: Chapter 18, Volume 1, Evolution of cotton breeding and in Chapter 1, Volume 2, Cotton fibres, picking, ginning, spinning and weaving. Other seed fibre which is also very well-known is coir (*Cocos nucifera* L.). Coir fibre is obtained from the husk surrounding the nut—fruit of the coconut plant. Coir is described in the Chapter 10, Volume 1 entitled Coir—coconut fibre. Another member of this group is kapok (*Ceiba pentandra* (L.) Gaertn.), coming from Southwest Asia and Indonesia, intensively explored for upholstery material, used for life buoys and belts, waistcoats, seat covers, and other application for life saving at the sea. Kapok is soft, very brittle, and inelastic. It will not retain moisture which is very important in using for sanitary and quite vermin-proof. Chemically it consists of about 64% of cellulose, 13% lignin and 23% of pentosanes.

Balsa fibre (*Ochroma pyramidale*) is grown in West Indies; the fibre is yellow under microscope, flattened, often much folded, the diameter of fibre is 25–50 μ and is mainly used for stuffing of mattresses and cushions.

Other representatives of this group are Kumbi (galgal) (*Cochlospermum gossypium* DC.), from India, applied for stuffing cushions; *Chorisia speciosa* from Brazil, *Beaumontia grandiflora* Wall., *Strophanthus* spp., milkweeds (*Asclepias* L.), *Calotropis/ Calotropis* Floss/Akund floss, also called, ak, madar, or mader, (‘Akund’), and Cattail Fibres (Typhaceae family) (Mauersberger, 1954).

20.2.1.1 *Calotropis gigantea*. Gentiles – *Asclepiadaceae*

This plant is from Asia and Africa; the fibre from the bast in the stalk is white and strong; it contains ca 77% of cellulose, usually applied for cordage. The fibre can also be obtained from seed pods of *C. gigantea* and other species of *Calotropis*. Seed fibre is also termed ‘yercum’ (Mauersberger, 1954).

Correlation between pod shape traits and its fibre quality of wild *C. gigantea* is described by Yumei Cui and the coworkers from Key Lab of Textile Science & Technology, Ministry of Education, Donghua University, Shanghai, China, and College of Textiles and Clothing, Xinjiang University, China, in the *Journal of Natural Fibers* (Cui et al., 2017).

Seed fibre in the *C. gigantea* pods is the most valuable component and can be developed into textile material. Cultivating high-yield and high-quality seed fibres for *C. gigantea* is one of the critical preconditions for its textile industrialization. In this chapter, pods picked from wild *C. gigantea* plants were used as the experimental materials. The fibre length distribution of the pods was studied. The pod characters (pod shape, single pod weight, fibre weight per pod and seed number per pod, etc.) and the crucial quality characteristics of seed fibres (length, diameter and strength) were tested. The correlations between them were analysed. The results show that the fibre lengths at the tip, the petiole and the middle part of the pods are significantly different: the fibres from the two sides of the middle part are the longest and more uniform. The pod shape is closely correlated with the fibre length, fineness and strength: longer pods have longer fibre length, smaller length variation and larger fibre diameter, and the percentage of low strength fibre (single fibre strength <1.2 cN) is significantly lower compared with smaller pods. Cultivating *C. gigantea* with large size and heavy pods will improve the seed fibre yield and the fibre quality (Yumei Cui et al., 2017).

20.2.2 Bast fibres

Bast fibres are obtained from the stem of various plants (McIntyre et al., 2002).

The family of bast plants is large and contains many species. The most known and already largely utilized bast plants are jute, ramie, kenaf, flax and hemp. These plants and derived fibres are presented in the following chapters of Volume 1, respectively: Chapter 3 Bast fibres: Jute, Chapter 4: Bast fibres: Ramie, Chapter 5: Bast fibres: Kenaf, Chapter 6 Bast fibres: Flax, Chapter 19: Developments in fibrous flax breeding and cultivation, Chapter 7: Fibre hemp. In Volume 2, jute is described in Chapter 2: Bast fibres: Processing of jute fibre and its applications and in Chapter 19: Environmental textiles from jute and coir.

To this family belong *Abroma augusta, Abutilon theophrasti* (Indian mallow, Tientsin 'jute', Chinese jute), *Acacia leucophloea* (piland bast), *Adansonia digitata* (baobab tree), *Apocynum cannabinum* (Indian 'hemp', dogbane), *Artocarpus elasticus* Reinw. ex Blume (kerbang, terep, and various other vernacular names), *Asclepias syriaca* L., *Bauhinia racemose* Lam. (malu creeper), *Brachystegia spiciformis* Benth., *Broussonetia papyrifera* (tapa cloth fibre, also kappa or masi), *C. gigantea* (yercum), *Clappertonia polyandra* K. Schum., (punga, a native name used in equatorial Africa to designate fibre from several plants), *Chonemorpha macrophylla* (bangi), *Commersonia fraseri* J.Gay (Blackfellow's 'hemp'; tie plant), *Couratari tauari, Cryptostegia grandiflora* (Roxb.) R. Br. (rubber vine), *Cytisus scoparius* (broom fibre, Spanish broom), *Daphnopsis guacacoa* Griseb., *Datisca cannabina* (Cretan 'hemp'), *Debregeasia hypoleuca* (Hochst.) Wedd., *Dirca palustris* (leather wood), *Dombeya buettneri* K. Schum., and *Datisca cannabina* L., *Epilobium angustifolium* and *Epilobium hirsutum* L. (willow herb), *Eupatorium gregaria* and *Eucalyptus gummifera, Ficus benghalensis* L. and *Ficus nekbudu* Warb. (banyan fibre), *California fremontia* (mountain leatherwood), *Girardinia palmata* (syn. *heterophylla),* (pha-pat, hypetye, and various native names), *Grewia occidentalis* L. (Kaffir 'hemp'), *Helicteres isora* L. (screw plant in Australia; puteran in Indonesia), *Hibiscus sabdariffa* L. (roselle, rosella, roselle 'hemp', and Java 'jute'), and within *Hibiscus,* there are known about 17 species (Mauersberger, 1954).

Honckenya ficifolia Willd. (called in Sierra Leone: napunti, bolo bolo, polepo etc.), *Lagetta lintearia* Lam. (lace bark, daguilla etc.), *Laportea gigas* Wedd (Rangoon 'hemp', stinging nettle), *Lavatera arborea* (tree mallow, also 'brotex'), *Lonchocarpus sericeus*, Malva, *Malachra Capita*, *Maoutia puya* (Puya 'hemp'), and 12 less-known cultivars (Mauersberger, 1954).

Moreover known is *Sida acuta* (broom plants) – cultivated in Mexico and tropical areas including Southeast Asia and Indonesia. The fibre is removed from the stalks by water or dew retting. The fibre cell is similar to that of genus *Hibiscus*. Chemical composition: moisture about 10%, ash 0.9%, lignin 14.8% and cellulose 70.2%. Used for cordage and course sacking. Furthermore, there is another list of bast fibres including 11 species (Mauersberger, 1954).

20.2.2.1 Urena bast fibre

It is obtained from just under the bark of the stalk of the bast plant Urena lobata, a member of the Malvaceae family. The name 'urena' is derived from the Malayalam word 'uram', which is used to designate the plant on the Malabar Coast of India. Urena fibre is utilized principally for the same purposes as jute and can be also mixed with jute or other fibres. Urena grows best in hot humid climates and in the soil especially rich in potash. The plant is perennial; it was mainly cultivated in Kongo and French Equatorial Africa. For highest quality of fibre, plants are harvested at full flowering state. Vegetation period takes from 120 to 150 days. In some favourable conditions, two harvesting per year can be achieved. After defoliating, the stalks are taken to retting area. The retting is very similar to that of jute; for retting, soft and moving water have to be used. Retting is complete after 9–12 days. After removal from the stalks fibre, strands are washed to remove impurities; next they are dried in the sun. Urena fibres, when properly prepared, are white, fine, lustrous, soft and flexible. Applications are similar to those of jute (Mauersberger, 1954).

20.2.2.2 Sunn fibre

Is a bast plant obtained from the *Crotalaria juncea* L. plant – an annual plant. Sunn fibres have been known from prehistoric time, mainly in India and Pakistan. The land preparation is similar to that of jute. Harvesting is done when the plant reaches flowering state 100–110 days. The plants are cut or pulled and remain on the field for 1 or 2 days after leaves are removed and then stalk are immersed in water for 4–6 days of retting and the fibres are stripped from the stalks by hand or mechanical method. Chemical composition of the raw sunn fibre is about 80% of cellulose, pectin 6.5%, and rest are water extract, fats and wax (Mauersberger, 1954).

20.2.2.3 Piassava fibre

This fibre belongs to hard bast fibres, derives from several palms, and can also be obtained from palm *Attalea funifera*. It grows in sandy soil of the region of South of Bahia in northern Brazil and also extends in two parts of Venezuela. African piassava comes from the palm *Raphia vinifera* (Mauersberger, 1954). Piassava fibre is also used as a reinforcement of polymer matrix (recycled polyethylene).

Morphological, structural, thermal and mechanical characterization of piassava fibres are described by Elzubair Amal and the team from Universidade Federal do Rio de Janeiro, Brazil.

The authors carried out extensive characterization of morphological, structural, thermal and mechanical properties of two species of piassava fibres, namely *Leopoldinia piassaba* (Amazon, Brazil) and *A. funifera* (Bahia, Brazil). Scanning electron microscopy (SEM) images revealed that each fibre is essentially a composite in which rigid cellulose microfibrils are embedded in a soft lignin and hemicellulose matrix. In addition, the tubular microfibrils are helically wound along the fibre axis forming ultimate hollow cells. An irregularly dimensioned mass, composed mainly of silica, appears on the surface of each fibre. The crystallinity of *L. piassaba* is found to be inferior to that of *A. funifera*. On the other hand, thermal analysis shows that both species start to decompose at approximately similar temperature (~265°C) and that the moisture content in each is approximately 4%–5%. The mechanical properties correlate well with the fibre diameter. Finally, the *L. piassava*, as compared with *A. funifera*, is softer and more ductile (Elzubair et al., 2007).

20.2.2.4 Attalea funifera *piassava fibres*

The topic properties and structure of *A. funifera* piassava fibres for composite reinforcement – a critical discussion – is presented by Sergio Neves Monteiro in the *Journal of Natural Fibers*.

Results on morphological, structural, thermal, tensile mechanical and chemical characterization of *A. funifera* piassava fibre have been presented by different research groups. In particular, a paper by Elzubair et al. (2007) showed apparent differences from values reported by previous works. These results needed to be complemented in terms of information required for the use of this natural fibre as a polymer composite reinforcement. In this work, a critical discussion on the differences reported so far is presented. A complementary information on the properties and structural characteristics of *A. funifera* fibre with findings related to its application as a composite reinforcement is also provided (Neves Monteiro, 2009).

20.2.2.5 *Hildegardia populifolia*

It is a species of flowering plant in the family Malvaceae, grown in India. Tensile properties of polycarbonate/polymethyl methacrylate blend coated natural fabric *H. populifolia* were tested by Jagadeesh D., and the team of the Department of Polymer Science and Technology, Sri Krishnadevaraya University, Anantapur, India, presented in the *Journal of Natural Fibers*.

The uniaxial natural fabric *H. populifolia* was coated with polycarbonate/polymethyl methacrylate blend, and its tensile strength, modulus and percentage of elongation at break were determined. The effect of alkali treatment on the tensile properties of the fabric was also studied. The tensile properties were enhanced by alkali treatment. The morphology of the fabric before and after coating was studied by microscopic technique (Jagadeesh et al., 2010).

20.2.2.6 Isora fibre

Isora fibre, morphology, chemical composition, surface modification, physical, mechanical and thermal properties and a potential natural reinforcement is presented by Lovely Mathew.

'Isora' is a bast fibre separated from the bark of *Helicteres isora* plant by retting process. Chemical constituents, morphology and physical properties of these fibres have been studied. Density and microscopic methods are used to determine the cross-sectional area and diameter of fibre bundles. Surface modification by alkali and silane treatment was tried. Tensile properties of the treated and untreated fibres were determined by density method. Average tensile strength of the fibre decreased and density increased to some extent on treatment with alkali and silane. The thermal characteristics, crystallinity index (CI), reactivity, and surface morphology of the untreated and treated fibres have been studied by thermogravimetric analysis (TGA), differential scanning calorimetric (DSC) analysis, differential thermogravimetric (DTG) analysis, WAXRD, Fourier transform infrared spectroscopy (FTIR) and SEM. SEM studies showed that fibre surface becomes rough on modification. Thermal analysis showed that chemical modification improves the thermal stability of the fibre. The strength of the fibre was theoretically calculated (Lovely et al., 2007).

20.2.2.7 Sida acuta *(broom plant)*

This is grown in Mexico and tropical areas, including Southeast Asia and Indonesia. Broom plants also grow up on hills and mountains and are common in many Mediterranean countries. The fibre is removed from the stalk by water or dew retting. The fibre is light coloured and harsh in feeling, used for cordage and coarse sacking (Mauersberger, 1954).

The acoustic characterization of broom fibres was carried out by Berardi and the team from Umberto Berardi Department of Architectural Science, Ryerson University, Toronto, Canada.

Natural materials are becoming a valid alternative to traditional synthetic ones for sound absorption applications. In this chapter, the acoustic properties of broom fibres are reported and discussed. The materials were cut and shredded to prepare several samples. Test specimens made of broom pieces with a diameter from 1.5 to 4 mm were prepared. The normal sound absorption for samples with a thickness of 60 mm, 80 and 120 mm was then measured. Results are particularly promising especially above 500Hz, as the samples consistently reported a normal sound absorption above 0.5. The increase in the sample thickness allowed obtaining higher sound absorption at lower frequencies, as generally expected for porous materials. The main limit of the investigated samples was the appearance of sound absorption resonances. These were more evident for the samples composed of pieces with larger diameter and may be attributed to the loose nature of the tested samples (Berardi, Iannace and Di Gabriele, 2017).

20.2.2.8 Sida cordifolia *(bast)*

Sida cordifolia L. ('ilima, flannel weed, bala, country mallow or heart-leaf sida) is a perennial subshrub of the mallow family Malvaceae, native to India. It has naturalized throughout the world, and is considered an invasive weed in Africa, Australia, the southern United States, Hawaiian Islands, New Guinea and French Polynesia. The specific name, *cordifolia*, refers to the heart-shaped leaf. *Sida* (broom plant), origin from Mexico and tropical areas, Southeast Asia and Indonesia (Wikipedia).

S. cordifolia is an erect perennial that reaches 50–200 cm (20–79 in) tall, with the entire plant covered with soft white felt-like hair that is responsible for one of its common names, 'flannel weed'. The stems are yellow-green, hairy, long and slender. The yellow-green leaves are oblong-ovate, covered with hairs and 3.5–7.5 cm (1.4–3.0 in) long by 2.5–6 cm (0.98–2.36 in) wide. The flowers are dark yellow, sometimes with a darker orange centre, with a hairy 5-lobed calyx and 5-lobed corolla (Wikipedia).

The fibre is removed from the stalk by water and dew retting. The fibres are similar to that of genus *Hibiscus*. Chemical composition: hemicellulose 17.6%, lignin 18.0 % and cellulose 69.5%, where alfa cellulose 68%, beta-cellulose 17% and gamma cellulose 15%. Strength of this fibre is about half of the sisal strength and 1/3 of abaca; application – cordage and sacking (Mauersberger, 1954).

Physicochemical, tensile and thermal characterization of new natural cellulosic fibres from the stems of *S. cordifolia* are described by P. Manimaran and the team of the Department of Mechanical Engineering, Kamaraj College of Engineering and Technology: Virudhunagar, Tamil Nadu, India.

This investigation deals with the newly identified *S. cordifolia* fibres characterized by chemical analysis, single fibre tensile test, TGA/DTG analysis (Thermogravimetric Analysis/Differential thermal Analysis), FTIR analysis, X-ray diffraction (XRD) and atomic force microscopy. The chemical constituents of *sida cordifolia* fibres are cellulose (69.52%), hemicellulose (17.63%) and lignin (18.02. %). The *sida cordifolia* fibres are thermally stable up to a temperature of 338.2°C evidenced by TGA analysis. The XRD confirmed that these fibres were rich in cellulose fraction with a CI of 56.92% (Manimaran et al., 2018a,b).

20.2.2.9 Cytisus scoparius *(bast)*

The common broom, Scotch broom, Spanish broom, is the bast fibre from the stalk and principally grows in South of Europe, which is utilized for making cordage (Mauersberger, 1954).

The history and perspective of Spanish broom *(Spartium junceum* L.) are presented in the survey prepared by Drago Katović and the team in the *Journal of Natural Fibers*

The survey is summarized as follows: the Spanish broom (*S. junceum* L.) plant is almost a forgotten textile raw material. The chapter describes procedures practised in the past and nowadays for obtaining fibres from this plant. New discoveries about its use as a component of special composite fibres as well as its probable benefits for the households of poorer limestone areas were made. Some chemical and physical properties of the fibres extracted from local plants are also reported (Katović et al., 2011).

20.2.2.10 Nettle bast fibre

It is obtained from three species of the stinging nettle: *Urtica dioica* L., *Urtica urens* L. and *Urtica pilulifera* L. (family Urticaceae). *Urtica dioica* is a perennial plant and the other two species are annual plants. Nettle is known as 'Swedish hemp' and it was cultivated in Sweden for the production of fibre used for making of cordage and sail cloth. It was also known in Germany and Russia. Italy and France used this in middle age. Cultivation of nettle is similar to cultivation of flax. Fibre extraction is usually performed by dew and water retting and then mechanical decortication (Mauersberger, 1954).

After retting, the obtained fibre is hackled and combed and then oiled to soften it. The fibre contains high cellulose level with traces of lignin on the surface (Mauersberger, 1954).

Tensile behaviour of nettle fibre composites exposed to various environments is described by Pramendra Kumar Bajpaiet and his team in the *Journal of Natural Fibers*.

Recently, natural fibre-reinforced biocomposites have received widespread attention because of their application in automotive, construction, industrial and other commercial markets. Their lightweight characteristics have direct positive impact on the fuel economy in automotive applications. The mechanical behaviour of natural fibre–based composites is affected by the different environmental conditions. Observing the application spectrum, there is an imminent need to investigate the effect of different operating environments on mechanical behaviour of biocomposites. The objective of the present research endeavour is to characterize the tensile behaviour of the nettle fibre–reinforced polypropylene composites when exposed to river water, diesel oil, freezing conditions, sunlight (UV) and soil for different exposure times (64–512 h). The experimental results show that the tensile strength of the developed biocomposite is affected by varying magnitudes when subjected to selected conditions for different exposure times (Bajpai et al., 2013).

An experimental investigation on mechanical and tribological properties of Himalayan nettle fibre composite is presented in the *Journal of Natural Fibers* by Mayank Pokhriyal and the team from the Department of Mechanical Engineering, G.B. Pant Engineering College, Pauri Garhwal Uttarakhand, India.

The main focus of the presented work was to explore mechanical and tribological properties of Himalayan nettle fibre and unsaturated polyester resin (GP) using hand layup method. Four composite laminates were prepared by adding the Himalayan nettle fibres by weight percentage (5%–20%). The various properties (e.g., tensile strength, hardness, impact strength and abrasion wear) were calculated for the composite laminates. It was found that the significant variation in properties was observed when fibre addition was in the range 15%. The change in the properties was negligible when fibre addition was increased from 15% by weight (Pokhriyal et al., 2018).

20.2.2.11 Hibiscus abelmoschus

Some botanical authorities classify this as *Abelmoschus esculentus* (okra). This is grown in tropical and subtropical areas and the plant is cultivated mainly for fruit (Mauersberger, 1954).

The structure and property of *Abelmoschus manihot* fibres were investigated by Chinese scientists and presented in the *Journal of Natural Fibers*.

Its transverse section was irregular polygon and the single fibre was cylindrical in shape. It belonged to cellulose I crystalline structure and its crystallinity was near to that of cotton and ramie. The breaking strength was lower than ramie fibres and the elongation at break was approximate with these fibres. The moisture regain was better than cotton. It can be dyed easily using high-temperature reactive red dyes and blue dyes. In addition, it had the higher thermal stability than cotton (Qu et al., 2016).

20.2.2.12 Okra fibre

The synthesis and applications of functional polymers from natural okra fibres for removal of Cu(II) ions from aqueous solution were investigated and described in the *Journal of Natural Fibers* by Amar Sigh Singha and Ashish Guleria from the Department of Chemistry, National Institute of Technology, Hamirpur, Himachal Pradesh, India.

Functional groups ($-COOCH_3$ and $-NH_2$) in natural okra fibre had been introduced by free-radical graft polymerization of methyl methacrylate (MMA) and acrylamide (AAm) from their binary solution using ascorbic acid/H_2O_2 as a redox initiator at a temperature of 45°C for 90 min. Factors affecting the grafting of lignocellulosic fibre such as feed molarity and comonomer composition were investigated. The grafted copolymers were characterized by FTIR, scanning electron microscopy EDX (SEM-EDX) and TGA. Functional groups were tested for their potential application in the removal of copper ions from aqueous solutions. Batch experiments were performed to evaluate the adsorption capacity of the maximum grafted cellulosic fibres towards the Cu(II) ions. Pseudo-first-order and pseudo-second-order kinetic models were studied to analyse the Cu^{2+} adsorption process, and results showed that the pseudo-second-order model was more suitable than the pseudo-first-order model. Furthermore, Langmuir, Freundlich, and Tempkin models were also applied to describe the adsorption isotherm of Cu^{2+} by different fibre adsorbents. The results indicate that the Langmuir model fitted the adsorption equilibrium data better than other two models. The maximum monolayer capacity qm was calculated using the Langmuir isotherm for raw, and MMA + AAm grafted fibre were 4.89 and 33.05 mg/g, respectively (Singha and Guleria, 2015).

20.2.2.13 Roselle (Hibiscus sabdariffa)

It is called roselle 'hemp', rosella and Java 'jute', probably native to West Africa, and grown in India, East Pakistan, Southeast Asia (e.g., in Egypt), Ceylon, Indonesia, Philippine Islands and West Indies. It has been grown on a small experimental scale in the United States too. The plant has general appearance similar to that of kenaf

(*Hibiscus cannabinus*); stalks and leaves of roselle are dark green to red and flowers are creamy white to yellow. Roselle is a perennial plant and in favourable climate, it is able to produce two crops annually. In most areas, including Indonesia, the fibre is removed from the stalk by retting, then stripping or beating to free fibre, and washing it in a manner similar to that used for jute. In Ceylon, bark with fibre adhering to it is stripped from the stalks in the field immediately after these are cut. These strips are then retted; after the retting is complete, the fibre is freed from the bark and shive by washing. This procedure requires less retting time and less expenditure of labour than if the whole stalk is retted. In some areas in West Indies, mechanical decortication is used to a very limited extend to remove fibre; considerable experimental work was being carried out in the US as well as in other countries to improve mechanical decorticators. Annual yield of fibre ranges from 600 to 2000 lb per acre (110−367 kg/ha), with the average ranging between 800 and 1200 lb (147−220 kg/ha) in most areas (Mauersberger, 1954).

Carefully prepared roselle fibre is creamy to silvery white, lustrous, and with moderately good strength. The book *Mathews' Textile Fibres* presented the table with the comparison of the breaking strength, as well as other selected physical factors, of roselle with that of kenaf, urena and jute. Length of the fibre strands is 3−5 ft (90−150 cm). Mendiola reported that the length of ultimate fibres ranged from 0.05 to 0.13 in. (1.27−3.3 mm) and averages 0.07 in. (1.78 mm) and that diameter ranges from 10 to 32 μ, with an average of 20 μ (Mauersberger, 1954).

Roselle fibres are used among others for the production of bast fibre (applied e.g. in diversified composites) and as an infusion, in which it may be known as carcadè (*Hibiscus* healthy tea) (Wikipedia).

A review of roselle fibre and its composites was prepared by R. Nadlene and other experts from the Department of Mechanical and Manufacturing Engineering, Universiti Putra Malaysia, Serdang, Selangor and Department of Material and Structure, Universiti Teknikal, Melaka, Malaysia, published in the in the *Journal of Natural Fibers*. Their achievements are the plant's history, physical description, cultural methods, and usages were explained to further understand this potential plant. Based on a literature review, it is clear that extensive research has been performed on the fibre extraction methods, morphology, properties and possible surface treatments of roselle to enhance its properties in the manufacturing of natural fibre−reinforced polymer composites; however, no review has covered the properties and applications of roselle fibres in detail. Therefore, a review was performed on roselle to determine its potential to enable a better selection of materials to produce composites for potential applications in the construction and building, automotive and aerospace industries (Nadlene et al., 2016).

20.2.2.14 Grewia occidentalis *(Kaffir 'hemp')*

It is grown in South Africa. The fibre removed from the stalk by retting is strong and light coloured. It is used for cordage and clothing fabrics (Mauersberger, 1954).

Preparation of unidirectional *Grewia serrulata* fibre – reinforced polyester composites and evaluation of tensile and flexural properties was studied and described by G.T. Mahesha, published in the *Journal of Natural Fibers*.

Partially biodegradable composites were prepared using bast fibres extracted from *G. serrulata* trees with unsaturated polyester resin. Different durations of fibre pretreatment and volume fraction variations were tried for fabricating the composites. Hand layup technique was used for the fabrication; tensile strength and flexural strength of the specimens were tested following ASTM standards. It is found that the mechanical properties of *G. serrulata* bast fibre–reinforced unsaturated polyester composites are superior in comparison with relevant properties of unsaturated polyester resin. Such natural fibre–reinforced composites are adaptable in applications such as automotive interiors, partition walls and false ceilings and ensure easy material disposal ways at the end of service life as raw materials used are mostly biodegradable (Mahesha et al., 2016).

Mechanical characterization and water ageing behaviour studies of *G. serrulata* bast fibre–reinforced thermoset composites was evaluated and presented by G. T. Mahesha, Shenoy B. Satish, and the team from the Department of Aeronautical & Automobile Engineering, Manipal Institute of Technology, Manipal University, Manipal, India, published in the *Journal of Natural Fibers*.

Natural fibre–reinforced polyester composites were prepared using bast fibres as the reinforcement, which were extracted from *G. serrulata* trees. Chemical treatments such as alkali pretreatment followed by permanganate treatment, acetylation and silane treatment were exclusively applied to modify the fibre surface. Hand layup technique with compression moulding was adapted for the fabrication of axially oriented fibre-reinforced composites. Tensile, flexural and impact strength properties of the specimens prepared were evaluated following the standard procedures. The SEM images of the fractured surfaces show improvement in compatibility between the chemically treated fibres and the matrix. It was found that the tensile and flexural strengths of chemically treated fibre–reinforced polyester composites have significantly improved when compared with untreated fibre–reinforced polyester specimens. The properties of 1200 h water aged specimens were found deteriorated considerably due to diffusion of water into the composite material system. The chemical treatment of fibres before fabrication of composites improves the resistance to water absorption tendencies (Mahesha et al., 2017).

Effect of *Grewia optiva/Bauhinia vahlii* fibres on physicomechanical and dry sliding wear behaviour of epoxy composites were investigated and described by Kumar, Sandeep and some other experts.

The physical, mechanical and sliding wear properties of *G. optiva* and hybrid *G. optiva/B. vahlii* fibres–reinforced epoxy composites were investigated. Results reveals that 6 wt.% hybrid *G. optiva/B. vahlii* fibre added composites resulted in highest physical (void content, water absorption) and mechanical (impact energy, hardness, tensile and flexural strength) properties. An L27 orthogonal array of Taguchi method was implemented to extract the optimal level of control factors. The optimal combination of control factors for achieving lowest wear rate was obtained as 4 wt.% fibre content, 2.5 m/s sliding velocity, 15 N normal load and 2000 m sliding distance (Kumar et al., 2019).

20.2.2.15 *Kydia calycina*

K. calycina plant – familia: Malvaceae, subfamilia: Malvoideae, tribus: *Hibisceae*, genus: *Kydia*, species: *calycina* (Wikipedia).

Extraction of unconventional bast fibres from *K. calycina* plant and their characterization were evaluated and described in the *Journal of Natural Fibers* by Tayyaba Fatma and Shahnaz Jahan.

The *K. calycina* fibres were screened out from various plant sources through the water retting method and then the time duration for extraction of fibres was optimized. After optimization, the extracted *K. calycina* fibres were compared with cotton, flax and jute fibres on the basis of physical and mechanical properties. The morphology of extracted fibres was also studied through SEM. The results showed that the *K. calycina* plant produces better-quality bast fibres with respect to plants such as jute and flax (Fatma and Jahan, 2018).

20.2.2.16 *Epipremnum aureum*

This is a species of flowering plant in the family of Araceae, native to Mo'orea. The species is a popular houseplant in temperate regions, but has also become naturalized in tropical and subtropical forests worldwide, including northern Australia, Southeast Asia, South Asia, the Pacific Islands and the West Indies, where it has caused severe ecological damage in some cases. The plant has a multitude of common names including golden pothos, Ceylon creeper, hunter's robe, ivy arum, money plant, silver vine, Solomon Islands ivy and taro vine. It is also called devil's vine or devil's ivy because it is almost impossible to kill and it stays green even when kept in the dark. It is sometimes mistakenly labelled as a Philodendron in plant stores. It is commonly known as money plant in many parts of the Indian subcontinent. It rarely flowers without artificial hormone supplements; the last known spontaneous flowering was reported in 1964. Evergreen vines are 20 m high with diameter up to 4 cm and are a source of bast fibre (Wikipedia).

Characterization of natural cellulosic fibre from *E. aureum* stem was presented by M. V. Maheshwaran and the team of Department of Mechanical Engineering, P.S.R. Engineering College, Sivakasi, Tamil Nadu, India, in the *Journal of Natural Fibers*.

The natural fibre *E. aureum* was extracted from its plant. *E. aureum* fibres (EAFs) were investigated by chemical analysis, FTIR, XRD, TGA, SEM, and single fibre tensile test. Chemical analysis, FTIR, and X-ray analysis evidenced that these fibres has 66.34% cellulose content with CI of 49.33%. The TGA reveals that EAFs can thermally withstand temperatures until 328.9°C. The morphology of the EAFs was observed by scanning electron microscope. It was established that the fibre can be utilized as reinforcement in polymer composites (Maheshwaran et al., 2018).

20.2.2.17 *Apocynum venetum*

Order: Gentianales, family: Apocynaceae, genus: *Apocynum,* species: *venetum.*

A. venetum, commonly known as sword-leaf dogbane, is a plant species in the dogbane family; it is poisonous but used as a source of fibre, medicine and nectar for production of honey. *A. venetum* leaves have been used in the traditional medicine for hypertension treatment. *A. venetum* fibres can be extracted from the *A. venetum* bast; these fibres possess the lustre of silk, smoothness of ramie, malleability of cashmere and the softness of cotton. *A. venetum* is considered to be native to a wide range in northern Asia and SE Europe (Wikipedia).

Investigation of the degradation of biorecalcitrance in *A. venetum* fibre biodegumming was evaluated and described by Jixian Gong and the team from China, published in the *Journal of Natural Fibers*.

Traditional degumming of bast fibres results in problems such as environmental pollution and inferior fibre quality. For these reasons, biodegumming was considered as a potential process for bast fibres. In biodegumming process, decomposition of noncellulosic components was the pivotal problem. The bast fibre of *A. venetum* was processed with combinatorial processing, containing physical, chemical and microbial effect. The structural biorecalcitrance was analysed at molecular, supermolecular and morphological level in each stage. The results show that main substance in recalcitrance of *A. venetum* bast is lignin and hemicellulose. And the protein and wax is also the constituent of biorecalcitrance. The recalcitrance in the bast tissue can be divided into three layers. The spatial structure and the distribution of the noncellulosic substances also contribute to the recalcitrance. This material provides directions for pretreatment design to render biomass more amenable to bioprocessing (Gong et al., 2019).

20.2.2.18 Kudzu

Kudzu, also called Japanese arrowroot, is a group of plants in the genus *Pueraria*, in the pea family Fabaceae, subfamily Faboideae. Native to much of Eastern Asia, Southeast Asia, and some Pacific islands. It is also called Kudzu 'hemp' (Mauersberger, 1954).

Performance analysis of kudzu fibre prepared by using combined steam explosion and chemical degumming was studied by Yan Song and other experts from China, presented in the *Journal of Natural Fibers*.

Kudzu is cosmopolitan, less costly, and is used as a source of fibre, food, medicine and fuel. However, it is not well-utilized as a kind of fibre because of the degumming problem. To take full advantage of the kudzu resource in the world and expand the application field of kudzu fibre, this study deals with preparation of kudzu fibre by combined steam explosion and chemical degumming. Kudzu fibres are characterized by SEM, FTIR and XRD. It is found that kudzu fibre after combined degumming has a clean surface with most nonfibrous composition removed and fibres dispersed; the cellulose I structure of kudzu cellulose did not change after treatment, but its crystallinity increased. These results suggest this combined degumming is feasible for preparing kudzu fibre. It will probably provide theoretical basis for subsequent utilization of kudzu fibres (Song et al., 2017).

20.2.2.19 *Salago,* Wikstroemia *spp.*

Salago, belonging to family Thymelaeceae, is a sturdy slow-growing shrub in primary and secondary forest throughout the Philippines at low and medium altitudes.

The inner bark is light-coloured and has a silky appearance and produces long and strong fibres. The fibre is used as an excellent material for the manufacture of currency paper, banknotes, check, documentary papers, stencils, handmade papers for art purposes, calligraphy papers and other paper materials where a certain degree of permanence, strength and durability is desired. Likewise, it is used for making rope, fishing lines and nets, clotheslines, sashes, strainers, wallets, colourful hats and other raw materials in making Japanese sliding doors (shoj), kimono and components for radio and microcomputers (http://www.mixph.com/techno-guide-on-growing-salago-fibre/).

Characterization of untreated and alkaline-treated salago fibres (genus *Wikstroemia* spp.) was revealed in the *Journal of Natural Fibers* by Mohammadhosein Pouriman and the team of the School of Chemical Engineering and Chemistry, Mapúa Institute of Technology, Intramuros, Manila, the Philippines.

Salago fibre (Genus *Wikstroemia* spp.) is a bast natural fibre native of the Philippines. This fibre is found all over this country and has various applications, e.g., as handmade paper, currency paper or banknotes. In this study, untreated and 5 wt% NaOH-treated salago fibres were characterized chemically, physically, thermally and morphologically for the first time. It was found that the treatment increases density and improves the mechanical properties of the fibre. FTIR analysis revealed stretching O—H and C—H alkane groups at 3330 cm^{-1} and 2918 cm^{-1}, respectively, and confirmed lignin reduction for the treated fibre. TGA showed that treated fibres possessed greater thermal stability. Scanning electron micrographs used to measure fibre diameters showed the rugged surfaces of untreated fibres and the glossier but more distorted surfaces of the treated ones (Pouriman et al., 2018).

20.2.3 Leaf fibres

Leaf fibres were obtained from the leaves of monocotyledonous plants. The fibres are in bundles. Leaf fibres are often named hard fibres. The leaves are utilized commercially as well as the full length bundles, obtained from these leaves. Main applications are various types of cordage and twines and can also be used in manufacturing woven fabrics. Fibres such as abacá, sisal and henequen, which occupy more than 80% of commercially produced leaf fibres, still have high commercial value. To this group belong, in addition to those mentioned before, *cantala,* palma *(Palma istle),* fique *(Furcraea macrophylla), Formium,* caroa, letona, Mauritius, cabuya *(Furcraea andina)* and other leaf fibres (Mauersberger, 1954).

To abacá is devoted a special chapter, Chapter 8. Abaca. *Musa* is one of two or three genera in the family Musaceae; it includes bananas and plantains. Around 70 species of

Musa are known, with a broad variety of uses. Although they grow as high as trees, banana and plantain plants are not woody and their apparent 'stem' is made up of the bases of the huge leaf stalks. Thus, they are technically gigantic herbs (Wikipedia).

20.2.3.1 Sisal

Among leaf fibres, an important role is played by sisal fibres, which can be obtained from the plant *Agave sisalana,* to which belong several varieties. Sisal plants have short, thick stems or trunks. Main producers of sisal are Brazil, Haiti, Venezuela, Indonesia, Kenya, Madagascar and Mozambique. Application of sisal fibres is mainly for ropes and twins for agriculture, marine, wrapping and other industrial purposes. Sisal grows best in warm, moist climate and well-drained soil. Fibre from the leaf is extracted by machine decortication. After decortication, fibres are washed and dried mainly in sun. Commercially it is utilized in the form of strands. The fibres have a very good strength and elongation properties (Wikipedia).

20.2.3.2 Henequen

This fibre is obtained from the leaves of *Agave fourcroydes* Lem. (native of Mexico, mainly the state Yucatan and also cultivated in Cuba). Henequen is cultivated in the system similar to sisal. Leaves are cut in the manner similar to this of sisal. First, the henequen leaves are harvested when the plant is 3–4 years old, and can be continued up to 15–16 years old plants. For extraction of fibre, decortication machinery is used. After decortication, the fibres are washed, dried in sun and brushed. Utilization of this fibre is in form of strands which are composed of bundles of individual fibres (Mauerberger, 1954).

Another leaf fibre from this family is *cantala,* obtained from *Agave cantala*, growing on Philippines, Indonesia, India and Mexico. Usage is the same as that of sisal and henequen (Mauerberger, 1954).

20.2.3.3 Arenga pinnata (syn. Arenga saccharifera)

It is an economically important feather palm native to tropical Asia, from eastern India to East Malaysia, Indonesia and the Philippines in the east. Common names include sugar palm, areng palm (also aren palm or arengga palm), black sugar palm and kaong palm, among other names (Wikipedia).

Toughness characteristic of *Arenga pinnata* fibre concrete was revealed by Hashim, Abdul Razak and T. Ferdiansyah, published in the *Journal of Natural Fibers.*

The source informed about the results of an investigation on the toughness characteristics of *A. pinnata* fibre. Lengths of the tested fibres are 15 mm, 25 mm and 35 mm, used in two volume fractions, i.e., 0.6% and 0.8%. The values of toughness, first crack deflection, first crack toughness, toughness indices and residual strength factors are presented for ages up to 180 days. The addition of *A. pinnata* fibre to concrete increased the toughness characteristic of concrete. This fibre is recommended to use in production in the cheap composite for building purposes (Razak and Ferdiansyah, 2005).

20.2.3.4 Palm (*Trachycarpus fortunei*)

The exploration on UV-blocking performance of lignin from palm (*Trachycarpus fortunei*) was investigated by Chinese researchers Zhang Tonghua, Wang, Yanlai and the team published in the *Journal of Natural Fibers*, and described as follows: to reveal the ultraviolet (UV)-blocking mechanism of palm fibre, alkali lignin was extracted from palm and bamboo fibre and its UV absorption spectrum was tested, followed by some structural characterization. Palm fibre lignin was found to have a higher UV absorbance. FTIR analysis revealed that the UV-absorbing functional groups in palm fibres include phenol OH, aliphatic OH and aromatic ring and triple bond structure. TGA and elemental analysis indicated high protein content and highly concentrated aromatic structures also contribute to the superior UV-absorption properties of palm fibres (Wang et al., 2019).

20.2.3.5 Borassus

Order: Arecales, family: Arecaceae, genus: *Borassus*.

Borassus flabellifer L., commonly known as doub palm, palmyra palm, tala palm, toddy palm, wine palm or ice apple, is native to the Indian subcontinent and Southeast Asia, including Nepal, India, Bangladesh, Sri Lanka, Cambodia, Laos, Burma, Thailand, Vietnam, Malaysia, Indonesia and the Philippines. It is reportedly naturalized in Pakistan, Socotra and parts of China. The fibre is used for cordage, fishing nets and coarse, and also in producing brushes (Wikipedia).

20.2.3.6 Bromelia

Family: Bromeliaceae, origin: Mexico, Argentina and Paraguay, Central America. Application: for belts, cordage, hammocks and course.

There are other leaf fibres such as from palma (*Samuela carmerosana*, Mexico); application: for cordage and bagging fabrics, and also for brush; *Yucca* species (Liliaceae family); *Nolina* spp.; *Hesperaloe funifera;* Palmaceae family including several members, which are utilized for textiles (Mauersberger, 1954).

20.2.3.7 Caroa fibre

It is obtained from the leaves of the plant *Neoglazovia variegate* (Bromoliaceae family) and grows mainly in Brazil. It is used for making cords, ropes, twine, paper, rayon, burlap in mixture with other fibres and cheap shoes (Mauersberger, 1954).

20.2.3.8 Pineapple fibre from Ananas comosus

Pineapple fibre is obtained from leaves of the plant *A. comosus* (L.) Merr., which produces well-known pineapple fruit. The name pineapple derives from the Spanish word piña, meaning cone-shaped. It belongs to Bromeliaceae family, indigenous to Brazil and Mexico. The leaves are 3–5 feet long. The fibre is removed from the leaves by combination of water retting and strapping, and then fibre is washed and dried. Yield of fibre is 2.5%–3.5% weight of green leaves. Main application is manufacturing twines, threads and fibre. In the Philippines, thin fabrics are produced, known as piña cloth (Mauersberger, 1954).

Research on the main constituents of *A. comosus* (pineapple) leaf fibre and its photooxidative degradation was conducted by M.D. Shamsul Alam and the team, from the Department of Applied Chemistry and Chemical Technology, Islamic University, Kushtia, Bangladesh, published in the *Journal of Natural Fibers*.

An estimation of main constituents of pineapple leaf fibre (PALF) by conventional method was attempted and the values of 74.44% α-cellulose, 13.39% hemicellulose, 7.12% lignin, 2.89% pectic matter and 0.58% aqueous extract were obtained. So, the main constituents of PALF are α-cellulose, hemicelluloses and lignin. The molecular weight of raw, bleached PALF and α-cellulose of PALF were 1.517 × 105, 1.303 × 105 and 1.225 × 105, respectively. Bleached PALF was dyed with two dyes, Direct Green 27 and Acid Orange 52. Maximum dye exhaustion was found to be 98.97% for Direct Green 27 under the conditions of 3% dye concentration, 5% electrolyte concentration and 70°C temperature, and 93.27% dye exhaustion was found for Acid Orange 52, where dyeing conditions were 4% dye concentration, 6% electrolyte concentration and 80°C temperature. Dyeing time was fixed for 60 min for both dyes. The colour fastness of PALF on exposure to sunlight has also been studied. It was observed that colour fastness of raw fibre is higher than bleached fibre. The degradation of PALF under sunlight was studied by tensile strength and molecular weight measurements. Bleached fibre exhibited lesser loss of tensile strength and molecular weight than raw fibre. Infrared spectra (IR) of PALF were measured for different exposure times under sunlight. The PALF containing higher α-cellulose has good characteristics, similar to materials of higher molecular weight. For this, it gives high tensile strength, more dyeability, and better colour fastness properties. Estimation of main constituents of *A. comosus* (pineapple) leaf fibre and its photooxidative degradation (Shamsul Alam et al., 2009).

Comparison of different methods to produce pineapple leaf fibres with steam explosion was described by Song, Yan and four experts from China, published in the *Journal of Natural Fibers*.

Pineapple leaf fibres are abundant but not completely utilized because of the production methods. In this research, the steam explosion process combined with traditional methods including steam explosion after retting, retting after steam explosion, steam explosion after scraping and hand-scraping only were employed, and they were characterized by XRD, SEM, gel permeation chromatography, FTIR and chemical composition and mechanical test. Results showed that the fibres obtained by steam explosion after scraping method contain the highest percentage of cellulose, lowest percentage of hemicellulose and lignin, also the best crystallinity and breaking tenacity with reasonable degree of polymerization (Song et al., 2019).

Characterization and comparative analysis on mechanical and acoustical properties of camellia sinensis/Ananas comosus/glass fibre hybrid polymer composites were presented by S. Gokulkumar and the team of researchers from India.

During last few decades, the hybrid composites emerged from natural fibre has brought incontestable advantages over synthetic materials. The environmental impact such as toxicity, nonbiodegradable and noise pollution has led to the identification of new class of fibres and also subsequent failure in critical stiffness and weight sensitivities of fibre-reinforced polymers has initiated magnification of mechanical toughness. In this present paper, fibre-reinforced composite materials are made with *Camellia sinensis – A. comosus* hybridized by glass fibre and epoxy resin matrix. The amalgamated laminates were fabricated by a compression moulding process with the composition of 40% wt. of reinforcing fibre and 60% wt. of epoxy matrix in a constant proportion. A hydrophilic content present inside natural fibres are alkalized by 5% of NaOH to modify as hydrophobic nature for intensifying bonding between fibre and matrix. These hybridized fibres are mechanically, chemically, morphologically and acoustically analysed to meet the requirements of mechanical and sound proofing properties. The results shows that 25% wt fraction of *Camellia sinensis* amalgamate with unidirectional fibre shows better acoustical absorption coefficient of 0.46 at 2000–6300 Hz frequency range and compared with existing material to establish the potential of using it in sound proofing acoustical applications (Gokulkumar et al. 2019).

20.2.3.9 Banana leaf fibre

These plants are perennial belonging to the family Musaceae and include large number of herbaceous plants. The name of the genus comes from the Antonio Musa, a physician of Octavius Augustus, first Romanian emperor (63–14 BC). *Musa paradisiaca* L. – in botanical order, it is *Iridales* belonging to the family Musaceae.

M. paradisiaca L., common name being banana fibre, grow in Far East and also in western hemisphere (Hawaiian Islands, Polynesia, etc.). The fibre from this plant is used by the natives for textile. Harvesting and fibre removal are carried out in a way similar to hand-stripping abaca or by primitive equipment. The clean fibre has high level of cellulose (more than 80%) (Mauersberger, 1954).

Natural fibres from plantain pseudostem (*M. paradisiaca*) for use in fibre-reinforced composites were studied and published in the *Journal of Natural Fibers* by Edith M. Cadena and the team of researchers.

Agricultural crops from plantain produce a significant amount of wastes and they are currently considered worthless. Accordingly, in this study, nonwood fibres from pseudostem of plantain plants were extracted through mechanical processing to be used as reinforcing material in polyester composites. Bio-based composites were obtained using a 4% wt. of lignocellulosic reinforcement and were prepared after the fibres underwent alkylation and acetylation treatments to enhance the compatibility of organic loads with the polyester matrix. The higher cellulose content of plantain fibres indicates that they can be used to reinforce composites with a polymeric matrix. The plantain fibres have bast fibre bundle of around 120 μm; single fibres of around 5 μm; and mesofibres with a diameter between 0.5 and 1 μm. The results showed that plantain fibres can be used as a filler material to obtain an alternative polymer composite. The flexural strength of composites (polyester with acetylated plantain fibres) was improved to 28% when the properties are compared with control composite (Cadena et al., 2017).

Evaluation of structural, tensile and thermal properties of banana fibres was studied and revealed by Raghavendra Subramanya, and the team from the Department of Mechanical Engineering, Acharya Institute of Technology, Bengaluru, Karnataka, India, published in the *Journal of Natural Fibers*.

Recognizing the importance of properties of the plant fibres for their use in polymeric composites and unavailability of data on tensile, thermal and other properties of banana fibres of Karnataka (India), a study was undertaken with the objective of determining some of these properties. This article presents determination of structural aspects of the fibres by FTIR and XRD techniques; obtaining stress-strain curves and tensile properties by tensile testing; finally, thermal properties by differential scanning calorimetry and TGA. FTIR spectrum of banana fibres showed aromatic character, whereas XRD results indicated the fibre to have mainly the cellulose I structure and CI of 52%. Thermal-degradation details of fibre samples were revealed by thermal studies. Stress-strain curves of banana fibres suggested their brittle nature with moderate values of tensile strength, but low percentage elongation. Weibull analysis of obtained tensile strength values revealed variation of 'characteristic strength' values from 3800 MPa at 99% reliability to 22,700 MPa with 0.01% reliability. Morphology studies revealed the number of defects along the length of the fibre, whereas fractured surface exhibited flat surface with intracellular fractures clearly indicating brittle nature of the fibre (Subramanya et al., 2017).

Banana pseudostem characterization and its fibre property evaluation on physical and bioextraction were revealed by J. S. Jayaprabha, and the team from the Environmental Technology Section, National Institute for Interdisciplinary Science and Technology, Council of Scientific and Industrial Research, Thiruvananthapuram, India, published in the *Journal of Natural Fibers*.

Their manuscript presents composition of different layers of banana pseudostem sheath and detailed characteristics of fibres from physical extraction and bioextraction. Cellulose, hemicellulose, lignin, pectin and ash were determined by chemical analysis of pseudostem sheaths. Lignocellulose constitutes about 60%–85% dry weight of the banana pseudostem sheath, in which cellulose accounts for about 50%. Except for the minor changes in hemicellulose, lignin, and ash contents, the physical extraction and bioextraction of fibres are comparable with their chemical and physical properties. Long bioextraction may affect the fibre qualities. SEM studies showed clean and smooth surface of bioextracted fibres unlike in physically extracted ones (Jayaprabha et al., 2011).

20.2.3.10 *Ensete ventricosum*

Commonly known as the Ethiopian banana, Abyssinian banana, false banana, enset or ensete, it is a herbaceous species of flowering plant in the banana family Musaceae. The domesticated form of the plant is only cultivated in Ethiopia, where it provides the staple food for approximately 20 million people. The name *E. ventricosum* was first published in 1948 in the Kew Bulletin, 1947, *p*. 101. Its synonyms include *Musa arnoldiana* De Wild., *Musa ventricosa* Welw. and *Musa ensete* J.F. Gmel. In its wild form, it is native to the eastern edge of the Great African Plateau, extending northwards from South Africa through Mozambique, Zimbabwe, Malawi, Kenya, Uganda and Tanzania to Ethiopia, and west to the Congo, being found in high rainfall forests on mountains, and along forested ravines and streams (Wikipedia).

Mechanical behaviour of *E. ventricosum* fibre under tension loading was tested by C. Mizera and the team from the Department of Mechanical Engineering, Faculty of Engineering, Czech University of Life Sciences Prague, Czech Republic, published in the *Journal of Natural Fibers*.

The mechanical behaviour of *E. ventricosum* fibres, expressed as the dependency of tension force in relation to the elongation, was determined for fibres with gauge length of 50 mm and transformed into general linear mathematical model describing dependency of tension stress and strain. The average rupture force (10.88 $\pm$ 1.11) N, rupture stress (390.33 $\pm$ 21.96) MPa, rupture strain (0.021 $\pm$ 2 $\cdot$ 10−5) and volume energy (3.39 $\pm$ 0.22) J m−3 were also determined at this conducted study. Using image analyses, it was determined that the cross-section of *E. ventricosum* fibre has a circle shape and its dimension was specified. From the study, it implies that *E. ventricosum* fibre, with respect to its mechanical behaviour under tension loading having regard to its density and to the fact that it is environment-friendly, biodegradable and recyclable, could be used as a perspective construction material of the future. Determined model of mechanical behaviour could be applied as background for further research focused on the *Ensete* fibre application (Mizera et al., 2017).

20.2.3.11 Sansevieria *fibre*

Another representative of leaf fibres is *Sansevieria* fibre, obtained from the leaves of Liliaceae family; originated from tropical Africa and India, now also produced in Yucatan (Mexico). Applications of this fibre include cordage, fish net, twines and coarse fabrics for bagging (Mauersberger, 1954).

Sansevieria trifasciata is a common perennial plant which freely grows and widely found in homes, parks and woodlands (Wikipedia).

Morphological, thermal and mechanical characterization of *S. trifasciata* fibres was revealed and published in the *Journal of Natural Fibers* by Samson Rwawiire from the Busitema University, Torro, Uganda and Blanka Tomkova from the University of Liberec, the Czech Republic.

In this research, the morphology was studied using scanning electron microscope and FTIR; thermal properties were studied using TGA and DSC analyses; mechanical behaviour was studied through tensile tests of *S. trifasciata* fibre obtained from Butaleja in Eastern Uganda. Findings show that the fibre has an irregular cross-sectional shape with lumens in the centre; the fibre diameter was between 80 and 120 μm. TGA tests showed that the fibre is stable below 200°C with maximum cellulose decomposition temperature of 315°C. DSC analysis showed that the fibre's crystallization temperature was 310.5°C and lignin decomposition temperature.

This study aims to isolate cellulose nanofibres from locally abundant pomelo fruit. The peels were subjected to alkali treatment (NaOH) with different concentrations and soaking times. Acid hydrolysis was also carried out to obtain an aqueous suspension of nanocellulose. The treated cellulose fibres were characterized by various methods.

Alkali treatment was able to remove noncellulosic constituents. Microscopy techniques yielded morphology of elementary fibres with a compact structure with both micro- and nano-scaled fibres. An improved onset of degradation was likely caused by an increase in crystallinity evidenced by X-ray diffractometry. Acid hydrolysis could successfully individualize cellulose nanofibres as observed by transmission electron microscopy (TEM) (Rwawiire and Tomkova, 2015).

20.2.3.12 Ficus *(peepal tree)*

Extraction and characterization of cellulose from pretreated *Ficus* L. (peepal tree) leaf fibres was revealed by K. Obi Reddy, and the team from the Department of Chemical Engineering Technology, Doornfontein Campus, University of Johannesburg, Johannesburg, South Africa, published in the *Journal of Natural Fibers*.

The authors carried out the physicochemical characterization of the cellulose extracted from the forestry residue of *Ficus* leaves using chemical method. Detailed chemical composition of the *Ficus* leaf fibres (both untreated and pretreated) and extracted cellulose was determined. The chemical analysis confirmed the sequential removal of the lignin and hemicellulose components from the *Ficus* leaves fibres. The morphology of the *Ficus* leaf fibres and extracted cellulose was investigated using SEM. Structural analysis was carried out by Fourier transform infrared and nuclear magnetic resonance spectroscopy. The studies indicated that on chemical treatments, the lignin and hemicellulose were removed extensively from the *Ficus* leaf fibres. XRD studies reveal that the extracted cellulose was more crystalline than the *Ficus* leaf fibres. The thermal stability of *Ficus* leaf fibres and extracted cellulose was investigated by TGA and the extracted cellulose had good thermal stability (Obi Reddy et al., 2016).

20.2.3.13 Agave

There are several plants from the *Agave* genus from which textile fibre is obtained, principally for domestic use from the country where grown (Mauersberger, 1954).

20.2.3.14 Letona

Letona fibres are produced from *Agave letonae* Trel. plant (sometime termed as Salvadorian 'henequen').

Main application of this fibre is production of bags for coffee bean. Others belonging to *Agavae* fibres are *Agave amaniensis* (Blue 'sisal'), *Agave cocui* (Venezuela), *Agave decipiens* (False 'sisal'), *Agave deweyana, Agave falcata, Agave heterocantha, Agave lespinassei, Agave lophantha, Agave lurida* and other six species.

Other leaf fibres are from plants of the genus *Furcraea*, related with *Agave* genus. This plant's origin is from tropical South America, Mexico and the west of India. To this family belongs Mauritius fibre, obtained from the leaves of *Furcraea gigantea*; the stalk of this plant is short, and the plant is cultivated mainly in Mauritius.

The yield of fibres is about 2% -2.5% of the weight of freshly cut leaves. Extraction of fibres is by decortication also after retting of the leaves. Other *Furcraea* fibres are

F. andina (cabuya), Ecuador, which can be harvested when the plant is 3–4 years old, and can be explored until the plants are 10–20 years old. Application of fibre: bags for rice, coffee, cacao, cotton bailing, threads and cordage (Mauersberger, 1954).

20.2.3.15 Agave vera-cruz

The study on characterization of *Agave vera-cruz* Mill. leaf fibre for textile applications – an exploratory investigation was published by M. Kanimozhi and N. Vasugi from the Department of Textiles & Clothing, Avinashilingam University, Coimbatore, India, in the *Journal of Natural Fibers*.

Their work demonstrates the textile potential of *Agave vera-cruz* L. fibre. It deals with the extraction of fibre from *Agave vera-cruz* Mill. leaves, analysing its chemical composition such as cellulose, lignin, wax, ash and moisture content. Physical properties such as fibre length, density and their mechanical behaviour in tensile tests were examined. Scanning electron microscopic study was conducted both in longitudinal and cross-sectional direction to identify the surface topology of the fibre. Infrared spectra and XRD of *Agave vera-cruz* Mill. fibres were recorded to determine the characteristic absorption bands with its degree of crystallinity. The chemical content estimation shows that cellulose (77.43%) was the major constituent present in the fibre. Physical properties reveal that *Agave vera-cruz* Mill. fibre has good prospective in future. SEM image depicts the presence of polygonal voids. FTIR and XRD assessment of the fibre reveals the presence of cellulose Iβ with CI of 60% (Kanimozhi and Vasugi, 2012).

20.2.3.16 Tunisian Agave americana L

Study of the mechanical properties of fibres extracted from Tunisian *A. americana* L. was conducted by Slah Msahli, and the team, published in the *Journal of Natural Fibers*.

A. americana L. fibres, the most abundant variety in Tunisia, have a quite important textile potential. This potential is demonstrated by studying the extraction of these fibres from leaves, their physical properties such us fineness or density and their mechanical behaviour in tensile tests. In this work, results of a mechanical behaviour study of fibres extracted from the *A. americana* L. plant are presented. These results deal with the principal and mechanical characteristics of these fibres which are the strain at break, the elasticity modulus and the rupture facture. These results permitted to situate these fibres, compared with the other textile fibres, as materials that can be used in technical applications such as reinforcing composites or geotextile. To understand the mechanical properties of these fibres, a correlation study between the properties already cited and the fine structure was performed. The obtained results showed that the mechanical properties of *A. americana* L. fibres are closely related to the individual fibres deformations and to the natural matrix (lignin and gums) that are linked to these elementary fibres (Msahli et al., 2015).

20.2.3.17 Cantala

It is a species of *Agave* (*A. cantala* Haw.), family Agavaceae (agaves), a tropical American plant, cultivated chiefly in warm regions of the Old World for its leaf fibres. The fibre of this plant is used for twine, rope and nets (https://www.thefreedictionary.com/cantala).

Cantala grows in the Philippines, Indonesia, India and other parts of Far East; it is native to Mexico. It was introduced to the Philippines by Spaniards in 1783. It is similar to henequen and is used for the same purposes as sisal and henequen, and sometimes for woven fabrics by natives, where the plants are grown. Cultivation and harvesting methods are similar to methods used for sisal and henequen. In the Philippines, water retting as well as mechanical decortication is used for extraction of fibres; often sea water is used for retting. *Cantala* fibre is long and has a fine diameter of fibre strands, similar to henequen (Mauersberger, 1954).

The influence of chemical treatments on *cantala* fibre properties and interfacial bonding of *cantala* fibre/recycled high-density polyethylene (rHDPE) was studied by Wijang Wisnu Raharjo, and the team of Mechanical Engineering Department, Sebelas Maret University, Surakarta, Indonesia; Mechanical Engineering Department, Brawijaya University, Malang, Indonesia, published in the *Journal of Natural Fibers*.

The influence of chemical treatments on the properties of *cantala* fibre as well as on the quality of the interfacial bonding of *cantala* fibre/rHDPE was investigated. The fibres were treated with alkali, silane and a combination of both. The results showed that the loss of hemicellulose and lignin after the alkali treatment, and the presence of a silane layer on the fibre surface after the silane or alkali-silane treatment, improved the thermal stability, surface energy and IFSS. The highest surface energy of 45.37 mN/m was obtained during the alkali treatment (NF12). The alkali-silane treatment with 0.75% wt of silane (NSF075) gave the highest thermal stability and IFSS value (Raharjo et al., 2018).

20.2.3.18 Furcraea *fibres*

These are the plants of the genus *Furcraea,* family Amaryllidaceae and are closely related to those of the *Agave* genus. Generally leaves are retted for 1–2 days before decortication, and then fibre is washed and dried. After drying, it may be brushed to soften and add lustre (procedure applied especially in case of Mauritius fibre obtained from leaves of the *F. gigantea*).

F. andina (cabuya), Ecuador, leaves are cut by hand and spines cut from edges. Fibre is removed by hand or with the aid of small decorticating machines. Principal uses are for bagging fabrics for rice, coffee, cacao and cotton baling, threads and cordage (Maeuersberger, 1954).

The study on evaluation of *F. andina* fibres as bacterial support for biological denitrification was conducted by J.J. Gallardo-Rodríguez and the team from the Department of Engineering, National University of Chimborazo, Riobamba, Ecuador; and from the Department of Chemical Engineering, Faculty of Engineering, University of Concepción, Bio-Bio, Chile, published in the *Journal of Natural Fibers*.

The aim of the research was to determine the performance of *F. andina* fibres as biofilter support. *F. andina* fibres could be a candidate to be used as support for bacterial immobilization to be applied, among others, in environmental processes such as bioremediation and biofiltration of wastewaters. Adhesion of *Pseudomonas* irreversibly occurred in lower times than inert supports such as polyethylene terephthalate demonstrating the advantages of using natural supports. Nitrate consumption rates obtained were comparable with rates reported in bibliography. A first-order kinetics was found for denitrification in the biofilm because of the diffusion of nitrate into the biofilm (Gallardo-Rodríguez et al., 2016).

20.2.3.19 Bagasse

The study on morphology of bagasse fibres obtained from the elaboration process of mezcal and effects on their tensile properties was evaluated by Carlos Inocencio Cortés-Martínez and Rey Fernando García-Méndez from the Department of Engineering, Instituto Politécnico Nacional, Oaxaca, Mexico, published in the *Journal of Natural Fibers*.

The morphology of the cross-section and longitudinal section of bagasse fibres of *Agave angustifolia* Haw from the elaboration process of mezcal were investigated and tensile tests were performed in function of the diameter (0.20–0.79 mm), gauge length (10, 15 and 20 mm) and strain rate (5–50 mm/min). The cross-section of the fibres is ribbon-like shape with dislocations, and the longitudinal section has mechanical damage in its surface. The ultimate tensile strength (14.83–86.51 MPa) and Young's modulus (0.20–1.26 GPa) are influenced by the diameter and the strain rate, whereas the strain at failure (16%–26%) is influenced by the gauge length. These results are discussed in the light of information on relationship between morphology and tensile properties of natural fibres and possible effects of the elaboration process of mezcal (Cortés-Martínez, García-Méndez, 2017).

20.2.3.20 Sugarcane bagasse

Influence of the length and the content of cellulose fibres obtained from sugarcane bagasse on the mechanical properties of fibre-reinforced mortar composites was studied and described by Montedo, Oscar Rubem; De Pellegrin, Marcelo; Acordi, Juliana, Brazil. Asbestos fibre can be substituted by cellulose fibre extracted from the sugarcane bagasse. The extraction of the fibres occurred using a heated water bath and a subsequent treatment with sodium hydroxide. This work aims to evaluate the effect of the cellulose fibre addition on mechanical properties of fibre-reinforced mortar composites. A 22 factorial design with a central point was used to evaluate the effect of fibre content and fibre length on the mechanical properties. The results demonstrated that the composites increased the flexural strength. The most expressive result was the modulus of elasticity because of the good cellulose fibre/cement interface (De Pellegrin et al., 2019).

Research on low-density polyethylene/sugarcane fibre composites from recycled polymer and treated fibre by steam explosion was conducted by Guilherme Luis Ribeiro, and the team, published in the *Journal of Natural Fibers*.

Low-density polyethylene (LDPE) has caused serious environmental damages, and it has been a challenge for waste management in large cities. The LDPE recycling as composites with natural fibres such as sugarcane fibres has been an interesting alternative to conciliate economic aspects with environmental benefits. In this study, SCF - sugarcane fibres has been treated by steam explosion and incorporated to low-density polyethylene waste (LDPEW) to generate LDPEW/SCF composites with higher fibre/polymer interaction and improved mechanical properties. LDPEW/SCF composites with fibres content until 20wt% showed mechanical properties with performance adequate for nobler applications when compared with the isolated recycled LDPEW (Ribeiro et al., 2019).

20.2.3.21 *Raffia palms (*Raphia*)*

Raphia is a genus of about 20 species of palms native to tropical regions of Africa, and especially Madagascar, with one species (*Raphia taedigera*) also occurring in Central and South America. *R. taedigera* is the source of raffia fibres, which are the veins of the leaves, and this species produces a fruit called 'brazilia pods', 'uxi nuts' or 'uxi pods'. They grow up to 16 m (52.5 ft) tall and are remarkable for their compound pinnate leaves, the longest in the plant kingdom; leaves of *Raphia regalis* up to 25 m (82.38 ft) long and 3 m (9.84 ft) wide are known. The plants are monocarpic, meaning that they flower once and then die after the seeds are mature. Some species have individual stems which die after fruiting, but have a root system which remains alive and sends up new stems which fruit (Wikipedia).

Palmaceae family is used for textile; fibres from the leaves of this palm are used for cordage, fishing nets and coarse fabrics. For commercial purposes, fibres from the most of these genera are used in brushes (Mauersberger, 1954).

R. vinifera grows in the valleys along the Atlantic coast of Central Africa, between Ecuador and 10° N. latitude (Mauersberger, 1954).

Investigation of physical and mechanical properties of *R. vinifera* fibres along the stem was carried out by a team of researchers from Cameroon and France, published in the *Journal of Natural Fibers*.

The study of the physical and mechanical properties of *R. vinifera* fibres along the stem was examined. In Central Africa, these fibres are already being used by craftsmen in the field of textile and decoration. The cross-section of these fibres is bigger at the base of the stem and grows from the periphery towards the centre on a transverse position. By the Archimedes method, the obtained bulk density was between 0.1288 g/cm^3 and 0.2368 g/cm^3, and in a transverse position, it increases from the centre towards the periphery. The dynamic tensile tests were used to estimate the Young modulus that lies among 0.88 and 7.9 GPa and the low values are at the base of the stem. The specific Young's modulus was deduced and varied from 7.33

GPa/(g/cm^3) to 66.00 GPa/(g/cm^3). Finally, through the strain, the microfibril angle was found and it is oscillated from 29.81 degrees to 48.65 degrees (Rodrigue et al., 2017).

20.2.3.22 Mesocarp fibres

The palm oil fruits are pressed, and the kernel is separated from the press cake (mesocarp fibres).

Study of nonlinear mechanical behaviour of oil palm mesocarp fibres was conducted by Suhaiza Hanim Hanipah, and a group of experts from the Department of Process and Food Engineering, Faculty of Engineering, Serdang, Selangor; and Faculty of Chemical Engineering, Universiti Teknologi Mara, Serdang, Selangor, Malaysia, published in the *Journal of Natural Fibers*.

This work investigates the nonlinear mechanical behaviour of oil palm mesocarp fibres (OPMF) using tensile tests, microstructure observation and finite element models. The micrograph images showed the fibre's surface with partly embedded silica bodies, whereas the cross-section contained cell wall structures. Viscoelastic behaviour was observed when the fibres were relaxed over time after being stretched, whereas the stress-strain curves from the cyclic tests indicated permanent set (plastic strain) due to the fibres' deformation. Finite element models were developed comprising single particles (2D and 3D) and 2D multiparticle geometries representing silica bodies embedded in a matrix representing the fibre. The modelling results suggested that silica bodies do not contribute much to the integrity of OPMF, highlighting the need to have a more complex model that considers cellular structures of the fibres and a constitutive relationship of cellulose, hemicelluloses and lignin (Hanipah et al., 2017).

20.2.3.23 Palm front fibres

Optimization of the strength properties of waste oil palm (*Elaeis guineensis*) front fibres – this problem was investigated by a group of researchers from the Universiti Sains Malaysia, Penang, Malaysia, and the Federal Institution of Industrial Research, Lagos in Nigeria, published in the *Journal of Natural Fibers*.

In this study, Box-Behnken design and response surface methodology was applied to study the effects of alkaline peroxide treatment variables (cooking time, sodium hydroxide concentrations, and hydrogen peroxide concentrations) on the pulp and paper strength properties (screened pulp yield, kappa number, tensile index, tear index, burst index), which are considered dependent variables or response variable of the extracted cellulose from oil palm (*E. guineensis*) fronds vascular bundle fibres. The results indicated that at the optimum operating conditions of 2.35% NaOH, 5.00% H_2O_2 and a 53.41-min cooking time, resulting maximum strength properties (tensile index, burst index, tear index) gave 9.92 Nm/g, 6.55 KPam^2/g and 6.22 mNm^2/g, respectively. The study revealed that alkaline peroxide pulped fibres of oil palm frond vascular bundles could be considered a suitable alternative for paper qualities with moderate strength requirements (Wahab Taiwo et al., 2017).

The use of residues from sustainable extraction of heart of palm in agglomerated panels was studied by a team of experts from Pontifícia Universidade Católica do Rio de Janeiro, Brazil, published in the *Journal of Natural Fibers*.

Residues obtained from the sustainable harvest of heart of palm from pejibaye (*Bactris gasipaes*) palms can be managed to produce long and chopped fibres. These fibrous residues can be used to manufacture agglomerated panels and also serve as reinforcement in polymer matrix composites. In this work, the characteristics of pejibaye fibrous residues are commented and the flexural mechanical properties of agglomerated panels made with these residues are determined and compared with the properties of conventional wood-based agglomerated panels. In addition, panels using both pejibaye residues and residues from coconut plantations were manufactured and their flexural mechanical behaviour was determined and compared with panels manufactured using only pejibaye residue. The results obtained show that the flexural mechanical properties of pejibaye agglomerated panels fall within the boundaries established by international standards for interior use (general applications – dry conditions). The hybrid coir/pejibaye panel showed the best set of properties, presenting a performance better than the agglomerated panels manufactured with only one kind of fibre (Quinayá et al., 2016).

Properties of natural fibres from the abaxial side of fireweed (*Gerbera delavayi*) leaf blade for manual spinning were studied by Xiaodan Xu, and a team from China, published in the *Journal of Natural Fibers*.

Natural plant fibres obtained from the abaxial side of fireweed *Chamaenerion* (*G. delavayi*) leaf blade were systematically studied. Fireweed fibres were manually dismantled from hot-bath–treated, refrigerated or untreated leaves and evaluated for different properties. In terms of the collection method, treatment with hot bath exhibited the highest efficiency. As for chemical composition, that of raw (untreated) fireweed fibres was similar to that of cotton, except for wax content that was four times higher in fireweed fibres that in cotton fibres. Fireweed fibre generally exhibited low values of length (6.5 mm), fineness (0.5 dtex), tenacity (0.7 cN/dtex), density (0.9 g/cm^3) and crystallinity (14.6%) but similar elongation at break and moisture regain to those of cotton. Furthermore, fireweed fibres showed a smooth surface, soft texture and water impermeability. After hot-bath treatment, fibres showed 20% lower wax component but higher thermal stability than raw fibres. The small size hindered the fibres from achieving the desirable standard of spinning and weaving by modern machine, but these unique cellulose fibres can be used for manual textile application without any chemical treatment (Xu et al., 2017).

20.2.3.24 Lotus plant

Light transmission by lotus plant stem and leaf tissue: natural textiles for optical fibres for optical fibres was studied by Masaya Goto, and a group of researchers from Japan, published in the *Journal of Natural Fibers*.

Application of plant stem for optical transmitting devices is considered. This research indicates a possibility of application of vascular bundles as natural textiles for optical fibres. Observations of lotus stem tissue were made. Laser beam transmission in lotus stem was confirmed. New functions of photoconduction through lotus stem and confinement of incident light on the leaf surface are discussed.

Lotus leaf surface is superhydrophobic. It is named as 'Lotus effect'. The hydrophobic function comes from fine surface asperity by papillae and thin plant oil. This surface asperity on the leaf consists of transparent tissue that can play the role of optical fibres. The optical fibres system can collect sunlight to effectively perform photosynthesis. Furthermore, light confinement of the surface can be carried out. The surface transparent fibre transmit light to the leaf surface. Lotus may carry out the photosynthesis in June/July during rainy season in Japan (Goto et al., 2017).

20.2.4 Grass fibres

20.2.4.1 Bamboo

Bamboo is a group of woody plants in the true grass family Poaceae (Wikipedia). More knowledge about bamboo can be gained from Chapter 9, Volume 1: Bamboo fibre.

Natural bamboo (*Neosinocalamus affinis keng)* fibre identification using FTIR and 2D −IR correlation spectroscopy was studied by Bailing Sun and a team from the Research Institute of Wood Industry, Chinese Academy of Forestry, Beijing, P.R. China, and Beijing Key lab. of Clothing, P.R. China, published in the *Journal of Natural Fibers*.

The aim of this research was to identify natural bamboo (*N. affinis Keng*) fibre through FTIR, second-derivative IR spectra and two-dimensional infrared correlation spectroscopy (2D-IR). Textile plant fibres, including natural bamboo, jute and flax fibres, were collected and isolated by hydrogen peroxide and glacial acetic acid becoming single fibres to avoid uncontrolled influence. All the chemically treated plant fibres showed generally similar IR spectral profiles. However, different numbers of peaks in the range of 1300−1200/cm and the variation in peak intensity at around 1059/cm were observed for bamboo fibre differentiation. A few differences in their second-derivative IR spectra between bamboo fibre and the other plant fibres also provided information for differentiation. Furthermore, the characteristic features of bamboo fibre were presented in 2D-IR correlation synchronous spectra in the range of 800−1200/cm and 1425−1750/cm, which were obviously distinguished from other investigated fibres. In addition, the results showed that the distinction between bamboo fibre and bamboo pulp fibre could be easily carried out using FTIR and 2D-IR spectroscopy because of their different chemical compositions and crystal lattice types of cellulose (I, II). Therefore, it is demonstrated that FTIR combined with 2D-IR correlation spectroscopy can become a new approach for natural bamboo fibre identification (Sun et al., 2015).

Analysis of mechanical behaviour of bamboo (*Dendrocalamus strictus*) by using finite element method (FEM) was studied by Navin Chand, and a team from India, published in the *Journal of Natural Fibers*.

Tensile strength of bamboo has been experimentally determined parallel and perpendicular to the fibre direction. Different properties are exhibited in two directions in bamboo because of the basic structural difference present in the two directions. Striking differences exist in the distribution of cells within one culm, both horizontally and vertically. Stress and strain values of bamboo under tensile loads are also determined by using the FEM software ABAQUS, and the failure load patterns have been generated and analysed. Flexural strength and deflection in bamboo determined experimentally matches closely with the FEM-generated values (Chand et al. 2008).

20.2.4.2 *Typha latifolia*

T. latifolia belongs to cattail fibres (*Typha C.*). The fibres are dull-grey coloured and are soft and fine with a diameter 15 μ. The fibre is use as a good insulating material for heat and sound. They are very elastic. *Pennisetum alopecuroides* belongs to family Poaceae and genus *Pennisetum.* The Chinese *Pennisetum,* Chinese fountain grass, dwarf fountain grass, foxtail fountain grass or swamp foxtail grass, is a species of perennial grass native to Asia and Australia (Mauersberger, 1954).

Paper properties of *T. latifolia, P. alopecuroides* and *A. americana* fibres and their effect as a substitute for kraft pulp fibres were studied by Hind El Omari, Ahmed Belfkira and François Brouillette from Canada and Morocco, published in the *Journal of Natural Fibers*.

The authors extracted and bleached fibres from three plants (*T. latifolia, A. americana* and *P. alopecuroides*). The obtained pulp contains more than 50% of fines and fibres with lengths not exceeding 1 mm. Generally, the amount of fines depends on fibre source, fibre morphology, pulping conditions and the extent of mechanical treatment. Handsheets were produced with plant fibres alone or mixed (in different percentages: 7.5%, 15%, 22.5% and 30%) with a commercial kraft pulp. The mechanical and optical properties of the sheets (breaking length, tear index, Young's modulus, whiteness and opacity) show that plants could be used as a substitute fibre supply for the paper industry. However, fines were hardly retained in the sheets. A model to explain the removal of fines is proposed (El Omari et al., 2017).

20.2.4.3 *Ampelodesmos*

Ampelodesmos is a genus of Mediterranean plants in the grass family, which is known by the common names stramma, Mauritania grass, rope grass, and diss grass. It is classified in its own tribe *Ampelodesmeae* within the grass subfamily *Pooideae* (Wikipedia).

20.2.4.4 Diss (*Ampelodesmos mauritanicus*)

Chemical and mechanical properties of Diss (*A. mauritanicus*) fibres were studied by M. El H. Bourahli and H. Osmani from Algeria, published in the *Journal of Natural Fibers*.

The natural Diss fibres (*A. mauritanicus*) were treated and characterized in terms of physicochemical performances and mechanical properties. The Diss fibres' main constituents with their weight proportions are cellulose (44.1%), hemicelluloses (27%), lignin (16.80%), extractives (9%) and ashes (3.1%). The measured specific density

(0.89 g/mL) decreases about 5% when alkaline treatment was used. The average elastic modulus, tensile strength and tensile elongation at failure are, respectively, 9.3 Gpa, 149 Mpa and 1.72% for a 20-mm gauge length. The degree of variability in fibre strength results at different gauge lengths was quantified by Weibull statistics. The Weibull modulus decreases from 2.96 to 2.73 and the tensile strength scale parameter decreases from 161 to 105 Mpa as the gauge length increases from 20 to 80 mm. Investigated were the alkali treatment effects on the mechanical properties and on the fibres structure. When treated with 5% NaOH during 2h, the Diss fibres tensile strength increased tremendously by about (37%). Longer processing time did not show any appreciable improvement. However, the tensile strength remains higher than that of untreated fibres (Bourahli and Osmani, 2013).

Uni- and bimodal Weibull distribution for analysing the tensile strength of Diss fibres were studied by M. E. H. Bourahli from the Institute of Optics and Precision Mechanics, University Ferhat Abbes Setif in Algeria, published in the *Journal of Natural Fibers*.

The work considered several statistical distribution functions with the aim to critically analyse the strength data of natural Diss fibres. Besides the Weibull two-parameter distributions, three-parameter and bimodal Weibull were used. Kolmogorov–Smirnov, Anderson-Darling and chi-square goodness-of-fit test were used to judge how accurate the theoretical Weibull probability density function (PDF) fits with tensile strength data.

Tensile test results indicated that the fibres have good tensile strength and Young's modulus. However, their fracture strength exhibits high scatters because of the presence of defects within the fibre and/or on the fibre surface. Microscopic observation revealed that the rupture of the fibres is governed by two types of defects, the rough surface of the fibre formed by several spines and the internal defects such as the defects of connection between fibrils. The statistical analysis of variability in strength showed that the results of the shape and scale parameters calculated using bimodal Weibull distribution give larger values compared with those using the single Weibull distribution. In addition, the strength data points fit well to the bimodal Weibull PDF curve. All three goodness-of-fit tests indicated also that the bimodal Weibull model gives a more accurate prediction of data strength compared with the unimodal Weibull model (Bourahli, 2018).

20.2.4.5 Mendong

Morphology, structure and mechanical properties of natural cellulose fibre from Mendong grass *(Fimbristylis globulosa* Retz.) were studied by Heru Suryanto, Eko Marsyahyo, Yudy Surya Irawan, Rudy Soenoko, published in the *Journal of Natural Fibers*.

This study deals with investigating the morphology, structure, and chemical properties of the Mendong fibres extracted from Mendong grass (*F. globulosa*) in the form of raw and treated fibre by alkali-included chemical content and functional group and to evaluate the strength and properties of Mendong fibres compared with other natural fibres. These studies explore the chemical properties of the fibre including fibre composition and functional group by FTIR, mechanical properties of fibre and the structural and morphological analysis of the fibre using SEM and XRD. The results showed that the chemical contents of Mendong fibres were 72.14% cellulose, 20.2% hemicellulose, 3.44% lignin, 4.2% extractive and moisture of 4.2%–5.2%. Mechanical properties of the fibre include a strong character with tensile strength of 452 MPa and modulus of 17 GPa. The structural properties of Mendong fibre such as crystallinity, crystalline index, microfibril angle and crystalline size were 70.17%, 58.6%, 22.9 degrees and 14.3 nm, respectively. This fibre has competitive advantages compared with other natural fibres and can be developed further as a potential reinforcement of polymer matrix composites (Suryanto et al., 2014).

20.2.4.6 *Imperata brasiliensis*

Characterization of a new lignocellulosic fibre from Brazil: *I. brasiliensis* (Brazilian satintail) as an alternative source for nanocellulose extraction was conducted by Kelly Cristina Coelho de Carvalho Benini, and a team from Fatigue and Aeronautical Materials Research Group, Department of Materials and Technology, UNESP-Univ. Estadual Paulista, Guaratinguetá, São Paulo, Brazil. The paper was published in the *Journal of Natural Fibers*.

The chemical, physical and thermal properties of a new lignocellulosic fibre from Brazil (*I. brasiliensis*) were examined by SEM, chemical composition, XRD, FTIR and TGA. Fibres were analysed aiming to compare the properties of its new natural resource with other lignocellulosic fibres used as a source of nanocellulose extraction. Microscopy analysis demonstrated that the bundle of fibres presented a variety of size and shapes, ranging between 25 and 500 μm, while a single fibre has a diameter of 5 μm. The chemical composition showed the presence of 37.7% of cellulose, 35% of hemicellulose and 14.3% of lignin. The total CI calculated using Segal method was of 36.6%. By TGA, it was possible to identify the degradation step of each primary component of lignocellulosic fibre and to observe that the onset degradation temperature was 157°C. With the results of ATR-FTIR technique, it was possible to estimate the CI, and the results exhibited good agreement with that calculated by XRD. Finally it was possible to conclude that fibres obtained from *I. brasiliensis* are suitable to be used as a resource for nanocellulose obtainment because it presents almost the same properties of other lignocellulosic fibres successfully used in literature for nanocellulose extraction (Coelho de Carvalho Benini et al., 2017).

20.2.4.7 *Heteropogon contortus* plant

It is a tropical, perennial tussock grass with a native distribution encompassing Southern Africa, Southern Asia, Northern Australia, Oceania and southwestern North America (Wikipedia).

Black spear grass is a clump-forming grass producing a clump of leaves up to 30 cm long and flowering stems 30–100 cm tall, up to 3 mm in diameter at the base. Originating in the tropics of Asia and Africa, it has spread to most other tropical areas. The plant has been grown in Hawaii and southwestern North America to provide erosion control and revegetation of degraded areas. The plant has often been grown as a pasture grass and is also utilized as a material for thatching and weaving, making coarse mats (http://tropical.theferns.info/viewtropical.php?id=Heteropogon+contortus).

Characterization of new natural cellulosic fibre from *H. contortus* plant was studied by N. Rajesh Jesudoss Hyness, and a team from India; the paper was published in the *Journal of Natural Fibers*.

This study was conducted to realize physicochemical properties of biofibre obtained from *H. contortus* plant. *H. contortus* fibres (HCFs) had cellulose (64.87 wt.%), hemicellulose (19.34 wt.%), lignin (13.56 wt.%) and low density (602 kg/m^3). The chemical functional group of HCFs was established by FTIR, thermal stability of the fibre up to 220°C discovered by TGA. Furthermore, the assets of HCFs proved that it can act as an excellent reinforcement material as a biocomposite. Finally, the tensile properties were carried out through single fibre tensile tests, such as tensile strength, tensile modulus and microfibrillar angle (Hyness et al., 2018).

20.2.4.8 *Carex meyeriana* fibre

Extraction and performance characterization of *C. meyeriana* fibre was studied by Chunhong Wang and a team from China. The paper was published in the *Journal of Natural Fibers*.

In this investigation, *C. meyeriana* fibre (CMF) was extracted by using the ultrasound alkali–oxygen technology in the research, and the performance characterization of CMF was revealed. Parameters such as ultrasonic and alkali–oxygen acting times, and the concentrations of NaOH and H_2O_2 were optimized by using the multitarget orthogonal method. The performance characteristics of CMF were investigated by means of SEM, UV, FTIR, XRD and other basic testing methods. The result revealed that the optimal parameters in the extraction process were ultrasonic irradiation for 70 min, soaking in the alkali–oxygen bath for 100 min and the concentrations of NaOH as 8 g/L and H_2O_2 as 12 g/L. After extraction, the characteristic absorption peaks of the noncellulose CMF in the FTIR spectra decreased significantly, and the crystallinity increased from 46.5% to 50.6%. Warmth retention and excellent antibacterial properties of the extracted CMF were observed. It was expected that CMF can be developed as a functional fibre in the future (Wang et al., 2016).

20.2.5 Bark

20.2.5.1 Barkcloth (*Ficus natalensis*)

F. natalensis is a tree in the family Moraceae. It is commonly known as the Natal fig, or 'Mutuba', to the locals. These trees are distributed from northeastern South Africa to Uganda and Kenya.

The bark of the tree is harvested, without harming the tree, to make barkcloth, an environment-friendly, renewable material. Skilled artisans incorporate this unique fabric into many modern uses, including fashion, accessories, housewares, interior design and art. The vision is to create sustainable jobs in East Africa by creating a global demand for barkcloth. Mutuba trees can be harvested annually for up to 40 years, yielding up to 200 m^2 of cloth individually (Wikipedia).

Static and dynamic mechanical properties of barkcloth (*F. natalensis*)-reinforced epoxy composites were studied by Samson Rwawiire from the Busitema University, Tororo, Uganda, and Blanka Tomkova from the University of Liberec, The Czech Republic. The paper was published in the *Journal of Natural Fibers*.

An exploratory investigation of barkcloth, a nonwoven material produced through a series of pummelling processes from mutuba tree in Uganda, is fronted as reinforcement for epoxy composite laminates. The fabric and composite morphology was investigated using SEM. The composite response to loading against temperature, time and frequency was investigated using dynamical mechanical analysis. The results show that the developed composites are stable with considerable tensile strength and bending rigidity thus providing material engineers with the possibility of applying the material for semistructural applications (Rwawiire and Tomkova, 2016).

20.2.5.2 *Ceiba pentandra*

C. pentandra is a tropical tree of the order Malvales and the family Malvaceae (previously separated in the family Bombacaceae), native to Mexico, Central America and the Caribbean, northern South America, and (as the variety *C. pentandra* var. *guineensis*) to tropical West Africa. A somewhat smaller variety is found throughout Southern Asia and the East Indies. *C. pentandra* fibres can be obtained from seed pods and bark.

Kapok is the most used common name for the tree and may also refer to the cotton-like fluff obtained from its seed pods. The tree is cultivated for the seed fibre, particularly in Southeast Asia and is also known as the Java cotton, Java kapok, silk-cotton, samauma or ceiba (Wikipedia).

C. pentandra is commonly obtained from Southwest Asia and Indonesia. Main application includes upholstery material, kapok in compressed condition in water can extract 36 times its weight, also is used as life buoys, belts and waistcoats (Mauersberger, 1954).

Physicochemical and thermal properties of *C. pentandra* bark fibre were studied by R. Kumar, N. and a team from India, and published in the *Journal of Natural Fibers*.

Owing to their low weight-to-high strength ratio and recyclable features, the natural fibres are the most potential choice in place of synthetic fibres and been used as reinforcement materials in polymer matrix composites. Characterization of *Ceiba pentandra* bark fibres (CPFs) was carried out with a help of chemical analysis, Fourier Transform-Infrared Analysis (FTIR), X-ray diffraction (XRD), thermogravimetric analysis, and Differential thermogravimetric analysis (DTG). CPFs contain 60.9% (w/w) of cellulose, 17.5% (w/w) of hemicellulose and 23.5% (w/w) of lignin. Besides, its density and CI are 682 kg m^{-3} and 57.94%, respectively. TGA and DTG analysis discovered that CPFs are thermally stable up to 342.1°C. Furthermore, all the resources of CPFs ensured that it can be an excellent alternative for synthetic fibres in polymer matrix composites (Kumar et al., 2018).

20.2.5.3 *Azadirachta indica*

A. indica, commonly known as neem, nimtree or Indian lilac, is a tree in the mahogany family Meliaceae. It is one of the two species in the genus *Azadirachta*, and is native to the Indian subcontinent, i.e., India, Nepal, Pakistan, Bangladesh, Sri Lanka and Maldives. It is typically grown in tropical and semitropical regions. Neem trees also grow in islands located in the southern part of Iran. Its fruits and seeds are the source of neem oil for, e.g., polymeric resins. Bark: neem bark contains 14% tannin, an amount similar to that in conventional tannin-yielding trees (such as *Acacia decurrens*). Moreover, it yields a strong, coarse fibre commonly woven into ropes in the villages of India (Wikipedia).

Physiochemical properties of new cellulosic fibres from *A. indica* plant were studied by P. Manimaran, and a group of experts from the Department of Mechanical Engineering of Kamaraj College of Engineering and Technology, Virudhunagar, Tamil Nadu, India. The paper was published in the *Journal of Natural Fibers*.

This research study was aimed at examining newly identified natural fibre from the bark of *A. indica*. The various properties were analysed and compared with other available bark fibres. The chemical composition of *A. indica* fibres (AIFs), high cellulose (68.42 wt.%) content and low lignin (13.58 wt.%) were discovered. The lower density of 740 kg/m^3 and CI of 65.04% were identified. The maximum peak temperature obtained was 321.2°C in DTG curve. Taken together, all the properties of AIFs indicated that they could be suitable to make green composites for various types of applications (Manimaran et al., 2018a,b).

20.2.5.4 *Dichrostachys cinerea*

D. cinerea, known as sicklebush, Bell mimosa, Chinese lantern tree or Kalahari Christmas tree (South Africa), is a legume of the genus *Dichrostachys* in the Fabaceae family.

Other common names include acacia Saint Domingue (French), el marabú (Cuba), 'Mpangara' (Shona), Kalahari-Weihnachtsbaum (German of former Southwest Africa), kéké or mimosa clochette (Réunion), Latin *cinereus* ('ashes'). It is native to Africa, Indian subcontinent and North Australia and introduced to the Caribbean and parts of Southeast Asia. In Ethiopia, the species is common in the Nechisar National Park.

The tree was brought to the Caribbean in the 19th century. In Cuba, where it is known as El Marabú or Marabou weed, it has become a serious invasive species problem, occupying about 4,900,000 acres (20,000 km^2) of agricultural land. Biomass is a source for renewable power generation (Wikipedia).

The study on characterization of new natural cellulosic fibre from the bark of *D. cinerea* was presented by P. G. Baskaran, and a team from the Department of Mechanical Engineering, Sri Venkateswara College of Engineering & Technology, Thirupachur, Thiruvallur, India, published in the *Journal of Natural Fibers*.

The aim of this investigation is to understand the physicochemical properties of fibres extracted from the bark of the *D. cinerea* plant. *D. cinerea* fibres (DCFs) has cellulose (72.4 wt.%), hemicellulose (13.08 wt.%), lignin (16.89 wt.%), density (1240 kg/m3), CI (57.82%) and tensile strength (873 ± 14 MPa). Besides the cellulose degradation of DCFs at 359.3 degrees vide by the thermogravimetric analysis and chemical groups are identified by Fourier transform analysis. Eventually the characterization results of DCFs strongly show the possibility of reinforcement in polymer matrices (Baskaran et al., 2018).

20.2.6 *Fruit fibres*

20.2.6.1 Borassus *and tamarind fruit fibres*

Research on *Borassus* and tamarind fruit fibres as reinforcement in cashew nut shell liquid-epoxy composites was carried out by Ullal Achutha Kini, and a team from India, published in the *Journal of Natural Fibers*.

In this study, *Borassus* and tamarind fruit fibres were extracted and their physical and mechanical properties such as diameter, density, tensile strength and interfacial adhesion strength (IAS) were experimentally determined. To study the effect of alkali treatment, both the fruit fibres were treated with 5% vol. sodium hydroxide solution for 0.5 h, 1 h, 2 h and 4 h durations. Morphological studies of untreated and alkali treated fibres by using scanning electron microscope SEM revealed the presence of the surface impurities on the untreated fibre, whereas the same were absent on the treated fibres. FTIR and XRD analysis also confirmed the elimination of amorphous hemicellulose of the fibres on treatment. *Borassus* fruit fine fibres and tamarind fruit fibres treated for 2 h exhibited better mechanical properties and improved IAS with cashew nut shell liquid-epoxy matrix (Kini et al., 2018).

20.2.6.2 *Pomelo fruit fibres*

Isolation of nanocellulose from pomelo fruit fibres by chemical treatments were studied by Niti Yongvanich from the Department of Materials Science and Engineering, Faculty of Engineering and Industrial Technology, Silpakorn University, Nakornpathom in Thailand, and published in the *Journal of Natural Fibers*.

This study aims to isolate cellulose nanofibres from locally abundant pomelo fruit. The peels were subjected to alkali treatment (NaOH) with different concentrations and soaking times. Acid hydrolysis was also carried out to obtain an aqueous suspension of nanocellulose. The treated cellulose fibres were characterized by various methods. Alkali treatment was able to remove noncellulosic constituents. Microscopy techniques yielded morphology of elementary fibres with a compact structure with both micro- and nano-scaled fibres. An improved onset of degradation was likely caused by an increase in crystallinity evidenced by X-ray diffractometry. Acid hydrolysis could successfully individualize cellulose nanofibres as observed by TEM (Yongvanich, 2015).

20.3 Conclusion

The global environmental situation connected with terrible pollution, danger for the future of our planet and growing population increased awareness among many world and governmental organizations, as well as NGOs to act towards limitation of this catastrophic situation. Most plastics and polymers, including man-made fibres and textiles, became serious hazard for life of lands and oceans. Residues of small polyesters and plastics decomposed and broken down by bacteria are now widely commented by many sources (Luong, 2017). For this reason, we notice the great interest in movement towards coming back to natural renewable resources with full satisfaction. Personally we are connected for almost 60 years with the research on the development of natural fibrous resources. In 50s, 60s and 70s of the 20th century, most of the research and industrial people were interested in man-made polymers, including man-made fibres. But now, in the 21st century, there are so many research centres, scientists and experts looking for the new sources of renewable fibrous materials. Finally, we can state '*Now is the end of nylon era*'. The excellent example of the change of public awareness is that the agenda of the United Nations – Food and Agriculture Organization of the United Nations – had proclaimed the year 2009 as the *International Year of Natural Fibres* and the entire world conducted several meetings, conferences, workshops and events devoted to the development of natural fibres. Another example is that the former Institute of Bast Fibres in Poznań, Poland, created in 1930, which published its research results dealing with fibrous plants and natural fibres in Polish, started at the beginning of 90s published in English. The Haworth Press took over this yearbook '*Włókna Naturalne. Natural Fibres*' and started publishing *Journal of Natural Fibers* in 2004. Finally, Taylor & Francis, who is currently the publisher of this journal, increases the frequency of issues' appearance to monthly frequency. This story confirms the growing tendency in the interest in natural materials including fibrous raw materials development. In this situation, looking for new natural fibrous raw materials and their novel, advanced applications is compatible with the needs of humanity and the environment, on top and in accordance with the requirements of a modern society. New fibres should be utilized in practice, especially in the regions where there is abundance, contributing to the development of these areas, especially rural ones and societies.

In two volumes of the second edition of the *Handbook of Natural Fibres*, published by Elsevier, there are chapters devoted to well-known natural fibres as well as to new emerging fibres such as bamboo, abaca and fibres from plants of arid regions of Africa, and also a chapter describing almost 70 characteristics of new natural fibres and their diversified applications. Summing up, taking into consideration the balance between man-made and natural fibres, there is satisfying competition from the view point of better environment-friendly sustainability, high quality, healthy influence and economy. We are looking for the brilliant future for new fibres characterized in this chapter.

References

Bajpai, P.K., Meena, D., Vatsa, S., Singh, I., 2013. Tensile behavior of nettle fibre composites exposed to various environments. J. Nat. Fibers 10 (3), 244–256. https://doi.org/10.1080/15440478.2013.791912.

Baskaran, P.G., Kathiresan, M., Senthamaraikannan, P., Saravanakumar, S.S., 2018. Characterization of new natural cellulosic fibre from the bark of Dichrostachys cinerea. J. Nat. Fibers 15 (1), 62–68. https://doi.org/10.1080/15440478.2017.1304314.

Berardi, U., Iannace, G., Di Gabriele, M., 2017. J. Nat. Fibers 14 (6), 858–863. https://doi.org/10.1080/15440478.2017.1279995.

Bourahli, M.E.H., 2018. Uni- and bimodal Weibull distribution for analyzing the tensile strength of Diss fibres. J. Nat. Fibers 15 (6), 843–852. https://doi.org/10.1080/15440478.2017.1371094.

Bourahli, M.E.H., Osmani, H., 2013. Chemical and mechanical properties of diss (*Ampelodesmos mauritanicus*) fibres. J. Nat. Fibers 10 (3), 219–232. https://doi.org/10.1080/15440478.2012.761115.

Cadena, E.M., Vélez, J.M., Santa, J.F., Otálvaro, V., 2017. Natural fibres from plantain pseudostem (Musa paradisiaca) for use in fibre-reinforced composites. J. Nat. Fibers 14 (5), 678–690. https://doi.org/10.1080/15440478.2016.1266295.

Chand, N., Shukla, M., Sharma, M.K., 2008. Analysis of mechanical behaviour of bamboo (dendrocalamus strictus) by using FEM. J. Nat. Fibers 5 (2), 127–137. https://doi.org/10.1080/15440470801928970.

Coelho de Carvalho Benini, K.C., Cornelis Voorwald, H.J., Hilário Cioffi, M.O., Milanese, A.C., Ornaghi Jr., H.L., 2017. Characterization of a new lignocellulosic fibre from Brazil: Imperata brasiliensis (Brazilian Satintail) as an alternative source for nanocellulose extraction. J. Nat. Fibers 14 (1), 112–125. https://doi.org/10.1080/15440478.2016.1167647.

Cortés-Martínez, C.I., García-Méndez, R.F., 2017. Morphology of the bagasse fibres obtained from the elaboration process of mezcal and effects on their tensile properties. J. Nat. Fibers 14 (2), 250–261. https://doi.org/10.1080/15440478.2016.1193089.

Cui, Y., Cheng, L., Shan, X., Fan, W., 2017. Correlation between pod shape traits and its fibre quality of wild *Calotropis gigantea*. J. Nat. Fibers 14 (1), 1–9. https://doi.org/10.1080/15440478.2015.1029197.

De Pellegrin M.Z., Acordi J., and Montedo O.R.K., Influence of the length and the content of cellulose fibres obtained from sugarcane bagasse on the mechanical properties of fibre-reinforced mortar composites, published online 18 June 2019, https://doi.org/10.1080/15440478.2019.1612311.

El Omari, H., Belfkira, A., Brouillette, F., 2017. Paper properties of *Typha latifolia, pennisetum alopecuroides*, and *agave americana* fibres and their effect as a substitute for kraft pulp fibres. J. Nat. Fibers 14 (3), 426–436. https://doi.org/10.1080/15440478.2016.1212766.

Elzubair, A., Bonelli, C.M.C., Suarez, J.C.M., Mano, E.B., 2007. Morphological, structural, thermal and mechanical characterization of piassava fibres. J. Nat. Fibers 4 (2), 13–31. https://doi.org/10.1300/J395v04n02_02.

Fatma, T., Jahan, S., 2018. Extraction of unconventional bast fibres from Kydia calycina plant and their characterisation. J. Nat. Fibers 15 (5), 697–706. https://doi.org/10.1080/15440478.2017.1354745.

Gallardo-Rodríguez, J.J., Ríos-Rivera, A., Felipe-Hernández, D., 2016. Evaluation of *Furcraea andina* fibres as bacterial support for biological denitrification. J. Nat. Fibers 13 (6), 691–701. https://doi.org/10.1080/15440478.2015.1130004.

Gong, J., Zhang, Q., Lou, J., Zhang, T., Li, H., Li, Z., Li, Q., Zhang, J., 2019. Investigation of the degradation of bio-recalcitrance in Apocynum venetum fibre biodegumming. J. Nat. Fibers 16 (1), 1–12. https://doi.org/10.1080/15440478.2017.1379043.

Gokulkumar, S., Ramaswamy, P., Thyla, L., Prabhu, S.S., 2019. Characterization and comparative analysis on mechanical and acoustical properties of camellia sinensis/ananas comosus/glass fibre hybrid polymer composites. J. Nat. Fibers (in press).

Goto, M., Wang, A., Goto, H., 2017. Light transmission by Lotus plant stem and leaf tissue: natural textiles for optical fibres. J. Nat. Fibers 14 (2), 262–270. https://doi.org/10.1080/15440478.2016.1193090.

Hanipah, S.H., Xiang, L.Y., Mohammed, M.A.P., Baharuddin, A.S., 2017. Study of non-linear mechanical behavior of oil palm mesocarp fibres. J. Nat. Fibers 14 (2), 153–165. https://doi.org/10.1080/15440478.2015.1093575.

Hyness, N.R.J., Vignesh, N.J., Senthamaraikannan, P., Saravanakumar, S.S., Sanjay, M.R., 2018. Characterization of new natural cellulosic fibre from heteropogon contortus plant. J. Nat. Fibers 15 (1), 146–153. https://doi.org/10.1080/15440478.2017.1321516.

Jagadeesh, D., Varada Rajulu, A., Guduri, B.R., 2010. J. Nat. Fibers 7 (2), 93–99. https://doi.org/10.1080/15440478.2010.481096.

Jayaprabha, J.S., Brahmakumar, M., Manilal, V.B., 2011. Banana pseudostem characterization and its fibre property evaluation on physical and bioextraction. J. Nat. Fibers 8 (3), 149–160. https://doi.org/10.1080/15440478.2011.601614.

Kanimozhi, M., Vasugi, N., 2012. Characterization of agave vera-cruz Mill leaf fibre for textile applications—an exploratory investigation. J. Nat. Fibers 9 (4), 219–228. https://doi.org/10.1080/15440478.2012.733526.

Katović, D., Katović, A., Krnčević, M., 2011. Spanish broom (*Spartium junceum L.*) - history and perspective. J. Nat. Fibers 8 (2), 81–98. https://doi.org/10.1080/15440478.2011.577277.

Kini, U.A., Nayak, S.Y., Heckadka, S.S., Thomas, L.G., Adarsh, S.P., Gupta, S., 2018. Borassus and tamarind fruit fibers as reinforcement in cashew nut shell liquid-epoxy composites. J. Nat. Fibers 15 (2), 204–218. https://doi.org/10.1080/15440478.2017.1323697.

Kozlowski R.M. and Muzyczek M., 3rd Quarter 2017. Chapter 11. Trends in natural ligno-cellulosic fibre functionalization and finishing processes, In: Ryszard M. and Kozlowski, M.M., (Eds.), Natural Fibres: Properties, Mechanical Behavior, Functionalization and Applications, Nova Publishers, USA, ISBN: 978-1-53612-071-4 (available online), 215–236.

Kumar, R., Hynes, N.R.J., Senthamaraikannan, P., Saravanakumar, S., Sanjay, M.R., 2018. Physicochemical and thermal properties of ceiba pentandra bark fibre. J. Nat. Fibers 15 (6), 822–829. https://doi.org/10.1080/15440478.2017.1369208.

Kumar S., Patel V.K., Mer, K.K.S., Gangil B., Singh, T., Fekete, G., May 20, 2019. Himalayan natural fiber-reinforced epoxy composites: effect of grewia optiva/bauhinia vahlii fibers on physico-mechanical and dry sliding wear behaviour, J. Nat. Fibers, https://doi.org/10.1080/15440478.2019.1612814.

Lovely, M., Joseph, K.U., Joseph, R., 2007. Chemical Composition, Surface Modification, Physical, Mechanical and Thermal Properties—A Potential Natural Reinforcement, J. Nat. Fibers 3 (4), 13–27. https://doi.org/10.1300/J395v03n04_02.

Luong, T., 2017. Microplastic Pollutants Highlights the Dangers Posed to the Environment and the Food Chain. Chemical & Engineering News, Posted on: February 21, Based on Christopher Blair Crawford, Brian Quinn. 2016. Microplastic Pollutants, first ed. Elsevier Science.

Mahesha, G.T., Satish, S.B., Vijaya, K.M., Bhat, K.S., 2016. Preparation of unidirectional Grewia serrulata fibre-reinforced polyester composites and evaluation of tensile and flexural properties. J. Nat. Fibers 13 (5), 547–554. https://doi.org/10.1080/15440478.2015.1081575.

Mahesha, G.T., Satish, S.B., Kini, M.V., Subrahmanya, B.K., 2017. Mechanical characterization and water ageing behavior studies of Grewia serrulata bast fibre reinforced thermoset composites. J. Nat. Fibers 14 (6), 788–800. https://doi.org/10.1080/15440478.2017.1279103.

Maheshwaran, M.V., Hyness, N.R.J., Senthamaraikannan, P., Saravanakumar, S.S., Sanjay, M.R., 2018. Characterization of natural cellulosic fibre from Epipremnum aureum stem. J. Nat. Fibers 15 (6), 789–798. https://doi.org/10.1080/15440478.2017.1364205.

Manimaran, P., Prithiviraj, M., Saravanakumar, S.S., Arthanarieswaran, V.P., Senthamaraikannan, P., 2018a. Physicochemical, tensile, and thermal characterization of new natural cellulosic fibres from the stems of Sida cordifolia. J. Nat. Fibers 15 (6), 860–869. https://doi.org/10.1080/15440478.2017.1376301.

Manimaran, P., Senthamaraikannan, P., Murugananthan, K., Sanjay, M.R., 2018b. Physicochemical properties of new cellulosic fibres from *Azadirachta indica* plant. J. Nat. Fibers 15 (1), 29–38. https://doi.org/10.1080/15440478.2017.1302388.

Mauersberger, H.R. (Ed.), 1954. Matthews' Textile Fibres: Their Physical, Microscopic, and Chemical Properties. John Wiley, Chapman & Hall, New York, London.

McIntyre, J.E., Denton, M.J., Daniels, P.N., 2002. Textile Terms and Definitions. The Textile Institute, Manchester.

Mizera, C., Herak, D., Hrabe, P., Muller, M. and Kabutey, A., 2017. Mechanical behavior of ensete ventricosum fibre under tension loading, J. Nat. Fibers 14 (2), 287–296, https://doi.org/ 10.1080/15440478.2016.1206500.

Msahli, S., Jaouadi, M., Sakli, F., Drean, J.-Y., 2015. Study of the mechanical properties of fibres extracted from Tunisian *Agave americana* L. J. Nat. Fibers 12 (6), 552–560. https://doi.org/10.1080/15440478.2014.984046.

Nadlene, R., Sapuan, S.M., Jawaid, M., Ishak, M.R., Yusriah, L., 2016. A review on roselle fibre and its composites. J. Nat. Fibers 13 (1), 10–41. https://doi.org/10.1080/15440478.2014.984052.

Neves Monteiro, S., 2009. J. Nat. Fibers 6 (2), 91–203. https://doi.org/10.1080/15440470902961128.

Pokhriyal, M., Prasad, L., Raturi, H.P., 2018. An experimental investigation on mechanical and tribological properties of Himalayan nettle fibre composite. J. Nat. Fibers 15 (5), 752–761. https://doi.org/10.1080/15440478.2017.1364202.

Pouriman, M., Caparanga, A.R., Ebrahimi, M., Dahresobh, A., 2018. Characterization of untreated and alkaline-treated Salago fibres (*genus wikstroemia* spp.). J. Nat. Fibers 15 (2), 296–307. https://doi.org/10.1080/15440478.2017.1329105.

Qu, L., Tian, M., Zhu, S., Guo, X., Zhang, Y., Zhang, X., Tang, X., Sun, K., 2016. Study on structure and property of abelmoschus manihot fibres. J. Nat. Fibers 13 (2), 215–226. https://doi.org/10.1080/15440478.2014.1004010.

Quinaya, D.C.P., da Silva, E.S., d'Almeida, J.R.M., 2016. On the use of residues from the sustainable extraction of heart of palm in agglomerated panels. J. Nat. Fibers 13 (2), 172–177. https://doi.org/10.1080/15440478.2014.1004009.

Raharjo, W.W., Soenoko, R., Irawan, Y.S., Suprapto, A., 2018. The influence of chemical treatments on cantala fibre properties and interfacial bonding of cantala fibre/recycled high density polyethylene (rHDPE). J. Nat. Fibers 15 (1), 98–111. https://doi.org/10.1080/15440478.2017.1321512.

Razak, H.A., Ferdiansyah, T., 2005. Toughness characteristics of Arenga pinnata fibre concrete. J. Nat. Fibers 2 (2), 89–103. https://doi.org/10.1300/J395v02n02_06.

Reddy, K.O., Maheswari, C.U., Muzenda, E., Shukla, M., Rajulu, A.V., 2016. Extraction and characterization of cellulose from pretreated Ficus (peepal tree) leaf fibres. J. Nat. Fibers 13 (1), 54–64. https://doi.org/10.1080/15440478.2014.984055.

Ribeiro, G.L., Gandara, M., Pinzón-Moreno, D.D., Saron, C., 2019. Low-density polyethylene/sugarcane fibre composites from recycled polymer and treated fibre by steam explosion. J. Nat. Fibers 16 (1), 13–24. https://doi.org/10.1080/15440478.2017.1379044.

Rodrigue, S.T.N., Ebénézer, N., Dieunedort, N., Didier, F., Fogue, M., Drean, J.-Y., Harzallah, O., 2017. Investigation of the physical and mechanical properties of raffia vinifera fibres along the stem. J. Nat. Fibers 14 (5), 621–633. https://doi.org/10.1080/15440478.2016.1250025.

Rwawiire, S., Tomkova, B., 2015. Morphological, thermal, and mechanical characterization of Sansevieria trifasciata fibres. J. Nat. Fibers 12 (3), 201–210. https://doi.org/10.1080/15440478.2014.914006.

Rwawiire, S., Tomkova, B., 2016. Static and dynamic mechanical properties of barkcloth (*Ficus natalensis*)-reinforced epoxy composite. J. Nat. Fibers 13 (2), 137–145. https://doi.org/10.1080/15440478.2014.984061.

Shamsul Alam, M.D., Arifuzzaman Khan, G.M., Abdur Razzaque, S.M., 2009. Estimation of main constituents of ananus comosus (pineapple) leaf fibre and its photo-oxidative degradation. J. Nat. Fibers 6 (2), 138–150. https://doi.org/10.1080/15440470902955369.

Singha, A.S., Guleria, A., 2015. Synthesis and applications of functional polymers from natural okra fibres for removal of Cu(II) ions from aqueous solution. J. Nat. Fibers 12 (6), 587–603. https://doi.org/10.1080/15440478.2014.984049.

Song, Y., Han, G., Li, M., Jiang, W., Li, X., Zhang, Y., Yu, R., 2017. Performance analysis of kudzu fibre prepared by using combined steam explosion and chemical degumming. J. Nat. Fibers 14 (6), 759–768. https://doi.org/10.1080/15440478.2016.1212770.

Song Y., Li X., Han G., Zhang Y. and Jiang W., May 8, 2019. Comparison of Different Methods to Produce Pineapple Leaf Fibers with Steam Explosion, J. of Nat. Fibers.https://doi.org/10.1080/15440478.2019.1612301.

Subramanya, R., Satyanarayana, K.G., Pilar, B.S., 2017. Evaluation of structural, tensile and thermal properties of banana fibres. J. Nat. Fibers 14 (4), 485–497. https://doi.org/10.1080/15440478.2016.1212771.

Sun, B., Huang, A., Wang, Y., Liu, J., 2015. Natural bamboo (neosinocalamus affinis keng) fibre identification using FT-IR and 2d-IR correlation spectroscopy. J. Nat. Fibers 12 (1), 1–11. https://doi.org/10.1080/15440478.2013.796907.

Suryanto, H., Marsyahyo, E., Irawan, Y.S., Soenoko, R., 2014. Morphology, structure, and mechanical properties of natural cellulose fibre from Mendong grass (Fimbristylis globulosa). J. Nat. Fibers 11 (4), 333–351. https://doi.org/10.1080/15440478.2013.879087.

Wahab Taiwo, O.F.A., Alkarkhi, Arniza Ghazali, A.F.M., Wan Daud, W., 2017. Optimization of the strength properties of waste oil palm (elaeis guineensis) fronds fibre. J. Nat. Fibers 14 (4), 551–563. https://doi.org/10.1080/15440478.2016.1215947.

Wang, C., Bai, S., Yue, X., Ren, Z., 2016. Extraction and performance characterization of the Carexmeyeriana fibre. J. Nat. Fibers 13 (6), 759–768. https://doi.org/10.1080/15440478.2015.1137528.

Xu, X., Zheng, W., Xu, J., Chen, L., 2017. Properties of natural fibres from the abaxial side of fireweed (*Gerbera delavayi*) leaf blade for manual spinning. J. Nat. Fibers 14 (1), 78–85. https://doi.org/10.1080/15440478.2016.1146644.

Yongvanich, N., 2015. Isolation of nanocellulose from pomelo fruit fibres by chemical treatments. J. Nat. Fibers 12 (4), 323–331. https://doi.org/10.1080/15440478.2014.920286.

Wang, Y., Xiao, X., Wang, S., Li, K., Jiang, Y., Ma, Y., Zhang, T., Ran, R., 2019. Exploration on UV-blocking performance of lignin from palm (trachycarpus fortunei) fibre. J. Nat. Fibers, published online May 07. https://doi.org/10.1080/15440478.2019.1612307.

Sources of further information and advice

Buitrago, B., Jaramillo, F., Gómez, M., 2015. Some properties of natural Fibres.Sisal, pineapple, and banana) in comparison to man-made technical fibres (aramide, glass, carbon). J. Nat. Fibers 12 (4), 357–367.

Franck, R.R. (Ed.), 2001. Silk, Mohair and Other Luxury Fibres. Woodhead Publishing, Cambridge.

Franck, R.R. (Ed.), 2005. Bast and Other Plant Fibres. Woodhead Publishing, Cambridge.

Kozłowski, R., Kozłowska, J., Rawluk, M., Barriga, J., 2004. Potential of lignocellulosic fibrous raw materials, their properties and diversified applications. Nonlinear Optics, Quantum Optics 31 (1–4), 61–89.

Kozlowski, R., Baraniecki, P., Barriga-Bedoya, J., 2005. Bast fibres (flax, hemp, jute, ramie, kenaf, abaca). Chapter 2. In: Blackburn, R.S. (Ed.), Biodegradable and Sustainable Fibres. Woodhead Publishing, Cambridge, pp. 36–110.

Kozłowski, R.M., Mackiewicz-Talarczyk, M., Barriga-Bedoya, J., 2010. 'Natural fibres production, processing, and application: inventory and future prospects', *Contemporary Science of Polymeric Materials*. ACS (Am. Chem. Soc.) Symp. Ser. 1061 (3), 41–51.

Kozłowski, R.M., Mackiewicz-Talarczyk, M., Muzyczek, M., Barriga-Bedoya, J., 2012. Future of natural fibres, their coexistence and competition with man-made fibres in 21st century. Mol. Cryst. Liq. Cryst. 556 (1), 200–222. Taylor & Francis, MS ID: 635962.

Index

Note: 'Page numbers followed by "*f*" indicate figures and "*t*" indicate tables'.

C

J

M

X

9780128183984